Instability and Plastic Collapse of Steel Structures

Instability and Plastic Collapse of Steel Structures

Proceedings of
The Michael R. Horne Conference
organised by
The University of Manchester
with co-sponsor
The Institution of Structural Engineers

Edited by
L. J. Morris

GRANADA
London Toronto Sydney New York

Granada Publishing Limited – Technical Books Division
Frogmore, St Albans, Herts AL2 2NF
and
36 Golden Square, London W1R 4AH
515 Madison Avenue, New York, NY 10022, USA
117 York Street, Sydney, NSW 2000, Australia
60 International Boulevard, Rexdale, Ontario R9W 6J2, Canada
61 Beach Road, Auckland, New Zealand

ISBN 0 246 12196 3

First published in Great Britain 1983 by Granada Publishing

British Library Cataloguing in Publication Data
Instability and plastic collapse of steel structures.
1. Buildings, Iron and steel—Congresses
2. Structural stability—Congresses
I. Morris, L. J.
624.1'821 TA684

Printed in Great Britain by Mackays of Chatham, Kent

Contents

5. DESIGN OF COLUMN MEMBERS

6. ANALYSIS OF RECTANGULAR PLATES

11. DESIGN OF TUBULAR CONNECTIONS AND CONSTRUCTION

Foreword

Professor Michael Horne has received many honours, as befits a leader in the engineering and academic worlds. He ran through all the degrees that Cambridge has to offer its working sons by 1956 when he took his Sc.D. degree. The great professional engineering institutions in this country have given him Premiums and medals; the Institution of Structural Engineers paid him the greatest compliment of all when it made him its President in 1980. He was admitted to the Fellowship of Engineering in 1978 and in 1981 he received a Royal Honour, being made an Officer of the Order of the British Empire, was elected to the Royal Society of London and received an honorary D.Sc degree from the University of Salford. However, I suspect that, in his heart, the honour that will touch him most is this collection of papers dedicated to him by distinguished workers throughout the world.

Michael Horne's own contribution to the theory of structures is monumental. He is the author of fifty-five books and papers, the co-author of forty-four more while his contribution to discussions containing results of original work amount to a further twenty. While he has enlarged our knowledge in other fields, his main work has been in instability, plastic behaviour and the design of steel structures. These are the subjects dealt with in this volume which contains the papers presented in his honour at the Conference held in Manchester, 20-22 September, 1983.

Baker

Preface

A Conference on the instability and plastic collapse of steel structures would not be expected to cover all the major aspects in which significant advances in structural theory, analytical techniques and applications to design have been made in the last forty years. The topic is sufficiently wide, however, for the papers received to have reflected the influence of a great number of those advances. Nevertheless, some papers were received which could not be fitted into the main theme of the conference, and it is much regretted that they had to be omitted.

The sixty papers that do appear in these Conference Proceedings have been divided into eleven subject groups that inevitably leave considerable overlap. One consequence of what has amounted to a significant improvement in our fundamental understanding of structural behaviour has been a certain integration and interrelationship between various fields of structural theory and application, and it could well be argued that too tight a classification into subject themes would in any case be undesirable.

The historical development of plastic theory as an important model of structural behaviour in teaching and research, and its establishment as an essential tool in present day approaches to design is clearly established. The idealised concept of plastic collapse is mirrored by the equally idealised concept of the elastic critical load. A unifying theme for most of the papers is their concern to bring together these two idealised aspects of structural behaviour, to understand their limitations as well as their usefulness, and to place them within the context of more elaborate and all-embracing analytical procedures.

In the field of bar structures, attention is paid to the particular elastic-plastic stability problems of complete structures, of columns and of compression members generally, of slender beams, and also the problem of local instability in thin-walled members including interaction with member stability. The production of analytical and design procedures for columns in building frames is still of central interest and is the subject of several papers - still providing a challenge to research workers and the producers of codes some fifty years after the work of the Steel Structures Research Committee had highlighted some of the complexities. The papers on the lateral stability of beams show an interest in advancing beyond the simplifying assumptions that cross-sections do not distort - revealing problems of analysis that only recent advances in computer-based numerical procedures have enabled us to tackle at all adequately.

The papers include evidence of the desire to attempt formal design procedures which will deal explicitly with all the limiting conditions which need to be satisfied if structures are to perform satisfactorily - involving not only plastic collapse and stability, but also considerations of minimum weight, shakedown, limiting stresses at working load and limitations on deflections. Different criteria dominate for different types and usages of structures - for example in the design of radio telescopes the imposition of severe deformation constraints necessitates an approach completely different from that of the design and analytical techniques used say for office blocks.

Problems concerning the local buckling of structural members are particularly to the fore in considering cold-formed sections and tubular members. Considerations of economy, convenience and appearance are amongst the factors that have led to an increased interest in these forms of construction, and a number of papers dealing with them are included.

The design of plated structures introduces stability problems that differ markedly from those mainly encountered in bar-type structures. Idealised plastic collapse and classical elastic buckling solutions play a less dominating role - mainly because of the significance of the changes in membrane stress distribution due to deflections that become significant in relation to plate thickness well before collapse takes place. Of considerable significance also can be the effect of initial plate panel imperfections and in some cases also the effect of residual stresses due to fabrication. Many of the papers are concerned with the effect of such factors on collapse load capacity. Numerical procedures exploiting computers - finite element and finite difference - have been used of recent years to carry out analyses hitherto impossible - and many of the papers show such applications. The production of new codes of practice - particularly the new British Code for Bridges - has stimulated much of the research, and still challenges comparisons to check further the validity of its clauses.

The papers presented at the Conference have been concerned predominately with the assessment of the strength and performance of structures under given loads, and with design procedures which use those assesssments to proportion structures, again for given load conditions. This is only part of the total process of design - a paper on the safety of structures performs a useful function in challenging our thinking about the way in which we as engineers meet our responsibilities for the safety of the structures we designed.

As editor of this volume of papers, I wish to express my thanks to all the authors - and to express the hope that the many ideas and analytical, computational and experimental results they contain may be a stimulus to further progress in our understanding of structural behaviour and its application to the more efficient and economic design of steel structures.

Linden J Morris
May 1983

Opening Address

Lord Baker of Windrush

The young which, by the most recent medical definition, means all those under the age of 80, do not always realise that the old are forced, on occasion, to adopt evasive tactics. When the invitation to open this Conference came to me, I calculated that, by 20th September, 1983, I would be more than half way through my 83rd year and might find the journey to Manchester and the excitement of meeting old friends too much for me. I could not bear the thought of being completely out of a function dedicated to the subject of plastic collapse and to honouring one of my most distinguished collaborators, so I decided to make sure of some small part by preparing this tape which can speak for me.

I consider myself one of the most fortunate of men. From my earliest days on airships I have been engaged on exciting work. As a result I have had no difficulty in attracting outstanding assistants. One of the first two men I recruited in 1931 for the Steel Structures Research Committee work, described in Vol I of *The Steel Skeleton* which I must mention since without that experience we would probably not have attacked the plastic problem so vigorously, was P.D. Holder. After taking his Ph.D. in 1935 he joined the Royal Air Force and became one of our most distinguished and gallant airmen, retiring in 1968 as Air Marshal Sir Paul Holder, K.B.E., C.B.,D.S.O.,D.F.C.

With that frustrating elastic design method out of my system, and after my meeting with Maier Leibnitz in 1936 convinced me that plastic behaviour was the key to rational design, I secured John Roderick, one of my Bristol graduates, to assist me. You will be familiar with his pioneer work, which sent me into the 1939-45 War well equipped. He is a key figure in my story, because he joined me again in Cambridge in 1944 and, as Michael Horne will confirm, as my right-hand man trained Michael and all our young team to be meticulous experimenters. Unfortunately for me, he left Cambridge relatively early for a Chair in Australia but there in the University of Sydney, he created a branch establishment, as it were, which has contributed much and added many distinguished workers to our field.

The War, and my position in the Ministry of Home Security, provided marvellous scope for early plastic design, but it did more for me, it showed what could be done with the backing of brilliant colleagues. I did my recruiting during the phoney-war period, when most people were confusedly marking time, so I was able to collect a most outstanding team, many of them architects. I found that with such colleagues the most one had to do was to state the problem, preside over the occasional design conference and then modestly accept the credit for success. I wish industry to-day could absorb this lesson. I understand that, particularly in manufacturing industry, the greatest mistake one can make is to be brighter than the boss. Let me impress upon you that this nation of ours will only recover its strength, its wealth, when you, in the university engineering departments, are able to attract a far greater proportion of brilliant pupils who, after their undergraduate education, you will pass to an eager industry.

With the War over, and my move to Cambridge, this pattern of brilliant colleagues repeated itself, though they were younger and needed preliminary training. This is where Michael Horne joins the scene in 1945. To my annoyance I cannot remember vividly our first meeting, though I have no doubt it was heralded by a penetrating gusty laugh, pealing up the dignified staircase which led to my room in Scroope House - a laugh which I am proud to say rang in my ears for 15 fruitful years. This lapse of memory is frustrating as I can remember clearly another meeting at just this time. This was with Bernard Neal, later to be a Professor at Swansea University College and to retire as Head of the Civil Engineering Department, Imperial College. I remember this, probably, because it was not so much me interviewing Bernard to see if he was bright enough to be admitted as a research student as Bernard interviewing me to see if I was bright enough to be his supervisor.

I must not take up too much time naming all that marvellous Cambridge team. Anyone interested will find a complete list in my Introduction to Vol II, *The Steel Skeleton*, though it carries down only to 1956. It therefore omits L. Gordon Johnson who joined us later and who, from his complete knowledge of orthodox steelwork practice, kept us straight when we embarked on the critical task of designing, for a somewhat doubting industry, a series of varied and important structures, including our own new premises in Cambridge, to demonstrate that the plastic design method was not only economical but was safe. Gordon went on to carry the message to the IDC Group and to join Ian Phillipps in the ranks of those distinguished industrialists produced by the team. We cannot include Robin Cole of Conder International as he left Cambridge before my time but he kept in such close touch with our work that he was almost a member.

I may have been living for nearly 40 years in a fool's paradise but, looking back, those pioneering days appear to me a joy - the hilarious excitement of testing full scale portal frames to collapse in the grounds of the Welding Institute at Abington, with dozens of undergraduate helpers and with someone, not Michael, laying odds on the result - the regular discussion meetings in the office, at one of which, after months of intellectual blockage, the rule for the mechanisms of collapse of a portal frame was revealed - a mass Eureka, if ever there was one - those remarkable sessions when, with Jacques Heyman and complete amity, we wrote the second volumne of *The Steel Skeleton*. At the centre of all this was the man we are honouring to-day, supported by this time by his marvellous calm wife Molly. One could see his powers developing as the months passed. The width of his intellectual interest was impressive and his industry was almost alarming. Fascinating new methods of analysis and the apparent intractability of stanchion behaviour and design alike occupied him. All problems attracted him. If one was toying with some new and possibly promising development, it was as well not to mention it if one hoped to pursue it at leisure, otherwise Michael would breeze in next morning with a complete solution or the announcement that the idea was fundamentally unsound.

Since then he has established his own distinguished band of workers and presided over a great school of Civil Engineering. Moreover he has not confined his energies to teaching and scientific work but has played a prominent part in furthering design as a member of the Merrison Committee on Box Girders and as a consultant. In addition to all this he has found the time and energy to devote to the organisation of the engineering profession, culminating in a year as President of the Institution of Structural Engineers.

This Conference, held to honour Michael Horne, will not disappoint him. The response to the call for papers has been excellent, yielding papers of high quality. Those whose work could not be accepted for lack of space will have the opportunity of joining in the discussions and making the Conference a success in strengthening the links between workers throughout the world and furthering our own understanding of the behaviour of steel structures, so leading to improvements in their design.

THE DEVELOPMENT OF PLASTIC THEORY 1936-48: SOME NOTES FOR A HISTORICAL SKETCH

J Heyman

Department of Engineering, University of Cambridge

John Baker (Professor J.F. Baker, Sir John Baker, Lord Baker of Windrush) was appointed to the Chair of Civil Engineering at Bristol University in 1933. However, he retained his post as Technical Officer to the Steel Structures Research Committee until 1936, the date of publication of that body's *Final Report* (the third), which had included 'Recommendations for Design' (1).

The Committee had been set up in 1929 to try to bring some order into the design of steel structures, and indeed one of their first acts was to patch together a 'Code of Practice' from the data that existed at that time; the Code was published in the *First Report* of 1931. This Code was meant to be a temporary expedient; however, it found favour at once with the London County Council, and it formed the basis of British Standard Specification No. 449, a Standard whose ghost, nearly sixty years later, is only now being exorcised.

Meanwhile the Committee engaged itself in a deep programme of research, both theoretical and experimental, including tests on actual buildings, and they rapidly concluded 'that the method of design inherent in the (1939) Code of Practice was almost entirely irrational and therefore incapable of refinement'. These are Baker's words (2), and they repay careful examination. The fact that a method is 'irrational', for example, does not imply that the method is necessarily unsafe. Indeed, the 'irrational' rules of the early BS 449 do, in many cases, give a safe method of design - but this is a conclusion based upon the theorems of the plastic theory, theorems not clearly articulated in 1931 (although perhaps unconsciously perceived), and not to be current for another twenty years. However this may be, it does seem that steelwork designers in the thirties and forties, and later, had an instinctive understanding of the safety of the elastic design rules of BS 449, and this perhaps helps to explain the long life of that Standard.

The Steel Structures Research Committee, however, was intent on producing a rational method of design. In this they succeeded; the *Recommendations* of 1936, based as they were on direct observation and theoretical analysis of the elastic behaviour of framed structures, gave a direct and economical method. Unfortunately there were two drawbacks. On the one hand the method was complex, or at least had not the utter simplicity of the (irrational) Code of

Practice of 1931, and on the other it was limited; as Baker remarked (3), it could deal only with the steel skeleton in its most regular form of vertical columns and horizontal beams.

In all this work the Committee showed itself aware of progress in other countries. The review of data in 1929, for example, presented information from the U.S., France, Germany, Spain and Belgium, together with summaries of Bye-Laws current in the Dominions (Australia, New Zealand, Canada and South Africa), and similarly the research papers in the three *Reports* do not neglect reference to foreign work. However, one finds what one seeks; the references are to elastic experiments, to elastic calculations, and *a fortiori* to elastic methods of design. There seems to be no hint in the work of the Steel Structures Research Committee that anything other than elastic methods could be used.

But work *had* been done on the inelastic behaviour of steel structures, and this work had its first impact on Baker in about 1936, when he met H. Maier-Leibnitz. In that year the second Congress of the International Association for Bridge and Structural Engineering was held in Berlin (Baker was to play host in 1952 in Cambridge to the fourth Congress of the Association). In the Berlin Congress one whole section, of eight papers, was devoted to plasticity in steel; if one of those papers might be singled out as a 'trigger' for Baker's later work, it is that in which Maier-Leibnitz reported his tests on the collapse of continuous beams (4). And if one statement, a minimal statement, should be made about Baker's contribution to structural design, it is not only that he realized, immediately and with absolute certainty, that the way ahead in the design of structural steelwork lay in the exploitation of plastic theory, but that he had also the drive and determination to develop that theory to the point where it has achieved world-wide acceptance.

Baker was joined in Bristol in 1936 by J.W. Roderick, a move made possible by a grant from the Institute of Welding, and the first early British experiments were made on the plastic collapse of steel frames. Small portal frames confirmed and extended the findings of Maier-Leibnitz that collapse loads were related only to the full plastic moments of the members of such simple framed structures, and that imperfections of one sort or another, which could alter markedly elastic behaviour, had little effect on ultimate strength.

This work went slowly, and was interrupted by World War II. However, the war gave rise to perhaps the most spectacular example of the application of plastic theory, certainly if measured by the weight of steel used - the Morrison shelter absorbed over a quarter of a million tonnes of steel (3,5). The intellectual basis of this design was both simple and elegant; the shelter, of the shape and size of a dining table under which the family could sleep, was required to squash down by no more than 12 inches if the house collapsed. The energy released in this collapse (design value 142,000 in lb) was absorbed in the rotation of a pair of plastic hinges, and the full plastic moment required was thus immediately calculable.

Towards the end of the war, in 1943, Baker was appointed to the Chair in Cambridge; Roderick arrived soon after, and from 1944 onwards Baker started building the team which helped him establish, once and for all, plastic theory as the basic tool of steelwork design. The team was large but continuously changing; some members stayed for two or three years and then moved on, while others remained as members of staff. Michael Rex Horne joined the team in 1945, and stayed for fifteen years before moving to Manchester. Baker was very much in control of the whole programme, and Roderick provided the firm discipline necessary for the day-to-day prosecution of the research.

A picture of the state of knowledge at about this time, that is, of the climate in which work was done in 1945-48, is given in the paper (6) by Roderick and Phillipps published in 1949. Theory and experiments are reported for the carrying capacity of simply-supported mild steel beams, and fifty-four references known to the Cambridge team are listed. Some of these stem from the 1936 Congress in Berlin, either directly or as secondary references; interestingly, W. Prager's 1933 paper (7) on the yield limit (written when he was 30) is noted, together with his 1948 survey (8) of stress-strain laws. The Cambridge work was built on the Bristol portal-frame tests, reported in 1938 (9), and was concerned on the one hand with further tests on beams and portals, and on the other with attempts to grapple with the problem of column design, culminating at this stage with the 1949 Royal Society paper by Baker, Horne and Roderick (10).

There is evident in this work of the 1940's a failure to uncouple problems in 'strength of materials' from those of 'theory of structures'. That is, it was known from Maier-Leibnitz (and in fact earlier) that, for the structure viewed as a whole, the collapse loads were insensitive both to practical imperfections and to the theoretical assumptions necessarily made to obtain elastic solutions. But it was not seen, at that time, how this information could be used to enrich and widen a more general theory of structures; indeed, the work was described within the limited (albeit widely limited) title of 'plastic theory of steel frames'. The investigations centred, therefore, on detailed local investigations: the formation of plastic hinges, the spread of plastic zones, the effect of strain hardening, and so on.

Thus the paper by Roderick and Phillipps was concerned to predict theoretically the load/deflexion curve for a simply-supported beam, and to correlate the predictions with experiment; for an accurate prediction, it was essential to use both an 'upper' and a 'lower' yield stress for the steel. Now both these are real quantities, but it passed almost without comment that the correct collapse load for the beam was always predicted, whatever value was taken for the upper yield stress. There was a curious failure to really grasp that it was the value of the full plastic moment that was of essential interest, and not an investigation of the partially plastic state.

Some central guiding precepts were lacking, and when the first complete experimental analysis of the single-bay rectangular portal frame (11) was published in 1950, the accompanying theory had been done on a mechanical basis, without the guidance of any general principles. The three basic modes (sidesway, beam collapse, and the combined mode) had been identified correctly earlier, but the analysis was done by statics (by drawing bending-moment diagrams) and had at most been confirmed by a 'real' work balance. The more sophisticated and illuminating concept of virtual work had not yet found its way into such analyses. All this was to change, slowly, after 1948; in that year Feinberg published (in Russian) the basic principles that were needed (12), but it was some little while before more than a handful of those in Europe and the U.S. knew of these.

In the mean time the simpler building frames, portals for example, whether rectangular or pitched, were in fact proving quite easy to analyse by statics, and by 1948 Baker and the Cambridge team felt ready to engage in practical design. It was in 1948 that a clause was inserted in BS 449 permitting the use of plastic methods; no guidance was given as to how the plastic method should be applied, but the value of 2 was specified for the collapse load factor. Once the clause had been inserted, plastic design was officially under way. Unofficially, the Cambridge team had had a trial launch, and had designed in 1947 a new pitched-roof laboratory block for the British Welding Research Association at Abington (and indeed the Welding Institute, as it now is, is a showcase for many examples of plastic steel design). But 1947 was a year early, and to satisfy the Local Authority calculations had to be submitted. Who better than Michael Horne to provide elastic calculations proving the safety of a structure that had been launched on a plastic slipway?

Baker had, indeed, in only twelve years, brought the work by 1948 to the point where it could be used in practice. Progress may have seemed slow, and it was interrupted by World War II, but it was an extraordinary achievement to have moved from an idea glimpsed in 1936 to an officially-permitted method in 1948. It is not to be-little this achievement to say that it was not yet fully understood by its progenitors, nor that others, some in the past and some contemporary, but unknown to the Cambridge team, had also obtained insights into the implications of plastic theory. On the formation of hinges, for example, Roderick and Phillipps knew of the work of Robertson and Cook (13) of 1913, and of Ewing (14) of 1899, but not of the exhaustive treatment by Saint-Venant (15) in 1864; this line of work can be traced back through Mariotte (16) in 1686 to Galileo (17) in 1638. Similarly, on the problem of the strength of beams, Mariotte knew that a fixed-ended beam under a transverse point load had twice the strength of a corresponding simply-supported beam; it was Navier (15) in 1826 who first clearly adulterated the pure spring of collapse analysis of structures with a polluting elastic theory. Even so, Kazinczy (18) in 1914, in Hungary, knew of and applied to design the plastic properties of fixed-ended beams.

But the plastic designer of 1948 (that is, a member of the Cambridge team, for plastic design had not yet spread far) was not, if he was honest, completely confident. It is true that he could design a grillage of intersecting plate girders by designing each span in turn, arranging each to collapse with a hinge at the centre and the two ends; the grillage of 57 redundancies (and this was before the days of computers) could be treated as a set of fixed-ended beams. But did the designer sometimes wake in the middle of the night and wonder whether there might be some other, more complex, pattern of hinges, which would lead to collapse at a load lower than he had calculated? Had he thought of every possible collapse mechanism?

There was, in fact, no 'lower-bound', no 'safe' theorem that was known to the Cambridge team, which would give the designer some comfort; the work could not yet be viewed against the wider background of a more general theory of plasticity. At this time, in 1947, Baker first met Prager, and each realized that there was much to be gained by an interchange of ideas. Baker had amassed immense experience of the plastic design and behaviour of steel frames; Prager knew of the mathematical theory of plasticity. There was concluded at once an arrangement to exchange post-doctoral members of the teams at Cambridge and at Brown University. Bernard Neal went to the U.S. in 1948 for a year, followed by the author in 1949 (and again in 1956); Paul Symonds came to Cambridge in 1949, and stayed for two years on the first of several visits. Baker lectured in the U.S. in 1949 on plastic design methods, and had a poor reception. However, the steel industry thought it just worthwhile to send Lynn Beedle to Cambridge from Lehigh, and he returned to build up a large research team and to prosecute wholeheartedly the practice of plastic design in the U.S. These visits were followed by many others on both sides.

Prager knew of Feinberg's principles (12); the proofs of these were supplied in 1949 by Greenberg (19) and were applied specifically to frames by Greenberg and Prager (20) in 1951. Feinberg's 'safe' principle is the one which enables the designer to sleep at night, being relieved of the necessity of trying to think of alternative mechanisms of collapse: if, for a given set of external loads acting on a structure, a set of internal forces can be found which satisfies the conditions of equilibrium, and for which the yield conditions are not violated (e.g. bending moments have values which do not exceed the full plastic values of the sections), then the structure is safe.

Such a principle had always been assumed by designers, consciously or not, at least before the advent of elastic theory persuaded them that they were doing something wrong if they were not calculating elastic stresses and strains. The stress-strain relations, and equations of compatibility, could be disregarded in the assessment of a structure's load-carrying capacity; the structure could be designed safely on an arbitrarily-assumed equilibrium state of stress. As always, if the designer made a 'good' choice for his arbitrary equilibrium state, then he would make an economical design; the quality of a designer is measured by the reasonableness of his assumptions.

It is the safe theorem also which justifies the use of BS 449 for design. The stresses calculated to satisfy the clauses of that Standard may be elastic, but they do represent a possible state of equilibrium for the structure. Thus a design based on those stresses will certainly be one capable of carrying the corresponding loads; the calculation, unfortunately, gives no indication of the margin of safety, of the load factor for the design. There is, moreover, an unfortunate consequence of too great a familiarity with BS 449, and that is that the designer may come to believe that he is calculating the 'actual' state of the structure (whereas the tests carried out by the Steel Structures Research Committee showed that the designer's calculations bear extremely little relation to the 'real' state of a building frame).

The Joint Committee which, nearly twenty years ago, produced recommendations (21) for the design of tall steel buildings braced against wind made no pretence that they were calculating the actual forces in a frame. Instead, they were fully aware of the two essential and distinct steps in the design of a steel frame: (a) the determination of a set of structural forces in equilibrium with the given loading, and (b) the design of an individual structural element (beam, column, connexion) to carry those forces. The two steps cannot in fact always be separated, but they are logically distinct. The Joint Committee recommended that each beam should be designed in isolation on the basis of plastic theory; the bending-moment diagram for each beam is therefore obtained at the same time that the beam is designed. By contrast, the bending moments in the columns are found by considering a limited substitute frame; each column length is then checked for stability, but the checking process is kept distinct from the problem of analysis of the structure.

Contact between Cambridge and Brown University enriched work at both centres. The U.S. became aware of a whole range of problems whose solutions were needed in the practical world; the Cambridge team could now view their achievements against a wider background, and could begin to see exactly what it was they had done. And that overview showed that Baker had achieved even more than perhaps he had intended. He had set out to devise a rational design method for framed structures made of ductile steel. What he had fashioned in the so-called plastic design method was in fact a completely new tool in the hands of the designer, which could be applied to a whole range of structures, and to structures made of any material that was, by usage, thought to be proper for structural use.

This universal design method has come to be called 'simple plastic theory'. Unfortunately, the real world is not always simple. The full comfort of the plastic theorems can only be felt if the designer knows that deflexions of his structure are small, if instability does not occur (either of the structure as a whole, or locally), if the material has adequate ductility, if the loads are static and not repeated, and so on. It is the solution of problems arising from considerations such as these that make plastic theory still a live subject.

Michael Horne was never content, even in the very early days, to work on the 'simple' problems only. One of the first interests of the Steel Structures Research Committee lay not only in the steelwork design for buildings, but also in the analysis of the loads to which they were subject. Twenty years later, in the late 40's, Horne was working on the statistical analysis of floor loading, on wind loads on structures, and on the effects of variable repeated loading.

Horne's contributions to the analysis of column behaviour has already been mentioned. This involvement with problems of stability, whether of columns or of plates, of beams or of whole structures, has continued through thirty years. He tackled head-on, and with success, the vexed and difficult problem of the effects of deflexions on the design of frames, particularly the single-storey pitched-roof portal on the one hand, and the multi-storey frame on the other.

In all this work he was not content either to stay within the confines of Academe. A project would be undertaken if it might be of use to practice; and if it were to be of use to practice, then monographs must be written, British Standards drafted, and committees attended. And having written the pamphlets to help the designer, then Horne could take his stand on the other side of the fence, and himself be responsible for some major applications in the real world of the plastic theory of design.

Horne has, in fact, played a part in almost every aspect of the history of plastic theory. He was an early and leading member of Baker's team which did the fundamental research; he has designed plastic structures so that he knows the designers' problems; he has written guides for those designers. And he has been a professor, teaching the principles of structural analysis and design, and writing books for all students of structural engineering, in which plastic theory forms an integral and essential part.

References

1. Steel Structures Research Committee, First Report, 1931; Second Report, 1934; Final Report, 1936, London, H.M.S.O.
2. Baker, J.F. The steel skeleton, vol. 1, Elastic behaviour and design, Cambridge, 1954.
3. Baker, J.F., Horne, M.R. and Heyman, J. The steel skeleton, vol. 2, Plastic behaviour and design, Cambridge, 1956.
4. Maier-Leibnitz, H. Test results, their interpretation and application, Second Congress, International Association for Bridge and Structural Engineering, Berlin, 1936.
5. Lord Baker of Windrush. Enterprise versus bureaucracy, Oxford (Pergamon Press), 1978.
6. Roderick, J.W. and Phillipps, I.H. Carrying capacity of simply-supported mild steel beams, Engineering Structures (Colston Papers), London (Butterworth), 1949.
7. Prager, W. Die Fliessgrenze bei behinderter Formänderung, Forschung auf dem Gebiete des Ingenieurwesens, vol. 4, 95, 1933.
8. Prager, W. The stress-strain laws of the mathematical theory of plasticity - a survey of recent progress, J. appl. Mech., vol. 15, 226, 1948.
9. Baker, J.F. and Roderick, J.W. An experimental investigation of the strength of seven portal frames, Trans. Inst. Weld., vol. 1, 1938.
10. Baker, J.F., Horne, M.R. and Roderick, J.W. The behaviour of continuous stanchions, Proc. Roy. Soc. A, vol. 198, 1949.
11. Baker, J.F. and Heyman, J. Tests on miniature portal frames, The Structural Engineer, vol. 28, 139, 1950.
12. Feinberg, S.M. The principle of limiting stress (in Russian), Prikladnaia Mat. Mekh., vol. 12, 63, 1948.

13. Robertson, A. and Cook, G. Transition from the elastic to the plastic state in mild steel, Proc. Roy. Soc. A, vol. 88, 482, 1913.

14. Ewing, J.A. The strength of materials, Cambridge, 1899.

15. Navier, C.L.M.H. Résumé des lecons données a l'Ecole des Ponts et Chaussées . . . , Paris, 1826. Third edition, with notes ana appendices by B. de Saint-Venant, Paris, 1864.

16. Mariotte, E. Traité du mouvement des eaux, Paris, 1686.

17. Galileo. Discorsi . . . intorno a due nuove scienze . . . , Leida, 1638.

18. Kazinczy, G. Experiments with clamped girders (in Hungarian), Betonszemele, vol. 2, 68, 83, 101, 1914.

19. Greenberg, H.J. Complementary minimum principles for an elastic-plastic material, Quart. Appl. Math., vol. 7, 85, 1949.

20. Greenberg, H.J. and Prager, W. On limit design of beams and frames, Proc. Amer. Soc. Civ. Engrs., vol. 77 (separate 59), 1951.

21. Joint Committee of the Institution of Structural Engineers and the Institute of Welding, Fully rigid multi-storey welded frames, London (The Institution of Structural Engineers), 1964.

THE FUNDAMENTAL THEOREMS OF PLASTIC THEORY OF STRUCTURES
- some teaching problems

F K Kong

Department of Civil Engineering, University of Newcastle upon Tyne

T M Charlton

Formerly Department of Engineering, University of Aberdeen

Synopsis

The paper discusses some of the long-standing problems associated with the fundamental theorems of plastic theory, which are often misunderstood and misquoted. It is based on the Authors' experience of structures teaching at several universities: Aberdeen, Belfast, Cambridge, Hong Kong, Leeds, Newcastle upon Tyne and Nottingham.

The presentation of the paper is illustrated with examples of misunderstanding and misquotation, including actual examples taken (anonymously) from current textbooks and research papers. An attempt is made to trace the cause of the problem and suggest a solution.

Introduction

It is now 33 years since the Institution of Civil Engineers published Horne's paper: 'Fundamental propositions in the plastic theory of structures' (1). Today, plastic design is firmly established in practice, and the plastic theory of structures is widely taught at universities and polytechnics. However, despite the prominent position which plastic theory now occupies in the structural engineering curriculum, the fundamental theorems are still frequently misunderstood and misquoted by students, including advanced students. Such misunderstanding could have adverse long-term effects on research and practice, not only in structural engineering, but also in other disciplines such as soil mechanics, and this International Conference in commemoration of Professor Horne's retirement would seem to be a fitting occasion to discuss some of the problems and trace the cause.

1. *THE UPPER BOUND THEOREM*

The upper bound theorem can be stated in different equivalent forms. For example, Horne (2) has stated that:

> 'If, for any assumed plastic mechanism, the external work done by the loads at a positive load factor λ is equal to the internal work at the plastic hinges, then λ is either equal to or greater than the load

factor at failure.'

Baker and Heyman (3), less formally, have stated that:

> 'If the guess (of the collapse mechanism) happens to be correct, then, of course, $\lambda = \lambda_c$; otherwise, the theorem states that the value of λ will always be greater than the true value λ_c, or at best equal to λ_c.'

where λ is the load factor resulting from the analysis of the guessed mechanism and λ_c is the true load factor at collapse of the structure.

Calladine (4) has expressed the theorem in a statement oriented to application:

> 'If an estimate of the plastic load of a body is made by equating internal rate of dissipation of energy to the rate at which external forces do work in ANY postulated mechanism of deformation of the body, the estimate will be either higher, or correct.'

It is important to note that none of the above statements makes any reference to a need for the assumed mechanism to be in a state of equilibrium. Horne's statement merely says 'any assumed mechanism'; Baker and Heyman use the term 'a guessed mechanism'; Calladine, in particular, stresses the word 'any' in 'ANY postulated mechanism'. Indeed, a study of the proof of the theorem in their books will quickly confirm that the equilibrium of the assumed mechanism is not a necessary condition for the upper bound theorem to be valid.

In the 12 to 14 years since the first appearance of the books mentioned above, plastic theory has been widely taught in civil and structural engineering degree courses, and many other books, research papers and higher degree theses have been published. Regrettably, a long-standing teaching problem never seems to have been overcome, in that the upper bound theorem is still frequently misunderstood and misquoted by students, including research students who have attended lecture courses on the subject and passed the examinations. The Authors believe that if a survey is carried out among structural engineering students, it will show that a large number of them incorrectly regard equilibrium as a necessary condition for the theorem to be valid. Occasionally, confusion is not restricted to students. For example, Fig 1 shows extracts taken from current textbooks and a research thesis; these extracts directly contradict the following explicit statements by Horne and others:

Horne (2): 'It should be noted that the postulation of a plastic mechanism ...... need not necessarily imply that a bending moment distribution in equilibrium with the external loads can exist for such a mechanism.'

Baker and Heyman (3): 'It should be noted that equilibrium need not be satisfied in (an upper bound) solution from an assumed mechanism; the mechanism could, for example, have extra hinges for which no possible equilibrium distribution of bending moments could be possible.'

Calladine (4): 'Note that in general the stresses (in a postulated mechanism of deformation) will NOT be in equilibrium.'

Baker and Heyman (3) have in addition displayed the fundamental theorems with particular clarity:

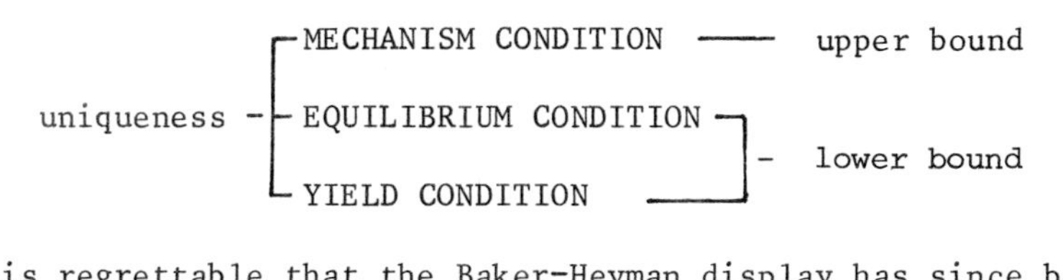

It is regrettable that the Baker-Heyman display has since been incorrectly modified by many others, Fig 1(c) being just one example. The upper bound theorem is a much more powerful tool to those who understand the irrelevancy of equilibrium than to those who do not, particularly if their activities extend beyond framed structures, e.g. to indentation problems (4). The difference between the incorrect (Fig 1) and the correct statements is therefore an important one, and should not be dismissed as 'little more than semantics'.

2. *THE LOWER BOUND THEOREM*

Horne's statement of the lower bound theorem is (2):

> 'If, at any load factor λ, it is possible to find a bending moment distribution in equilibrium with the applied loads and everywhere satisfying the yield condition, then λ is either equal to or less than the load factor at failure.'

The Authors feel that two points need to be emphasized in class teaching. First, the bending moment distribution need not be the actual distribution at the load factor λ; it must only be in equilibrium with the external loads at that load factor (3). Secondly, the word 'everywhere' needs stressing, not only in relation to the yield condition but also to the equilibrium condition, as otherwise gross mistakes might be made when the student's activities extend beyond framed structures. Consider, for example, Fig 2, which shows the stress distribution in the 'truss model' for a deep beam, proposed for a 'theoretical lower bound on the actual collapse load'. In this figure, the elements in the regions A and B are each subjected to a compressive stress in the direction of the respective member axis and to zero stress in the other two principal directions. For such a stress distribution, neither the elements in the loading region C nor those in the support region D can possibly be in a state of equilibrium (5); therefore, the value of the load calculated from such a model cannot be accepted as a lower bound on the true collapse load.

3. *THE UNIQUENESS THEOREM*

> 'If, at any load factor λ, a bending moment distribution can be found which satisfies the three conditions of equilibrium, mechanism, and yield, then that load factor is the collapse load factor λ_p.' (2)

As Horne (2) points out, the uniqueness theorem does NOT state that the bending moment distribution at collapse is itself unique. Neither does it state that the collapse mechanism is unique; indeed, specific examples of possible bending moment distributions and collapse mechanisms under the same collapse load have been given by Horne and others (2, 3, 6). In this connection, the statements in Figs 3 and 4 can be misleading when used in class-room teaching. In Fig 3, the reference to 'the resulting bending moment diagram' is likely to mislead students into thinking that the bending moment distribution is itself unique; similarly, in Fig 4, the reference to 'the correct mechanism' is likely to mislead students into thinking that the collapse mechanism is unique.

4. *CONCLUDING REMARKS*

The misunderstanding among students could have been started by the teachers' desire to teach well. In their eagerness to present an easy-to-understand account of the plastic theory of structures, they rephrased the fundamental theorems in simple terms, accepting a loss of precision as a reasonable price to pay. The loss of precision led to misunderstanding among some students; in due course the students became teachers and writers of textbooks, and the misunderstanding was further propagated.

An effective way to check this process is to insist on the students being taught the proofs of the theorems - currently, it is usual for the theorems to be merely stated but not proved, the emphasis being on the applications to the solution of particular problems. However, there is a difficulty here in that the proofs cannot be easily taught to students who do not possess a good understanding of the Principle of Virtual Work. Here probably lies the root cause of the problem: though the Principle of Virtual Work occupies a central position not only in the plastic theory of structures but in structural analysis generally (2, 3, 4, 6, 7, 8), it is seldom given the emphasis it deserves except perhaps in the more prestigious engineering schools. Thus it would seem that in the root cause of the problem lies its solution - the proper teaching of the all-important Principle of Virtual Work in the first place.

References

1. Horne, M.R. 'Fundamental propositions in the plastic theory of structures.' Journal of the Institution of Civil Engineers, vol. 34, 1950, p. 174.

2. Horne, M.R. 'Plastic theory of structures.' First Edition, Nelson, London, 1971. (Second Edition, Pergamon Press, Oxford, 1979)

3. Baker, Sir John and Heyman, J. 'Plastic design of frames, vol. 1: Fundamentals.' Cambridge University Press, 1969.

4. Calladine, C.R. 'Engineering plasticity.' Pergamon Press, Oxford, 1969.

5. Kong, F.K. and Kubik, L.A. Discussion of 'Collapse load of deep reinforced concrete beams.' Magazine of Concrete Research, vol. 29, 1977, p. 42.

6. Neal, B.G. 'The plastic methods of structural analysis.' Chapman and Hall, London, 1963. (Third Edition, 1977)

7. Charlton, T.M. 'Energy principles in applied statics.' Blackie, London, 1959.

8. Kong, F.K., Prentis, J.M. and Charlton, T.M. 'Principle of virtual work for a general deformable body - a simple proof.' The Structural Engineer, vol. 61A, 1983. (Scheduled for publication in June 1983.)

If a collapse mechanism can be found such that the associated moments satisfy

(i) the equilibrium condition, and
(ii) the mechanism condition,

the mechanism is kinematically sufficient, and the corresponding load system is greater than or equal to the true collapse load

(a) Extract from a 1977 textbook

If, in a structure subjected to loading defined by a positive load factor λ, a bending moment distribution satisfying the equilibrium and mechanism conditions can be found, then λ is greater than or equal to the collapse load factor λ_c.

(b) Extract from a 1981 textbook

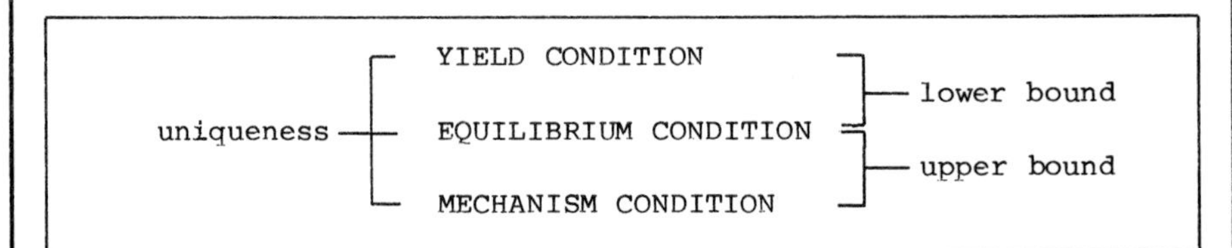

(c) Extract from a 1981 textbook

If a system of hinges can be described which transforms a structure into a mechanism, and the structure is still in equilibrium, then the corresponding load is either greater than or equal to the true collapse load of the structure.

(d) Extract from a 1979 thesis

Fig 1 Equilibrium is often (incorrectly) stated as a necessary condition for the upper bound theorem to be valid

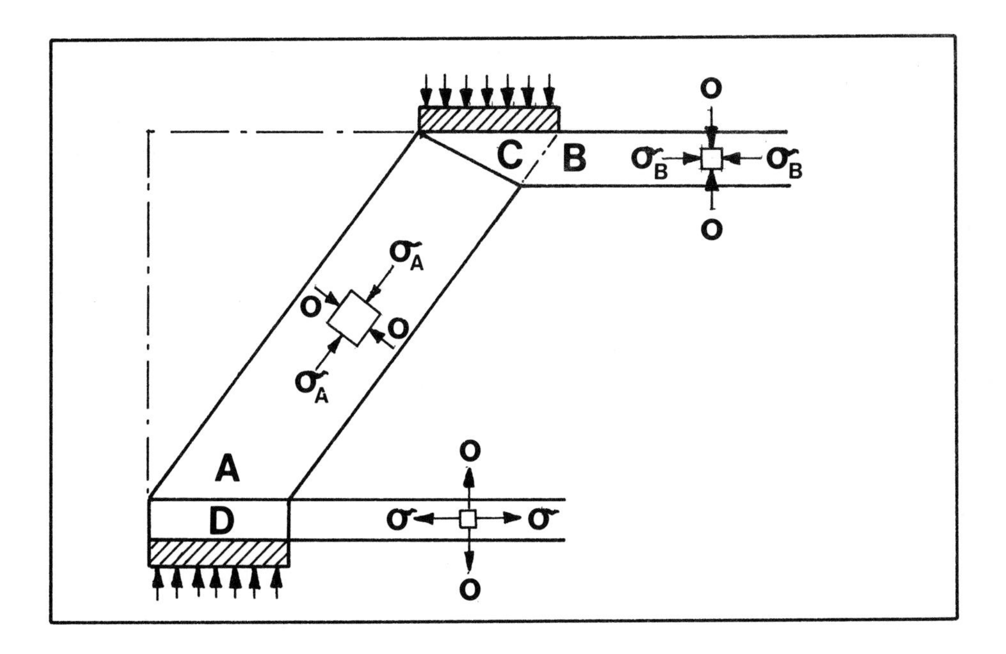

Fig 2 Extract from a recent research paper - an incorrect application of the lower bound theorem

UNIQUENESS THEOREM

If a structure is subjected to loading, defined by a positive load factor λ, such that the resulting bending moment distribution satisfies the three collapse conditions, then λ equals λ_c. It is impossible to obtain a bending moment distribution, at any other load factor, which satisfies all three conditions simultaneously.

Fig 3 Extract from a 1981 textbook

It can be proved that this is always true; i.e. of all the mechanisms that can be formed, all but the correct mechanism of collapse correspond to loads larger than the ultimate load the structure can support.

Fig 4 Extract from a 1972 textbook

SOME ASPECTS OF STRUCTURAL COMPUTING: 1943-1983

R. K. Livesley.

Department of Engineering, Cambridge University.

Synopsis

This paper is not intended as a comprehensive state-of-the-art review of computing in Structural Engineering. It is merely a personal account of some of the landmarks in the history of the subject, with special emphasis on the early work carried out at Manchester.

1. THE FIRST DECADE: 1943-1953

The first electronic computer, the ENIAC, was commissioned in 1946. Three years later the first British computer, the EDSAC, began operation at Cambridge, to be followed after a further two years by the Ferranti Mark 1 at Manchester.

These dates make a reference to 1943 seem like poetic licence. However, practical developments require theoretical foundations. That year saw the publication of a paper by Courant [1] which foreshadowed what we now know as the finite-element method, and it must have been during the same period that Gabriel Kron, an electrical engineer, wrote a paper [2] which laid the foundations of matrix methods of skeletal structural analysis. Mention should also be made here of Dantzig's development of the simplex algorithm, published a few years later. That algorithm provides structural theory with a formal solution to problems of plastic limit analysis and design.

It was on the EDSAC that Bennett [3] carried out the first structural calculations to be done on a computer - at least in this country. The importance of this work does not stem from its being a 'first performance'. Nor was it impressive in terms of the size of the problem solved, since the EDSAC had an extremely small store - smaller, indeed, than many present-day micro-computers. What made Bennett's work significant was the fact that the size limitations were imposed by computer hardware, *not* by restrictions within the algorithm used. Indeed, the matrix techniques used in that first program, based on Kron's original paper, survive with very little change in many of the structural analysis programs in use today.

The commissioning of the Ferranti Mark 1 at Manchester in 1951 provided research workers with something modern computer-users take for granted - auxiliary storage. Again, a tenth of a megabyte with an access time of 30 ms may seem small by modern standards, but it allowed the construction of a program which could handle frameworks (admittedly only plane ones) which were real engineering problems rather than mere academic exercises. The first commercial use of this program took place early in 1953 (Livesley & Charlton [4]) with the analysis of the power-station framework shown in Fig. 1.

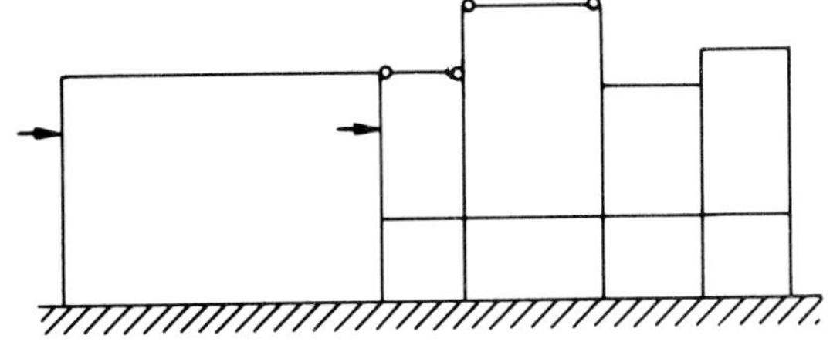

Fig. 1. One of the first frames to be analysed on a computer.

Because of the small working store of the Mark 1 a sub-structuring technique had to be used, a complete analysis taking about 15 minutes. It is perhaps of interest to note that there were significant differences between the computer analyses and manual monent-distribution calculations carried out by the Consultant Engineer. At first it seemed as though one set of results must be wrong, and each analyst naturally hoped the fault lay with the other. However, the discrepancies were eventually traced to the fact that the moment-distribution calculations ignored the axial compression of the structural members which was, of course, included automatically in the computer analysis. Thus for the first time a computer calculation provided a means of investigating the validity of the assumptions of conventional analysis.

Similar work to that described above was certainly carried out in the U.S.A. at about the same time (see Clough [5]), but it was done within industry and commercial reasons apparently hampered its publication in journals circulating in this country.

2. *THE SECOND DECADE: 1953-1963*

The ten years from 1953 to 1963 saw rapid changes in many aspects of computing. Hardware became more reliable, programming less of a black art, more programmers were trained and more programs written. It was the period of transition from experimental prototype to accepted professional tool.

2.1 Developments in hardware

In 1953 computing in Great Britain was centred on the computers at Cambridge, Manchester and the National Physical Laboratory. Ten years later there were upwards of 500 machines in this country and well over 3000 in the U.S.A. Only a small proportion of these were installed primarily for scientific or technical work - the majority were acquired for pay-roll calculations, stock control and other commercial purposes. However, the use of computers for commercial work was of indirect benefit to the scientific research worker, since it emphasised the need for reliability. It also encouraged the introduction of input/output devices designed specifically for use with computers. These were considerably faster and more convenient than the standard punched-card and telegraphic units used on the early machines.

The typical large machine of 1963 had fast and reliable input/output devices, a reasonable amount of working storage, with backing storage probably taking the form of magnetic tape. It might even have had a rudimentary graphical display. It was, however, essentially a single-user machine: time-sharing was still to come.

2.2 Developments in programming

In the early 50's the shortage of programmers was even more severe than the shortage of computers. Programming as a full-time professional activity hardly existed, all the early computer users having received their formal education in other disciplines in pre-computer days. Indeed, this variety of experience among the users contributed significantly to the creative excitement of those early years.

It is hardly necessary to say that at first all programs were local to the particular machines for which they had been written. Programming was entirely in machine code and low computing speed and small storage capacity made program efficiency essential. The small repertoire of machine-code instructions gave ample scope for ingenuity and each computer tended to have its group of local 'experts', who knew from experience all the idiosyncracies of their machine and regarded it with a proprietary affection usually reserved for boats and vintage cars. The intricacies of machine-code programming were capable of giving great delight to those programmers without a dead-line for the production of results. However, it was very quickly realised that computers not only needed to be made faster, more reliable and with larger stores: if they were ever to become a normal tool of industry they needed to be made much easier to program.

Several techniques for making programs easier to write were developed in the early 1950's. Indirect addressing, address modification, instruction labels, floating point hardware - all these reduced the labour of coding and hence the incidence of mistakes. But the real break-through came with the introduction of symbolic programming languages. The idea of writing a program in normal English and mathematical symbols, with the computer itself doing the translation into its own internal code, dates from the early 50's and perhaps from before that. Indeed, it probably occurred to Babbage himself. But it was not until computers had sufficient storage to hold the necessary compilers that the idea became a practical possibility. The Fortran programming language became available in the U.S.A. in about 1956, while similar languages known as 'Autocodes' were developed for British computers at about the same time. At first certain of the 'experts' treated these languages with some degree of scorn, regarding them as only suitable for novices trying to solve simple problems, but gradually they were adopted by beginner and expert alike.

The introduction of high-level languages not only made computers more accessible. It also made them less parochial. Program exchange became a real possibility, though to this day there are still problems when transferring any large program to a different type of machine.

2.3 Developments in programs for structural analysis

The first structural programs were concerned with the analysis of skeletal structures according to simple linear-elastic theory. With this problem 'solved', at least in principle, research workers looked for new problems to tackle. In the period under consideration developments followed four main lines.

The first of these had as its goal the analysis of really complex structures - structures with several hundreds or even thousands of joints and members. The central problem here is the efficient handling and manipulation of the stiffness or flexibility matrix of a complex structure. These matrices, although they may be large, are always sparse and can usually be made banded about the leading diagonal. Although this is a problem in 'pure' numerical analysis rather than structural theory, few numerical analysts seemed to be aware of it during the 50's and many of the early contributions came from engineers.

The second involved the development of analysis programs which used more exact (and therefore more complicated) models of structural behaviour. The early programs for frame analysis based on linear small-deflection elastic theory were extended to include the effects of axial forces, bowing, finite deformations and monotonically increasing plasticity. A group at Manchester was particularly active in this area, largely due to the enthusiasm of the late Professor Merchant [6][7]. Work also started on more difficult non-conservative structural problems including creep, plastic strain reversal and shakedown.

While most of this computing work on non-linear structural behaviour was research-oriented, a practical requirement for extensive non-linear analyses arose with the design and erection calculations for the Forth bridge. Much of the computing work for this bridge was done in Manchester (see Brotton et al. [8]) and the same program was later used for the Severn, Bosphorus and Humber bridges. A few years later the introduction of commercial television generated a requiremement for analysis programs for guyed masts. These twc types of structure are similar in the sense that the non-linearity comes from the use of flexible cables. In each case the non-linearities were dealt with by an iterative approach.

The third area of development concerned programs for the stress analysis of continua. Even the earliest aeronautical stress-analysis programs included the effect of shear-carrying panels (see Turner et al. [9]), and this period saw very rapid developments in the finite-element method. It is worth noting that at that time the method was regarded simply as a technique of stress analysis, with small pieces of material being assembled by the analyst in a quasi-physical manner, rather than as a general mathematical technique for solving elliptic partial differential equations. As with the developments in sparse-matrix techniques referred to earlier, many of the important mathematical contributions came from engineers rather than professional numerical analysts.

The fourth area of work concerned what at that time was known as 'The Plastic Theory', though it is now often given the more general title of 'limit analysis'. The first presentations of the theory occured in the late 40's and early 50's, and it was natural at that time to emphasise the superiority of the new approach over conventional elastic analysis by comparing the amounts of manual calculation required. As with elastic analysis, however, the best way of solving a limit analysis problem on a computer is not simply a programmed version of the manual approach. It was soon realised that the calculation of a plastic collapse load is essentially a problem in linear programming. Programs for collapse analysis (or for the equivalent problem of plastic minimum-weight design) were produced in the mid-50's by the present Author [10] and by Heyman & Prager [11].

2.4 The impact of computers on the Engineering Profession

The ten years after 1953 saw the gradual adoption of computers by the Structural and Civil Engineering professions.

The introduction of computers for structural calculations was most rapid in the aircraft industry, where the need to save weight in increasingly complex airframes led to a demand for more accurate prediction of stresses. The concentration of the aircraft industry into a relatively small number of large companies also helped in this development, since these organizations were of sufficient size to be able to install their own computers and train programmers from among their own staff.

The civil engineering industry also began to use computers, but at a rather slower rate. A large proportion of the calculations in civil engineering are done in the offices of consulting engineers, and at that time the majority of these firms were too small to justify either the purchase of a computer or the engagement of full-time computing staff. It was possible to obtain a certain amount of time on computers in Universities and research establishments, but the staff of such organizations were usually too busy to do more than give an outsider a brief introduction to the mysteries of programming, after which the novice was left to sink or swim. The conventionally-educated structural engineer was perhaps worse off than most in coming to terms with computers. For in addition to having to learn a lot of new computing concepts he was likely to be told that he had to adopt a new approach to structural analysis based on matrix algebra - an area of mathematics taught to few engineers in pre-computer days.

This growth of professional interest was marked by the organization of conferences devoted entirely to the subject of structural computing. The first such conference in this country was at Southampton in 1960, two years after the very successful ASCE conference of 1958. The Institutions also began to take notice of the subject, the Institution of Structural Engineers producing a report [12] and establishing a register of computer programs. The first text-books, too, appeared at about this time.

While public and professional interest was stimulated by the analysis of prestigious structures such as the Forth Bridge and the Sydney Opera House, a more mundane requirement for ordinary linear elastic frame analysis was growing within the construction industry, and computer service bureaux began to appear to meet this need. Initially these provided only programs and computing time, but gradually they acquired professional staff capable of providing the full consultancy services available today.

3. *THE LAST TWENTY YEARS: 1963-1983*

By 1960 it was no longer unusual for a computer to be used for structural calculations. The boom in the construction industry due to the development of motorways and North-Sea oil, the expansion of research in the Universities - these produced a rapid growth in computing activity, with an accompanying increase in the flow of published material.

3.1 The big program suites arrive

The early 60's saw the advent of solid-state electronics: valves were replaced by transistors, which in their turn were replaced by integrated circuits. With the spread of cheap reliable computing power came a growth in the size of computer programs. The programs of the 50's had been personal creations - programs which, if written today, would comprise only a few hundred Fortran statements. The 60's and 70's saw the birth of program suites such as ICES, NASTRAN, GENESYS and PAFEC - all of them the work of groups of programmers and involving many man-years of programming effort. The production of such a suite is a difficult management problem and the final product represents a large financial investment. However, management is a somewhat anonymous activity and no attempt will be made here to pick out individual names. The suites themselves are well documented by their producers.

3.2 Research activity: Structural Analysis

The frame analysis programs based on elastic beam theory, which had seemed such an advance in the early 50's, were clearly inadequate for the analysis of box-girder bridges and North-Sea oil rigs. Finite-element modelling provided the answer and research interest moved, to some extent, from line elements (i.e. beams and columns) to plates, shells and three-dimensional solid elements. One of the most significant advances in this area was the development, by Irons and others, of iso-parametric elements. Finite-element theory, which had begun as a technique of continuum elasticity, began to be used to model fluid flow, electromagnetic fields and non-linear materials such as soils. Continuum plasticity problems were studied both as general elastic-plastic problems and as problems in limit analysis.

The early static analysis programs were also extended to cover dynamic effects. Designers learnt through experience that both suspension bridges and television masts were susceptible to collapse from wind-excited vibrations, and the non-linear programs mentioned earlier were extended to give information about behaviour under dynamic loads (see, for example, Iwegbue & Brotton [13]). A similar extension occured in programs for the analysis of aircraft structures, while the designers of off-shore drilling platforms were conscious from the start of the need for an accurate prediction of the dynamic response of structures to random time-dependent loading.

The last twenty years has also seen continuing development in the other areas mentioned in section 2.3. The large program suites mentioned in section 3.1 require sophisticated schemes for handling sparse matrices, and mention must be made here of Irons' frontal solution method [14] for solving linear systems of algebraic equations. A great deal of work has also been done on the improvement of solution efficiency by band-width or front-width minimization (see

Millar [15]). Similar work is still required, however, to give programs for plastic limit analysis the same sort of capacity as those now available for elastic analysis.

3.3 Research activity: Structural Design

The success of computers as tools of analysis naturally led to an early consideration of their role in the overall process of design. Most of the work which has been done falls into one of the following three categories.

The first, optimum design, sees the central problem as the determination of an optimum set of design parameters and seeks to obtain values of those parameters by a formal numerical procedure. While the philosophy of this approach goes back to the work of Maxwell and Michell on the optimum layout of pin-jointed trusses, the numerical procedures of mathematical programming come from the post-war years. These procedures are more general than those of linear programming and allow optimization problems to be solved where (a) the constraints and/or merit function are non-linear, or (b) the design parameters can take only discrete values. However, the associated computations are much more complex than in the simple linear case, and the number of variables which can be handled is correspondingly smaller.

In the structural field this approach had early successes with the development of the programs for the plastic minimum-weight design of frames mentioned in section 2.3, which were essentially programs using variations of Dantzig's simplex algorithm to solve the linear programming problem. However, the introduction of inconvenient aspects of reality, such as displacement constraints and stability, into the simple plastic-theory model changes the basic mathematical problem to one of non-linear programming, with a dramatic reduction in the size of problem which can be handled. Difficulties at least as severe arise when account is taken of the limited range of steel sections available to a designer, since this transforms the optimization problem to one of integer programming.

A considerable amount of work has been done at Manchester on various aspects of optimum design during the last 20 years, largely because of Professor Horne's interest in the topic. Much of this has been concerned with conventional skeletal building frames, where the design parameters are clearly defined. Noteworthy early papers were produced by Majid & Anderson [16], Davies [17], Horne & Morris [18] and Toakley [19]. A good general account of the work done up to 1973 is given by Gallagher & Zienkiewicz [20].

Programs for optimum design have been used effectively in the design of simple structures, such as overhead gantries for railway electrification schemes, which have to be manufactured in large numbers. However, there are relatively few design problems which are sufficiently well-defined for all the decisions to be left to a computer and certainly many where it is impossible to define a quantifiable measure of merit. The advent of time-sharing and computer graphics in the mid-60's stimulated a second approach - interactive design.

Interactive design philosophy allows the search for the 'best' design to be carried out in the traditional intuitive manner. It places the human designer at the centre and sees the computer as providing an extension of the designer's powers of memory, visual imagination and analysis. The development of the light-pen in the 60's was followed by the appearance of many interactive design programs for geometrical layout in the structural, architectural and electrical fields. Such programs often incorporate optimization routines, but these operate under the control of the program user. For example, in large-scale finite-element analyses the most efficient way of producing a satisfactory well-graded mesh is for the computer to generate a first approximation which is then modified by the user on an interactive graphical display.

Just as the advent of the light-pen brought about an interest in interactive design, so the development of high-resolution plotters and fast printers encouraged a third approach, in which the computer is given a clerical role, carrying out the relatively mundane operations associated with conventional detailed design. Many programs now exist for the detailing of structural steelwork and reinforced concrete, the production of bills of quantities, bar-bending schedules, etc. While such programs may make choices which are, in a sense, 'design decisions', these choices are usually at a relatively low level - for example, ensuring that two reinforcing rods do not interfere, or checking the requirements of a standard code of practice.

For a general introduction to the philosophy of interactive design see Leckie et al. [21]. Recent work at Manchester on structural detailing has been described by Mills and Brotton [22].

The organization of a general computer-based system for structural design poses many non-trivial questions. For example, what is the best way to store the information defining the structure being designed so that modifications can be made subsequently by the designer in a natural and convenient way? How can this information be converted to data suitable for input to an analysis program? How are the results of an analysis best presented to the designer? How should designs be archived? These questions lead the programmer into areas distinct from structural engineering - computer graphics, data base design and management, the syntax of command languages - even the production of tapes for numerically-controlled machine tools.

3.4 The continuing need for special-purpose programs

A large proportion of the computing needs of structural designers are now met by the general-purpose program suites mentioned in section 3.1. However, there are still structures with unusual features, or unusual design requirements, for which special-purpose programs are needed. An interesting example of such a structure is the steerable radio-telescope.

The supporting structure of a radio-telescope has to maintain the parabolic shape of the primary reflecting surface for all positions of the telescope axis. The accuracy with which the shape must be maintained increases as the wavelength of the signals which are to be observed decreases, and in recent years the requirements for observations at ever shorter wavelengths has reached the point at which the deformation of the telescope structure under its own weight becomes an important design constraint.

For small dishes a feasible solution to the problem is to make the structure so stiff that deformations are negligible. For dishes above about 10m this solution becomes impractical, and in a number of recent designs the principle of *homology* has been used. This approach seeks to make the self-weight deformation such that the shape of the dish remains parabolic (though changes in focal length and axis are permissible). In practice this design goal cannot be realised completely and the designer must be content with minimising some measure (the r.m.s. average, for example) of the deviations from true parabolic form which occur under the various operating conditions.

The central component of this optimizing procedure is the elastic deformation analysis of the telescope dish. This structure usually consists of a central 'core', which connects it to the elevation mounting points, surrounded by radial ribs. The central core has, at most, simple reflective symmetry, but the the rest of the dish has a high degree of cyclic symmetry and it is important to take advantage of this feature in the analysis.

The homology principle has been used in the design of the British millimetre-wave radio-telescope, which is to be installed in Hawaii. The complexitiy of the design, and the number of design parameters, is such that a fully-automatic optimum design program is impractical. Interactive design is possible, however, and a program has been written at Cambridge by Smithers [23] which runs on the SERC interactive computing facilty. In an interactive program it is necessary for the designer to receive information about the extent to which a trial design fails to satisfy the homology criterion, and the graphical presentation of a clear picture of the discrepancies is seen as an important feature of the program.

4. *IN RETROSPECT*

Twenty years ago it looked as though the future of structural computing would be dominated by large multi-access machines and programs written by experts - a future in which designers might merely be taught how to operate terminals. It was not foreseen that solid-state technology would make computing available to all. Although the physical size (and degree of reliability) of the computer may be different, the schoolboy learning to use a present-day micro-computer still shares something of the excitement and frustration of computing in its early days.

References

1. Courant, R. 'Variational methods for the solution of problems of equilibrium and vibration.' Bull. Am. Math. Soc., 49, p 1, 1943.

2. Kron, G. 'Tensorial analysis and equivalent circuits of elastic structures.' J. Franklin Inst., 238, p 400, 1944.

3. Bennett, J.M. Ph.D. Thesis, Cambridge University, 1952.

4. Livesley, R.K. and Charlton, T.M. 'The use of a digital computer with particular reference to the analysis of structures.' Trans. N.E. Cst. Inst. Engrs. & Shipbldrs., 71, p 76, 1955.

5. Clough, R.W. 'Use of modern computers in structural engineering.' Proc. ASCE, 84, ST3, p 1636, 1958.

6. Merchant, W. 'Critical loads of tall building frames.' The Struct. Engr., 33, p 84, 1955.

7. Merchant, W. and Brotton, D.M. 'A generalised method of analysis of elastic plane frames.' 7th Congress, Int. Assoc. for Bridge and Struct. Engng., August 1964.

8. Brotton, D.M., Williamson, N.W. and Millar, M.A. 'Solution of suspension bridge problems by digital computer.' The Struct. Engr., 41, p 121 and p 213, 1963.

9. Turner, M.J., Clough, R.W., Martin, H.C. and Topp, L.J. 'Stiffness and deflexion analysis of complex structures.' J. Aero. Sci., 23, p 805, 1956.

10. Livesley, R.K. 'The automatic design of structural frames.' Quart. J. Mech. & Appl. Maths, 9, p 257, 1956.

11. Heyman, J. and Prager, W. 'Automatic minimum-weight design of steel frames.' J. Franklin Inst., 266, p 339, 1958.

12. 'The use of digital computers in structural engineering.' Inst. Struct. Engrs., London, 1962, rev. 1967.

13. Iwegbue, I.E. and Brotton, D.M. 'A numerical integration procedure for computing the flutter speeds of suspension bridges in erection conditions.' Proc. Inst. Civ. Engrs., 63, p 785, 1977.

14. Irons, B.M. 'A frontal solution program.' Int. J. Num. Meth. Eng., 2, p 5, 1970.

15. Millar, M.A. Ph.D. Thesis, Manchester University, 1974.

16. Majid, K.I. and Anderson, D. 'Elastic-plastic design of sway frames by computer.' Proc. Inst. Civ. Engrs., 41, p 705, 1969.

17. Davies, J.M. 'Variable repeated loading and the plastic design of structures.' The Struct. Engr., 48, p 181, 1970.

18. Horne, M.R. and Morris, L.J. 'Optimum design of multi-storey rigid frames.' Chapter 14 of Ref. 20.

19. Toakley, A.R. 'Optimum design using available sections.' Proc. ASCE, 94, ST5, p 1219, 1968.

20. Gallagher, R.H. and Zienkiewicz, O.C. (Ed.) 'Optimum Structural Design.' Wiley, 1973.

21. Leckie, F.A., Butlin, G.A. and Platts, M.J. 'The use of interactive graphics in engineering design.' Chapter 17 of Ref. 20.

22. Mills, P. and Brotton, D.M. 'Computer-aided detailing of reinforced concrete structures.' The Struct. Engr., 57A, p 19, 1979.

23. Smithers, T. Ph.D. Thesis, Cambridge University, 1981.

THE SAFETY OF STRUCTURES

D I Blockley

Department of Civil Engineering, University of Bristol

Synopsis

Limit principles are central to plastic theory. In the past they have perhaps been somewhat submerged in the apparent accuracy of elastic theory. The interaction diagram, which shows limits of plastic collapse, is used to demonstrate the central principle of limit state design.

The fundamental problems of reliability theory and the probability measure are discussed and it is suggested that a concept of responsibility should replace that of reliability in modern technology. Finally techniques of computer based artificial intelligence for the management of structural safety are very briefly outlined.

Introduction

"In a multi storey building every floor could be packed solid with people while a gale blew and the roof was piled with frozen snow, but the probability of this happening is so negligible that no engineer would think of assuming it as the working load condition. In catering for safety in such a case the engineer uses either his own or, when using a code of practice, other people's judgement in neglecting certain loading conditions which are theoretically possible. A discussion of the problem in terms of the theory of probability is, however, capable of leading to useful conclusions."

This quotation comes not from the introduction to some recent treatise on structural safety but from a classic treatise on plastic theory, The Steel Skeleton. It demonstrates that the authors, and in particular Michael Horne, had an early interest in structural safety. It is often forgotten that there are at least three stages in the analysis of a structure; the analysis of the loads; the analysis of the response of the structure to those loads, and the analysis of the acceptability or otherwise of that response. Clearly the response analysis is crucial, quite satisfactory structures can be built by making crude assumptions about loads and safety factors, whereas an incorrect understanding of response can quickly lead to disasters. Michael Horne has been involved with the development of powerful methods of response analysis but he has conjectured from time to time about the use of probability theory in the analysis of loads and safety (1, 2, 3). It is perhaps significant that his early interest did not blossom into a large scale research effort.

1. *CURRENT METHODS*

No-one will disagree that it is necessary but not sufficient that a structure is safe. This is because the word safe is vague and imprecise. As soon as an attempt is made to be more precise, possible strategies have to be listed and there is then inevitable disagreement. Should we use permissible stresses, ultimate loads or limit states? Clearly any assessment of safety is based on descriptions of behaviour using theoretical models (occasionally physical models) which vary in quality. What is more, the other categories of hazard, such as external random hazards (fires, floods, explosions, vehicle impact) and human error must not be forgotton. The latter for example is extraordinarily difficult to quantify and is normally dealt with by trying to ensure proper professional conduct. We must note however, that many structural failures are actually due to human error. It is argued in this paper that individual and collective responsibility is a central concept for professional conduct.

2. *SYSTEM AND PARAMETER UNCERTAINTY*

It is necessary for the discussion which follows to classify uncertainty in structural calculations into two types. System uncertainty is that which is due to the lack of dependability of a theoretical model when used to describe the behaviour of a proposed structure, assuming a precisely defined set of parameters. Parameter uncertainty is that which is due to the lack of dependability of parameter values, assuming the model is precise.

Thus, for example, the uncertainty associated with a prediction of the deflection of a beam is the uncertainty of the parameter values W, E, I, ℓ and the modelling of the actual problem with some simple one-to-one relationship such as $\delta = W\ell^3/KEI$. System uncertainty arised from two sources. Firstly, the theoretical model may be a good one when tested under precisely controlled laboratory conditions. However, the conditions in the actual built structure are different and the effects of these differences on the theoretical results are not precisely known or even understood. Secondly the model, even when tested in laboratory conditions, may not be a good one. The effects of differences between the actual structure and the model may be even less well understood than the first case.

3. *PLASTIC THEORY*

Elastic methods of analysis can lead, in the unwary, to a false impression of exactness and precision. There is no concept of limiting conditions or upper and lower bounds to solutions to problems. Deterministic relationships and specified parameter values are used to produce single answer solutions. Of course most designers know that there is uncertainty in the estimates used and that a lot of engineering judgement must be used in interpreting the figures, but nevertheless elastic methods do not lead naturally to the concept of bounded solutions.

The fundamental theorems of plastic collapse, however do bring our attention back to this important idea. The Safe Theorem shows why elastic solutions are safe whilst the Unsafe Theorem enables upper bounds to be calculated. If all of the bounds can be included in a solution then the region of all possible safe solutions can be identified in an interaction diagram. For example Fig. 2 (Appendix) is just such a diagram for the structure of Fig. 1. As there are only two variables, the safe region is an area, whereas for three variables it is a volume and for more than three variables it is known as a hypervolume. This example illustrates how the safe region is bounded by what are actually limit state boundary conditions. For the ultimate limit state, in this example, there are the descriptions of the three plastic collapse mechanisms. The safe region is the set of all points where the strength effect is greater than the load effect. For the serviceability limit state, the bounding or limiting equations are of a similar form except they refer, in this example, to limiting all of the elastic moments to the first yield moment and the elastic deflections to some specified limiting deflection. These equations are plotted in Fig. 3 and bound the set of all serviceable loads. Clearly in a practical example all limit state equations should be plotted including those concerning the foundations.

4. *RELIABILITY THEORY*

Reliability theory is concerned with calculating the probability that the actual values of load will lie within the safe region. Intuitively many engineers seem to mistrust these techniques. It may be that they see it as being of little relevance, it may be that the language of probability is foreign to them. Nevertheless there is a growing recognition that it has something to offer, and it has been used to help decide rationally on possible partial factor values.

Reliability theory can be used to deal with parameter uncertainty but it cannot cope well with system uncertainty. It does not therefore represent an ultimate method to which we can all aspire: the reasons are profound. Certain of the parameters are the familiar intersubjective phenomena (such as deflection, strain) which may be identified clearly and measured dependably in repeatable experiments in a controlled situation. Reliability theory as currently formulated can cope well with random uncertainty in these parameter values. At the other extreme some of the parameters involve concepts which are difficult to identify clearly and measure dependably in repeatable experiments (workmanship, site control). These latter concepts and relations between them are vague imprecise or fuzzy. Thus whilst it is possible to obtain one-to-one mappings or relations (causal) for isolable and measurable parameters in a laboratory it is not possible to do it for these vague concepts. It is therefore necessary to develop a mathematics which can handle this imprecision.

If one is to use probability as a measure of safety and risk the most profound question which must be answered is this; what is the probability that the model on which the prediction is based is true? Indeed what is the probability that the model enables the deduction of true propositions? It has been shown (4) that the answers to both these questions are that the probabilities are zero. It is arguable that the difficulty here is our concept of truth. It is suggested

(5) that we must think about dependable theories rather than true theories. Of course, a true theory is a dependable theory but a dependable theory may not be true. A dependable theory will have a high truth content, which means that the theory enables the deduction of many propositions which do correspond to the facts (are true) - but not always - there will be situations in which the theory leads to propositions which contradict the facts. Perhaps the major lesson we can take from physics and the philosophy of science is that our models (and our measurements) are part of a complex hierarchy of theories all of the theories are uncertain and if not already shown to be false, it is only a matter of time before thay are falsified. The most far reaching recent example is how Einstein replaced Newton. However the latter's theory is quite dependable enough for structural engineers. Many, true propositions can be deduced from Newton's Laws but nevertheless it is a false theory and therefore has a probability of being true of zero. As engineers (and scientists) we are interested not in highly probably theories but in theories with high truth content, which are highly tested, well corroborated and highly dependable. It is important to distinguish clearly between the meaning of the word probable as used in everyday language and its meaning through the calculus of probability theory. This has been discussed elsewhere (6, 7).

The whole problem of dealing with system uncertainty is complex. I believe a more fruitful way of tackling it, as well as the problem of controlling human error, is to model it using many-to-many relations on the possibility space with hierarchically ordered and logically related propositions which restrict the possibilities in a way analagous to the interaction diagram mentioned earlier. This will be described briefly in the last section of the paper.

5. *RESPONSIBILITY*

If the engineer cannot even rely on science to provide the 'truth' the whole problem seems to become overwhelmingly difficult. Settle (8) has tried to resolve this by suggesting another criterion which is in fact adopted by engineering practice. His suggestion is a development of Popper's philosophy of trial and error (criticism). He argues that the notion of the reliability of a hypothesis should be replaced by the notion of a responsibility to act on the hypothesis. The taking of responsibility implies not that one has earned the right to be right or even nearly right, but that one has taken what precautions one can reasonably be expected to take against being wrong. The responsible engineer is not expected to be right every time but he is definitely expected never to make childish or lay mistakes.

Engineering is decision making on the basis of information of varying dependability. A decision may be viewed as a suspension of criticism for a moment. Good decision making is not a static process, however, and criticism of the consequences must therefore be continued after the moment of the decision.

Responsibility, it is argued, is a more basic concept then reliability. It points to the role of the individual engineer and his duty to work with care and diligence. It is used in law where for example under tort, the standard of care is that of a reasonable practitioner. Ultimately, the judgement of what is a reasonable decision must be made by the peer group (5) of the decision maker and their judgement in turn must depend on their values and ethics. For structural and civil engineers the major peer groups must be the Institutions, but measurements of peer group opinion must be documents such as codes of practice and other design guides.

The argument presented here seems to have degenerated from the scientific and deterministic precision of Newtonian mechanics to such vague concepts as responsibility, reasonableness and ethics. This may be so but nevertheless the argument does point out the fundamental responsibility of the individual to be diligent and careful and to take part in the development of the values of the peer groups of which he is part. It helps resolve the tension between so-called scientific precision and engineering practice; it points the way to more research on methods by which the quality of infomation can be judged and managed more effectively. It points to the concept of controlling or managing safety rather than calculating some absolute measure of safety. The ideas described in the next section are being developed for just this purpose.

6. *FUZZY INFERENCE*

In previous work using fuzzy logic (5) to analyse structural failure it has been suggested that it may be possible to distill conclusions from past successes and failures about the safety of a planned project. A fuzzy relational inference language (FRIL) is being developed and used for this purposes and has been described elsewhere (9). No more than a flavour of the method can be given here.

Imagine we have evidence from three existing and reasonably successful structures (*E1, E2, E3*), from three failures (*F1, F2, F3*) and two alternative design proposals (*D1, D2*). We wish to set up evidence in a computer using FRIL so that we may ask various important questions. In this very simple example the evidence is in the form of notional probabilities of failure and degrees of proneness to failure through human error. As this is an illustrative example two further designs will be included (*D3, D4*) which have extreme values of these measures. All the information is stored in base relations, the names of which are reasonably self explanatory if *LS* stands for Limit State, *HE* for human error, *EX* for existing structures, *F* for failure, *DS* for proposed designs. The base relations are

EX-LS	Name	-log p_f	χ
	E1	*5*	*1*
	E2	*6*	*1*
	E3	*7*	*1*
DS-LS			
	D1	*4*	*1*
	D2	*7*	*1*
	D3	*0*	*1*
	D4	*10*	*1*
F-LS	*F1*	*0*	*1*
	F2	*6*	*1*
	F3	*0*	*1*

EX-HS	Name	Proneness	χ
	E1	*0.1*	*1*
	E2	*0.2*	*1*
	E3	*0.4*	*1*
DS-HE			
	D1	*0.1*	*1*
	D2	*0.4*	*1*
	D3	*1*	*1*
	D4	*0*	*1*
F-HE	*F1*	*0.7*	*1*
	F2	*0.9*	*1*
	F3	*0.3*	*1*

The degree of proneness to failure through human error is measured on a scale 0,1 and 0 represents no proneness to failure (*D4*) and 1 represents certain failure (*D3*). χ is a fuzzy membership level.

Three so called re-write rules and four set theoretic base relations will be used in the example.

The first re-write rule expresses the fact that for a design to be safe, it should be similar to existing designs and it should not be the case that it is similar to the failures.

DS-SAFE (x) ← *EX-DS-SIM (x)* and not [*F-DS-SIM (x)*]

The relation *DS-SAFE* is a list of designs which are safe and is obtained by logically connecting the two relations on the right-hand side of the backwards arrow, which should be read as 'if'.

The relation expressing similarity between the designs and the existing structures needs also to be re-written.

EX-DS-SIM (x) ← *DS-LS (x, a)* and *EX-LS (y, b)* and

SIM-PROB (a, b)

and *DS-HE (x, c)* and *EX-HE (y, d)* and *SIM-PRON (c, d)*

This states that a relation containing a list of names of designs (*x*) which are similar to existing structures is made up of the list of names from the base relations *DS-LS* and *F-LS* which have similar notional probabilities *SIM-PROB* and names from *DS-HE* and *EX-HE* which have similar degrees of proneness to failure.

Also in like manner

F-DS-SIM (x) ← *DS-LS (x, e)* and *F-LS (y, f)* and *GT-PROB(e, f)*

and *DS-HE (x, g)* and *F-HE (h, i)* and *GT-PRON (g, i)*

The last base relations to be defined are

$$SIM\text{-}PROB\ (a, b) = [\ (a, b)\ |\ \chi = 1 - \frac{|a-b|}{10}\ ]$$

which states that *SIM-PROB* is a relation consisting of a set of points (a, b) with a fuzzy membership level of $1 - \frac{|a-b|}{10}$

$$SIM\text{-}PRON\ (a, b) = [\ (a, b)\ |\ \chi = 1\ |a-b|\ ]$$

$$GT\text{-}PROB\ (a, b) = [\ (a, b)\ |\ \chi = (1 - \frac{(a-b)}{10}) \wedge 1\ ]$$

$$GT\text{-}PRON\ (a, b) = [\ (a, b)\ |\ \chi = (1 - (b-a)) \wedge 1\ ]$$

The last two base relations express a one sided similarity so that designs are only considered to be disimilar to failures if they are somewhat safer. Once this knowledge base is set up there are different questions which may be asked, using a specially formulated WHICH query. Only the answers are given here.

1 Which Designs are safe?

WHICH [*x*, *DS-SAFE (x)*] produces

DS-SAFE	NAME	χ
	D1	*0.4*
	D2	*0.5*
	D4	*0.6*

2 How prone are safe designs to human error?

WHICH [*(x, y), DS-SAFE (x)* and *DS-HE (x, y)*]

produces

NAME	PRONENESS	χ
D1	*0.1*	*.4*
D2	*0.4*	*.5*
D4	*0*	*.6*

3 How safe in the limit state are designs which are not prone to human error and are similar to existing structures?

WHICH [*(x, y), DS-LS (x, y)* and *DS-HE (x, w)* and not *PRONE (w)* and *EX-DS-SIM (x)*]

Where not *PRONE*

w	χ
0	*1*
0.5	*0.5*
1	*0*

produces

NAME	$-\log_{10}p_f$	χ
D1	*4*	*.9*
D2	*7*	*.6*
D4	*10*	*.6*

The answers to the questions are themselves fuzzy relations, which indicate the uncertainty of the answer. FRIL is one of the developments in fuzzy expert systems. An expert system is a computer knowledge base which can respond to simple questions and can update the relations contained within the light of new knowledge (learn). The system has exciting possibilities for the future.

7. *CONCLUSION*

Absolute safety cannot be guaranteed: nor even can a measure of uncertainty such as a probability of failure be guaranteed. Instead of striving to calculate some absolute measure it is conjectured that the concept of a responsibility to act on an uncertain hypothesis should be used. The reasonableness of any decisions made is determined by the peer group. This is effectively what happens in practice and many of the tensions between researcher and practitioner may be resolved by recognising this fundamental philosophical argument. Much more research effort should be concentrated on the control of safety by a more effective management of information. Techniques of the present information technology revolution must be exploited.

References

1 Baker, J.F., Horne, M.R., Heyman, J., "The Steel Skeleton" Cambridge University Press, 1956.

2 Horne, M.R., "The Variation of Mean Floor Loads with Area", Engineering, Vol. 171, 1951.

3 CIRIA, "Structural Codes - the Rationalisation of Safety and Serviceability Factors" Proceedings of the Seminar, London, 1976.

4 Blockley, D.I., "A Probability Paradox" ASCE Eng. Mech. Div., Dec. 1980. Vol. 106, No. EM 6.

5 Blockley, D.I., "The Nature of Structural Design and Safety", Ellis Horwood, 1980.

6 Lakatos, I., "Mathematics, Science and Epistemology" Philosophical Papers No. 2, Cambridge University Press, 1978.

7 Cohen, L.J., "The Probable and the Provable", Clarendon Press Oxford, 1977.

8 Settle, T.W. "Scientists: Priests of Pseudo-Certainty or Prophets of Enquiry?", Science Forum, Vol. 2, No.3, 1969, pp. 21-4.

9 Blockley, D.I., Pilsworth B.W., Baldwin, J.F., "Structural Safety as Inferred from a Fuzzy Relational Knowledge Base", ICASP-4, Florence, June 1983.

Appendix

Consider the portal frame of Fig. 1 with two idealised loads W_C, representing a dead loading and H_C representing a sway loading. For the plastic collapse mechanisms shown, the solutions are well known. For example for mechanism *(1)*.

$$M_p = \frac{h_1}{h_1 + h_2} \left(\frac{V\ell}{8} + \frac{Hh_2}{2} \right) \qquad (1)$$

Express the loads as characteristic values x partial factors λ. For the ultimate limit state the load effect (in this case the bending moment of equation *(1)*) must be less than the strength effect $(\sigma_y \times Z_p)$, the yield stress x the plastic modulus.

If a 457 x 152 x 60 UB of steel to BS4360 Grade 43 is used and the numerical values for h_1, h_2, ℓ, V_C, H_C are substitued this inequality for mechanism *(1)* becomes:-

$$0.536\lambda_V + 0.047\lambda_H \leq 1$$

and for mechanism *(2)* $\quad 0.369\lambda_V + 0.18\lambda_H \leq 1$

and *(3)* $\quad 0.238\lambda_H \leq 1$

These inequalities can be plotted as an interaction diagram Fig. 2 and the safe values of λ are shown. Clearly for a complete example all conceivable ultimate limit states should be included in the diagram. Of course, for simple plastic collapse, the Uniqueness Theorem applies, but as discussed in the paper, it is never certain that all possible ultimate limit states can be identified.

A similar diagram can be plotted for the serviceability limit states Fig. 3. In this case the limiting inequalities are that the elastic moments should not exceed some values (first yield in this example) and that specified elastic deflections should not be exceeded (span/200 for vertical deflection, height/200 for sway deflection, in this example). Again a safe region of permissible values can be identified. For a complete design, more limitstates (e.g. geotechnical) and more variables can be included, when as stated in the paper the safe region becomes a hypervolume.

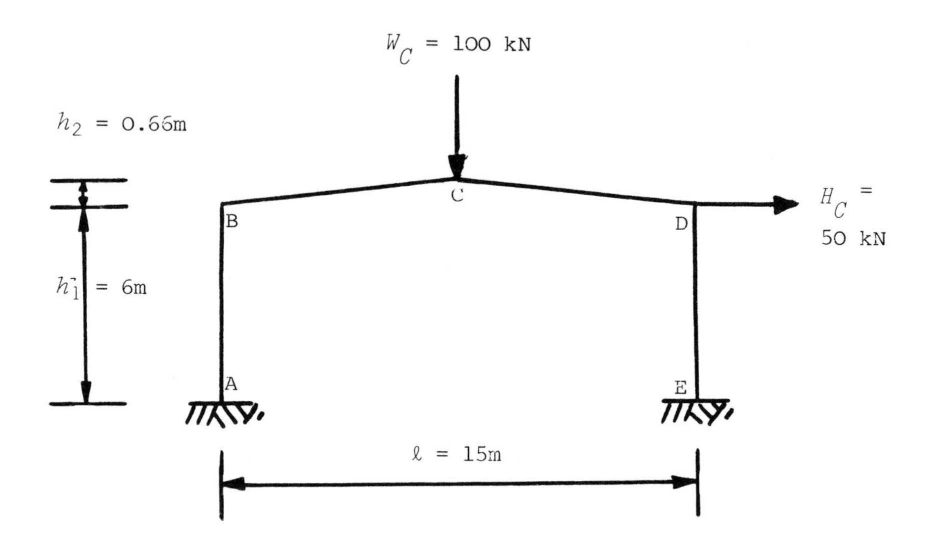

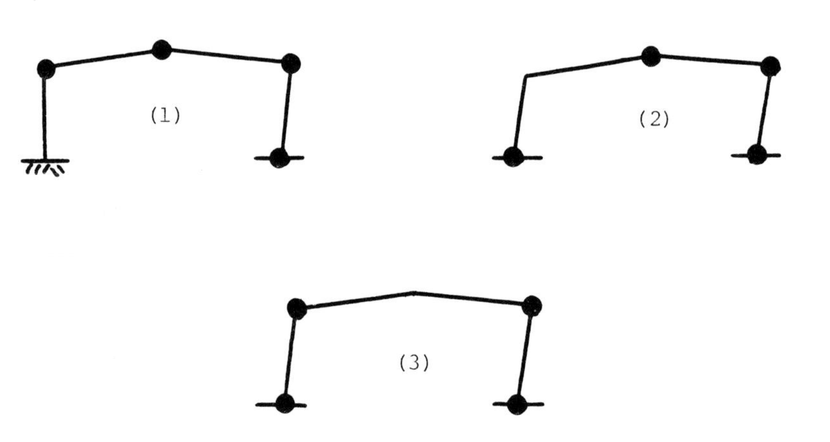

Fig. 1 Portal Frame and Collapse Mechanisms

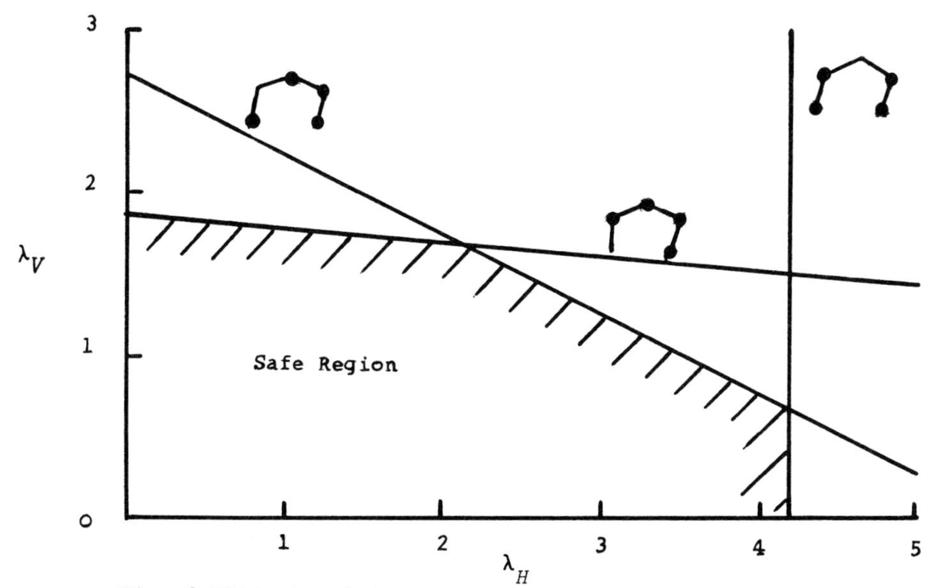

Fig. 2 Ultimate Limit State

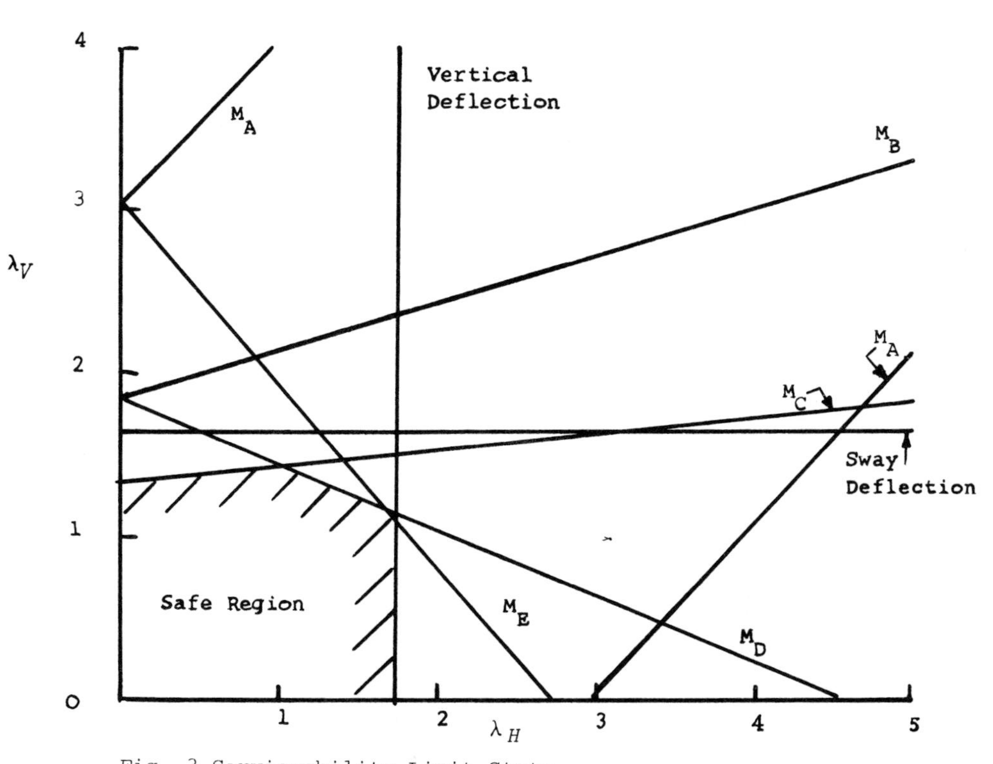

Fig. 3 Serviceability Limit State

THE PRACTICAL SIGNIFICANCE OF THE ELASTIC CRITICAL LOAD IN THE DESIGN OF FRAMES

L K STEVENS

Department of Civil Engineering University of Melbourne

Synopsis

The development of the recognition of the significance of the elastic critical load of frames is reviewed and the available methods for determining the critical condition are briefly discussed. Means by which the effects of elastic instability are considered for the allowable stress and plastic methods of design of steel and concrete frames are described. The need to consider torsional modes of instability for certain configurations is illustrated and the possible effects of flexible connections are discussed.

Notation

P	load
P_C	plastic collapse load
P_{EC}	elastic critical load
P_F	failure load
P_W	working load
P_Y	yield load
α	ratio of P_{EC} to P_W
λ_c	elastic critical load factor
λ_p	rigid plastic load factor
Δ	displacement

Introduction

The elastic critical load condition of a frame may be defined as that condition at which the elastic displacements, as predicted by small displacement theory, become undefined (1). This condition is of considerable practical significance since the approach to the critical load of a frame will almost always be associated with an increasingly rapid growth of displacements and stresses. The most significant critical load for many framed structures is the primary side-sway mode and considerable research effort has been directed towards analysis and design problems associated with this mode.

The possible importance of the elastic critical load has been recognised for many years although it was seldom of practical design significance for early framed structures which were of moderate height with substantial bracing and often with massive cladding. The P-Δ effect was identified as an amplifying factor in the design of mill frame buildings as early as 1932 by Ketchum (2) and effects of changes in geometry in special structures, such as slender arches, were analysed by re-iterative methods; that proposed by Newmark (3) being a typical and efficient approach.

The rapid development of the steel-framed office building from 1930 onwards gave an impetus to further study of the elastic critical load as a possible design condition. Increase in building height, elimination of massive cladding, architectural fashion for clear spaces unencumbered by bracing, use of higher strength steels and reduction in member sizes due to plastic design methods; all served to bring the elastic critical load into consideration as a possible design limit state.

Although there appear to be no reported failures directly traceable to the elastic instability of frames which have been designed to satisfy normal structural criteria, the author is aware of at least one building into which extra bracing was inserted after construction because of concern over P-Δ effects associated with a low elastic critical load. Several other designs are known to have been modified before final design when the effects of a low elastic critical load had been recognised. There are however many other systems, particularly in scaffolding and temporary falsework supporting structures, where inadequate bracing or low lateral stiffness has led to collapse through the development of sidesway instability originating in the elastic range.

1. *THE DETERMINATION OF THE ELASTIC CRITICAL LOAD*

Elastic critical loads may be determined by a number of "exact" and approximate methods, these being conveniently subdivided into undefined displacements, equivalent length, and energy methods.

1.1 *Undefined displacements*

If the stiffness matrix is set up to relate loads to displacements, the condition of undefined displacements occurs when the stiffness becomes zero (1). This is evidenced in the stiffness matrix by the eigen values or by the conditions when the determinant of the coefficients of the stiffness matrix becomes zero. Extraction of eigen values or evaluation of determinants is a tedious and rather impractical task by manual methods and many ingenious methods have been developed to enable manual solutions to be obtained. Such methods include substitute equivalent frames and sub-assemblage analysis while the moment distribution method has been modified for the effects of axial load through use of stability functions (1). The critical condition for modified moment distribution occurs when the moments cease to converge. This is sometimes a difficult condition to identify precisely and a more convenient approach was developed particularly for sub-assemblages in which stiffness distribution replaced moment distribution with the somewhat more readily identified zero stiffness condition providing the criterion for the critical load (4).

Other approximate methods have been proposed by which second order effects are evaluated and corresponding equivalent load systems calculated. These are then reapplied to the frame and the critical condition estimated from the amplification factor so obtained. This method has the advantage that first order linear analysis may be used without the need to use stability functions. Typical of such methods is that developed by Roberts (5) who has applied the concept for the determination of primary side sway critical loads for multi-storey, multi-bay frames.

Modern computer programmes are now available which take into account the axial load effects through a finite element approach and which extract the eigen values corresponding to all designated degrees of freedom including three dimensional behaviour. Programmes such as ANSYS (25) are highly accurate, convenient and relatively cheap to use, thus avoiding much of the tedious manual analysis previously needed for the estimation of elastic critical loads.

1.2 Equivalent length

The concept of equivalent length of a column with specified end conditions is well established for simple frames. Wood (4) extended this concept in a masterly manner through the use of sub-assemblages and stiffness distribution to cover the case of practical multi-storey frames. His three-part paper published in 1974 must be regarded as a landmark in the development of frame instability design consciousness, both for the material presented and for the discussion which it generated. Comprehensive design charts were presented which were applicable to both steel and concrete framed systems.

1.3 Energy methods

At the elastic critical load the stiffness becomes zero and small variations of displacement occur without change in the total energy of the system; the gain of strain energy stored in the members being equal to the loss of potential energy of the loads. This method is "exact" if the correct displaced shape is chosen to calculate the variations in energy, but will provide an over-estimate for other assumptions of displaced shapes.

This method has been applied by Appeltauer and Barta (6) and was further developed by Stevens (7) who used the deflected shape developed under lateral wind loading to provide an approximation to the primary side-sway mode. This lateral deflected shape was calculated by first order analysis and formed part of the usual information required for design. Estimates of the approximate upper and lower bounds were provided and it was suggested that an accuracy to within 10% of the correct value could be obtained through this simple approximate method of analysis.

The energy method has also been developed by Anderson (8) for steel frames and by McGregor (22) for concrete frames. Horne (9) simplified the approach adopted by Stevens (7) very significantly by concentrating attention on to the storey having the greatest side sway displacement, or sway index, produced by lateral loads equal to 1% of the vertical loads. The accuracy for this approach was shown to be within 20% of the correct value and was always an under-estimate.

The energy methods provide simple and readily understood physical models for determining approximate estimates of critical loads and are particularly valuable as a design tool when initial feasibility studies are being made.

2. SIGNIFICANCE OF ELASTIC CRITICAL LOAD

There are several limit states which are of practical significance in structural behaviour.

The initial yield load , P_Y ,is a reliable and usually conservative lower bound estimate of failure when based on simple first order analyses provided that second order effects are insignificant. Full second order analysis become tedious manually but are readily performed by computer (25).

The simple plastic theory collapse load, P_C , is a reliable upper bound estimate in the absence of strain hardening. It is simple to determine but efforts to predict the actual failure load P_F by taking into account second order effects in a complete analysis are impractical by manual methods and are very expensive in computer time for large frameworks.

The elastic critical load, P_{EC} , is an upper bound to collapse which may often be greatly in excess of the true values for practical structures. It is, however, a fairly simple quantity to estimate by approximate methods.

Changes of geometry effects arising from the primary displacements have been used to produce reduced estimates of collapse loads in arches by Stevens (10) and portal frames by Heyman (11) but these methods breakdown for more complex multi-storey, multi-bay frames. Stevens (12) attempted to apply this approach to the design of multi-storey frames by introducing sway displacement limitations in each storey at the design stage. The P-Δ effects were taken into account in proportioning members and amplification allowance was conservative if the displacements were everywhere less than the limiting values. The method proved to be cumbersome but introduced the concept of controlling the instability effects in the preliminary design stage of a first order analysis. Majid and Okdeh (27) have recently applied this concept in a more practical manner by designing primarily to satisfy sidesway limits and have confirmed sidesway displacements as the usually predominant criterion for many frames.

In the elastic range of behaviour it has been shown by Horne (1) that first order stresses and deflections in the primary mode are amplified, as the elastic critical load is approached, by the factor $1/(1-P/P_{EC})$. By applying this factor it is relatively simple to obtain a reasonably accurate estimate of the initial yield condition modified for second order effects provided that the buckling mode is similar in form to the displaced shape induced by the loading. This modified value of initial yield is, however, still a lower bound to the actual failure load P_F and, since the amplification factor only applies to elastic behaviour, it is not applicable for predicting post elastic behaviour as a collapse mechanism is developing.

Merchant (13) from consideration of the Rankine formula for strut behaviour and from observations of test results on frames, sought to obtain a simple relationship between the actual failure load P_F , the simple plastic theory collapse load P_C and the elastic critical load P_{EC} , the latter two quantities being upper bounds for very stocky and very slender systems respectively. He concluded that a simple interaction relationship provided the best result and postulated that the collapse load would be estimated with reasonable accuracy from the expression:

$$\frac{1}{P_F} = \frac{1}{P_C} + \frac{1}{P_{EC}} \qquad (1)$$

Horne (14) in a typically ingenious fashion, later demonstrated the validity of the expression for frames, provided that the primary buckling mode corresponded to the lateral sidesway mode under lateral loading and that all plastic hinges formed simultaneously at the collapse condition. He also concluded that the effect of strain hardening in increasing the value of P_C would normally tend to cancel the reduction associated with the other assumptions.

Since this interaction expression was first presented, much effort has been applied in testing its accuracy and general acceptance has been gained for its application in a wide range of pratical circumstances.

It must be emphasised that the expression relies on the coincidence of the side sway mode and the primary buckling mode, a requirement which was dramatically demonstrated by studies made by Lu (15) who carried out tests and analyses on symmetrical frames subjected to symmetrical vertical load systems without significant lateral loads. Results lay well above the straight line predicted by Merchant and earlier versions of the Australian steel design code (16) relied upon these results to justify neglect of second order effects if the elastic critical load was greater than three times the plastic collapse load. Donnelly and Stevens (18) analysed a range of frames by second order non linear analysis and showed that, in the presence of significant lateral load, there could be unacceptably large reductions in failure loads due to sidesway instability effects. The code provisions have now been extended to require instability effects to be considered for all frames over three storeys (17).

Many of the frames for which instability effects have been calculated, either in the elastic or non-elastic range, have been arbitrarily chosen to demonstrate a particular characteristic. Wood (4) and Donnelly and Stevens (18) pointed out that some of these examples were unrealistic in that they did not satisfy normal loading stress or deflexion criteria. In order to obtain an overall appreciation of where excessively low elastic critical loads might be encountered, Stevens (7) designed a large number of frames using approximate anaylses with normal criteria for stress and deflexion as specified in the then current British Standard for steel buildings (19). The elastic critical load was estimated by the energy method described above and compared with the allowable stress design load in the form of a ratio

$$\alpha = P_{EC}/P_W$$

A typical set of results is given in Fig. 1 where α is shown for a framing system with varying storey heights and numbers of bays designed to satisfy acceptable stress and deflection criteria. These results demonstrate that lateral deflexion limits will control for narrow frames of few bays, with stress limits under combined vertical and wind load controlling for wider frames. In this latter range, taller frames show smaller elastic critical loads than lower frames. In very wide frames of many bays, the controlling criterion is stress due to vertical load and the elastic critical load remains constant; as extra bays are added there are corresponding increases in vertical load and lateral stiffness.

It was shown that very tall and very wide frames designed without attention to overall instability would have low elastic critical loads leading to excessive amplication factors and reductions in failure loads. At the time the paper was written (1967) no guidance was available on desirable levels of P_{EC} and it was suggested that a value of ten times the design working load should be adopted. It was also suggested that a satisfactory method of ensuring that this was achieved would be to satisfy a sway deflection limit for a given percentage of the total vertical load applied as a notional lateral load.

3. *CURRENT PROVISIONS FOR DESIGN OF FRAMES*

Many of the current provisions for the design of frames make use of the Merchant interaction formula to determine the possibility of instability effects and also to make allowance for them.

For example, the draft 1982 British Steel Structures Code (20) classifies frames as either "non-sway" or "sway" frames according to their lateral displacement when subjected to a notional lateral load of 0.5% of the factored vertical load. Non-sway frames do not require further stability analyses but sway frames are required to be designed taking into account the amplification of stresses and deflexions for allowable-stress elastic design, and for the effect on plastic collapse loads when the ultimate limit state is considered. In the later case, the analysis may be made by either a full elastic-plastic sway analysis or by a simplified procedure based on Merchant's formula. If the simplified procedure, the elastic critical load factor, λ_c must be not less than 4.6 times the ultimate limit state design load and the permitted hinge locations are restricted to the beams and columns bases. In $4.6 \leqslant \lambda_c < 10$, the required rigid plastic load factor or the ultimate load must be not less than $0.9\ \lambda_c/(\lambda_c - 1)$ while for $\lambda_c \geqslant 10$ the effect of instability may be neglected.

Allowances are also made for increased stiffness due to concrete encasement of columns and for the effects of cladding on the estimated value of the elastic critical load.

Other design methods (21) including those for concrete frames (22) rely upon methods based on second order analysis, effective lengths of columns , P-Δ effects as determined by interative methods, amplified stresses, and limitations of storey drift under a notional lateral load which is a specified proportion of the vertical load. The application of the Merchant formula to the prediction of reasonably accurate failure loads of large reinforced concrete frames is not usually practical since collapse loads are not reliably predictable from simple plastic theory when significant redistribution of moments is required to produce a complete mechanism . The formula is, however, applicable for lower bound solutions provided that the uncertanties of estimation of Young's modulus for concrete are recognised.

4. *THREE DIMENSIONAL STRUCTURES AND TORSIONAL INSTABILITY*

Almost all studies of frame instability appear to have concentrated on planar systems but the possibility of instability occuring in any of the modes associated with all degrees freedom cannot be neglected (23). The torsional mode of dynamic response is well recognised in seismic design and it may also be a potential source of danger for elastic instability as the simple example in Fig. 2 illustrates. The single storey frame is assumed to have a diaphragm at the first floor level which ensures a fully braced action in that horizontal plane. The lateral and torsional restraint is provided solelyby the inner core structure of rigidly jointed beams and columns. All other connections are pinned. By applying the simple energy approach to the calculation of the elastic critical loads, the torsional mode with rotation about 0 is about 29% of the elastic critical loads for modes with displacements soley in the x or y directions. This occurs because the extreme points such as E,F,G,H are subjected to much greater movement in the torsional mode with a corresponding greater loss of potential energy, than is experienced in the sidesway mode.

This reduction would not occur if the stiffness is provided at the perimeter rather than concentrated at the core; it is therefore a matter of primary concern at the design stage to ensure that all possible modes are considered and that an appropriate configuration is adopted.

5. *FLEXIBLE CONNECTIONS*

The rigidity of a connection may be less than that of the members connected due to a deliberate design feature such as bolted end plates or through deterioration of rigidity by non-elastic action. This latter effect may occur in steel structures by local buckling or alternating loading into the non-elastic range which may produce marked reduction in the tangent modulus (24), or by cracking and crushing of concrete encasement (4).

In order to demonstrate the effect of reduction in connection rigidity the frame shown in Fig. 3 was analysed with varying values of rigidity in the sections immediately adjacent to the connections. This analysis was accomplished using the ANSYS (25) program to give the primary side sway instability loads with the results presented in Fig. 4. The reduction in elastic critical load is initially small as the connection rigidity is decreased, the reduction only becoming significant in the range which might be expected, for example, in very flexible bolted end-plate connections which may develop the full strength of the beam after considerable non-linear behaviour, or where alternating plasticity may occur. The application of a reduced elastic critical load value should therefore be considered in circumstances where the full rigidity of connections may not be achieved. It is also interesting to note that substantial local increases in joint rigidity are effective in raising the elastic critical load significantly.

6. *CONCLUSIONS*

The above review indicates, with regrettably many omissions, something of the wide range of research carried out into the effects of frame instability. Johnson (26) questioned whether, in the light of the final outcome, the work has been, or ever can be cost effective, but dismissed short term cost effectiveness in research as an inappropriate criterion. The fact that simple design techniques have been developed and that the range of potentially dangerous situations has been delineated and shown to be of limited significance to many practical structures is reward enough for the effort expended since frames can now be designed with both confidence and economy.

References

1. Horne, M. R. and Merchant W. 'The Stability of Frames.' Comm. and Intern Library. Gr. Brit. 1965.

2. Ketchum, M.S. 'Design of steel mill buildings.' McGraw Hill. N.Y. 1932.

3. Newmark, N.M. 'Numerical procedure for compulsory deflections, moments and buckling loads.' Trans. Am. Soc. Civ. Eng. Vol. 108 (1953).

4. Wood, R. H. 'Effective lengths of columns in Multi Storey Buildings, Parts I,II, and III.' Structural Engineers, July, Aug., Sept. 1974 Vol. 52.

5. Roberts, T.M. 'Second order effects and elastic critical loads of plane multi-storey embraced frames.' Structural Engineer Vo. 59A No.4 April 1981.

6. Appeltauer, J.W. and Barta, T.A. 'Critical Loads of plane frames.' Conc. and Construct. Engrng No. 59 Aug. 1964.

7. Stevens, L.K. 'Elastic stability of practical multi-storey frame.' Proc. Inst. Civ. Eng. Vol 36 Jan 1967.

8. Anderson, D. 'Simple calculation of elastic critical loads for unbraced multi-storey steel frames.' Structural Engineer Vo. 58A, No. 8, Aug. 1980.

9. Horne, M.R. 'An approximate methods for calculating the elastic critical loads of multi storey plane frames.' Structural Engineer Vol. 53, No.6, June 1975.

10. Stevens, L.K. 'Carrying capacity of mild-steel arches.' Proc. I.C.E. Vol. 6, March 1957.

11. Heyman, J. 'Plastic design of pitched roof portal frames.' Proc. I.C.E. Vol. 8, October 1957.

12. Stevens, L.K. 'Control of stability by limitation of deformations.' Proc. I.C.E. Vol. 28, July 1964.

13. Merchant, W. 'Critical loads of tall building frames.' Struct. Eng. Vol. 33, March 1955.

14. Horne, M.R. 'Elastic-plastic failure loads of plane frames.' Proc. Roy. Soc. A274, 343, 1963.

15. Lu, L.W. 'In elastic buckling of steel frames.' Proc. ASCE. ST6 Vol. 91, Dec. 1965.

16. Standards Association of Australian. SAA Steel Structures Code AS CA1 -1968.

17. Standards Association of Australia. SAA Steel Structures Code AS 1250-1981.

18. Donnelly, M.C and Stevens, L. K. 'Stability of practical multi-storey frames.' Inst. Eng. Aust. Metal Structures Conference, Sydney, November 1972.

19. British Standard 449. 'The use of structural steel in building.' BS1 London, 1959.

20. Draft British Steel Structures Code BS5950. 'The use of steelwork in building.' BSl London, 1982.

21. Structural Stability Research Council. 'Guide to stability design criteria for metal structures.' 3rd Edition, N.Y., July 1977.

22. McGregor, J.G. and Hage, S.E. 'Stability analysis and design of concrete frames.' Proc. A.S.C.E., ST 10, Vol. 103, Oct. 1977.

23. Adams, P.F. 'Stability of three dimensional building frames.' State of art report Committee 16 ASCE - 1ABSE Joint Committee on Tall Buildings, Lehigh 1972.

24. Pavlovic, M.N. and Stevens, L.K. 'The effect of prior flexural prestrain on the stability of structural steel columns'. Engineering Structures Vol. 3 No.2, April 1981.

25. ANSYS 'Engineering Analysis System Users' Manual.' Rev. 4.0, Feb. 1982, Swanson Analysis Systems, Houston, Penn., USA.

26. Johnson, R. P. 'Discussion on effective lengths of columns in multi storey buildings.' Struct. Eng., June 1975, No.6, Vol. 53, p. 237.

27. Majid, K.I. and Okdeh, S. 'Limit design of sway frames.' Struct. Eng. Vol. 60B, No.4, Dec. 1982.

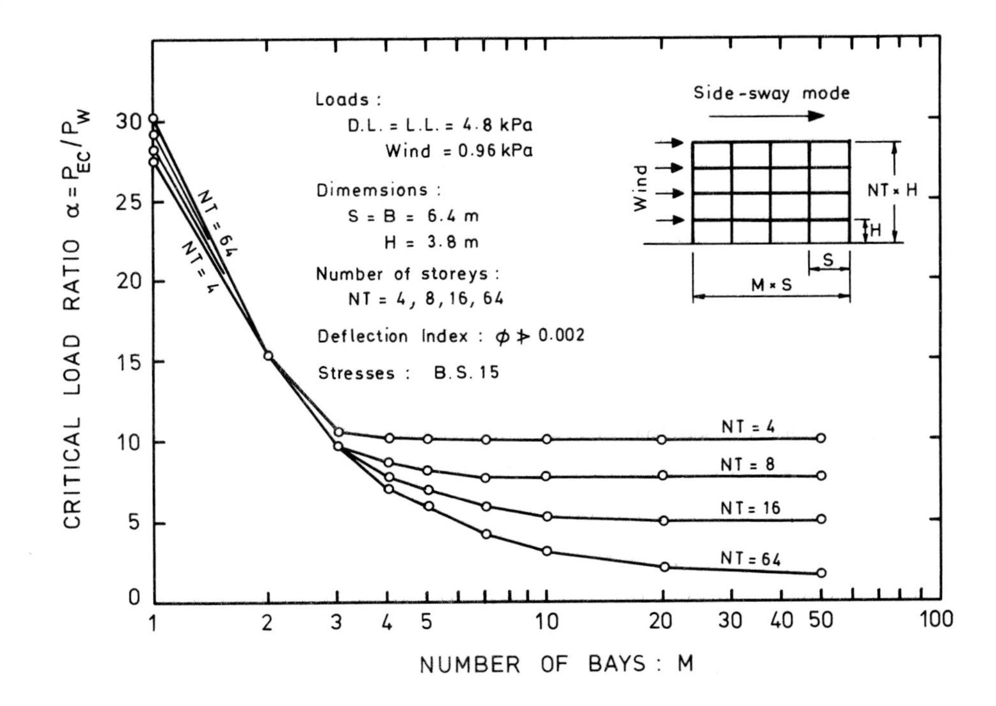

Fig 1 Variation of critical load ratio with number of bays and storeys - practical stress and deformation criteria satisfied

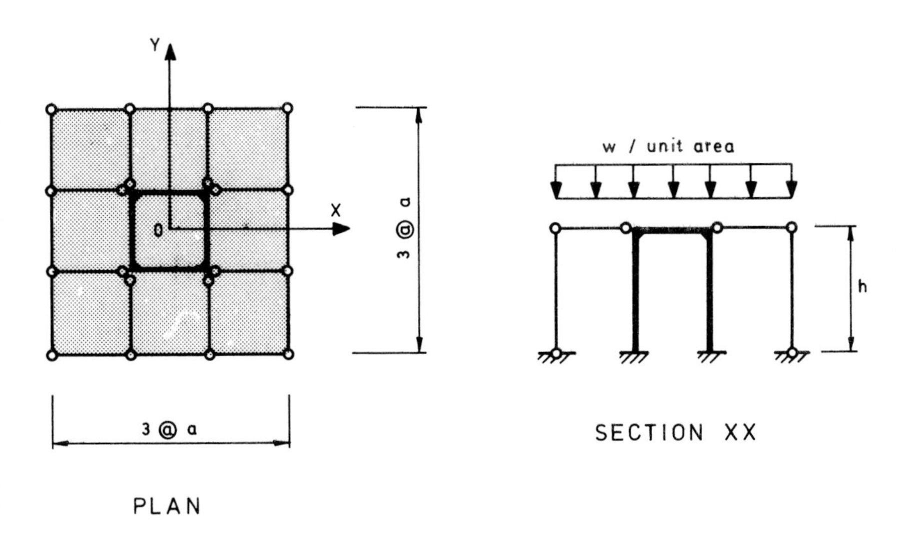

Fig 2 Three dimensional instability model

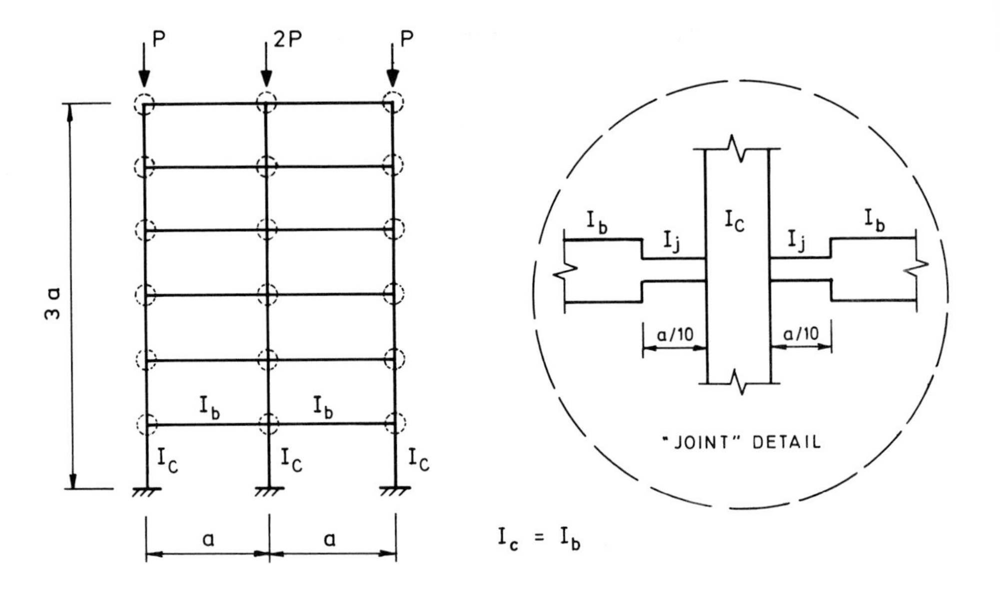

Fig 3 Model for investigating joint flexibility

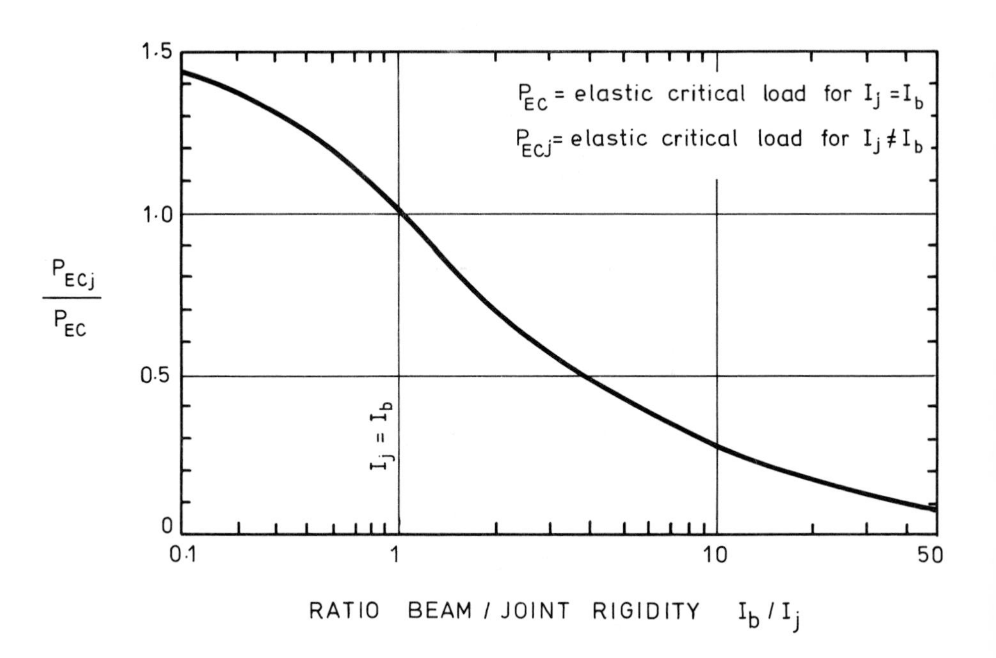

Fig 4 Effect of joint flexibility on elastic critical load

A NEW MULTI-CURVE INTERACTION VERSION OF THE MERCHANT-RANKINE APPROACH

H Scholz

Department of Civil Engineering, University of the Witwatersrand, Johannesburg

Synopsis

Simplified interaction methods to check overall stability in plastically designed sway frames have recently been adopted by various design documents. However, the use of these rules is severely limited by many restrictions. Furthermore, within their range of applicability, in many instances, inaccurate results are obtained. In addition, the incorporation of P-Delta effects other than that from lateral loadings and the recognition of residual stresses is not readily possible. It is demonstrated that these shortcomings can be avoided by using a multi-curve interaction approach.

Notation

E	Young's modulus
f_P	stress at onset of yield
f_Y	yield stress
I	second moment of area
K_C	column stiffness
L	member length
M	elastic second-order bending moments
P	axial member force
P_C	elastic buckling load
P_F	failure load
P_P	rigid-plastic collapse load
r	radius of gyration
S_B	bracing stiffness
$\bar{s}$	relative bracing stiffness
y	distance from centroid of section to extreme fibre
α	reduction factor defined in eqn (1)
λ	slenderness ratio of member in actual frame

λ_ℓ slenderness ratio of member in *limiting frame*

$\lambda_{H0,4}$ equivalent slenderness ratio related to $0,4\ P_C$

Introduction

A new multi-curve interaction method has been developed for the approximate elasto-plastic analysis of unbraced and partially-braced frameworks by Scholz (1). In essence, the method constitutes a refinement and extension of the Merchant-Rankine formula (2,3). A modified version of the latter by Wood (4) has been adopted by the European Convention for Constructional Steelwork, ECCS, (5) and appears in the draft for the new British Steel Design Code, B20, (6). The use of the modified Merchant-Rankine rule is limited in many ways, furthermore, inaccurate results are often obtained. It could be shown that these shortcomings are not applicable to the new multi-curve interaction method and that a much better correlation with rigorous solutions is achieved (1).

In this paper the new method is initially used to analyse a simple framework subjected to combined vertical and lateral loading. Subsequently, it is demonstrated that this procedure can also incorporate other P-Delta conditions, as well as the effects of partial bracing, semi-rigid joints and residual stresses.

1. FUNDAMENTAL PRINCIPLES

The fundamental modus operandi of the procedure is embodied in the interaction graph of Fig 1. To analyse a frame in accordance with these curves the factor α and the two ratios $\alpha P_C/P_P$ and $(\alpha P_C/P_P)_\ell$ need to be known, where P_P represents the rigid-plastic collapse load and P_C the elastic buckling load of the framework. The factor α is related to the P-Delta effects. Each curve of Fig 1 represents a certain family of frames characterised by the same parameter $(\alpha P_C/P_P)_\ell$. This ratio belongs to the so-called *limiting frame* of the frame family, a frame that just reaches first yielding at a critical section when subjected to the same relative loading condition as that of the actual frame but in magnitude related by the factor α to its own elastic buckling load or to its own geometry, as appropriate.

For a given frame the ratio $(\alpha P_C/P_P)_\ell$ identifies the specific failure curve from the array of possible curves of Fig 1, whereas the slope $\alpha P_C/P_P$ through the origin enables the location of the actual structure on the selected curve. With the elastic buckling load a function of the lateral frame stiffness, the factor α related to the elastic P-Delta effects and first yielding varying with load and residual stresses, it is shown that it is in principle possible to account for all these conditions in the actual frame by accepting an interaction approach between completely elastic behaviour and the unaffected structure, for which the failure load equals P_P.

The curve shape in Fig 1, which is empirical, is based on typical load related failure curves of frames with a changing slenderness. They have been compared with rigorous analytical results and laboratory tests.

1.1 Factor α and limiting slenderness ratio λ_ℓ

The factor α is equal to $1,0$ for completely symmetrical structures and loading conditions and reduces to less than *1,0* in accordance with eqn (1) for any deviation from full symmetry.

$$\alpha = \frac{0,4}{1 - 0,6\left(\frac{700 - \lambda_{H0,4}}{700}\right)^{3}} \tag{1}$$

The term $\lambda_{H0,4}$ is representative of the non-symmetrical conditions and takes the form of a slenderness ratio which becomes >0 for unsymmetrical loading conditions or geometry. It is evaluated from eqn (2) using elastic second-order moments and axial forces from loading corresponding to $0,4P_C$. In addition, f_P is set equal to $250\ N/mm^2$ and y/r equal to $1,0$. The smallest ratio $\lambda/\lambda_{H0,4}$ found for the structure is significant.

$$\lambda_{\ell} = \frac{y}{r}\,\frac{M}{I}\,\frac{L}{f_P}\left[\frac{1}{2} + \frac{1}{2}\sqrt{1 + \frac{4f_P\,PI}{(\frac{y}{r}M)^2}}\right] \tag{2}$$

For the general case, eqn (2) is evaluated twice, first for $\alpha = 0,4$ to obtain $\lambda_{H0,4}$ to calculate the final value for α from eqn (1), then for the final value for α, which will give λ_{ℓ}. For the latter computation the appropriate stress at the onset of yield and the prevailing ratio y/r need to be substituted.

The term λ_{ℓ} is defined as the *limiting slenderness ratio* of a *limiting frame*; equivalent to the *limiting column* in conventional strut design. This *limiting frame* signifies the transition from completely elastic failure to inelastic failure. Actual frames of a specific frame group and their *limiting frame* are related by eqn (3).

$$\left(\frac{\alpha P_C}{P_P}\right)_{\ell} = R\,\frac{\alpha P_C}{P_P}\,\frac{\lambda}{\lambda_{\ell}} \tag{3}$$

The factor R recognises that the reduction due to axial forces to the fully-plastic moment capacity of the section will be different for the actual frame and the *limiting frame*. The value R can easily be estimated, however, $R = 1$ will often give satisfactory results. For the structure as a whole the lowest ratio λ/λ_{ℓ} is significant.

2. *FRAME SUBJECTED TO COMBINED LOADING*

The procedure described above is now applied to the basic two-storey steel frame with the proportionate loading as shown in Fig 2. The effects of lateral bracings and settlements will be examined later. The ratio of elastic buckling load to rigid-plastic collapse load was calculated as *3,3*. The small variation of the fully-plastic moment due to axial load has been neglected for the example.

Initially, elastic second-order forces are required for horizontal loading proportionate to a vertical load equal to $0,4P_C$. These are given in Fig 3. For this analysis the approximate method by Rutenberg (7) was adopted. This technique and similar procedures substitute a modified first-order approach for a rigorous elastic second-order analysis. It has been found that such approximations are perfectly adequate for use in the multi-curve interaction concept.

The largest bending moment $(0,1011P_C L)$ is significant in finding $\lambda_{H0,4}$ from eqn (2).

$$\lambda_{H0,4} = 203$$

Equation (1) can then be solved to obtain α.

$$\alpha = 0,51$$

Subjecting the structure to horizontal and vertical loading related to αP_C will give the elastic second-order bending moments shown in Fig 4. The largest bending moment $(0,223P_C L)$ is used to obtain λ_ℓ from eqn (2). The term λ_ℓ defines the geometry of the *limiting frame*.

$$\lambda_\ell = 531$$

It is now possible to determine the ratio of elastic failure load to plastic collapse load of the *limiting frame* from eqn (3).

$$(\alpha P_C/P_P)_\ell = 0,32$$

The term $(\alpha P_C/P_P)_\ell$ is used to identify the relevant curve of Fig 1 and the ratio $\alpha P_C/P_P=1,68$ is entered in the same figure as a line through the origin. The resulting intersection point defines the magnitude of the failure load as a function of the rigid-plastic collapse load on the left vertical axis.

$$P_F/P_P = 0,79$$

3. OTHER P-DELTA EFFECTS

It is here argued that additional P-Delta effects such as due to non-uniform temperature, differential settlement, column-axial shortening or an initial-storey eccentricity can be considered in just the same way as the strength loss associated with applied horizontal loading. This is demonstrated for the frame of Fig 2 using the same horizontal and vertical loading as before but adding the influence of an imposed differential settlement of $L/100$. Elastic second-order analyses for the settlement $L/100$ and proportionate horizontal loading related to $0,4P_C$ and, subsequently, for the settlement $L/100$, proportionate horizontal loading and vertical loading related to αP_C is required. The salient results are given below:

$\alpha = 0,4$: $\rightarrow max M_{H+S} = 0,1116\ P_C L$

Equation (2) : $\lambda_{H0,4} = 212$

Equation (1) : $\alpha = 0,50 \rightarrow max M_{H+S+V} = 0,227\ P_C L$

Equation (2) : $\lambda_\ell = 540$

Equation (3) : $(\alpha P_C/P_P)_\ell = 0,31$

Figure 1 : $P_F/P_P = 0,78$

Compared with the case without settlement only a marginal additional loss in strength is predicted.

4. *SEMI-RIGID JOINTS AND PARTIAL-LATERAL BRACING*

4.1 Semi-rigid joints

The effect of semi-rigid joints can readily be incorporated in the presented method. Cheong Siat Moy (8) has demonstrated that flexible joints can be accounted for by replacing the original beam stiffness by a reduced equivalent elastic value. This will affect the elastic buckling load and the second-order elastic analyses for horizontal loading and the P-Delta effects. For the vertical load analysis the column stiffnesses should be adjusted accordingly.

Using the multi-curve interaction method for the frame of Fig 2 and allowing for an equivalent beam stiffness equal to *1/2* of that of the fully-rigid connection leads to the following results considering vertical loading and a proportionate horizontal load as in section 2. The relevant ratio P_C/P_P now equals *2,63*.

$\alpha = 0{,}4$: $\rightarrow max M_H = 0{,}115\ P_C L$

Equation (2) : $\lambda_{H0,4} = 183$

Equation (1) : $\alpha = 0{,}53 \rightarrow max M_{H+V} = 0{,}2088\ P_C L$

Equation (2) : $\lambda_\ell = 400$

Equation (3) : $(\alpha P_C/P_P)_\ell = 0{,}35$

Figure 1 : $P_F/P_P = 0{,}76$

4.2 Partial-lateral bracing

The introduction of a linear elastic lateral restraint can be used to account for lateral bracing or cladding. The presence of such a restraint will reduce the sway and thus the P-Delta effects compared with the bare frame.

The frame of Fig 2 is here analysed for vertical and proportionate horizontal loading as in section 2 allowing for a dimensionless bracing restraint of $\bar{s} = \Sigma(S_B L^2)/(\Sigma EK_C) = 2$. The revised ratio P_C/P_P equals *4,26*; other salient results are given below:

$\alpha = 0{,}4$: $\rightarrow max M_H = 0{,}0794\ P_C L$

Equation (2) : $\lambda_{H0,4} = 208$

Equation (1) : $\alpha = 0{,}51$ $max M_{H+V} = 0{,}1849\ P_C L$

Equation (2) : $\lambda_\ell = 568$

Equation (3) : $(\alpha P_C/P_P)_\ell = 0{,}38$

Figure 1 : $P_F/P_P = 0{,}84$

5. *RESIDUAL STRESSES*

Residual stresses such as from welding or cooling-down will introduce earlier yielding of the section and thus aggravate the sway losses. It is here argued that this behaviour can be reflected by the introduction of a lower stress at the onset of yield, f_P, which appears in eqn (2) for the computation of the *limiting slenderness ratio*.

Effectively, for the multi-curve interaction method this stress change will linearly modify the *limiting slenderness ratio* compared with an otherwise identical structure and leads to the selection of a lower curve in Fig 1.

Assuming a residual stress equivalent to $0,4f_Y$ for the problem examined in section 2 will yield the following results:

$$(\alpha P_C/P_P)_\ell = 0,32 \times 0,6 = 0,19$$

$$\alpha P_C/P_P = 1,68$$

$$P_F/P_P = 0,75$$

6. *FRAME SUBJECTED TO VERTICAL LOADING*

To complete the investigation of the basic frame of Fig 2 and in order to further demonstrate the simplicity of the new multi-curve interaction method, the frame of section 2 is finally analysed for pure vertical loadings.

For this completely symmetrical structure and loading the factor α is equal to $1,0$, obviating the need to evaluate eqn (1). Moreover, since a first-order elastic analysis is very close to a second-order approach for multi-storey frames, only a direct conventional plane frame analysis for vertical loading related to P_C is required to identify the smallest ratio λ/λ_ℓ for the problem.

For the rigid-plastic collapse load a beam mechanism is now applicable with the ratio $P_C/P_P = 2,35$. The relevant parameters and results of the multi-curve interaction method for this case are then as follows:

$\alpha = 1,0$: $\rightarrow max M_V = 0,143\ P_C L$

Equation (2) : $\lambda_\ell = 341$

Equation (3) : $(P_C/P_P)_\ell = 0,69$

Figure 1 : $P_F/P_P = 0,94$

7. *CONCLUSIONS*

It has been shown in principle that it is possible by adopting the recently developed multi-curve interaction method to allow for the common secondary effects encountered in multi-storey steel frames, including those resulting from partial bracing, residual stresses and the use of flexible connections. It has been argued that in each case a specific *limiting frame* can be identified and that the actual framework and its failure load can be found between a structure unaffected by secondary effects and the corresponding *limiting frame*, for which all effects and failure are completely elastic. Suitable boundary and identification parameters have been defined. The required elastic second-order analyses can readily be replaced by one of the many available modified first-order procedures.

References

1. Scholz, H. 'A new Multi-Curve Interaction Method for the Plastic Analysis and Design of Unbraced and Partially-Braced Frames.' PhD Thesis, University of the Witwatersrand, Johannesburg, 1981.

2. Merchant, W. 'The Failure Load of Rigid Jointed Frameworks as influenced by Stability.' The Structural Engineer, Vol.32,1954, pp.185-190.

3. Rankine, W.J.M. 'Useful Rules and Tables.' London, 1866

4. Wood, R.H. 'Effective Lengths of Columns in Multi-Storey Buildings.' Building Research Establishment (BRE) - Current Paper, CP 85/74, 1974.

5. ECCS 'European Recommendations for Steel Constructions.' Vol.11, Recommendations, European Convention for Constructional Steelwork (ECCS), 1976.

6. BSI 'Draft Standard Specification for the Structural use of Steelwork in Buildings.' British Standards Institution (BSI), London, 1977.

7. Rutenberg, A. 'A Direct P-Delta Analysis Using Standard Plane Frame Computer Programs.' Computers and Structures, Vol.14, No.1-2, pp.97-102, 1981.

8. Cheong Siat Moy, F. 'Consideration of Secondary Effects in Frame Design.' Journal of the Structural Division, Proceedings, ASCE, Vo.ST10, pp.2005-2019, Oct.1977.

Acknowledgement

I wish to acknowledge the partial support received from the South African Council for Scientific and Industrial Research (CSIR) and the South African Institute of Steel Construction (SAISC).

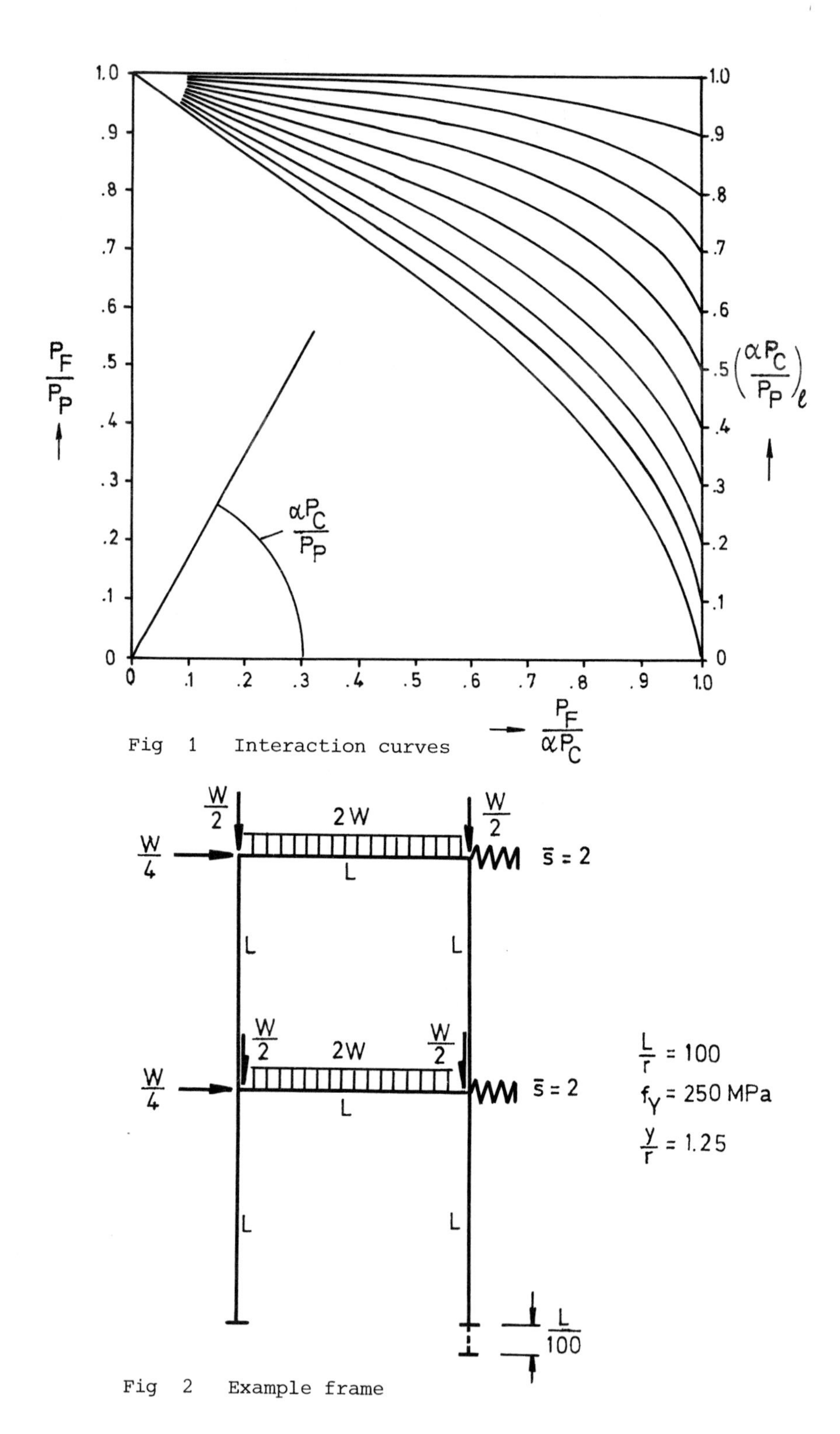

Fig 1 Interaction curves

Fig 2 Example frame

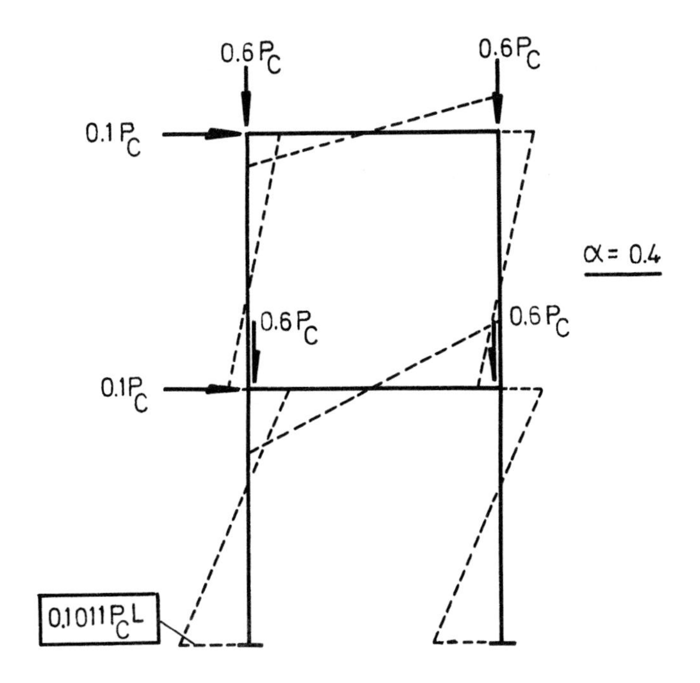

Fig 3 Elastic second-order moments due to horizontal load

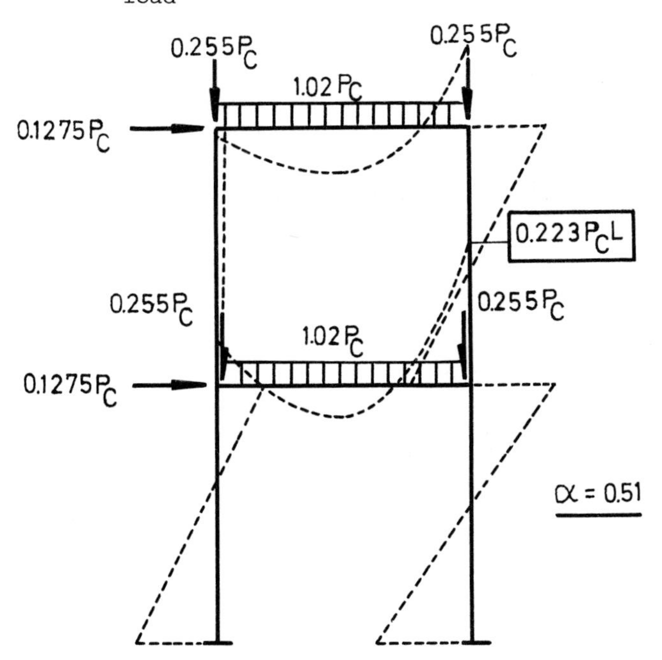

Fig 4 Elastic second-order moments due to horizontal and vertical loading

ELASTIC POST-BUCKLING RESPONSE OF PLANE FRAMES

H. B. Harrison

School of Civil and Mining Engineering, University of Sydney.

Synopsis

The elastic critical load is an important parameter in predicting a frame's load-carrying capacity. It is also important to know the nature of the post-buckling response since the influence of imperfections on strength will be less severe when the post-buckling behaviour is stable than when it is unstable.

For some classes of elements and frames, it is possible to program a discrete element method of second-order analysis which takes into account large deformations and yet can be processed by minicomputer and even microcomputer systems.

The method is explained briefly and a description given of its application to a range of circular and parabolic arches, ring and portal frames. The geometries can be determined for which there is a transition from stable to unstable post-buckling response. The introduction of imperfections into the frame geometry or its loading makes it possible to understand the equilibrium bifurcation states that may exist in an arch or portal frame.

Introduction

The elastic critical load is an important parameter in estimating the frame's ultimate load-carrying capacity. From it and the plastic failure load factor one may obtain some estimate of the load carrying capacity of a frame, making use of the Rankine type, interaction formula in the manner suggested by Horne and Merchant (1). The problem of how to compute both the elastic critical load and the plastic failure load has been haunting the structural engineer since the sixties when it became clear that linear-elastic analysis could be programmed for automatic solution by computer. What had been derided as both an irrational and time-consuming method of stress analysis became suddenly popular again when associated with the developing electronic computer.

The generality of the matrix stiffness method of frame analysis is quite remarkable and follows from the two primary assumptions upon which it is based. The first is that the material should be Hookean and the second is that the deformations should be small so that they may be neglected in formulating the equations of equilibrium and compatibility. It is unfortunate that the irrationality of the linear-elastic method also stems from these same two assumptions. If we regard the prediction of the maximum static strength of a framed structure as the primary objective of any analysis, then it is evident that while the material may sometimes be Hookean when failure is imminent, the deformations seldom will be negligibly small.

Second-order methods of frame analysis by computer require skill on the part of the analyst as the methods are rarely fool-proof. The same can be said for critical load calculations. In this paper a numerical method of second-order analysis is used which is very powerful but which lacks the generality of the first-order stiffness method. It can only be applied to restricted classes of frames. While the assumption is made that the material is Hookean, the deformations can be large. By adhering to the simple Hookean concept for the material and concentrating upon accurately allowing for deformations in both equilibrium and compatibility expressions, an efficient technique is developed that suits an interactive computer system. Having such a numerical tool permits the study of the static load response of a variety of frames, rings and arches and makes possible the study of bifurcation buckling modes, the post-buckling behaviour and the influence of imperfections.

1. *DISCRETE ELEMENT METHOD*

The method suits line structures forming a single closed loop from one support to another. It is iterative and convergence may be fast or slow depending upon the shape and line density of the resulting diagram. It might take several hours of computation to complete a typical load-deformation plot to the stage where one is certain that no significant curve branches or bifurcations have been overlooked.

The method is explained in (2). The structure is subdivided automatically into a number of straight elements (usually 50 is sufficient). One guesses the three reactions and/or displacements at one abutment and sets in motion a chain calculation across the frame to the other abutment where three errors in reactions and/or displacements will certainly exist. Since each element is very short, simple first-order flexibilities may be used to solve for deformations and stress-resultants across it. Having computed three miscloses from three initial guesses, Newtonian iteration is used to improve the guessed values in a semi-automatic manner. The role of the computer is mainly a book-keeping one and the rates-of-change of error with respect to each initial variable are evaluated numerically by successive chain analyses.

2. *ARCH BEHAVIOUR*

An arch is used to transfer load less by flexure and more by the development of thrust. Fuss predicted 200 years ago that a parabolic rib under a uniform distribution of load would be free of moment and shear. In theory one can approach the ideal of a fully stressed condition. Flexible structures carrying large axial forces are prime candidates for elastic instability and the arch rib is no exception. Unlike the simple Euler strut with uniform axial force, the rib thrust in a uniformly loaded arch will vary from a minimum at the crown to a maximum at the abutments.

Circular arches subjected to a single concentrated load were described by the author in (3). Fig. 1 shows a plot of load against central deflection for a low-rise circular arch. The critical load occurs at the initiation of asymmetric deformations and the post-buckling behaviour is clearly unstable. Figures 1 and 2 show a reduction of nearly 30 per cent in the critical bifurcation load when the area of section is changed by a factor of 10. While this is a large variation in area, one may note that published values of critical loads for arches (4) are based upon the assumption of infinite axial stiffness. Fig. 1 shows the advantages of a second-order analysis over a critical load analysis. With a falling bifurcation equilibrium line, the arch will be sensitive to imperfections. This is confirmed by the results for a off-centre load in figures 3 and 4 which also shed some light on the significance of bifurcation lines. These occur only under conditions of symmetry in both the arch and its loading.

Figures 1 to 4 show that a point on an equilibrium bifurcation line represents a state when sway can occur in one direction or the other and so symmetry must exist for the concept of a critical load to have any validity. Unsymmetrical deformations may occur in one of two sway modes at a bifurcation point.

The curves in Fig. 5 have been computed for centrally loaded circular arches (5) and demonstrate the influence of the rise-to-span ratio on both the maximum load-carrying capacity and the nature of the post-buckling behaviour. Four pin-based arches were studied and the rise values for each were chosen so that the angle subtended at the centre of the arch curvature would be 50, 90, 140 and 180 degrees; the span being constant. Published critical loads for arches of these shapes agree closely with both the maximum and bifurcation loads for each arch in Fig. 5. It is interesting to see that the post-buckling behaviour is highly stable for the semi-circular arch and becomes very unstable for the flattest arch.

Studies of parabolic arches in (6) involved loading taken as uniformly distributed from the left-abutment over varying lengths approaching the full span. Only one geometry was considered; namely where the rise was a quarter of the span. For both fixed and pinned bases the post-buckling behaviour was unstable and involved an equilibrium bifurcation with an asymmetric sway mode well below the symmetric buckling load. The second-order curves shown in Fig. 6 show the effect of introducing an imperfection in the form of a premature curtailment in the length of the loading. The studies showed that as the proportion of the span covered decreased, the maximum supportable intensity decreased and then rose but the trend clearly demonstrated that little more than half the total load to cause buckling if spread over the whole span could be safely carried over half the span.

3. CLOSED RINGS

The discrete element method was applied (7) to ring frames subjected to radial loading of a hydraulic nature. Five ring frames were considered with the first being circular and the remainder being polygons of various geometries but all having the same perimeter as the circular specimen. They could then be considered as imperfect forms of the circular case.

The results are summarised in Fig. 7 where it is clear that the post-buckling behaviour is stable in all cases. An imperfection was introduced into the loading by varying the intensity from the top to the bottom of the ring. The vertical symmetry of the hydraulically loaded ring frame enabled analysis to be done by considering only a half frame. Hence the arch program could be called upon with a few modifications made in the load generation and closure error routines. The curves in Fig. 7 agree well with the lowest buckling load as predicted by Levy's formula. Provided one chose the appropriate initial values, there was no difficulty in generating solutions corresponding to the higher buckling modes.

4. PORTAL FRAMES

Portal frames are linear arches whose geometry is defined by the three variables- span, rise and eaves height. The discrete element method (2) can process any type of arch whether regular or irregular in shape and whether uniform or non-uniform in section. In studying the post-buckling behaviour of symmetric portals of varying geometry, it seemed sensible to use as a vehicle a uniform frame with span and section properties similar to those adopted in the earlier studies of circular and parabolic arches.

Altogether some 18 portals have been analysed, nine with pinned and nine with fixed bases. The pitch angles chosen were 15, 7.5 and zero degrees and with a constant spans of 10000, three values of eaves height were chosen, namely 3500, 2500 and 1500. The load condition was that with a symmetrical and uniformly distributed vertical load covering the whole span. The results have been shown in figures 8 - 13 in the form of load-deflection curves where the deformation chosen is the vertical downwards movement of the apex. Not shown are the results of many additional analyses carried out with a distributed loading slightly curtailed in length to detect the location on the curves of points corresponding to equilibrium bifurcation. Such a strategy follows from the earlier experience demonstrated in figures 1-4. For the symmetric situation in Fig. 1 it is possible and in fact likely that a step-by-step numerical solution may skip completely the bifurcation line. Such an error may not be important for the arch in Fig. 2 but it would clearly be disastrous for that in Fig. 1. The existence of a bifurcation line is clearly indicated by the results plotted in Fig. 3 for the case where a slight imperfection is introduced into the symmetry of the loading.

In all nine portals with fixed bases, no equilibrium bifurcation lines were detected despite a careful search using a distributed load displaying slight asymmetry. It would appear that the only possible deformation modes for these frames are symmetrical ones. The curves in Fig. 8 for the low and medium rise frames show an unstable post-buckling curve but the buckling mode in both cases involved symmetrical deformations. The highest rise frame in Fig. 8 demonstrates a post-buckling curve that is marginally stable. As the pitch angle decreases, the critical load concept becomes irrelevant as can be seen in figures 9 and 10. Rectangular frames with fixed bases are studied in Fig. 10 and there are no clearly defined critical loads states. These frames are beginning to resemble beams with elastic end restraints and in-plane buckling is hardly relevant.

The load-deflection curves for the pin-based frames are shown in figures 11-13 and they differ markedly from the fixed-based curves. The frames with a 15 degree pitch in Fig. 11 all show bifurcation curves with the high-rise frames being stable in the post-buckling range. The 7.5 degree frames in Fig. 12 have post-buckling curves of somewhat lower slope and the frame with the smallest eaves height shows no bifurcation line. The zero pitch or rectangular pinned frames in Fig. 13 all have well defined maximum loads but only the highest rise frame demonstrates a bifurcation curve.

Calculated elastic critical loads have been compared with the bifurcation and maximum loads produced by the discrete element method. The critical loads were calculated using a program developed by Hancock from one used in another context (8). Columns and rafters were subdivided into two elements and the distribution of axial force determined from a first-order stiffness analysis. The results are summarised in Table 1. There is good agreement between the alternative predictions of load factors at the onset of asymmetric buckling but no agreement at all with the maximum loads corresponding with symmetric deformations.

5. *CONCLUSIONS*

The discrete element shooting method is an effective tool in the analysis and design of arches, rings and portals. Earlier studies of arches and rings have been summarised and discussed. The portal frame results have been set out in some detail. In assessing the significance of the results for eighteen portal frames, the picture presented is a complex one as indeed it would be if results were being presented for eighteen laboratory tests. Frames with pinned bases behave very differently from those whose bases are fixed and on the whole it is seen that the frames with pinned bases and appreciable pitch angles tend to display asymmetric buckling modes with stable post-buckling behaviour.

The implications of the second-order studies for low-rise arches of parabolic or circular shape are important because the case of uniform loading is a very practical one and with unstable post-buckling bifurcation curves very much in evidence, designers should be aware of the need for second-order analyses and the availability of appropriate programs. The situation is different with portal frames where the structure is primarily a flexural one and plastic failure will dominate the in-plane behaviour. The elastic critical load factors could be expected to be many times the plastic load factors and large errors in the predictions of critical loads can be tolerated. In any event the discrete element method can be extended to deal with complex moment-curvature relationships as has been shown by Bridge (9).

The agreement in Table 1 between the load factors at the onset of unsymmetrical deformations in pin-based frames confers relevance on eigenvalue calculations based essentially on small deformation theory. There remains the problem of explaining the significance of eigenvalue calculations for the fixed base frames where there is no correlation with the results of the second-order large deformation analyses.

References

1. Horne, M.R. 'Plastic Theory of Structures.' Pergamon Press, Oxford, 1979.
2. Harrison, H.B. 'Structural Analysis and Design: Some Minicomputer Methods.' Pergamon Press, Oxford, 1980.
3. Harrison, H.B. 'Post-buckling behaviour of elastic circular arches.' Proceedings of the Institution of Civil Engineers, Part 2, June 1978.
4. Austin, W.J. and Ross, T.J. 'Elastic buckling of arches under symmetrical loading.' Journal of the Structural Division, ASCE, vol. 102, no. ST5, May 1976.
5. Harrison, H.B. 'Post-buckled arch behaviour and the influence of imperfections.' Proceedings of 7th Australasian Conference on the Mechanics of Structures and Materials, University of Western Australia, May 1980.
6. Harrison, H.B. 'In-plane stability of parabolic arches.' Journal of the Structural Division, ASCE, vol. 108, no. ST1, January 1982.
7. Harrison, H.B. 'Large deformation analysis of submerged ring frames.' Journal of the Engineering Mechanics Division, ASCE, vol. 105, no. EM5, October 1979.
8. Hancock, G.J. 'Local, distortional and lateral buckling of I-beams.' Journal of the Structural Division, ASCE, vol. 104, no. ST11, November 1978.
9. Bridge, R.Q. 'Large deflection analysis of portal frames.' Civil Engineering Transactions, Institution of Engineers, Australia, vol. CE21, no. 2, 1979.

TABLE 1. Summary of Analyses of 18 Portal Frames.

Frame no.	Geometrical details pitch	bases	eaves	Refer Fig.	Second-order values L.F.(asy)	L.F.(sym)	Eigen-values
1	15	fixed	3500	8	-	0.76	1.348
2	15	fixed	2500	8	-	0.72	1.640
3	15	fixed	1500	8	-	0.82	1.582
4	7.5	fixed	3500	9	-	-	1.395
5	7.5	fixed	2500	9	-	-	1.669
6	7.5	fixed	1500	9	-	-	1.408
7	0	fixed	3500	10	-	-	1.448
8	0	fixed	2500	10	-	-	1.491
9	0	fixed	1500	10	-	-	0.948
10	15	pinned	3500	11	0.35	0.87	0.366
11	15	pinned	2500	11	0.54	0.80	0.548
12	15	pinned	1500	11	0.76	0.76	0.830
13	7.5	pinned	3500	12	0.37	0.75	0.373
14	7.5	pinned	2500	12	0.57	0.63	0.562
15	7.5	pinned	1500	12	-	0.50	0.838
16	0	pinned	3500	13	0.39	0.61	0.376
17	0	pinned	2500	13	-	0.47	0.569
18	0	pinned	1500	13	-	0.31	0.828

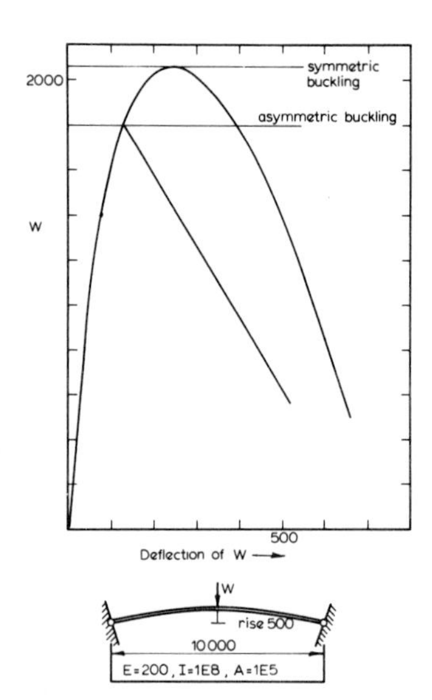

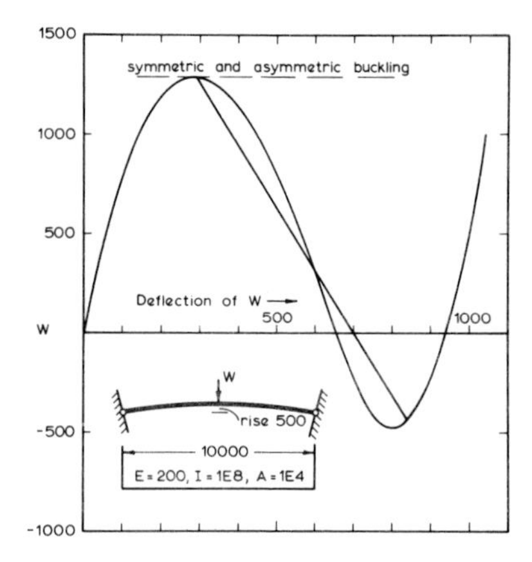

Fig. 1 Load-deflection curves for a low-rise arch.

Fig. 2 Load-deflection curves for a low-rise arch with reduced axial stiffness.

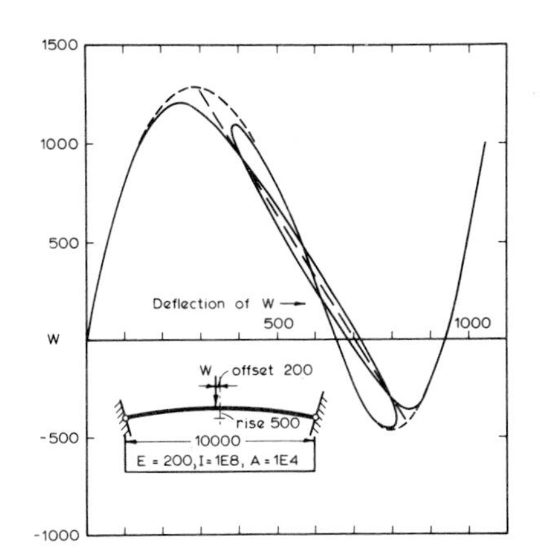

Fig. 3 Load-deflection curves for a low-rise arch under near central point load.

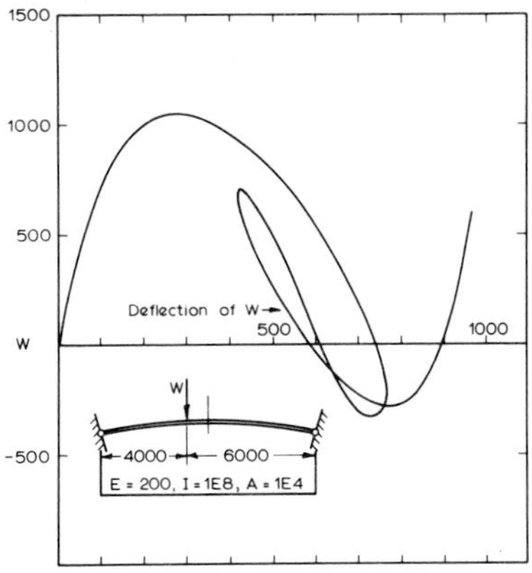

Fig. 4 Load-deflection curve for a low-rise arch under off-central point load.

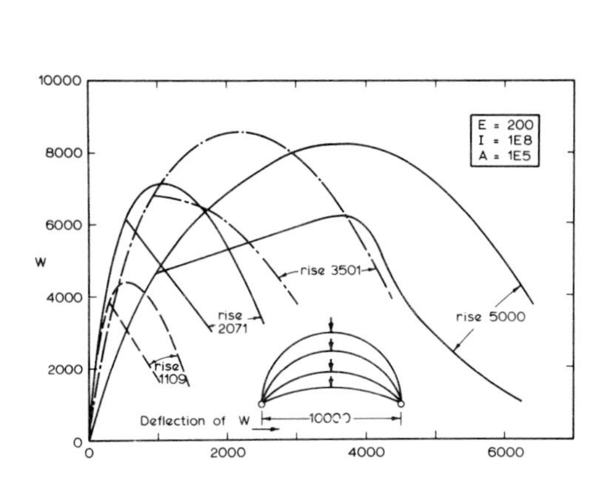

Fig. 5 Load-deformation curves for centrally loaded circular arches of varying rise.

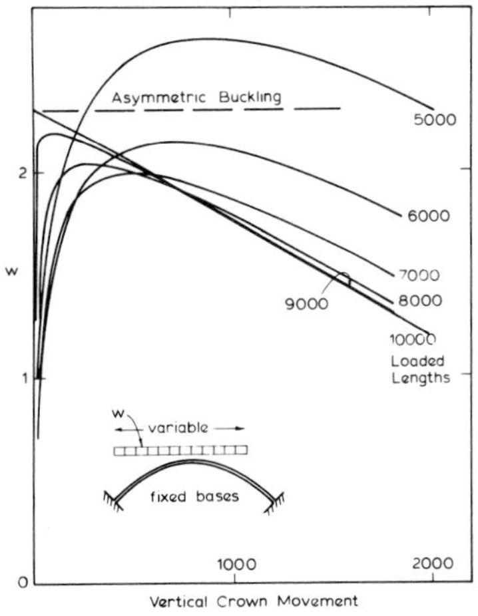

Fig. 6 Load-deflection curves for parabolic arches carrying distributed loading of varying length.

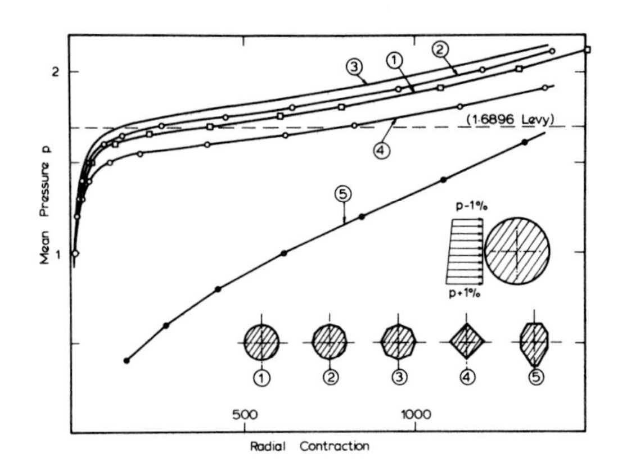

Fig. 7 Load-deformation curves for various ring frames under nonuniform pressure.

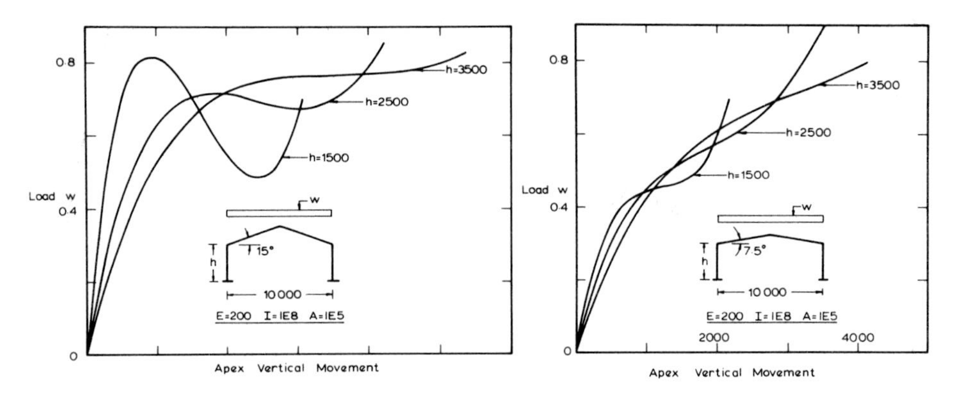

Fig. 8 Load-deformation curves for fixed base portals with 15 degree pitch.

Fig. 9 Load-deformation curves for fixed base portals with 7.5 degree pitch.

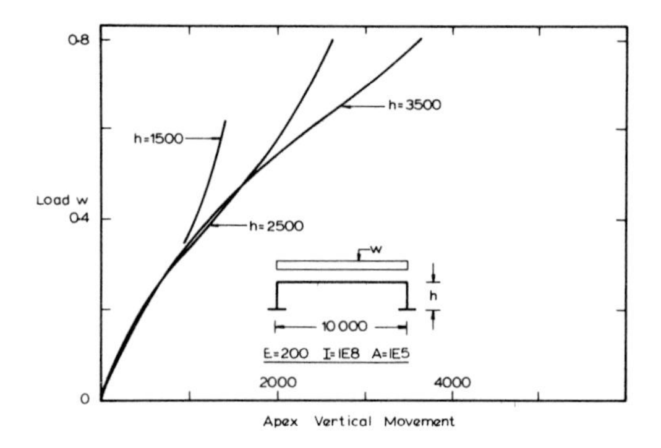

Fig. 10 Load-deformation curves for fixed base portals with zero degree pitch.

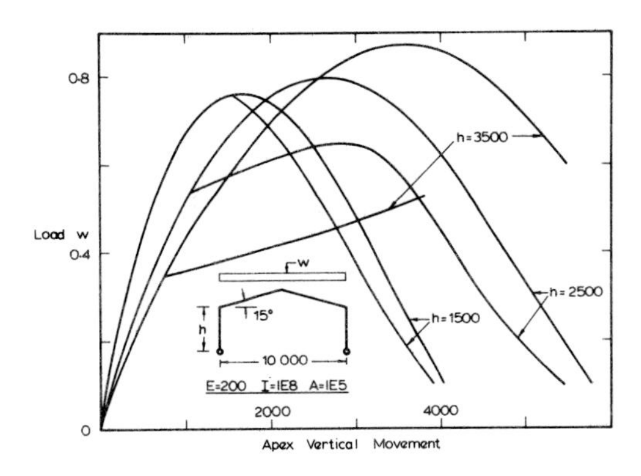

Fig. 11 Load-deformation curves for pinned base portals with 15 degree pitch.

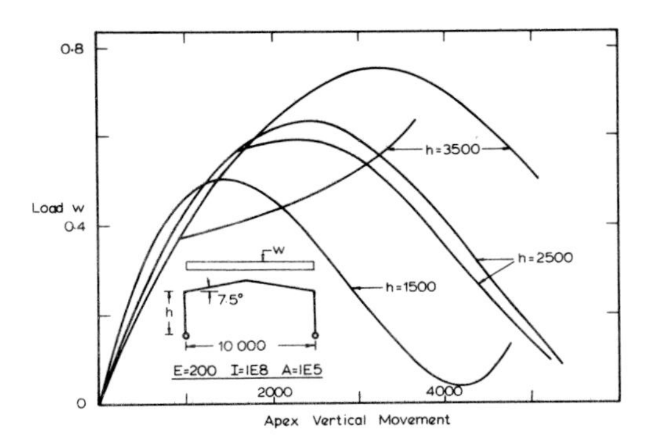

Fig. 12 Load-deformation curves for pinned base portals with 7.5 degree pitch.

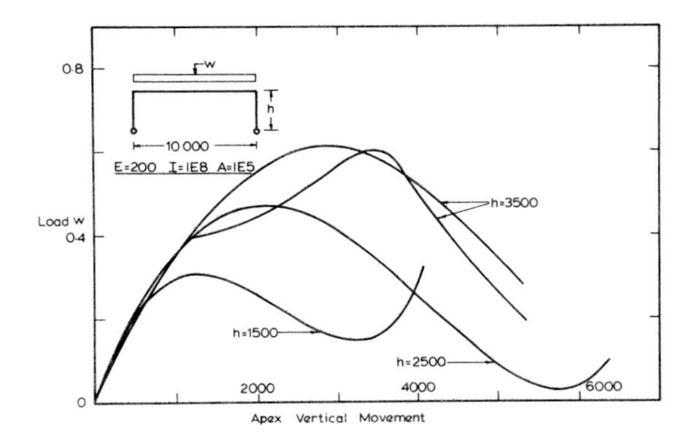

Fig. 13 Load-deformation curves for pinned base portals with zero degree pitch.

THE STABILITY OF IMPERFECT STRUCTURAL SYSTEMS REINFORCED BY VERTICAL TRUSSES.

E. G. Galoussis

School of Technology Democritus Univ. of Thrace. Xanthi - Greece.

Synopsis

The non-linear behaviour of frames consisting of columns and beams stiffened by diagonal bracings is examined. Some unavoidable imperfections are taken into account and a critical state, still in the elastic range, is investigated. The reduction of the critical load is shown in diagrams. Finally, some experimental results are given, for an easy comparison, which proved that the theoretical results and the experimental ones are in very close agreement.

Notation

A	cross-sectional area of bracing member
α_o	initial lack of straightness of diagonals
E	Young's Modulus
ℓ	length of bracing member
P	axial force of bracing member
P_E	Euler load of bracing member
Q_{res}	shear resistance of the structure
Q_{ext}	applied horizontal external force
r	radius of gyration of the diagonal
S'_1, S'_2	tangent stiffness of substructures with bar in tension and compression respectively.
W	load in columns
W_{cr}	critical load of frame
$\rho=\frac{P}{P_E}$	stability ratio
ϕ	slope of columns
ϕ_o	initial value of ϕ

Introduction

The design of eccentrically braced frames are being discussed extensively due to their flexibility and ability to absorb horizontal forces. Lastly Popov (1), Popov and Roeder (2) et al, gave some instructions for designing such tall flexible buildings and also some buildings designed and constructed according to the recommendations concerning eccentric bracings as referred to by Merovich, Nicoletti, Hartle (3) and Libby (4).

Some imperfections such as section or material variations along the slender diagonals, initial curvature, eccentric connections etc. of the bars appear very often in the structures. The recommendation proposed for a check of the deflections of the structures by an elastic analysis in determining whether they are acceptable is not safe.

In the paper the reduction of the critical load is investigated with respect to the presence of imperfections in the diagonals and the out-of-plumb position of the structure or the application of external horizontal loads.

1. *THE BAR IN COMPRESSION AND TENSION*

To incorporate in the calculations all the imperfections referred to, the member can be considered to have an initial half sine-wave curvature as shown in Fig. 1 and given by

$$Y_o = \alpha_o \sin\left(\frac{\pi x}{\ell}\right) \qquad \ldots\ldots(1)$$

Then the total deformation δ is given by

$$\delta = \frac{P\ell}{AE}\left[1 + \frac{1}{4}\left(\frac{\alpha_o}{r}\right)^2 \frac{2+\rho}{(1+\rho)^2}\right] \qquad \ldots\ldots(2)$$

The stability ratio is positive when the axial force is tensile. Thus the effective stiffness of the member changes with the initial curvature and the stability ratio.

When sufficient axial force is applied to the member, full yielding will occur at one cross-section; for a member loaded in tension the member can elongate for no change in force. For a member in compression the full yield will occur at midlength forming a plastic hinge at that point. The end displacement δ_ρ, shown in Fig. 1, is given by Smith (5) as

$$\delta_\rho = \ell\,(1 + \cos\theta) + \frac{P\ell}{AE}\cos^2\theta \qquad \ldots\ldots(3)$$

where $\theta = \arcsin\left(\frac{2e^*}{\ell}\right)$

The behaviour of a member can be described by equations (1) to (3) to give the non-linear elastic and inelastic response of an initially curved members as shown in Fig. 2 (curves EDO, OABC).

2. *THE MODELLING OF THE STRUCTURE*

Some typical eccentric bracings are shown in Fig. 3a. Because of the assumptions above, the member-structure interaction can be described solely in terms of axial force/deflection, equilibrium and compatibility relationships. A typical relationship between the non-linear response of a diagonal in compression and of the adjoining structure to which the member is attached is shown in Fig. 2. Curves OGA, OHB are the responses of the adjoining structure without the diagonal, subject to a given external load and the application of P in place of the member.

Neglecting the non-linear effect of the bar in tension the response of the adjoining structure is described by a line GA or HB. At point A equilibrium and compatibility exist between the diagonal and its adjoining structure. It is obvious that point B is an unstable position since a slight increase in the axial deformation cannot find any other equilibrium state with greater force. Point B depends also on the imperfections of the diagonal. In Fig. 4 the behaviour of the system is described in different coordinates. So the curve OACB is the response of the structure in a system of coordinates Q_{res} - ϕ_c and the straight line DEAB gives the effect of the loads in the coincident systems of coordinates Q_{ext} - ϕ. The slope of the last straight line is the magnitude of the external load. The line DEAB can be defined by point D when an initial slope ϕ_o exists, or by point E, when an additional external horizontal load Q_{ext} is applied to the perfect structure. Points A and B are points of equilibrium, stable and unstable respectively.

Finally, points C (or C') define a critical state of the structure because in any slight increase in the external load or in the deformation of the frame no equilibrium exists. It has been proved by Dooley, Galoussis (6), Galoussis - Panagiotopoulos (7) and Galoussis (8) that the response of the model of Fig. 5 is similar to that of a panel of the frame. The constant c, of the springs is calculated by the stiffness of the joining beams. The horizontal member D is an ideal stiff bar and the dimensions are those of a panel of the structure.

3. *THE TANGENT STIFFNESS AND THE CRITICAL LOAD*

It is apparent (see Fig. 4) that the tangent DC of the curve OACB is the critical load for the critical state described. So, if S'_1, S'_2 are the tangent stiffnesses of ADB and BEC we get

$$W_{cr} = S'_1 + S'_2 \qquad \text{.............................(4)}$$

Considering the stiffness of the joining beam, for a constant c of the spring we have

$$S'_1 = S_1 = \frac{b^2 h}{\ell^2 \ (c + \ell/AE} \qquad \text{.......................(5)}$$

The secant stiffness S_2 of the subsystem ADB can be easily calculated taking into account equation (2), by

$$S_2 = \frac{b^2 h/\ell^2}{c + \frac{\ell}{AE}\left[1 + (\frac{\alpha_o}{2r})^2 \quad \frac{\rho(2-\rho)}{(1-\rho)^2}\right]} \qquad \text{.............(6)}$$

and the tangent stiffness is given by

$$\frac{1}{S'_2} = \frac{1}{S_2} + \rho \frac{d}{d\rho} (\frac{1}{S_2})$$

or

$$S_2' = \frac{b^2 h/\ell^2}{c + \frac{1}{AE}\left[1 + (\frac{\alpha_o}{2r})^2 \frac{1}{(1-\rho)^3}\right]} \qquad (7)$$

Equations (4), (5) and (7) permit the calculation of the critical load W_{cr}. The equations referred to were used to predict the critical load for the experiments.

4. EXPERIMENTAL WORK

Special attention was paid in designing the specimen, so that any undesirable effect of the imperfections due to manufacture were minimized. The specimen is shown in Fig. 6. Two nominally identical structures were manufactured, the uprights 3" x 3" x 1/4" (75 mm x 76 mm x 64 mm) mild steel tubes welded and bolted via gusset plates to cross-members of similar section. Each was a very stiff rigid-jointed portal frame whose failure load by sidesway in its own plane, by buckling or by material failure was considerably higher than any load to which it was to be subjected. The diagonal bracings of the upper and lower frames were 1/2" (12.7 cm) bright steel solid round bars of length between pin-ends of 39.4" (1.0 m) giving a slenderness ratio of 317.5. The ends of each bar strut were threaded left and right-hand respectively, so that the initial slope of the uprights could be varied. Each such threaded end was screwed into an end-fitting comprising a commercially-manufactured spherical bearing. Independent trial showed appreciable friction, so each spherical bearing's shaft was fitted with roller bearings. With these added no appreciable friction remained. The tests of the structure described above were carried out with struts of initial bow 1/300 x the length.

To simulate the flexibility of the eccentric bracing in tension, the tension brace used for the experiments consisted of two 1/2"(12.7 mm) bars, with a piece of structural channel section joining them elastically as shown in Fig. 7. By using different lengths of channel S_1 could be varied. The channel used was calibrated by direct test before being inserted into the test specimen and the constant C_1 of the spring of Fig. 5 was determined. Unfortunately such an arrangement could not be applied to the compression member and the prediction of the critical load was made for a value of C_2 = 0 (equation (7)). Load was applied by four hydraulic jacks, one at each of the top corners, each fitted with a load cell. The whole assembly was held in an Avery 150 ton testing machine whose recorded load checked that indicated by the four load cells. It had the advantage in giving warning of approaching instability by the wavering of the pointer on the load indicator. Tests to correlate the critical load, the initial slope and the initially curved eccentric braces were carried out. The results are shown in Fig. 8.

It is perhaps necessary to mention that the very low stiffnesses used were in order to have the tangent stiffness $S_1' = S_1$ of the same order as S_2' when the brace was at a high fraction of its Euler load.

CONCLUSION

It has been shown that some imperfections such as the initial curvature of members and the out-of-plumb erection of tall building frames stiffened by vertical trusses cannot be neglected. A critical state has been defined by considering the combination of those imperfections. The critical load is given by a simple formula. Horizontal forces due to wind or earthquake can also lead the frames to a similar critical state. The effect of the flexibility due to eccentric connections of the diagonals causes deterioration in the critical state, reducing the value of the critical load. The occurence of bolt slippage magnifies the sidesway as well which causes a further reduction of the critical load. Experiments verified this critical state and the influence of the imperfections. Since the aforementioned imperfections are of a random nature only a stochastic investigation can result in useful conclusions based on the deterministic formulae given above.

Acknowledgement

The author would like to extend his gratitude to Professor D.M.Brotton and Dr. J. F. Dooley for their encouragement and their support in performing the experiments in the "Merchant Laboratories" at U.M.I.S.T. during a years visit.

References

1. Popov, E.P. 'An update on Eccentric Seismic Bracing'. Engineering Journal, AISC, 3rd Quarter, 1980.

2. Roeder, C.W. - Popov, E.P. 'Design of an Eccentrically Braced Steel Frame'. Engineering Journal, AISC, 3rd Quarter, 1978.

3. Merovich, A.T. - Nicoletti, J.P. - Hartle, E. 'Eccentric Bracing in Tall Buildings'. Journal of the Structural Div. ASCE, ST3, Sept. 1982.

4. Libby, J. R. 'Eccentrically Braced Frame Construction - A Case History'. Engineering Journal, AISC, 4th Quarter, 1981.

5. Smith, E.A. - Smith, G.D. 'Collapse Analysis of Space Trusses'. Proceedings of Symposium on Long Span Roof Structures, ASCE, Oct. 1981.

6. Dooley, J.F. - Galoussis, E.G. 'Critical State of Structures'. Mechanics Research Communications. Vol.9, 1982.

7. Galoussis, E.E. - Panagiotopoulos, P.D. 'Über durch Imperfectionen verusachtes Versagen von Fachwerkkonstructionen-eine Stochastiche Näherung'. Der Stahlbau, Vol. 50, Dec. 1981.

8. Galoussis, E.G. 'On a Critical State of Braced Structures Caused by Imperfections'. The Sino-American Symposium, Sept. 1982.

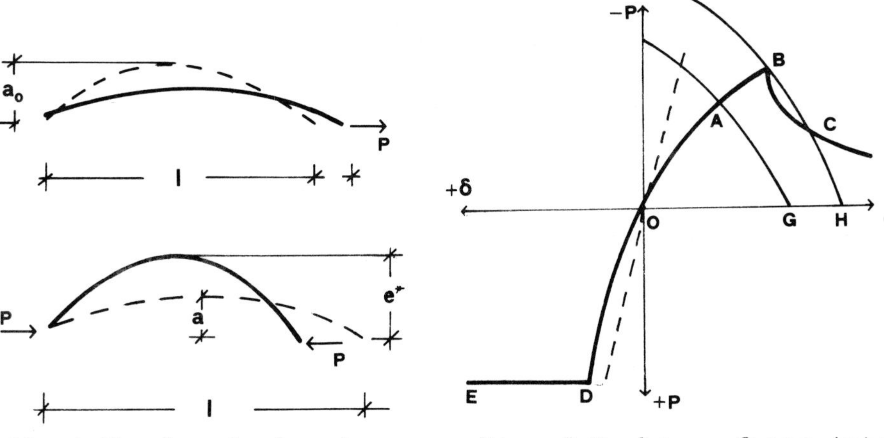

Fig. 1 The bar in tension and compression.

Fig. 2 Member - frame interaction.

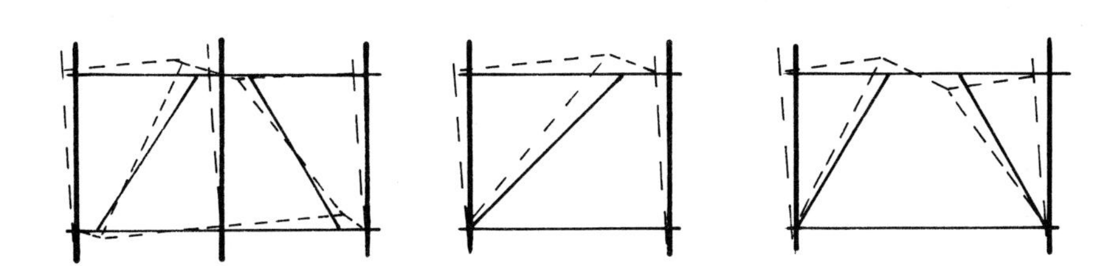

Fig. 3 Panels with eccentric bracings.

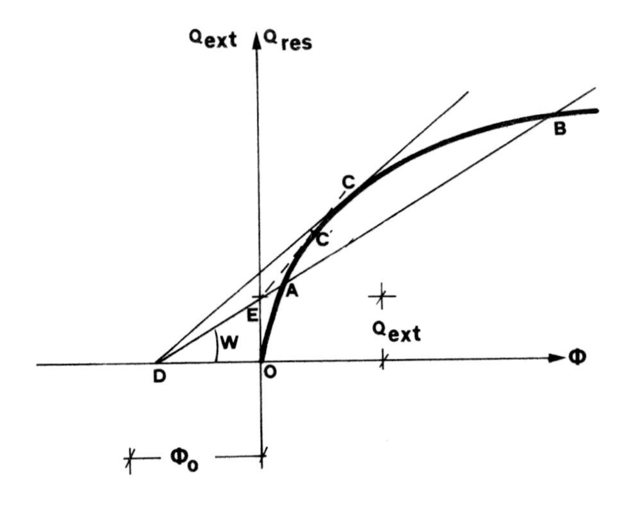

Fig. 4 Q-Φ curve

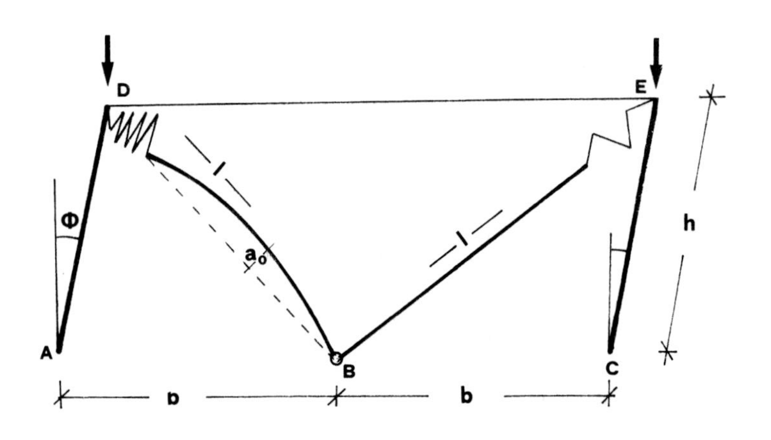

Fig. 5 The model used

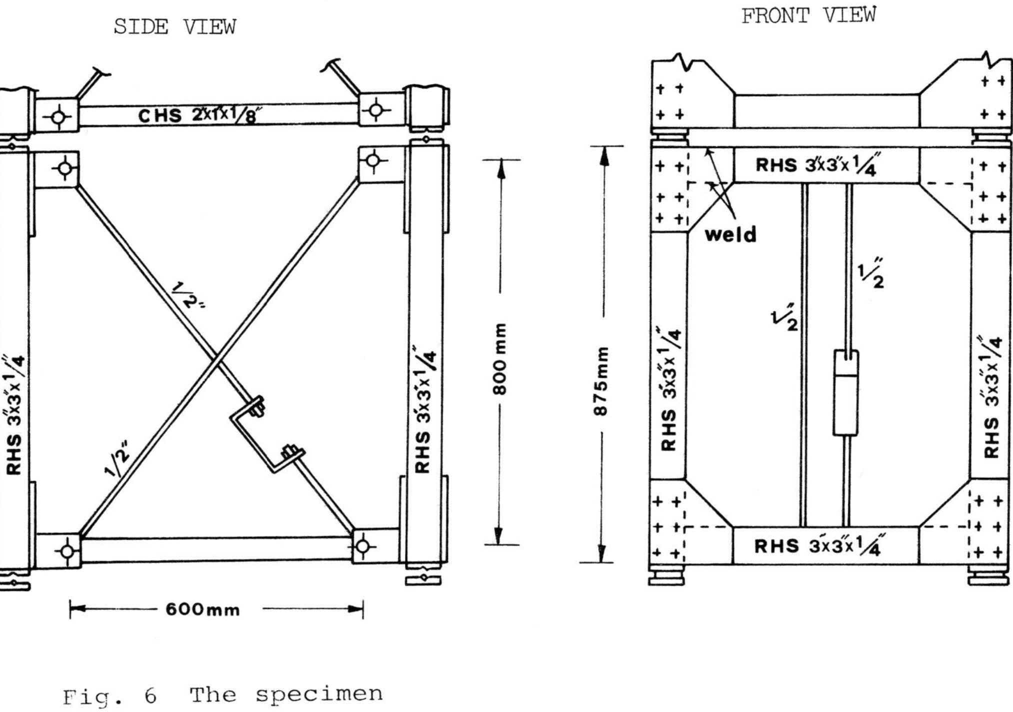

Fig. 6 The specimen

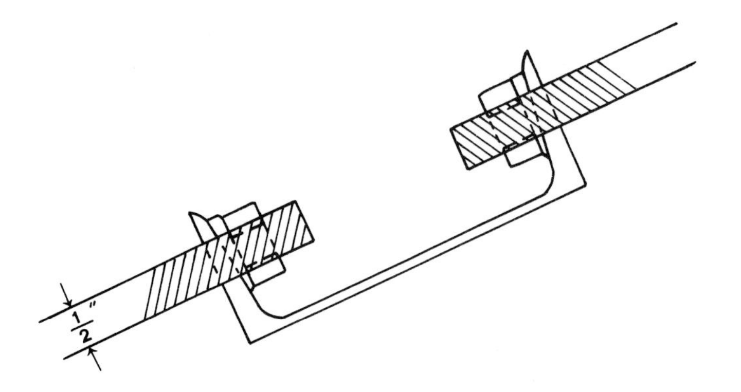

Fig. 7 Detail of the bar in tension

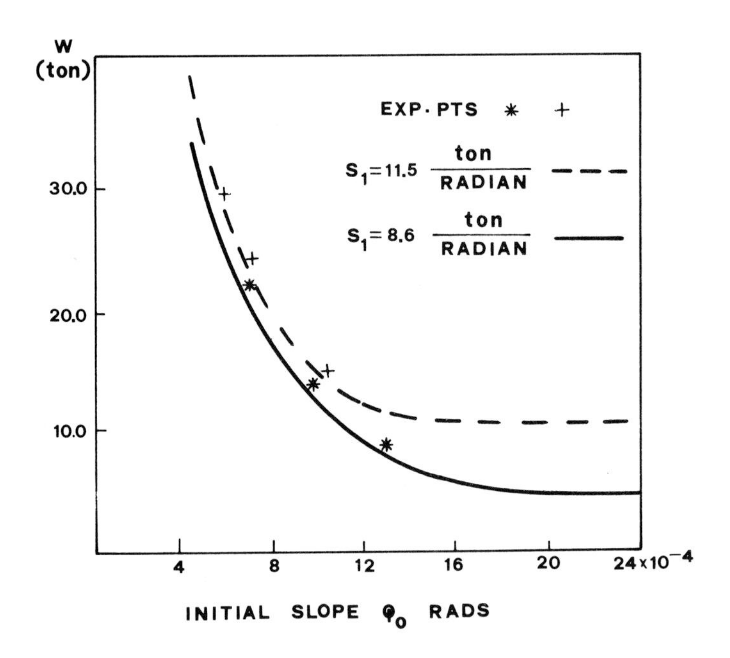

Fig. 8 $W_{cr} - \Phi_o - S_1$ curves

ELASTIC STABILITY OF BUILDING STRUCTURES

I.A. MacLeod and J. Marshall

Department of Civil Engineering, University of Strathclyde

Synopsis

A simply method of assessing the elastic critical sway mode load of a building is developed. This approach follows from Horne's method for sway frames but is based on overall stiffness. Comparison is made with computer based estimates of critical load for a shear wall and frame system.

Introduction

Elastic critical load is normally considered with respect to single columns or single frames. However, the elastic collapse mode of a complete building involves most if not all of the support structure and therefore in any test of elastic stability it seems reasonable that more than just localised effects should be considered. Columns tend to be considered as part of shear mode frames such that their storey height behaviour can be considered in isolation. If there is a significant bending mode component in the lateral deformation of the system then a storey height approach is invalid and distribution of lateral load and consideration of sway stability must be treated with respect to the total height of the structure. Likewise the no-sway stability of part of a building depends on the lateral support provided by the other frames and walls.

A realistic evaluation of the elastic critical load of a building must therefore take into account the complete structure.

1. BEHAVIOUR OF BUILDING STRUCTURES

The sway stability of a building is mainly governed by its lateral stiffness. The parts of the structure which provide the lateral stiffness are normally discussed in terms of two basic modes of deformation, i.e. shear mode and bending mode. Figure 1(a) shows typical lateral deflection shapes for shear mode and bending mode deformation with a uniformly distributed lateral load. The shear mode shape is a second order curve and the bending mode is a fourth order curve.

It is normal to identify parts of the structure which resist lateral load and to classify them according to whether they are shear mode or bending mode units. In fact all such units will have a combination of both modes but it is convenient for simplified analysis to assume that only one mode is present in each case. Thus, for example, a shear wall would normally be considered as bending unit (although it will have a shear mode component due to shear deformation or to the effect of openings if present). Similarly a rectangular rigid jointed frame would be assumed to deform only in a shear mode (neglecting the bending mode component due to axial deformation of the columns).

A common model for a complete building structure is to sum the stiffness of all shear mode units to make one shear mode frame and to connect this to a bending mode unit which is the sum of all bending mode stiffness - Fig. 2. This simplified model is used here to study the effect of varying the proportion of shear to bending mode stiffness on elastic stability.

The combined system has a lateral mode of deformation which is between the two extremes. A typical shape for the lateral deflection of such a system is shown in Fig. 1. Note that the curve is fairly flat.

First of all we discuss the stability of units having only one mode of deformation.

1.1 Stability with shear mode deformation

Shear mode deformation is characterized by the relationship:

$$V = C\, dv/dx \tag{1}$$

where V is the applied shear
C is the shear stiffness (shear per radian)
v is the lateral deflection.

Consider the element of a shear cantilever with axial load P shown in Fig. 3. If this element suffers a shear strain dv/dx then the applied load will have a horizontal component Pdv/dx. The restoring shear will be Cdv/dx (eqn (1)). Thus for unstable equilibrium:

Overturning shear = Restoring shear

i.e. $$P_{cs}\, dv/dx = C\, dv/dx$$

$$P_{cs} = C \tag{2}$$

where P_{cs} is the critical load with shear mode deformation.

Denoting the storey height stiffness of a frame as k i.e. k is the lateral load at the top of a storey to cause unit sway deformation over the storey height, then (approximately)

$$C = kh \tag{3}$$

where h is the storey height.

Thus using eqn (2)

$$P_{cs} = kh \tag{4}$$

Note that with shear mode deformation the elastic critical load appears to be independent of the total height of the building. This is not entirely correct (as discussed later) but is used for design of columns on a storey height basis.

To get eqn (3) it is assumed that the storey height deformation is straight. To allow for the fact that it is not straight but in double curvature in a rectangular sway frame, eqn (4) can be modified to (1,2,3):

$$P_{cs} = 0.9\,kh \tag{5}$$

1.2 Stability with pure bending mode deformation

The elastic critical load for a pure bending mode cantilever with a top point load is given by the well-known relationship:

$$P_{cb} = \frac{4\pi^2 EI}{h^2} \tag{6}$$

where P_{cb} is the critical load with pure bending mode deformation.

A uniform bending cantilever with uniformly distributed vertical loading has an elastic critical load (4).

$$P_{cb} = \frac{7.81\,EI}{h^2} \tag{7}$$

It should be noted that the elastic critical load of a bending mode cantilever is a function of its overall height and thus any such calculation based on a storey height of a shear wall is not valid.

2. SIMPLIFIED METHOD FOR CALCULATING THE ELASTIC CRITICAL LOAD OF A BUILDING STRUCTURE

In order to make a simplified assessment of elastic critical load the following assumptions are made:

(1) The critical load is a sway mode

(2) The lateral distribution of the vertical load does not affect the critical load.

(3) The distribution of vertical load is uniform with height, i.e. the vertical applied load at each storey level is the same.

(4) The shape of the lateral deflection curve under a uniformly distributed lateral load is a straight line.

(5) The distribution of lateral restoring force is uniformly distributed

i.e. we are assuming that a lateral uniformly distributed load causes a straight line displacement which is also the shape of the primary lateral buckling mode. Since the deformation mode of the complete structure must lie between the shear and bending limits, a straight line deflected shape is reasonable - see Fig. 1. Suppose that a system which conforms to this ideal is given a top lateral defelction Δ - Fig. 4. Then for unstable equilibrium in this position

Overturning moment = Restoring moment

i.e.

$$\frac{P_{cr}\Delta}{2} = \frac{WH}{2} \qquad (8)$$

where P_{cr} is the total vertical load to cause buckling
W is the total lateral restoring load
H is the total height

Defining $K = W/\Delta$

where K is the lateral top stiffness under a uniformly distributed load and substituting into (8) gives:

$$P_{cr} = KH \qquad (9)$$

i.e. KH is an approximation to the elastic critical load of the system.

This expression is reasonably accurate for a bending cantilever under uniformly distributed vertical load since in this case:

$$P_{cr} = KH = (8EI/H^3)H = 8EI/H^2 \qquad (10)$$

Comparing this with eqn (7) shows that eqn (9) overestimates the critical load by about 2% in this case.

For a shear mode frame, using eqn (1) gives:

$$dv/dx = V/C$$

Under a uniformly distributed lateral load w kN/m, $V = wx$ (x is the distance down from the top of the frame)

$$dv/dx = wx/kh$$

$$\therefore \Delta_{top} = \int_0^H \frac{wx}{kh}\,dx = \frac{wH^2}{2kh} = \frac{WH}{2kh}$$

where W is the total lateral load,

$$K = W/\Delta_{top} = 2\,kh/H$$

$$\therefore P_{cr} = KH = 2\,kh \qquad (11)$$

i.e. eqn (11) appears to be rather inaccurate with pure shear mode deformation in comparison with eqn (4). This is discussed later.

3. *COMPARISON WITH OTHER SOLUTIONS*

To check the validity of eqn (9) a shear wall and frame system of the type shown in Fig. 2 was used. The following properties were used:

E = 28 N/mm² ; 10 stories at 3.0 m
Frame Beams : I = 3.375 E-4, m⁴, A = 0.115, m², span = 3.0 m
Frame Columns : I = 6.75 E-4 m⁴, A = 0.09 m²
Wall : I varies as shown in Table 1

Connecting links : $I = 0$ $A = 0.115\,m^2$ span = $3.0\,m$
Shear deflection of the members was neglected.

The vertical load on the structure was distributed evenly with height, i.e. the same load was applied at each storey level.

The second moment of area of the wall was varied to alter the relative stiffness of the walls. α is defined as the ratio of the top stiffness of the wall to the top stiffness of the frame. The top stiffness in each case was calculated by a linear elastic frame analysis using a lateral load system having the same form as the vertical load, i.e. a unit point lateral load was applied at each storey level. The top stiffness is then the total lateral load divided by the top deflection.

Table 1 gives predictions of the elastic critical loads by the following methods:

(1) *Stability function analysis.* Livesley's stability functions (5) were used in a program in which the load factor was increased until the determinant of the structural stiffness matrix passed through zero. This type of analysis is described by Majid (6).

For this analysis the total vertical load was distributed to the wall and the frame in the proportions shown in Table 1. This distribution ensures sway mode instability in each case. The effect of the lateral distribution of vertical load is discussed later.

(2) *Horne's Method* (1) in which the estimate of elastic critical load is based on the storey height stiffness of the system under a uniformly distributed lateral load. In Table 1 the critical storey for instability is noted in each case. The 0.9 correction factor noted in Section 1.1 was not applied to the results quoted in Table 1.

(3) *Unit Shear Method.* An alternative to Horne's method is to apply a unit shear to the top of the structure and estimate the critical load at each storey as the reciprocal of the storey height sway angle. Results for this are also given in Table 1 with the position of the critical storey noted.

(4) Equation (9).

Also shown in Table 1 are the critical loads for sway stability of the walls and the frame separately and the no sway critical load for the frame. These were calculated using the stability function analysis.

From the results given in Table 1 the following observations are made:

(1) For the combined wall and frame system used, eqn (9) gives predictions of the sway critical load to a satisfactory degree of accuracy. The frame part of the model used has a significant bending mode component due to column axial deformation. Therefore, even the $\alpha = 0.22$ case does not represent a dominant shear mode situation. If the frame column axial stiffness was increased so that the axial deformation became negligible then one would expect eqn (9) to significantly overestimate the critical load as discussed in Section 2. Equation (9) might also give less satisfactory results with non uniform structures,e.g. where the lateral stiffness is decreased at the base for architectural reasons.

(2) Horne's Method is better than the unit shear method but is less accurate than use of eqn (9) for the cases considered. This accuracy is due to the fact that in the cases considered bending mode deformation either dominates or is at least significant. Accuracy of Horne's Method is likely to improve with dominant shear mode systems.

4. *EFFECT OF LATERAL DISTRIBUTION OF VERTICAL LOAD*

To investigate the effect of the distribution of load between the wall and the frame the $\alpha = 11$ and the $\alpha = 0.22$ cases were reanalysed with variation of load distribution as noted in Table 2. The elastic critical loads given in Table 2 are those from a stability function analysis. In the $\alpha = 0.22$ case the critical load is in a sway mode and the distribution of vertical load does not significantly affect the prediction of critical load. With $\alpha = 11$ the reverse is true. This is because with the high α value a no-sway mode can develop. The transition from no-sway to sway instability appears to occur near the load ratio 0.75 : 0.25. The sway critical load is of the order of 450 MN and the no-sway critical load for the frame is 116 MN. Thus no-sway instability may occur if the frame under consideration is effectively braced by the rest of the structure and is relatively heavily loaded. To be braced the stiffness of the rest of the structure must be significantly stiffer than the no-sway frame (probably of the order of ten times stiffer).

5. *DISCUSSION*

The most important factor in sway instability of a building which deforms in a bending mode appears to be its overall lateral stiffness. To accurately estimate such a stiffness is particularly difficult since cladding and non loadbearing partitions are significant and are not easy to include in an analytical model. Also if the structure is in reinforced concrete then such factors as cracking, creep and shrinkage can be equally difficult to model. However, given a structure with defined elastic properties it is nowadays relatively easy to estimate stiffness. A computer frame model, of which that at Fig. 2 is the simplest form, can be solved. The stiffness of the model of Fig. 2 can in fact be estimated to a satisfactory degree of accuracy by hand methods (7). Accepting the constraints on accuracy imposed when creating an elastic model, the elastic critical sway mode load of a building structure can be estimated by a relatively simple analysis.

6. *CONCLUSION*

While elastic instability of a normal building structure is extremely unlikely, it is worthwhile in design to have some measure of how much the actual vertical load differs from the critical value. This paper points the way towards a relatively simple method of doing this. A tentative procedure is as follows:

(1) Calculate the top stiffness of the complete structure either using the model of Fig. 2 or by other means. Use eqn (9) to estimate the elastic critical load of the system.

(2) If the system has a dominant shear mode, e.g. if the stiffness of the shear mode components is greater than the bending mode components then the use of Horne's Method may be more accurate than eqn (9).

(3) Check each frame and wall of the system to ensure that the local load to that frame will not cause no-sway instability.

While the above procedure needs further development before it could be used with confidence in design it does demonstrate the important feature of a dominant bending mode system (a common situation in practice) that consideration of storey height behaviour does not reflect true behaviour.

References

1. Horne, M.R. 'An approximate method for calculating the elastic critical loads of multi-storey plane frames.' The Structural Engineer, vol. 53, No. 6, 242/248, June 1975.

2. Stevens, L.K. 'Elastic stability of practical multistorey frames.' Proceedings Institution of Civil Engineers, vol. 36, 99/117, Jan. 1967.

3. Rosenblueth, E. 'Slenderness effects in buildings.' Journal Struct. Div., ASCE, vol. 91, No. ST1, 229/252, February 1965.

4. Timoshenko, P. and Gere, J.M. 'Theory of elastic stability.' 2nd Ed., McGraw-Hill, 1961.

5. Livesley, R.K. 'Matrix method of structural analysis.' Pergamon 1975.

6. Majid, K.I. 'Non linear structures.' Butterworths, London, 1972.

7. MacLeod, I.A. 'Shear wall-frame interaction - a design aid with commentary.' Portland Cement Association, U.S.A., 1970.

I WALL m^4			1.93	0.772	0.386	0.193	0.0772	0.0386
$\alpha = \frac{\text{Wall Stiffness}}{\text{Frame Stiffness}}$			11.0	4.41	2.2	1.1	0.44	0.22
Elastic Critical loads in MN	*Frame and Wall System*	*Stability Function Analysis*	449	206	123	83.0	56.8	47.0
		Equation (9)	489	216	130	87.5	61.0	51.5
		Horne's Method (critical storey)	351 (10)	164 (10)	102 (9)	70.8 (8)	50.5 (6)	42.6 (5)
		Unit Shear Method (critical storey)	319 (5)	147 (5)	88.7 (5)	59.8 (5)	41.8 (4)	35.1 (4)
	Frame only sway	*Stability Function Analysis*	32.5	32.5	32.5	32.5	32.5	32.5
	Frame only no sway		116	116	116	116	116	116
	Wall only sway		412	163	82.0	41.0	16.0	8.0
Ratio $\frac{\text{Vertical load on Wall}}{\text{Vertical load on Frame}}$			5.0	2.0	1.0	0.5	0.2	0.1

TABLE 1 ESTIMATES OF ELASTIC CRITICAL LOAD

$\alpha = \dfrac{\text{Wall Stiffness}}{\text{Frame Stiffness}}$		11.0	0.22	
Vertical load ratio Wall : Frame	0 : 1	119	46.7	*Elastic critical loads in MN*
	0.25 : 0.75	158	46.3	
	0.5 : 0.5	238	47.7	
	0.75 : 0.25	448	48.0	
	1 : 0	452	48.2	

TABLE 2 EFFECT OF DISTRIBUTION OF VERTICAL LOADS

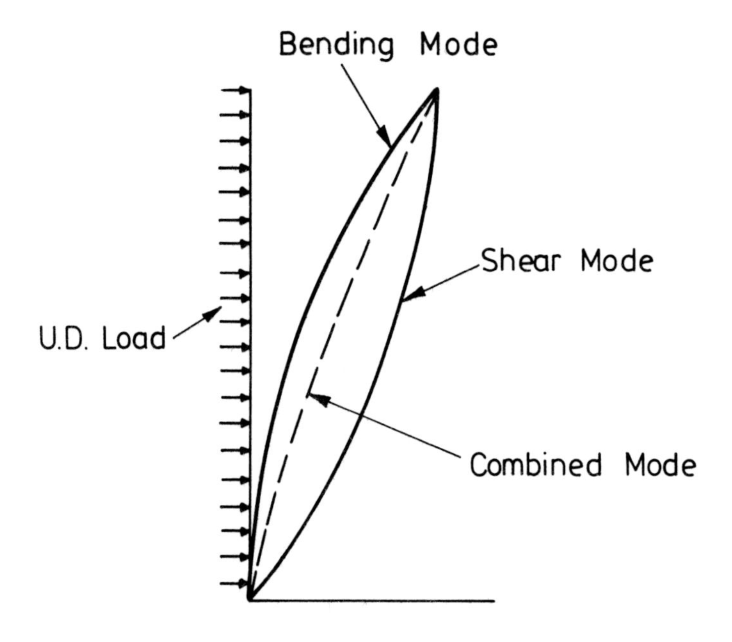

Fig. 1 Lateral deflection modes

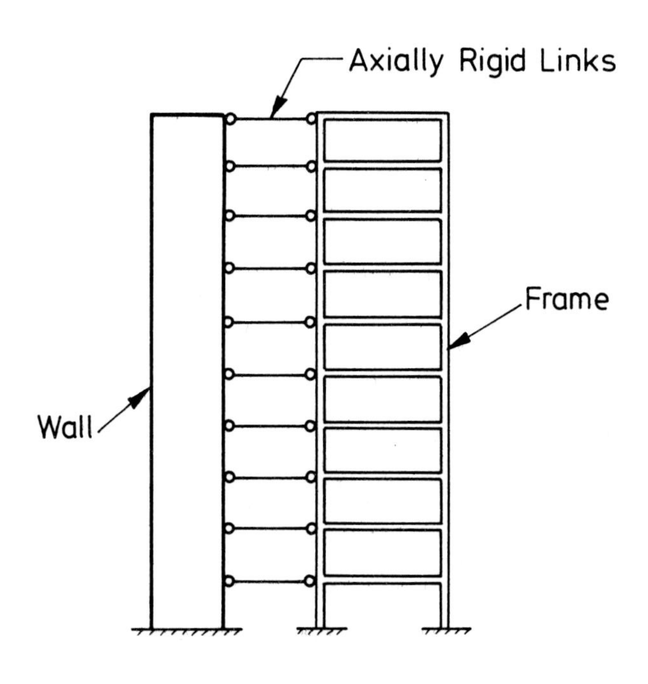

Fig. 2 Wall and frame model

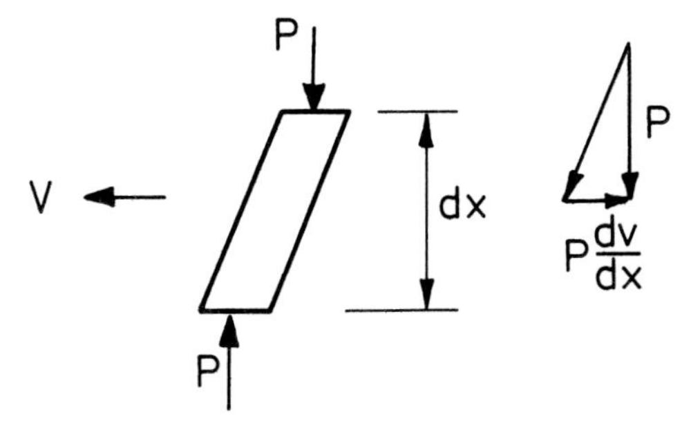

Fig. 3 Loading in shear mode element

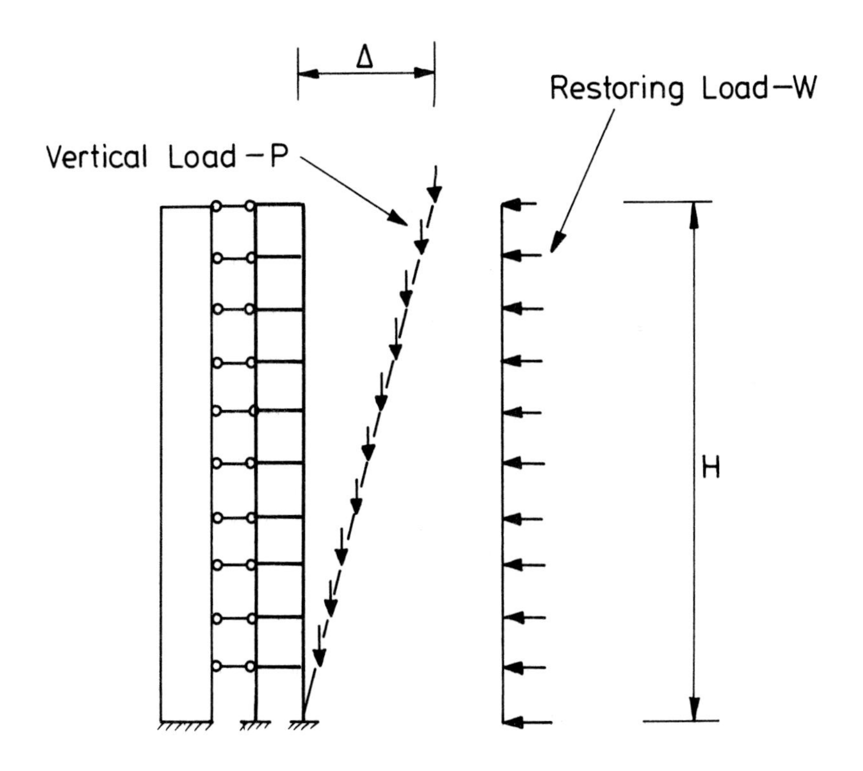

Fig. 4 Applied and restoring loads

EFFICIENT SOLUTION OF PLASTIC COLLAPSE ANALYSIS AND OPTIMUM DESIGN OF STEEL BAR STRUCTURES WITH ACCOUNT TAKEN OF INSTABILITY.

Nguyen Dang Hung, Ch Massonnet

Department of Civil Engineering, University of Liège.

Synopsis

The paper presents the basic principle and the method for working out a computer program for processing both limit analysis and limit design of frame structures under proportional loading as well as variable loading (shakedown analysis and shakedown design).

The automatic search of independent mechanisms and the dual aspects of different formulations are examined. For example, in the case of limit analysis, the kinematical formulation with rotation rates as variables is used but, by considering the duality of the linear programming method, the distribution of the moments of all sections are obtained directly from the primal calculation. In the case of limit design, the statical formulation with moments as variables is adopted but, by duality, the Foulkes' mechanism is obtained directly from the primal calculation.

Due to lack of space, present paper covers in detail only the solution of the analysis problem by the kinematic method. The other developments referred to hereabove may be found in References (11) to (15).

Notation

A_r	matrix of connection of elongation
A	reduced matrix of connection of elongation
B_r	matrix of deflexion connection
B	reduced matrix of deflexion connection
C^T	independent mechanism matrix
C	matrix of independent equilibrium equations
D	boolean connection matrix of the plastic spans
E	unit matrix, elasticity modulus
G, resp H	matrix involved in the Gauss-Jordan treatment of matrix A (displacements resp. rotations)
I	moment of inertia of the bar

K	rigidity of the bar
R	matrix of rotation kinematic connection
R_r	réduced matrix of kinematic connection of rotation
S	condensation matrix
$H_i V_i$	components of the generalized force applied to hinge i
N_p	limit normal force
A_a	web area
M_p	limit plastic moment
V_p	limit shear force
S_x	static moments of the half section
M	current moment
N	current normal force
V	limit design constraint matrix. - Current shear force
$\overline{N}$	reduced collapse stress
W	limit analysis constraint matrix
x_i, y_i	coordinates of hinge i in the structural axes
u_i, v_i	displacements of hinge i in the structural axes
n_b	number of bars
n_r	number of plastic hinges
l_k	length of bar k
n	number of nodes. - vector of normal forces
n_1	number of type 1 hinges
n_2	number of type 2 hinges
n_3	number of support hinges
n_4	number of type 4 hinges
q_r	vector of hinge displacements
d	vector of bar elongations
b	vector of bar deflexions
s	vector of absolute rotations of the bars
q	vector of node displacements
g_r	vector of generalized forces conjugated to q_r
t	vector of shear forces
g	vector of forces conjugated to q
z	vector of dependent displacements
$\overline{w}$	vector of independent displacements
m	vector of hinge moments
$\overline{m}$	vector of limit plastic moments

r	vector of the relative rotation velocities
x	primal variables (kinematic)
y	dual variables (static)
l	span length vector
α_k	angle of bar k with the x-axis
δ_k	elongation of bar k
β_k	deflexion of bar k
ϕ	total plastic dissipation of the structure
λ	kinematically admissible load multiplier
η	artificial variable for the simplex algorithm
σ_p	yield stress

Introduction

According to the Project of EUROCODE 3 (Steel Structures), the choice of the appropriate analysis methods for steel bar structures depends on the cross-sectional properties with reference to the limit state requirements.

According to the limit state conditions based on the b/t ratios of the plates composing the cross sections, the method of analysis can be

a) plastic without deformation restriction ;
b) plastic with deformation restriction ;
c) elastic with yield stress in the extreme fibres, etc...

The corresponding limit b/t ratios given in EUROCODE 3 define various types of sections called respectively :

case a) plastic sections ;
case b) compact sections ;
case c) non-compact sections, etc...

It is understood that the sections of the bar structures envisaged in present paper comply with Table 2 of the project of EUROCODE 3, which defines *plastic sections*.
This being said, there is an urgent need for an efficient computer automatic method taking account of all secondary effects :

effects on N and V on M_p capacity ;
existence of discrete series of rolled profiles
second order (stability) effects.

In the book of the senior author (1),(pp.107-133), as well as in all classical treatises of Structural Plasticity (3), (4), (5), it is shown that all problems of limit analysis and optimum limit design can be reduced to linear programs. However, there are two dual approaches to the problem :

1) the first approach is based on the technique of combining independent mechanisms due to Neal and Symonds (2), (3). This method was developed by Baltus in an internal Report (8), then published by Massonnet and Anslijn (9) and Massonnet (10).
2) the static approach taking as auxiliarly unknowns the redundants of the structure (6).

Present paper is based on the first approach for analysis and on the second for optimum design. Its basic tool is the automatic generation of a complete set of independent mechanisms, first presented in (7). Nguyen Dan Hung and de Saxce have improved this technique, that will be presented shortly in the next sections 3 and 4.

In subsequent writings ((11),(12),(13),(14),(15)) Nguyen Dang Hung has generalized the procedure in various directions (e.g. he has studied its application to Shake Down Analysis and Optimum Design, to reinforced concrete structures, to a better approach of the instability phenomena by taking account the $(P-\Delta)$ effects, etc...

In this paper, we restrict ourselves to insist on some decisive aspects of the automatic calculations and practical implementations. These are :

1) the efficient choice between the statical and kinematical formulation which leads to a minimum number of variables and limits considerably the dimension of the problem ;
2) the automatic construction of the characteristic matrices, mainly the independent equilibrium-compatible matrix ;
3) the direct calculations of dual variables, which allows to obtain all the necessary results during a simple procedure ;
4) the realistic verifications of the stability conditions imposed by European norms for steel structures.

These aspects are taken into account during the realization of a multipurpose computer program named CEPAO, written in FORTRAN IV, available in the Department of Structural Mechanics and Stability of Constructions, University of Liège. This package supplied with some facilities like interactive input data, graphic display of output results, is running now on IBM, VAX and some other mini-computer like WANG, HP, etc...

Due to lack of place, we shall assume that the basic theorems of Limit and Shakedown Analysis as well as of Optimum Limit Design are known to the reader. For the establishment of these theorems, see e.g. (1), (2), (4), (5).

1. BASIC ASSUMPTIONS OF PLASTIC ANALYSIS AND DESIGN

The object of plastic analysis is to determine the collapse load of a structure when its plastic capacities are known. Concerning the optimal plastic design, one has to minimize some merit function (usually the total weight) when the loads and the geometric dimensions of the structure are specified. The following assumptions are generally made :

- the strains are small ;
- the loading is quasi-static ;
- the material behaviour is rigid perfectly plastic ;
- the flexural action is predominant. Shear, torsion, axial forces are negligible or taken into account only a posteriori ;
- the geometric effects of stability intervene only after the plastic collapse state ;
- the plastic hinges are concentrated at critical sections ;
- there exists a continuous series of members ;
- the weight of a member is proportional to the plastic capacity.

2. CONNECTION PROBLEMS

2.1 General

Consider a plane structure formed of n_b bars numbered from 1 to n_b, n_r critical sections or plastic hinges numbered from 1 to n_r.
We will distinguish three types of critical sections in the structure. The critical sections where there is no joint mechanism belong to the type 1. The first critical section at node where joint mechanism exists belongs to the type 2. The rest of critical sections at the same node belongs to the type 4. The critical sections which constitute the supports (pinned supports fixed supports belong to the type 3.

Let x_i, y_i (Fig 1) the cartesian coordinates of critical section i, let u_i, v_i the displacement components at critical section i, let l_k the length of the element k :

$$l_k = [(x_j - x_i)^2 + (y_j - y_i)^2]^{1/2}, i,j \; [1,n_r] \; ,k \quad [1,n_b]$$

The direction of this element is defined by the angle α_k :

$$\cos \alpha_k = \frac{x_j - x_i}{l_k} \; , \; \sin \alpha_k = \frac{y_j - y_i}{l_k}$$

2.2 Kinematic connection.

Under the action of the loads, the nodes are displaced. The bar k undergoes :

1) an elongation (relative displacement of its two ends along the axis of the bar):

$$\delta_k = \cos \alpha_k u_j + \sin \alpha_k v_j - \cos \alpha_k u_i - \sin \alpha_k v_i \tag{2}$$

2) a lateral deflexion (relative displacement of the two ends in the direction perpendicular to the bar) :

$$\beta_k = - \sin \alpha_k u_j + \cos \alpha_k v_j + \sin \alpha_k u_i - \cos \alpha_k v_i \tag{3}$$

This deflexion is accompanied by a rotation of the bar

$$w_k = \beta_k / l_k \tag{4}$$

Consider, now, for the structure under study :

a) The vector "displacements of the movable nodes of types 1 and 2":

$$q^T = [u_1 v_1, \ldots, u_n v_n] \tag{5}$$

b) the vector "elongations of the bars"

$$d^T = [\delta_1, \ldots, \delta_{n_b}]$$

c) the vector "deflexion of the bars"

$$b^T = [\beta_1, \ldots, \beta_{n_b}]$$

d) the vector "absolute rotations of the bars"

$$s^T = [w_1, \ldots, w_{n_b}]$$

Equations (2), (3) and (4) can be written in matrix form as follows:

$$d = A\,q \qquad (6)$$

$$b = B\,q \qquad (7)$$

$$s = R\,q \qquad (8)$$

where :

A is a matrix of dimensions $n_b \times 2(n-n_3)$, called matrix of kinematic connection of elongation ;

B is a matrix of dimensions $n_b \times 2(n-n_3)$, called matrix of kinematic connection of deflection ;

R is a matrix of dimensions $n_b \times 2(n-n_3)$, called matrix of kinematic connection of rotation.

Considering (2), (3) and (4), it appears that the terms in row number k of A, B, R, are zero, except

$$A(k,2j-1)=\cos\alpha_k;\ A(k_1 2i-1)=-\cos\alpha_k;\ A(k_1 2j)=\sin\alpha_k;\ A(k_1 2i)=-\sin\alpha_k \qquad (9)$$

$$B(k,2j-1)=-\sin\alpha_k; B(k,2i-1)=\sin\alpha_k;\ B(k,2j)=\cos\alpha_k;\ B(k,2i)=-\cos\alpha_k \qquad (10)$$

$$R(k,2j-1)=-\sin\alpha_k/l_k; R(k,2i-1)=\sin\alpha_k/l_k; R(k,,2j)=\cos\alpha_k/l_k;$$
$$R(k,2i) = -\cos\alpha_k/l_k \qquad (11)$$

2.3 *Static connection*

Suppose, now, that, at each hinge number i, a concentrated force be applied whose components in global axes are H_i, V_i.

We may now introduce the vector of the generalized forces conjugated to the displacement vector q_r :

$$g^T = [H_1\ V_1, \ldots, H_n\ V_n]$$

These forces induce in each bar k a normal force N_k and a shear force T_k. These internal forces of the bar elements of the general structure form the following vectors :

$n^T = |N_1, \ldots, N_{n_b}|$: vector of the normal forces conjugated to the vector of elongations, d ;

$t^T = |T_1, \ldots, T_{n_b}|$: vector of the shear forces conjugated to the vector of the deflexions, b ;

Let us write now the equation of virtual work of the general structure :

$$n^T \delta d + t^T \delta b = g^T \delta q \qquad (12)$$

Replacing the variations δd, δb by their respective values defined in (6) and (7) gives :

$$(n^T A + t^T B)\ \delta q = g^T \delta q \qquad (13)$$

Identifying to the two members of this equation, we obtain :

$$g = A^T n + B^T t \qquad (14)$$

3. *AUTOMATIC SEARCH OF THE INDEPENDENT MECHANISMS OR OF THE INDEPENDENT EQUILIBRIUM EQUATIONS OF THE STRUCTURE*

3.1 *Deflexion mechanisms*

The deflexion mechanics are also called displacement mechanisms, because they are connected to the displacements of the nodes ; these are the mechanisms which produce non zero work. The panel and beam mechanisms belong to this category.

As the bars are supposed to deform by bending only, these mechanisms are characterized by no elongation of any bar. According to equation (6), this gives

$$A\, q = 0 \tag{15}$$

This is the relation connecting the node displacements of the structure in a flexural mechanism. If vector q is decomposed in two partial vectors z and $\overline{w}$, we have :

$$q^T = |z^T\ \overline{w}^T| \tag{16}$$

(after eventual permutation of the components) where

z (of dimension n_b is the vector of the dependent displacements ;
$\overline{w}$ (of dimension $2(n-n_3)-n_b)$ is the vector of the independent displacements.

Condition (22) may be written explicitely

$$|A_z\ A_w| \begin{vmatrix} z \\ \overline{w} \end{vmatrix} = 0 \tag{17}$$

where :
A_z is square with dimensions n_b x n_b ;
A_w is rectangular with dimensions n_b x $(2n-2n_3-n_b)$.
The structure being fixed against any rigid movement, A_z may be inverted, so that :

$$z = -\, A_z^{-1}\, A_w\, \overline{w} = G\, \overline{w} \tag{18}$$

with

$$G = -\, A_z^{-1}\, A_w \tag{19}$$

On the other hand, the decomposition (14) yields the decomposition of the dual vector

$$g^T = |h^T\ \overline{e}^T| \tag{20}$$

which yields itself the following decomposition of relation (14)

$$\begin{vmatrix} h \\ \overline{e} \end{vmatrix} = \begin{vmatrix} A_z^T & B_z^T \\ A_w^T & B_w^T \end{vmatrix} \begin{vmatrix} n \\ t \end{vmatrix}$$

whence we obtain :

$$n = A_z^{-T}\, [h - B_z^T\, t] \tag{21}$$

$$\overline{e} = -\, G^T\, [h - B_z^T\, t]\ + B_w^T\, t \tag{22}$$

Equation (21) enables to compute the normal forces in the bars as

soon as the shear forces are known. Equation (22) leads to the independent equations of flexural equilibrium as follows :
Let M_i be the bending moment at hinge i of the structure. The general moment vector is then

$$m^T = |M_1, \ldots, M_{n_r}|$$

The shear force T_k in bar k is connected to the moments at the ends of this bar by the equilibrium equation

$$T = \frac{M_i - M_j}{l_k} \qquad (23)$$

In matrix form, the general vector of shear forces is connected to the moment vector by means of a matrix U which is easily deduced by (23) :

$$t = U\,m \qquad r = U^t b \qquad (24)$$

Replacing t in (22) by its value (24), one finds the final equilibrium equations corresponding to the independent deflexion mechanisms

$$e_1 = \bar{e} + G^T h = C_1 m \qquad (25)$$

with

$$C_1 = (G^T B_z^T + B_w^T)\,U \qquad (26)$$

The dimensions of C_1 are obviously $(2n-2n_3-n_b) \times n_r$.

The kinematic relation, dual of (25), which connects the relative rotations of the hinges

$$r^T = |\theta_1, \ldots, \theta_{n_r}|$$

(which are conjugate to the bending moments), to the independent displacements w, is obtained by writing the complementary virtual work equation as follows :

$$\delta m^T r = \delta h^T z + \delta\bar{e}^T\bar{w} = (\delta h^T G + \delta\bar{e}^T)\bar{w} = \delta\bar{e}_1^T\,\bar{w}.$$

Replacing δe_1^T by its value (25) and identifying both members of the equation gives

$$r = C_1^T\,\bar{w}. \qquad (27)$$

This shows clearly that C_1^T is the matrix of the independent displacement mechanisms.

C_1^T could also have been found directly by computing the relative rotations of the hinges by means of the absolute rotations of the bars defined by (8) and the independent displacement $\bar{w}$ defined by (18).

Practically, the inversion process (17), (18) is produced automatically by the Gauss-Jordan elimination method, whose algorithm is available on any computer.

3.2. Rotation mechanisms

In addition to the deflexion mechanisms, we must introduce in C_1 the rotation mechanisms which correspond to the node equilibrium equations. We have an equation of this type in all nodes where a hinge type 2 exists. These equations do not involve any work and connect the moments at type 2 hinges to the moments at type 4 hinges.

$$e_2 = 0 = C_2 m \tag{28}$$

The coefficients of C_2 are defined to a common factor and their signs correspond to the customary sign convention. Practically, it is convenient to adopt as common factor the smallest term (in absolute value) of C_1, in order to have a good conditioning of the complete matrix C :

$$C = \begin{vmatrix} C_1 \\ C_2 \end{vmatrix}$$

The equation dual of (28) is

$$r = C_2^T w_2 , \tag{29}$$

where w_2 is the dual displacement vector of the zero e_2 forces. If we set

$$e^T = |e_1^T \quad 0| \tag{30}$$

$$w^T = | \bar{w}, w_2 |$$

we can rewrite (32), (34), (35) and (36) in the following dual forms:

$$e = C\, m \tag{31}$$

$$r = C^T w \tag{32}$$

Note that the dimension of matrix C is $(2n_1 + 3n_2 - n_b) \times n_r$.

As the member of independent mechanisms of a structure n_h times statically indeterminate and having n_r critical sections is

$$n_m = n_r - n_h \tag{33}$$

it is imperative to verify the following relation

$$n_r - n_h = 2n_1 + 3n_2 - n_b \tag{34}$$

for any admissible structure.

3.3 Example

To illustrate preceding formulation, we consider the simple portal frame of Fig 2 which has 7 critical sections $(n_r=7)$ and 4 bars $(n_b = 4)$.

As the degree of redundancy of the structure is $n_h=3$, we must find $n_m = n_r = n_h = 4$ independent mechanisms. Using the numbering of hinges and bars indicated on fig.2, we have

$$n_1 = 1, \quad n_2 = 2, \; n_3 = 2, \; n_4 = 2.$$

The number of nodes is $n = n_1 + n_2 + n_3 = 5$.

Equation (34) is verified.

The connexion matrices A, B, of equ.(4), (7), are respectively

$$
A = \begin{vmatrix} 0 & 0 & 0 & 1 & 0 & 0 \\ 1 & 0 & -1 & 0 & 0 & 0 \\ -1 & 0 & 0 & 0 & 1 & 0 \\ 0 & 0 & 0 & 0 & 0 & 1 \end{vmatrix} \begin{matrix} \text{bar 1} \\ \text{bar 2} \\ \text{bar 3} \\ \text{bar 4} \end{matrix} \qquad B = \begin{vmatrix} 0 & 0 & -1 & 0 & 0 & 0 \\ 0 & 1 & 0 & -1 & 0 & 0 \\ 0 & -1 & 0 & 0 & 0 & 1 \\ 0 & 0 & 0 & 0 & -1 & 0 \end{vmatrix} \begin{matrix} \text{bar 1} \\ \text{bar 2} \\ \text{bar 3} \\ \text{bar 4} \end{matrix}
$$

(columns of A: $u_1 \; v_1 \; u_2 \; v_2 \; u_3 \; v_3$)

Considering the form of matrix A, it is visible that the decomposition (17) is possible with :

$$
z^T = |v_2 \quad u_2 \quad u_3 \quad v_3| \quad ; \quad \bar{w}^T = |u_1 \quad v_1|
$$

We find easily :

$$
A_z = \begin{vmatrix} 1 & 0 & 0 & 0 \\ 0 & -1 & 0 & 0 \\ 0 & 0 & 1 & 0 \\ 0 & 0 & 0 & 1 \end{vmatrix} = A_z^{-1}
$$

$$
A_w = \begin{vmatrix} 0 & 0 \\ 1 & 0 \\ -1 & 0 \\ 0 & 0 \end{vmatrix} , \quad G = -A_z^{-1} A_w = \begin{vmatrix} 0 & 0 \\ 1 & 0 \\ 1 & 0 \\ 0 & 0 \end{vmatrix}
$$

$$
B_z = \begin{vmatrix} 0 & -1 & 0 & 0 \\ -1 & 0 & 0 & 0 \\ 0 & 0 & 0 & 1 \\ 0 & 0 & -1 & 0 \end{vmatrix} \qquad B_w = \begin{vmatrix} 0 & 0 \\ 0 & 1 \\ 0 & -1 \\ 0 & 0 \end{vmatrix}
$$

so that displacements u_1, v_1, are the only independent variables.
Equations (18) are simply :

$$
v_2 = v_3 = 0 \quad ; \quad u_2 = u_1 \quad , \quad u_3 = u_1
$$

The dual decomposition (20) being :

$$
h^T = |V_2 \quad H_2 \quad H_3 \quad V_3| \quad ; \quad \bar{e}^T = |H_1 \quad V_1|
$$

The external forces defined by (25) are :

$$
e_1 = \begin{vmatrix} \bar{H}_1 \\ \bar{V}_1 \end{vmatrix} = \begin{vmatrix} H_1 \\ V_1 \end{vmatrix} + \begin{vmatrix} 0 & 1 & 1 & 0 \\ 0 & 0 & 0 & 0 \end{vmatrix} \begin{vmatrix} V_2 \\ H_2 \\ H_3 \\ V_3 \end{vmatrix} = \begin{vmatrix} H_1 + H_2 + H_3 \\ V_1 \end{vmatrix}
$$

Figure 2 shows that :

$$
H_1 = H_3 = 0 \quad ; \quad H_2 = P \quad ; \quad V_1 = - P.
$$

This gives the left-hand member of equation (25) :

$$e_1 = P\begin{vmatrix} 1 \\ -1 \end{vmatrix}$$

To find matrix C_1 defined by (26), we must first find matrix U defined by (24). One finds easily

$$U = \frac{1}{l}\begin{vmatrix} 0 & -1 & 0 & 1 & 0 & 0 & 0 \\ -1 & 0 & 0 & 0 & 0 & 1 & 0 \\ 1 & 0 & 0 & 0 & 0 & 0 & -1 \\ 0 & 0 & 1 & 0 & -1 & 0 & 0 \end{vmatrix}$$

so that :

$$C_1 = \frac{1}{l}\begin{vmatrix} 0 & 1 & -1 & -1 & 1 & 0 & 0 \\ -2 & 0 & 0 & 0 & 0 & 1 & 1 \end{vmatrix}$$

and we obtain finally the independent deflexion equilibrium equations:

$$Pl = -M_4 + M_2 - M_3 + M_5$$
$$Pl = 2M_1 - M_6 - M_7$$

The first equation corresponds to a panel mechanism, and the second to a beam mechanism.

We are now faced with the rotation mechanism ; there are two of them because we have the two corresponding equilibrium equations :

$$M_2 - M_6 = 0$$

$$M_7 - M_3 = 0$$

The complete rotation matrix defined by (30), (31), is then

$$C = \frac{1}{l}\begin{vmatrix} 0 & 1 & -1 & -1 & 1 & 0 & 0 \\ -2 & 0 & 0 & 0 & 0 & 1 & 1 \\ 0 & 1 & 0 & 0 & 0 & -1 & 0 \\ 0 & 0 & -1 & 0 & 0 & 0 & 1 \end{vmatrix} \qquad (35)$$

4. *KINEMATIC FORMULATION OF THE PROBLEM OF PLASTIC LIMIT ANALYSIS WITH PROPORTIONAL LOADING*

4.1 General

We call proportional or simple loading a loading governed by a single parameter λ, called load multiplier. The kinematic formulation of the analysis problem is based on the fundamental kinematic theorem, also called upper bound theorem (1), (2), (3), which states that *the true multiplier is the smallest among all kinematically admissible multipliers* λ_+ *corresponding to licit mechanisms*. We call "licit" at the plastic point of view any mechanism compatible with the topology of the structure and which induces a positive dissipation in all its yielded sections.
Let

$$\bar{m}^T = |M_{p1}, M_{p2}, \ldots, M_{pn}|$$

be the vector of the plastic moment capacities of the critical sections ; the total plastic dissipation of the structure is

$$\phi = \bar{m}^T |\dot{r}|$$

where $\dot{r}$ is the vector of the relative rotation velocities of the critical sections. The equation of virtual powers for the general licit mechanism reads :

$$\lambda g^T \dot{q} = \bar{m}^T |\dot{r}| \qquad (36)$$

Taking account of (16), (18), (20), (25) and (30), the left-hand member of (36) may be expressed as a function of the independent displacement velocities :

$$\lambda e^T \dot{w} = \bar{m}^T |\dot{r}| \qquad (37)$$

The kinematically admissible rotations must satisfy (32). In these conditions, the kinematic formulation of the problem of Limit Analysis reduces to the following linear program :

To find $\dot{r}$, $\dot{w}$

such that

$$\phi = \bar{m}^T |\dot{r}| = \textit{minimum} \qquad (38)$$

with the conditions

$$\dot{r} = C^T \dot{w}$$

and $\quad e^T \dot{w} = \xi = \textit{constant}.$

The limit multiplier will then be :

$$\lambda = \phi/\xi$$

The formulation (38) is similar to the formulation adopted in (6), (7), (12), (1). As the simplex method generally utilized for solving the linear program accepts only positive variables, the following variables are often introduced :

$$\dot{\theta}_i = \dot{\theta}_i^+ - \dot{\theta}_i^- \quad ; \quad |\dot{\theta}_i| = \dot{\theta}_i^+ + \dot{\theta}_i^- \qquad (39)$$

$$\dot{w}_k = \dot{w}_k^+ - \dot{w}_k^- \quad ; \quad |\dot{w}_k| = \dot{w}_k^+ + \dot{w}_k^- \qquad (40)$$

in which case the function to minimize becomes

$$\phi = \sum_{i=1}^{n_r} (M_{pi}^+ \dot{\theta}_i^+ + M_{pi}^- \dot{\theta}_i^-) \qquad (41)$$

In the CEPAO program, we adopt the following change of variables in order to minimize their number. We choose a displacement velocity $\dot{w}_o$ such that the new unknowns :

$$\dot{w}_k' = \dot{w}_k + \dot{w}_o \qquad (42)$$

are always positive. The estimate of $\dot{w}_o$ is based on the relative values of the rotation matrix C. This translation transforms the problem (41) in an equivalent problem, which has the following convenient matrix form :

To find

$$\dot{r}_+ , \dot{r}_- , \dot{w}'$$

such that

$$\phi = \bar{m}_+^T \dot{r}_+ + \bar{m}_-^T \dot{r}_- = minimum \qquad (43)$$

with the conditions

$$- \dot{r}_+ + \dot{r}_- + C^T \dot{w}' = C^T \dot{w}_o$$

and

$$e^T \dot{w}' = e^T \dot{w}_o + \xi = \xi'$$

The shadow prices d_+, d_-, are obtained (17) by applying the simplex of the primal problem (49). Problem(56) gives the moments in all hinges :

$$\text{if } \theta_i = \theta_i^+, \text{ then } M_i = M_i^+ = M_{pi}^+ - d_i^+ \qquad (44)$$

$$\text{if } \theta_i = - \theta_i^-, \text{ then } M_i = - M_i^- = d_i^- - M_{pi}^-$$

The details of the calculations on the computer are given in (8). As soon as the moment distribution in the hinges is known, the shear forces in the bars are computed by formula (24), then the normal forces by formula (21). To obtain the collapse mechanism, one uses the solution of the primal problem (43), then eqn. (42), to obtain the independent displacement velocities. The other components of the displacement field are obtained by (18).

5. *GENERAL FLOW CHART OF PROGRAM CEPAO (see Fig 3)*

6. *ADDITIONAL CONSIDERATIONS*

For what regards the optimum (minimum weight) limit design, linear programming furnishes the maximum theoretical plastic moments of the spans. The program CEPAO contains a data bank of the available rolled profiles (HEA, HEB, IPE) in which the computer selects automatically the smallest one which has the plastic strength at least equal to that required.

As soon as the choice of the profiles is finished, the program undertakes the systematic verification effect of V and N on the plastic moment capacity, as well as of the stability criteria proposed by ECCS (European Construction for Constructional Steelwork) (16). These criteria concern plate buckling (of the walls of the profiles), and lateral torsional buckling of the beams and columns).

If any of these criteria is violated, and this at any place of the structure, the program searches automatically in the list of rolled profiles mentioned above the smallest one which complies with all stability requirements.

At the end of the optimization process, CEPAO executes a limit analysis of the structure found with these revised profiles. The limit load multiplier obtained gives a supplementary estimate of the safety of the frame against plastic collapse.

This connection "après coup" for the additional effects and the fact that the general stability of the frame is not controlled in CEPAO may be considered as a weakness of the program in its present state of development. This is the reason why work is under way at Liège to obtain automatically (by linear programming) the elastoplastic displacements at impending collapse. This will enable the program to take account of the (P-Δ) effects and, if desired, to control if the displacements at impending collapse are not excessive.

Some final remarks about the main characteristics and the *efficiency of the CEAPO program.*

The input of data in CEPAO is identical to that of any classical program of elastic analysis. The preprocessor of CEPAO determines *automatically* the critical hinges necessary to the computations. The results are printed with complete numbering system. The CEPAO has also a postprocessor which enables to draft (on a Benson plotter) the shape of the structure, its loading, the M, N, V diagrams, the optimal mechanism and the distribution of optimal profiles. The program written in FORTRAN is presently operational for any bar structure not exceeding 60 critical sections, 40 independents mechanisms, 40 bars, 20 loading cases, 30 unknown profiles and 20 technological conditions.

To give some idea of the efficiency of CEPAO within the place available in this paper, we shall only mention the two storey, two bay pitched roof portal frame represented at Fig 4. For a detailed presentation of the results involving all data mentioned above, the reader is referred to (11). The total CPU computation time for two optimization processes plus the control analysis on the IBM 370-158 computer of University of Liège was 17.3 seconds.

References

1. Massonnet, Ch. and Save, M. 'Plastic analysis and design of structures, Vol. 1 : Structures depending on one parameter' Third french edition, 637 pp., ed.:B. Nélissen, rue Denis Lecocq, 22, 4000 Liège, Belgium.

2. Neal, B.G. and Symonds, P.S. 'The rapid calculation of the plastic collapse load for a framed structure'. Proc. Inst. Civil Engrs. London, pp. 58-100, 1952.

3. Neal, B.G. 'The plastic methods of structural analysis'. Chapman and Hall, London, 1956.

4. Martin, D.B. 'Plasticity, Fundamentals and general results.' MIT Press Cambridge, Massachusetts, Londres, 1975.

5. Majid, K.I. 'Non-linear structures'. Londres, Butterworths, 1978.

6. Lescouarc'h, Y. 'Development of the plastic computation methods of bar structures (in french)'. Ann. I.T.B.T.P. Suppl. to N°326, March 1975.

7. Sander, M.L.G. 'Etude sur le calcul automatique de la charge limite plastique'. Travail de fin d'études, Laboratoire de Mécanique des Matériaux et de Statique des Constructions, Université de Liège, 1971-1972.

8. Baltus, R. 'Calcul automatique de la charge de ruine plastique des structures planes formées de barres. Travail de fin d'études, Laboratoire de Mécanique des Matériaux et Statique des Construc-

tions. Université de Liège, 1973-1974.

9. Massonnet, Ch. and Anslijn, R. 'Optimal plastic design of plane bar structures (in french), in the book : 'Les méthodes d'optimisation dans la construction'. Ed. Eyrolles, Paris, pp. 125-161, 1975.

10. Massonnet, Ch. 'Développement d'un programme automatique de dimensionnement plastique des structures planes en acier à l'aide de la programmation linéaire.' Proceedings of the Symposium on plastic analysis of structures, Roumanie, septembre 1972.

11. Nguyen Dang Hung and de Saxce, G. 'Plastic analysis and design of bar structures including stability'(in french). Construction Métallique, N° 3, pp. 15-38, 1981.

12. Nguyen Dang Hung 'An automatic program for rigid-plastic and elastic-plastic analysis and optimization of framed structures.' Proceedings of the Workshop on "The application of the Mathematical programming method to structural analysis and design. To appear in the International Journal of Engineering Structures (J. Munro, editor).

13. Nguyen Dang Hung 'Further aspects of analysis and optimization of structures and proportional and variable loadings'. Proceedings of the EUROMECH-Colloquium 164 "Optimization methods in structural design , held at University of Siegen (Fed. Rep. of Germany) on October 12-14, 1982. (to be published).

14. Nguyen Dang Hung, de Saxce, G., Lemaire, E., Villers, P. 'User's guide of CEPAO-82'. Internal Report N°127 of the Laboratory of Mechanics and Materials and Statics of Structures, Liège University, November 1982.

15. Nguyen Dang Hung 'An automatic method for the elastoplastic displacements at impending collapse - Laboratory of Mechanics of Materials and Theory of Structures, Liège University, Internal Report N° 125, November 1982.

16. E.C.C.S. 'European Recommendations for Steel Constructions' E.C.C.S. - EG.77-2E, March 1978.

17. Beale, E.M.L. 'Mathematical programming in practice'. John Wiley and Sons, New-York, 1968.

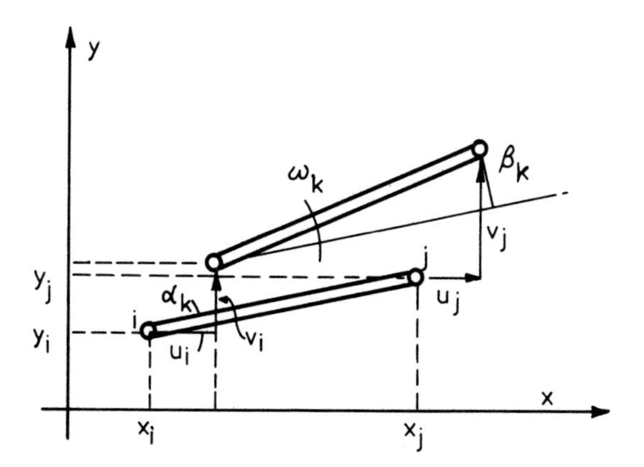

Fig 1 Bar element

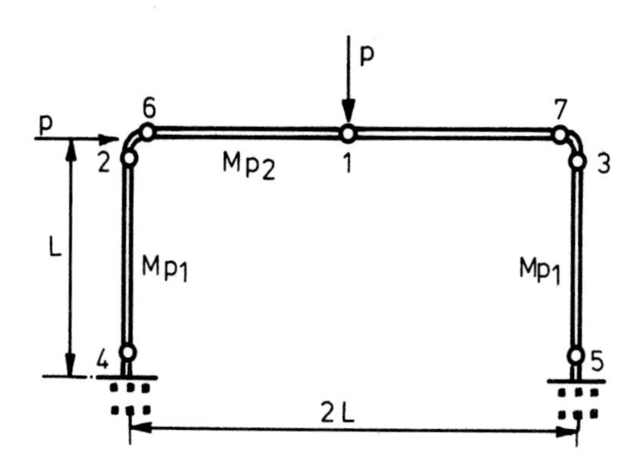

Fig 2 Simple rectangular frame

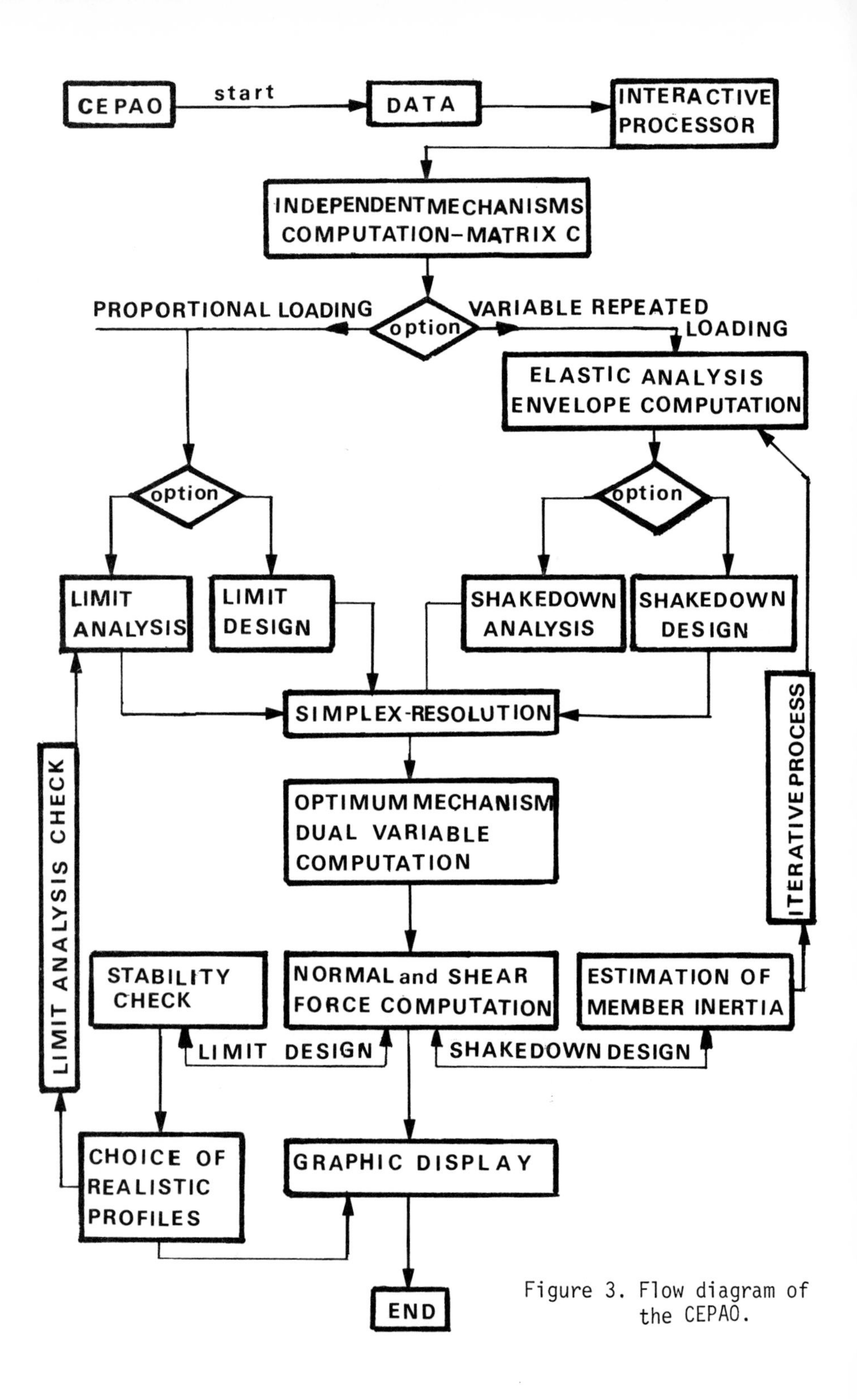

Figure 3. Flow diagram of the CEPAO.

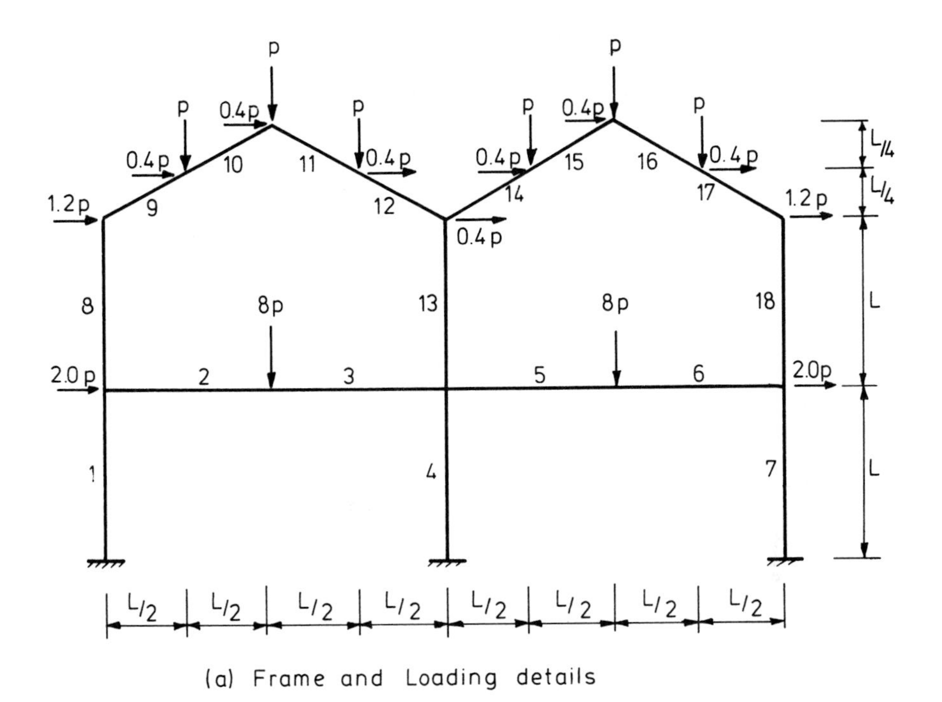

(a) Frame and Loading details

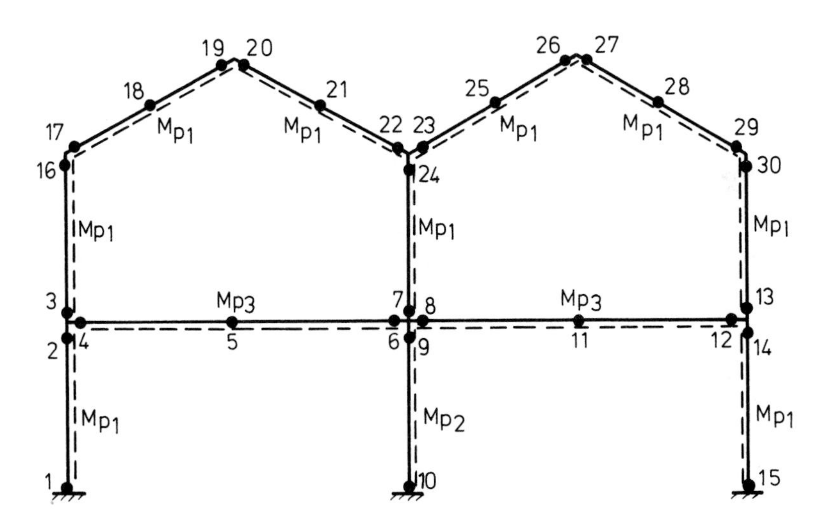

$M_{p1} = M_p$; $M_{p2} = M_{p3} = 2M_p$

(b) Possible hinge locations

Fig 4 Two bay, two storey pitched roof portal frame

AN ELASTO-PLASTIC HAND METHOD FOR UNBRACED RIGID-JOINTED STEEL FRAMES.

D. Anderson and T.S. Lok

Department of Engineering, University of Warwick

Synopsis

A simple method is presented for the estimation of second-order elasto-plastic collapse loads for unbraced rigid-jointed plane steel frames. The proposal requires no additional parameters apart from those necessary for the well-known Merchant-Rankine formula. Results are compared with the modified formula due to Wood. The proposed method gives closer agreement to accurate computer results for bare frames than the modified formula. Suggestions are made for the inclusion of strain-hardening and cladding stiffness where appropriate.

Notation

λ	load factor; collapse load predicted by proposed method
λ_c	lowest elastic critical load
λ_{det}	deteriorated critical load
λ_f	collapse load from second-order elasto-plastic computer analysis
λ_{MR}	collapse load predicted by Merchant-Rankine formula
λ_p	rigid-plastic collapse load
λ_{WMR}	collapse load predicted by modified Merchant-Rankine formula

Introduction

Unbraced steel frames provide a convenient construction for relatively low multi-storey buildings with open interior layouts. Such frames have traditionally been designed under combined loading by the 'simple' method which, although based on contradictory assumptions, leads to designs which are regarded as adequate and is easy to apply(1).

In contrast, the plastic design of sway frames is a much more complicated task, due to the need to give a fuller consideration to instability effects. For this reason, computer-based methods are often the most appropriate (e.g. 2,3). However, if suitable software is not readily available to a designer, then the use of a relationship such as the Merchant-Rankine formula (4) provides an attractive alternative. Such methods are also required when checking designs provided by computer. According to the Merchant-Rankine formula, the load level at failure, λ_{MR}, is given by

$$1/\lambda_{MR} = 1/\lambda_p + 1/\lambda_c \qquad (1)$$

where λ_p is the load level at rigid-plastic collapse and λ_c denotes the lowest elastic critical load. In the past, verification of the formula has relied on elasto-plastic analyses of small frames and tests on model structures (e.g. 5,6), which showed eqn. (1) to be generally conservative. Recently, an extensive parametric study has been carried out by the authors on frames up to 10 storeys in height and 5 bays in width (7). The study confirmed the conservative nature of the formula when realistic combinations of horizontal and vertical load were applied. Indeed, λ_{MR} could be below the load at which the second-order computer analysis (8) showed the first plastic hinge to form.

Wood (9) has proposed a modified relationship to make some allowance for strain-hardening and stray composite action. Denoting the resulting collapse load by λ_{WMR}

$$1/\lambda_{WMR} = 0.9/\lambda_p + 1/\lambda_c, \text{ with } \lambda_{WMR} \not> \lambda_p \qquad (2)$$

This form of the equation will also offset the tendency to underestimate the load level at failure. It has been recommended (10, 11) that eqn. (2) should be used only when rigid-plastic collapse is by a combined mechanism, and the authors' studies showed that λ_{WMR} varied between 97% and 104% of the computer result when this restriction was observed. In practice, though, the designer will frequently wish to analyse a trial set of sections which already satisfy criteria such as adequate stiffness at working load, and a combined mechanism cannot therefore be guaranteed. When the rigid-plastic collapse mode was unrestricted, the study showed that λ_{WMR} could now exceed the computer result by as much as 7%. It was argued that this is acceptable because eqn. (2) strictly applies only to clad buildings, whilst the computer analyses were on bare frames and strain-hardening was neglected.

It is recognised, though, that some engineers will prefer not to rely on strain-hardening and cladding to ensure adequate strength, and also that certain kinds of structure have minimal composite action. The authors have therefore sought an expression for the collapse load which will retain the simplicity of the Merchant-Rankine approach, but will provide closer agreement with the computer analyses on bare frames, irrespective of the shape of the rigid-plastic collapse mode. Whereas Wood used a single factor of 0.9 to allow for strain-hardening and composite action (as well as the conservative tendency of eqn. (1)), the authors intend that after further research, such effects be included as optional items to enhance the basic strength of the frame at the designer's discretion.

1. DETERIORATION OF STIFFNESS

The parametric showed that eqn. (2) tended to significantly exceed the results of the computer analyses when λ_c/λ_p was relatively high. This is not surprising, as eqn. (2) enables λ_{WMR} to be taken as λ_p once λ_c/λ_p reaches 10.0. A compromise factor of say 0.95, substituted in eqn. (2), would therefore be unsatisfactory because this would cause an approximately constant percentage reduction, no matter what the value of λ_c/λ_p.

The expression for the collapse load proposed below has been obtained by using the order of hinge formation shown by computer to calculate the deterioration of frame stiffness under increasing load, λ. A typical result is shown in Fig. 1, the ratio of the deteriorated critical load, λ_{det}, to λ_c being plotted against λ/λ_p. λ_{det} only changes when a new plastic hinge forms, but these points have been joined by a discontinuous line in Fig. 1 to represent a gradual reduction in stiffness. The expression for the collapse load has then been obtained by seeking a smooth curve to fit this result. The following has been adopted

$$\frac{\lambda}{\lambda_p} = \left[1 - \frac{0.4\lambda_p}{\lambda_c}\right]\left[1 - \left(\frac{\lambda_{det}}{\lambda_c}\right)^2\right] \qquad (3)$$

As collapse occurs when the rising load factor λ equals the deteriorated critical load λ_{det} (12), the failure load is found by solving the quadratic

$$\frac{\lambda}{\lambda_p} = \left[1 - \frac{0.4\lambda_p}{\lambda_c}\right]\left[1 - \left(\frac{\lambda}{\lambda_c}\right)^2\right] \qquad (4)$$

for λ. The factor of 0.4 has been chosen to give close agreement between λ from eqn. 4 and failure loads given by computer, λ_f, but it can be seen from Fig. 1 that the expression also gives an approximate representation of the deterioration of stiffness once hinges begin to form.

2. COMPARISON WITH PARAMETRIC STUDIES

The parametric study used to examine the Merchant-Rankine formulae also demonstrates the accuracy of eqn. (4). Full details of the frames are given in (7) and will not be repeated here. It is worth noting, though, that they all satisfied a limit of (height/300) on sway under unfactored wind load, and the minimum beam sections were governed by the need to sustain vertical loading to a higher level than that required under combined loading. Partial safety factors were taken from the 1977 British draft code (10), and the design loads for combined loading were taken as corresponding to a load factor of unity. Two different bay widths were considered, with one storey height. Loading generally consisted of extreme values of the ratio of vertical to wind load, although some intermediate values were also considered.

The results are summarised in Tables 1 and 2, the failure loads given by the proposed equation being denoted by λ and the computer results by λ_f. The rigid-plastic collapse mode is indicated by a letter, a beam-type mechanism by B, column sway by S and a combined mechanism by C. The first figure in the frame identification refers to the number of storeys, and the second to the number of bays, with W and N referring to wide or narrow bay widths. Comparisons show that the proposed expression exceeds the computer results by a maximum of only 3%, compared to 7% by the Wood-Merchant-Rankine formula. Equations (2) and (4) both give maximum underestimates of 8%, but this arises in a frame for which λ_c/λ_p is just 3.19. It has been proposed that eqn. (2) should only be applied when λ_c/λ_p is at least 4.0. Observing this restriction for both equations, eqn. (2) gives a maximum underestimate of 3%, and the proposed expression 5%.

3. *FURTHER COMPARISONS*

Other frames from the literature provide further comparisons between eqns. (2) and (4), and second-order computer analysis. The results are given in Table 3, the load factors now being multiples of the working loads. The proposed expression gives a maximum deviation of + 4% from the computer results, whereas eqn. (2) gives + 9%. In defence of the Wood-Merchant-Rankine formula, it should be noted that both the second and sixth examples had sway deflections at working load which exceeded the usual limit of (height/300); indeed the roof beam of the second example in Table 3 was simply-supported to enable small column sections to be used. However, it can be seen that the proposed equation was able to deal with these difficult cases in a significantly more satisfactory manner.

4. *APPLICATIONS IN PRACTICE*

The comparisons given above have shown the proposed expression being used as an analysis tool, with λ determined by solution of eqn. (4), once λ_c and λ_p are known. As accurate comparisons were required to validate the expression, λ_c and λ_p were determined from suitable computer programs (8).

In practical design, the most convenient procedure is to take λ as the specified load level for collapse. Using design (i.e. factored) loads, λ will therefore be unity. The elastic critical load λ_c for a trial design can be very easily determined to good accuracy from charts (9, 11). The minimum value required the rigid-plastic collapse load can be then found by solving eqn. (4) for λ_p. In this way an exact calculation of the rigid-plastic load of a trial design can often be avoided, use being made instead of the lower-bound theorem.

With $\lambda = 1.0$, eqn. (4) can be re-arranged to give λ_p

$$0.4\lambda_p^2 - \lambda_c.\lambda_p + \lambda_c^3 / (\lambda_c^2 - 1) = 0 \qquad (5)$$

This expression is shown graphically in Fig. 2, safe designs being above the solid curve. Equations (1) and (2) are also plotted, showing that the proposed method gives results which are very similar to eqn. (2) for low values of λ_c/λ_p, where the latter equation has been found to be particularly successful. The parametric study (7) indicated that as λ_c/λ_p increases, a more conservative result than eqn. (2) is required. Figure 2 demonstrates that the proposed method successfully provides this by requiring higher minimum values of λ_p in order to attain the design load.

5. *CONCLUSION*

A simple expression has been presented for collapse loads of plane unbraced multi-storey steel frames under combined loading. Studies on 38 examples have shown the calculated value to lie between 92% and 104% of the figures given by second-order elasto-plastic analysis. The lower limit rises to 95% if frames for which λ_c/λ_p is less than 4.0 are excluded. The authors believe that the method will therefore prove acceptable to designers who do not wish to rely on strain-hardening and stray composite action to offset the higher collapse loads that can be predicted by the modified Merchant-Rankine formula. The proposed method should always be used instead of the latter whenever cladding is minimal. The method is easy to apply.

It is hoped that further research will enable to beneficial effects of strain-hardening and cladding stiffness to be included in the proposed method by using enhanced values of λ_p and λ_c where these are justified by accurate analysis.

Acknowledgement

This work has been supported by the Science and Engineering Research Council and the Building Research Establishment (BRE) in the form of a CASE research studentship. The authors gratefully acknowledge this support, and the permission of the Director, BRE, to publish. The opinions expressed are those of the authors. They also wish to thank Dr. R.H. Wood for his encouragement to undertake this work.

References

1. 'Steel Designers' Manual', Crosby Lockwood Staples, London, 1972, Chapter 31.

2. Majid, K.I. and Anderson, D. 'Elastic-plastic design of sway frames by computer'. Proceedings of the Institution of Civil Engineers, vol. 41, 1968, Dec.

3. Horne, M.R. and Morris, L.J. 'Optimum design of multi-storey rigid frames'. Optimum Structural Design: Theory and Applications, ed. Gallagher, R.H. and Zienkiewicz, O.C., Wiley, London, 1973.

4. Merchant, W. 'The failure loads of rigid jointed frameworks as influenced by stability'. The Structural Engineer, vol.32, 1954, July.

5. Horne, M.R. 'Elastic-plastic failure loads of plane frames'. Proceedings of the Royal Society, A, vol. 274, 1963, July.

6. Low, M.W. 'Some model tests on multi-storey rigid steel frames'. Proceedings of the Institution of Civil Engineers, vol.13, 1959, July.

7. Anderson, D. and Lok, T.S. 'Design studies on multi-storey steel unbraced frames'. The Structural Engineer, Part B, to be published.

8. Majid, K.I. and Anderson, D. 'The computer analysis of large multi-storey frames structures'. The Structural Engineer, vol. 46, 1968, Nov.

9. Wood, R.H. 'Effective lengths of columns in multi-storey buildings'. The Structural Engineer, vol.52, 1974, July-Sept.

10. 'Draft Standard Specification for the Structural Use of Steelwork in Building Part 1 : Simple Construction and Continuous Construction', British Standards Institution, London, Documents 77/13908DC and 79/13966DC, 1977 and 1979.

11. 'European Recommendations for Steel Construction', European Convention for Constructional Steelwork, 1978, March.

12. Wood, R.H. 'The stability of tall buildings'. Proceedings of the Institution of Civil Engineers, vol.11, 1958, Sept.

13. Anderson, D. and Islam, M.A. 'Design of unbraced multi-storey steel frames under combined loading'. Proceedings of the Institution of Civil Engineers, vol.69, Part 2, 1980, March.

14. Anderson, D. 'Investigations into the design of plane structural frames'. Ph.D. Thesis, University of Manchester, 1969.

Frame	Min. vertical : max. wind					Max. vertical : min. wind				
	$\frac{\lambda_c}{\lambda_p}$	λ_{WMR}	λ	λ_f	Mec	$\frac{\lambda_c}{\lambda_p}$	λ_{WMR}	λ	λ_f	Mec
4 x 2W	9.15	1.14	1.08	1.07	B	5.75	1.04	1.02	1.04	C
4 x 2N	11.4	1.34	1.28	1.30	B	6.56	1.17	1.14	1.20	B
4 x 3W						5.25	1.04	1.02	1.04	S
4 x 3N						5.37	1.10	1.08	1.08	C
4 x 4W						4.93	1.04	1.02	1.05	S
4 x 4N						3.34	0.88	0.87	0.92	S
4 x 5W	5.34	1.01	0.99	1.04	C	4.78	1.03	1.01	1.05	S
4 x 5N	5.34	1.16	1.14	1.13	C	3.19	0.89	0.89	0.96	S
7 x 2W	12.8	1.15	1.11	1.10	B	5.66	1.02	1.00	1.01	C
7 x 2N	16.6	1.42	1.38	1.39	B	7.98	1.20	1.15	1.15	B
7 x 3W						4.74	0.89	0.88	0.89	C
7 x 3N						5.30	1.07	1.05	1.04	S
7 x 4W						4.48	0.96	0.95	0.97	S
7 x 4N						4.11	0.98	0.97	0.99	S
7 x 5W	5.06	0.96	0.94	0.95	C	4.36	0.96	0.95	0.99	S
7 x 5N	5.78	1.24	1.21	1.26	B	3.98	1.00	0.99	1.03	S
10 x 2W	15.3	1.15	1.11	1.11	B	6.02	1.05	1.02	1.01	C
10 x 2N	14.9	1.42	1.37	1.40	B	8.07	1.18	1.13	1.10	S
10 x 3W						4.67	0.96	0.94	0.96	C
10 x 3N						6.19	1.04	1.01	1.02	S
10 x 4W	6.57	1.06	1.03	1.04	C	3.87	0.89	0.88	0.91	S
10 x 4N	8.23	1.31	1.25	1.24	B	4.56	1.00	0.99	1.02	S

Table 1 Parametric studies under extreme loading

Frame	Vert load	Wind load	$\frac{\lambda_c}{\lambda_p}$	λ_{WMR}	λ	λ_f	Mec
4 x 2W	Max.	Inter.	5.82	1.03	1.01	1.03	C
4 x 3W	Max.	Max.	5.72	0.97	0.94	0.94	S
4 x 5W	Max.	Max.	4.88	1.02	1.00	1.03	S
7 x 2W	Max.	Inter.	6.15	1.03	1.00	1.00	C
7 x 5W	Max.	Max.	4.63	0.91	0.89	0.89	S
10 x 2W	Max.	Inter.	8.07	1.12	1.08	1.05	B
10 x 3W	Max.	Max.	6.41	1.08	1.05	1.04	B
10 x 2N	Max.	Inter.	5.78	1.08	1.05	1.06	S

Table 2 Parametric studies under various loadings

Ref.	Frame	$\frac{\lambda_c}{\lambda_p}$	λ_{WMR}	λ	λ_f
2	4 x 1	13.9	1.56	1.51	1.49
14	4 x 1	7.71	1.55	1.49	1.43
14	8 x 2	5.06	1.51	1.48	1.48
12	4 x 1	6.01	2.01	1.96	1.91
7	6 x 1	5.02	1.59	1.56	1.54
13	6 x 1	10.3	1.63	1.55	1.49
13	15 x 3	3.49	1.37	1.37	1.38
7	6 x 3	4.18	1.70	1.67	1.64

Table 3 Other comparisons

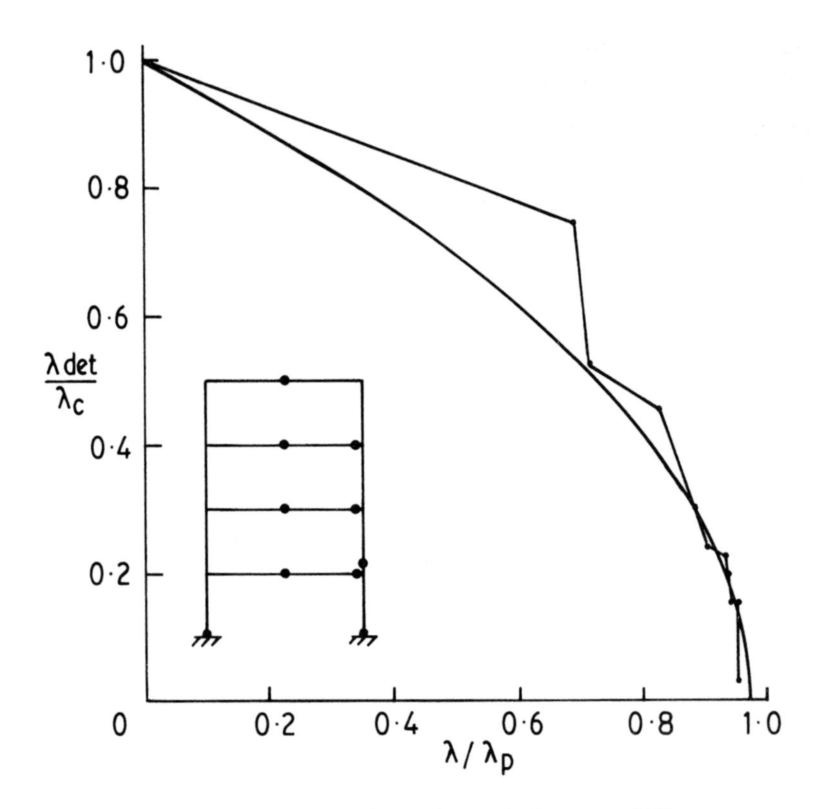

Fig. 1 Deterioration of frame stiffness

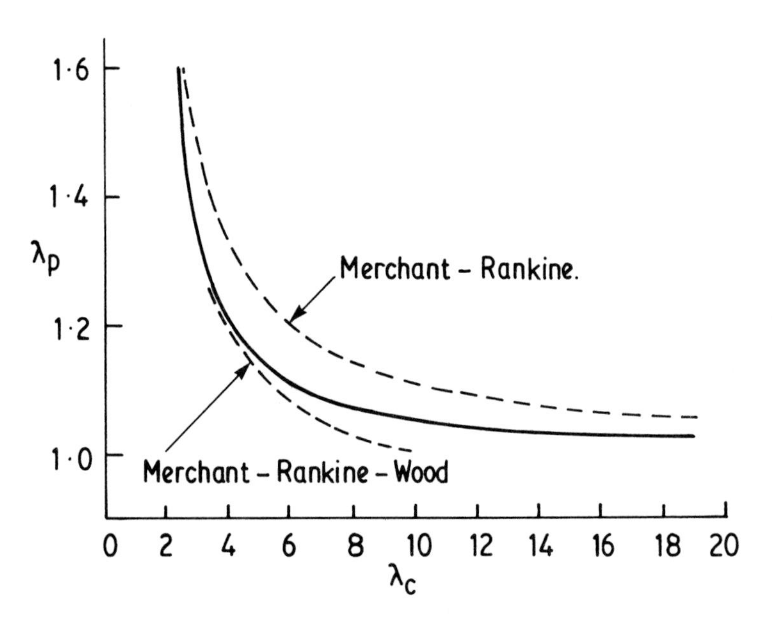

Fig. 2 Design curve for $\lambda = 1.0$

ANALYSIS OF ELASTIC-PLASTIC FRAMES AT LARGE DISPLACEMENT RANGE ACCOUNTING FOR FINITE SPREAD OF YIELDING ZONES

U. Andreaus and P. D'Asdia

Istituto di Scienza delle Costruzioni, Faculty of Engineering, University of Rome

Synopsis

In the present paper a method of analysis is described which allows for the carrying out of the elasto-plastic analysis of plane frame structures under the assumption both reversible and irreversible behaviour of material, accounting or not for the effect of geometrical non-linearities, and using piece-wise linear yield conditions at any desired degree of discretization in the space of the active stress resultants (axial force, shear force and bending moment).

Furthermore the presented procedure allows the assumption of concentrated plastic hinges to be abandoned thus accounting for the influence of the effective extension of plastic zones on the structural behaviour.

Introduction

Elasto-plastic analysis procedures have been cast under the assumption of both geometrically linear and non-linear structural behaviour (1,2,3). In the former case the validity of the results should be assessed by a subsequent check of the displacement magnitude (4), in the latter one the deformation history influences the following equilibrium path of the structure. In both cases a correct evaluation of the actual deflections seems to be of paramount importance, including the influence of the finite extension of plastic zones, which may produce significant effects for the structural scheme under investigation (5).

It is well known that complex frames are unlikely to collapse in a regular mechanism; on the contrary, elastic-plastic instability occurs when the number of plastic hinges does not yet suffice to transform at least a portion of the frame into a mechanism. It is precisely in this situation where neglecting the effect of the finite extension of plastic zones leads to an overestimate of the structural stiffness and hence to an underestimate of deflections and hence to an overestimate of the collapse load. In fact the classical assumption of concentrated plastic hinges preserves the full elastic rigidity of the single cross-section until full plastification has been reached, whereas considering the spread of plastic zones from hinges allows the correct evaluation not only of the greater flexibility of member portions near effective plastic hinges, but also the greater flexibility of those structural members in which no cross-section reaches full plastification.

The proposed method, by using the independent elastic-plastic kinematical compatibility equations, restricts the problem size to within not more than twice the number of the redundant unknowns in the complete elastic frame, independently of the nodal degrees of freedom, and requires only one factorization of the matrix governing the problem when no local unloading occurs.

The presented procedure allows for carrying out the elasto-plastic analysis of plane frame structures under the assumption of both reversible and irreversible behaviour of material, accounting or not for the effect of geometrical non-linearities by means of 'fictitious' forces according to the matrix force method for large deflection analysis, also tracing its possible descending branches, and using piece-wise linear yield conditions at any desired degree of discretization in the space of the active stress resultants (axial force, shear force and bending moment) without increasing the problem size.

Furthermore, the proposed algorithm allows a restriction of the additional computational effort in the case of local unloading, inasmuch that it requires the performance of a new factorization of a part of the matrix governing the problem of a size small with respect to the total size of the matrix.

At fairly greater computational cost it is possible to drop the assumption of concentrated plastic hinges and to account for the influence of the effective extension of plastic zones on the structural behaviour.

1. *INCREMENTAL PROCEDURE*

In this section only governing relations and the fundamental steps of the incremental procedure to be performed cyclically are presented, while the improvements adopted to account for finite spread of yielding zones and geometrical non-linearities are given in the following sections.

Step 1

According to a standard procedure the equilibrium matrix $\underline{H}$ of the structure is assembled and decomposed into a non-singular square matrix $\underline{Z}_0$ (equilibrium matrix of the auxiliary statically determinate system) and the matrix $\underline{Z}$ of influence coefficients of the redundant unknowns $\underline{x}$.

Employing the virtual work principle supplies the independent elastic-plastic compatibility equations, which reduce to the following form when the frame is entirely elastic:

$$\underline{Z}^T \underline{F} \underline{Z} \underline{x} = - \underline{Z}^T \underline{e}_0^d - \alpha \underline{Z}^T \underline{e}_0^1 \qquad (1)$$

where $\underline{F}$ is the block-diagonal matrix of member elastic flexibilities and $\underline{e}_0^d$ and $\underline{e}_0^1$ are vectors of natural elastic deformations due respectively to the dead and live loads acting on the auxiliary statically determinate system, the last term being affected by the positive factor α.

Factorizing the matrix $\underline{Z}^T \underline{F} \underline{Z}$[1] gives the vector $\underline{x}$ as the sum of a constant term and of a term depending on the factor α.

Step 2

The two terms of $\underline{x}$ allow the evaluation of the corresponding constant and variable (according to α) parts of the vector $\underline{s}$ (which collects the stress resultants at the critical cross-sections of the frame) and of the nodal displacement vector $\underline{d}$.

Step 3

The first level of α at which plastic flow occurs and its increments between the onsets of two subsequent plastic hinges are obtained by equating to zero at least one of the (previously negative) components of the vector of plastic potentials

$$\underline{\Psi} = \underline{N}^T \underline{s} - \underline{k} \leq \underline{0} \tag{2}$$

where $\underline{N}$ is the block-diagonal matrix whose columns are outward unit normals to the piece-wise linear yield modes, and $\underline{k}$ is the vector of the generalized plastic capacities.

Step 4

A different type of check is performed[2] according to the irreversible or reversible behaviour assumed for the material.

In the first case, if the non-negativity condition on the plastic multiplier increments $\Delta\underline{\lambda} = \Delta\alpha\underline{\lambda}^{(2)}$ is satisfied, allowance must be made in the virtual work equations for hinge discontinuities and fulfilment of the yield condition(s) must be enforced for the stress state at the active mode(s), identified in the step 3 by the vanishing components of $\underline{\Psi}$. Thus eqns (1) take the following incremental form:

$$\begin{aligned} \underline{Z}^T \underline{F} \underline{Z} \Delta \underline{x} + \underline{Z}^T \underline{E} \underline{N} \Delta \lambda &= - \Delta \alpha \underline{Z}^T \underline{e}_0^1 \\ \underline{N}^T \underline{E} \underline{Z} \Delta \underline{x} &= - \Delta \alpha \underline{N}^T \underline{s}_0^1 \end{aligned} \tag{3}$$

where $\underline{E}$ is a block-diagonal matrix of statics transformation which essentially self-equilibrates the generalized stresses over the

---

(1) The symmetry of the matrix governing the problem, preserved during the whole procedure, allows the use of standard techniques thus saving computer storage and time.

(2) This step can be skipped in the first cycle, when the structure is still entirely elastic.

(3) The superimposed dot denotes derivation with respect to the load parameter α.

elements, and $\underline{s}_0^1$ is the part of $\underline{s}$ due to the live loads acting on the auxiliary statically determinate system. Therefore, the onset of subsequent plastic hinges is accounted for by adding to system (3), the corresponding yield equations and unknown plastic multiplier increments.

In the case of irreversible behaviour, if the non-negativity condition on $\underline{\Delta\lambda}$ is not satisfied, local unloading occurs and the corresponding equations and unknowns in the set (3), added in previous cycles, must be dropped and $\Delta\alpha$ becomes zero

In the second case (reversible behaviour), if the non-negativity condition on the current plastic multipliers $\underline{\lambda} + \Delta\alpha\dot{\underline{\lambda}}$ is satisfied, the same equations and unknowns as in the first case are added to the set (3). If the condition $\underline{\lambda} + \Delta\alpha\dot{\underline{\lambda}} \geqslant \underline{0}$ is not satisfied, the loading increment $\Delta\alpha$ is reduced as much as required to satisfy it, and the corresponding equations and unknowns are suppressed in the set (3) as in the first case.

Step 5

Update the vectors of redundant unknowns, stress resultants elastic and plastic deformations and nodal displacements, by adding to them the corresponding increments calculated by multiplying the relevant rates by the loading increment $\Delta\alpha$, evaluated at the steps 3 and 4.

Step 6

Factorize the part of the system (3) matrix not yet manipulated; in other words, factorize

- the last column(s) and row(s), if no local unloading occurs;
- the part of the matrix constituted by the columns and rows following the dropped one(s) because of its (their) unloading and reversal in the elastic range.

When the system (3) becomes nearly singular (incipient collapse), normalizing the vector $\dot{\underline{\lambda}}$ yields the collapse mechanism and the incremental procedure is stopped. Otherwise solving eqns (3) gives the vectors $\dot{\underline{\lambda}}$ and $\dot{\underline{x}}$ which allow the evaluation of the rates $\dot{\underline{s}}$ and $\dot{\underline{d}}$.

Return to perform step 3.

2. *INFLUENCE OF FINITE EXTENSION OF PLASTIC ZONES*

The finite spread of yielding zones from plastic hinges as the external loads are monotonically increasing leads to a continuous reduction in member stiffness.

Such flexibility variation can be accounted for at the desired degree of accuracy by subdividing the single loading step, defined by two subsequent full plastifications of critical cross-sections, into a number of intermediate steps, where the local flexibility is updated for those members exhibiting significant changes in spread of yielded material.

In this paper two methods are presented which allow the accomplishment of outlined procedure by relating along each member the stiffness changes due to the spread of yielding to the stress resultant distribution rate obtained by a polynomial interpolation of the relevant components of $\underline{\dot{s}}$.

A preliminary study is required in order to find the apropriate relationships between the stress state and the elastic zone size of the single cross-section of given shape by using equilibrium equations between internal and external stresses.

In the present paper I-shaped cross-section, widely adopted in engineering practice, have been considered under axial force N and bending moment M and the above mentioned relations can be formulated as follows:

$$(N, M, h_u, h_l) = 0 \tag{4}$$

where h_u and h_l denote the distances from the centroid of the upper and lower bounds of the elastic zone. Equation (4) represents analytically a set of non-linear relations describing the cross-sectional behaviour in the different cases of flange and/or web plastification; the change from one case to another, at a finite number of critical cross-sections along the member, places a restraint on the length of the loading step when direct iterative solution of eqns (4) is performed during the procedure and allows the evaluation of the reduced axial and flexural stiffness distribution along the member by interpolating the relevant values at the critical cross-sections.

Alternatively, eqns (4) can be used to generate a priori piece-wise linear domains in the space of the active stress resultants at prefixed percentages of the axial and flexural full elastic stiffness, which allow a reduction of the loading step so that one of the critical cross-sections moves to the subsequent domain. The reduction in member stiffness is then performed by interpolation as in the case of the direct use of eqns (4).

An interesting feature of the use of constant stiffness domains is that groups of cross-sections of the same shape and different dimensions can be described through a single set of domains. In the present paper the European standard profiles IPE and HEB have been investigated, and the relevant domains are shown in Figs 1 and 2 respectively, in the space of non-dimensional active stress resultants $n = N/N_y$ and $m = M/M_y$, where N_y and M_y are the full plastic values. Figures 1a and 2a show the axial stiffness domains, while Figs 1b and 2b represent the flexural stiffness domains. It is worth noticing that the proposed 16-sided domains present a maximum scattering of about 7%. with respect to the exact non-linear domains of any cross-section, while the maximum difference between the IPE and HEB domains attains about 10%.

On the basis of the above mentioned criteria the incremental procedure presented in Section 1 is modified as follows.

Step 3

The maximum admissible loading increment $\Delta\alpha$ is restrained not only by equating to zero at least one component of the plastic potential vector, eqns (2), but also either by solving eqns (4) at all critical cross-sections or by equating to zero at least one component of a set of conditions similar to eqns (2), which describe the constant stiffness domains, defined by different values of $\underline{N}$ and $\underline{k}$ (see Figs 1 and 2).

Step 4

Under the assumption of irreversible behaviour of material, if the non-negative condition on $\underline{\lambda}$ is satisfied, no changes are required with respect to the procedure of Section 1, if $\Delta\alpha$ has been determined by eqns (2), otherwise it is necessary to update member flexibilities in matrix $\underline{F}$ as well as the corresponding terms of the natural elastic deformation vector $\underline{e}_0$, and then to go to step 6, where the system (3) is factorized up from the first changed term and then solved. On the contrary, if the non-negative condition on $\underline{\lambda}$ is not satisfied (local unloading), drop as previously the corresponding equation and unknown in system (3), assign their fully elastic value to the terms of the flexibility matrix $\underline{F}$ influenced by the unloading, and to step 6, where the system (3) is factorized up from the first changed term.

3. *INFLUENCE OF GEOMETRICAL NON-LINEARITIES*

Geometrical non-linearities are accounted for in the framework of the matrix force method, the equilibrium matrix $\underline{H}$ and the matrix $\underline{Z}$ of influence coefficients of the redundant unknowns are not changed during the incremental procedure, while additional loads, the so-called 'fictitious' forces, are introduced in order to account for the influence of axial force on bending moment and deflection distribution along each member. Such fictitious forces for each member are evaluated by multiplying by the current natural deformations the difference between the linear elastic stiffness matrix and the non-linear stiffness matrix, which is evaluated by means of the stability functions when the finite extension of plastic zones is neglected (axial and flexural stiffness are constant along the member) or otherwise is calculated by numerical integration of the differential equation of the beam element. Elemental fictitious forces are then assembled in global coordinates and added to the variable part of the external loads at the step 4 of the incremental procedure of Section 1. In fact, determining the equilibrium path of the structure between the onset of two subsequent plastic hinges now requires the performance of iteratve cycles involving step 3, that part of step 4 concerning the aforementioned assembling of the known quantities and the last part of step 6, where the solution of eqns (3) is made with respect to the known quantity vector, without performing a new factorization is updated by improving the evaluation of the additional loads on the basis of the stress state provided by the preceding cycle. The iteration is stopped at step 4, when the difference between two subsequent evaluations of the loading increment and of the stress state is less than a prescribed tolerance. This procedure usually converges adequately in few iterations. So far step 4 is completed by modifying eqns (3) as already shown in Section 1 or 2, then step 5 is performed and the next loading increment is looked for.

The adopted method alows for a correct evaluation of the buckling load in the elastic-plastic range before the plastic collapse mechanism occurs, both at the onset of a plastic hinge (or at a step-wise stiffness variation) and during the intermediate equilibrium path. Moreover, it is possible to evaluate the descending branch of the equilibrium path until the structure transforms into a mechanism.

4. *NUMERICAL APPLICATION*

The four-bay, three-storey frame shown in Fig 3 has been analysed under monotonically increasing loads, which have the reference values (α = 1) given in the same figure.

The structural behaviour has been investigated accounting or not for the finite spread of yielding zones adopting the yield domains of Figs 1 and 2, and accounting or not for geometrical non-linearities.

Moreover, in order to compare the influence of different plastic domains, the analysis has been performed at small and large displacement range, under the assumption of concentrated plastic hinges using the yield surface shown in Fig 4, where T denotes the shear force.

The six equilibrium paths are represented in Fig 5, where δ is the horizontal displacement of the joint acted upon by the force F_3.

Acknowledgements

The authors are indebted to Dr. F. Iannozzi for his contribution in the development of the computer program and in the analysis of results.

References

1. Hodge, Jr. P.G. 'Computer solutions of plasticity problems.' Proc. Int. Symp. on Foundations of Plasticity, Warsaw, 1972, pp. 261-286.

2. Zienkiewicz, O.C., Vallipan, S. and King, I.P. 'Elastoplastic solutions of engineering problems, initial stress, finite element approach.' Int. J. Num. Eng., Vol. 1, 1969, pp. 75-100.

3. Argyris, J.H. and Sharpf, D. 'Methods of elastoplastic analysis.' J. Appl. Mech. and Phy., (ZAMP), Vol. 23, 1972.

4. Andreaus, U. and D'Asdia, P. 'Displacement analysis in elastic-plastic frames at plastic collapse.' Accepted for publication in Comp. Meth. in Appl. Mech. and Eng.

5. Andreaus, U. and Sawczuk, A. 'Deflection of elastic-plastic frames at finite spread of yielding zones.' Comp. Mech. in Appl. Mech. and Eng. (in press).

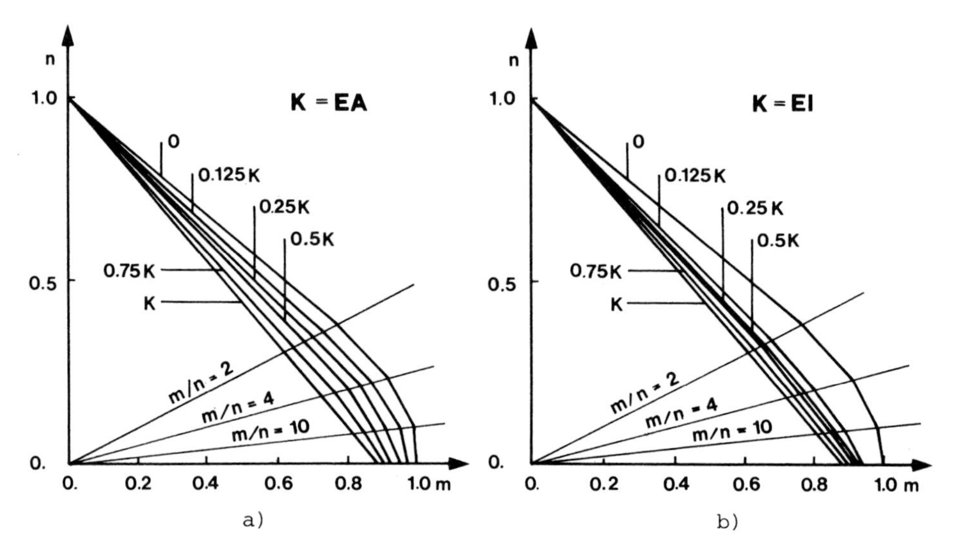
n
1.0
0.5
0.
K = EA
0
0.125K
0.25K
0.5K
0.75K
K
m/n = 2
m/n = 4
m/n = 10
0.
0.2
0.4
0.6
0.8
1.0 m
a)
n
1.0
0.5
0.
K = EI
0
0.125K
0.25K
0.5K
0.75K
K
m/n = 2
m/n = 4
m/n = 10
0.
0.2
0.4
0.6
0.8
1.0 m
b)

Fig 1

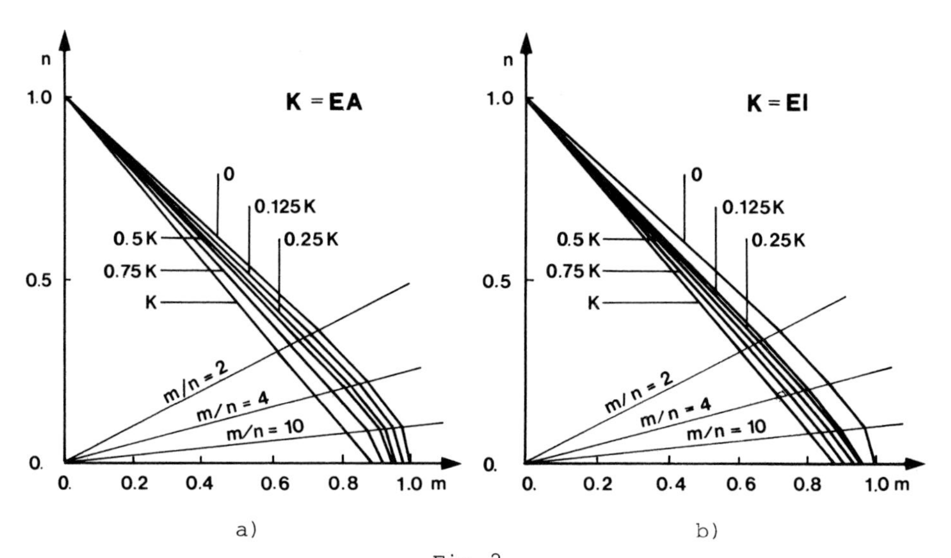
n
1.0
0.5
0.
K = EA
0
0.125K
0.25K
0.5K
0.75K
K
m/n = 2
m/n = 4
m/n = 10
0.
0.2
0.4
0.6
0.8
1.0 m
a)
n
1.0
0.5
0.
K = EI
0
0.125K
0.25K
0.5K
0.75K
K
m/n = 2
m/n = 4
m/n = 10
0.
0.2
0.4
0.6
0.8
1.0 m
b)

Fig 2

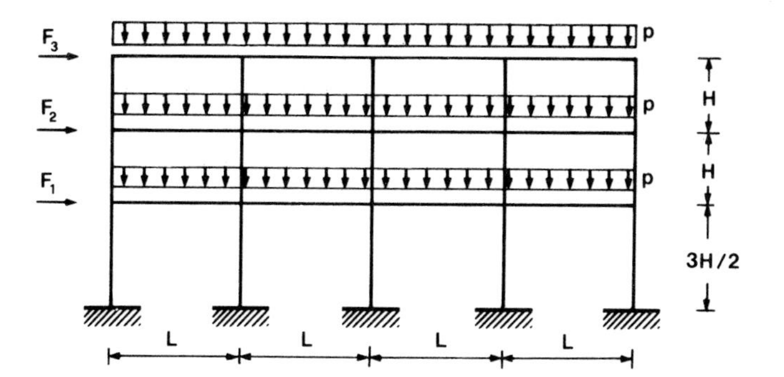

Fig 3

L = 6,00m; H = 3,00m.
Columns (HEB 180):
Ny = 148,88t; Ty = 17,085 t; My = 10,99tm.
Beams (IPE 300):
Ny = 122,66t; Ty = 26,158t; My = 14,318tm.
p = 2,700t/m; F_1 = 3,705t; F_2 = 2,964t; F_3 = 1,482t.

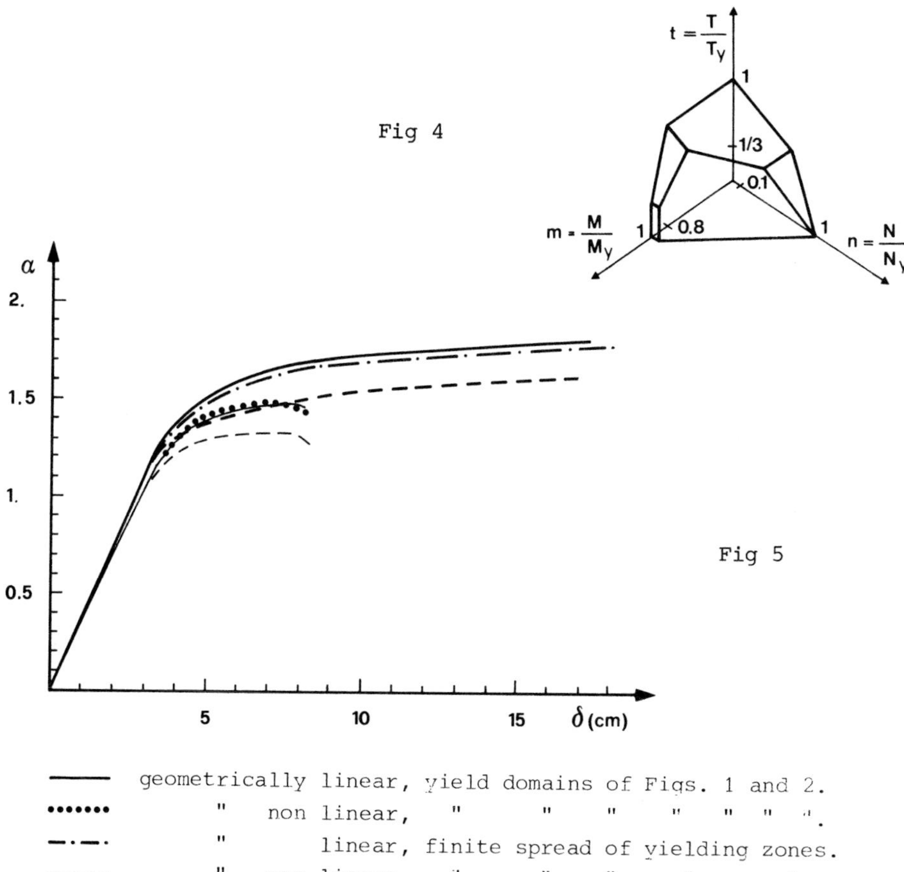

Fig 4

Fig 5

————	geometrically linear, yield domains of Figs. 1 and 2.
••••••••	" non linear, " " " " " " ".
—·—·	" linear, finite spread of yielding zones.
————	" non linear, " " " " ".
- - - -	" linear, yield domain of Fig. 4.
– – – –	" non linear, " " " " ".

INSTABILITY AND PLASTIC COLLAPSE OF PLANE FRAMES WITH GENERALIZED YIELD FUNCTION

G J Creus, P L Torres and A G Groehs

Department of Civil Engineering, Universidade Federal do Rio Grande do Sul, Porto Alegre, Brasil

Synopsis

A numerical method for the analysis of elastoplastic plane frames in the presence of finite displacements is proposed. Finite displacements effect is taken into account using the nonlinear tangent stiffness matrix. Generalized plastic hinges (that account for shear and normal force effects) are concentrated at the critical sections. The method has been implemented into a computer code with problem oriented language. Several examples are presented that indicate good agreement with known solutions.

Introduction

This paper describes a numerical procedure for the study of plane frames allowing:

i elastic (small and large displacements) analyses and the determination of buckling loads

ii elastoplastic (small and large displacements) analises considering the effect of normal and shear forces with the determination of limit and instability loads.

The effect of large displacements may be important in elastoplastic frames (1, 2). These effects may of course be studied analytically, but such analyses are complex except for the simplest situations. The present work describes a numerical method to solve these problems via a general incremental matrix procedure.

In the present work finite displacements effects are taken into account by means of the nonlinear tangent stiffness matrix, following Yang (3) for the elastic case. Then, the procedure is extended to the elastoplastic case defining a general yield criterion that takes into account the effects of moment and shear and normal forces. Use of this yield criterion leads to an elastoplastic relationship that may be looked upon as representing a generalized plastic hinge, in the sense that it reduces to the classical hinge concept in special situations. On the other hand, the procedure has the advantage of allowing a consistent modelling of material and section properties, including hardening effects, through the choice of specific yield functions. Moreover, when the large displacements formulation is used, an instability check is performed as a part of the elastoplastic analysis. The procedure has been implemented into a computer code using a problem oriented language.

The effect of finite displacements and normal stresses on the behavior of elastoplastic plane frames have been studied in pionnering works by Horne (1) and Onat (2, 4). A numerical procedure, considering classical plastic hinges and small displacements was presented by Wang (5). Numerical elastoplastic analysis of arches with consideration of normal force and small displacements effects have been presented by Cohn and Rohman (6); their work uses a transfer matrix approach that is of somewhat restricted application. Tranberg et al (7) studied large displacements without consideration of shear and normal force effects. Mc Ivor et al (8) presented a complete analysis; their interest was the consideration of thin walled beam structures.

1. *INCREMENTAL ELASTIC ANALYSIS*

The elastic analysis is performed writing for a bar element, Fig. 1:

$$\dot{\underset{\sim}{F}} = \dot{\underset{\sim}{K}}\ \dot{\underset{\sim}{U}}^{e} \tag{1}$$

where $\dot{\underset{\sim}{F}}^{T} = \{\dot{F}_{xi}, \dot{F}_{yi}, \dot{M}_{zi}, \dot{F}_{xj}, \dot{F}_{yj}, \dot{M}_{zj}\}$

and $\dot{U}^{eT} = \{\dot{u}^{e}_{i}, \dot{v}^{e}_{i}, \dot{R}^{e}_{wi}, \dot{u}^{e}_{j}, \dot{v}^{e}_{j}, \dot{R}^{e}_{wj}\}$

indicate rates of the vectors of load and (elastic) displacements and $\underset{\sim}{K}$ is the incremental tangent stiffness matrix. The superindex T is used for the transpose of a matrix and i and j indicate the left and right ends of the element respectively. In the present analysis, an updated reference configuration is used. Thus, a new frame of reference is established for the deformed element at the end of each load increment and F, U^{e} and K are updated accordingly. As it is well known, the stiffness matrix K must include the terms of the linear stiffness matrix plus the initial stress (or geometric) matrix; see for example Yang (3), whose procedure we follow on this subject.

2. *INCREMENTAL ELASTOPLASTIC ANALYSIS*

2.1 *Generalized yield condition*

The yield condition used is written in the form (see for example Hodge (9), Drucker (10)):

$$Y = \left(\frac{F_x}{N_p}\right)^{\alpha} + \left(\frac{F_y}{Q_p}\right)^{\beta} + \frac{|M_z|}{M_p} - 1 - q = 0 \quad (2)$$

where N_p, Q_p, M_p are respectively the plastic values of normal force, shear force and moment for the given section. This yield function is fairly general, as it represents the behavior of many simmetrical sections with adequate choices of α and β. The state variable q models a form of isotropic hardening; other forms may be adopted, without major changes in the formulation below. The influence of the displacement of the plastic hinges because of hardening (11) may be approximately accounted for through the yield function. For the example in this paper we shall be working with uniform rectangular sections of height h and width b and a perfectly plastic material with yield strength σ_e. Thus (*)

$$\begin{aligned} &\alpha = 2; \quad \beta = 2 \\ &M_p = 0.25\ \sigma_e\ b\ h^2; \quad N_p = \sigma_e\ b\ h; \quad Q_p = 0.5\ \sigma_e\ b\ h \\ &q = 0 \end{aligned} \quad (3)$$

2.2 *Elastoplastic matrix*

Using the yield function (2) and the normality rule for the plastic rates of displacement $\dot{\underset{\sim}{U}}^p = \dot{\underset{\sim}{U}} - \dot{\underset{\sim}{U}}^e$

$$\dot{\underset{\sim}{U}}^p = \Lambda^{\cdot} \left\{\frac{\partial Y}{\partial \underset{\sim}{F}}\right\} ; \quad \Lambda^{\cdot} \geq 0 \quad (4)$$

and following the general procedure (as outlined for example in Zienkiewicz (12) for the continuum) we obtain

$$\dot{\underset{\sim}{F}} = \underset{\sim}{K}_p\ \dot{\underset{\sim}{U}} \quad (5)$$

where
$$\underset{\sim}{K}_p = \underset{\sim}{K} - \frac{\underset{\sim}{K} \left\{\frac{\partial Y}{\partial \underset{\sim}{F}}\right\} \left\{\frac{\partial Y}{\partial \underset{\sim}{F}}\right\}^T \underset{\sim}{K}}{A + \left\{\frac{\partial Y}{\partial \underset{\sim}{F}}\right\}^T \underset{\sim}{K} \left\{\frac{\partial Y}{\partial \underset{\sim}{F}}\right\}} \quad (6)$$

and
$$A = \left\{\frac{\partial Y}{\partial q}\right\} \left\{\frac{\partial q}{\partial \underset{\sim}{U}_p}\right\} \left\{\frac{\partial Y}{\partial \underset{\sim}{F}}\right\} \quad (7)$$

When a plastic hinge is formed only at end i (**) of the

(*) $\beta=2$ is in the safe side compared with $\beta=4$ recommended by Drucker (10). The same is true for the coefficient 0.5 in Q_p. The effect of shear stresses in the examples studied was very small.

(**) In fact, hinges are considered immediately to the right of node i and immediately to the left of node j. Adequate programming avoids formation of superfluous hinges (joint mechanisms).

bar, we have

$$\{\frac{\partial Y}{\partial \tilde{F}}\} = \{\{\frac{\partial Y}{\partial \tilde{F}}{}_i, \tilde{0}\} \tag{8}$$

Thus, writing $\tilde{K}$ in the partitioned form

$$\tilde{K} = \begin{vmatrix} \tilde{K}_{ii} & \tilde{K}_{ij} \\ \tilde{K}_{ji} & \tilde{K}_{jj} \end{vmatrix} \tag{9}$$

eqn (6) reduces to

$$\tilde{K}_{pi} = \tilde{K} - \frac{\begin{vmatrix} \tilde{K}_{ii} \\ \tilde{K}_{ji} \end{vmatrix} \{\frac{\partial Y}{\partial \tilde{F}}\}_i \{\frac{\partial Y}{\partial \tilde{F}}\}_i^T |\tilde{K}_{ii} \ \tilde{K}_{ij}|}{A + \{\frac{\partial Y}{\partial \tilde{F}}\}_i \tilde{K}_{ii} \{\frac{\partial Y}{\partial \tilde{F}}\}_i} \tag{10}$$

A similar form may be found for a hinge at end j. From eqn (2) we have

$$\{\frac{\partial Y}{\partial \tilde{F}}\}_i^T = \{\frac{2N}{N_p^2}, \frac{2Q}{Q_p^2}, \frac{1}{M_p} \ sgn \ (M_z)\}_i \tag{11}$$

The numerical analysis is performed basically as in the elastic case. At each stage the stiffness matrix is updated including the changes due to plastic action in the active hinges. The load is applied in increments controlled in such a way that eqn (2) is satisfied approximately, with the load point $\{F_x, F_y, M_z\}$ remaining always close to the yield surface. In the examples shown, this condition is established as

$$1 - \delta \leq (\frac{F_x}{N_p})^2 + (\frac{F_y}{Q_p})^2 + \frac{|M_z|}{M_p} \leq 1$$

with $\delta = 0.01$

3. EXAMPLES

Several examples were run to check the program. The results of Yang (3) on finite elastic displacements were reproduced. There were also analized several examples from Hodge (9) concerning rigid-plastic frames. The conditions of classical limit analysis were approximated taking high values for Q_p, N_p and the elasticity modulus E.Then, the matrix (10) naturally transforms into the matrix of an element with a moment discontinuity at end i, as used in Wang (5). These examples may be found in (13) together with more details about the computational procedure. In the following we analize a few selected examples.

3.1 Free-clamped column

This is a classical example frequently used to check geometrically nonlinear solution procedures. Figure 2 shows the results from the program for a uniform column axially loaded with initial excentricity e=L/400, where L is column length. These results are compared with the theoretical solution for e=0 (see for example Timoshenko (14)). The discretization of the column into five elements is shown in the insert.

3.2 *Two story frame*

The two story frame shown in Fig 3 was analized by Horne (1), whose results are taken for comparison. This is an interesting example that allows a comparison among different types of analysis: linear elastic (line A); geometrically nonlinear elastic (B), elastoplastic without (C) and with (D) consideration of changes in geometry. For this structure we have: L=400 in; area A=3.2 in^2; moment of inertia I=51.2 in^4; elasticity modulus $E=30x10^6$ lb/in^2; M_p=464 Kips.in, Q_p=29 Kips. N_p=58 Kips; vertical loads have the value λP_p and horizontal forces the value 0.25 λP_p, where P_p is the load that produces rigid-plastic collapse. The value of normal and shear stresses remain comparatively small. Thus, our results are quite close to those of Horne, as could be expected for a slender frame.

3.3 *Beam with fully constrained ends*

Haythornwaite (15) studied the case of a beam with ends both clamped and prevented from moving axially, Fig 4. For a central load P, the classical limit value is $P_p=8\ M_p/L$; this value corresponds either to the absence of axial constrains or to the use of the initial geometry in the analysis. Taking into account the deformed configuration, Haythornwaite obtained a theoretical solution for a rigid plastic material that satisfies eqn (2) with q=0, $Q_p \rightarrow \infty$. This solution is indicated by line A in Fig. 4. Haythornwaite indicates some experiments that support this result. This is a good example because the stress state at the hinges goes all the way from simple flexure to pure tension, as the beam becomes a two bar truss. The program described in this paper gave the results indicated by the small circles in Fig 4. The material was considered elastic-perfectly plastic with the constants shown in the figure. Two elements were used to model the half beam.

4. CONCLUDING REMARKS

The approximate analysis described is an extension of the well known method of plastic hinges. It uses a generalized yield condition that allows the consideration of shear and normal force effects and accounts for finite displacements through an incremental procedure with tangent stiffness matrix. The examples shown, with fairly large displacements and (in one case) a stress state varying from almost pure flexure to almost pure tension indicate a satisfactory performance at a reduced computational cost.
Deliberately, the authors checked the program against well

known classical problems. But the computational procedure is quite general and may be used with complex plane frames.

References

1. Horne, M.R. and Merchant, W. "The Stability of Frames', Pergamon Press, Oxford, 1965.
2. Onat, E.T. 'On Certain Second Order Effects in the Limit Design of Frames', J.Aeronaut. Sci. 22, 681-684 (1955).
3. Yang, T.Y. 'Matrix Displacement Solution to Elastica Problems of Beams and Frames', Int. J. Solids Structures, 9, 829-842 (1973).
4. Onat, E.T. and Prager, W. 'Limit Analysis of Arches', J.Mech. Phys. Solids, 1, 77-89 (1953).
5. Wang, C.K. 'General Computer Program for Limit Analysis', Proc. Am. Soc. Civil Engnrs., J. Struct. Div., 89, No. ST6, 101-117 (1963).
6. Cohn, M.Z. and Rohman, M.A. 'Analysis up to Collapse of Elastoplastic Arches', Computers and Structures, 6, 511-517 (1976).
7. Tranberg, W., Swannell, P. and Meek, J.L., 'Frame collapse using tangent stiffness', Journal ASCE, vol. 102, No. ST3, March, 1976.
8. Mc Ivor, I.K., Wineman, A.S. and Wang, H.C. 'Plastic Collapse of General Frames', Int. J. Solids Structures, 13, 197-210 (1977).
9. Hodge, P.G. Jr. 'Plastic Analysis of Structures', Mc Graw-Hill, New York (1959).
10. Drucker, D.C. 'The Effect of Shear on the Plastic Bending of Beams', J.Appl. Mechanics, 23, 509-514 (1956).
11. Horne, M.R. and Medland, I.C. 'Collapse loads of steel frameworks allowing for the effect of strain hardening', Proc. Inst. Civ. Engrs, Paper 6825, 1965.
12. Zienkiewicz, O.C. 'The Finite Element Method, 3rd. Edn., Mc Graw-Hill, London (1977).
13. Torres, P.L. 'Nonlinear Analysis of Elastoplastic Frames', M.Sc. Thesis (in preparation), Universidade Federal do Rio Grande do Sul, Porto Alegre, RS, Brasil (1983).
14. Timoshenko, S. and Gere, J.M. 'Theory of Elastic Stability'. Second edition, Mc Graw Hill, New York, 1961.
15. Haythornthwaite, R.M. 'Beams with Full end Fixity, Engineering, 183, 110-112 (1957).

Acknowledgement

Financial assistance from CNPQ and FINEP are gratefully acknowledged.

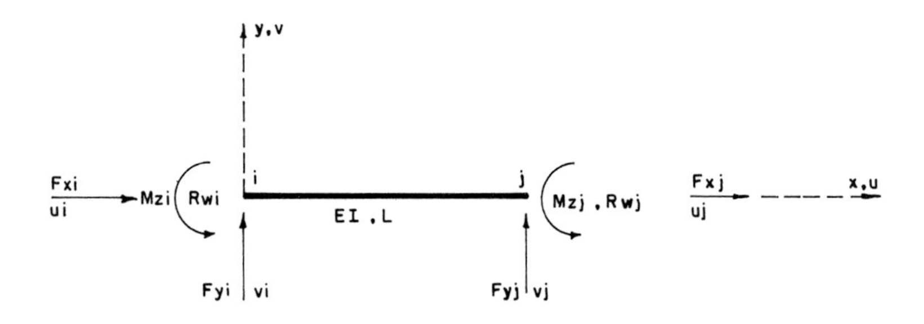

Fig 1

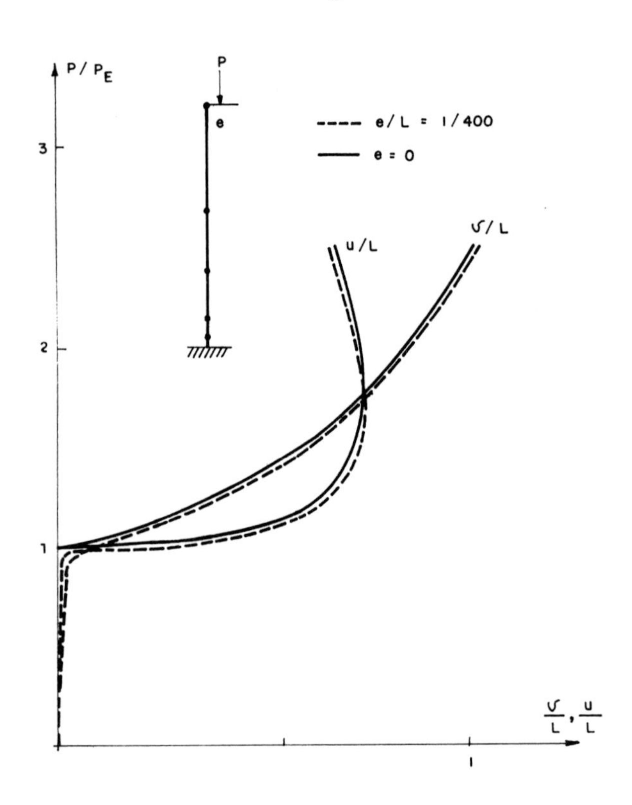

Fig 2

Two recent studies by Grierson et al (3,4) considered thin-walled structures and planar frameworks, and developed a minimum weight design method whereby performance criteria are satisfied *simultaneously* at both the service-load and ultimate-load levels specified for the design. Specifically, for proportional static loading and first-order behaviour, the design method ensures acceptable elastic stresses and displacements under service loads while, at the same time, ensuring adequate post-elastic strength reserve of the structure under ultimate loads. Stated otherwise, a serviceable design is found cognizant of the margin of safety against failure in a plastic mechanism mode. The design method involves an iterative process having the following essential feasures: (i) for a given design (e.g., the initial 'trial' design), sensitivity analysis techniques are employed to approximate the performance constraints as linear functions of the member sizing variables; (ii) optimization techniques are applied to find an improved (i.e., lower weight) design; (iii) the performance constraints are updated for the next weight optimization; (iv) the process is repeated until weight convergence occurs after a number of design stages.

To date, the design method described in the foregoing has been applied considering the member sizes as *continuous* variables to the design. The present study extends the method to design for *discrete* member sections. Specifically, for planar frameworks comprised of axial and/or flexural members of any cross-section type (e.g., WF, T, Double-Angle, etc.). For axial members, no restrictions are placed on the specified sets of discrete sections from which the design cross-sections are to be selected. For flexural members, however, the sets of discrete sections specified for the design are restricted in the sense that all sections in each set have the same constant shape. This latter restriction facilitates a reasonable and convenient means to update the stiffness and strength properties of a flexural member as its discrete section changes over the design history that produces the minimum weight structure.

The notation used in the paper is as follows: an underscored 'bar' (_) denotes a matrix or vector; a superimposed 'tilde' (~) denotes the transpose of a matrix or vector; a superimposed 'dot' (˙) denotes rate quantities.

1. *THE DESIGN PROBLEM*

All loads are static and the specified service loads are proportionally related to the specified ultimate loads. The framework is discretized into an assemblage of n prismatic members. For each member, the type of cross-section is specified a priori (i.e., WF or T, etc.) and a corresponding set of discrete sections from which its design cross-section is to be selected is identified. The design variable for each member i is its cross-section area a_i. Expressed in its general form, the minimum weight design problem is

$$\textit{Minimize:} \quad \sum_{i=1}^{n} w_i a_i \tag{1a}$$

$$\textit{Subject to:} \quad \underset{\sim}{\delta}_j \leq \delta_j \leq \hat{\delta}_j \qquad (j=1,2,\ldots,d) \tag{1b}$$

$$\underset{\wedge}{\sigma}_k \leq \sigma_k \leq \hat{\sigma}_k \qquad (k=1,2,\ldots,s) \qquad (1c)$$

$$\underset{\wedge}{\alpha}_m \leq \alpha_m \leq \hat{\alpha}_m \qquad (m=1,2,\ldots,p) \qquad (1d)$$

$$a_i \in A_i \qquad (i=1,2,\ldots,n) \qquad (1e)$$

Equation (1a) defines the weight of the structure (w_i is the weight coefficient for member i); eqns (1b) define the d service-load constraints on displacements δ_j (quantities with under- and super-imposed 'hat' ^ denote specified lower- and upper-bounds, respectively); eqns (1c) define the s service-load constraints on stresses σ_k; eqns (1d) define the p ultimate-load constraints on plastic collapse-load factors α_m; eqns (1e) require each member cross-section area a_i to belong to a specified set of discrete sections $A_i \equiv \{a_1, a_2, ..\}_i$.

2. *EXPLICIT PERFORMANCE CONSTRAINTS*

2.1 *Service-Load Constraints*

Since elastic displacements and stresses vary inversely with the sizing variables a_i, corresponding first-order constraints of 'good' quality are achieved by formulating them as explicit functions of the *reciprocal* sizing variables

$$x_i = 1/a_i \qquad (i=1,2,\ldots,n) \qquad (2)$$

First-order Taylor's series expansions and elastic sensitivity analysis based on 'virtual-load' techniques are employed to formulate each of the displacement and stress constraints eqns (1b) and (1c) as the 'explicit-linear' constraints

$$\underset{\wedge}{\delta}_j \leq \sum_{i=1}^{n} d^o_{ij} x_i \leq \hat{\delta}_j \; ; \qquad \underset{\wedge}{\sigma}_k \leq \sum_{i=1}^{n} s^o_{ik} x_i \leq \hat{\sigma}_k \qquad (3a,b)$$

where the superscript ^{o} denotes quantities evaluated for the current design stage (e.g., the initial 'trial' design), and the x_i are the variables to the next design stage.

The displacement and stress sensitivity coefficients (i.e., gradients) d^o_{ij} and s^o_{ik} in eqns (3) are evaluated as

$$d^o_{ij} = \frac{1}{x^o_i} (\tilde{\underline{u}}_j \underline{K}_i \underline{u})^o \; ; \qquad s^o_{ik} = \frac{1}{x^o_i} (\tilde{\underline{u}}_k \underline{K}_i \underline{u})^o \qquad (4a,b)$$

where $\underline{K}_i$ is the global-axis stiffness matrix for member i, and the vectors $\underline{u}$, $\underline{u}_j$ and $\underline{u}_k$ of nodal displacements for the structure are found from finite-element elastic analysis as

$$\underline{u} = \underline{K}^{-1}\underline{P} \; ; \qquad \underline{u}_j = \underline{K}^{-1}\underline{b}_j \; ; \qquad \underline{u}_k = \underline{K}^{-1}\underline{t}_k \qquad (5a,b,c)$$

in which $\underline{K} = \sum_{i=1}^{n} \underline{K}_i$ is the structure stiffness matrix for the current design stage, $\underline{u}$ is the vector of nodal displacements associated with (each) vector $\underline{P}$ of applied service loads, and $\underline{u}_j$ and $\underline{u}_k$ are vectors of 'virtual' nodal displacements associated with the vectors $\underline{b}_j$ and $\underline{t}_k$ of 'virtual' loads.

The 'virtual-load' vectors in eqns (5b,c) are identified from

$$\delta_j = \tilde{\underline{b}}_j \underline{u} \; ; \qquad \sigma_k = \tilde{\underline{t}}_k \underline{u} \tag{6a,b}$$

That is, from eqn (6a), $\underline{b}_j$ is a specified vector that identifies the particular nodel displacement δ_j as being of concern to the design (e.g., if $\underline{b}_j = [1, 0, 0, \ldots, 0]$ then $\delta_j = u_1$). Similarly, from eqn (6b), $\underline{t}_k$ is row k of the global-axis stress matrix for the member associated with the stress σ_k that is of concern to the design. Once specified, the vector $\underline{b}_j$ remains constant for the design. As well, the vector $\underline{t}_k$ is invariant for the design of trusses comprised of axial members alone (in this case, $\underline{t}_k$ is row k of the topological matrix for the truss multiplied by Young's modulus for the material and divided by the length of member k). For flexural members, however, the vector $\underline{t}_k$ is a function of the neutral-axis position for the member cross-section associated with the stress σ_k. As such, since the neutral-axis position changes from stage to stage, the vector $\underline{t}_k$ must be updated over the design history that produces the minimum weight structure. Moreover, regardless of the type of member, each member stiffness matrix $\underline{K}_i$ must also be updated over the design history. The means to update the structure stiffness matrix $\underline{K}$ and, when necessary, the virtual-load vector $\underline{t}_k$ are discussed in the following.

The stiffness matrix for prismatic member i of a planar framework can be expressed as

$$\underline{K}_i = \underline{K}_i^A a_i + \underline{K}_i^B I_i \tag{7}$$

where $\underline{K}_i^A$ and $\underline{K}_i^B$ are constant matrices that correspond to the axial and bending stiffness properties of the member, respectively, and I_i is the cross-section moment of inertia. For a flexural member i, it is assumed that all sections in the specified set of discrete sections from which its design cross-section is to be selected have the same constant shape. As such, the relationship between the cross-section area and the moment of inertia for flexural member i can be expressed as

$$I_i = k_i a_i^2 \tag{8}$$

and the position of the neutral axis y_i (the maximum distance to the extreme fibres of the cross-section) can be expressed in terms of the cross-section area as

$$y_i = n_i a_i^{0.5} \tag{9}$$

where the constants k_i and n_i depend only on the shape of the cross-section and, therefore, apply for all sections in the discrete set specified for member i. (Formulae for k_i and n_i are given by Chiu (5) for a range of cross-section types; e.g., WF, T, etc.). To facilitate the weight optimization for each design stage, eqn (8) is approximated as the linear expression

$$I_i = k_i (a_i^o) a_i = k_i^* a_i \tag{10}$$

where a_i^o is the known cross-section area of member i for the current design, a_i is the sizing variable to the weight optimization for the current design stage, and the constant k_i^* is such that eqns (8) and (10) give the same value for the moment of inertia I_i when $a_i = a_i^o$. Therefore, from eqns (7) and (10), the stiffness matrix $\underline{K}_i$ for member i can be linearly related to the cross-section area a_i as

$$\underline{K}_i = \underline{K}_i^A a_i + \underline{K}_i^B k_i^* a_i = \underline{K}_i^* a_i \tag{11}$$

where $\underline{K}_i^* = \underline{K}_i^A + \underline{K}_i^B k_i^*$ is a constant matrix for the current design stage. Note that $\underline{K}_i^* = \underline{K}_i^A$ for a member under axial stress alone, while $\underline{K}_i^* = \underline{K}_i^B k_i^*$ for a member under bending stress alone.

The structure stiffness matrix $\underline{K}$ and, when necessary, the virtual-load vector $\underline{t}_k$ are updated for each design stage through eqns (9), (10) and (11). Then, elastic analysis is conducted to formulate the service-load constraints eqns (3) for the next weight optimization.

2.2 *Ultimate-Load Constraints*

To maintain consistency with the service-load constraints eqns (3), the ultimate-load constraints are also formulated in terms of the reciprocal sizing variables x_i defined by eqn (2). To this end, first-order Taylor's series expansions and sensitivity analysis techniques are employed to formulate the load-factor constraints eqns (1d) as the 'explicit-linear' constraints

$$\underset{\wedge}{\alpha}_m - 2\alpha_m^o \leq \sum_{i=1}^{n} p_{im}^o x_i \leq \hat{\alpha}_m - 2\alpha_m^o \qquad (m=1,2,\ldots,p) \tag{12}$$

where, again, the superscript ^{o} denotes quantities evaluated for the current design stage, and the x_i are the variables to the next design stage.

The load factor α_m^o in eqn (12) defines the load level at which plastic-collapse mechanism m forms for the current design. It is evaluated using a 'finite-incremental' plastic analysis technique due to Franchi (6) as

$$\alpha_m^o = (\underline{\tilde{R}}\,\underline{\dot{\lambda}}_m)^o = \sum_{i=1}^{n} (\underline{\tilde{R}}_i \underline{\dot{\lambda}}_{im})^o \tag{13}$$

where $\underline{R}$ is the vector of plastic capacities for all members of the structure (the subvector $\underline{R}_i$ refers to that for member i) and $\underline{\dot{\lambda}}_m$ is the vector of member plastic deformation-rates associated with the collapse mechanism m (the subvector $\underline{\dot{\lambda}}_{im}$ refers to that for member i).

The load-factor sensitivity coefficient (i.e., gradient) p_{im}^o in eqn (12) is evaluated as

$$p_{im}^o = -\frac{1}{x_i^o} (\underline{\tilde{R}}_i \underline{\dot{\lambda}}_{im})^o \tag{14}$$

Equation (14) derives, in part, from the fact that for fixed structure topology the vector $\underline{\dot{\lambda}}_m$ of plastic deformation-rates characterizing collapse mechanism m is *invariant* with changes in the design (i.e., $\partial\underline{\dot{\lambda}}_m/\partial x_i = 0$).

To facilitate efficient plastic analysis for each design stage, a piecewiselinear (PWL) yield condition is adopted to govern plastic behaviour at the two end-sections j and k of each member i. As such, the vector $\underline{R}_i$ of plastic capacities for each member i can be expressed as

$$\tilde{\underline{R}}_i = [\tilde{\underline{R}}_j;\ \tilde{\underline{R}}_k] \tag{15}$$

where the components of the subvectors $\underline{R}_j = \underline{R}_k$ are functions of the 'principal' axial and/or bending plastic capacities N_p and M_p for the member. For a member under axial or bending stress alone, $\tilde{\underline{R}}_i = [N_p;\ N_p]$ or $\tilde{\underline{R}}_i = [M_p,\ M_p;\ M_p,\ M_p]$, respectively. For a member under combined axial and bending stresses, $\tilde{\underline{R}}_j = \tilde{\underline{R}}_k = [r_1, r_2, \ldots]$ where each component r_ℓ is the orthogonal distance from the origin of the stress space to a particular linear yield surface of the PWL yield condition for the cross-section. Moreover, each r_ℓ is a constant function of the principal plastic capacities N_p and M_p for all cross-sections having the same shape. (Formulae for r_ℓ are given by Chiu (5) for a range of cross-section types and PWL yield conditions).

The principal axial plastic capacity N_{pi} of each member i is linearly related to the cross-section area a_i as

$$N_{pi} = \sigma_y\ a_i \tag{16}$$

where σ_y is the material yield stress. For all members i having the same cross-section shape, the principal bending plastic capacity M_{pi} is related to a_i as

$$M_{pi} = m_i\ a_i^{1.5} \tag{17}$$

where the constant m_i depends only on the shape of the cross-section and, therefore, applies for all sections in the specified set of discrete sections from which the design cross-section for member i is to be selected. (Formulae for m_i are given by Chiu (5) for a range of cross-section types). To facilitate the weight optimization for each design stage, eqn (17) is approximated as the linear expression

$$M_{pi} = m_i (a_i^o)^{0.5} a_i = m_i^*\ a_i \tag{18}$$

where a_i^o is the known cross-section area of member i for the current design, a_i is the sizing variable to the weight optimization for the current design stage, and the constant m_i^* is such that eqns (17) and (18) give the same value for the plastic moment capacity M_{pi} when $a_i = a_i^o$. Therefore, from eqns (15) and (16) and/or (18), the vector of plastic capacities $\underline{R}_i$ for member i can be linearly related to the cross-section area a_i as

$$\tilde{\underline{R}}_i = [\tilde{\underline{R}}_j^*;\ \tilde{\underline{R}}_k^*] a_i = \tilde{\underline{R}}_i^*\ a_i \tag{19}$$

where $\tilde{\underline{R}}_i^*$ is a constant vector for the current design stage. Note that $\tilde{\underline{R}}_i^* = [\sigma_y;\ \sigma_y]$ for a member under axial stress alone, while $\tilde{\underline{R}}_i^* = [m_i^*,\ m_i^*;\ m_i^*,\ m_i^*]$ for a member under bending stress alone. For a member under combined axial and bending stresses, $\tilde{\underline{R}}_j^* = \tilde{\underline{R}}_k^* = [r_1^*,\ r_2^*,\ \ldots]$ where each component r_ℓ^* is a fixed function of the constants σ_y and m_i^*.

The vector $\underline{R}$ of plastic capacities for the members of the structure is updated for each design stage through eqns (17), (18) and (19). Then, plastic analysis is conducted to formulate the ultimate-load constraints eqns (12) for the next weight optimization.

A discussion is in order at this point concerning the specific set of p plastic collapse mechanisms that the ultimate-load constraints eqns (12) actually refer to for each design stage. In fact, the number p of ultimate-load constraints progressively increases from stage to stage as different collapse mechanisms become 'critical' to the design (i.e., the constraint set for each load case is progressively augmented to account for any 'new' mechanism that forms at the lowest load level for any given design stage). As such, each collapse mechanism m referred to by eqns (12) is either a new 'critical' mechanism for the current design, or it is a mechanism that has been accounted for by the weight optimization of the previous design stage. In the former case, the load factor α_m and the vector $\dot{\underline{\lambda}}_m$ of plastic deformation-rates for the mechanism are determined directly by the plastic analysis conducted for the current design stage. The corresponding ultimate-load constraint is then formulated through eqns (14) and (12) and added to the constraint set for the next weight optimization. For the latter case, however, the vector $\dot{\underline{\lambda}}_m$ is already known from a plastic analysis conducted at a previous design stage (recall that $\dot{\underline{\lambda}}_m$ is invariant with changes in the design). As such, the load factor α_m is updated through eqn (13) directly (i.e., without need for further plastic analysis). Thereafter, the corresponding ultimate load constraint is updated through eqns (14) and (12) for the next weight optimization.

3. *THE EXPLICIT DESIGN PROBLEM*

From eqns (2), (3) and (12), the minimum weight design problem eqns (1) is expressed explicitly in terms of reciprocal sizing variables x_i as

$$\textit{Minimize:} \quad \sum_{i=1}^{n} w_i/x_i \tag{20a}$$

$$\textit{Subject to:} \quad \underset{\wedge}{\delta}_j \leq \sum_{i=1}^{n} d_{ij}^o x_i \leq \hat{\delta}_j \qquad (j=1,2,\ldots,d) \tag{20b}$$

$$\underset{\wedge}{\sigma}_k \leq \sum_{i=1}^{n} s_{ik}^o x_i \leq \hat{\sigma}_k \qquad (k=1,2,\ldots,s) \tag{20c}$$

$$\underset{\wedge}{\alpha}_m - 2\alpha_m^o \leq \sum_{i=1}^{n} p_{im}^o x_i \leq \hat{\alpha}_m - 2\alpha_m^o \qquad (m=1,2,\ldots,p) \tag{20d}$$

$$x_i \in X_i \tag{20e}$$

where the components of each discrete set $X_i \equiv \{x_1, x_2, \ldots\}_i$ in eqn (20e) are the reciprocals of the cross-section areas comprising the corresponding discrete set A_i in eqn (1e).

The discrete weight optimization problem eqns (20) is solved for each design stage using an efficient 'generalized optimality criteria technique' due to Fleury (7).

After each design stage, the sensitivity coefficients d_{ij}^o, s_{ik}^o and p_{im}^o are updated, if necessary new ultimate-load constraints are added to the constraint set, and the weight optimization is repeated to find an improved (lower weight) design. The iterative process terminates with the final design when there is no change in the structure weight from one design stage to the next.

A STRUctural SYnthesis computer code named STRUSY-2 has been developed by Lee (8) to implement the described design method. The code also incorporates a number of additional features that enhance computational efficiency, stabilize the convergence of the design history and ensure fabrication and architectural requirements.

4. *EXAMPLE APPLICATION*

The industrial steel-mill building in Fig 1 is to be designed accounting for the three independent load cases noted in Fig 2; load P = 44.4 KN (10 kips) defines the service-load level. The 7 web members in the roof-truss experience axial stress alone and may be designed for any cross-section shape. The 26 beam, column and chord members in the framing system experience combined axial and bending stresses, and their design cross-sections are each specified to have the constant WF shape shown in Fig 2. The design cross-section area for each of the 33 members is to be selected from the discrete set of 50 area values given in Fig 2.

Service-load displacement constraints of ±38 mm (1.5 in) are imposed on the horizontal translations of nodes 2,3,7,9,11 and 13 for load case 2. Service-load stress constraints of $\pm 0.67\sigma_y$ are imposed at the end-sections of each member for all three load cases. (There are 90 stress constraints in total.) Ultimate-load plastic collapse constraints impose a lower bound of 2.5 on the values of the collapse load factors for all plastic mechanisms forming for load cases 1 and 2; i.e., the structure is designed to withstand a 150% overload beyond the service-load level (load factors are not bounded from above). To satisfy symmetry and member-continuity requirements, design-variable-linking is employed to reduce the number of independent sizing variables from 33 to 12 (the 12 linked groups of members are indicated in fig 1).

The STRUSY-2 computer code due to Lee (8) is applied to conduct the minimum weight design. Starting with an arbitrary 'trial' structure weighing 9618.52 kg, the 'minimum-weight' structure weighing 4737.2 kg is found after eight design stages. The 'optimum' cross-section areas for the 12 groups of members are shown in Fig 3 along with with iteration history for the design process. Eight 'critical' plastic collapse modes (numbered 1,2,...,8 in Fig 3) became active at various stages of the design history. Two displacement constraints, one stress constraint and three plastic-collapse constraints are 'active' at the final design stage. A number of other constraints are potentially active.

5. *CONCLUDING REMARKS*

Through the coordinated use of elastic analysis, plastic analysis, approximation concepts and optimization techniques, the described design method provides an efficient and effective means to optimally size practical engineering frameworks using discrete member sections while simultaneously satisfying performance criteria imposed at both the service and ultimate load levels. For example, the steel-mill building design presented herein required only 80 seconds to complete using the STRUSY-2 computer code on an IBM 4341 computer (University of Waterloo).

For service-load constraints alone, the STRUSY-2 code due to Lee (8) can also be applied for commercially available standard sections (for which the 'constant-shape' assumption employed herein does not apply). Work is presently underway to extend this capability to design under ultimate-load constraints as well.

References

1. Emkin, L.Z. and Litle, W.A. 'Storywise Plastic Design of Multistory Steel Frames.' ASCE, J. of Str. Div., Vol. 98, No. ST1, January, 1972.
2. Horne, M.R. and Morris, L.J. 'Optimum Design of Multi-Storey Rigid Frames.' Proceedings, Int. Sym. on Computers in Optimization of Structural Design, Univ. of Wales, Swansea, January, 1972.
3. Grierson, D.E. and Schmit, L.A., Jr. 'Synthesis under Service and Ultimate Performance Constraints.' J. of Comp. and Structures, Vol. 15, No. 4, 1982.
4. Grierson, D.E. and Chiu, T.C.W. 'Synthesis of Frameworks under Multilevel Performance Constraints.' NASA CP-2245, Sym. on Advances and Trends in Structural and Solid Mechanics, Washington, D.C., October, 1982.
5. Chiu, T.C.W. 'Structural Synthesis of Skeletal Frameworks under Service and Ultimate Performance Constraints.' M.A.Sc. Thesis, Univ. of Waterloo, Canada, April, 1982.
6. Franchi, A. 'STRUPL-ANALYSIS: Fundamentals for a General Software System.' Ph.D. Thesis, Univ. of Waterloo, Canada, 1977.
7. Fleury, C. 'Structural Weight Optimization by Dual Methods of Convex Programming.' Int. J. Num. Meth. in Engng., Vol. 14, 1979.
8. Lee, W.H. 'Optimal Structural Synthesis of Frameworks using Discrete and Commerically Available Standard Sections.' M.A.Sc. Thesis, Univ. of Waterloo, Canada, May, 1983.

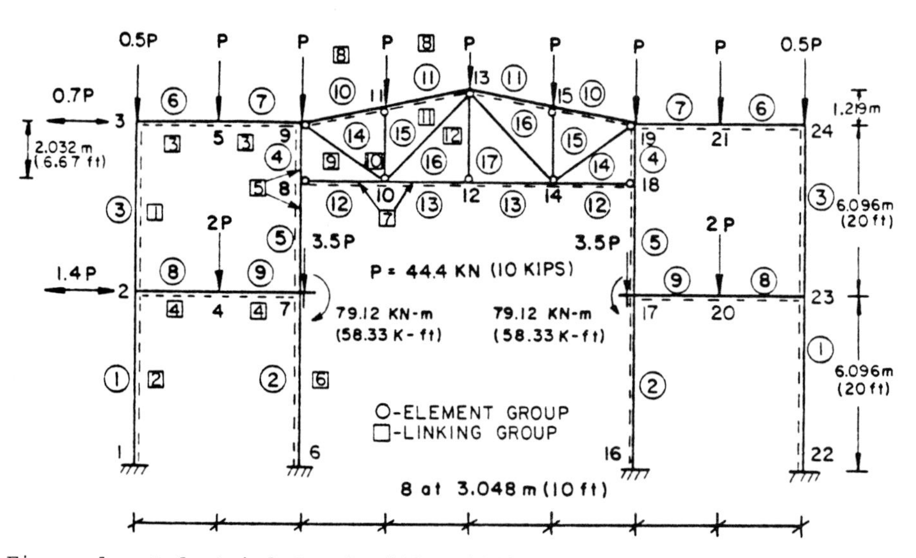

Figure 1: Industrial Steel-Mill Building

Design Load Cases

CASE	LOAD COMBINATIONS
1	ROOF + FLOOR + CRANE LOADS
2	CASE 1 + WIND LOAD FROM LEFT
3	CASE 1 + WIND LOAD FROM RIGHT

Cross-Section Areas (mm^2) Available for Design

500.	3871.	7097.	10323.	13548.
968.	4194.	7419.	10645.	13871.
1290.	4516.	7742.	10968.	14194.
1613.	4839.	8064.	11290.	14516.
1935.	5161.	8387.	11613.	14839.
2258.	5484.	8710.	11935.	15161.
2581.	5806.	9032.	12258.	15484.
2903.	6129.	9355.	12581.	15806.
3226.	6452.	9677.	12903.	16129
3548.	6774.	10000.	13226.	16452.

- ALL WEB MEMBERS ARE TRUSS MEMBERS
- TOP AND BOTTOM CHORD ELEMENTS, BEAM AND COLUMN ELEMENTS ARE WF SECTIONS

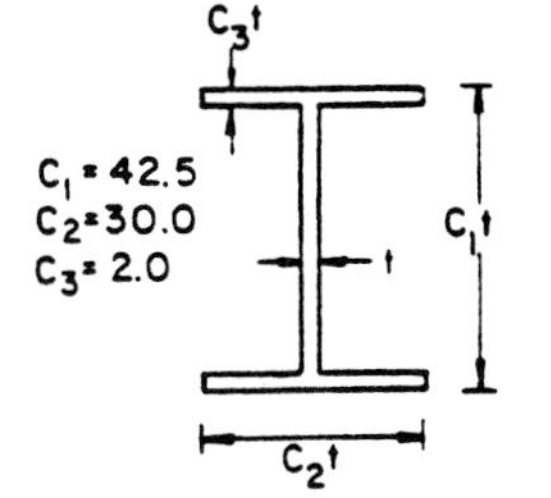

Cross-Section Data

Figure 2: Data for Steel-Mill Building Design

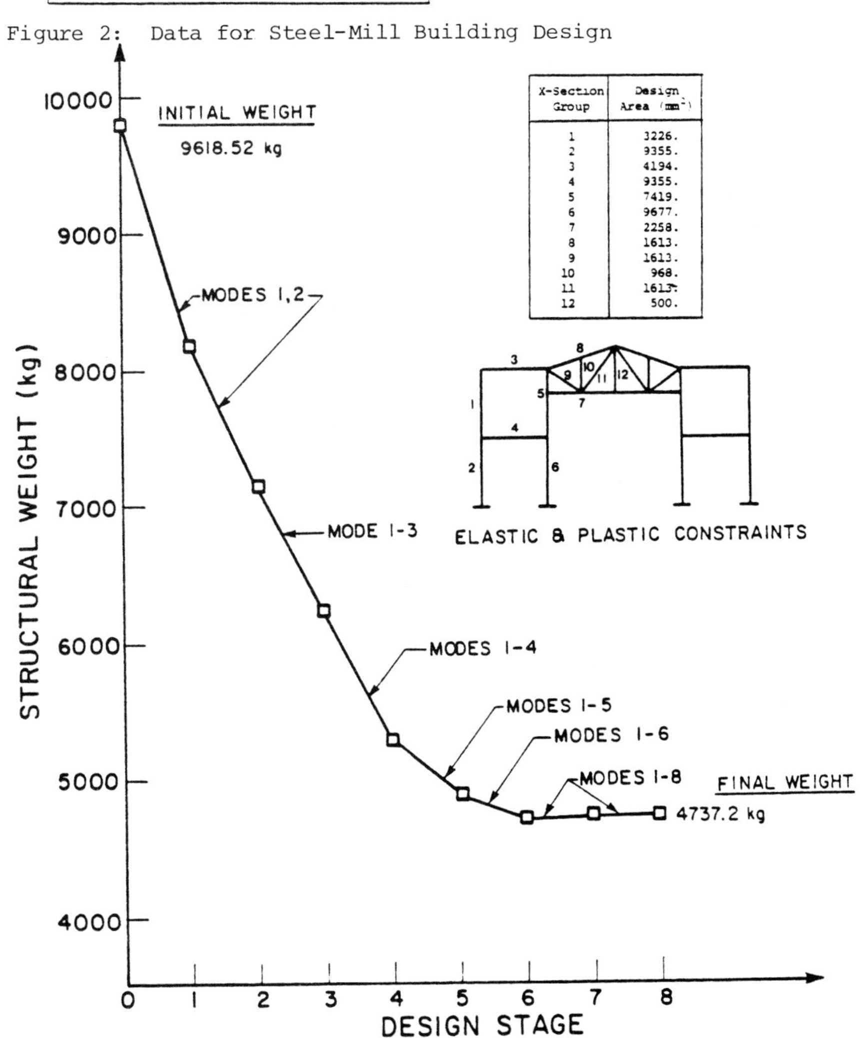

X-Section Group	Design Area (mm^2)
1	3226.
2	9355.
3	4194.
4	9355.
5	7419.
6	9677.
7	2258.
8	1613.
9	1613.
10	968.
11	1613.
12	500.

Figure 3: Steel-Mill Building Design History

DESIGN OF STEEL FRAMES WITH INTERACTIVE COMPUTER GRAPHICS

C. I. Pesquera and W. McGuire

Department of Structural Engineering, Cornell University

Synopsis

This paper summarizes the analysis and design tools that have been implemented in an integrated design system with interactive computer graphics and demonstrates some of their capabilities. The design system allows the definition, analysis, and design of two or three-dimensional steel frames of arbitrary geometry. Provisions are included to re-size the members of a structure, either automatically or interactively, such that both serviceability and ultimate limits states are satisfied. Second order geometric effects as well as material nonlinearities may be included in the analytical model to permit prediction of the response of the structure up to ultimate loads.

Notation - Strength Design

F	nodal forces
p	ratio of axial load to squash load
m_z	ratio of strong axis bending moment to corresponding plastic moment
λ_a, λ_b	reference load factors
λ_d	target load factor

Notation - Stiffness Design

A	area of a member
C	constant that relates moment of inertia to weight of a member (= $\rho A/I_z$, where ρ is a specific weight)
E	Young's modulus
I	set of active members
I_z	moment of inertia
I_{zm}	minimum acceptable moment of inertia
J	set of active constraints
K	iteration number, member number

l member length

M_z bending moment due to real loads

$\bar{M}_z$ bending moment due to virtual loads

P axial load due to real loads

$\bar{P}$ axial load due to virtual loads

r displacement at a constrained degree of freedom

r_u displacement limit

α ratio of moment of inertia to area

ρ step size factor

λ Lagrange multipliers

Γ maximum ratio of actual displacement to specified displacement limit

Introduction

The authors and associates at Cornell have been conducting research aimed at the advancement of practical, economical methods for the nonlinear analysis and design of three dimensional steel framed structures. Under development is an interactive computer graphics, computer aided design system that will place at the command of the user a battery of linear and nonlinear analytical tools that he may use either independently or in conjunction with established design equations, which are also part of the system.

The project is a comprehensive one. This paper is focused on one portion of it: the interactive tools for improving strength and stiffness characteristics of trial designs of planar frames. Before describing these tools the principles underlying the development will be sketched. A more general description of the system is presented in Reference (1).

1. GENERAL PRINCIPLES

Fundamental to this work is the idea that interactive graphics in an advanced minicomputer environment is a vehicle that will enable engineers to use modern design specifications and methods of analysis most efficiently and effectively.

The type of specification most amenable to computerization is the modern limit states or load and resistance factor specification. In the present system, provision is made for the interactive use of the major design equations of Reference (2). It is anticipated that formulas from Reference (3) will also be utilized.

Figure 1 is a flow chart that illustrates one way in which the assembled interactive computer graphics capability may be used. The first step is the definition of the structure. The next group of steps is an iteractive procedure for satisfying serviceability and ultimate limit states. For each load combination the structure is analyzed and its response is evaluated. If necessary, a new design is generated, that is, member properties are revised using the results of the existing analysis. A new analysis is then made and the process is repeated until a satisfactory design is obtained. In some cases the designer may find it necessary to modify the structural configuration or add new members. In this event, these changes are made in the structure definition component, as illustrated on the chart.

The method suggested considers strength and serviceability limit states separately. It is assumed that one of these states will control the major features of the design and that only minor changes will be required when the second is considered.

2. *ANALYSIS AND DESIGN SYSTEM*

The system consists of a number of modules served by a common database. In a typical problem, the structure geometry, boundary conditions, member properties, and loads are specified in the problem definition module, the first component of the system, (see Reference 4). In the second component, the load combination for the limit state under consideration is defined, the type of analysis to be used is selected, and the analysis is performed. Tools are provided to evaluate and, if necessary, to modify the current design using a member-by-member approach based on satisfying code type equations. Also, an iterative procedure has been implemented to modify member sizes so that the displacement limits are not exceeded. The third component of this system contains provisions for aiding the interpretation of the behavior of the structure through the graphical display of information, and for assisting in the generation of a new design when the full nonlinear design procedure for strength is being used.

The basic analysis program is a modification of a stiffness method program developed by Orbison (Ref. 5). It employs a straight, prismatic, twelve degree-of-freedom beam-column element. Rigid connections are assumed. The program includes provisions for the following types of analysis: 1) linear elastic, 2) second order elastic, 3) first order elastic plastic, and 4) full geometric and material nonlinear. Second order effects are modeled with a geometric stiffness matrix and updated reference geometry. Material nonlinearity is incorporated by permitting plastification of end cross sections under combinations of axial force and bending that satisfy a specified yield criterion. The criterion currently used is a continuous polynomial approximation of an interaction diagram for steel wide flange sections developed in Reference (6). Elastic unloading of previously plastified cross sections is accommodated.

The system is a flexible one. It can be used in a number of ways, depending upon the designer's needs. Two different design methodologies have been explored. The first is based on first or second order elastic analysis, with code type equations used to evaluate the design. The second is based on full nonlinear analysis, with the structure's performance expected to satisfy or exceed some user-specified criteria. Both strength and stiffness requirements must be considered in each method. As an illustration of the capability of the system, two of the procedures used in the second method in the search for efficient solutions to the requirements of strength and stiffness will be described.

3. *STRENGTH DESIGN*

For each iteration of the strength design process a full nonlinear analysis is required. The analysis can be peformed iteractively, with the user having the ability to monitor the response of the structure as the results are generated, or through a batch process in which control of the analysis is predefined. Figure 2 shows the type of information that is displayed at the end of each load step during an interactive analysis. It includes the location of plastic hinges, a graph of the lateral displacement of the top story versus a characteristic load parameter, analysis statistics, and various analysis control options. After each step, nodal displacements, member end forces, and plastic hinge locations are saved. This information is then available to the third component of the system, which is dedicated to the review of response and to the support of the re-design procedures used when full nonlinear analysis is the design basis.

After the analysis of a trial design, it is judged by its resistance to gravity loads and to combined gravity and lateral loads. Failure load factors such as the 1.7 for gravity loads and 1.3 for combined loads that are used in Part 2 of the AISC Specification are prescribed. For the improvement of a trial design, the system accomodates the traditional approach in which the designer relies upon his experience to select new sizes for members that appear either too weak or too strong. This "manual" approach is facilitated by graphic displays and interactive member modification provisions. It works well for simple structures, but for structures of large size a more systematic re-design approach is also needed. The procedure incorporated in the system follows a method developed by Majid and Anderson in Reference (7). In this method beams are required to remain elastic up to service load and columns up to collapse load. An interpolation procedure is applied to the analytical results to obtain estimates of the member sizes required to satisfy these requirements.

In principle, the method involves the use of linear interpolation or extrapolation to calculate a vector of nodal forces on an element at a target load from nodal forces determined in the analysis at two other loads. Thus:

$$\{F_d\} = \{F_a\} + \{\{F_b\} - \{F_a\}\} \left(\frac{\lambda_d - \lambda_a}{\lambda_b - \lambda_a}\right) \qquad (1)$$

where $\{F_a\}$ and $\{F_b\}$ are the nodal forces at the load factors λ_a and λ_b selected as the basis for interpolation. These factors are selected automatically by the program following the recommendations given in Reference (7). They depend on the way the member is responding to load and may vary from member to member. The factor, λ_d called the target load factor, is specified by the user. Common values for λ_d, would be 1.0 for beam elements and the collapse load factor for column elements. If it develops that initially specified values of λ_d result in unsatisfactory overall performance they may be modified interactively by the user in subsequent iterations.

Following the establishment of λ_a, λ_b, and λ_d, the computer selects the lightest acceptable section to develop a plastic hinge under the nodal forces $\{F_d\}$. Sections selected in this way are then used in the next analysis.

The procedure has been modified in several ways to take advantage of the interactive environment. For example, curves such a those in Fig. 3 are displayed. They aid in the selection of reference loads in cases in which the automated procedure fails to predict reasonable design forces $\{F_d\}$, and they help in monitoring changes in the behavior as sections are modified. Cross section plasticity is calculated through use of the following two dimensional version of Orbison's yield surface for wide flange members (Ref. 6).

$$1.15p^2 + m_z^2 + 3.67p^2m_z^2 = 1.0 \qquad (2)$$

In the section search, the applied forces are taken from $\{F_d\}$ and the section resistances from the system database. Any section selected by the program can be interactively modified by the user if desired (Ref. 1).

4. *STIFFNESS DESIGN*

A trial design, or even one which fully satisfies all strength criteria, may still deflect excessively under service loads. To handle such situations, an optimiztion procedure has been incorporated in the system to modify the design so that prescribed drift limits or other displacement constraints are satisfied with a minimum increase in weight. The procedure employs member moments of inertia as the basic design variables and follows established optimality criteria concepts and techniques (Ref. 8, 9, 10, 11, 12).

The algorithm used starts with consideration of the basic optimization problems: to find a vector of design variables $\{I_z\}$ so that

$$W = \{C\}^T\{I_z\} \quad \text{-->} \ min. \qquad (3)$$

Subject to:

$$\{r\} - \{r_u\} \leq 0 \qquad \text{Displacement limits}$$

$$\{I_{zm}\} - \{I_z\} \leq 0 \qquad \text{Size limits}$$

If a strength design precedes the stiffness design the minimum moments of interia, I_{zm}, are taken as those obtained in that process. Consideration is also given to user imposed depth limits and to grouping, that is, to the use of the same shape for several members. Only displacement constraints are explicitly considered in the optimization algorithm however. Depth and grouping constraints are treated as "side" constraints. They are dealt with through the concept of active and passive members in which members governed by side constraints become inactive or "passive" (Ref. 8).

For optimum design the Kuhn-Tucker conditions must also be satisfied. These conditions are

$$\frac{\partial W}{\partial I_{zi}} + \sum_{j=1}^{J} \lambda_j \frac{\partial r_j}{\partial I_{zi}} = 0 \qquad i = 1, \ldots, I$$

$$\lambda_j \geq 0 \qquad j = 1, \ldots, J \tag{4}$$

where J is the set of active constraints, I is the set of active members, and λ_j are the Lagrange multipliers.

The optimality criterion method used employs the following recurrence relationship to generate a new design vector based on the current design vector and the Kuhn-Tucker conditions (Ref. 9):

$$I_{zi}^{(k+1)} = (T_i^{\,k})^{1/\beta}\, I_{zi}^{\,k} \tag{5}$$

where β is a step size factor, and k is the iteration number. T_i is obtained as follows: At the optimum, the condition satisfied by each active member is, from eqn (4)

$$1 = \frac{-\sum_{j=1}^{J} \lambda_j \frac{\partial r_j}{\partial I_{zi}}}{\frac{\partial W}{\partial I_{zi}}} \tag{6}$$

the values of T_i are obtained by evaluating the right hand side of this equation which, prior to convergence, will not be unity. First however, Lagrange multipliers associated with each active constraint and the gradients of the active constraints must be determined. To calculate the Lagrange multipliers it is necessary to transform eqn (4) into a determined set since, in general, the number of active constraints and the number of active members are not the same. A transformation that minimizes the error in the Lagrange multipliers in a least squares sense is used to convert this equation into a symmetric system of J unknowns (Refs. 10, 11). In matrix form, the transformed equation is

$$-[r']\{W'\} = [r']^T[r']\{\lambda\} \tag{7}$$

To obtain r_i', the gradient of an active constraint, the displacement r_i is first expressed as a function of the internal forces and member properties through the principle of virtual work:

$$r_i = \sum_{j=1}^{m} \frac{1}{I_{zj}} \left[\frac{P_j \bar{P}_{ji} \alpha_j}{E} + \int_0^{\ell_j} \frac{M_{zj}(x)\bar{M}_{zji}(x)dx}{E}\right] \tag{8}$$

where, l_j is the member length, $\bar{P}_{ji}$ and $\bar{M}_{zji}$ are the axial load and the bending moment in member j due to a unit load at the location of displacement i, P_j and M_{zj} are the axial load and the bending moment in member j due to the real loads, and α_j is the ratio of the moment of inertia to the area of the member. The derivative of r_i with respect to the design variable of member k is

$$\frac{\partial r_i}{\partial I_{zk}} = \frac{-B_{ki}}{I_{zk}^2} + \sum_{j=1}^{m} \frac{\partial B_{ji}}{\partial I_{zk}} \frac{1}{I_{zj}} \tag{9}$$

where B_{ji} is the term inside the bracket of eqn (8). It has been shown that, based on virtual work principles, the second term of eqn (9) is identically zero (Ref. 12). It therefore becomes

$$\frac{\partial r_i}{\partial I_{zk}} = -\frac{B_{ki}}{I_{zk}^2} \tag{10}$$

The design algorithm used in employing the above relationships can now be summarized as follows:

1) Displacement limits are prescribed by specifying the maximum displacement of one or more degrees of freedom. In wind resistant design, the horizontal degree of freedom of one node at each level is constrained to satisfy a prescribed drift limit.

2) The trial design, or one based on strength requirements, is used as the initial design for the optimization process. The α_i factor for each member is computed.

3) A linear elastic service load analysis of the structure is performed. The ratio of the actual displacement to the specified displacement limit is computed for each active constraint. The maximum ratio, Γ, is then used to scale the design variables so that the most violated constraint becomes an equality. The design and displacement vectors are updated by using this scale factor as follows:

$$\begin{aligned} \{\bar{I}_z\} &= \Gamma\{I_z\} \\ \{\bar{r}\} &= \frac{1}{\Gamma}\{r\} \end{aligned} \tag{11}$$

4) The weight monitor curve is updated. For the second and all other subsequent iterations it is checked for convergence. If a prescribed weight convergence criterion is satisfied, standard sections which most closely satisfy the current designed vector (assumed to be continous up to this point) are selected and the process is stopped.

5) If convergence has not been obtained, virtual forces in the members due to unit loads at the location of the active displacement constraints are calculated. Coefficients B_{ji} *and the gradients of* the active constraints are determined from equations (9) and (10).

6) Lagrange multipliers are calculated using eqn (7). If a multiplier is negative, its associated constraint becomes inactive and a new set of Lagrange multipliers is calculated. This process is repeated until all the Lagrange multipliers are positive.

7) The continuous design vector $\{I_z\}$ is updated using the recurrence relationship, eqn (5), in which it is assumed that α retains its initial value for each member.

8) Size and grouping constraints are checked. If the design of a member is governed by either of these criteria that member becomes inactive. If there is a change in the active/inactive list the process is repeated starting at step 6, otherwise it is a resumed at step 3.

Figure 4 shows the type of information that is displayed during the stiffness design process. It includes the weight/iteration curve, the set of active constraints, the deflected shape of the structure, and various options to control the process.

5. *EXAMPLE*

A planar example frame is shown in Fig. 5. This example was used in Reference (11) in which it was designed to satisfy both stress and displacement constraints. The design criteria used here are also presented in Fig. 5. The frame is first designed to resist gravity and combined loads. The minimum increase in member sizes required for the satisfaction of the drift limits is then obtained. In non-linear analysis each beam was subdivided into three segments and distributed loads were treated as concentrated loads at the internal nodes. During the design iterations, collapse was reached at load load factors less than the minimum specified for both loading conditions even when individual members became plastic at load factors greater than or equal to the initial target load factors. In order to obtain satisfactory collapse load factors for the overall structure, the target load factor for hinge formation in the beams was increased from its original value of 1.0 to 1.1 for combined loads and to 1.5 for gravity loads. The weight of the structure and the sections obtained in the strength and stiffness designs are given in Table 1. It can be seen that it was necessary to increase the size of some members in order to obtain adequate stiffness, but no changes were required in the upper stories.

6. *SUMMARY*

This paper has presented some of the reasons for using interactive computer graphics in the design of steel frames. A design system with the capability for satisfying serviceability and ultimate limit states efficiently has been described. An example that demonstrates the application of two of the weight reduction features of the system has been presented. It is believed that this type of system can facilitate conventional design and that it also provides a medium for the practical use of advanced methods of analyses in design whenever they are advantageous.

Acknowledgements

This work was supported by the National Science Foundation under Grant No. CEE-8117028. The authors wish to thank John F. Abel and Donald P. Greenburg for their invaluable contributions to the research. The original analysis routines were developed by James G. Orbison. His contribution is gratefully acknowledged. Thanks are also due to Professor Shlomo Ginsburg for his guidance in the use of optimality criteria methods for stiffness design.

References

1. McGuire, W., and Pesquera, C. I., "Interactive Computer-Graphics in Steel Analysis/Design - A Progress Report", Presented at The National Engineering Conference, American Institute of Steel Construction, Chicago, Illinois, March 11, 1982.

2. Steel Structures for Buildings - Limit States Design, CBA Standard s16.1-1974, Canadian Standards Association, Rexdale, Ontario, Canada, 1976.

3. Tentative Load and Resistance Factor Design Specification, the American Institute of Steel Construction, Chicago, Illinois, scheduled for issue 1983.

4. Pesquera, C. I., McGuire, W., and Abel, J. F., "Interactive Graphical Preprocessing of Three-Dimensional Framed Structures," Computers and Structures, Vol. 17, pp. 1-12, 1983.

5. Orbison, J. G., "Nonlinear Static Analysis of Three-Dimensional Steel Frames", doctoral dissertation, Cornell University, Ithaca, N.Y., May, 1982.

6. Orbison, J. G., McGuire, W., and Abel, J. F., "Yield Surface Applications in Nonlinear Steel Frame Analysis", Computer Methods in Applied Mechanics and Engineering", Vol. 33, 1982, pp. 557-573.

7. Majid, K. I., and Anderson, D., "Elasto-Plastic Design of Sway Frames by Computer", Proceedings of the Institute of Civil Engineers, Vol. 41, December, 1968, pp. 705-729.

8. Gellatly, R. A., and Berke, L., "The Use of Optimality Criteria in automated Structural Design", Proceedings of the Third Conference on Matrix Methods in Structural Mechanics, AFFDL-TR-71-160, December, 1971.

9. Kirsh, U., Optimum Structurl Design, McGraw Hill, Inc., 1981.

10. Jennings, A., Matrix Computation for Engineers and Scientists, John Wiley and Sons, 1977.

11. Tabak, E. I., and Wright, P. M., "Optimality Criteria Method for Building Frames", Journal of the Structural Division , ASCE, Vol. 107, No. ST7, July, 1981, pp. 1327-1342.

12. Berke, L., "Convergence Behavior of Optimality Criteria Based Iterative Procedures", USAF AFFDL-TM-72-1-FBR, January 1972.

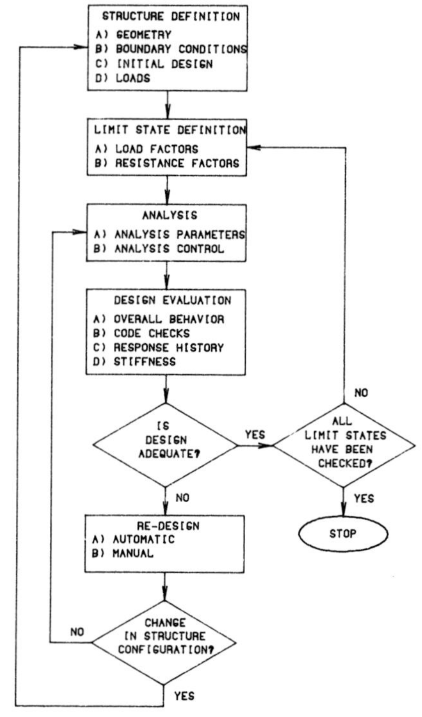

Fig. 1 Flow chart of a possible use of the design system.

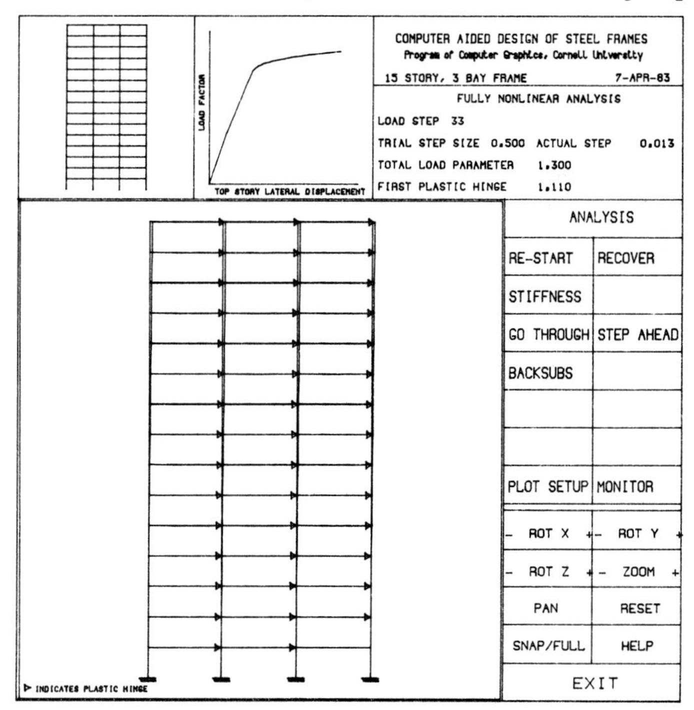

Fig 2 Display layout during an interactive analysis.

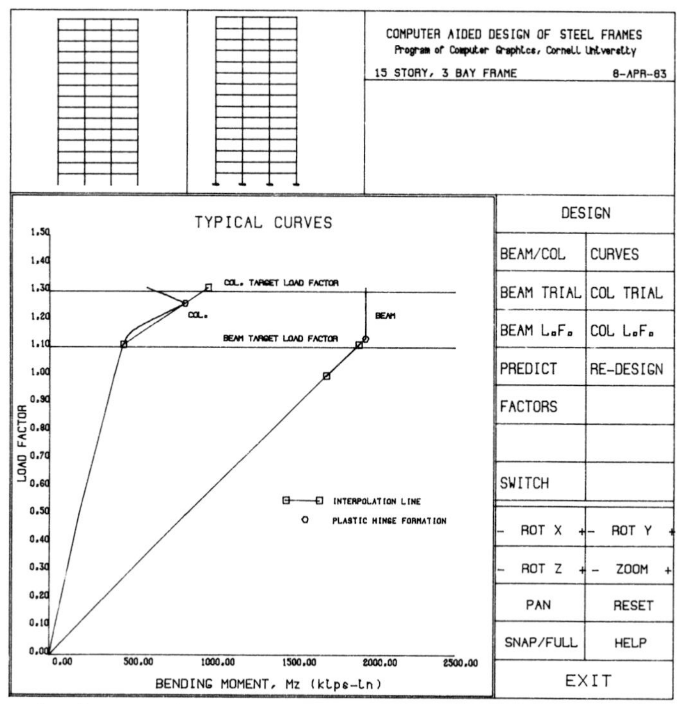

Fig. 3 Display used for the specification of target load factors and study of the response of selected sections.

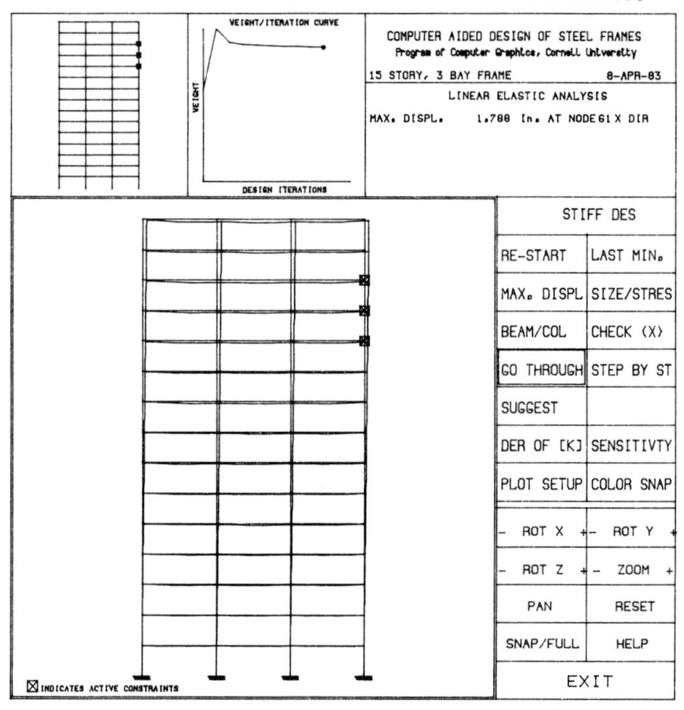

Fig. 4 Information displayed during the stiffness design.

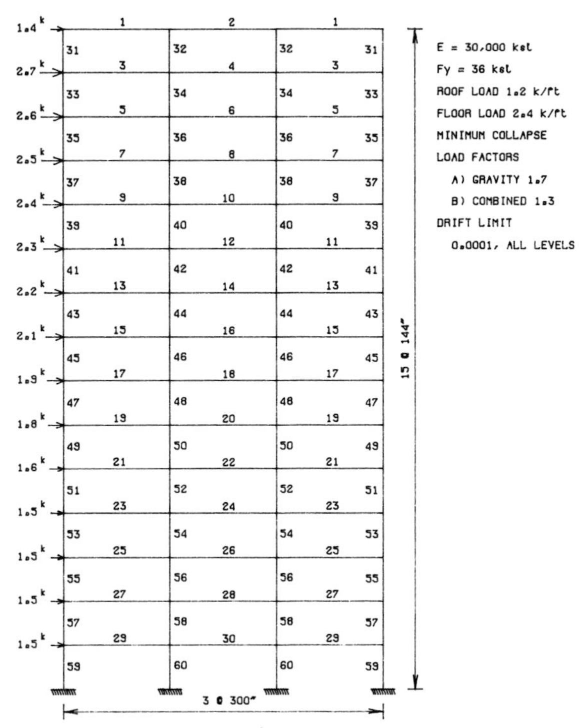

Fig. 5 Geometry, loads, material data, and design criteria for example structure. Member numbers refer to cross sections in Table 1 below.

STRENGTH DESIGN			
1	W12X 22	31	W14X 22
2	W12X 22	32	W6 X 9
3	W16X 31	33	W12X 26
4	W18X 35	34	W8 X 18
5	W16X 31	35	W12X 30
6	W18X 35	36	W8 X 28
7	W16X 31	37	W14X 34
8	W18X 35	38	W14X 38
9	W14X 34	39	W14X 43
10	W18X 35	40	W14X 48
11	W14X 34	41	W14X 43
12	W18X 35	42	W12X 58
13	W14X 34	43	W14X 48
14	W18X 35	44	W14X 68
15	W18X 35	45	W14X 53
16	W18X 35	46	W12X 79
17	W18X 35	47	W12X 58
18	W18X 35	48	W10X 88
19	W18X 35	49	W14X 61
20	W18X 35	50	W14X 99
21	W18X 35	51	W14X 68
22	W18X 35	52	W14X 109
23	W18X 35	53	W12X 72
24	W18X 35	54	W14X 120
25	W18X 35	55	W14X 74
26	W18X 35	56	W14X 132
27	W18X 35	57	W14X 82
28	W18X 35	58	W14X 145
29	W18X 35	59	W14X 82
30	W18X 35	60	W12X 152
38.6 TONS			

STIFFNESS DESIGN			
1	W12X 22	31	W14X 22
2	W12X 22	32	W6 X 9
3	W16X 31	33	W12X 26
4	W18X 35	34	W12X 16
5	W16X 31	35	W12X 30
6	W18X 35	36	W14X 26
7	W16X 31	37	W14X 38
8	W16X 31	38	W14X 38
9	W14X 48	39	W16X 45
10	W16X 31	40	W14X 53
11	W16X 50	41	W14X 74
12	W16X 36	42	W12X 65
13	W16X 57	43	W14X 68
14	W14X 53	44	W14X 68
15	W18X 71	45	W14X 99
16	W16X 45	46	W12X 79
17	W18X 71	47	W12X 120
18	W16X 45	48	W10X 88
19	W18X 71	49	W12X 152
20	W16X 40	50	W14X 99
21	W16X 77	51	W12X 170
22	W18X 50	52	W14X 109
23	W18X 65	53	W12X 152
24	W18X 50	54	W14X 120
25	W18X 60	55	W12X 190
26	W18X 55	56	W14X 132
27	W16X 67	57	W14X 176
28	W18X 50	58	W14X 145
29	W18X 46	59	W14X 193
30	W14X 38	60	W12X 152
54.8 TONS			

Table 1 Steel shapes and weight (short tons) obtained from the strength and stiffness designs.

USE OF THE ELASTIC EFFECTIVE LENGTH FOR STABILITY CHECKS OF COLUMNS AND CONSEQUENCES FOR CHECKS ON BEAMS IN BRACED FRAMES

H H Snijder, F S K Bijlaard, J W B Stark

Institute TNO for Building Materials and Building Structures (IBBC) of the Netherlands Organisation for Applied Scientific Research (TNO)

Synopsis

In general, checking for stability of columns in braced frames is based on effective lengths. The formulae for checking stability are, however, derived for individual pin-ended columns where the effective length is equal to the system length. This paper deals with the above mentioned problem. Requirements necessary for the beams of a simple braced frame are sought, when the column is checked for stability using the elastic effective length.

Notation

A	cross-sectional area
b	beam length
E	Young's modulus
e^*	imperfection parameter
F	concentrated load
F_A	column capacity based on $\ell_k = \ell_{eff}$
F_B	column capacity based on $\ell_k = \ell_{sys}$
F_E	Euler buckling load
h	column length
I	moment of inertia
k	rotational stiffness
ℓ_{eff}	elastic effective length
ℓ_k	buckling length
ℓ_{sys}	system length
M_1	end-moment with the smallest absolute value
M_2	end-moment with the largest absolute value
M	absolute value of M_2
M_p	plastic moment of a column section

$M_{p,red}$ reduced plastic moment

M_r restraining moment

N axial load

N_p squash load

q uniformly distributed load

W_e elastic section modulus

W_p plastic section modulus

w displacement

w_o imperfection

$\hat{w}_o$ amplitude of imperfection

x, y rectangular coordinates

β equivalent moment factor

δ displacement

λ slenderness

λ_{eff} elastic effective slenderness

λ_{sys} system slenderness

μ ratio between Euler buckling load and axial force

σ_r yield stress

ω buckling coefficient

Introduction

According to most regulations the stability of a braced frame is checked using the first order elastic force distribution. Stability check of the total braced frame is replaced by stability checks of the individual columns. The columns are 'cut' out of the frame and the resultant pin-ended columns, with the imposed bending moments and axial forces, are checked for stability (Fig 1 and (1)). The problem of the stability of a braced frame is thus simplified to that of an individual column.

Non-linear analyses of such columns have led to interaction curves that describe the collapse of columns under combined bending and compression. Interaction formulae are derived from these interaction curves. Equation (1) gives the ECCS (2) formula:

$$\frac{N}{A} + \frac{\mu}{\mu - 1} \frac{\beta M + Ne^*}{W_e} \leqslant \sigma_r \qquad (1)$$

where μ and e^* are based on the buckling length

$$\beta = 0.6 + 0.4 \frac{M_1}{M_2} \geqslant 0.4 \qquad (2)$$

Equation (1) can be modified as below:

$$\frac{\omega N}{A} + \frac{\mu}{\mu - 1} \frac{\beta M}{W_p} \leqslant \sigma_r \qquad (3)$$

where ω and μ are based on the buckling length and β is as given in eqn (2).

Equation (3) can be written as follows:

$$\frac{\omega N}{N_p} + \frac{\mu}{\mu - 1}\frac{\beta M}{M_p} \leqslant 1 \qquad (4)$$

The β factor accounts for the shape of the moment distribution and is a measure of the magnitude of the bending moment in a critical section of the column in the deformed situation (3). In fact, the stability check of the individual column is reduced to a check on the yield stress for a cross-section with an axial force and a characteristic moment.

1 PROBLEM DESCRIPTION

The buckling length concept is used in eqns (1-4). When the individual column approach is consistently applied, the buckling length should be taken equal to the system length (Fig 1). In actual fact, the columns form part of the braced frame and it can be shown that the use of the system length as buckling length is conservative in many cases. In the extreme case of a column in conjunction with a beam with infinite stiffness and no bending moments, the elastic carrying capacity is four times that of a pin-ended column (Fig 2). When the end moments, obtained by linear elastic moment distribution, are small, the use of the individual column approach with $\ell_k = \ell_{sys}$ gives conservative results. This is also shown by geometric and material non-linear analyses of braced frames with initially imperfect columns (Fig 3). These analyses are carried out with the finite element method (f.e.m.) using computer program DIANA of IBBC-TNO (4).

The imperfections in the columns are so determined that a pin-ended axially loaded column attains collapse at the load carrying capacity resulting from the ECCS buckling curves using $\ell_k = \ell_{sys}$.

The f.e.m. analyses show that the interaction formula, using $\ell_k = \ell_{eff}$, gives a good approximation to the load carrying capacity of the frame, if the collapse of the columns characterizes the frame behaviour (Fig 3). Therefore, checks on column stability in braced frames are acceptable on the basis of $\ell_k = \ell_{eff}$ (2). This, however has consequences for checking of the beams. The moment distribution at collapse is different from the moment distribution using the first order elastic analysis. At collapse the sum of the bending moments on the beam is greater than $1/8\ qb^2$ (Fig 4). This is caused by the top restraining moment M_r that has to stabilise the column. At the top of the column, the bending moment can even change sign (Fig 5). Figures 4 and 5 show that knowledge of the moment distribution at collapse is indispensible for checking of the beams. In view of this, it is clear that checking the beams on the basis of the first order moment distribution is insufficient.

In (5) it is suggested that the first order elastic moment distribution should be magnified by multiplication with the amplification factor of the column (Fig 6b). This approach does not satisfy equilibrium. In addition, a possible change of sign of the moment at the top of the column is not brought into account. Therefore, other methods are discussed in this paper.

Only by shifting the first order elastic bending moment diagram of the beam is it possible to satisfy equilibrium and to account for a possible change in the sign of the moment (Fig 6c). The bending moment diagram is shifted by a value of M_r. The restraining moment M_r is the moment carried over from the column to the beam when the column achieves collapse. In other words, M_r is the moment that the beam should provide in order to stabilise the column. The restraining moment M_r, that the beam has to offer to the column at the moment of column collapse, will be attempted to be determined in a simple manner.

2 *DETERMINATION OF THE RESTRAINING MOMENT* M_r

First, a simple braced frame is considered for determining the restraining moment. For a single storey, single bay braced frame with concentrated loads on the columns, two methods are described for determining the restraining moment.

2.1 Method I: The individual column

The column capacity for frame A (Fig 7a) is F_A, based on $\ell_k = \ell_{eff}$. If the column is isolated from the frame ($\ell_k = \ell_{sys}$) it can only carry load F_B (Figs 7b and 7c). The carrying capacity of the individual column can be increased to F_A which is greater than F_B by the application of a restraining moment M_r (Fig 7d). The value of M_r, required to increase the carrying capacity, has to be determined. The magnitude of the restraining moment decides the strength requirement for the beam. The restraining moment M_r from Fig 7d has been calculated. The following assumptions are made for the calculations:

- The sections possess a bilinear moment-curvature diagram, which is a reasonable approximation for I-sections.
- The yield criterion used (6) is as given in eqn (5).

$$\begin{aligned} N &< 0.15\ N_p \qquad & M_{p,red} &= M_p \\ N &\geqslant 0.15\ N_p \qquad & M_{p,red} &= 1.18\ M_p\ (1 - \frac{N}{N_p}) \end{aligned} \tag{5}$$

- The calculations are geometric non-linear computations based upon equilibrium in the deformed situation (7).
- The columns have parabolic initial deformations which include geometrical imperfections and residual stresses.

$$w_o = -\frac{4\ \hat{w}_o}{h^2}\ (-\ x^2 + xh) \tag{6}$$

- For all calculations, the ECCS buckling curve b is used (2).

2.1.1 Procedure for determining the initial imperfection $\hat{w}_o$

For a pin-ended column with a parabolic initial imperfection, the differential eqn (7) can be derived.

$$\frac{d^2 w}{dx^2} + \alpha^2 w = -\frac{8\ \hat{w}_o}{h^2} \quad \text{with } \alpha^2 = \frac{F}{EI} \tag{7}$$

for $x = o$, $w = o$
for $x = h$, $w = o$

F is taken equal to F_B, on the basis of $\ell_k = \ell_{sys}$. Then $\hat{w}_o$ is determined so that the column just attains collapse. A plastic hinge then occurs in the middle of the column.

2.1.2 Procedure for determining the required restraining moment M_r if the column in loaded with F_A

From the column of Fig 7d, the differential eqn (8) can be formulated.

$$\frac{d^2w}{dx^2} + \alpha^2 w = -\frac{8\,\hat{w}_o}{h^2} - \frac{M_r\,x}{EI\,h} \quad \text{with} \quad \alpha^2 = \frac{F}{EI} \tag{8}$$

for $x = o$, $w = o$
for $x = h$, $w = o$

F is taken equal to F_A, on the basis of $\ell_k = \ell_{eff}$. The restraining moment M_r is then determined so that the column just attains collapse. A plastic hinge occurs somewhere in the column.

2.1.3 The results of method I

The Figs 8a and 8b give the results for two system slendernesses. The solid line gives the necessary restraining moment as a function of the amount of partial restraint of the column. The dashed line gives the reduced plastic moment as a function of the amount of partial restraint of the column. Both figures show that with increase in the amount of partial restraint of the column, the restraining moment increases. Because F_A also increases, the reduced plastic moment decreases.

Figure 8a is characteristic for system slendernesses $\lambda_{sys} < 100$. To the left of the intersection point in Fig 8a, the necessary restraining moment is greater than the reduced plastic moment. Therefore, the required restraining moment cannot be achieved at the top of the column. In this area, method I cannot be used with $\ell_k = \ell_{eff}$ for checking the column.

Figure 8b is characteristic for system slenderness $\lambda_{sys} \geqslant 100$ and in comparison to Fig 8a is discontinued before an intersection point is achieved. This is because the model in Fig 7d cannot carry loads greater than the Euler buckling load based on $\ell_k = \ell_{eff}$. Also see Fig 9.

For areas in Fig 8 where the restraining moment is smaller than the reduced plastic moment, $\ell_k = \ell_{eff}$ may be used for checking the column stability. The strength requirement necessary for the beam can be obtained from Fig 8. Because of compatibility requirements at the column/beam junctions, a stiffness requirement for the beams can be derived. This stiffness requirement is, however, not discussed further in this paper.

The individual column approach for $\lambda_{sys} \geqslant 100$ gives a relatively large range where checks on column stability cannot be carried out using $\ell_k = \ell_{eff}$. It is shown with a f.e.m. analysis, that the load carrying capacity calculated with the interaction formula, using

$\ell_k = \ell_{eff}$, underestimates the load carrying capacity (Fig 10). The bound $F_A = F_E$ in Fig 8b has no physical meaning but is a result of method I, the individual column approach. Therefore method II has been developed.

2.2 *Method II: The column with rotational end restraint*

The column of Fig 7a is schematised as given in Fig 11. The restraining moment is replaced by an elastic rotational spring at the top of the column. The moment M_r at the top of the column will again be calculated. The assumptions given in 2.1 are again valid. The procedure for determining the initial imperfection $\hat{w}_o$ is as described in 2.1.1.

2.2.1 *Procedure for determining the required restraining moment M_r when the column is loaded with F_A*

The differential eqn (9) for the column in Fig 11 can be derived.

$$\frac{d^2 w}{dx^2} + \alpha^2 w = - \frac{8\,\hat{w}_o}{h^2} - \frac{M_r\,x}{EI\,h} \quad \text{with } \alpha^2 = \frac{F}{EI} \tag{9}$$

for $x = o$, $w = o$

for $x = h$, $w = o$ and $M_r = k\left(\frac{d(w + w_o)}{dx}\right)_r$

F is taken equal to F_A on the basis of $\ell_k = \ell_{eff}$. It is then determined whether the yield criterion is exceeded somewhere in the column. If not, F_A is supported elastically by the column. If the yield criterion is exceeded, then a value of F smaller than F_A is determined so that the column remains elastic. The restraining moment M_r is determined from eqn (9).

2.2.2 *The results of method II*

The Figs 12a and 12b give the results for two system slendernesses. The strength requirement for the beam can be obtained from these figures. Because compatibility between the top of the column and the end of the beam is included in the calculations, a stiffness requirement for the beam is no longer necessary. According to method II, F_A can be supported elastically for a large range of elastic effective slendernesses. For those parts of Fig 12 to the left of the intersection points, the restraining moments cannot be achieved at the top of the column. Therefore, checks on column stability cannot be made using $\ell_k = \ell_{eff}$. To the right of the intersection point, the field moment exceeds the reduced plastic moment for small system slendernesses. However, this violation results in a reduction of the load carrying capacity (see 2.2.1) by less than *1%*. If this violation is ignored, the check on column stability to the right of the intersection points in Fig 12 can be carried out with $\ell_k = \ell_{eff}$, if the beam can provide the necessary restraining moment.

All the results for method II are summarised in Fig 13. In the shaded area, the use of $\ell_k = \ell_{eff}$ is allowed and the required restraining moment can be read off. For example, for the slenderness $\lambda_{eff} = 100$ and $\lambda_{sys} = 120$, the restraining moment $M_r = 0.32\ M_p$. To avoid the use of graphs to determine the restraining moment, a formula has been derived. On linearisation of the curves in Fig 13, the following expression can be obtained for M_r.

$$\frac{M_r}{M_p} = f(\lambda_{sys})\ (\lambda_{sys} - \lambda_{eff}) \tag{10}$$

$$\text{with } \lambda_{sys} < 110:\ f(\lambda_{sys}) = 1.3x10^{-4}\ \lambda_{sys} + 2.6x10^{-3}$$
$$\lambda_{sys} > 110:\ f(\lambda_{sys}) = 1.7x10^{-2} \tag{11}$$

When the column is not completely loaded by its carrying capacity F_A, based upon $\ell_k = \ell_{eff}$, the beam does not have to provide the full restraining moment according to eqn (10). For practical checks, eqn (12) should be used. This relationship accounts for the reduced load on the column.

$$\frac{M_r}{M_p} = f(\lambda_{sys})\ (\lambda_{sys} - \lambda_{eff})\ \frac{(F_E - F_A)}{(F_E - N)}\ \frac{N}{F_A} \tag{12}$$

using eqn (11) and F_E on the basis of $\ell_k = \ell_{eff}$.

3 FURTHER RESEARCH

Further investigations are necessary on braced frames with intermediate loads on the beam and on columns in braced frames with two rotational end restraints. For these cases, approximation formulae for the restraining moment can be derived in the same way as given before.

4 CONCLUSIONS

- Design formulae, derived for checking the stability of the individual columns where $\ell_k = \ell_{sys}$, are commonly used with $\ell_k = \ell_{eff}$ for braced frames.
- On the basis of a parameter study, it is concluded that the design formulae can be used with the elastic effective length $(\ell_k = \ell_{eff})$ if the beams can provide the necessary restraining moment.
- So far, there are no design formulae available for the necessary restraining moment.
- In this study, an approximation formula for the necessary restraining moment has been derived (12) for a one storey, one bay braced frame, with concentrated loads on the columns.
- Two methods to determine the necessary restraining moment have been used. A method based on the individual column approach, is not effective for geometrically non-linear problems. Therefore, a second method has been developed, using a rotational spring at the top of the column. This rotational spring has two functions. It generates the restraining moment M_r and accounts for the elastic boundary condition. The geometrical non-linearity is therefore correctly taken into account.

References

1. ECCS, Second International Colloquium on Stability, Tokyo-Liege-Washington. Introductory Report, 2nd edition, Chapter 7, Beam-Columns.
2. ECCS-CECM-EKS, European Recommendations for Steel Construction, March 1978.
3. Austin, W.J. 'Strength and design of metal beam-columns.' Proc. ASCE, Journ. Struct. Div., vol. 87, no. ST 4, April 1961.
4. Borst, R. de, Kusters, G.M.A., Nauta, P., Witte, F.C. de. 'DIANA - A comprehensive, but flexible finite element system.' Paper presented to the 4th International Seminar on Finite Element Systems, July 6-8th, 1983, Southampton, England.
5. Roik, K., Kindmann, R. 'Das Ersatzstabverfahren - Tragsicherheitsnachweise für Stabwerke bei einachsiger Biegung und Normalkraft.' Der Stahlbau 5/1982, pp. 137-145.
6. ASCE, Plastic Design in Steel. 2nd edition, Manual and reports on engineering practice of the ASCE, no. 41, New York, 1971.
7. Timoshenko, S.P., Gere, J.M. 'Theory of Elastic Stability.' 2nd Edition, McGraw-Hill, 1961.

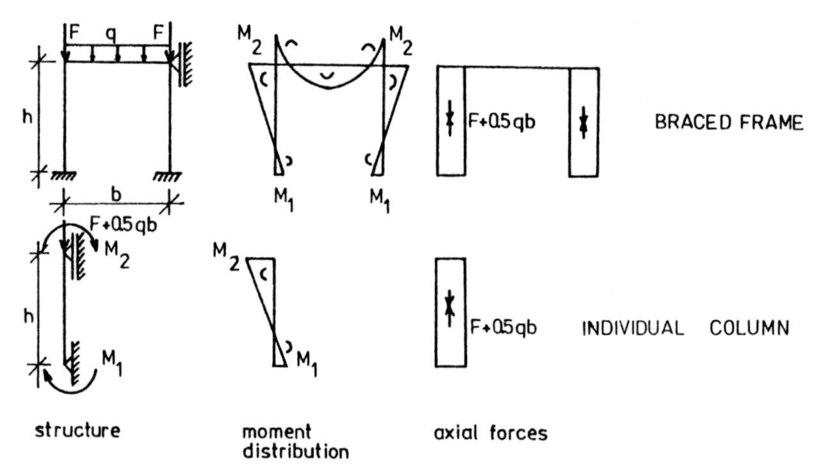

Fig. 1 The individual column approach.

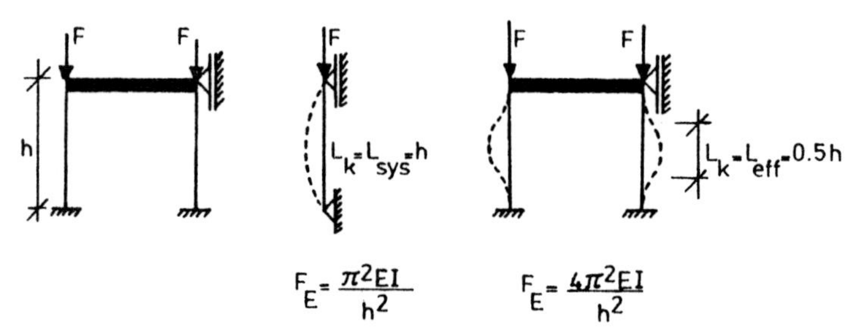

Fig. 2 Braced frame where $\ell_k = \ell_{sys}$ is conservative.

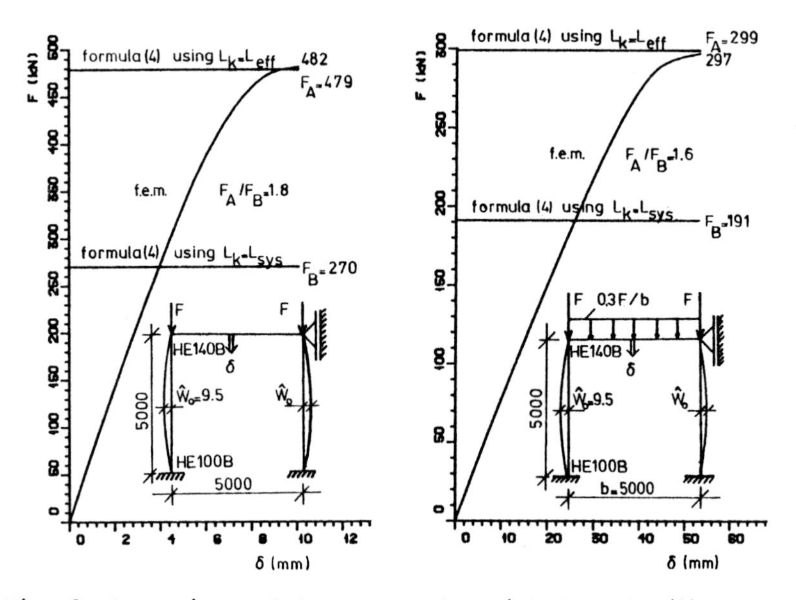

Fig. 3 Comparison of f.e.m. results with formula (4).

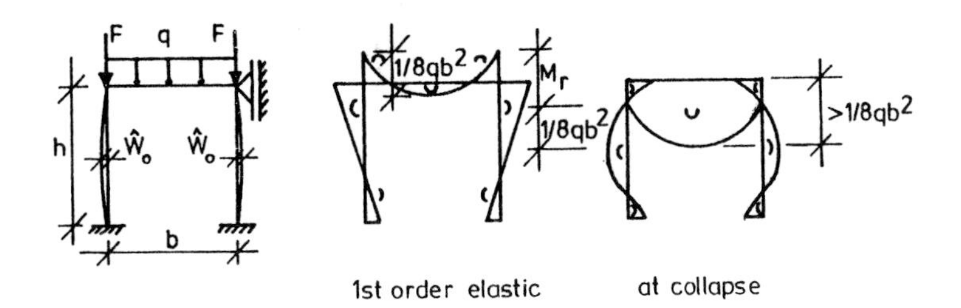

Fig. 4 Comparison of bending moment diagrams.

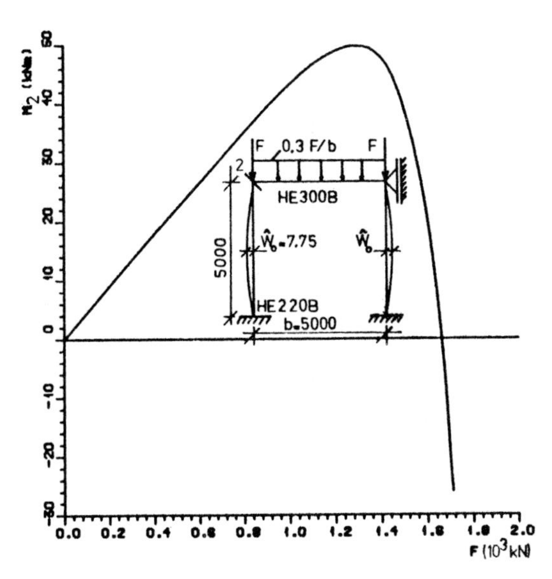

Fig. 5 Change of sign of bending moment at the top of the column.

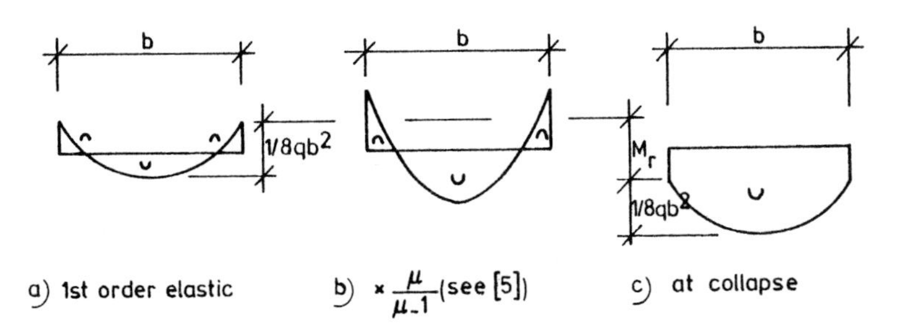

Fig. 6 Bending moment diagrams of the beam.

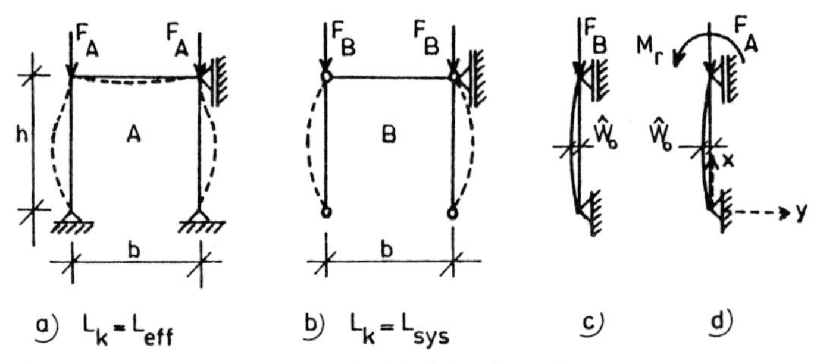

Fig. 7 Method I: The individual column.

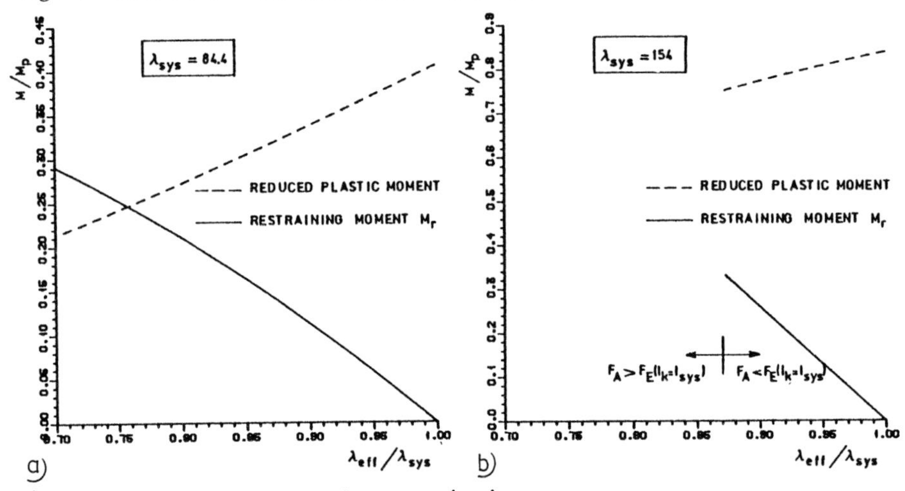

Fig. 8 Results for two characteristic system slendernesses (method I).

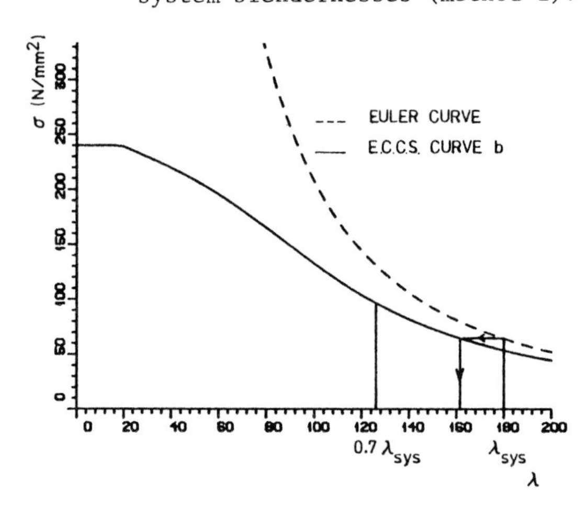

Fig. 9 F_A cannot exceed F_E ($\ell_k = \ell_{sys}$).

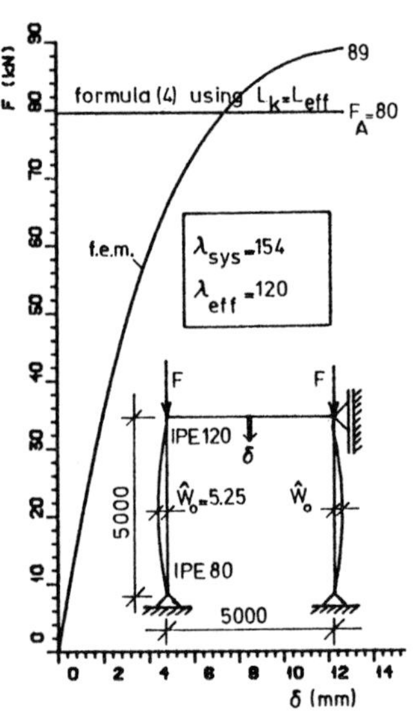

Fig. 10 Comparison of f.e.m. results with formula (4).

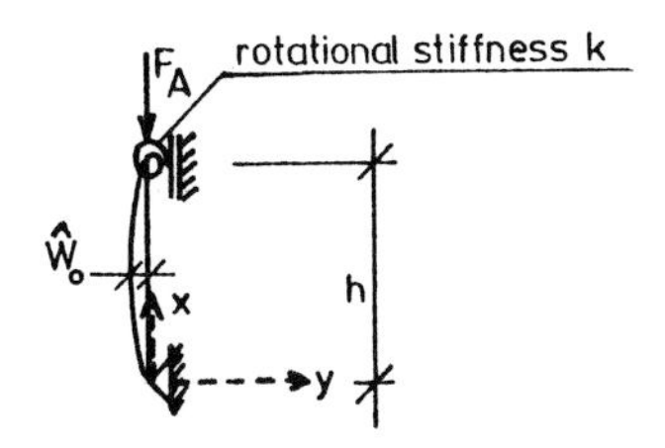

Fig. 11 Method II: The column with rotational end restraint.

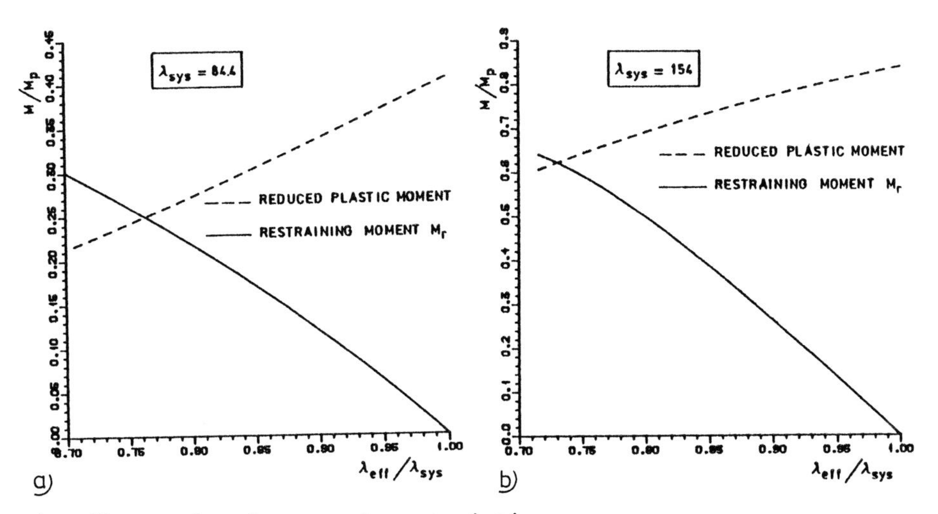

Fig. 12 Results for two characteristic system slendernesses (method II).

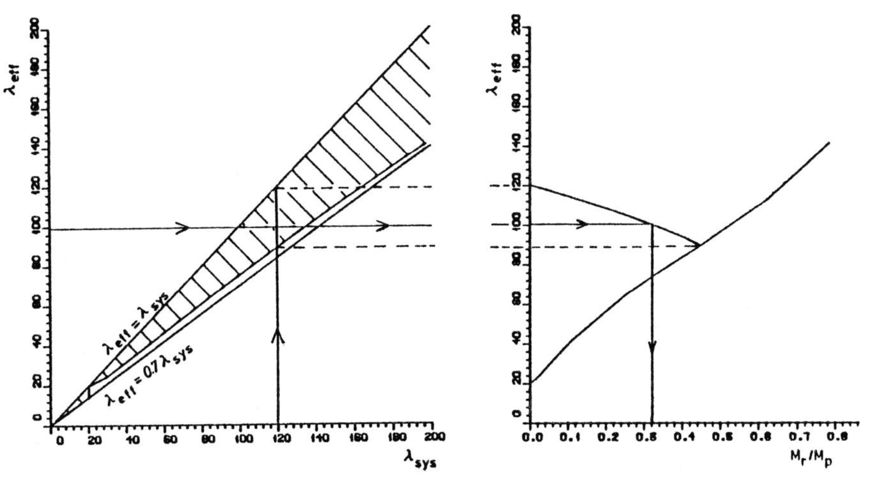

Fig. 13 Summary of the results of method II.

ADAPTING THE SIMPLEX METHOD TO PLASTIC DESIGN

Alan Jennings

Department of Civil Engineering, Queen's University, Belfast

Synopsis

A modified simplex technique is described for the plastic design of plane frames subject to one or more loading cases. The method differs from Livesley's in so far as the equilibrium equations are constructed using redundants and involving a smaller set of initial equations.

Introduction

The number of times elastic frame analyses have been programmed for computers must run into hundreds or thousands. However in contrast there have been relatively few programmes written using the plastic theory. Furthermore the most appropriate techniques for computer implementations have not been extensively investigated.

One method of obtaining collapse loads investigated by Jennings and Majid (1) is to perform an elastic-plastic analysis in which a series of linear analyses are performed with a plastic hinge being introduced after each such analysis. The collapse condition may be recognised by monitoring when the stiffness matrix of the structure including plastic hinges loses its positive definite property. This approach has the advantage of yielding the deflections at the various stages of loading and also permits buckling effects to be introduced. However to use this technique as a design tool by repeatedly calling the elastic-plastic analysis for different trial designs, in common with many structural optimisation schemes, is so expensive in computing time that it is not the type of method to use on present day micro-computers.

On the other hand one of the advantages of the simple plastic theory is that, since an analysis reduces to one of linear programming, it is easy to include a function optimization if it is linear in the variables. Thus, with very little extra complication or computing time, it is possible to obtain a minimum weight design assuming a linear relationship between weight and plastic moment. This technique has been investigated by Jennings (2) and Livesley (3) and computer implementations have been developed by Livesley (4,5).

Criticism may be levelled against the use of such a program in a batch processing mode. For instance:

a) The designer should be trying to minimise cost rather than weight.

b) The variation of plastic moment versus weight is not linear, and is indeed stepped if only certain prefabricated sections are to be used.

c) The designer does not have the degree of control over the design process which he would like.

However with interactive computing these criticisms largely disappear. Since it is possible to allow the user to include restrictions on the minimum plastic moment of particular members and to use the grouping facility to be described later, the designer is able to supply a significant degree of control over the design process. For fast interactive response with micro-computers it is important that programs are developed which are as efficient as possible.

1. *LOWER BOUND EQUATIONS*

The method to be described here is based on the equilibrium equations for the structure using redundant forces. Example 1 (Fig 1a) shows a continuous beam problem first solved by Livesley (3). With redundants R_1 and R_2 specified as the hogging moments over the supports, the sagging moments at the five critical points throughout the frame are given by the equations

$$\left.\begin{aligned} M_1 &= 3 - \tfrac{1}{2}R_1 \\ M_2 &= -R_1 \\ M_3 &= -R_1 \\ M_4 &= 1 - \tfrac{1}{2}R_1 - \tfrac{1}{2}R_2 \\ M_5 &= -R_2 \end{aligned}\right\} \qquad (1)$$

From the lower bound theorem of plastic collapse, (Baker and Heyman (6)), if any choice of redundants leads to a set of bending moments such that

$$|M_i| \leq \bar{H}_i \qquad (2)$$

where $\bar{H}_i$ is the fully plastic moment at position i, then the collapse load will not be exceeded. Therefore, for any particular set of redundants, a choice of plastic moments satisfying equation (2) yields a safe design. If the designer wishes to select sizes for some members which are different from others, then the moment points are arranged in groups so that all members in one group shall have the same plastic moment. If point i is contained in group r, then

$$\bar{H}_i = H_r \qquad (3)$$

and with a linear relationship between plastic moment and weight, the weight of a given design is proportional to the parameter

$$w = \sum \ell_r H_r \qquad (4)$$

where ℓ_r is the total length of the members associated with group r and summation takes place over all the groups.

One method of achieving the optimum design is to compute w for an arbitrary choice of redundants and then to vary the redundants in such a way that w is reduced until a minimum is achieved. However it is also possible to solve these equations by linear programming techniques such as the simplex method for which standard programmes are available. The fact that a linear programming formulation is possible establishes that the minimum is unique.

2. *A SIMPLEX METHOD FORMULATION*

Consider example 1 for the case in which the two spans of the beam to be specified in different groups. A simplex method formulation requires the minimisation of the parameter

$$w = 2H_1 + 4H_2 \tag{5}$$

subject to the constraints

$$\left.\begin{aligned}
H_1 &\geq 3 - \tfrac{1}{2}R_1 + \tfrac{1}{2}\bar{R}_1 \\
H_1 &\geq -3 + \tfrac{1}{2}R_1 - \tfrac{1}{2}\bar{R}_1 \\
H_1 &\geq -R_1 + \bar{R}_1 \\
H_1 &\geq R_1 - \bar{R}_1 \\
H_2 &\geq -R_1 + \bar{R}_1 \\
H_2 &\geq \bar{R}_1 - \bar{R}_1 \\
H_2 &\geq 1 - \tfrac{1}{2}R_1 + \tfrac{1}{2}\bar{R}_1 - \tfrac{1}{2}R_2 + \tfrac{1}{2}\bar{R}_2 \\
H_2 &\geq -1 + \tfrac{1}{2}R_1 - \tfrac{1}{2}\bar{R}_1 + \tfrac{1}{2}R_2 - \tfrac{1}{2}\bar{R}_2 \\
H_2 &\geq -R_2 + \bar{R}_2 \\
H_2 &\geq R_2 - \bar{R}_2
\end{aligned}\right\} \tag{6}$$

In order to satisfy the normal linear programming constraint that all the variables are non-negative, extra variables $\bar{R}_i$ have been introduced which take the place of R_i when it is negative. A minimisation by the simplex method requires the inclusion of slack variables to convert the conditions to equalities resulting in, for this example, 10 equations in 16 constrained variables (excluding the weight function itself). A standard solution by the simplex method involves sequences of steps firstly to obtain a feasible solution and then to convert that to the optimum solution. Each step has similarities with the elimination of a column in the Gauss-Jordan method of solving simultaneous equations but the number of steps required cannot be predicted in advance.

3. *A MODIFIED SIMPLEX METHOD*

The above formulation using the standard simplex method can be improved in the following three ways if a special program is written:

a) By retaining bending moment equations, each equation can be used to represent two conditions.
b) Only one variable needs to be allocated for each redundant.
c) Advantage can be taken of the fact that a feasible solution can always be obtained to initiate the minimisation.

A tableau is formed from the coefficients of the bending moment equations (eqns (1) for example 1) with the redundants considered as 'non-basic' variables and the following three modifications:

a) An extra equation is included for each group specifying the minimum plastic moment for that group. (For example 1, $M_6 = M_7 = 0$ has been assumed where M_6 and M_7 are in groups 1 and 2 respectively).

b) The bending moment in each group having the greatest constant term is recorded as the critical moment. If necessary the sign of the whole equation is changed so that the constant term is positive.

c) The slack variables associated with the critical moments are specified as extra non-basic variables. The column associated with each slack variable is initially null except for a unit term in the row of its associated bending moment.

The initial tableau for example 1 is therefore as shown in Table 1. Its first column represents the values of the bending moments at the current trial design in which the non-basic variables are zero, and each of the other columns indicate what happens to these bending moments if the particular non-basic variable is modified by a unit amount. The weight function, which is shown below the tableau is a linear combination of the critical rows. The first term in this equation is the value of w for the current trial design and the other terms indicate its possible movement with changes in the non-basic variables. This function is used to decide which non-basic variable to move by identifying which has the element of largest modulus, discounting positive coefficients associated with slack variables (because these variables cannot adopt negative values). For the example this method indicated that R_1 should be increased. By examining columns 1 and 2 of the tableau it is possible to ascertain that the weight function remains valid until $|M_3| = |M_4|$. Following simplex strategy, a slack variable assiciated with M_3 replaces R_1 in the set of non-basic variables giving the modified tableau shown in Table 2. The next stage is achieved by increasing R_2 until $|M_5|$ becomes equal to $|M_4|$ giving the stage 3 tableau shown in Table 3.

Because the redundants have all been eliminated and the slack variables have all positive coefficients in the weight function, the minimum weight design has been found with a tableau one quarter the size of the full simplex tableau, but still making use of the slack variable concept to expedite the search. Figures 1(b), (c) and (d) show the bending moment diagrams for the three stages of the optimisation.

4. *GENERALISATION OF TABLEAU OPERATIONS*

Let the unit term followed by the non-basic variables be represented by the vector x and let the tableau correspond to matrix A. If variable x_k is being increased and M_i is critical for a particular group in which M_j is non-critical, then $|M_j|$ will also become critical with $M_i = M_j$ or $M_i = -M_j$ at

$$\left.\begin{aligned} x_k &= (a_{j1} - a_{i1})/(a_{ik} - a_{jk}) \\ \text{or}\quad x_k &= -(a_{j1} + a_{i1})/(a_{ik} + a_{jk}) \end{aligned}\right\} \qquad (7)$$

if in each case the denominator is negative. The determination of which moment becomes critical next is therefore obtained by examining all the non-critical moments to find which has the least value of x_k (discounting those for which the denominator is positive). If the moment M_j becomes critical next at a negative value then the sign of all elements on row j are changed so that it represents $-M_j$.

The elimination of variable x_k from the tableau, if it is a redundant which is increasing, is implemented by modifying elements in all columns other than k according to the algorithm

$$\bar{a}_{\ell m} = a_{\ell m} - \frac{a_{\ell k}(a_{im} - a_{jm})}{(a_{ik} - a_{jk})} \qquad (8)$$

The coefficients for the new slack variable s_j which replaces x_k can then be computed according to

$$\left.\begin{aligned} \bar{a}_{\ell k} &= \frac{a_{\ell k}}{a_{ik} - a_{jk}} \qquad (\ell \neq j) \\ \text{and} \quad \bar{a}_{jk} &= \frac{a_{jk}}{a_{ik} - a_{jk}} + 1 \end{aligned}\right\} \qquad (9)$$

If the variable x_k is a redundant which is decreasing, the sign of all coefficients on column k of the tableau are changed so that x_k becomes an increasing variable and equations (8) and (9) apply. The other possibility is that x_k is the slack variable s_t associated with a critical moment M_t. In this case eqn (8) is implemented and then 1 is subtracted from a_{tk} before eqn (9) is implemented. This subtraction effectively transfers the variable s_t to the left-hand side where it is implicitly retained as a basic variable associated with the moment M_t. It is also necessary to strike this moment off the list of critical moments and, if necessary, adjust the specification of which moments contribute to the weight function.

5. *A PLANE FRAME PROGRAM*

For the case of a portal frame in which the redundants are chosen as the forces and moment acting at one of the supports, the initial tableau is very easily constructed. With external forces and redundants designated as shown in Fig 2 the bending moment at station i is given by

$$\left.\begin{aligned} M_i &= a_o + a_1 R_1 + a_2 R_2 + a_3 R_3 \\ \text{where} \quad a_o &= \sum_{j=i+1}^{s} [(y_i - y_j) H_j + (x_j - x_i) V_j] \\ a_1 &= y_i - y_s \\ a_2 &= x_s - x_i \\ a_3 &= 1 \end{aligned}\right\} \qquad (10)$$

The coefficients $a_o, \ldots, a_3$ for all the moment points, forming the core of the initial tableau can therefore be automatically constructed very easlily from a list of coordinates of the moment points and a list of applied forces. To facilitate the design of more complex frames, loop data is input which comprises a list of the moment points in each loop of the structure in cyclic order. Three redundants are introduced in each loop, and the form of the initial tableau is equal to $(n_m + n_g)(1 + 3n_\ell + n_g)$ where n_m, n_g and n_ℓ are the numbers of bending moment stations, groups and loops respectively.

A frame program using the modified simplex algorithm has been implemented in UCSD PASCAL on Apple II and ITT 2020 micro-computers and compared with a previous implementation using a standard linear programming package using equations of the form shown by eqns (6). Table 4 shows the resulting tableau sizes and computing times for the six problems illustrated in Fig 3. This indicates a storage reduction factor of between ¼ and 1/10 as compared with the standard tableau with similar reductions in computing time.

6. *THE USE OF THE MINIMUM MOMENT FACILITY*

The minimum moment facility (giving rise to the moments M_6 and M_7 in Table 3) has been included for the following two reasons:

a) There are certain frames in which more than one group is specified for which the above algorithm might produce a fallacious negative plastic moment without this facility. Example 2 would be such a frame if the left-hand column were considerably lengthened.

b) A designer may wish to specify a minimum plastic moment. For instance, the result for example 4 indicated a minimum weight structure having plastic moments of *135* and *280* kNm for the columns and beam respectively. If the plastic moment of the first available section above *135* is *152* kNm, it would be possible to perform a re-design with *152* kNm specified as the minimum column plastic moment to see if that reduces the plastic moment required for the beam.

7. *MULTIPLE LOADING CASES*

If there is only one group, multiple loading cases may be minimised separately and the largest plastic moment required for any of the cases chosen as the best design. However, if several groups are to be considered, it is necessary to include all the loading cases in the same minimisation. Thus the weight for example 8 shown in Fig 4, in which the columns and rafters are to be in separate groups, may be minimised with $n_m = 32$, $n_\ell = 3$ and $n_g = 2$ giving a tableau of size *22 x 9*. This is *2½* times the size of the tableau required for the first loading case on its own).

8. *COMPARISON WITH LIVESLEY'S METHOD*

The above formulation differs from that of Livelsey (5) in so far as the equilibrium equations are constructed using redundants rather than by using joint equilibrium equations. As a result the basic equations are of a similar order to Livesley's compact equations after Gauss-Jordan elimination. By comparison the sizes of Livesley's initial tableaux for examples 3, 5, 6 and 7 are between two and four times larger. The advantage of Livesley's formulation is that loop data does not need to be input.

Acknowledgements

The author would like to thank W.P. McIlfatrick and R.G. Kerr who programmed and tested the linear programming and the modified simplex frame algorithms.

References

1. Jennings, A. and Majid, K. 'An elastic-plastic analysis by computers for framed structures loaded up to collapse'. The Structural Engineer, vol. 43, 1965.
2. Jennings, A. MSc Thesis, Manchester University, 1954.
3. Livesley, R.K. 'The automatic design of structural frames' Quart. J. Mech. and Applied Math., vol. 9, 1956.
4. Livesley, R.K. 'A compact FORTRAN sequence for limit analysis', Int. J. for Num. Methods in Engineering, vol. 5, 1973.
5. Livesley, R.K. 'Matrix Methods of Structural Analysis', 2nd Ed., Pergammon, Oxford, 1975.
6. Baker, J.F. and Heyman, J. 'Plastic Design of Frames'. Vol. 1. 'Fundamentals', Camb. Univ. Press, 1969.

Table 1 Stage 1 tableau for example 1

		Constant	R_1	R_2	s_1	s_4	
Group 1	M_1	3	-1/2	0	1	0	← critical
	M_2	0	-1	0	0	0	
Group 2	M_3	0	-1	0	0	0	
	M_4	1	-1/2	-1/2	0	1	← critical
	M_5	0	0	-1	0	0	
Group 1	M_6	0	0	0	0	0	
Group 2	M_7	0	0	0	0	0	
$w = 2\|M_1\| + 4\|M_4\|$		10	-3	-2	2	4	

Table 2 Stage 2 tableau for example 1

		Constant	s_3	R_2	s_1	s_4	
Group 1	M_1	8/3	1/3	1/6	1	-1/3	← critical
	M_2	-2/3	2/3	1/3	0	-2/3	
Group 2	$-M_3$	2/3	1/3	-1/3	0	2/3	← critical
	M_4	2/3	1/3	-1/3	0	2/3	← critical
	M_5	0	0	-1	0	0	
Group 1	M_6	0	0	0	0	0	
Group 2	M_7	0	0	0	0	0	
$w = 2\|M_1\| + 4\|M_4\|$		8	2	-1/3	2	2	

Table 3 Stage 3 tableau for example 1

		Constant	s_2	s_5	s_1	s_4	
Group 1	M_1	11/4	3/8	-1/8	1	-1/4	← critical
	M_2	-1/2	3/4	-1/4	0	-1/2	
Group 2	$-M_3$	1/2	1/4	1/4	0	1/2	← critical
	M_4	1/2	1/4	1/4	0	1/2	← critical
	$-M_5$	1/2	1/4	1/4	0	1/2	← critical
Group 1	M_6	0	0	0	0	0	
Group 2	M_7	0	0	0	0	0	
$w = 2\|M_1\| + 4\|M_4\|$		15/2	7/4	3/4	2	3/2	

Table 4 Comparison of modified simplex with standard linear programming methods

Example	Tableau size		Computing time (secs)	
	Standard LP	Modified simplex	Standard LP	Modified simplex
2	18 x 21	12 x 7	45	10
3	10 x 16	6 x 5	15	3
4	18 x 20	7 x 6	35	8
5	26 x 23	13 x 5	80	7
6	16 x 29	9 x 8	90	15
7	20 x 33	12 x 8	110	17

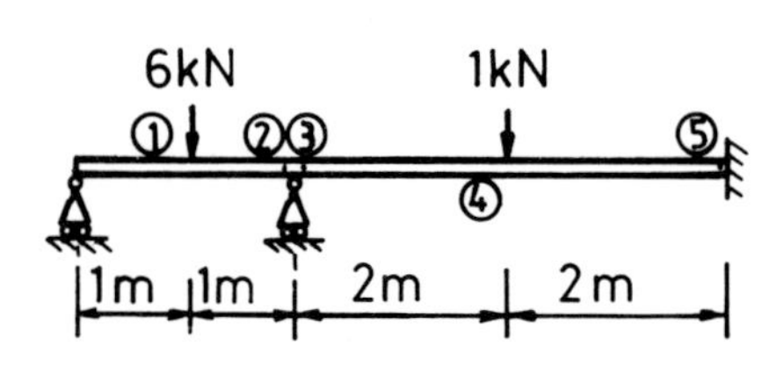

(a) General arrangement

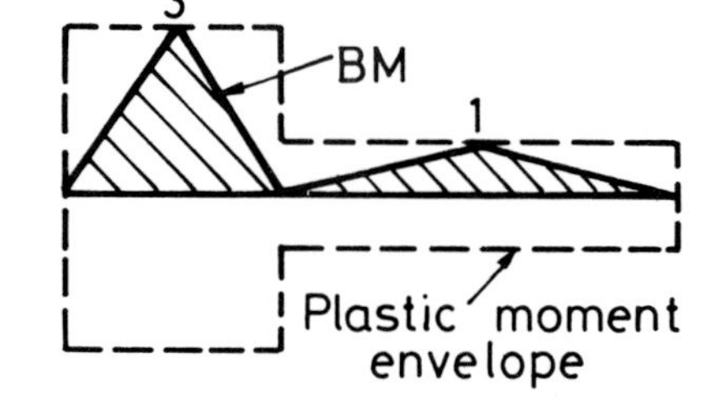

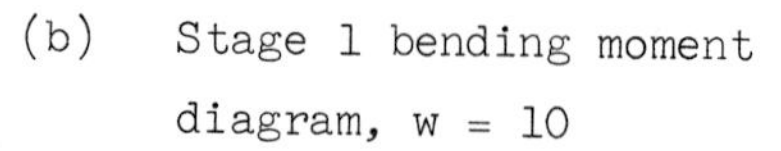

(b) Stage 1 bending moment diagram, w = 10

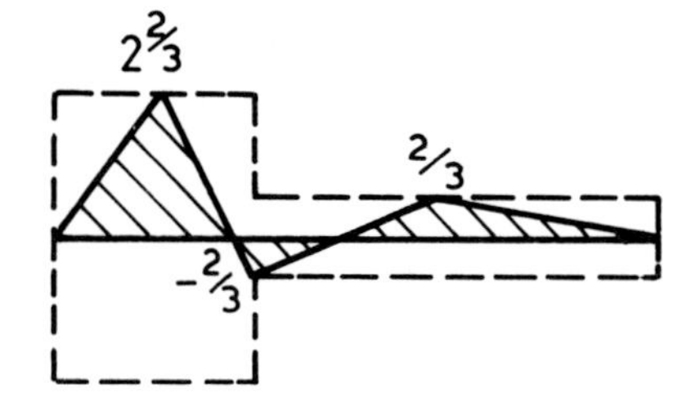

(c) Stage 2 bending moment diagram, w = 8

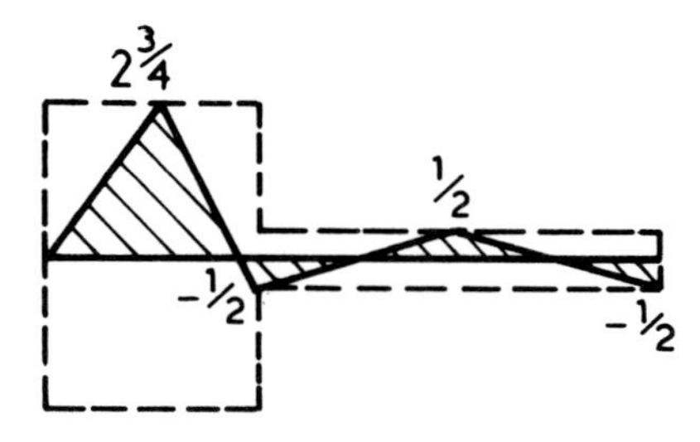

(d) Stage 3 bending moment diagram, w = 7.5

Fig. 1. A cantilever beam optimisation (example 1)

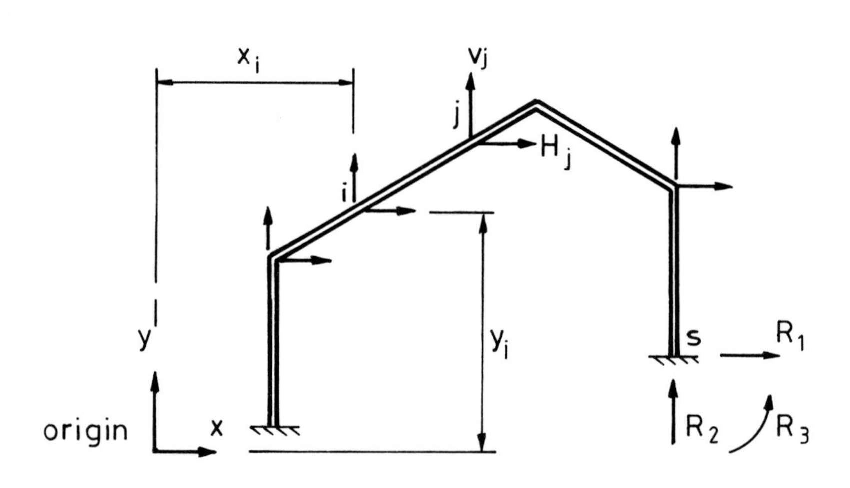

Fig. 2. General arrangement for a portal frame

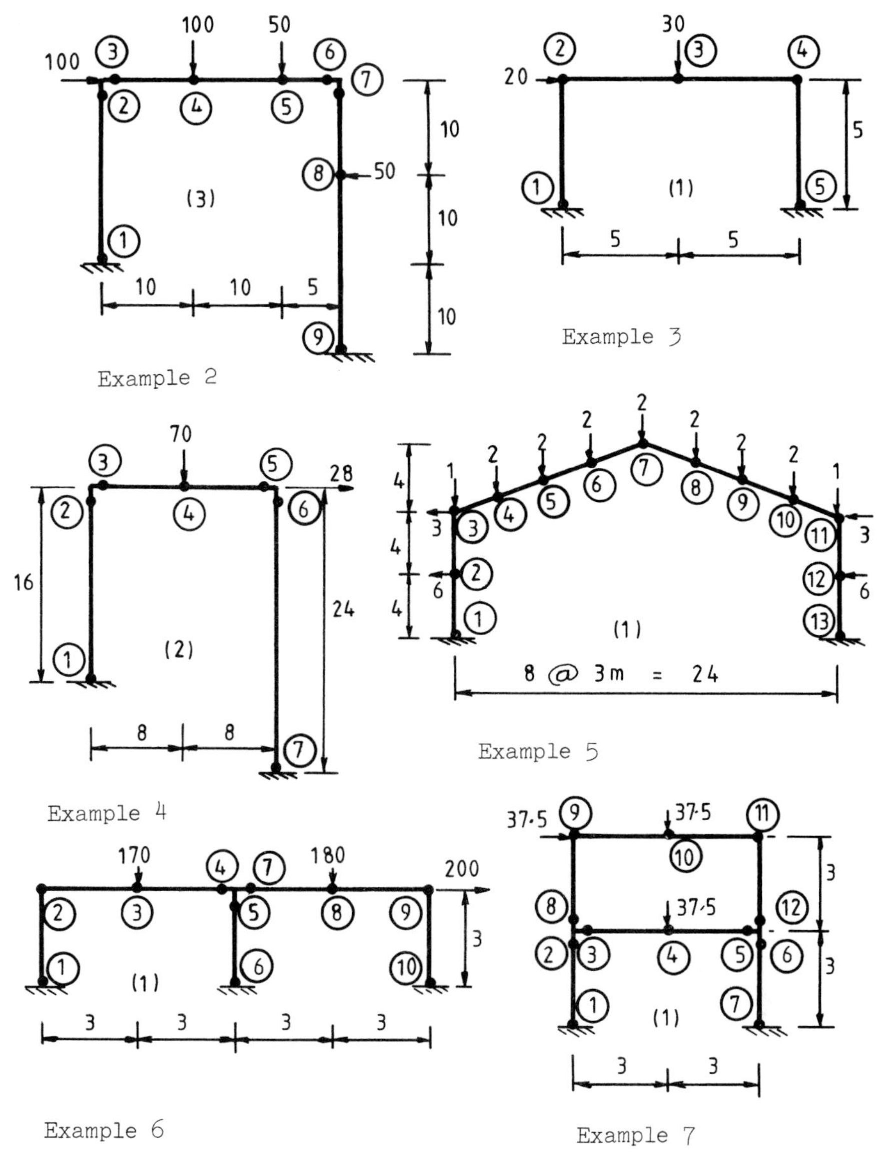

Fig. 3. Plane frame examples No. of groups given in brackets.

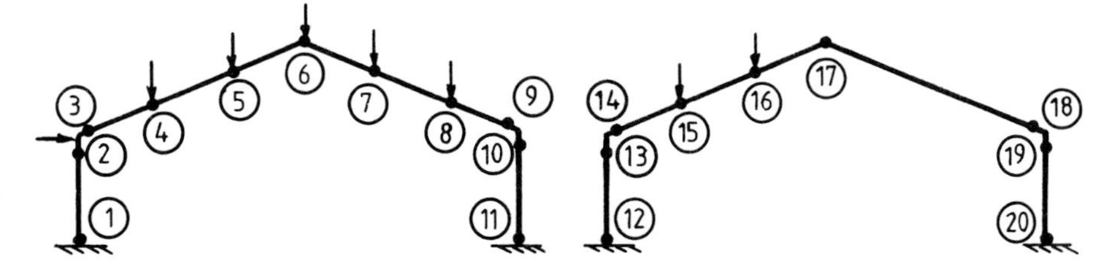

Fig. 4. Frame with two loading cases

A BUILT-IN SEMI CIRCULAR ARCH UNDER A CENTRAL CONCENTRATED LOAD

H.S.Harung, M.A. Millar, M.G. Moore

Department of Civil and Structural Engineering, UMIST.

Synopsis

An existing non-linear elastic-plastic bar element computer program is used to model the in-plane pre- and post-mechanism behaviour of a semi-circular arch of constant cross-section under a single central concentrated load. The results are compared with the rigid-plastic solution presented by Gill (1) which illustrates a number of important structural principles associated with large deflections and plastic behaviour. Return of elasticity and reversal of plasticity occur at nodes prior to the mechanism collapse. The phenomenon of a plastic zone travelling along the arch profile is effectively modelled during the post-mechanism behaviour.

It is suggested that the example could be incorporated in a set of benchmark problems to be used to evaluate non-linear elastic plastic bar element programs.

Introduction

The in-plane behaviour of a built-in semi-circular arch of constant cross-section subjected to a central concentrated load is a classical problem in structural analysis. Elastic theory of arches has been summarised by Austin (2) and by the American Structural Stability Research Council (3) which was formerly the Column Research Council.

Gill (1) treated the problem as a teaching exercise in rigid-plastic behaviour in the post mechanism regions, using a pocket calculator for his calculations. Modern finite element computer programs provide the facility for the arch to be analysed using a large number of discrete elasto-plastic elements as described by Lee and Murphy (4) and Komatsu and Sakimoto (5) who include three-dimensional behaviour. Spread of plasticity both along the length and through the depth of the arch cross-section can be observed, but Bernoulli's theory of linear strain distribution is still used and change of geometry is sometimes neglected. For such detailed analysis where elastic unloading of plastic regions is a possibility all the plastic deformations of each element have to be stored and the requirements in terms of computation time and storage become extensive as noted by Jain (6). In this paper the circular arch has been modelled using a change of geometry bar-element plane frame analysis computer program by Harung and Millar (7) which has been upgraded and maintained by Moore. The bar element is straight and of constant cross-section. Plastic hinges are only allowed to form at the element end nodes, no allowance is made for partial plasticity prior to the plastic moment of resistance being attained but facilities exist for the plastic moment of resistance to be reduced to account for the interaction with axial strains. The effects of axial plastic strains are also included but strain hardening is neglected. (The use of this mathematical model will in general produce a lower bound to the pre- and post-mechanism behaviour).

The general computer program was used to investigate the behaviour of the semi-circular arch with a range of flexural stiffnesses varying from a very flexible cross-section, with no elastic limit to the stresses, to a flexurally stiff cross-section with realistic plastic limits.

1. *MATHEMATICAL MODELLING OF THE ARCH*

The semi-circular arch is represented by 36 linear two-dimensional elements each subtending a 5 degree angle at the centre of the circle as shown in Fig 1. A concentrated vertical load, W, is applied at the centre of the arch, point C, which is free to move asymmetrically. A vertical axial spring element is attached to the centre of the arch to control the deflection when the stiffness of the arch becomes small or negative. The plastic moment-axial load interaction used in the particular analyses is linear but the computer program has facilities for defining second order interaction curves.

Four examples have been analysed using the non-dimensional properties given in Table 1. All examples are carried out with a constant spring stiffness with a value greater than $10EI/R^3$. In the elastic range the applied load increment/spring stiffness is constant at a value of $0.05R$ when the non-linear convergence is rapid, but if there is difficulty in convergence this increment is reduced. Also when plasticity is detected the load increment is limited to the next discrete plastic discontinuity.

2. *BEHAVIOUR OF THE ARCH*

The four non-linear examples investigated are summarised in Table 1 as follows:-

1) Elastic analysis of a very flexible arch.

2) Elastic-plastic analysis of a similar very flexible arch.

3) Elastic-plastic analysis of a flexible arch.

4) Elastic-plastic analysis of a stiff arch.

Example	1	2	3	4
EI/M_pR	0	0.4	4	40
Approx. T/R	0	1/800	1/80	1/8
W_eR/M_p	-	3.9	6.1	6.5
δ_e/R	-	0.29	0.0195	0.0019
$W_eR^3/(\delta_eEI)$	-	33	78	84
W_mR/M_p	0	3.9	7.9	9.1
δ_m/R	0.5	0.29	0.062	0.0061
$W_mR^3/(\delta_mEI)$	21	34	31	37
$W_mR^2/(EI)$	10.7	9.8	1.94	0.22

Table 1. Arch behaviour up to first limit load.

δ is the vertical deflection of node C.

T is the thickness of the arch assuming a rectangular cross-section.

The values listed in Table 1 give the critical non-dimensional values for the cross-sectional properties, the displacements and the applied load in terms of the flexural modulus, EI, and the plastic moment of resistance, M_p.

Suffix e indicates the limit of elasticity.

Suffix m indicates the first maximum load.

2.1 Elastic very flexible arch

The series of curves, shown in Fig 2a, give the deformed shape of half the arch at various load levels. The non-dimensional load deflection curve is shown in Fig 3. Without the restraining spring element the arch would snap-through from point X_1 to point Y_1. The arch was not restrained against lateral displacement but the behaviour remained symmetrical throughout. It is clear that a relatively small number of straight bar elements provide an effective model for frames subjected to gross deformations.

2.2 Elastic-plastic very flexible arch

The series of curves, shown in Fig 2b, give the deformed shape of the half arch at various load levels. The deformations after the first hinge has formed at the centre point C exhibit a slight asymmetrical component which is attributed to the way the axial plastic deformations are included in the program. As the maximum lateral displacement of node C is only 0.005R the plot has been restricted to half the arch. The non-dimensional load-deflection curve presented in Fig 3 shows clearly that snap-through occurs at the formation of the central hinge. The controlled displacement analysis indicates that the moment at point C reduces prior to the formation of any further hinges, and that when the arch returns to the all elastic state it is actually stiffer than the arch without the plastic yielding. This is indicated by the shaded area on Fig 3. A further snap-through of the structure occurred immediately after the last deformation plot on Fig 2b. Ultimately all arches will tend towards the triangular shape but the introduction of plasticity hastens this behaviour. The load-deflection response of example 3 is also shown on Fig 3 for comparison.

2.3 Elastic-plastic flexible arch

The arch properties for example 3 as shown in Table 1 correspond to a realistic slender arch of constant cross-section. The behaviour is almost linear up to the formation of the first plastic hinge at C since the displacements are too small for there to be any significant change of geometry effect or any tendency to buckle.

The plastic behaviour of this example exhibits a number of interesting features which are presented graphically in Fig 5 and are described sequentially in the following paragraphs. The non-dimensional load-deflection response is given in Fig 3 and is compared with the stiff and the very flexible arches of examples 4 and 2 respectively.

After the formation of the central plastic hinge under the load, hinges form virtually simultaneously at supports A and A'. A very slight component of the asymmetrical mode is excited by the axial plastic strain that accompanies the plastic rotation at C. As with example 2 this is due to the frame being modelled such that the central hinge forms at the end of the specified element rather than at the node. In all the examples presented here the hinge is assumed to form in the element to the right of centre C which causes the arch to sway from right to left. As the load, W, increases, a further hinge forms at the left hand side mid-point B and hinge A' immediately ceases to rotate and therefore returns to its elastic state and at a slightly higher load a further hinge forms at B' thus creating a plastic sway mechanism with plastic active hinges at ABCB'. As the mechanism deforms an unrestrained arch would snap-through from X_3 to Y_3. As node C moves down, node B is forced outwards whilst the location of node B remains relatively static, but when C reaches the horizontal line BB', nodes B and B' are subsequently pulled inwards and the hinge at A starts to unload elastically. The plastic deformation at A is locked into the structure as was the previous small rotation at A'.

Further movement of C below the line BB' induces tension in the arch between BC and CB' which reduces the moment at B and B' and increases it at the next node down towards the supports. This behaviour is the start of the phenomenon known as the travelling hinge. Similar behaviour continues down both lower portions of the arch until the central node is below the springing line AA'. The plastic deformation at the end of each segment is locked into the structure once the elasticity returns. The lower segments have been plastically deformed into a shape of greater curvature than the original circle.

As the centre of the arch moves below the springing line it is clear that the distance AC increases and plastic hinges, in the opposite direction to those previously present, form at A and A'. Soon after the formation of these hinges at the supports the plastic hinge at C returns to its elastic state and subsequent plastic travelling hinges move sequentially from C to B and B' Due to the initial sway of the frame which produced a maximum horizontal displacement at C of approximately 0.06R, the behaviour is not absolutely symmetrical. The plastic discontinuities tend to move alternatively between the left and the right side of the arch. After the relieving of the large plastic sets at B and then B' the travelling hinges continue their path towards the supports. Small but definite kinks are left at B and B' at this stage but they are less pronounced than those in the rigid plastic analysis presented by Gill (1). Later plastic behaviour spreads back towards C as a continuous plastic length. The Vee shape at node C subsequently starts to open again. The behaviour has been followed over most of the range from the semi-circular arch to the triangular deformation. In the closing stages of the analysis a large number of nodes had become plastic.

2.4 Elastic-plastic stiff arch

The load-deformation behaviour of an arch 10 times the elastic stiffness of the flexible arch described in Section 2.3 was investigated using the same computer modelling. The plastic properties were kept constant.

In the load-deflection plot presented in Fig.4 it is not possible to differentiate between the response of the stiff arch and the rigid-plastic behaviour predicted by Gill (1) over the whole of the range of the post-buckled behaviour of the stiff arch. The effects of axial load reduction on the plastic moment and elasticity caused a reduction in the mechanism peak non-dimensional load from 9.6 to 9.1.

A detailed study of the complete frame deformations shows that the sway mechanism as predicted by Gill was the most critical. The sway deformation was similar in nature to the flexible arch but the maximum lateral displacement increased from $0.06R$ to approximately $0.1R$. Comparisons of typical deformed states of the flexible, the stiff and the symmetrical rigid plastic analyses are presented in Fig 6.

3. CONCLUSIONS

It is shown that the predominant behaviour of an arch subjected to a central point load is due to the change of geometry effect and that only in the case of very flexible arches is the post-buckled behaviour significantly different from the rigid plastic mathematical model.

Convergence of non-linear problems will always be difficult when a large number of variables are changing simultaneously. In the arch analyses presented the behaviour of the series of plastic links forming the straightened segments of the arch was traced using the cable stiffness formulation developed by Johnson and Brotton (8).

It is proposed that the four examples presented in the paper should be included in a benchmark test for evaluating elastic-plastic plane frame computer programs.

References

1. Gill, S.S. 'Large deflection rigid-plastic analysis of a built-in semi-circular arch'. Int.J.Mech.Eng. Ed. October 1976, pp 339-355.

2. Austin, W.J. 'In plane bending and buckling of arches'. J Struc D ASCE, vol. 97, ST5, May 1971, pp 1575-92.

3. Johnston, B.G.ed. 'Guide to stability design criteria for metal structures'. Structural Stability Research Council, 3rd edition, John Wiley & Son, New York, 1976, p465.

4. Lee, L.H.N. and Murphy, L.M. 'Inelastic buckling of shallow arches'. J Eng Mech Div ASCE, vol. 94, EM1, February 1968, pp 225-239.

References (Continued)

5. Komatsu, S. and Sakimoto, T. 'Ultimate load carrying capacity of steel arches'. J Struc Div ASCE, vol. 103, ST12, December 1977, pp 2323-36.

6. Jain, A.K. Discussion on 'Model for mild steel inelastic frame analysis' by Santhanam, T.K. J. Struct Div ASCE, vol. 105, ST11, November 1979, pp 2480-1.

7. Harung, H.S. and Millar, M.A. 'General failure analysis of skeletal plane frames'. J Struc Div ASCE, vol. 99, ST6, 1973, pp 1051-1074.

8. Johnson, D. and Brotton, D.M. 'A finite deflection analysis for space structures' from 'Space Structures' International Conference on space structures at University of Surrey, Blackwell, Oxford, 1967, pp 244-255.

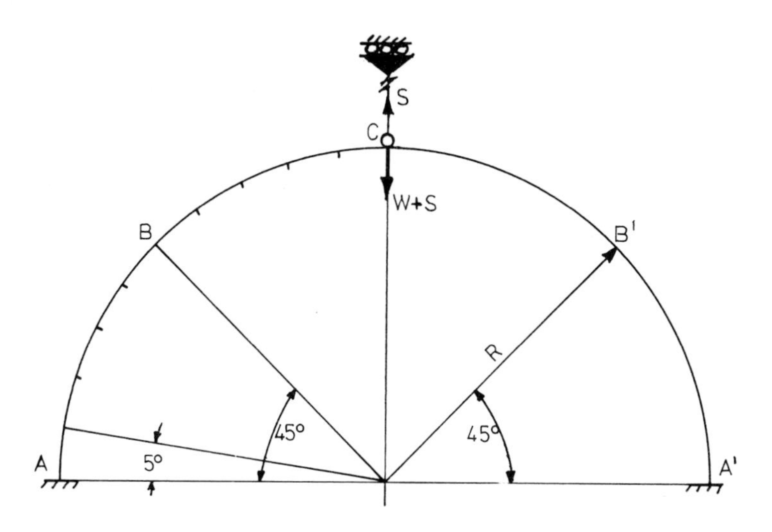

Fig. 1 Arrangement of elements of arch

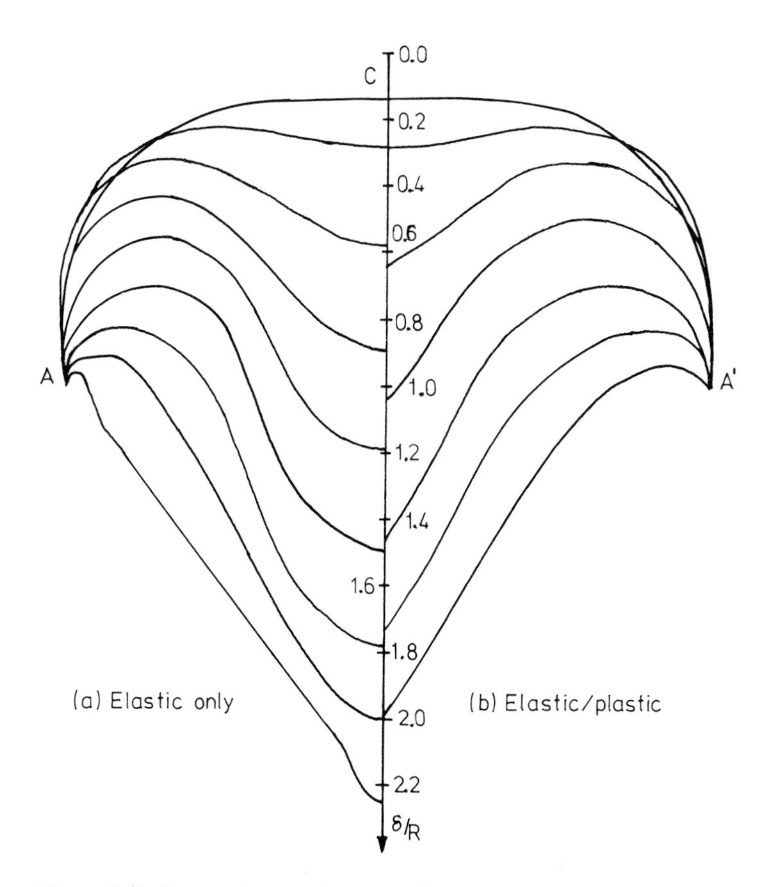

Fig. 2 Deformations of very flexible arches

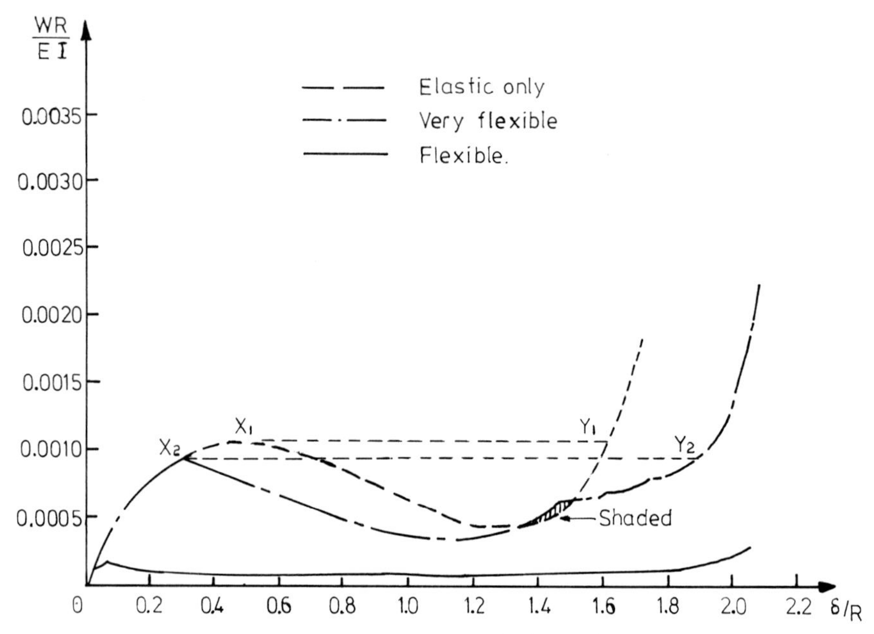

Fig. 3 Non-dimensional elastic load-deflection graphs

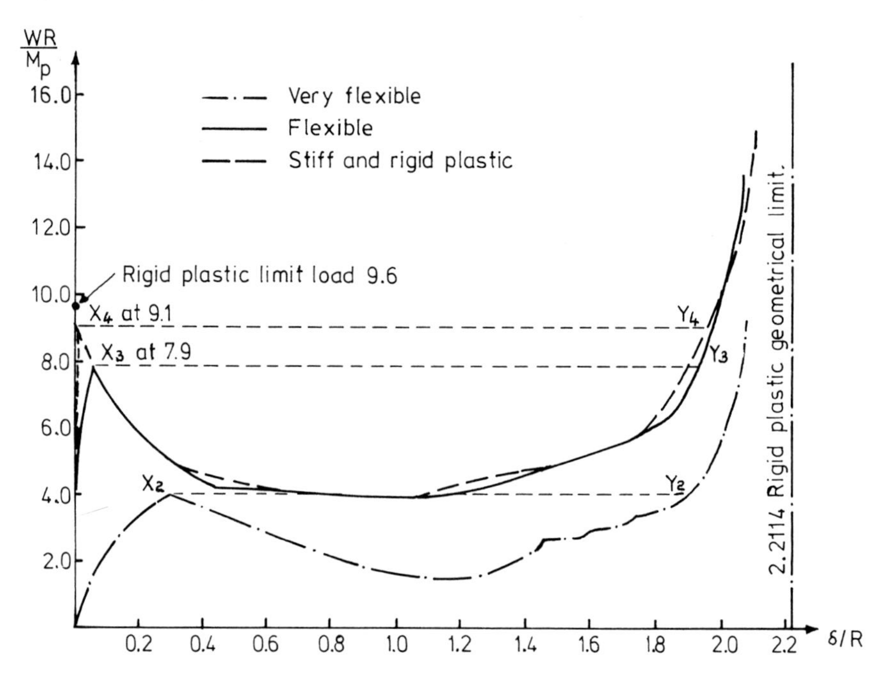

Fig. 4 Non-dimensional plastic load-deflection graphs

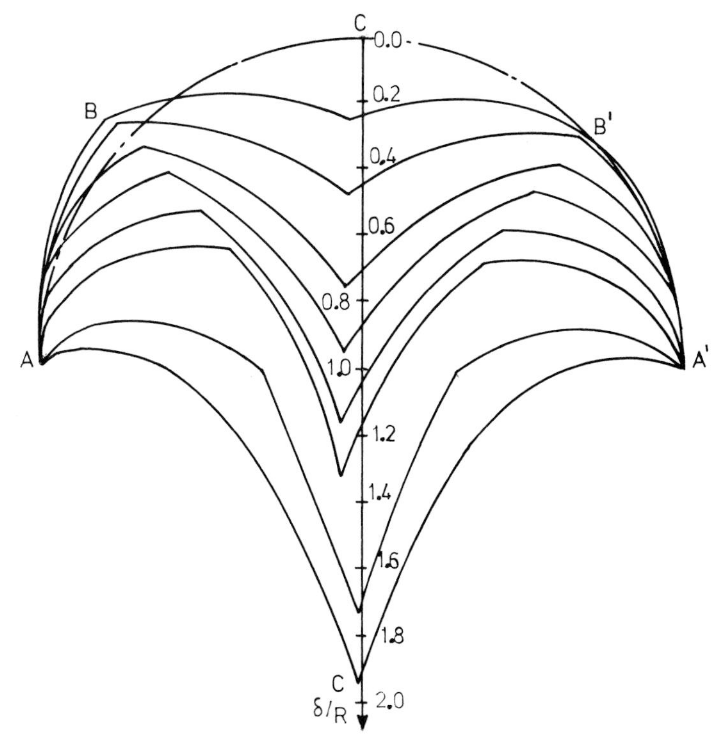

Fig. 5 Deformations of elastic-plastic flexible arch

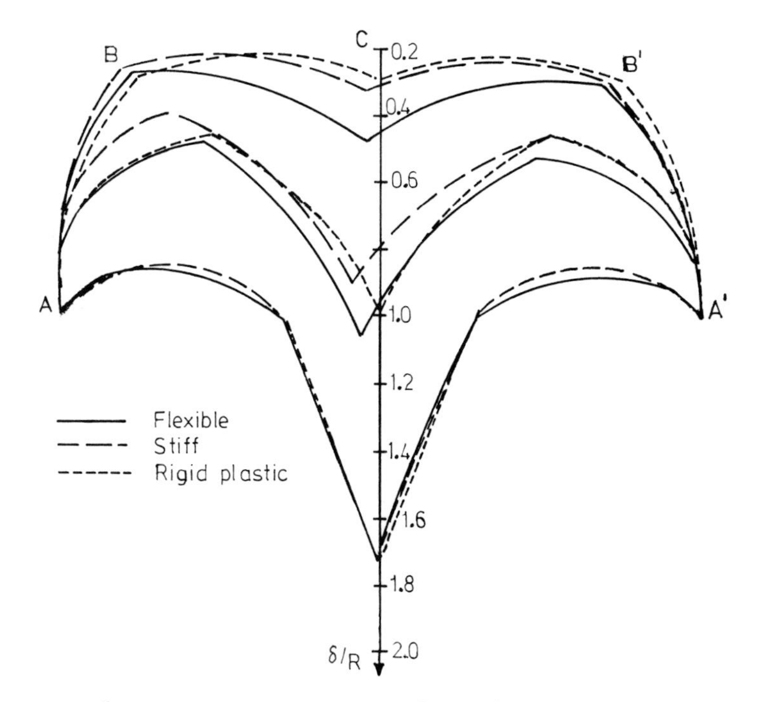

Fig. 6 Comparisons of arch deformations

THE APPLICATION OF SHAKEDOWN THEORY TO THE DESIGN OF STEEL STRUCTURES

P. Grundy.

Department of Civil Engineering, Monash University.

Summary

The limit states of incremental collapse and alternating plasticity, not usually important in the design of building frames, can be the most significant criteria for the design of structures, especially grids, for moving loads. They are also the appropriate criteria for structures subjected to variable repeated loading where designers consider plastic design inappropriate. Studies of grids show that wide margins exist between initial yield (where present design approaches usually stop), the incremental collapse or alternating plasticity limit, and plastic collapse limit. Using the shakedown criterion by an upper bound method easy to use, an efficient design can be achieved in which residual deflections and ductility requirements are limited.

Introduction

The condition of shakedown applies to a structure, when, for a given domain of variable repeated loading, after an initial period of adaptation through plastic deformation, it sustains all loads without increments of plastic strain occurring. The condition of non-shakedown or inadaptation exists when plastic strain increments are repeated indefinitely with variable repeated load within a domain. This inadaptation can be of two types. The first is incremental collapse (IC), in which the structure becomes unserviceable through the incremental growth of a collapse mechanism with the repetition of a variable load program. The second is alternating plasticity (AP), in which alternating tensile and compressive plastic strains occur at specific locations within the structure, causing local damage which may be fatal for the whole structure. Local damage as observed for AP is also possible at plastic hinges during IC.

As a design limit state, shakedown is therefore a serviceability limit state for deflections, and an ultimate limit state for collapse as a mechanism or through local fracture [1,2]. Even as a serviceability limit state, however, the condition of incremental collapse - which, after all, requires the word "collapse" to describe it - is more serious than an ordinary deflection criterion, since it describes a condition of cumulative and irreversible damage.

The shakedown limit lies between the elastic limit, when first yield of the material theoretically occurs, and the collapse load limit, when the structure fails under the application of a steady load. The elastic limit is of theoretical interest. In practice it cannot be known since the initial stresses in a structure cannot be known. The shakedown limit and collapse load limit can be found correctly, since they are independent of the initial stress in the structure.

If the elastic limit and the collapse load are close together, it is unlikely that the shakedown limit is a controlling parameter in the determination of the reliability or capacity of the structure. The shakedown limit in this case will be close to the collapse load. Many cycles of load might be required for the damage through IC or AP to become significant, whereas only one load event is required at a slightly higher level for the structure to collapse altogether. Under such circumstances the single overload causing collapse is more likely to occur than the multimple overloads of slightly lower magnitude to cause incremental collapse or alternating plasticity. This was the conclusion reached by Horne in relation to conventional building frames [3].

For the shakedown criterion to be important in the design of a structure certain features not found in ordinary portal frame buildings must be present. These are:

- a large margin between elastic limit and collapse
- significantly different alternative loads, and
- many repetitions of major loads.

Those structures for which designers have resisted plastic design methods for fear of fatigue damage or unserviceable plastic deformations are the more obvious candidates for useful application of shakedown theory. Bridge deck systems, offshore structures and cranes fall into this group. Of course, shakedown theory has an established position in the design of containments of nuclear reactors and pressure vessels generally, but this discussion is limited to steel frameworks.

The United States Steel Corporation has formulated design procedures for continuous bridge girders based upon the shakedown criterion, known as Autostress Design [4]. The procedure concentrates on the longitudinal stringers and considers the build up of residual stress with load, using a stress history approach. The behaviour of the cross girders is not emphasised, although the studies below show that they are important. Whether cross girders are used or not, the correct assessment of load distribution by elastic analysis is of paramount importance.

The starting point for shakedown analysis is the correct evaluation of elastic behaviour under load. In all but the simplest of structural designs an elastic analysis is necessary. In many design situations the elastic analysis is also regrettably the end of all investigation of structural behaviour. The step to calculate the ultimate load capacity is not taken. Depending on the susceptibility of the structure either to collapse from a single excessive load or to non-shakedown type of damage in service, one of these ultimate limit states should be checked. It is not perhaps realised that the effort to calculate the shakedown limit once the elastic analysis has been performed is about the same as, or even less than, that required for the static collapse load.

1. THEORY

Upper and lower bound theorems apply to the shakedown limit as to the collapse limit. The lower bound or static theorem was first enunciated by Melan [5]. It can be stated as follows:

> For a structure of elastoplastic material subjected to a variable repeated load in a given load domain, shakedown will occur if the total stress $\sigma(t)$ at any time t at every point in the structure satisfies

$$\sigma(t) > \sigma_Y^-, \tag{1a}$$

$$\sigma(t) < \sigma_Y^+, \tag{1b}$$

where $$\sigma(t) = \sigma_E(t) + \sigma_R. \tag{2}$$

> $\sigma_E(t)$ is the elastic stress due to the load at time t, ignoring yield limitations,
>
> σ_Y^+ and σ_Y^- are the tensile and compressive yield stresses of the material, and
>
> σ_R is any statically admissible self-equilibrating residual stress.

To find λ_S, the shakedown limit load factor, it is necessary to write

$$\sigma(t) = \lambda\sigma_E(t) + \sigma_R, \tag{3}$$

and the analysis consists of finding σ_R which maximises λ in eqs (1).

The upper bound theorem was first enuciated by Koiter [6]. It can be stated as follows:

> For a structure of elastoplastic material subjected to a variable repeated load in a given load domain shakedown will not occur if there exists a kinematically admissible displacement rate with a corresponding strain rate $\dot{\varepsilon}^+$ and $\dot{\varepsilon}^-$ for maximum and minimum strain rates respectively such that

$$\int_V \sigma_{E\,max}\,\dot{\varepsilon}^+\,dV + \int_V \sigma_{E\,min}\,\dot{\varepsilon}^-\,dV > \int_V \sigma_Y^+\,\dot{\varepsilon}^+\,dV + \int_V \sigma_Y^-\,\dot{\varepsilon}^-\,dV \tag{4}$$

> where $\sigma_{E\,max}$ and $\sigma_{E\,min}$ are the maximum and minimum repeated elastic stresses due to the loads.

To find λ_S by the upper bound theorem it is necessary to apply λ to σ_E in eq (4) and then to find the displacement rate (or corresponding collapse mechanism which minimises λ in eq (4).

In applying either theorem, if either the positive or negative bound is acting in difference parts of the structure for different load cases, then the non-shakedown condition is IC. If the positive and negative bounds are active at the same point for different load cases then the non-shakedown conditions is AP. The AP limit may be independently stated as follows:

$$\sigma_{E\ max} - \sigma_{E\ min} < \sigma_Y^+ - \sigma_Y^- \qquad (5)$$

However, this statement is not independent of (1) unless stresses are replaced by stress resultants.

If the model of the structure can be reduced from a continuum to a discrete system where the stresses or stress resultants are evaluated only at the critical locations where plastic flow might occur, the calculation of λ_S can be reduced to a mathematical program, which is in fact a linear program in one-dimensional stress situations [7].

2. *SHAKEDOWN OF BARS*

In the analysis of frames and grids stress resultants are evaluated in critical locations where peak values of bending moments are obtained and where plastic hinges could form part of a collapse mechanism. The relevant features of the well-known moment-curvature relationship of bars, with consequent stresses, are summarised in Fig 1. The element, assumed to be initially stress free at O, reaches yield stress in the extreme fibres at A, develops plastic regions in the top and bottom at B, and remains in a state of residual stress at C with total axial force and bending moment zero.

From this behaviour it is evident that even a statically determinate frame which is subjected to a load program which takes parts of the structure beyond initial yield, M_Y, but not up to full plastic moment, M_P, will reach a shakedown condition.

The unloading path of Fig 1 is linear over the range BD. This represents the elastic moment range which can be sustained without AP. It is always less than the plastic moment range. In the case of a symmetric cross-section with a material which has tensile and compressive yield strengths of equal magnitude, shakedown theorems can be restated in terms of moments.

The lower bound theorem for frames can be stated:

> For a frame subjected to a variable repeated load within a given load domain, shakedown will occur if, at every point in the structure,
>
> $$M_{min} < M_P^+, \qquad (6a)$$
>
> $$M_{min} > M_P^-, \qquad (6b)$$
>
> and $M_{E\ max} - M_{E\ min} < 2M_Y \qquad (6c)$
>
> where M_{max} and M_{min} are the maximum and minimum values of the moment $M(t)$ occurring during the repeated load program,
>
> $$M(t) = M_E(t) + M_R, \qquad (7)$$
>
> $M_E(t)$ is the elastic moment in response to the load, ignoring yield limitations and
>
> M_R is any statically admissible residual moment.

The lower bound approach was adopted by Neal and Symonds [8] for the analysis of frames.

Similarly the upper bound theorem for frames can be stated:

> For a frame subjected to a variable repeated load in a given load domain shakedown will not occur if a kinematically admissible collapse mechanism exists, with associated discrete plastic rotations, θ_i, such that

$$\Sigma M^{+}_{i\,E\,max}\,\theta^{+}_{i} + \Sigma M^{-}_{j\,E\,min}\,\theta^{-}_{j} > \Sigma M^{+}_{iP}\,\theta^{+}_{i} + \Sigma M^{-}_{iP}\,\theta^{-}_{j}, \qquad (8a)$$

or for any i, $M_{i\,max} - M_{i\,min} > 2\,M_Y$. (8b)

These statements are limited to frames with bars having only one significant stress resultant, in this case, bending moment about one axis. The yield criterion is linear. In fact there are six stress resultants. If two or more of these are significant, for example, bending moment about one axis, torsion and axial force, then the yield surface is curved, requiring non-linear expressions for shakedown between the elastic limit and the plastic limit in a position dependent upon the range of the independently varying axial force and bending moment [9], and the incremental collapse mechanism includes both plastic curvature and plastic extension. As Fig 2 shows, the elastic limit surface tends to be made up by hyperplanes whereas the limit surface is generally curved and convex.

3. *ANALYSIS OF FRAMES*

A single example will demonstrate the ease of estimating the shakedown limit for frames. Consider the pitched portal frame of Fig 3. An elastic analysis for all load cases will yield an envelope of maximum and minimum bending moments as in Fig 3(a). There is a single internal redundancy with the corresponding bending moment as shown in Fig 3(b). A typical collapse mechanism is shown in Fig 3(c).

A lower bound analysis would be achieved by applying the appropriate load factor to the elastic moment envelope and then searching for a magnitude of residual moment to satisfy eqs (6a) and (6b) elsewhere.

Note that this procedure is even easier than lower bound plastic collapse analysis, which would require a different residual moment for each load case. The starting point for either analysis is the elastic moment envelope, which the analyst usually requires for design checking anyway.

An upper bound analysis would be achieved by applying the appropriate load factor to the elastic moment envelope as before and then using eq (8a) to establish the value of the load factor for various trial collapse mechanisms such as the one illustrated in Fig 3(c). A plastic collapse analysis would be achieved if the moment envelope were replaced by the elastic moments for each load case being considered. Again this would entail more effort on the part of the analyst, although the direct application of the work equations to the collapse mechanism would be quicker if an elastic analysis was not required.

4. *ANALYSIS OF GRIDS*

Shakedown analysis is relevant to the analysis of grids more often than it is to frames because moving concentrated loads with a consequent greater range of local effects are more likely to occur, for example, in bridge decks or offshore platforms. Torsionless grids were studied experimentally and theoretically by Spencer [10,11]. The grids were simple supported on four sides or on two opposite sides with the other sides free. The tests established that the theoretical shakedown limit was indeed the point beyond which incremental collapse took place. The rate of growth of the collapse mechanism was approximately proportional to the amount by which the load factor exceeded the shakedown limit load factor. This also conforms with simple theoretical models of incremental collapse behaviour [12,13].

Two examples will illustrate torsionless grid behaviour. Figure 4 shows a one way grid with three crossbeams. Figure 5 shows a one way grid continuous over two equal spans, with a single crossbeam at each midspan. The single concentrated load, P, is distributed directly to the adjacent stringers for analysis. The load domain is narrower than the grid so that the external stringers carry a maximum of $3/4P$ while the adjacent interior stringer carries the remaining $1/4P$. The relative properties of stringers and crossbeams are given in Fig 4.

An elastic analysis was performed for each grid with the load(s) positioned on the designated nodes, and the envelope of maximum and minimum elastic moments was determined. An inspection of the envelope located the critical nodes for elastic limit load, P_Y, and alternating plasticity limit load, P_{AP}. An upper bound analysis using the likely mechanisms shown in Figs 4 and 5 was carried out to establish the shakedown limit load, P_{IC}, and the collapse load, P_C. For the two span grid the results apply equally to either span and the results are given for only one. The results are given in Table 1, where a form factor of 1.15 is assumed.

From Table 1 it is evident that the margin between initial yield and collapse is large for grids. The more redundant two span grid has the greater margin. Efficient design for the moving live load condition should exploit this post elastic range. Adopting the IC criterion instead of the initial yield criterion would allow an additional 70% of load for the one span grid and an additional 88% for the two span grid.

5. *PLASTIC DEFORMATION AT SHAKEDOWN*

Concern is sometimes expressed at the amount of plastic deformation which must occur before shakedown is achieved, just as for plastic

Table 1. Grid Load Capacity

Grid Type	One Span	Two Span
Initial Yield, $P_y\ L/M_p$, at Nodes e,h	5.43	5.93
Alternating Plasticity, $P_{AP}\ L/M_p$, at e,h	9.27	10.97
Incremental Collapse, $P_{IC}\ L/M_p$, Mechanism	7.71, I	9.12, A or I
Plastic Collapse, $P_C\ L/M_p$, Mechanism	9.14, F	13.13, R
P_Y/P_C	0.595	0.444
P_{IC}/P_C	0.845	0.695
P_{IC}/P_Y	1.705	1.880

collapse. The designer's concern for adequate ductility up to plastic collapse is increased for repeated loads up to incremental collapse. In either case the simple elastoplastic model of the moment-curvature relationship leads to singularities in strain and curvature at plastic hinges. These do not occur in practice because of the knee in the actual moment-curvature relationship and subsequent strain hardening. However, even with the elastoplastic model it can be shown that the plastic deformations are adequately contained [14]. In the case of moving loads on a grid a single passage of the maximum load is sometime sufficient for complete plastic adaptation.

Consider a beam continuous over two equal spans subjected to a strain of equally spaced moving concentrated loads. The plastic collapse load is $5.83\ M_P/L$ and the elastic limit load is $4.82\ M_P/L$, assuming a form factor of unity. The shakedown limit load is $5.74\ M_P/L$ for a single load and $4.93\ M_P/L$ for two loads spaced $0.845\ L$ apart [15]. The effect of passing a single load in excess of the elastic limit load across the two spans is shown in Fig 6. The envelope of maximum positive bending moment is traced under the load point as it moves. When this moment reaches M_P it cannot increase further and some plastic curvature takes place, leading to the development of residual moment.

If $m_R M_P$ is the residual moment at the centre support, it can be shown that, during plastic deformation under moving load,

$$\frac{\partial m_R}{\partial x} = -\frac{m_R}{x} - \frac{PL}{M_p}\left(\frac{1}{x} - \frac{5}{2} + x^2\right) \qquad (9)$$

provided that $\frac{\partial m_R}{\partial x} < 0$, where xL is the distance from the side of centre of the load. α, the ratio of plastic curvature to M/EI, is given by

$$\alpha = \frac{2}{3x}\ \frac{\partial m_R}{\partial x} \qquad (10)$$

Figure 6(a) shows the moment envelope for $P = 5.4\ M_P/L$. Plastic curvature develops over the range $0.286 < x < 0.41$, commencing with $\alpha \simeq 12$, falling tangentially to zero. Plastic curvature distributions for various loads are shown in Fig 6(b).

It is noted that a plastic hinge does not form at the centre support unless the load exceeds the shakedown limit. Of course, an unfavourable initial moment arising from the substantial settlement of a side support - a small one would be favourable - would lead to some plastic hinge rotation. In this simple example, shakedown is achieved by one passage of the load.

In practice it is impossible to predict residual deflections after shakedown, since initial stresses are not known. Bounds on predictions are progressively widened as the load factor approaches the shakedown limit [16]. For loads beyond the shakedown limit following an ordered cycle, deflection increments appear to be linearly proportional to the excess [12,13]. Interaction between IC and AP has not been found significant in some cases studied [13]. Studies to date suggest that, at worst, no more ductility is required in the application of shakedown theory than is required for simple plastic collapse theory.

6. *EFFECT OF STRAIN HARDENING*

In the case of elastic-strain hardening material, even Prager kinematic hardening [13], adaptation against IC will take place at any load level. However, AP will occur. During ordered load cycles the effective residual stress will fluctuate between limits which become more widely spaced with increasing load.

When the material has elastic-plastic-strain hardening characteristics it becomes more difficult to find bounds. An alternative is to perform an incremental analysis keeping a record of the spread of the strain hardening zones [11]. The attractive simplicity of elastoplastic theory has been lost, but the results underscore the conservative nature of the elastoplastic solutions. The studies show that the residual deflection at shakedown - which always occurs - is far more sensitive to the size of the yield plateau than to the strain hardening modulus. The possibility exists of using a simplified model with a locking material akin to the rigid-plastic-rigid model used by Horne [17].

7. *EXTENSION OF THEORY TO BEAM AND SLAB SYSTEMS*

The comprehensive application of shakedown theory to grids requires an extension to include the slab or deck which usually acts compositely with the beam in practice. Yield line theory of slabs needs to be adapted to shakedown analysis. This has been done [18,19], although further experimental study of shakedown of reinforced concrete slabs is required, and the numerical procedures are at present formidable. It is possible that the upper bound approach outlined above can be adapted to the combined grid and plate to bring the shakedown criterion within the reach or ordinary design [2].

8. *CONCLUSIONS*

For structures, especially grids, subjected to repeated moving loads, there is a strong case for the adoption of the shakedown criterion as the appropriate ultimate limit state for design.

The economy available in pursuing a shakedown limit approach instead of the present elastic limit approach normally used in designing for variable repeated loads can be substantial.

Given the customary elastic analysis of structures the step to an upper bound estimate of the shakedown limit is not difficult.

The ductility requirements for the establishment of shakedown are generally less than or equal to those for legitimate use of plastic collapse. With moving loads the plastic deformation required for adaptation is not excessive and it is generally distributed through the structure rather than concentrated at hinges.

Acknowledgement

The grid analyses were performed by M. Sivakumar, research student, using program GENGRID by N. Sneath, lecturer, in the Department of Civil Engineering, Monash University.

References

1. Grundy, P. 'Shakedown as a limit state.' 6th Australasian Conf. on Mechanics of Structures and Materials, 1-6, Christchurch, N.Z., (1977).

2. Grundy, P. 'Bridge Design Criteria.' Metal Structures Conf. I.E. Aust., 73-77, Adelaide, 1976.

3. Horne, M.R. 'Effect of variable repeated loads in the plastic theory of structures,' Engg. Struct. Acad. Press Inc., NY, 141-151, (1949).

4. Carskaddan, P.S. 'Autostress design of highway bridges. Phase 1: Design Procedure and Example Design.' USS Tech, Rep., March 8, 1976.

5. Melan, E. 'Theorie statisch unbestimmter Systeme aus ideal-plastischen Baustoff.' Sitzungsber. Akad. Wiss. Wien. 145: 195-218, (1936).

6. Koiter, W.T. 'A new general theorem on shakedown of elastic-plastic structures.' Proc. Konk. Ned. Akad, Wet. B59: 24-34, (1956).

7 Maier, G. 'Shakedown Analysis, Engineering Plasticity by Mathematical Programming.' NATO-ASI, Waterloo, 107-134 (1977).

8. Neal, B.G. and Symonds, P.S. 'A method for calculating the failure load for a frames structure subjected to fluctuating loads.' J. I.C.E., 35, : 186-197, (1950-51).

9. Grundy, P. 'Shakedown of bars in bending and tension.' J. Eng. Mech. Div., ASCE, 95, EM3, 519-529, (June 1969).

10. Grundy, P. and Spencer, W.J. 'Shakedown of Elastoplastic Grids.' 5th Australasian Conf. on Mechanics of Structures & Materials, 219-236, Melbourne, (1975).

11. Spencer, W.J. 'Toward limit state design - upper bound solutions for plastic collapse and incremental collapse of structures'. 7th Australasian Conf. on Mechanics of Structures and Materials, Perth, (1980).

12. Guralnick, S.A. 'Incremental collapse under conditions of partial unloading.' Publ. IABSE, 33, II, 69-84, (1973).

13. Tin Loi, F. and Grundy, P. 'Deflection stability of work-hardening structures.' J. Struct. Mech., 6(3), 331-347 (1978).

14. Eyre, D.G. and Galambos, T.V. 'Shakedown of Grids'. J. Struct. Div. ASCE, 99, ST10, (Oct. 1973).

15. Grundy, P. 'Shakedown under moving loads.' Civ. Engg. Trans. I.E. Aust., CE13, (April, 1971).

16. Brzezinski, R. and König, J.A. J. Struct. Mech., 2, 211 (1973)

17. Horne, M.R. 'Instability and the plastic theory of structures.' Trans. Engng. Inst. Canada. 4, (1960).

18. Alwis, W.A.M. and Grundy, P. 'Shakedown of plates under moving loads.' 7th Australasian Conf. on Mechanics of Structures and Materials, Perth (1980).

19. Alwis, W.A.M. and Grundy, P., 'Shakedown analysis of plates' and 'Capacity of rectangular plates under moving load'. For publication. (Int. J. Mech. Sci.).

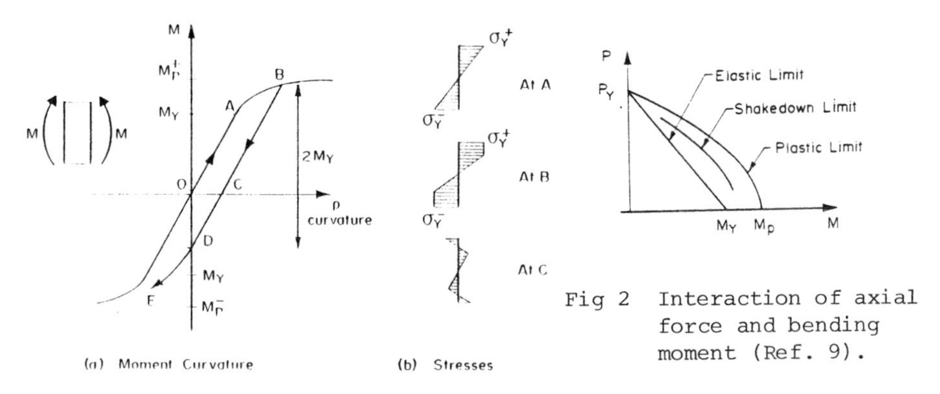

Fig 2 Interaction of axial force and bending moment (Ref. 9).

Fig 1 Moment-Curvature-Stress Relationships

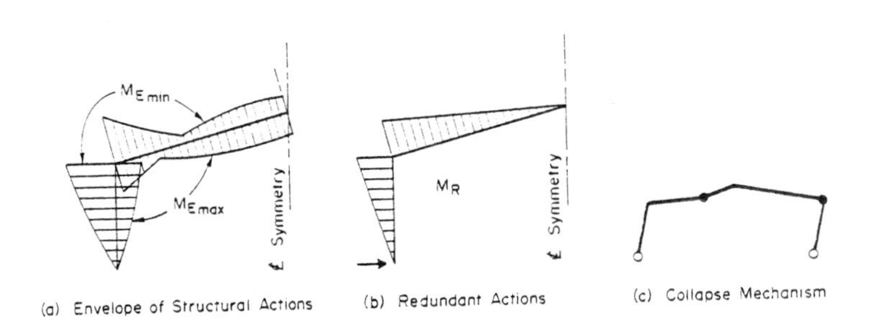

Fig 3 Shakedown Analysis of a Portal Frame.

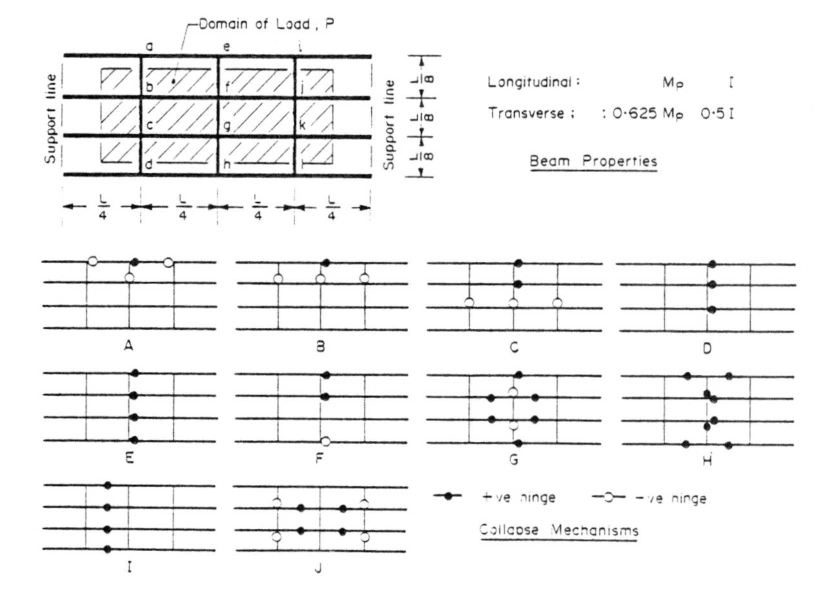

Fig 4 One Way Grid.

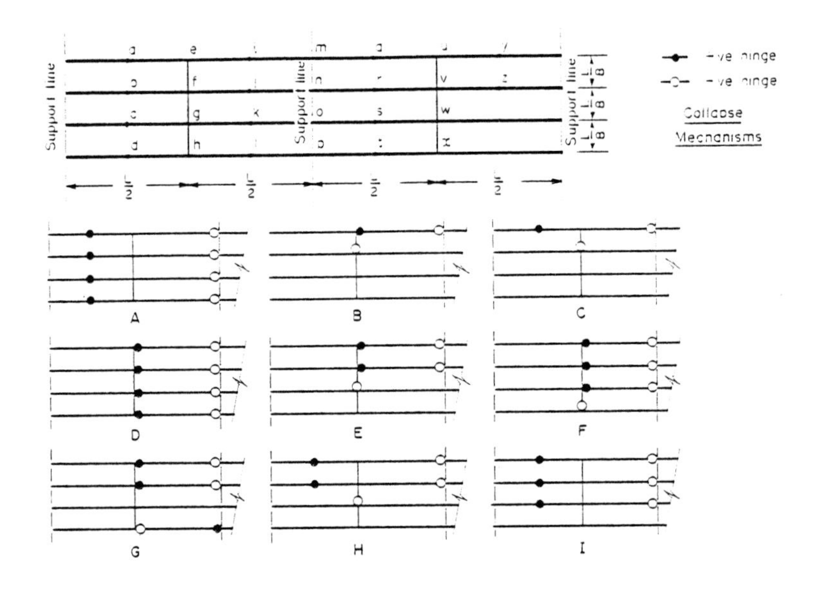

Fig 5 Two Span Grid.

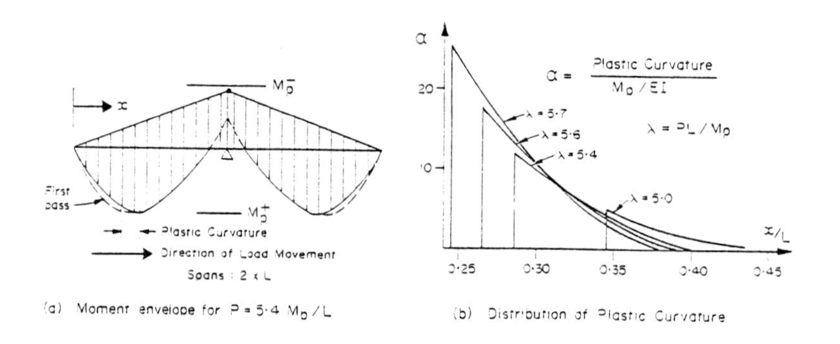

Fig 6 Adaptation of Two-span Beam to Moving Load.

CORRELATION STUDY OF ULTIMATE LOAD CAPACITIES OF SPACE TRUSSES

L C Schmidt

Department of Civil and Mining Engineering, University of Wollongong

P R Morgan, B M Gregg

Department of Civil Engineering, University of Melbourne

Synopsis

The correlation between the experimental and theoretically derived load carrying capacities of twenty space truss tests has been investigated. Both model and full scale systems are considered. The study has furnished alternative empirical corrections to the theoretical methods for the close estimation of the ultimate strength of the class of space trusses studied.

Notation

R ratio of the degree of indeterminacy to the total number of truss members and constraints

S ratio of slenderness ratios:

for $(\ell/r)/(\ell/r)_T \geq 1$, $S = (\ell/r)/(\ell/r)_T$

$(\ell/r)/(\ell/r)_T < 1$, $S = (\ell/r)_T/(\ell/r$

ℓ pin-ended or fixed-ended length of a critical compressive element as appropriate to the truss system investigated

r radius of gyration of strut cross section

T subscript denoting transition - length strut, when yield stress of material equals the Euler critical stress

Introduction

To the practising engineer the design of space trusses of the type indicated in Fig 1 is considered a relatively simple matter nowadays as computers can be used to analyse readily any complex network of members in the elastic range. Few, if any, inelastic analyses are carried out to find the ultimate strength, as it is generally considered that an enormous reserve of strength exists beyond the load at which the first member buckling occurs. Experimental evidence by Schmidt et al (1) suggests that where compressive chord collapse initiates the inelastic behaviour the reserve of strength does not exist, owing to the brittle type of failure of the compressive elements that are usual in such structures. As illustrated by other tests of Schmidt et al (2), only in the case where tensile chord yielding occurs is there an increasing load capacity beyond yielding of the first tensile chord.

The present paper attempts to correlate experimental results with theoretical results derived from relatively simple analyses. Empirical relationships are derived, and these allow corrections to be applied to the results of the simple theoretical analyses. The truss systems investigated herein are plate-like with two layers of chords, and are so designed that initial failure occurs by either compressive chord collapse or tensile chord yield; web failure is excluded.

A survey was made of tests to destruction reported by others, but the only result that could be used in the current investigation was that published by Mezzina et al (4). Even in this case a series of control tests on compressive members had to be carried out in Melbourne in order to determine basic member information for the theoretical analyses.

The present paper considers theoretical results derived from analyses that take into account the post-buckling behaviour of compressive elements and yield of tensile elements. Correlation studies are made and conclusions are then drawn.

1. *THEORETICAL ANALYSES*

In the work discussed herein, attention is drawn to two significant load capacities in the theoretical analyses. The first occurs at the buckling load of the first critical compressive member, or the first yield of a tensile member, and is referred to as a Point A analysis. Such an analysis is linear and is the usual analysis procedure carried out by designers. The second significant load capacity is considered to be that at which successive compressive chord failures furnish complete lines of collapsed members, after which a significant loss in truss load capacity and stiffness occurs. Alternatively, with mixes of yielding tensile chords and collapsing compressive chords, Point B can be considered to be the load at which the stiffness matrix of the structure is about to become singular, or the load at which significant reductions in truss stiffness occur.

In these non-linear analyses, the post-buckling characteristics of the struts are assumed to be known, either for the pin-ended or fixed-ended condition. The choice between the two is determined by the jointing system used. The significant loads derived from the theoretical analyses carried out for the range of trusses investigated are listed in Table 2.

2. *TEST RESULTS*

To the 12 truss tests reported earlier (Schmidt et al (3)) a further eight have been added making 20 tests in all. Data available for each of these further tests are a full experimental collapse load test plus the simplified collapse analyses performed for comparison purposes. Truss layouts are given in Fig 1 or the earlier paper. Table 1 provides details of all trusses included in this study. The tests reported now include trusses which have four corner supports, a truss which is statically determinate, and the full scale welded tubular steel truss tested in Italy (Mezzina et al (4)). Table 2 summarises the significant points derived from the test results.

3. *COMMENTARY AND CONCLUSIONS*

Comments on tests 1 to 12 are already available (Schmidt et al (3)) including the load-deflection plot for each truss. In referring to these earlier graphs the truss load capacities A and B have been derived from the simplified analyses referred to in Section 1 herein, and are given in Table 2, columns 5 and 6. In general it may be seen from Table 2, columns 8 and 9, that the truss load capacity denoted by Point B does not always overestimate the measured ultimate load capacity, but where it does the overestimate is generally less than that given by Point A.

For the purposes of the present study, the theoretical Point A load capacities for truss tests 10 and 11 have been taken as those at Point B owing to the extremely small load capacities at which tensile chord yield first occurred. However, the tests have been excluded from one set of regression analyses, as seen in Table 3, case (4).

Trusses 13 to 17 were identical in form to trusses 6,7 and 8 and the test results for trusses 13 to 17 (seen in Fig 2) were similar to those for trusses 6,7 and 8. The Points A provided similar and quite large overestimates of test capacity (19 to 25% as seen in Table 2, column 8) but the Points B provided a much closer estimate (within 5% as seen in Table 2, column 9). Truss 18 is a special case. It is a statically determinate system and in this case failure of the first member (Point A) and of incipient truss collapse (Point B) correspond, as seen in Fig 3.

Truss 19 was supported on four points only but was otherwise similar to trusses 13 to 17. Fig 4 presents the test and theoretical curves of load versus deflection and it is seen that Point A is a good prediction, but Point B is even better. Truss 20 was supported at four intermediate points (see Fig 1) and yielded the test and theoretical results given in Fig 5. This truss was a full scale test performed in Italy (Mezzina et al (4)) on a tubular steel truss with concentric welded joints preventing slip but offering little flexural resistance. This truss was designed to permit some tensile yielding before compressive elements became critical and controlled the behaviour. Points A and B were good estimators of the load capacity (Table 2, columns 8 and 9).

In all cases the theoretical analyses used member characteristics derived from control tests; it could be expected that experimental results would be closely predicted. Close agreement was not always achieved, however, as can be seen in Table 2, columns 8 and 9. Reasons for this discrepancy appear to be initial forces due to tolerancing of joints and members, the possibility of joint slip, joint fixity or member continuity, joint eccentricity, whether tension or compression controls the design, and the degree of statical indeterminacy.

A summary and correlation study of the results are given in Figs 6,7 and 8 and Table 3. This work has been an extension of that presented earlier (Schmidt et al (3)) and is based upon the same empirical parameter R/S, which reflects the degree of indeterminacy. Fig 6 presents a plot of R/S against the percentage by which the Point A analysis for each truss overestimates the test results, and presents a linear regression line for all results. The coefficient of correlation for these results, 0.63, is significant at the 95% level, and the sample gives 95% confidence that the population coefficient of correlation (of all possible truss tests) lies between 0.26 and 0.84. Confidence limits at the 95% level were then computed, and the confidence band is shown on Fig 6.

Considering the factors special to different trusses, specific test results were progressively excluded from the regression, correlation and confidence limit calculations. Fig 7 presents these regression lines but for clarity omits the confidence bands. Table 3 summarises these results. Similar calculation techniques were applied to the Point B theoretical predictions of truss capacity given in columns 6 and 9 of Table 2. In all of the cases analysed (see column 1 of Table 3) the sample coefficients of correlation (given in column 5 of Table 3) did not permit rejection (at the 95% significance level) of the hypothesis that the population coefficient of correlation was zero. In other words, for the Point B analyses the percentage overestimate of truss capacity is independent of the parameter R/S. To present the data for the Point B analyses a simpler approach has therefore been adopted. The results are presented in Table 3 columns 5 to 8. For each case (column 1, Table 3) the sample mean and standard deviation are computed (columns 6 and 7), as well as the range which included 95% of all tests (column 8). The range has been included on Fig 8.

Table 3 presents the statistical analyses for the Point A and Point B results for all tests and then for the four selected reduced samples (see column 1). The reduced samples were selected on the basis of excluding trusses with clear physical differences in truss construction or structural behaviour or both. Of the exclusions tested only one yielded significant improvement in predictive ability over case (1) that uses all tests; this was case (4) referred to in Table 3, column 1. The other cases ((2), (3) and (5)) were as reliable as case (1) as predictors. Thus for all of cases (1), (2), (3) and (5) it is accurate enough to apply the same rules. It would appear sensible to use the results achieved from the full sample (case (1)) of 20 truss tests. To obtain a reasonable estimate of the actual truss capacity either of the following approaches could be adopted.

Based on Point A: Reduce the Point A load capacity by the percentage defined for case (1) in column 2 of Table 3, using the rule: percentage reduction = 164 R/S + 1. For 95% of trusses and over a wide range of R/S values the real load capacity of the truss should be within $\pm$ 5% of the predicted value, because of the confidence bands given in Fig 6.

Based on Point B: The actual capacity should be (for 95% of trusses) between 83% and 125% of the Point B value, or even (for 70% of trusses) between 93% and 115% of the Point B value.

Case (4) excluded the four trusses where extensive tensile yielding occurred. The 16 trusses included in this sample were designed so that compressive member buckling controlled the behaviour. More reliable predictions can be made for this case. Using a Point A analysis the percentage reduction may be computed as (190 R/S). This reduced figure should be within $\pm$4% of the actual truss capacity for a wide range of R/S values. A Point B analysis should provide an answer between 85% and 119% of the actual capacity for 95% of trusses, or between 94% and 111% of capacity for 70% of trusses.

A Point A analysis, used in unmodified form for the case where compressive chord collapse occurs first, is seen to overestimate truss strength significantly, and is therefore unsafe. Conventional elastic analysis, therefore, is considered to be inadequate for highly redundant systems with brittle-type behaviour of the compressive chords. However, such a Point A analysis, combined with the corrective function derived from the appropriate regression line (column 2, Table 3) enables a good estimate of truss ultimate strength to be found that is satisfactory for competitive design purposes.

The load at first buckling, Point A, is seen to overestimate truss carrying capacity to an increasing extent as the degree of indeterminacy increases, as the truss load capacity is strongly dependent on *R/S*. This increase is attributed to the sensitivity to imperfections of trusses with brittle-type behaviour of the compressive elements. The correlation test indicated the significance of the factor *R/S*. The increased pool of results allows the observation to be made more strongly than was possible earlier by Schmidt et al (3). However, the Point B analysis did not correlate significantly with the value of *R/S*. But the Point B analysis can be directly related to the ultimate strength of the truss with a reasonably small margin of error, regardless of whether compressive chord buckling or tensile chord yield occurs initially. Moreover, the Point B analysis furnishes some idea of the inelastic range of deformation, which is unavailable from a Point A analysis.

References

1. Schmidt, L.C., Morgan, P.R. and Clarkson, J.A. 'Space Trusses with Brittle-Type Strut Buckling.' Journal of the Structural Division, ASCE, vol. 102, No. ST7, July 1976.

2. Schmidt, L.C., Morgan, P.R. and Clarkson, J.A. 'Space Truss Design in Inelastic Range.' Journal of the Structural Division, ASCE, vol. 104, No. ST12, December 1978.

3. Schmidt, L.C., Morgan, P.R. and Hanaor, A. 'Ultimate Load Testing of Space Trusses.' Journal of the Structural Division, ASCE, vol. 108, No. ST6, June 1982.

4. Mezzina, M., Prete, G. and Tosto, A. 'Automatic and Experimental Analysis of a Space Grid in Elastoplastic Behaviour.' 2nd International Conference on Space Structures, University of Surrey, Guildford, England, 1975.

TABLE 1. Truss Details (Y = yes; N = no)

Test No.	Member Details (mm) D = outside dimension R means solid rod C, S mean Circular or Square tube			Member Material Steel or Alumn.	Joint Detail P, pinned C, Continuous	Joint Eccentricity (intentional)	Joint Slip Possible	Primary Inelastic Behaviour C, Compressive T, Tensile	Nominal Slenderness Ratio of Comp. Chords	Transition $^{\ell}/_{r} = \pi\sqrt{\frac{E}{\sigma_y}}$ *	Overall Truss Dimensions (m)
(1)	Compr. Chord (2)	Web (3)	Tensile Chord (4)	(5)	(6)	(7)	(8)	(9)	(10)	(11)	(12)
1 2	C All members equal D = 12.7 t = 1.52	C	C	Al.	P P	N N	Y N	C C	76 76	47 47	0.915
3 4	C All members equal D = 33.8 t = 3.3	C	C	St.	P P	N N	Y Y	C C	125 125	98 98	4.116
5	R D = 6.4	R D = 4.9	R D = 6.4	St.	C	Y	N	C	192	66	2.439
6 7 8	C All members equal D = 12.7 t = 1.52	C	C	Al.	P P P	N N N	Y Y Y	C C C	76 76 76	47 47 47	1.829
9	C D = 48.3 t = 4.1	C Webs and tensile chords D = 1.33 and t = .13	C	St.	P	N	Y	T → C	87	98	9.604
10 11	C Comp. chords and webs D = 12.7 and t = 1.52	C	R D = 1.6	Al/St.	P P	N N	Y Y	T T	76 76	47 47	1.829
12	S D = 25.4 t = 3.3	C D = 26.9 t = 3.3	S D = 25.4 t = 3.3	St.	C	Y	Y	C	139	63	8.537
13 14 15 16 17	C All members equal D = 12.7 t = 1.52	C	C	Al.	P P P P P	N N N N N	Y Y Y Y Y	C C C C C	76 76 76 76 76	46 46 46 46 46	1.829
18	No comp. chord	C Web and tensile chord D = 12.7 C = 1.52	C	Al.	P	N	Y	C			1.829
19	C All members equal D = 12.7 t = 1.52	C	C	Al.	P	N	Y	C	76	47	1.829
20	C All members equal D = 60.3 t = 2.9	C	C	St.	P	N	N	T → C	74	79	9.000

* Young's Modulus values adopted were 67 and 200 GPa for Aluminium and Steel respectively.

TABLE 2 - SUMMARY OF RESULTS

Test No. (1)	Type of Truss (2)	Truss* Layout (3)	Fig. No for Load/Deflection Curve (4)	Truss Load Capacities (kN) Theoretical A (5)	Theoretical B (6)	Expt. (7)	Percentage over-estimate A (8)	B (9)	R** (see text) (10)	S (see text) (11)	R/S (12)
1	Triodetic	SOS	3	27	27	24.4	10	10	$4/_{67}$ = .06	1.62	.037
2	Triodetic	SOS	3	27	27	26.3	3	3	$4/_{67}$ = .06	1.62	.037
3	Bamford	SOS	4	173	129	130	25	0	$12/_{87}$ = .14	1.28	.109
4	Bamford	SOS	4	173	129	111	36	14	$12/_{87}$ = .14	1.28	.109
5	Ecc.joint welded	SOS	5	17.2	12.6	15.0	13	-19	$^{112}/547$= .20	$\frac{\frac{1}{2}\times 192}{66} = 1.45$	.138
6	Triodetic	SOS	6	44.8	36.8	36.5	19	1	$60/_{315}$= .19	1.62	.117
7	Triodetic	SOS	6	44.8	36.8	34.3	23	7	$60/_{315}$= .19	1.62	.117
8	Triodetic	SOS	6	44.8	36.8	33.0	26	10	$60/_{315}$= .19	1.62	.117
9	Bamford	SOD	7	397	430	365	8	15	$47/_{347}$= .14	1.13	.124
10	Triodetic	SOS	8	25.4	25.4	21.8	14	14	$60/_{315}$= .19	1.62	.117
11	Triodetic	SOS	8	25.4	25.4	17.8	30	30	$60/_{315}$= .19	1.62	.117
12	Ecc.joint bolted	SOS	10	445	335	280	37	16	$80/_{419}$= .19	$\frac{\frac{1}{2}\times 139}{63} = 1.10$	.173
13	Triodetic	SOS	2	48.4	37.5	36.0	26	4	$60/_{315}$= .19	1.65	.115
14	Triodetic	SOS	2	48.4	37.5	39.0	19	-4	$60/_{315}$= .19	1.65	.115
15	Triodetic	SOS	2	48.4	37.5	38.8	20	-4	$60/_{315}$= .19	1.65	.115
16	Triodetic	SOS	2	48.4	37.5	39.0	19	-4	$60/_{315}$= .19	1.65	.115
17	Triodetic	SOS	2	48.4	37.5	39.2	19	-5	$60/_{315}$= .19	1.65	.115
18	Triodetic	SOS	3	12.9	12.9	12.9	0	0	$0/_{255}$= 0		0
19	Triodetic	SOS	4	24.7	21.[illegible]	22.2	10	-2	$40/_{295}$= .14	1.62	.086
20	Conc.joint welded	SOS	5	314	273	294	6	-8	$40/_{295}$= .14	1.07	.131

For tests 1-12 the Fig. Numbers are those in the earlier paper (Schmidt et al, 1982)

* SOS = square-on-square; SOD = Square on diagonal.

** R is written as a ratio; Numerator the degree of indeterminacy, denominator the total number of members plus support constraints

TABLE 3 - TRUSS TEST STATISTICS - ANALYSIS POINTS A AND B

Tests Excluded (No. in sample)	POINT A			POINT B (Percentage Overestimate Statistics)			
	Linear Regression eqt.* (%)	Sample coefficient of correlation	95% confidence range on population coeff.of corr.	Sample coefficient of correlation	Mean (%)	Standard deviation, σ (%)	Range likely to include 95% of all possible trusses (%) **
Column No.→(1) Case No.↓	(2)	(3)	(4)	(5)	(6)	(7)	(8)
None (n = 20) (1)	$164 \frac{R}{S} + 1$	0.63	0.26) 0.84)	0.06	4	10.9	4 ± 21 (- 17 to + 25)
Welded trusses 2,5 and 20 - no joint slip (n = 17) (2)	$184 \frac{R}{S} + 1$	0.72	0.36) 0.89)	0.24	6	9.7	6 ± 19 (- 13 to + 25)
Trusses 5 and 12 where the fixed-end strut characteristic was used (n = 18) (3)	$158 \frac{R}{S} + 2$	0.59	0.17) 0.83)	0.06	4.5	9.6	5 ± 19 (- 14 to + 24)
Trusses 9, 10, 11 and 20 with extensive tensile yield (n = 16) (4)	$190 \frac{R}{S}$	0.79	0.48) 0.92)	0.00	1.7	8.6	2 ± 17 (- 15 to + 19)
Trusses 2, 5, 12 and 20, welded or using fixed-end strut characteristic (n = 16) (5)	$169 \frac{R}{S} + 2$	0.64	0.21) 0.86)	0.12	5.4	9.7	5 ± 19 (- 14 to + 24)

* Percentage overestimate = slope of regression line $\times \frac{R}{S}$ + intercept on percentage overestimate axis (see Figures 6 and 7 for clarification)

** 1.96σ on each side of the mean, assuming approximate normality of population of tests.

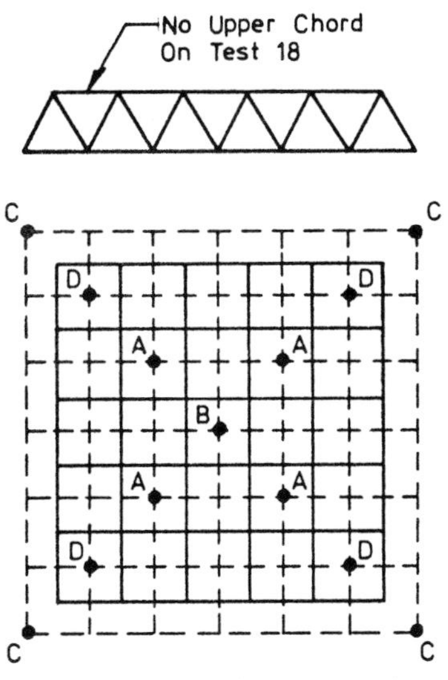

Fig 1 Double layer space truss.

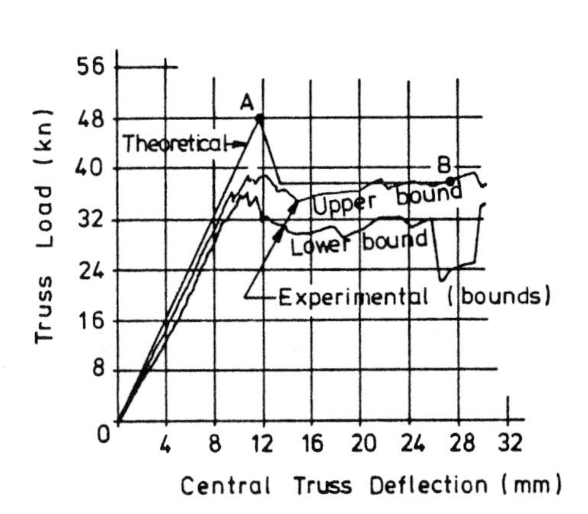

Fig 2 Results of tests and analysis for trusses 13 to 17.

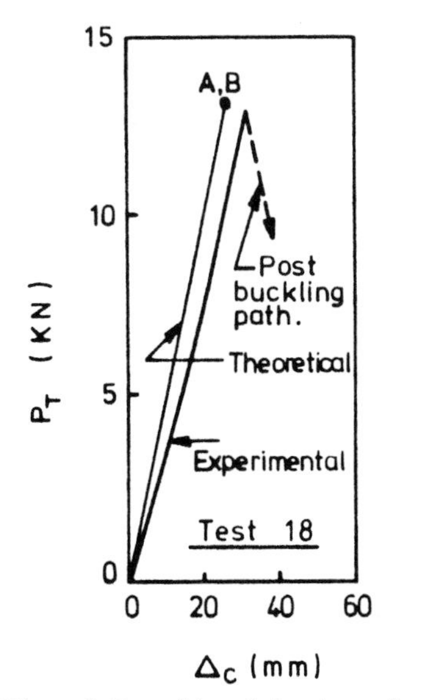

Fig 3 Result of test and analysis for truss 18.

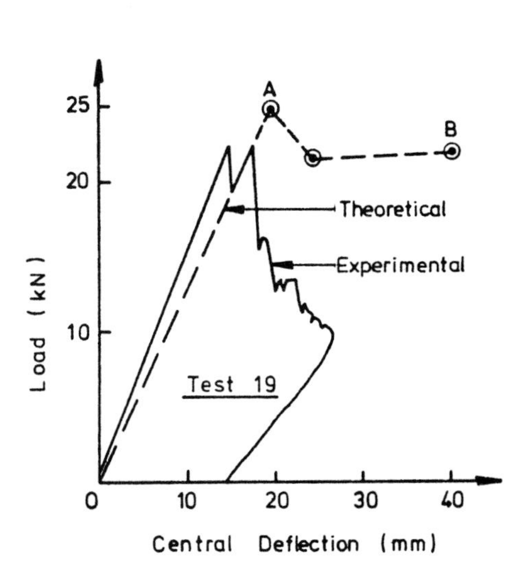

Fig 4 Result of test and analysis for truss 19.

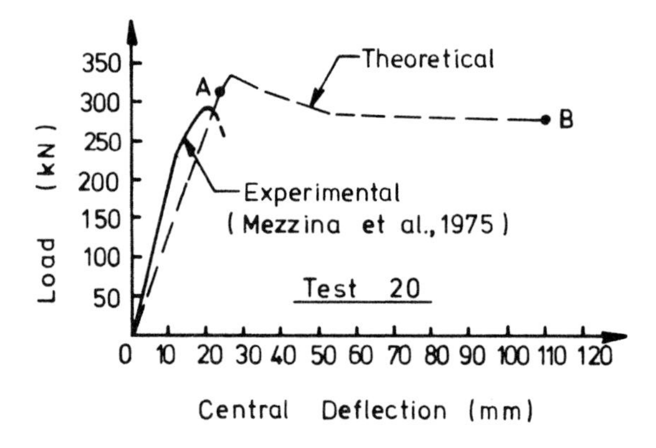

Fig 5 Result of test and analysis for truss 20 (Mezzina et al).

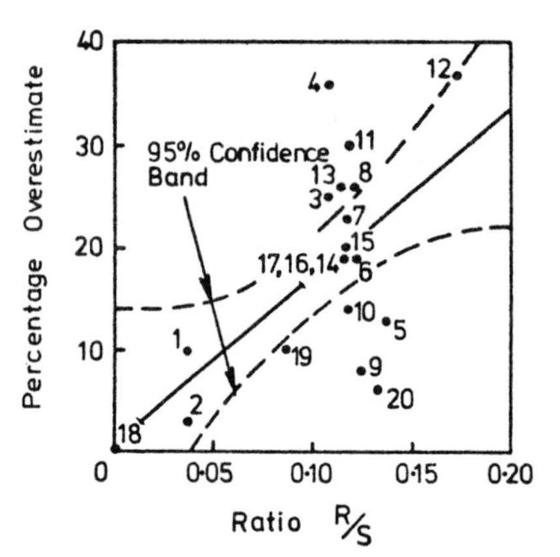

Fig 6 Regression line and 95% confidence limits for all truss tests, based on Point A analyses.

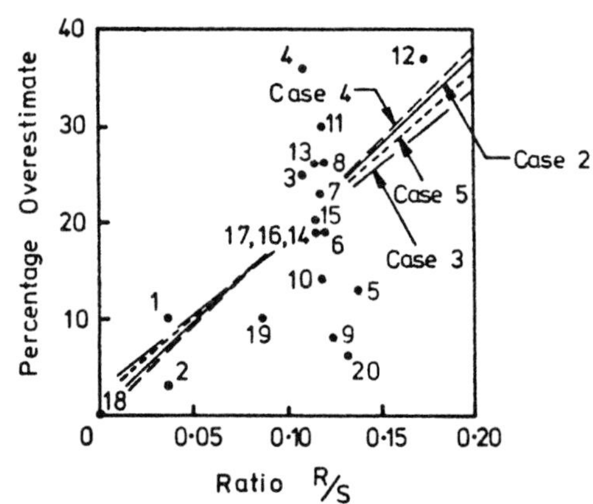

Fig 7 Regression lines after excluding various truss results, based on Point A analyses.

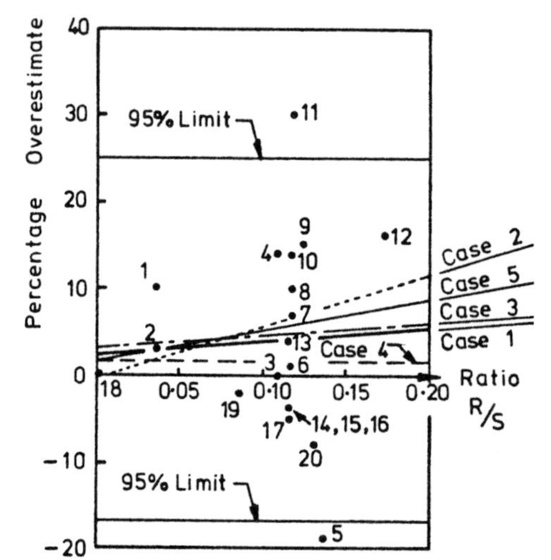

Fig 8 Regression lines after excluding various truss results, based on Point B analyses.

FAILURE OF COMPLETE BUILDING STRUCTURES

K Majid

Department of Civil and Structural Engineering, University College, Cardiff

T Celik

Department of Civil Engineering, Karadeniz Technical University, Trabzon, Turkey

Synopsis

In this paper the elastic-plastic method of analysis of frames is extended to cover the case of complete building structures consisting of a number of plane frames together with a grillage of floor slabs and shear walls. Particular attention is given to the case of non-proportional loading of the frames and the buckling of the slabs. The latter is dealt with in a manner proposed by Horne.

Notation

$\underline{a}$ vector of horizontal deflections of the shear walls.

$\underline{F}$ influence coefficient matrix for the frames.

$\underline{f}$ the vector of loads transmitted to the frames.

$\underline{f}_n$ the new forces transmitted to the frames.

$\underline{G}$ influence coefficient matrix for the grillage.

$\underline{g}$ the vector of loads transmitted to the grillage.

k stiffness.

M, M_{cr} moment and elastic critical moment.

$\underline{P}$ applied wind load vector.

$\underline{V}$ vector of vertical loads.

$\underline{w}$ the vector of loads applied to the shear walls.

λ_{cri} the load factor at the discontinuous stage i.

$\delta\lambda$ an increment of λ.

ϕ rotational restraining rate.

Θ rotation.

Other symbols are defined in the text.

Introduction

The plastic theory (1,2 etc) suggested that frames loaded proportionally collapsed by the development of a mechanism. Later this was developed to cope with the formation of hinges, one at a time, leading to a premechanism failure (3,4 etc). It was considered that the axial load in the members reduced the carrying capacity of a frame. This approach also considered proportional loading.

Here, the elasto-plastic method is developed to deal with complete structures. This will reveal that each individual frame will not sustain the loads proportionally. However, the approach of tracing the load-deflection history of a structure by changing its stiffness as proposed by the pioneers of the elastic-plastic theory, including Horne, is kept intact.

A simple complete building structure consists of a grillage system of floor slabs and parallel shear walls, together with parallel multi-storey skeletal frames as shown in figure 1. Grillages may be manufactured out of reinforced concrete. However, for testing purposes, these may be made out of perspex (5). Thus they may buckle during testing. To cope with this, methods suggested by Horne (6) and by Trahair (7) will be used. The outline of the method is given in the Appendix.

1. *NON-PROPORTIONAL LOADING*

The analysis of a structure reveals that because the stiffness of its components change differently, the amount of external wind loads transmitted to any frame changes non-proportionally. The applied wind load vector $\underline{P}$ at the junctions is divided into a vector $\underline{f}$ transmitted to the frames and another vector $\underline{g}$ transmitted to the grillage, thus

$$\underline{P} = \underline{f} + \underline{g} \tag{1}$$

If the influence coefficient matrix $\underline{G}$ expresses the horizontal displacements of the grillage at each junction due to unit loads applied one at a time, at various junctions, then the vector of horizontal deflections is given by $\underline{G}\ \underline{g} + \underline{a}$. Here $\underline{a}$ gives the horizontal deflections of the junctions when loads $\underline{w}$, shown in figure 1, are applied to the shear walls. Similarly the deflections at the floor levels in the frames are given by $\underline{F}\ \underline{f}$. The influence coefficient matrix $\underline{F}$ gives the horizontal deflections of the frames due to unit loads, again applied one at a time, at the various junctions (8,9). Compatibility gives :

$$\underline{G}\ \underline{g} + \underline{a} = \underline{F}\ \underline{f} \tag{2}$$

Hence $$\underline{f} = (\underline{G} + \underline{F})^{-1}(\underline{G}\ \underline{P} + \underline{a}) \tag{3}$$

Thus at each stage the loads transmitted to the grillage and to each frame can be calculated. The loads $\underline{V}$ are assumed to be carried by the frames as demonstrated by Creasy (10). These and the horizontal loads $\underline{P}$ are assumed to increase proportionally. As the loads are increased, a total of k-1 discontinuous changes (such as plastic hinges) can develop at load factors $\lambda_{cr1}, \lambda_{cr2}, \ldots\ldots, \lambda_{cr(k-1)}$.

The next, i.e. the kth change, may take place at the unknown load factor λ_{crk}. If the initial vertical load acting at a joint m is $\underline{V}_m$, then the element of the load vector corresponding to this load is $\lambda_{crk}\, V_m$. The horizontal wind load on the shear wall also increases proportionally and at stage k the load vector acting on these walls is $\lambda_{crk}\, \underline{W}$.

Everytime equation (2.A), see Appendix 1, are solved the forces $\underline{f}_n$ transmitted to a frame are calculated. For horizontal forces f_1, f_2, .., ..., f_{k-1}, f_k transmitted to junction ij of a frame, at stages 1 to k, the total load acting is :

$$\ell_{ijk} = (f_1 - f_2)\,\lambda_{cr1} + \ldots\ldots + (f_{k-1} - f_k)\,\lambda_{cr(k-1} + f_k\,\lambda_{crk} \qquad (4)$$

In equation (4) f_k is known but λ_{crk} has to be found iteratively. To do this λ_{crk} is replaced by a specified value $\delta\lambda$ and the load ℓ^*_{ijk} acting at ij is calculated from equation (4). The usual procedure (9) is then used to find λ_{crk}. Because the loading is non-proportional, the possibility of plastic hinges in steel frames becoming inactive increases. These are treated in the manner proposed by Davies (11).

2. *LATERAL BUCKLING OF SLABS*

A number of structures were tested by Onen (5) in which the grillage was made out of perspex sheets that buckled. These can be analysed by combining the methods proposed by Horne (6) and Trahair (7). The slab shown in Figure 2 is simply supported about its minor axis and is subject to unequal end moments M and M'. Horne dealt with the case of an I-beam loaded and supported similarly. Neglecting the flanges of such a beam, Horne's method, applied to this slab, gives the elastic critical moment M_{cr} as :

$$M_{cr}^2 = (F + F'\gamma)\, M_E^2 \qquad (5)$$

where M_E is the critical moment for equal and opposite end moments given by (12) :

$$M_E = \pi\sqrt{EI\ GJ}/L \qquad (6)$$

where GJ is the torsional rigidity and I is the second moment of area about the minor axes. In equation (5) γ is the warping rigidity factor which can be neglected in a narrow rectangular section. Equations (5) and (6) give :

$$M_{cr} = \pi\sqrt{F}\ .\ \sqrt{EI\ GJ}/L \qquad (7)$$

For equal and opposite moments for a slab with end restraints, Trahair's method gives the critical moment as :

$$|M_{cr}| = \delta\sqrt{EI\ GJ}/L \qquad (8)$$

where δ is a function of the rotational restraining rate ϕ, given by:

$$\phi = k\Theta/(k\Theta + 2\ EI/L) \qquad (9)$$

Here k is the rotational stiffness of the supports. For a panel connected to others it is assumed that the supporting panels are pinned at their far ends as in Figure 3. For the slab uv, the rotational restraints at u and v are :

$$k\,\Theta_u = \frac{3EI_1}{L_1} + \frac{3EI_2}{L_2} + \frac{3EI_3}{L_3} \tag{10}$$

$$k\,\Theta_v = \frac{3EI_4}{L_4} + \frac{3EI_5}{L_5} \tag{11}$$

In a beam restrained elastically while subjected to unequal moments, equations (7) and (8) combined give :

$$M = \delta\sqrt{F}\,\sqrt{EI\,GJ}/L \tag{12}$$

When $\delta = 6.34$ the end restraints are infinite. When the slab is subject to equal and opposite moments with no end restraints, equation (12) reduces to (7) with $\delta = \pi$. Finally when $F = 1$ and the slab is subject to equal and opposite moments, equation (12) reduced to equation (8). Values of F and δ can be represented as :

$$\sqrt{F} = 1.765-1.015\xi+0.245\xi^2-0.082\xi^3-0.02\xi^4+0.318\xi^5-0.212\xi^6 \tag{13}$$

$$\delta = 3.14+1.542\phi-3.127\phi^2+21.681\phi^3-53.97\phi^4+69.65\phi^5-41.912\phi^6+9.337\phi^7 \tag{14}$$

where $\xi = M'\ M$. Values of F versus ξ where tabulated by Horne. For a rectangular thin plate Patel et al. (13) gave :

$$GJ = (1 - 0.6302\ t/d)\ dt^3\ G/3 \tag{15}$$

where t is the thickness and d is the depth of the plate. When predicting the panel buckling load M, M' and ξ are unknown. To overcome this it was assumed that between one discontinuity stage and the next the ratio ξ does not change. This is reasonable because the interval between one critical stage and the next is small. Furthermore Horne showed that $\sqrt{F}$ is not very sensitive to changes in ξ. Perspex plates can also crack because it is brittle. This is dealt with in the manner given by Celik (14).

3. *COMPARISON WITH EXPERIMENTAL RESULTS*

Sixteen structures with reinforced concrete grillages and seven with perspex grillages were tested by Celik (14) and Onen (5). Six of Celik's structure had intermediate steel frames, four had no frames and the rest had reinforced concrete frames. Onen's perspex structures all had intermediate steel frames. Details of these together with the experimental procedures are given elsewhere (12,15). The largest discrepancy between the theoretical and the experimental failure load P of all Celik's structure was *16.51%*. In the analyses, the reinforced concrete frames and also the grillages were represented with bilinear or trilinear moment-curvature diagrams. The manner of representing the concrete behaviour did not change the results significantly. The largest discrepancy between the present theoretical results and the experimental results obtained by Onen was *24.5%*. The behaviour of one of the structures is reported here. In Figure 4 the load deflection graphs obtained by four different methods are shown for Onen's structure 1. Curve 1 is obtained by the present method. Curve 2 is experimental while curve 3 was obtained theoretically by Onen. Finally curve 4 shows the behaviour of one of the bare frames.

The present method predicts that the first floor slabs spanning F_1, F_2 and F_3 would buckle first at a factor of *0.96*. In fact Onen reported that experimentally, at a load factor of *0.98*, these slabs buckled and fractured. This is marked as *GB1* on the theoretical curve 1 and as *GB, GC* on curve 2. As the analysis continued 13 plastic hinges developed in the frames but collapse happened when the second floor slabs spanning frames 1 to 3 buckled. Figures 5, 6 and 7 show the portion of the wind load transmitted to the frames and to the grillage at various stages of loading. These indicate that neither the frames nor the grillage were sustaining the loads proportionally.

4. *ANALYSES OF A SIX STOREY STRUCTURE*

A comparison of the results indicated that the computer program produced, can trace the behaviour of structures with sufficient accuracy. The program was therefore used to analyse a number of full scale complete practical structures. One of these is reported here. The six storey structure shown in Figure 8 has one *304.8 mm* thick reinforced concrete shear wall, six *152.4 mm* thick slabs and two steel frames, with two unequal bays. The floor loads are shown in Figure 9 together with the analytical load-deflection diagrams. This structure was analysed twice. The first one neglected the effect of composite action between the floor slabs and the steel beams. The second included this effect in the manner given in reference 15. The results for both analyses are shown in Figure 9. In the first analysis the structure collapsed at a load factor of *1.778* with 34 hinges. When hinge 9 developed, hinges 1, 2 and 3 became inactive. Hinge 8 also became inactive when hinge 11 developed. In the analysis considering the effect of composite action 8 hinges developed in the frames, followed by a diagonal crack at a factor of *1.784* in the second storey of the shear wall. This was followed by the formation of hinges 10 to 16. At this stage hinges 1, 2, 3, 8, 11 and 13 all became inactive. With hinge 17 at $\lambda = 1.894$ hinges 7 and 12 became inactive. Hinge 12, however, was soon reactivated at $\lambda = 1.897$. Failure took place with hinge 19 at $\lambda_F = 1.974$, which is 10% higher than that obtained neglecting the effect of composite action.

5. *CONCLUSIONS*

The results obtained experimentally and theoretically indicate that the failure of complete structures is due to a combined effect of crack developments, hinge formations and frame instability. This cannot be simplified to that of a plane frame analysis.

References

1. Baker, J.F., Horne, M.R. and Heyman, J. 'The steel skeleton'. Vol. II, Cambridge, 1956.
2. Horne, M.R. 'Plastic theory of structures'.
3. Horne, M.R. Proc.Royal Soc. A., Vol. 274, 1963.
4. Jennings, A. and Majid, K.I. The Struct.Engr., Vol. 43, Dec. 1965.
5. Onen, Y.H. Ph.D. Thesis, University of Aston, 1973.
6. Horne, M.R. 'The flexural-torsional buckling of members of symmetrical I-sections under combined thrust and unequal terminal moments'. Quart.Jour.of App.Maths., VII, Pt. 4, 1954.
7. Trahair, N.S. 'Stability of I.beams with elastic end restraints'. The Journal of the Institution of Engineers, Australia, June 1965.
8. Majid, K.I. and Croxton, P.C.L. 'Wind analysis of complete building structures by influence coefficients'. Proc.Inst.Civ. Engrs., Vol. 47, Oct. 1970.
9. Majid, K.I. and Onen, Y.H. 'The elasto-plastic failure load analysis of complete building structures'. Proc.Inst.Civ.Engrs., Vol. 55, Sept. 1973.
10. Creasy, L.R. 'Stability of modern buildings'. The Struct.Engr., No. 1, Vol. 50, Jan. 1972.
11. Davies, J.M. 'The response of frameworks to static and variable repeated loading in the elastic-plastic range'. The Struct.Engr., No. 8, Vol. 44, Aug. 1966.
12. Timoshenko, S.P. 'Theory of elastic stability'. McGraw-Hill, 1936.
13. Venkataraman, B. and Patel, S.A. 'Structural mechanics'. McGraw-Hill, 1970.
14. Celik, T. 'Elastic-plastic analysis of complete structures with shear walls and frames'. Ph.D. Thesis, University of Aston, 1977.
15. Majid, K.I. 'The effect of composite action on the elasto-plastic analysis of complete building structures'. Proc.Finite Element Methods in Engineering, New South Wales, 1974.
16. Majid, K.I. and Celik, T. 'Bending, shear and torsion in deep rectangular reinforced concrete panels'. Proc.Inst.Civ.Engrs., Part 2, Sept. 1979.

APPENDIX

Progress Towards Failure

1. Matrix displacement method is used to calculate the matrices $\underline{G}$ and $\underline{F}$. Equations (3) then (1) are used to calculate the force vectors $\underline{f}$ on each frame and the vector $\underline{g}$ on the grillage. If the external loads are now multiplied by a load factor λ, the total vertical load at a joint in the frame will become λV. The external initial wind load P_{ij} is divided between the frame i and the slab at floor level j, thus

$$P_{ij} = f_{ij} + g_{ij} \qquad (1.A)$$

When the load factor is λ, junction ij of the frame is subject to a horizontal load f_{ij} while the grillage is subject to g_{ij}.

2a. Each steel frame is now analysed elasto-plastically (4) to obtain the load factor λ at which a plastic hinge develops.

2b. Each R.C. frame is analysed to obtain the factor λ at which a discontinuity is reached on the moment-curvature curve of any member. A tri-linear idealization (14) is used to represent the M-C relationship. As the loads $\underline{f}$ are different from frame to frame, the load factor for the formation of a hinge is also different from one frame to another. The lowest load factor λ_{FL} that may cause either a hinge in a steel member or a change in the EI of an R.C. section, is selected as the one that causes a discontinuity in the frames.

3. The grillage is analysed under factored loads $\lambda_G \underline{g}$ and $\lambda_G \underline{w}$. This is to calculate λ_{GC} at which a discontinuity occurs in the M-C curve of one of the panels. The lowest factor λ_{GB} at which a panel buckles is also calculated.

4. The next discontinuous stage is obtained by calculating λ_{cr}. This is the lowest λ_{FL}, λ_{GC} or λ_{GB}. At the kth discontinuity this load factor is λ_{crk}.

5. If $\lambda_{cr} = \lambda_{FL}$ the frame in question alters basically. It becomes more flexible and incapable of sustaining its share of the transmitted loads. Part of these have therefore to be transferred to the slabs and transmitted to the other frames and the shear walls. If $\lambda_{cr} = \lambda_{GB}$ then one of the panels buckles. Finally if $\lambda_{cr} = \lambda_{GC}$ using flexural rigidity of the panel in question is reduced in its M-C curve. It is possible for a panel to fail as a deep beam under combined bending, shear and torsion (16). In that case the stiffness of that panel is made nominal.

6. Once λ_{cr} is determined, the member forces and deflections are calculated for this load factor.

7. The influence coefficient matrices $\underline{F}$ and $\underline{G}$ are thus changing continuously. At each new stage the new influence coefficient matrices $\underline{F}_n$ or $\underline{G}_n$ is prepared. The inverse matrix transformation (3) is carried out to calculate the new forces $\underline{f}_n$. Equations (3) are now in the form :

$$\underline{f}_n = (\underline{G}_b + \underline{F}_n)^{-1} \; (\underline{G}_n \underline{P} + \underline{a}_n) \qquad (2.A)$$

The new matrix $\underline{G}_n$ is reconstructed only when the stiffness of a panel is altered while $\underline{F}_n$ is reconstructed after every alteration in the stiffness of a frame.

8. Once each λ_{cr} is calculated the search is continued for the next one and the process is terminated when the sway deflection at the structure increases considerably.

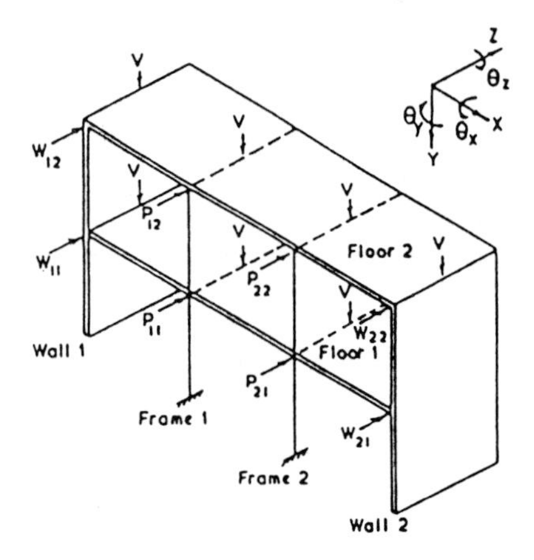

Fig 1 A complete structure - loading and sign convention

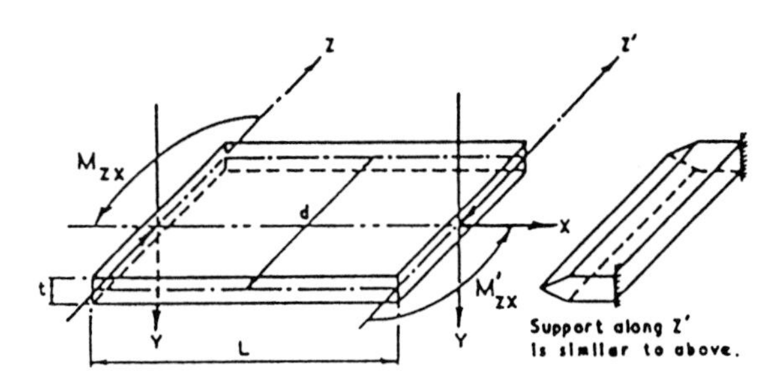

Fig 2 Panel with no lateral end restraints and subject to unequal end moments

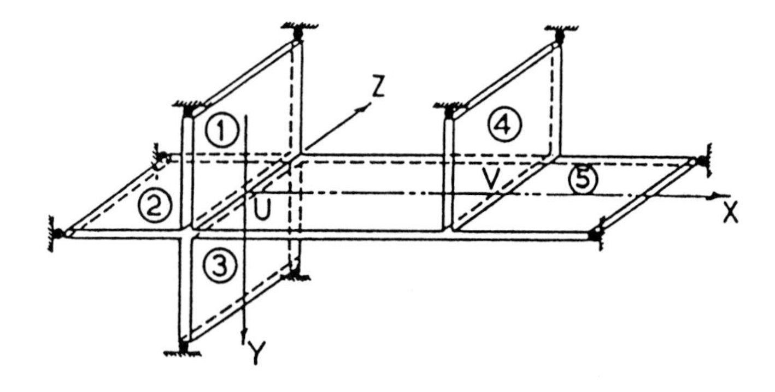

Fig 3 A grillage member connected to the joints I & J - the assumption of simply supported joining members at far ends

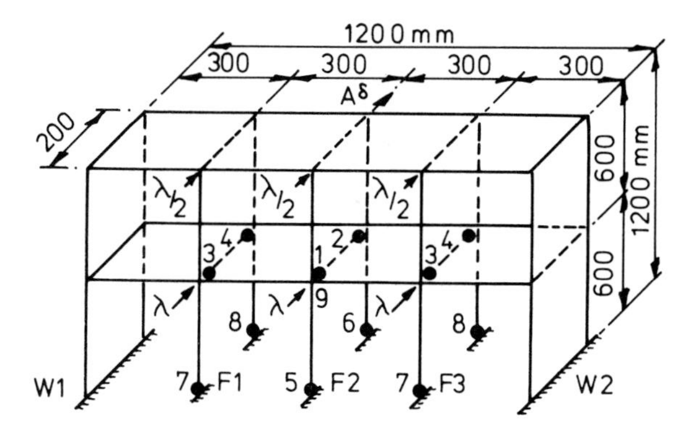

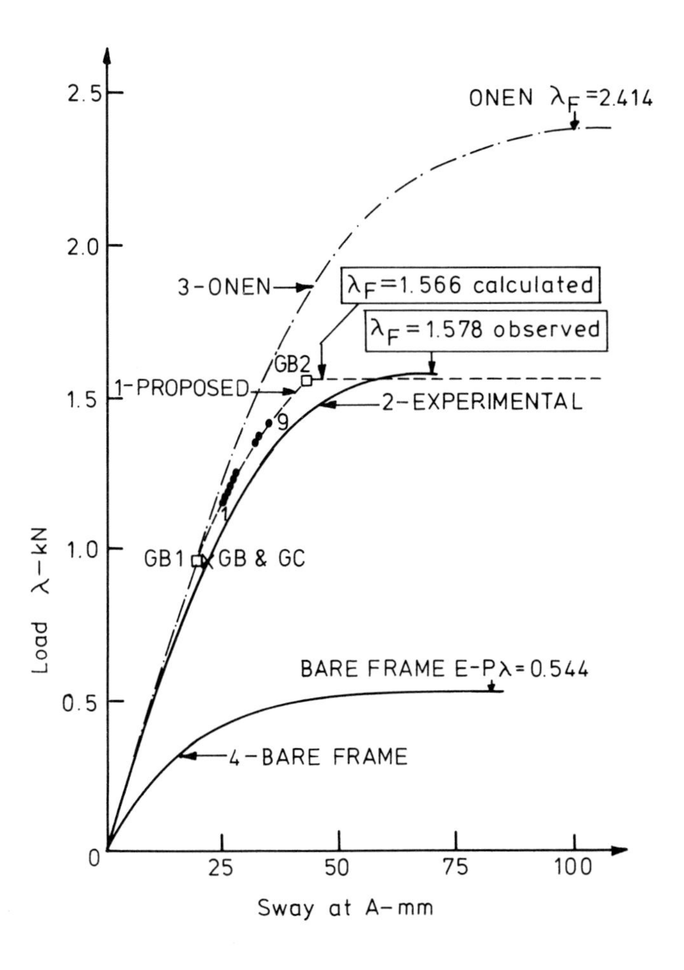

Fig 4 Load deflection graphs for Onen's first two storey structure

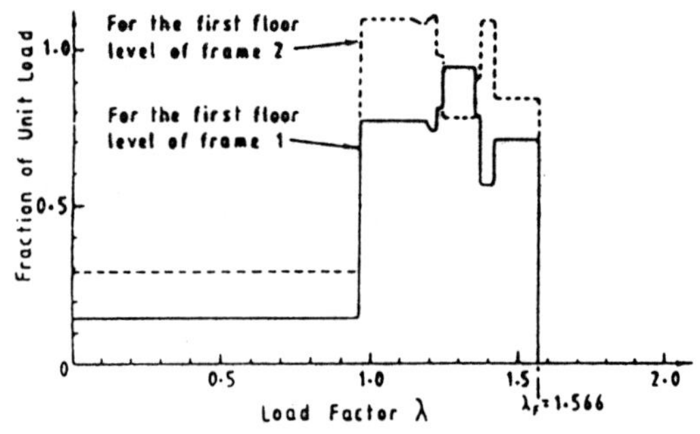

Fig 5 Variation of the fraction of a unit load transmitted to frames 1 and 2 of Onen's first two storey structure

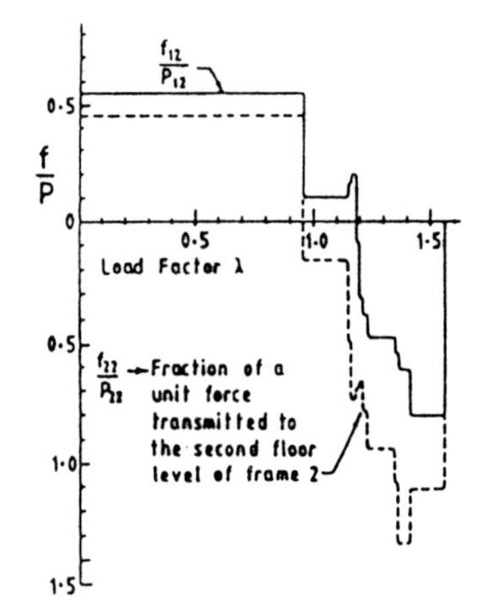

Fig 6 Variation of loads transmitted to the second floor levels of the frames

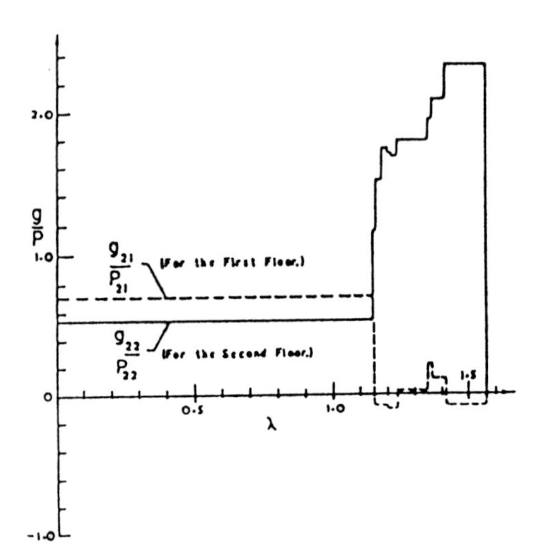

Fig 7 Variation of loads transmitted to the grillage

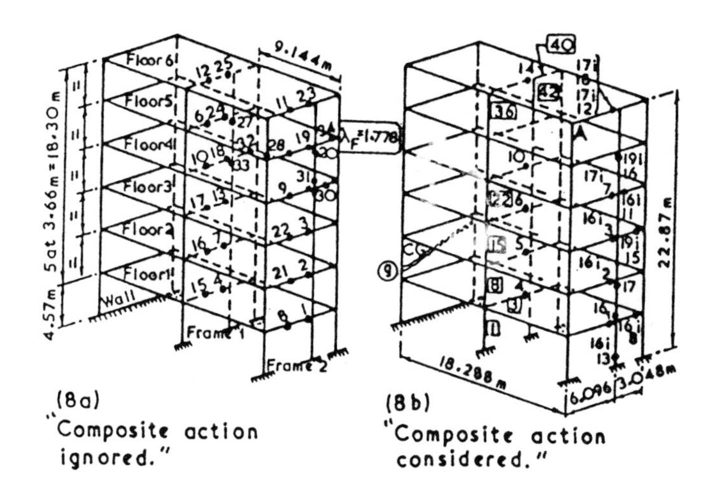

Fig 8 Six storey structure

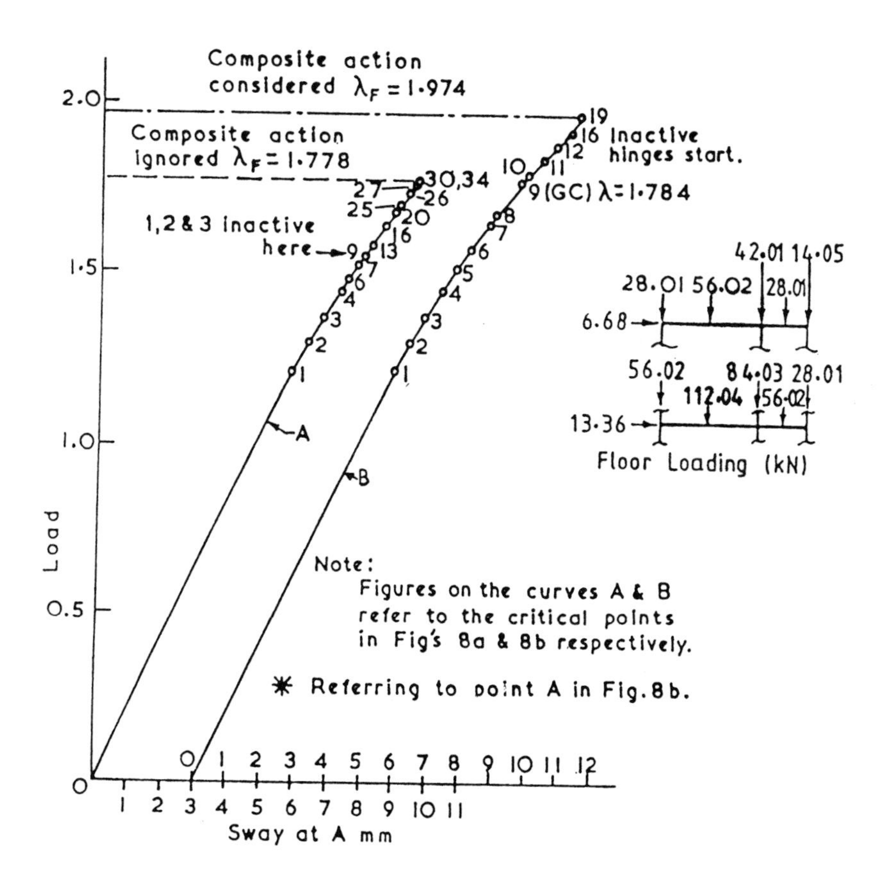

Fig 9 Load deflection diagrams

THE SEARCH FOR A UNIFIED COLUMN DESIGN

R.H. Wood

University of Warwick and W.J. Marshall (Consulting Engrs.) London

Synopsis

A short history of the main features of various attempts at column design in the elasto-plastic range is given. Starting with Horne's original classification of end conditions and his two particular methods, special attention is given to various proposals for simplificaton for codes of practice, and the part which the variable-stiffness concept could play in future unification.

Notation

A	area
C	elastic critical load factor N_{cr}/N_E
d	depth of section
e	eccentricity
e^*	eccentricity allowing for initial stresses
E	Young's modulus
I	moment of inertia
I_p	polar moment of inertia
GJ	torsional rigidity
k	Hardy Cross nominal distribution coefficient
K	stiffness I/L
ℓ	effective length
L	length
M	moment
M_p	plastic moment
M_{pN}	reduced for axial load
N	axial load

$\bar{N}$ axial load/squash load = $N/A\sigma_Y$

N_E Euler load

R reduction factor due to plastic zones

t flange thickness

T torsional constant (Horne)

Z section modulus

Z_p plastic modulus

α equivalent single curvature moment ratio

β ratio of end moments: +1(S/C) : -1(D/C)

μ magnification factor for end moments due to axial load

σ stress

σ_a axial stress

σ_Y yield stress

λ load factor

λ_p rigid plastic

λ_{cr} elastic critical

κ curvature

Suffices

a	axial	E	Euler
b	beam	R	reduced
c	column	Y	yield
cr	critical		
e	elastic		
p	plastic		
u, l	upper and lower		
x, y	axes		

Other symbols as introduced in text.

1. *INTRODUCTION TO THE PROBLEM OF COLUMN DESIGN*

Few engineers and research workers would dispute that column design is probably the most difficult in the whole of structural engineering. Non linearities (and even discontinuities) are involved from yielding, buckling (local, member and frame instability), properties of joints (semi-rigid and rigid), limited ductility, and, additionally with cased or reinforced concrete columns, cracking and crushing. It is impossible for the Author to do justice to sixty years of research in a short paper, but it is hoped that the important innovations of various authors can be highlighted.

Nowadays a tour-de-force on computers is possible but it is only half the problem. Equally important, and more demanding, is the *simplification* for Codes of Practice, for we face the relentless competition of 'simple' design, with its downright intuitive handling of 'pin' joints. In what follows, simple presentation of research must never be forgotten.

The author was fortunate (as D.S.I.R. representative) to be involved in the first active committee after the war, the so-called 'F.E.1' committee, run jointly by the British Welding Research Association and Cambridge University, chaired by Prof.J.F.Baker. It was he who seized upon the column problem as being the key to progress, and to Dr. Horne fell the task of presenting the first column designs in the plastic range: always remembering that the problem of frame instability loomed ahead, with Merchant and the Author heavily involved on that side. Because of this the first designs of columns were intended for sway-braced frames or portal frames.

2. *HORNE'S PIONEER COLUMN DESIGNS FOR PLASTIC FRAME DESIGN*

2.1 $P_x P_y$ *design*

Although it was realised that end-conditions for columns graduated from elastic to plastic, nevertheless Horne found it convenient (Fig 1) to separate the adjoining *beam* conditions into plastic (P), elastic (E), or zero-moment (O–pinned); of the nine permutations - more if the top is not the same as the bottom - Horne singled out the three most important cases. The $P_x P_y$ case was the one requiring an urgent general solution (1,2). The impressive feature of this design method was the first time that proper control of torsional stability was achieved; and the derivation of an effective Single Curvature (S/C) moment, αM_x, with terminal major axis moments M_x and βM_x, such that the combined buckling-plus-torsion criterion is now

$$\frac{N}{N_{Ey}} + \left(\frac{\alpha M_x}{M_E}\right)^2 = 1 \qquad (1)$$

the elastic critical moment for a beam being M_E, and where, for Codes of Practice, α may be simplified to

$$\begin{aligned} \alpha &= 0.6 + 0.4\beta \qquad \text{when } \beta > -0.5 \\ \alpha &= 0.4 \qquad \text{when } \beta < -0.5 \end{aligned} \qquad (2)$$

It should carefully be noted that plastic beams (P_y) mean that the effective length ratio, ℓ/L, must be unity; thus

$$M_E = \frac{\pi}{L}\sqrt{\left\{EI_y GJ\left[1 - \frac{\sigma_a I_p}{GJ} + \frac{N_E(d-t)^2}{4GJ}\right]\right\}} \qquad (3)$$

In this first column design, Horne balanced the negative Wagner term against the positive warping term, omitting both for simplicity. Also fron eqn (1) it was shown that if the axial stress σ_a is increased to

$$\sigma_a' = \sigma_a + \frac{(\alpha M/Z_x)^2}{T} \qquad (4)$$

where T is Horne's torsional constant ($AGJa_x^2/I_x^2$; $a_x = d/2$ with symmetvrical sections), then σ_a' must be used to give an increased value of $(N/N_{Ey})'$ for minor axis buckling for use with the well known magnification factor,

$$\mu = 1/(1 - N/N_E) \qquad \text{generally} \qquad (5)$$

This same magnification factor operated on the stresses due to initial curvature. Less obvious was Horne's use of the equivalent S/C moment factor α, originally for torsion, also for minor axis bending without torsion, but this he justified by trials which showed small errors on the unsafe side, and tolerable errors on the safe side. This device has since been used in many Codes of Practice. The final basic test was of the form

$$\sigma_a + \mu_x \sigma_x + \mu_y \sigma_y < \sigma \qquad (6)$$

where $\sigma_x = \alpha_x M_x / Z_x$; $\sigma_y = \alpha_y M_y / Z_y$

and σ is a reduced value of σ_Y on account of initial curvature.

The use of α_x and α_y enforces a separate check for yield at the ends of the stanchion. Horne's original symbols have here been changed to show better the relation with future codes of practice. It is noted that this is strict elastic design, which later (3) appeared under the somewhat misleading title of *Plastic design of columns*. However, this gives maximum protection to constancy of yield moments in a mechanism of collapse. Also it was easy (1) to produce charts for graphical evaluation of eqn (6).

2.2 $P_x O_y$ design

If the troublesome minor-axis moments could be removed then something less conservative then eqn (6) could be attempted. Thus came Horne's second column design (2,4) which could allow even full plastic hinges (at one end only; $\beta = -1$ excepted). The vital factor in this theory is the reduction of EI_y, involved in both N_E and M_E in eqn (1), to $0.7EI_y$, to account for the fact that, if yielding is just reached near mid-height, then partial plastic zones are restricted to one end. (This is a particular early example of a generalised variable-stiffness technique later developed by the Author). Subsequently (4) the advent of Universal Column sections made it advantageous to take account of the positive warping term in eqn (3), and to issue very simple design charts (3), typified in Fig 2, for a succession of T values. The condition of no-yield-at-mid-height plus full-plasticity-at-one-end prevents the attainment of $\beta = +1$, symmetrical S/C, even at $N=0$. Members of the FE1 committee noticed that this clashed with Neal's work (5) on laterally unsupported beams. By making the column curves fit Neal's results at $N=0$, and noting that tests had shown that the squash load could be reached if $N/N_{Ey} < 0.1$ approximately, Horne deduced a limited-stockiness line, below which smaller L/r_y values could tolerate any value of β at full plasticity.

Designers could go right ahead with Portal Frame design, but for multi-storey frames, 'patterned' loading presented a problem of saying what the end moments would actually be (Fig 3). The worst single-curvature component of live-loading can produce column moments which are easily evaluated by simple moment distribution, putting $K_b = 0$ for all *plastic* beams, but it leaves some beams elastic. Thus the condition $E_x E_y$ is nearer the truth, particularly if it is remembered that floors on both sides of one of the major beams have to be loaded, and this pattern then leaves the minor-axis beams elastic. Another feature is that it does not always pay to use P_xO_y. Apart from the problem of S/C, any trial increase of section automatically assumes an increase in acting plastic moment: if the conditions approach E_xO_y, possible in Fig 3(a), then the $0.7EI_y$ assumption might favour a switch to Horne's fully-elastic column design, in spite of its label P_xP_y.

In the correspondence (6) on Horne's first method, Campus and Massonnet indicated that they had independently reached an alternative equivalent S/C moment formula: and that....."*their buckling formula.... was simpler than .....[Horne's] formula*". Horne replied that....."*their approach was.....a purely empirical interaction curve*". Horne's analytical approach did give ...."*better agreement with [their] test results, at the cost of somewhat greater complexity*". Their 'empirical' formula was of the form (for x-axis loading only):

$$\frac{\sigma_a}{\sigma_c} + \frac{A\alpha_x M_x}{\sigma_Y Z_x(1 - N/N_{Ex})} = 1 \tag{7}$$

where σ_c = allowable stress for axially loaded columns
A = sectional constant for wide-flanges

As we shall see, this was the start of more extensive interaction formula in later codes.

3. *THE JOINT COMMITTEE ON RIGID MULTI-STOREY FRAMES (1957-1971)*

This committee (I.Struct.E. and I.Welding) was set up to recommend designs for multi-storey frames. It tackled braced frames only, and its final report (7) has well documented clauses for both plastically and elastically designed beams, with unique tables for the permissible span/depth ratios, to suit deflection/span = 1/360 even in the presence of elastic end-restraint, prepared by Horne and Wood. Once again the column problem was paramount. From the start experienced designers on the committee had insisted that effective lengths down to $0.7L$ or less must be possible if the minor-axis beams were designed to remain elastic, if only to compete with 'simple' design. Without realising it fully, they had plunged themselves into the P_xE_y problem, what Horne (1) had described as ....."*of great potential importance in the evolution of a design method....[where] ....the minor-axis beams....stabilise the stanchions ....[but].... this case is intractable analytically*". The only known instant graphical way of obtaining maximum column moments (stability functions and all) was by Wood's nomograms (8,9). These frightened the designers, but led to a slightly less accurate method based on a maximum-moment-multiplier (μ) interaction-plot between the end-moments designers would calculate, ignoring N/N_E , and the value of N/N_{cr}, Figs 4(a)(b). Since the k-values are Hardy Cross distribution coefficients this is purely graphical design. Other charts were given given for top storeys and for stresses due to initial curvature σ_{ic}. The design criterion was simply

$$\sigma_a + M_x/Z_x + \mu M_y/Z_y + \sigma_{ic} < \sigma_Y \tag{8}$$

This method is more accurate than the combined use of α and the multiplier $1/(1-N/N_{cr})$. Moreover since the D/C magnification factor only slightly exceeds unity, this method favours the partial D/C moments which often occur, particularly as the *committee deliberately chose to ignore greater D/C moments if the patterned loading were switched around to suit. Consequently this was partial-plastic design with restraining beams.* It was certainly the lightest column design then known - the nearest to $P_x E_y$ yet achieved.

But it was not enough. Calibration exercises showed insufficient total saving in weight versus 'simple' design, and it now becomes important to see why. 'Simple' column designs are based on

$$\frac{\sigma_a}{\sigma_c} + \frac{(V_x e_{ox}/Z_x + V_y e_{oy}/Z_y)}{\sigma_{bc}} < 1 \qquad (9)$$

where σ_c = permissible column stress based on an effective length ℓ.

σ_{bc} = permissible *beam* stress, for torsional buckling.

e = nominal eccentricity $\nless$ $d/2 + 100$ mm.

V^o = out-of-balance beam reaction.

Not only is there no moment magnification factor at all, nor even patterned loading, but all moments are treated as if D/C, and small moments at that; moreover this *pin-joint* method uses $\ell/L = 0.7$ with restraining beams, designed to be plastic with *pinned* ends!

Either

(a) there can not be much justification save for composite action,

or (b) rigid-frame designs deserve some reduction in design load factors by comparison,

but (c) rigid frame design must be super-competitive to survive.

4. *THE VARIABLE-STIFFNESS APPROACH TO $P_x E_y$ DESIGN*

The above situation was so serious that the author had already commenced a somewhat revolutionary approach whilst the Committee's report was in preparation. The basic trouble was obvious - too often an all-plastic or all-elastic approach - not enough partial plasticity. The author had noted Dr.A.R.Gent's observation that torsional buckling often starts when about half the section has been plasticized; also Horne's $0.7EI_y$ factor mentioned earlier; above all the gradual appearance of plastic hinges in frame instability cases, till collapse took place at the *deteriorated critical load* (9). The idea emerged of an overall continuous deterioration of EI_y (and EI_x), expressed as a variable $R.EI_y$, for example. If, intuitively, we write

$$R = \frac{\text{Stiffness Reduction factor}}{\text{due to moments predominantly}} \cdot \begin{matrix}\text{Stiffness reduction}\\ \text{factor due to axial}\\ \text{load predominantly}\end{matrix} \qquad (10)$$

then the true buckling load becomes

$$(N_{cr})_{det} = R.N_{cr} = R.\frac{N_{cr}}{N_E}.N_E = R.C_R.N_E \qquad (11)$$

where C_R is read directly from appropriate exact critical load charts like Fig 4(b), on the understanding that the Hardy Cross coefficients k_R are now calculated using $R.K_c$ instead of K_c. The choice of R is not critical, because if the column loses stiffness the degree of restraint, and therefore C_R, increases as long as end conditions refer to E_y. In view of the elastic beams remaining in Figs 3(a)(b), it was never expected that any buckling would take place first about the major-axis, and in all the BRS - BISRA tests (10) every column failed about the minor-axis (in spite of attempts to make them do otherwise). We invent a notation R_{xy}, R_{yy}, which is the reduction (overall) of EI_y due to X-X and Y-Y bending respectively.

The value of R_{xy} is decided by the necessity of preventing full plastic hinges forming at the ends, lest the elastic restraint be destroyed. Therefore we limit the allowable moment M_{ax} to the one-flange-plastic condition; thus approximately

$$M_{ax} = 1.08 Z_x \sigma_Y (1-\overline{N}) \qquad \text{and} \qquad \overline{N} = N/A_s \sigma_Y \tag{12}$$

With a small axial load, and with $\beta = 1$, if $M = M_{ax}$ the whole of the compression flange will go plastic (cf. Gent's suggestion) and shortly after that additional minor-axis plastic zones will certainly cause torsional buckling, suggesting $R_{xy} = 0.4$ (cf. Horne 0.7 with one plastic hinge only at one end). A corresponding value of 0.8 was suggested for D/C. For all possible moments a simple linear formula

$$R_{xy} = 1 - \frac{M_x}{M_{ax}}(0.4 + 0.2\beta) \tag{13}$$

is conservative, for only 8% reduction in M_x below M_{ax} would cause R_{xy} to revert to unity in reality.

As $\overline{N}$ increases the deterioration of R will not be linear, but will accelerate rapidly as $\overline{N}$ approaches unity. At first a correction factor $(1-\overline{N}^2)$ was suggested. However when $M_x = 0$, $R_{xy} = 1$, and with pinned ends $C_R = 1$ also, so we required a deterioration function on account of $\overline{N}$ only which will agree with simple design. It was found that $(1-\overline{N}^2)$ was satisfactory, so that generally

$$R_{xy} = (\text{value in eqn (13)}).(1-\overline{N}^2) \tag{14}$$

Later Dr.B.W.Young (cf.) made the ingenious suggestion that all his column curves A,B,C,D, for different initial stresses, could be inserted instead as polynomial functions of $\overline{N}$ (Appendix 5 of Ref 10). The factor $(1-\overline{N}^2)$ therefore allows for initial curvature and existing stresses.

It was noticed during the BRS - BISRA tests that buckling took the form of final *unwrapping* (2) about the minor axis, with a stationary value of $\overline{N}$ for a whole variety of end M_y values, passing through zero and changing sign. It is normally a waste of time calculating M_y, there being no exact value, the product $R.C_R$ remaining constant in eqn (11) during buckling. The bold step was taken

(a) to ignore M_y normally, since R_{xy} conservatively allows already for plastic zones associated with reversal of bending.

(b) for exceptional cases of heavy Y-Y loading, to investigate values of R_{yy}.

(c) to design Y-Y beams elastically, continuous but merely on column props.

The R_{yy} problem was solved by dimensional analysis, after fitting tangent beam lines to moment-rotation plots of columns obtained on the BRS differential analyser (9), also the Cambridge-Bristol model tests (2,9) and Imperial College tests (supplied by Dr. Gent)(10). An important parameter, against which to plot R_{yy}, took the form:

$$\frac{FEM}{\sigma_Y Z_y} \cdot \frac{K_{cy}}{\Sigma K_{by}} = \frac{(nominal\ beam\ rotation}{(nominal\ column\ rotation\ at\ yield)} \qquad (15)$$

where FEM is the out-of-balance moment of loads on Y-Y beams of stiffness ΣK_{by}. Design charts were given for R_{yy} for all types of Y-Y loading. Note that there is no actual calculation of M_y anywhere, except that values of (15), top and bottom, may be used to denote graduations between S/C and D/C.

The final design chart (entirely graphical), Fig 5, is based on

$$R = (least\ of\ R_{xy}\ or\ R_{yy}).(1 - \overline{N}^2) \qquad (16)$$

in combination with critical load charts such as Fig 4(b). L/r_y is obtained by multiplying $\surd[N_{collapse}/N_E]$ by [constant/$\surd(\overline{N})$] where the constant is 91.0, 76.9, 69.4 for grades 43, 50, 55 of steel. This method is partly intuitive (R values), partly exact (C_R values), but was extensively checked against the BRS/BISRA full-scale tests (10). It is the only column design so far which can consistently save steel (on the columns) versus competitive pin-joint design (eqn 9), and the latter has never been subjected to such controlled tests.

Any future computer-based improvements of this method must take careful note of the simple graphical checks on torsional instability (11) of *restrained continuous* columns. The Author first armed himself with a whole series of critical load charts, such as Fig 6, for combinations of M_x/M_E, β_x (from $+1$ to -1), N/N_{Ey}, obtained by Finite-Difference methods. Then by stiffness-distribution methods (12), the combined axial-torsional critical load was found for three continuous column lengths, each with its proper M_x values to suit Fig 3 and with beam restraints, showing the power of such methods. It was then discovered that Horne's eqn (1) could be generalised to modify eqn (11), with almost identical accuracy:

$$N_{collapse} = R.C_R.N_{Ey}[1 - (\frac{\alpha M_x/M_E}{\surd R.C_R})^2] \qquad (17)$$

and a slenderness-correction factor for torsion follows from the quantity in brackets, easily obtained from a straight-line nomogram (Fig 15, Ref 11). Beam restraint increases greatly the permissible L/r_y. (Note also that R affects the warping term directly if required). For most practical cases the reduction of L/r_y for torsion is only a few per cent.

5. *COLUMN DESIGN METHODS BASED ON INTERACTION FORMULAE*

5.1 *Young's Cambridge method*

There appeared, at about the same time, a fresh Cambridge attempt, headed by Young (13), Dwight et al., to improve on the $P_x P_y$ designs, by computer-integration in the presence of initial stresses, rated as important by the Lehigh school (14), and not included in Horne's method (cf.). A very useful start was the tabulation, Fig 7, of terminal column moments for major-axis failure only, giving obvious rise to a *stocky column concept*. This start avoids the jump from plastic-end-moments to elastic-end-moments implied in Horne's two methods, and has all gradations. Basically Young's approach depends on the intuitive product of four correction factors:

$$M_x = \sigma_Y Z_p (K_x . K_T . K_Y . K_H) \tag{18}$$

where K_x = factor for major-axis effects only (the ordinate of Fig 7).

K_T = torsional factor largely based on Dibley's work on beams.

K_Y = factor for minor axis buckling, even if $M_y \neq 0$.

K_H = factor for *biaxial* plastic hinge end-moments in the presence of both M_x and M_y.

The last two coefficients are more complicated, but the method will repay study for two reasons:

(a) as a more competitive design method than Horne's $P_x P_y$ method (but in the discussion Horne thought it would be unreliable for large minor axis moments).

(b) As a cut off for the Author's $P_x E_y$ designs, with very stocky columns which could generate the full plastic moment instead of M_{ax} (eqn 12).

This method does not appear to have been developed for restrained columns and is therefore not amenable to $P_x E_y$, or $E_x E_y$ conditions. Minor axis moments are uncertain if connected to elastic beams, in view of K_Y and K_H.

5.2 *The Continental (ECCS) and Lehigh designs*

These are taken together (14,15,16) since they seem to indicate very similar interaction formulae of the type:

$$\overline{N} + \left[\frac{1}{1-N/N_{crx}}\right]\frac{\Theta_x \beta_x M_x}{M_{px}} + \left[\frac{1}{1-N/N_{cry}}\right]\frac{\beta_y M_y + Ne^*_y}{M_{py}} \leqslant 1 \tag{19}$$

where $M_{px} = \sigma_Y Z_{px}$ and $M_{py} = \sigma_Y Z_{py}$

Θ = torsional constant and e^*_y = eccentricity for appropriate column curves (pinned with initial stresses). In addition the end moments are subject to a biaxial plastic interaction formula for M_x and M_y which is a simplified form of Young's K_H factor.

It may seem odd that elastic critical load multipliers can be combined with maximum plastic moments: and that such formulae are any criterion of buckling at all. However there is considerable justification about the strong axis (or weak axis if taken separately) by appeal to the Merchant-Rankine formula (see Fig 8).

Let the rigid-plastic collapse load be N_c for an applied (S/C) moment $N(e+e^*)$.

Then
$$N_c = \frac{M_{pN}}{e+e^*} \qquad (20)$$

and the M-R formula gives

$$\frac{N}{N_c} + \frac{N}{N_E} = 1 \qquad (21)$$

leading to

$$\frac{N(e+e^*)}{(1-N/N_E).M_{pN}} = 1 \qquad (22)$$

Provided M_{pN} is conservatively put at

$$M_{pN} = M_p(1-\overline{N}) \qquad (23)$$

then combining eqns (22) and (23) gives

$$\overline{N} + \frac{1}{(1-N/N_E)}.\frac{N(e+e^*)}{M_p} = 1 \qquad (24)$$

which is a 'justification' of (19) at least for P_xP_y conditions, i.e. identifying N_E with M_x or M_y separately. But for P_xE_y conditions there is a partial breakdown of the analogy. For major-axis bending, Fig 8(c), torsional buckling will be associated with N_{cry}, and provided Θ includes for this all is well (as it is in eqn 17): however, for minor axis bending, Fig 8(b), recalling that such beams are designed to stay elastic, the M-R formula has them *rigid-plastic* (for N_c) implying $N(e_y+e^*_y)$ virtually disappears for column design, leaving $\overline{N} + (N/N_{cry}) = 1.$ This is strongly reminiscent of the Author's variable-stiffness approach and explains why the ECCS formula is much more conservative for P_xE_y conditions than for P_xP_y or P_xO_y; and why it may be in difficulty competing with the over-optimistic 'simple' design (eqn 9). However eqn (19) is perhaps our best *generalised* method of the moment, since there are restrictions placed on the use of eqn (11), e.g. it is only valid for I sections which have the same unsupported length in X and Y directions, and where I_x is noticeably greater than I_y.

6. *POSSIBLE FUTURE DEVELOPMENTS OF THE VARIABLE STIFFNESS APPROACH*

6.1 *Frame instability*

This review would be incomplete without some mention of a frame-instability check. For a simplified check we fall back on the M-R formula working out λ_{cr} either by Horne's computerised deflection approach (17), or manually using the substitute Grinter Frame with stiffness distribution or by graphs (12). It is important never to use the M-R formula without first checking that sway/(storey height) does not exceed 1/300 at working conditions, which can be done graphically (18). An exceptionally good account of this whole graphical process is given by Massonnet and Save (16), with worked examples. Following the Author's suggestion a modified version (12) of the M-R formula is gaining in popularity, viz.

$$\lambda_{collapse} = \frac{\lambda_p}{(0.9+\lambda_p/\lambda_c)} \quad \text{provided} \quad 4 < \lambda_c/\lambda_p < 10 \qquad (26)$$

to allow for strain hardening and stray composite action. Dr.D.Anderson and the Author are investigating the use of a continuous function for determination of stiffness of plastic beams (with λ) to improve on the M-R formula.

6.2 *Indications from the latest reinforced concrete column research*

This involves cracking and limited-strain-crushing of the concrete in addition to yielding of steel, with two F values, and a stress-strain curve for concrete. The first problem was the design of R.C. columns failing either by attainment of ultimate moment or by buckling. Two basic requirements were performed on computers by integration:

(a) Moment rotation curves for $\beta = 1, 0, -1$.

(b) Charts of M-κ (curvature) properties on which the *cross-sectional* values of R were expressed as $R = M/EI\kappa$.

Such charts, Fig 9(a), are novel and very informative. Thus overall values of R for the whole column at any $\overline{N}$ value must lie between known limiting values, read off a horizontal line, and were evaluated as an approximate function of $\overline{N}$, shown in Fig 10(b) for 1% and 6% reinforcement. For patterned loading, Fig 3(a), the designer selects a trial value of $\overline{N}$, and, *by moment-distribution ignoring stability functions,* finds reduced end moments M, βM using $R.K_c$. Moment magnification factors can be given based on β and an increased value of $N/R.N_E$. Simpler procedures however were obtained by the author and Dr.I.May (University of Warwick) using a non-dimensional plot of M/M_{ult} versus $N/R.N_{cr}$, obtained graphically using $R.K_c$. A typical chart for $\overline{N} = 0.6$, Fig 10, shows acceptably small scatter versus exact computer results. These predict failure by ultimate moment somewhere in the column. A surprising feature is that charts for other $\overline{N}$ values are remarkably similar, and one such chart should suffice. They are conservative for in-plane buckling with still more more slender columns.

Evidently these extended ideas of variable stiffness would be even simpler if applied to steelwork. Already applied to both in-plane and out-of-plane buckling, with torsion, and moment-control without buckling, they may well hold the key to unification of all column designs; especially with recent revivals (20) of semi-rigid design where column design must indeed be competitive.

References

1. Horne,M.R. 'The stanchion problem in frame structures designed according to ultimate carrying capacity'. Proc.I.C.E., Vol 5, Pt 1, April 1956, 105-146.

2. Baker,J.F., Horne,M.R. and Heyman,J. 'The Steel Skeleton - Plastic Behaviour'. Vol II, C.U.P., 1956.

3. Horne,M.R. 'The plastic design of columns'. BCSA Publication No. 23, 1964.

4. Horne,M.R. 'Safe loads on I-section columns in structures designed by plastic theory'. Proc. I.C.E., Sept 1964, 137.

5. Neal,B.G. 'The lateral instability of yielded mild steel beams of rectangular cross section'. Phil. Trans (A), Vol 242, 1950, 197-242.

6. Correspondence on Ref 1., Proc I.C.E., Vol 5, Aug 1956, 558-571.

7. Joint Committee's 2nd Report 'Fully-rigid multi-storey welded steel frames'. I.Struct.E. and Welding Inst., 1971

8. Wood,R.H. 'A derivation of maximum stanchion moments in multi-storey frames by means of nomograms'. The Structural Engr, Vol 31, 1953, 316.

9. Wood,R.H. 'The stability of tall buildings' Proc. I.C.E. Vol 11, Sept 1958, 69-102.

10. Wood,R.H. 'A new approach to column design'. H.M.S.O. 1973

11. Wood,R.H. and Chakrabarti,B. 'Torsional buckling at the limit state of collapse in braced multi-storey I-section steel column design'. J.Strain Analysis, Vol 12, No.3, 1977, 233-250 (I.Mech.E).

12. Wood,R.H. 'Effective lengths of columns in multi-storey buildings'. The Structural Engr., Vol 52, 1974, (in 3 parts Nos. 7,8,9).

13. Young,B.W. 'Steel column design'. The Structural Engr, Vol 51, Sept 1973, 323-336

14. Driscoll,G.C., Beedle,L.S. et al. 'Plastic design of multi-storey frames'. Lecture Notes, Fritz Eng. Lab., Lehigh Report No. 273.20, 1965.

15. European Convention for Constructional Steelwork 'Recommendations for steel constructions Vol 1' Report ECCS-EG-76-IE, 1976

16. Massonnet,Ch. and Save,M. 'Calul plastique des constructions'. Vol 1, 3rd Edition - Nelissen B - Liege.

17. Horne,M.R. 'An approximate method of calculating elastic critical loads of multi-storey frames'. The Structural Engr, Vol 53, June 1975.

18. Wood,R.H. and Roberts,E.H. 'A graphical method of predicting sideway in the design of multi-storey buildings'. Proc. I.C.E., Pt.2, Vol 59, 1975, 353-372.

19. Wood,R.H. and Shaw,M.R. 'Developments in the variable-stiffness approach to reinforced concrete column design'. Mag. Conc. Res., Vol 31, Sept 1979, 127-141.

20. 'Joints in Structural Steelwork'. Proc.Int.Conf., Teesside Polytechnic, 1981, Pentech Press, London.

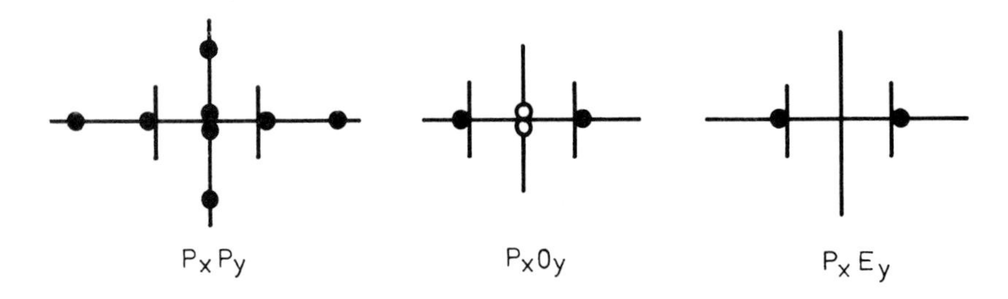

Fig 1 Horne's boundary conditions for column design

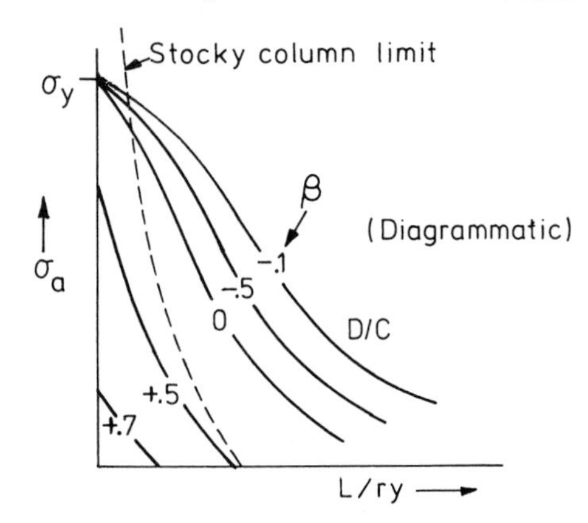

Fig 2 Horne's P_xO_y design: Plastic hinge at end only

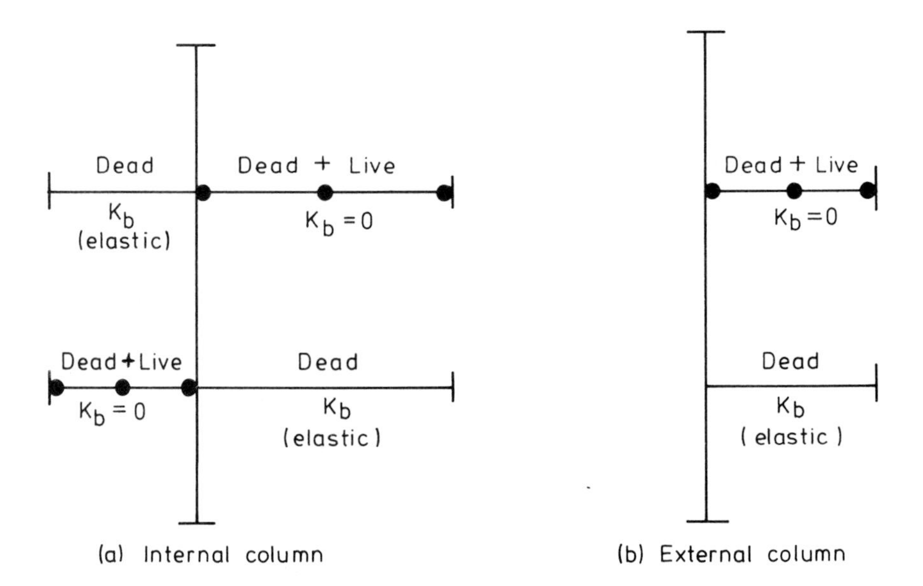

Fig 3 Patterned loading, and Joint Committee's limited frame

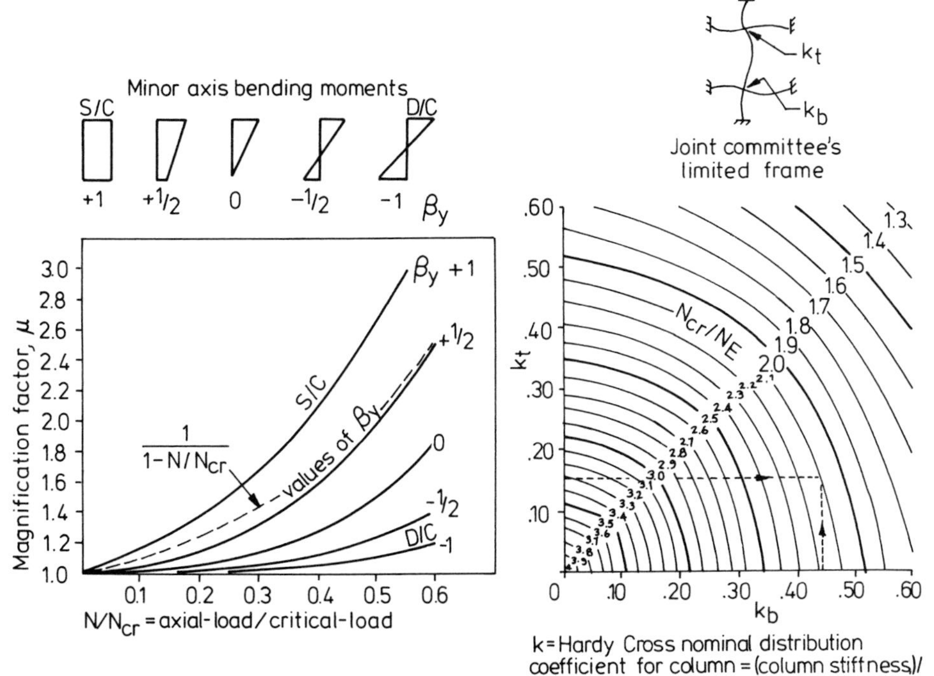

(a) Magnification factor, μ, for minor axis bending (from Joint Committee's 2nd report

(b) Values of C(N_{cr}/N_E) for a restrained continuous column.

Fig 4 Basis of Joint Committee's design method

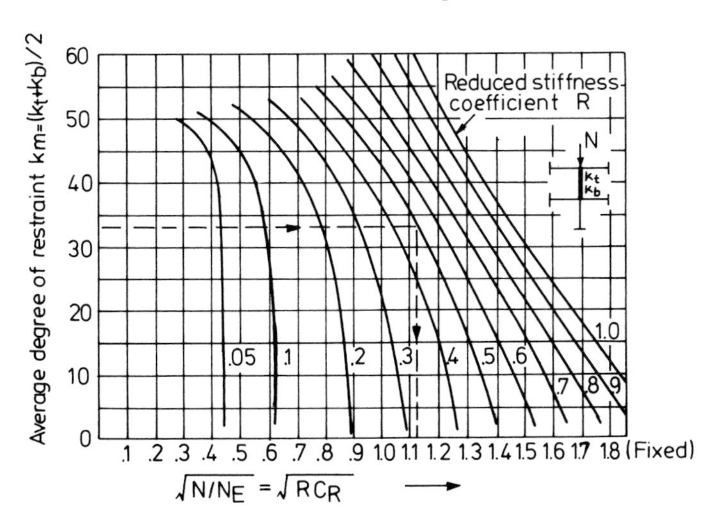

Fig 5 Direct design chart for continuous columns ignoring torsion, for the variable stiffness method

Similar charts are available for top and bottom storeys.

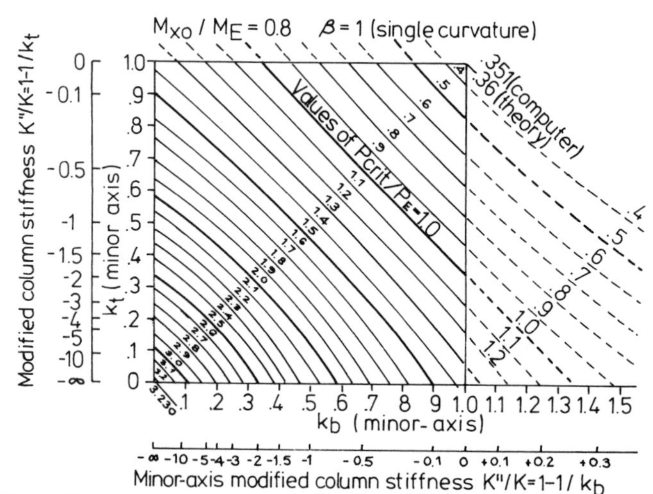

Fig 6 Elastic critical buckling loads (including torsion) for restrained column

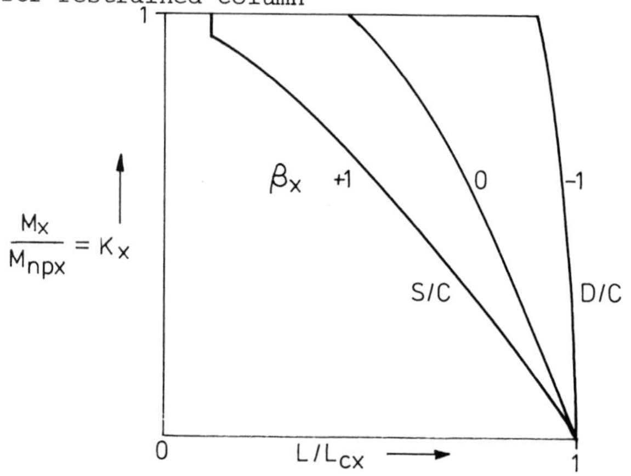

Fig 7 Universal column sections - major axis bending only
L_{cx} = limiting column length under axial load only

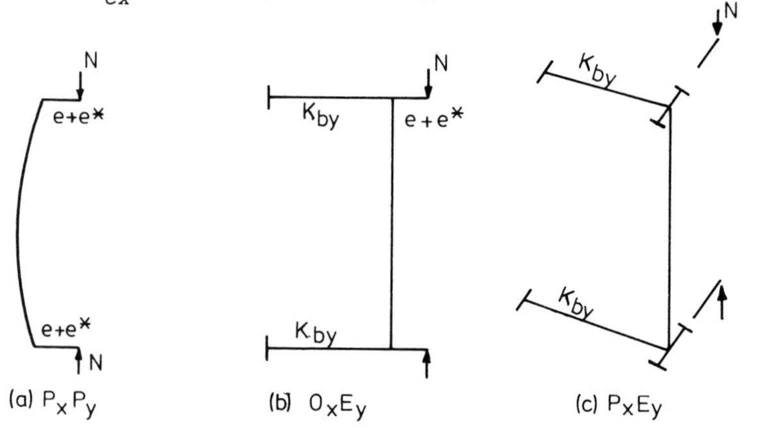

Fig 8 Models for Merchant-Rankine analogy

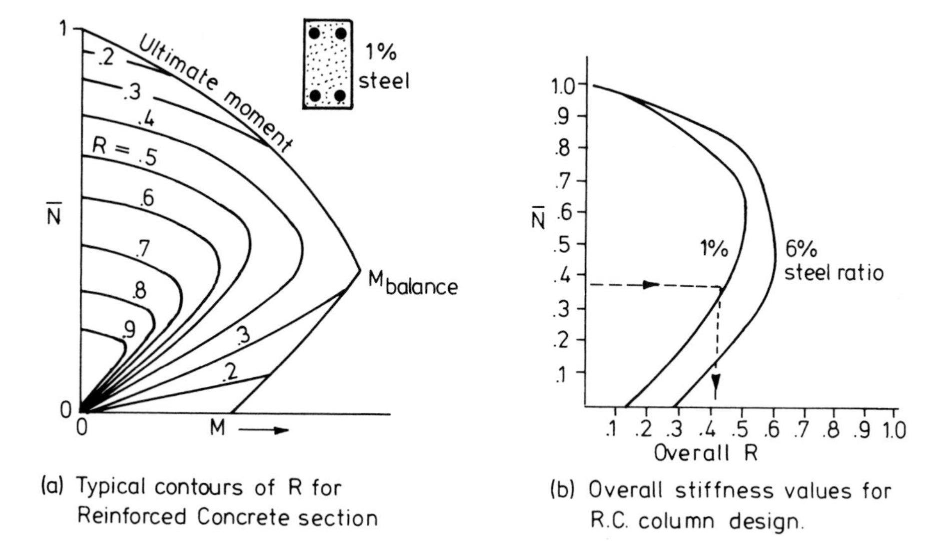

Fig 9 Variable-stiffness method and reinforced concrete columns

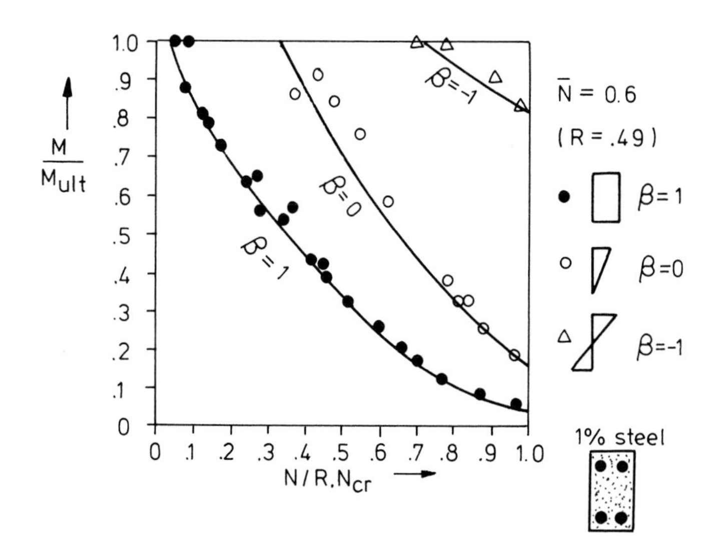

Fig 10 Typical design for r.c. columns (uniaxial bending)

INFLUENCE OF LOADING PATTERNS ON COLUMN DESIGN IN MULTI-STOREY RIGID-JOINTED STEEL FRAMES

Y.F. Yau, D.A. Hart, P.A. Kirby and D.A. Nethercot

Department of Civil and Structural Engineering, University of Sheffield

Synopsis

This paper describes some of the findings of a research project into loading patterns on multi-storey rigid-jointed frames. It shows that the well-known chequerboard patterns and the typical loading patterns of the draft steel code B/20 may significantly underestimate design moments and forces. Further, simplified frames, such as continuous beams and subframes, when used to determine design moments are shown to give erratic results. Two new types of loading pattern are introduced and are shown to more closely estimate the worst design forces in regular frames. In terms of member size, the use of simplified patterns may lead to the selection of sections up to two sizes lighter than those required for the worst possible loading pattern.

Introduction

In the terminology of the new U.K. Steel Code (1) steel frame structures may be classified into two main types: 'simple construction' for which the joints between members are assumed to be incapable of transmitting significant moments and 'continuous construction' for which the joints provide full continuity. Design of continuous steel structures seeks to exploit the better load sharing between different components as a means of justifying the greater fabrication costs of the joints. Thus the design forces for an individual member depend on both the other members of the structure and the pattern of loading assumed. It is therefore important to be able to correctly identify which of all the possible load arrangements is the 'most severe' for every component i.e. subjects that member to forces which take it nearest to its capacity. For all but the simplest of structures a rigorous treatment of this problem would involve consideration of a large number of separate load cases. It is therefore hardly surprising that several simplified load patterns have evolved whose aim is to approximate this worst case for different situations e.g. beam span moments, column end moments, etc.. The basis for these will be reviewed later in the paper.

The ready availability nowadays of matrix stiffness programs enables the linear elastic analysis of plane frames to be easily and efficiently conducted. It is therefore possible, in principle, to analyse a frame under any arrangement of loading and, using the principle of superposition, to combine the results of individual load cases. It was because of this facility that some doubt has recently been cast upon the ability of the accepted simplified patterns to closely estimate the worst conditions. Further study (2) has also suggested that approximate methods of analysis based on consideration of a portion of the structure e.g. sub-frame, continuous beam etc., significantly underestimate the design moments.

It is accepted that the likelihood of many types of frame structure actually being subjected to the exact arrangement of loading leading to the most severe design conditions for any member is small, (3). However, in certain cases where live loads are high relative to dead loads and where parts of the structure may be required to carry anything between full and zero live load e.g. a building housing liquid storage vats, such a possibility must be considered. It was as a result of discussions with an engineering department concerned with the design of this form of structure that the present study was undertaken. Although it deals with unbraced frames subject to vertical gravity load only, the extension to consider the interaction with lateral wind loading is currently in progress.

1. BASIS FOR DERIVATION OF CHEQUERBOARD PATTERNS

The traditional chequerboard loading patterns for multi-storey frames were derived from a consideration of the deflected shapes of members (4), as indicated in the regular frame of Fig 1.

With live load on span CD only, the distortions of the frame members are largest in and immediately adjacent to the loaded span, and decrease rapidly with increasing distance from the load. However, the loading of Fig.1 does not give the maximum possible sagging moment in CD. This is produced by loading all the spans which have been deflected downwards by the load on CD, leading to the well-known chequerboard pattern of Fig 2.

Incidentally, this pattern also produces the maximum hogging moments in the unloaded spans, and the maximum column end moments assuming that all columns are bent into single curvature.

By applying similar reasoning it is possible to construct other theoretical chequerboard patterns, e.g. for the maximum moment at the end of a beam. However, these patterns are more difficult to identify and hence, errors are more likely to occur in their construction. And also, many more patterns are required since a different pattern must be found for each joint in the frame. Any non-regularity of frames also renders this intuitive approach less reliable.

The draft steel code, B/20 (7) attempts to reduce the number of patterns required for a full frame analysis by proposing a set of typical loading patterns, the use of which, reduces the number of load cases requiring consideration to eight for beam design, plus four further patterns for each storey of the frame for column design.

2. METHOD FOR DETERMINING ACCURATE PATTERNS

The most appropriate method for determining the member forces in a complete multi-storey frame is the matrix stiffness approach. Here the designer assumes a "worst" loading pattern for the frame and a computer program calculates the member forces throughout the frame due to the loading pattern. By making a simple modification to this method so that it operates in an influence co-efficient manner it is possible to detect the worst load pattern for every design condition for every member in the frame. The method operates as follows:-

1. Load each beam of the frame in turn with, for example, a udl.
2. Analyse the frame under each of these load cases, and determine the influence co-efficients.
3. Combine the influence co-efficients to find the maximum possible value of each design force, moment, etc.. This requires the accumulation of all appropriate terms involving positive influence co-efficients. A check on the values used enables the corresponding load pattern to be identified.

The patterns thus derived almost invariably differ from the assumed chequerboard patterns.

3. *COMPARISON OF RESULTS OF ACCURATE AND CHEQUERBOARD PATTERNS*

Research conducted at the University of Sheffield (2) has compared the moments found from accurate and chequerboard patterns of live load with those from a consideration of continuous beam and subframe simplifications, for a number of different frames. Some surprising and perhaps disturbing findings have emerged. An example is the frame of Fig 3; the results for some typical locations are shown in Table 1.

For this example, differences of over 20% in beam support moments occurred at several locations when chequerboard loading was applied to the full frame, or when only a part of the frame was considered, as indicated in Fig 4. Moreover, these differences were almost always on the unconservative side i.e. actual moments were underestimated. Even larger discepancies (up to 40%) were obtained in some other frames which contained differing bay widths and storey heights. Column moments, in particular, were grossly underestimated (by up to 60% in certain cases). The results in the final column of Table 1 were produced using an alternative easily constructed loading pattern (5) which appears to give design moments close to those derived from the influence co-efficient approach for reasonably dimensioned frames.

The foregoing work has been confined to analysis i.e. the determination of design forces for individual members. However, it is recognized that the percentage discrepancies presented above may not be fully transmitted into a change of member size, hence, the problem now to be addressed is one of design, i.e. choosing a member whose capacity is just sufficient to withstand these design forces.

4. *APPLICATION OF ACCURATE PATTERNS TO THE DESIGN OF COLUMNS*

The basis for column design to BS449 (8) is the linear interaction formula of eqn 1, which is more conveniently written in load terms instead of stresses for the present work.

$$\frac{P}{P_o} + \frac{M}{M_o} \not> 1.0 \qquad (1)$$

In which P = axial load in member
P_o = axial strength of member in the absence of any moment
M = moment in member
and M_o = design moment of member in the absence of any axial load

Loading each individual beam in turn and considering the criterion

$$\frac{P}{P_o} + \frac{M}{M_o} > 0.0 \qquad (2)$$

it is possible to establish a loading pattern which produces the worst load case for the design of a particular column. The pattern thus derived may involve up to three-quarters of the beams in a particular frame being loaded. However, the contribution to the right-hand side of eqn 1 of many of the beams is less than 1% of the total. Thus, by changing the criterion of eqn 2 to

$$\frac{P}{P_o} + \frac{M}{M_o} > 0.01 \qquad (3)$$

it has been possible to establish much simpler patterns (see Fig.6) giving results of at least 90% of the maximum.

Four such patterns exist for each column, corresponding to compressive axial load coincident with moments at the top or base of the column which may be either clockwise or anti-clockwise in direction. The technique described above seeks to identify which loaded beams contribute significantly to the left-hand side of the BS449 interaction formula. As an example the results for Column 17 of the frame in Fig 3 are given in Tables 2 and 3. The two tables illustrate that the simple patterns for the column interaction problem can lead to results amounting to some ninety per cent of the maximum value i.e. disregarding minor contributions still gives reasonable results. An analysis of the magnitudes by which loaded beams contribute to the column interaction pattern has identified four types of contributing beam (see Table 4).

The patterns shown in Fig 6 therefore produce combinations of member forces (P and M) which effectively control the design of the column. The identification marks (I-IV) on the beams refer to Table 4 and the member numbers to Tables 2 and 3.

5. *THE CONCEPT OF A "DESIGNED FRAME"*

In the preliminary design of a multi-storey frame, a designer must determine the member sizes for his initial trial design in order that an analysis of member forces can be conducted. Also, he must choose perhaps arbitrarily, the effective length factors for beams and columns. Additionally, he must decide what loading patterns to apply. Moreover, it is unlikely that two independent designers, working on the same frame, would always make the same selections.

For the comparison of different types of patterns applied to multistorey frames such a state of affairs is unsatisfactory, and there is a need for a common basis of comparison. Such a basis is the concept of a "Designed Frame" whose member sizes are selected by computer analysis based on specified effective length factors and using BCSA universal beam and column section tables in which the members are arranged in order of increasing weight. The computer procedure is as follows:-

1. Select initial member size for the frame.
2. Determine design forces for every member by using an accurate pattern analysis.
3. Find the lightest section which can carry the design loads.
4. Check whether all member sizes are the same as those in the previous frame. If not repeat the above until this condition is satisfied. If true, then a "Designed Frame" has been identified.

Two observations on this procedure are, firstly the choice of member sizes for the initial frame is not important, but obviously less computing time will be required if the initial frame is close to the final frame. Secondly, the choice of effective length factors (provided that they are reasonable) and hence the choice of member sizes has no major discernable effect on the accurate patterns which control member proportions.

Using a "Designed Frame" there are two methods by which the results from different pattern types may be compared i.e. by member sizes or by the direct comparison of member forces. In terms of design the former method is the more realistic since ultimately it is the member sizes which will determine what forces exist in the structure.

Using the first method, the simplified typical patterns of B/20 have been found to lead to a requirement for section sizes which are up to two sizes lighter than those in the "Designed Frames" considered. Thus, if these lighter sections were actually used in a frame, then overstressing could occur. The degree of overstressing will be dependent upon two factors: (i) The difference in load carrying capacities of adjacent sections in the BCSA tables, and (ii) on the amount of spare load capacity in the member i.e. the difference between the left- and right-hand sides of Eq 1. It is in fact possible for there to be no overstressing.

6. COMPARISON OF COLUMN LOADS FOR VARIOUS PATTERNS

Using a "Designed Frame" with the geometry of the frame in Fig 3, the accurate interaction pattern results for columns 16, 17 and 18 have been found and are presented here in Table 5. Note that it is the column nearest to the edge of the frame, i.e. Column 16, which carries the greatest loads.

Comparing the values in the last column of Table 5 with those of the simplified patterns in Table 6 shows that the axial loads determined from the patterns of Fig 6 are in close agreement with the accurate pattern results, but moments are underestimated by about 25%.

The B/20 patterns identify as a typical worst load case, a pattern which leads to higher moments than the accurate result, but these are coincident with significantly smaller axial loads, thus leading to a less severe design requirement.

It is possible to conceive the typical patterns for column design in B/20 as a combination of the two simple patterns shown in Fig 6, which approximate the worst condition of combined axial load and moment at the top and base of a column respectively. However, these simple patterns are allowed to 'overlap' in B/20, thus reducing their effectiveness. By loading beams which are one bay removed from the column under consideration, the Draft Code patterns have significantly increased the moment, but have reduced the axial load by a greater degree, hence, the combined result is a less severe load case, as illustrated in Table 6. Additionally, in using the B/20 patterns it is implied that beams which are more than one storey below the base of the column being considered should be loaded. However, it has been shown that these beams give negligible contributions to the patterns.

7. *COMPARISON OF SIX DIFFERENT SUBFRAME ANALYSES USED FOR COLUMN DESIGN*

A common manual method for determining column end moments in multistorey frames is the use of subframe simplifications.

Many different suggestions have been made as to the properties of the subframe which should be used (6), hence from the two basic subframes of Fig. 8 it is possible to form the six different subframes ((i) to (vi)) described in Table 7. The results from these six subframes have been evaluated with respect to the "Designed Frame" mentioned in the last section. The assumption that a member possesses only half of its relative stiffness originates from an assumed deflected shape of a frame in which certain members are bent into double curvature, giving points of contraflexure at their mid-lengths (6). Consider any one of columns 16 to 20 of the "Designed Frame", which has an applied live gravity udl of intensity 30kN/m only. Dividing the total load on the supported area by the number of columns the design axial load in these columns may be estimated as 450kN.

Table 8 shows that the different subframe types estimate the column end moments as being in the range 9.85 to 22.44 kN.m. Further comparison, with the accurate pattern results in Table 5 (16.98 to 17.47 kN.m) better emphasizes the erratic nature of results from subframe analyses. It should be noted that the subframe analysis method takes no account of the position of a column within a given storey; thus the results in Table 7 apply equally to any one of columns 16 to 20 in the "Designed Frame", however as Table 5 shows, different moments and axial loads exist for the worst accurate load cases of these columns due to their respective positions within the storey. The American Concrete Institute has produced a Table of Co-efficient Values which makes an attempt to take the position of a column into account for these subframe simplifications. This refinement has not been included in Table 8.

8. *CONCLUSIONS*

This paper has shown that the well-known chequerboard patterns and the typical loading patterns of B/20 may significantly underestimate design forces in individual members of rigid-jointed frames. Further, simplified frames, such as subframes and continuous beams, when used to determine design moments, have been shown to give erratic results. Two new types of loading pattern have been introduced which are shown to more closely estimate the accurate results for regular frames. In terms of member size, the use of B/20 typical patterns has been shown to lead to the selection of sections up to two sizes lighter than those required with an accurate pattern analysis.

It is recognized that the presence of dead load will reduce the discrepancies found between results from the accurate patterns and the simplified patterns, and also between different idealisations of the structure. However, despite their variability the results from these approaches have produced satisfactory structures. Thus, it seems unreasonable to penalize the more thorough and rational approach i.e. an accurate pattern analysis, which requires the selection of heavier members to carry the greater design forces. Therefore, it would appear appropriate to permit a percentage reduction in live design loads for frames analysed using an accurate pattern approach.

Acknowledgements

The authors would like to thank the Design Systems Group of I.C.I. Engineering Department for their financial support and assistance during this research project. In particular thanks are due to Mr. D. Button of I.C.I. for his identification of the problem and its industrial implications.

References

1. British Standards Institution, BS5950. 'The Structural Use of Steelwork in Buildings, Part 1 : Simple Construction and Continuous Construction', to be published.

2. Yau, Y.F., Kirby, P.A. and Nethercot, D.A. 'Loading Patterns for Multistorey Frames', paper in preparation.

3. Beeby, A.W. 'A Proposal for Changes to the Basis for the Design of Slabs', Cement and Concrete Association Technical Report No. 547.

4. Winter, G. and Nilson, A.H. 'Design of Concrete Structures', 8th edition, McGraw-Hill, New York, 1973.

5. Yau, Y.F. 'Loading Patterns for Multistorey Frames', thesis presented to the University of Sheffield in 1982 in partial fulfillment of requirements for the degree of M.Eng.

6. Kong, F.K. and Evans, R.H. 'Reinforced and Prestressed Concrete', 2nd edition, Nelson, London, 1980.

7. British Standards Institution, 'Draft Standard Specification B/20 for the Structural Use of Steelwork in Building, Part 1 : Simple Constructuion and Continuous Construction', 1977.

8. British Standards Institution, BS449. 'The Structural Use of Steelwork in Buildings', 1969.

TABLE 1 - LEFT SUPPORT MOMENTS FOR BEAMS OF FIG.3

(Moment Values in kN.m - Values in brackets are percentage errors with reference to the accurate pattern. Negative sign indicates underestimates)

		APPROXIMATIONS			
Beam	Accurate Pattern	Chequerboard Pattern	Continuous Beam	Subframe	Corridor Pattern
B7	32.2	25.7(-20.2%)	23.5(-27.0%)	23.1(-28.3%)	31.8(-1.2%)
B8	41.1	37.7(- 8.3%)	36.3(-11.7%)	34.4(-16.3%)	40.9(-0.5%)
B9	43.0	39.3(- 8.6%)	35.9(-16.5%)	34.4(-20.0%)	42.8(-0.5%)
B13	30.8	24.5(-20.5%)	23.5(-23.7%)	23.1(-25.0%)	30.3(-1.6%)
B14	40.9	35.7(-12.7%)	36.3(-11.2%)	34.4(-15.9%)	40.5(-1.0%)
B15	42.3	36.5(-13.7%)	35.9(-15.1%)	34.4(-18.7%)	42.1(-0.5%)
B19	30.1	24.4(-17.9%)	23.5(-21.9%)	23.1(-23.3%)	29.2(-3.0%)

TABLE 2 - INTERACTION RESULTS FOR COLUMN 17 (TOP)

Beam Loaded	$\frac{P}{P_o}$	$\frac{M}{M_o}$	$\frac{P}{P_o} + \frac{M}{M_o}$	Beam Value Percentage	Cumulative Percentage
B15	0.06230	0.11897	0.18127	35.63	35.63
B 3	0.06255	0.00966	0.07221	14.19	49.82
B 8	0.06106	0.00869	0.06975	13.71	63.53
B 2	0.06255	-0.00988	0.05270	10.36	73.89
B 9	0.06109	-0.00888	0.05222	10.26	84.15
B21	0.00204	0.04753	0.04549	8.94	93.09
B10	-0.00013	0.01045	0.01058	2.08	95.17
etc.			<0.01	<1.00	
TOTAL			0.50879		

TABLE 3 - INTERACTION RESULTS FOR COLUMN 17 (BASE)

Beam Loaded	$\frac{P}{P_o}$	$\frac{M}{M_o}$	$\frac{P}{P_o} + \frac{M}{M_o}$	Beam Value Percentage	Cumulative Percentage
B20	-0.00207	0.11752	0.11545	21.60	21.60
B14	0.06232	0.04828	0.11060	20.69	42.29
B 2	0.06258	0.00718	0.06976	13.05	55.34
B 8	0.06106	0.00048	0.06154	11.52	66.86
B 9	0.06109	-0.00024	0.06085	11.38	78.24
B 3	0.06255	-0.00695	0.05560	10.40	88.64
B15	0.06230	-0.04828	0.01402	2.62	91.26
etc.			<0.01	<1.00	
TOTAL			0.53498		

TABLE 4 - TYPES OF CONTRIBUTING BEAM

Beam Type	Axial Load Term	Moment Term
(I)	High Positive	High Positive
(II)	Negligible	High Positve
(III)	High Positive	Negative
(IV)	High POsitive	Small

TABLE 5 - ACCURATE PATTERN RESULTS

Column	P(kN)	M(kN.m)	P/P_o	M/M_o	P/P_o+M/M_o
16(accurate)	495.47	17.47	0.709	0.236	0.945
17(accurate)	473.84	16.98	0.678	0.229	0.907
18(accurate)	473.20	17.03	0.677	0.230	0.907

TABLE 6 - RESULTS FOR SIMPLIFIED PATTERNS

Column(pattn)	P(kN)	M(kN.m)	P/P_o	M/M_o	P/P_o+M/M_o
16(Fig.6)	493.05	13.28	0.706	0.179	0.885
17(Fig.6)	478.01	14.30	0.684	0.193	0.877
18(Fig.6)	473.25	11.90	0.677	0.161	0.838
16(B/20)	404.29	20.46	0.579	0.276	0.858
17(B/20)	372.32	20.70	0.533	0.279	0.812
18(B/20)	375.71	20.68	0.538	0.279	0.817

TABLE 7 - SUBFRAME TYPES

Subframe Type	Subframe Drawing	Members with Full Stiffness	Members with Half Stiffness
(i)	A	All members	——
(ii)	A	All columns	All beams
(iii)	B	All members	——
(iv)	B	All columns	All beams
(v)	B	All beams, C2	Columns C1, C3
(vi)	B	Column C2	All beams,Columns C1, C3

TABLE 8 - RESULTS FROM SUBFRAMES

Subframe	P(kN)	M(kN.m)	P/P_o	M/M_o	P/P_o+M/M_o
(i)	450	9.85	0.644	0.133	0.777
(ii)	450	14.97	0.644	0.202	0.846
(iii)	450	13.70	0.644	0.185	0.829
(iv)	450	20.11	0.644	0.271	0.915
(v)	450	14.79	0.644	0.200	0.844
(vi)	450	22.44	0.644	0.303	0.947

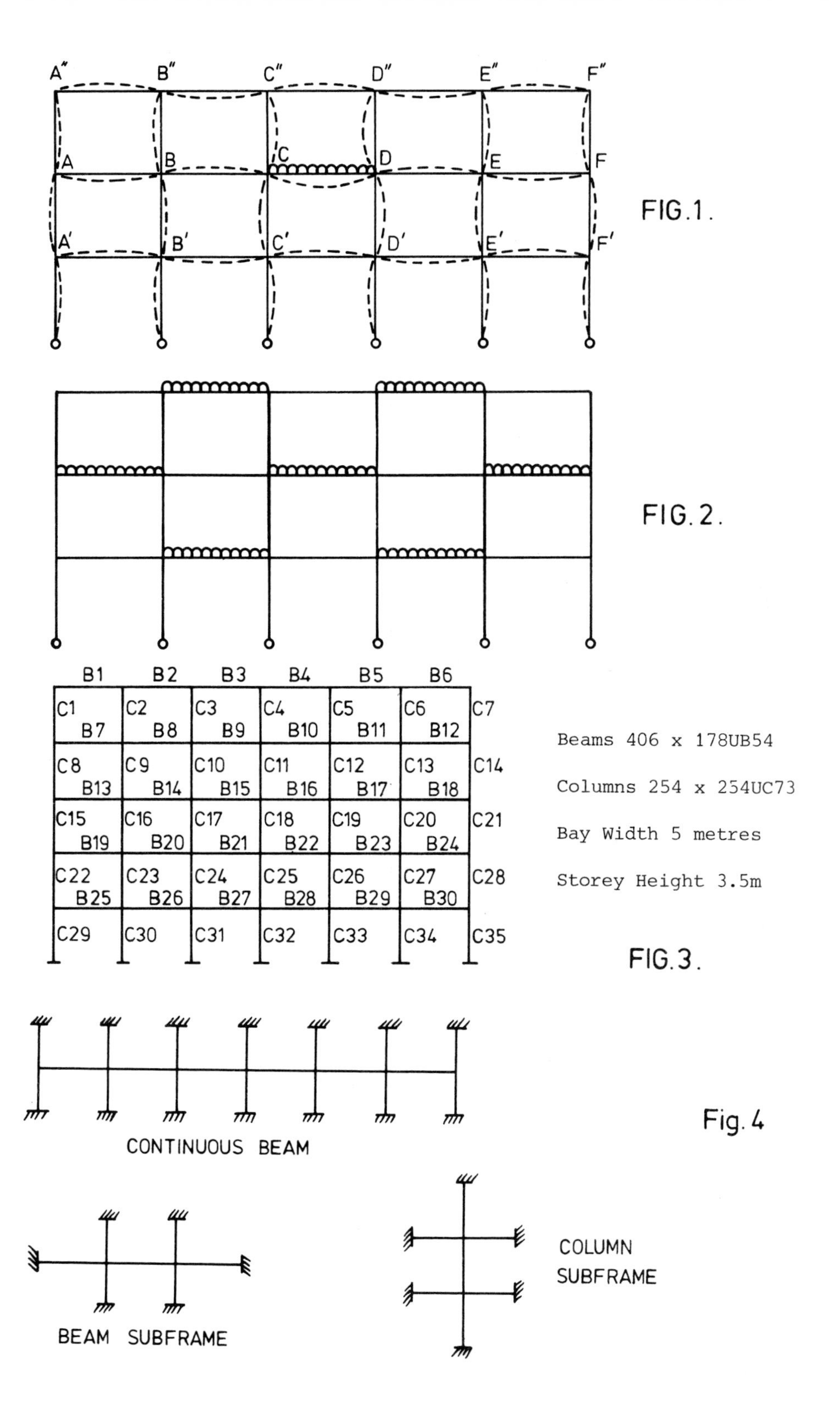
A"
B"
C"
D"
E"
F"
A
B
C
D
E
F
A'
B'
C'
D'
E'
F'
FIG.1.
FIG.2.
B1 B2 B3 B4 B5 B6
C1 C2 C3 C4 C5 C6 C7
B7 B8 B9 B10 B11 B12
C8 C9 C10 C11 C12 C13 C14
B13 B14 B15 B16 B17 B18
C15 C16 C17 C18 C19 C20 C21
B19 B20 B21 B22 B23 B24
C22 C23 C24 C25 C26 C27 C28
B25 B26 B27 B28 B29 B30
C29 C30 C31 C32 C33 C34 C35
Beams 406 x 178UB54
Columns 254 x 254UC73
Bay Width 5 metres
Storey Height 3.5m
FIG.3.
CONTINUOUS BEAM
Fig. 4
BEAM SUBFRAME
COLUMN
SUBFRAME

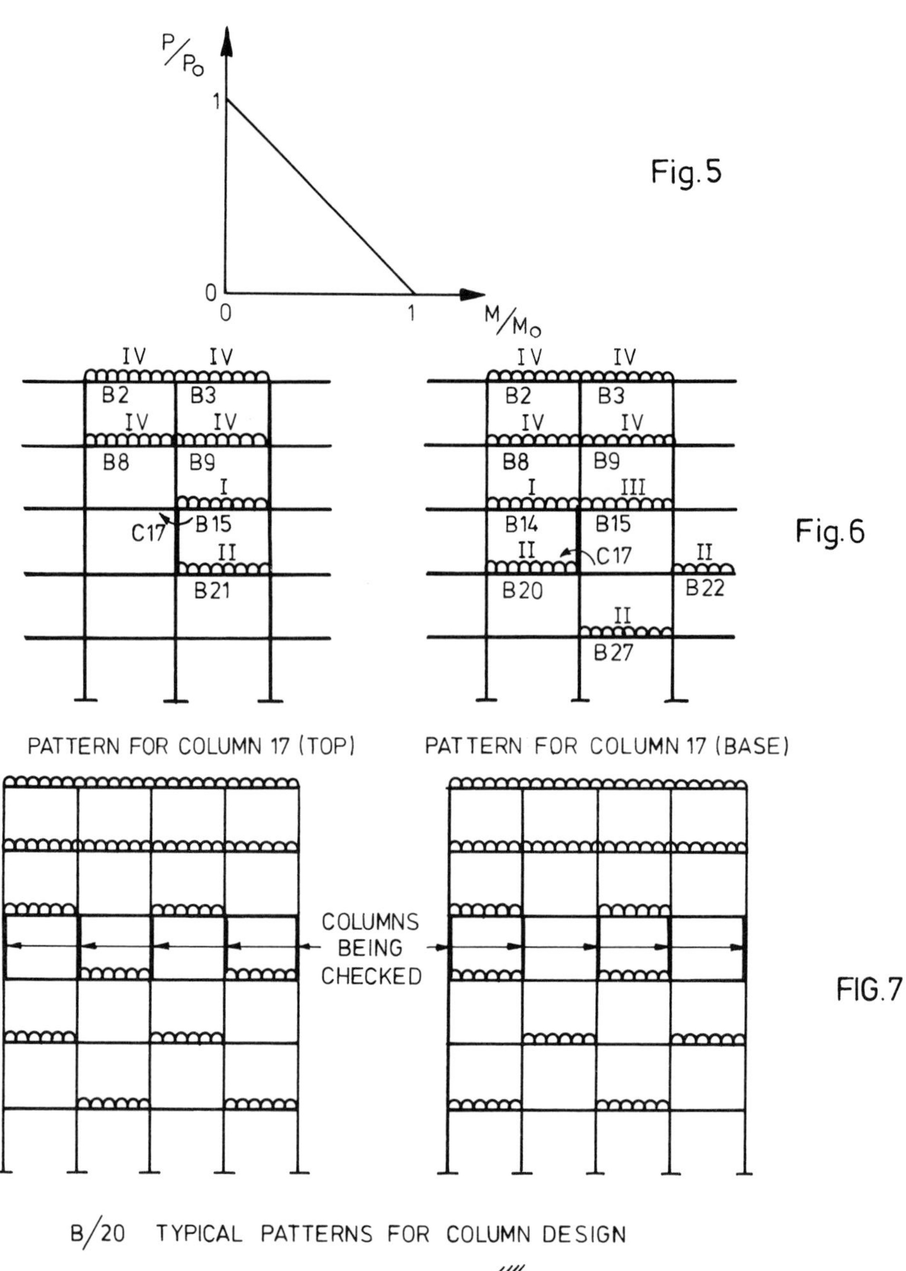

B/20 TYPICAL PATTERNS FOR COLUMN DESIGN

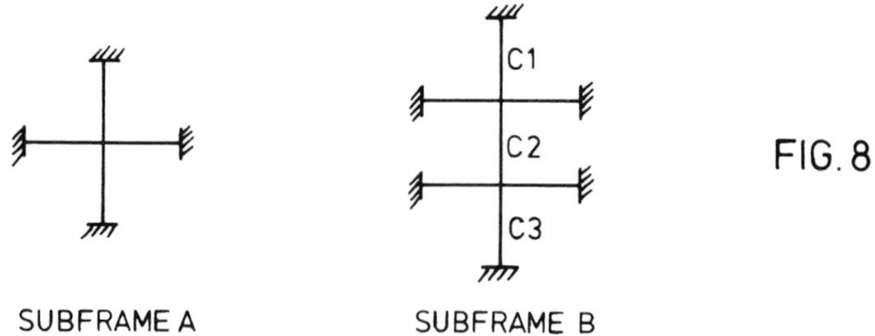

COMPUTER ANALYSIS OF COLUMNS

W B Cranston

Cement and Concrete Association, London

Synopsis

The use of computers over the last 30 years is reviewed. Details of programs used to analyse a particular column treated originally by hand by Horne are given. The cheapness and power of the currently available home micro-computers is demonstrated. Some speculations about the effects on education and practice are introduced.

Introduction

Computers have been used from their earliest days to analyse columns in the inelastic range. Eickhoff (1) used the EDSAC machine at Cambridge in the early 1950's to analyse rectangular mild steel columns. The author (2) used the DEUCE computer at Glasgow in 1960-61. Applications in the intervening years have dealt with reinforced concrete columns (3), biaxial bending in steel columns (4) and more recently biaxial bending in concrete columns (5), (6), (7).

In this paper the computer analysis of a rectangular steel column originally tested by Baker and Roderick (8) is considered. It was analysed initially by hand by Horne (9). The method of analysis has been programmed for 6 different computers and Horne's particular column has been analysed on 4 of them. At the time of writing the analysis is being mounted on a Jupiter ACE micro-computer (cheekily named after the ACE which was the prototype for the DEUCE computer mentioned above).

Comparisons are drawn between computer costs, ease of writing and debugging programs, and flexibility of programs in use.

It is suggested that radical changes to design procedures and to training must be made now that computers are - to quote the Jupiter Ace manual - "available to all".

1. THE COLUMN ANALYSED

The test arrangement is shown in Fig 1. The column cross-section is of mild steel with a yield stress of 19 T/in^2 (295 N/mm^2). In the experiment the beam loads were applied first and held constant as the column load was increased to failure. Rotations at the ends and centre of the column were measured. The behaviour was virtually symmetrical with the two end rotations being within or close to 10% of one another until very near maximum load.

2. *THE ORIGINAL HAND ANALYSIS*

The column was assumed to be symmetrical and thus only one half was analysed. Five different zones of behaviour arise, the most complex being where steel originally plastic in tension or compression is subject to strain reversal and unloading of the fibres occurs. Numerical integration of the appropriate - and highly complex - differential equations was carried out and the analysis took a total of six months continuous effort.

3. *METHOD OF COMPUTER ANALYSIS*

The system analysed is shown in Fig 2. The column length is divided into segments. The analysis is based on calculating curvatures at the division points between segments. Solutions are found as the deflection at a given control point along the column is increased in stages.

At the start of a given stage assumptions are made for the deflected shape, end rotation and load factor. Forces corresponding to these assumptions are calculated at each division point and using an iterative method the curvatures corresponding to these forces are calculated in turn. The curvatures are integrated to give a calculated deflected shape which is compared with that assumed. The original assumptions are corrected and the procedure repeated until the calculated and assumed shapes agree.

The cross-sections are divided into strips (Fig 3) and a strain history is maintained for each strip which updates at each stage the maximum strain in each element, once it is strained beyond the elastic limit.

The method has been developed and improved as follows :

1964 : Integration procedure changed to deal with unequal segments. Up to this time equal segments were used requiring a large number of segments for some analyses.

Rigid end gusset lengths introduced to allow consideration of joints.

1970 : Analysis extended (10) to include non-linear analysis of framing beams. For the systems illustrated in Fig 2 the relation between restraint and end rotation has to be worked out and input as data to the program.

1982 : Analysis extended to cover biaxial bending (7) and refinement introduced to allow number of elements in cross-section to be reduced for some accuracy.

1983 : Convergence procedures improved.

These improvements reduce the amount of calculation for a given problem to rather less than half that needed for an analysis in 1959.

4. THE ANALYSIS

Symmetry was assumed, as for the hand analysis. The idealization of the half-column for the SIRIUS program in 1964 is shown in Fig 3. An elasto-plastic stress-strain curve was used. An initial trial run with 10 strips/cross-section produced convergence problems under high curvatures since strain profiles across the section gave plasticity in all elements. This was overcome by using narrower strips for the elastic 'core' of the highly stressed cross-sections.

The idealization for one of the more recent analyses is shown in Fig. 4. The convergence difficulties in the early analysis with ten strips have been avoided by giving the plastic plateau of the stress-strain curve a very slight upward slope.

5. COMPARISON OF RESULTS

It will be seen from Figure 5 that there is very good agreement between analysis and experiment in respect of end and central rotations. In Fig 6 comparisons are made with the plastic zones established by Horne and in Fig 7 curvatures are compared. Again agreement is good. A closer spacing of division points in the analysis in the quarter point region would have improved agreement.

6. GENERAL COMPARISONS

The computer details are given in Table 1, and program details in Table 2. Significant trends can be distinguished, some of which are mentioned below.

The costs of analysis are now 1/1000 of what they were in 1959 and 1/100 of what they were in 1972. The author remarked in 1972 (page 23 of reference 3) that rigorous computer analysis of reinforced concrete columns was not an economical design option. This is no longer true.

Turning now to speed, it is estimated that the Jupiter ACE will be around 10 times as fast as the DEUCE in 1959. This is being achieved by using the FORTH language which gives a large gain in speed at the expense of around double the programming effort. Trading up to a more expensive micro at £2,500 with a BASIC compiler and efficient floating point arithmetic routines would give a speed increase of between 100 and 1000.

The program mounting time has reduced by a factor of around 10.

The question of debugging difficulty is also worth comment. Using the Telefile Services T85 computer for instance, log on time was several hours and a lot of computer time was expended on tracing faults. The BASIC language is very little used on this machine and syntax errors were identified by the cryptic remark "BAD STMT". After diagnosis the complete program line had to be typed anew.

The syntax checking on the ZX81 BASIC is by contrast excellent as also is the facility to bring down a line of program for editing. Also it is possible on the ZX81 to interrupt the program at any stage, manually print out variables, modify variables, modify program, and then continue at the interruption point. In fairness it should be added that similar facilities are available using PASCAL language on the T85. The ZX81, aimed at the home market, is an exceptionally 'user friendly' machine and has been shown to be superior in this respect even to the IBM personal computer - as well as making a strong showing against it in speed (11).

An additional feature incorporated in the latest micro-computer programs was to incorporate details of the cross-sections and stress strain curve directly into BASIC instruction lines. The program has in this way been applied to several different series of concrete columns. To adapt for application to Horne's special column required modifications to only 30 program lines.

7. *DISCUSSION AND SUGGESTIONS*

The information and comparisons given above confirm the enormous increase in computing power now easily and cheaply available. How is this affecting and going to affect structural design ? With the pace of development still growing only some speculations can be ventured.

What is certain is that with the expenditure of £10,000 a design office can have the power to tackle the sort of problem outlined in this paper on an interactive basis. This expenditure, capitalised over a number of years, represents only a few percent increase in costs for even a small practice. So nearly all practices will have this sort of computing power available on hand within a year or so.

A clue as to how this power could be used can be found in experience with platforms designed for the offshore industry. The large sums of money available for design have meant that these structures have been automatically analysed by very expensive finite element techniques. More complicated analysis has been applied, because "it is here".

Naesje (12) produces an interesting case history in regard to such a platform where a finite element analysis was used and amidst the large volume of output a high stress concentration was not identified. When the platform was submerged severe cracking and leakage occurred leading to large repair costs. It seems fair to assume that the engineer concerned had not thought through the general behaviour of the structure and in simple terms had "left it to the analysis".

Positive action is necessary to avoid this sort of misuse of the enormously increased available power now available. Some suggestions relating to training and education are made below :

a) It is suggested that much less time need now be expended on teaching techniques of analysis.

b) Some of the time saved should be expended on developing a deeper understanding of actual structural behaviour, preferably by actual tests on materials and structural components, although suitable video material could be just as effective.

c) Some of the time should be expended on training in communication and recording skills. It is of the utmost importance that programs written for computers should be written in a readable way and supported by annotated listings and appropriate guidance for users.

Finally, it is suggested that since structural design is very much a non-standard operation there will be many occasions where it will be of great benefit to a user to be able to modify or extend the program being used. Software where the listing is not available to the user precludes this. The commercial and ethical problems introduced by this dilemma need to be tackled.

Acknowledgements

This paper is published by permission of Dr G Somerville, Director of Research and Development at the Cement and Concrete Association. Thanks are due to Dr D Flower of the Operations Department who improved and mounted the program on the C&CA's HP3000 computer, and to the author's son, Andrew, who assisted in the work using the ZX81 and ACE computers.

References

1. Eickhoff, K.G. 'The plastic behaviour of columns and beams.' PhD Thesis presented to Cambridge University, 1955.

2. Cranston, W.B. 'Restrained metal columns.' PhD Thesis presented to Glasgow University, 1963.

3. Cranston, W.B. 'Analysis and design of reinforced concrete columns.' Research Report 41.020, Cement and Concrete Association, April 1967.

4. Sugimoto, H. and Chen, W.F. 'Small end restraint effects on strength of H-columns.' Journal of Structural Division, ASCE, Vol 108, No. ST3, March 1982, pp. 661-681.

5. Menegotto, M. and Pinto, P.E. 'Slender RC compressed members in biaxial bending.' Proceedings of the American Society of Civil Engineers, Journal of the Structural Division, Vol. 103, No. ST3, March 1977, pp. 587-605.

6. Al-Noury, S.F. and Chen, W.F. 'Finite segment method for biaxially loaded RC columns.' Journal of the Structural Division ASCE, Vol 108, No. ST4, April 1982, pp. 780-799.

7. Cranston, W.B. 'Analysis of slender biaxially loaded restrained columns.' Presented to CEB Commission III, Buckling, Munich, October 1982.

8. Baker, J.F. and Roderick, J.W. 'The behaviour of stanchions bent in double curvature.' Welding Research, Vol 2, 1948.

9. Baker, J.F., Horne, M.R. and Heyman, J. 'The steel skeleton.' Cambridge University Press, 1956, Chapter 14.

10. Sturrock, R.D. Private communication.

11. Babsky, D 'ZX81 v IBM PC!' Which Micro, April 1983, pp 34-36.

12. Naesje, K. Dahl, P.C. and Muksnes, J. 'Structural damage in an offshore concrete platform, the cause and the cure.' FIP Notes, September/October 1978, pp 8-17.

TABLE 1 - COMPUTER DETAILS

Computer and date	Store details (bytes)	Capital cost	Mainten. and support costs	% availab. to indiv. user	Remarks
DEUCE	Mercury delay lines, 1k + Drum, 32k	£ 40,000	Not known	5%	Used 'hands on'
SIRIUS 1964	Nickel delay lines, 35k	£ 10,000	£15/hr	10%	Used 'hands on'
1903A 1969	Core, 96k + Disks	£225,000	£30/hr	N/A	Batch mode
ZX81 1982	ROM, 8k RAM, 16k	£250**	Negligible	100%	Author's property
HP3000 1982	Core,1000k + Disks	£200,000	£50/hr	50%	Time-shared 20 users
T85* 1983	Core,800k + Disks	£250,000	£100/hr	95%	Time-shared 50 users
ACE 1983	ROM, 8k RAM, 24k	£250**	Negligible	100%	Author's property

* Made available by TELEFILE SERVICES, Slough (SIGMA compatible)
** Includes cost of television, tape recorder, printer

TABLE 2 - PROGRAM DETAILS

Computer and date	Time to mount & debug (days)	Language used	Run-time on computer (hours)	Reliability	Debugging difficulty	Cost of analysis 1983 prices
DEUCE 1959	100*	Machine Code	{20}	50%	High	{£2,000}
SIRIUS 1964	100*	Machine	5	85%	Medium	£750
1903A 1964	25*	ALGOL	{2}	95%	High	£100
ZX81 1982	10*	BASIC Interp.	8	100%	Low	{£2}**
HP3000 1982	5	BASIC Compiled	0.10	100%	Medium	£10
T85 1983	2	BASIC Compiled	0.02	100%	High	£50
ACE 1983	{20}*	FORTH (Threaded)	{2}	100%	Low	{£2}**

* Includes learning of language
Indicates estimated value
** Major component of estimated cost is to cover printing

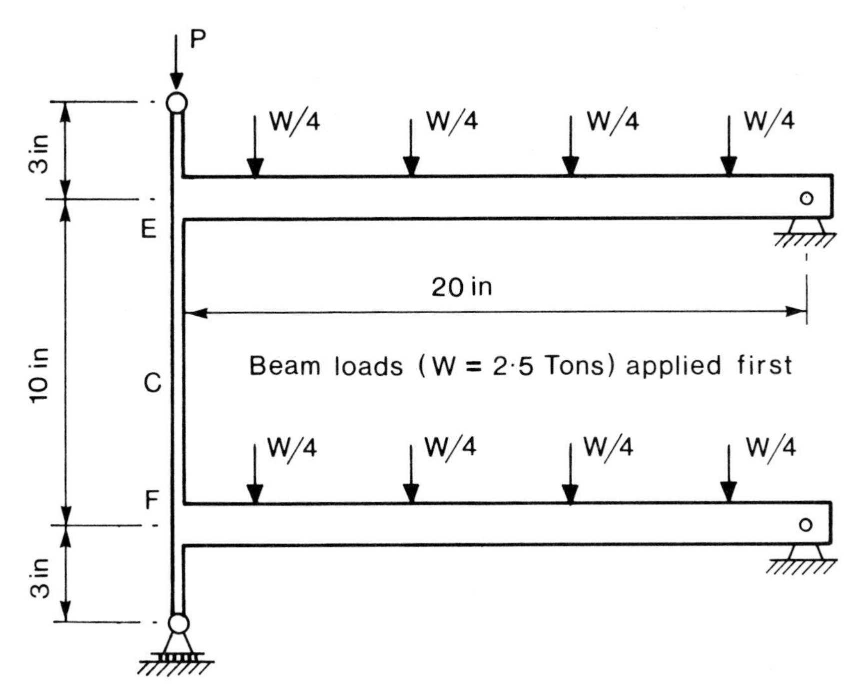

Fig 1 Column tested

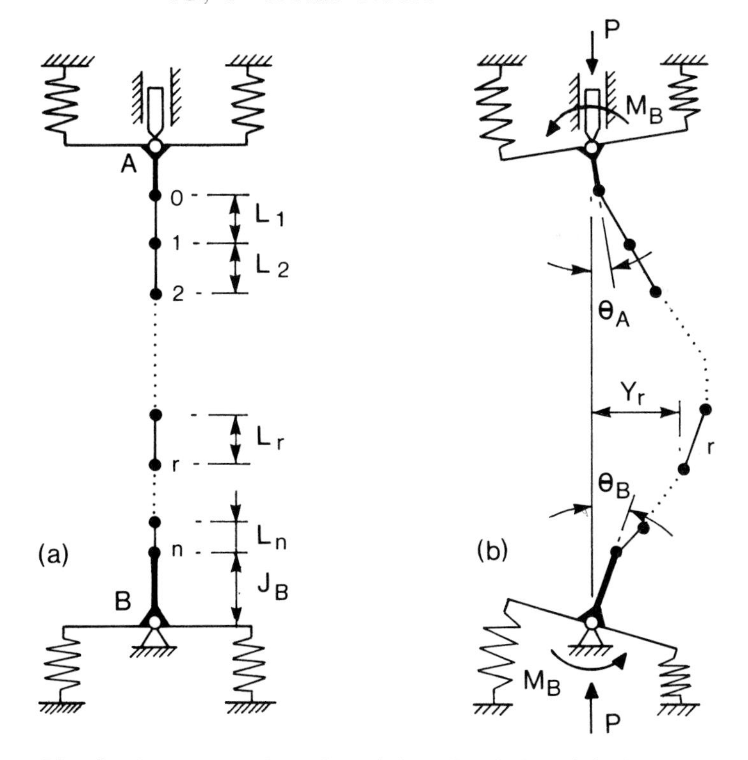

Fig 2 System analysed ; (a) unloaded ; (b) loaded

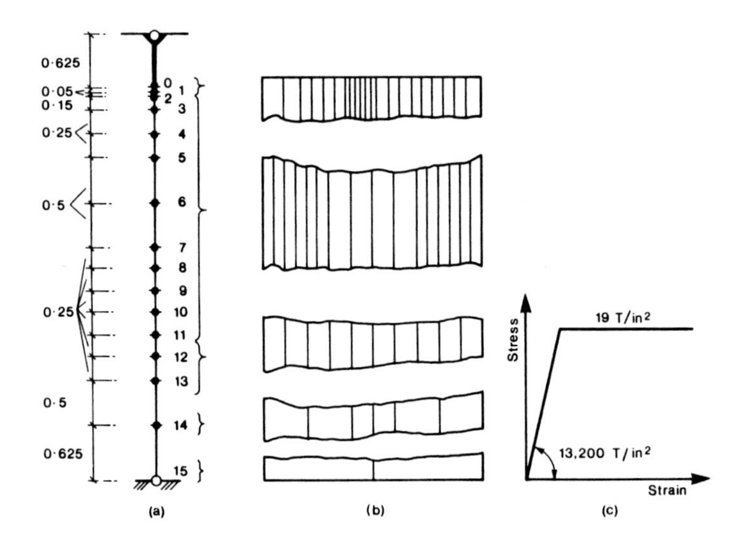

Fig 3 Idealization for SIRIUS analysis

(a) segments
(b) strips
(c) stress-strain

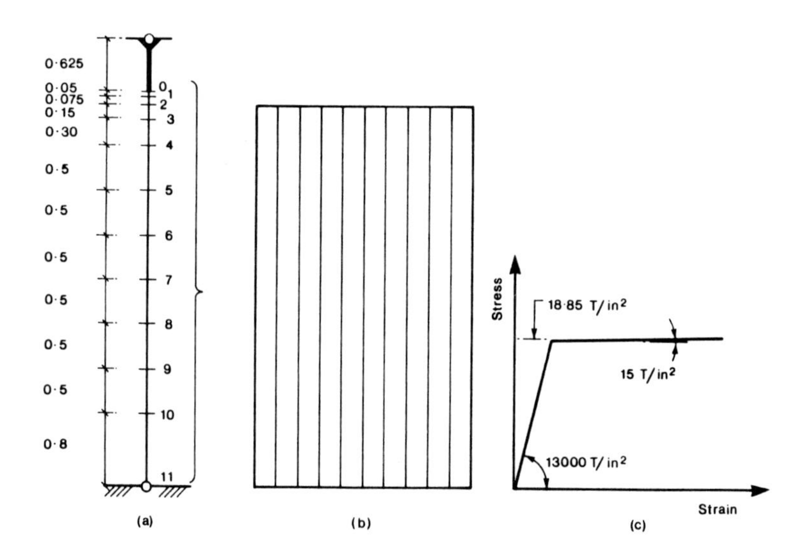

Fig 4 Idealization for ZX81 analysis

(a) segments
(b) strips
(c) stress-strain

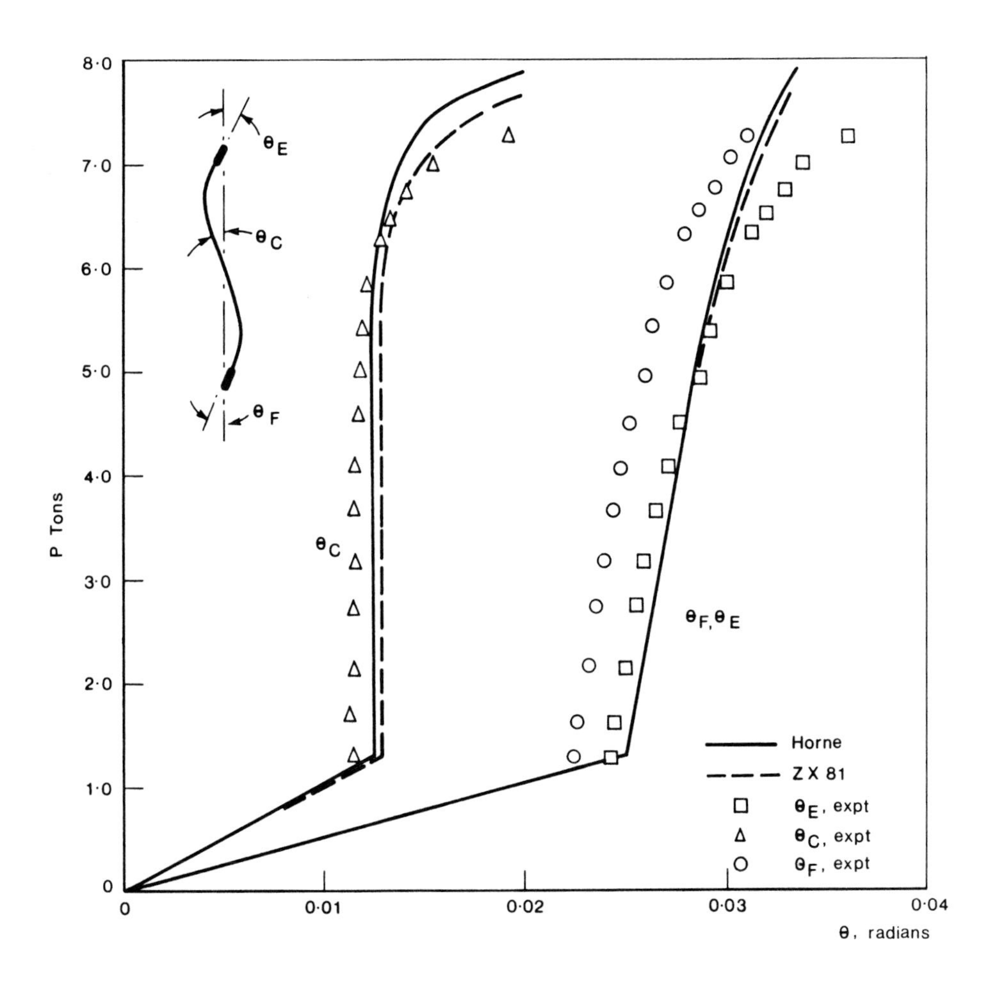

Fig 5 Load - slope diagrams

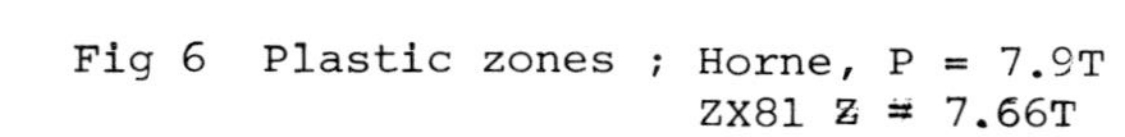

Fig 6 Plastic zones ; Horne, P = 7.9T
ZX81 Z ≠ 7.66T

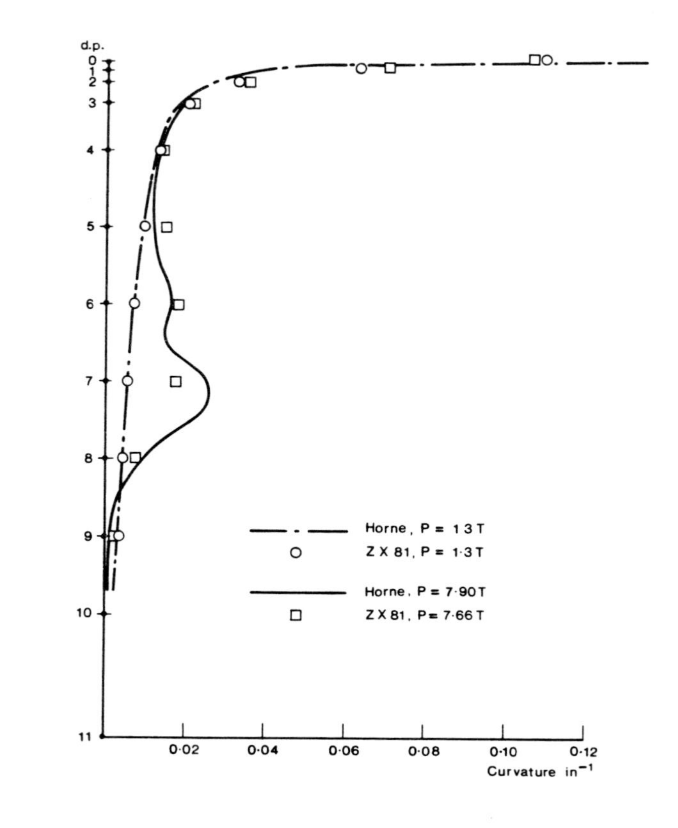

Fig 7 Curvature distribution along column

STABILITY OF SLENDER COLUMNS WITH MINOR AXIS RESTRAINT AND END PLASTICITY

D C Stringer

AMCA International Ltd., Ottawa, Canada

Synopsis

A theoretical and experimental study of the stability of slender I-section columns, loaded plastically in double-curvature about the major axis and restrained by elastic minor axis beams, is presented. The study investigates the extent to which the restraint remains effective as plastic hinges, at the column ends, participate in a collapse mechanism in the plane of the major axis. Design charts are presented to predict the minor axis restraint required by the column if it is to sustain plastic hinge action without instability.

Notation

A	cross sectional area of column
A_f, A_w	area of flange and web of column
b, d	width of flange and distance between flange centroids
E, G	modulus of elasticity, modulus of rigidity
F	shear force at ends of column
I_y	second moment of area of column about minor axis
I_{by}	second moment of area of minor axis beam
K	St. Venant torsion constant
K_b, K_c	ratios I_{by}/L_y, I_y/L
L, L_y	lengths of column and minor axis beam
M_{px}, M_{py}	full plastic moment in column in uniaxial bending in absence of axial load
M_f, M_y	full plastic moment of flanges and web of column for uniaxial bending about major axis
M_{xo}, M_{yo}	terminal bending moments in column about major and minor axes in presence of axial load
M_o	terminal bending moment in column about minor axis in absence of axial load

n	ratio P/P_p
P	axial load in column
P_E	Euler load of pin-ended column
P_{crit}	elastic critical load of column
P_p	squash load of column
Q	ratio of minor axis beam to column stiffness K_b/K_c
Q_{min}	minimum ratio of above as determined by analysis
R	ratio M_o/M_{py}
R_c	rotation capacity of column
r_1, r_2	non-dimensional factors defining position of neutral axis in flanges
r_x, r_y	radii of gyration of column about major and minor axes
To	longitudinal torque at ends of column
u	deflection of column about minor axis
w	non-dimensional factor defining position of neutral axis in web
z	distance along longitudinal axis of column
Γ	warping constant of column
θ	major axis hinge rotation of column
θ_1, θ_2	minor axis hinge rotation for 'compression' and 'tension' flange respectively
θ_e, θ_p	elastic and plastic end rotation about major axis of col.
ϕ	twist of column about longitudinal axis
ρ	ratio P/P_E
ρ_{crit}	ratio P_{crit}/P_E

Introduction

After considering various alternative procedures, a Joint Committee of the Institution of Structural Engineers and the Institute of Welding (1) recommended that columns in fully-rigid, non-sway frames be designed according to the P_xE_y loading condition. With this design procedure, major axis beams are designed in accordance with plastic theory while minor axis beams are designed to remain elastic under factored loads, thus providing rotational end restraint to the columns. The most critical loading pattern for column instability was taken to be the pattern which produces the largest component of single-curvature (S/C) major axis bending. The Joint Committee considered it inadvisable to allow plasticity at mid-height of the columns without precise control of the extent of plastic zones. Consequently, the basis of column design was to allow the extreme fibre stress near mid-height to just reach yield.

If the loading pattern were switched from S/C to double-curvature (D/C), however, larger major axis bending moments would be induced in the column and end plasticity would probably occur. A collapse mechanism involving plastic hinges at the column ends is shown in Fig. 1. Such a mechanism is more likely to take place in the case of external columns as the major axis beam bending moments must then be sustained solely by the column. The question which arises concerns the effect of end plasticity on the beam/column continuity about the minor axis. A simplified analysis by Horne (2) predicted that the loss of restraint due to end plasticity would result in sudden collapse of the column when the axial load was sufficiently above its Euler load. To further investigate the problem, a program of structural testing was conducted and an improved analytical model was developed at the University of Manchester (3), the results of which are described in this paper.

1. *EXPERIMENTAL PROGRAM*

1.1 *Test Variables*

To determine the conditions under which end plasticity leads to subsequent instability of a column, seventeen column tests were carried out with the variables selected as follows:

a) Axial load ratio $\rho = P/P_E$ (ρ = 1.5, 2.0, 2.4)

b) Minor axis beam stiffness ratio $Q = K_b/K_c$ (Q = 7.0, 3.0, 1.0)

c) Minor axis beam loading ratio $R = M_o/M_{py}$ (R = 0.0, 0.05)

d) Load path - two load paths were investigated:

   i) constant axial load accompanied by increasing beam load
   ii) constant beam load accompanied by increasing axial load

The slenderness ratios of the test specimens were kept constant at L/r_x = 50 and L/r_y = 150 for all tests.

1.2 *Test Apparatus and Loading Procedure*

The test assemblage is shown schematically in Fig. 2. Axial load was applied through crossed knife-edges. Since the knife-edges were not coincident with the intersection of the axes of the column and the beams, the effect of the applied axial force was to decrease the stiffness of the minor axis beams. This second-order effect was taken into account when comparing experimental and theoretical results. The deformation in three dimensions of the test specimen necessitated special supports at the remote end of the minor axis beams to avoid the introduction of parasitic forces. All loads were applied by hydraulic jacks and were measured by load cells. Instrumentation also consisted of transducers to measure column displacement, twist and end rotation, together with strain gauges for the computation of minor axis beam moments.

The load procedure for the majority of the tests involved the application of pre-selected axial load and minor axis beam load followed by the progressive increase of major axis beam load. After plasticity had developed at the column ends, further increments of D/C major axis deformation were applied to the column until failure occurred. Failure was defined to be either the point at which the column could no longer sustain its axial load or the point at which the major axis moments had fallen 20% from their maximum value.

1.3 Test Results

Of primary interest was the amount of major axis end rotation which could take place prior to the instability of the column. Following Galambos and Lay (4) rotation capacity is defined as: $Rc = (\theta_p - \theta_e)/\theta_e$; the ratio of the plastic to the elastic end rotation at a moment of 0.95 $Mmax$. Test results are summarized in Table 1. Typical load-deformation curves are shown in Fig. 4. Important factors influencing the behaviour of the restrained column were the ratio $\rho/\rho crit$ and the minor axis beam loading. When the axial load was close to the elastic critical load, failure of the column occurred almost immediately upon formation of end plasticity. Loads on the minor axis beams further reduced the rotation capacity. For values of $\rho/\rho crit$ less than approximately 0.6, the column gained considerable benefit from the restraint offered by the minor axis beams and the moment-rotation curves about the major axis were flat-topped.

A typical pattern of plastic zones in the column is shown in Fig. 3. Initially, yield was confined to the 'compression' flange at each end of the column. Later, part of the 'tension' flange yielded, but plasticity never penetrated the full depth of the section. Finally, due to flexure about the minor axis, yielding occurred in the flange tips near mid-height of the column (Fig. 3). At this stage, the major axis moments had significantly decreased and the pre-selected axial load could not be maintained.

2. ANALYTICAL MODEL

2.1 The Biaxial Plastic Hinge

An analytical investigation was carried out, based on the idealization that the restrained column remains elastic except at the ends where biaxial plastic hinges or complex hinges form. This assumption may be justified due to the high moment gradient. Several patterns of complex hinge may occur in an I-section depending on the relative magnitude of axial load, bending about the two axes and warping. It is also probable that, during the history of loading, a complex hinge may change from one pattern to another. Fig. 5 shows one pattern of biaxial hinge where the position of the neutral axes is defined by non-dimensional quantities r_1, r_2 and w in terms of which equilibrium equations may be written expressing axial force, major and minor axis moments:

$$n = \frac{1}{A}\,[(r_1 - r_2)\,Af + wA_w] \qquad (1)$$

$$M_{xo} = \frac{(r_1 + r_2)}{2}\,Mf + (1 - w^2)\,M_w \qquad (2)$$

$$M_{yo} = M_{10} + M_{20} = (2 - r_1^2 - r_2^2)\,\frac{M_{py}}{2} \qquad (3)$$

where M_{10} and M_{20} are the flange moments about the minor axis for the 'compression' and 'tension' flanges, respectively.

Values of r_1, r_2 and w must satisfy strain compatibility at the junctions of the flanges and the web (Fig. 6) and are related by the plastic discontinuity angles:

$$\Delta\theta_1 = \frac{1 + w}{r_1} \frac{d}{b} \Delta\theta \qquad (4)$$

$$\Delta\theta_2 = \frac{1 - w}{r_2} \frac{d}{b} \Delta\theta \qquad (5)$$

2.2 *Differential Equations for Minor Axis Flexure & Torsion*

The deflected shape of a I-section column subjected to D/C major axis bending, S/C minor axis bending and axial thrust is shown in Fig. 7. It is assumed that displacements in the YZ plane are negligible. The differential equations for minor axis bending and torsion of the column may be written as:

$$EI_y \frac{d^2u}{dz^2} - M_{xo} \left(1 - \frac{2z}{L}\right)\phi + Pu - M_{yo} = 0 \qquad (6)$$

$$E\Gamma \frac{d^4\phi}{dz^4} - (GK - Pr_o^2) \frac{d^2\phi}{dz^2} - M_{xo} \left(1 - \frac{2z}{L}\right) \frac{d^2u}{dz^2} = 0 \qquad (7)$$

To simultaneously solve the second and fourth order differential equations the following boundary conditions are required:

$$u(o) = \phi(o) = 0 \qquad (9)$$

$$\phi(L/2) = du/dz\,(L/2) = d^2\phi/dz^2\,(L/2) = 0 \qquad (10)$$

$$d\phi/dz(o) = 0 \text{ (prior to hinge formation)} \qquad (11)$$

$$d^2\phi/dz^2(o) = -2\,(M_{10} - M_{20})/EI_yd \text{ (after hinge formation)} \qquad (12)$$

The finite difference method was used to solve the equations, the major axis end moment being increased until the end forces were consistent with those necessary for the formation of a biaxial plastic hinge.

At the stage of the solution where a complex hinge had just formed, the angles of discontinuity θ_1, θ_2 and θ are zero. Further deformation of the column will produce plastic flow at the column ends necessitating a change in the boundary conditions. Subsequent points on the load-deformation curve were obtained by increasing the plastic deformation angles and iterating the solution of the differential equations until strain compatibility at the plastic hinge was satisfied. Moment-rotation curves from the analytical model, illustrating the effect of variables ρ, Q and R, are shown in Fig. 8. A comparison between the experimental results and theoretical prediction is given in Fig. 4. In general, the theoretical predictions of major axis end moment were smaller than the measured values because the analytical model does not account for strain hardening nor the "gusset plate" effect (the stiffened region at the beam to column interface).

3. *MINIMUM RESTRAINT REQUIREMENTS*

Design charts were developed from which the required minor axis restraint can be obtained for a given column subjected to end plasticity from D/C major axis bending and axial load. The derivation of the charts is based on the elastic critical load of a 'deteriorated' column in which hypothetical structural hinges in alternate flanges replace the plastic zones which actually form in a column of this type (Fig. 9). The numerical results from the analysis are presented in the form of minimum restraint curves for various D/t values as shown in Fig. 10. To ensure a reasonably flat-topped M-θ curve, however, more than the theoretical minimum restraint is required. From an examination of the test results, a suitable empirical relationship between the required stiffness ratio Q and the theoretical minimum stiffness ratio Q_{min} is given by:

$$\frac{1}{Q+1} = \frac{1}{Q_{min}+1} - 0.15 \qquad (13)$$

The required stiffness ratio Q is limited to external columns subjected to symmetrical D/C major axis bending with minor axis beams either unloaded or symmetrically loaded about the axis of the column. The axial load is assumed to be such that the neutral axis for major axis bending is inside the 'tension' flange.

4. *CONCLUSIONS*

A significant factor influencing the stability of restrained columns with end plasticity is the proximity of the axial load to the elastic critical load. Test results and the theoretical model indicated that if $\rho/\rho_{crit} < 0.60$ and if there is no significant bending moment about the minor axis, the moment-rotation curve of the columns would be flat-topped. The minimum minor axis restraint required for a given column size and axial load may be computed with the use of theoretical design charts such as the one illustrated in Fig. 10. To ensure stability, however, while providing rotation capacity at terminal plastic hinges, the minimum restraint should be increased by a factor (eqn. 13) established from test results.

References

1. "Fully Rigid Multi-Storey Welded Steel Frames", Second Report of the Joint Committee of the Institution of Structural Engineers and the Institute of Welding, 1971.
2. Horne, M.R., "Failure Loads of Biaxially Loaded I-Section Columns Restrained About the Minor Axis." Engineering Plasticity, ed. by J. Heyman and F.A. Leckie, Cambridge University Press, 1968.
3. Stringer, D.C., "The Elastic-Plastic Behaviour of Restrained Columns." Ph.D Thesis, University of Manchester, Feb. 1972.
4. Galambos, T.V., Lay, M.G., "End-Moment End-Rotation Characteristics of Beam-Columns." Report No. 205A.35, Fritz Engrg. Lab., Lehigh University, Bethlehem, Pa., 1962.

Test No.	Q'	R	ρ	ρ'_{crit}	$\frac{\rho}{\rho'_{crit}}$	$\frac{Mxo}{Mpx}$	R_c
A1	7.13	0.0	1.45	3.32	0.436	0.95	4.7
A2	7.13	0.052	1.45	3.32	0.436	0.88	4.8
A3	6.98	0.0	1.93	3.32	0.581	0.68	4.1
A4	6.98	0.054	1.93	3.32	0.581	0.63	2.3
A5	6.86	0.0	2.33	3.32	0.703	0.62	3.7
B1	2.06	0.0	1.58	2.37	0.667	0.83	5.1
B2	2.10	0.042	1.45	2.37	0.612	0.97	4.2
B3	1.95	0.0	1.93	2.37	0.815	0.77	1.3
B4	1.95	0.041	1.93	2.37	0.815	0.65	0.6
B5	1.83	0.0	2.33	2.37	0.981	0.52	1.5
B6	1.95	0.0	1.93	2.37	0.815	0.73	0.7
B7	1.95	0.030	1.93	2.37	0.815	0.53	0.5
C1	0.60	0.0	1.45	1.59	0.913	0.93	2.1
C2	0.60	0.036	1.45	1.59	0.913	0.80	0.8

Table 1: Test Results

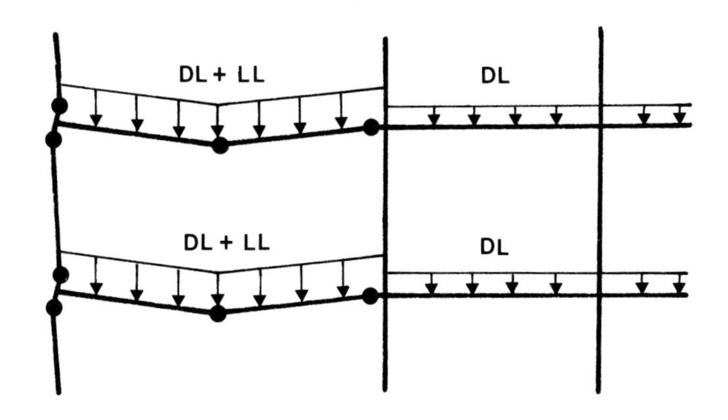

Fig. 1 Collapse Mechanism involving Plastic Hinges in External Columns

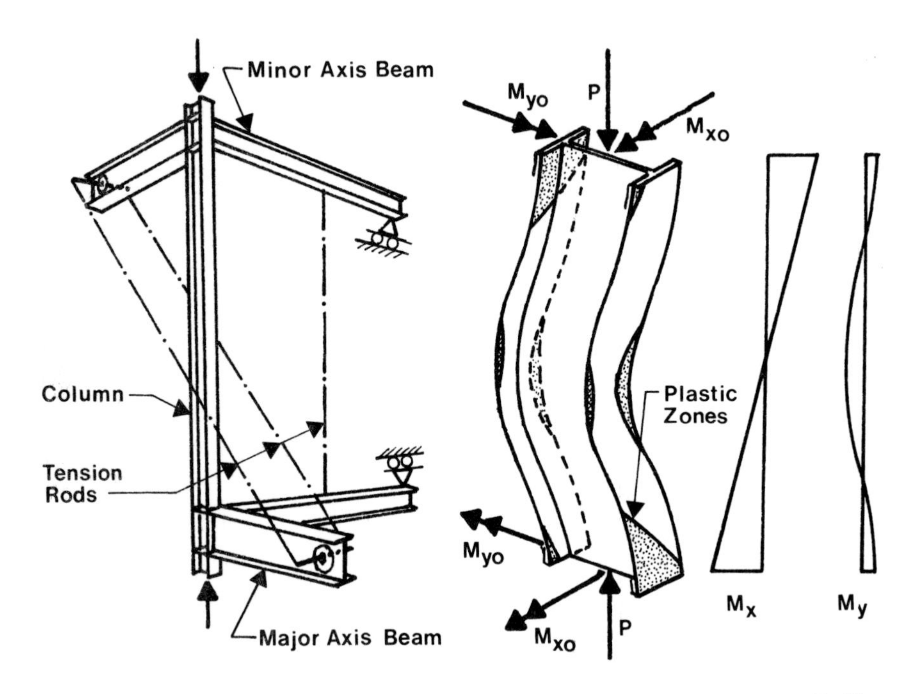

Fig. 2 Test Assemblage

Fig. 3 Plastic Zones at Failure

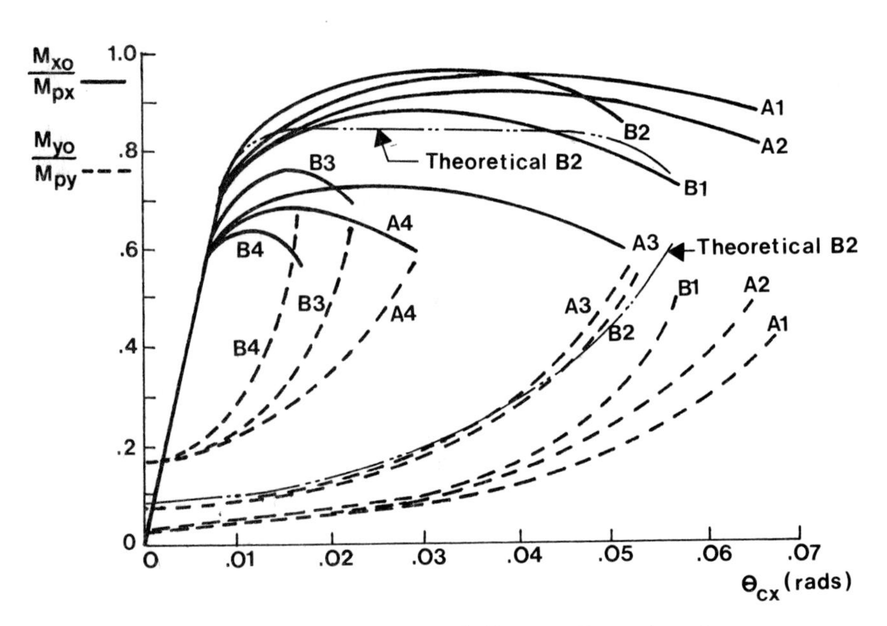

Fig. 4 Experimental Moment-Rotation Curves

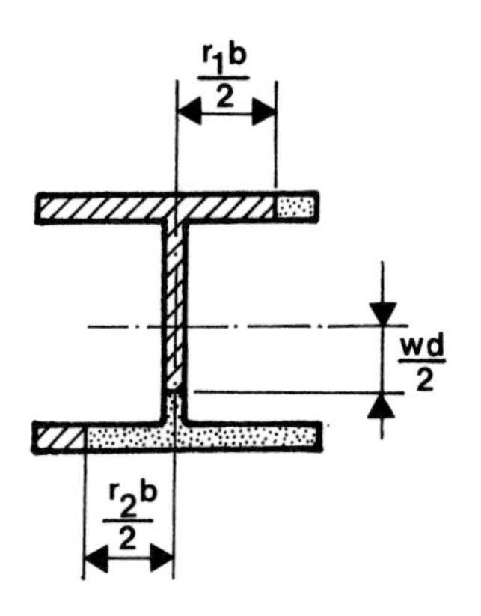

Fig. 5 Biaxial Plastic Hinge

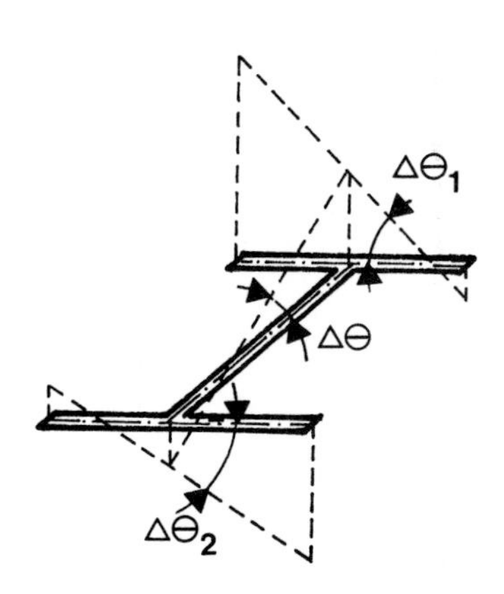

Fig. 6 Plastic Discontinuity Angles at Biaxial Hinge

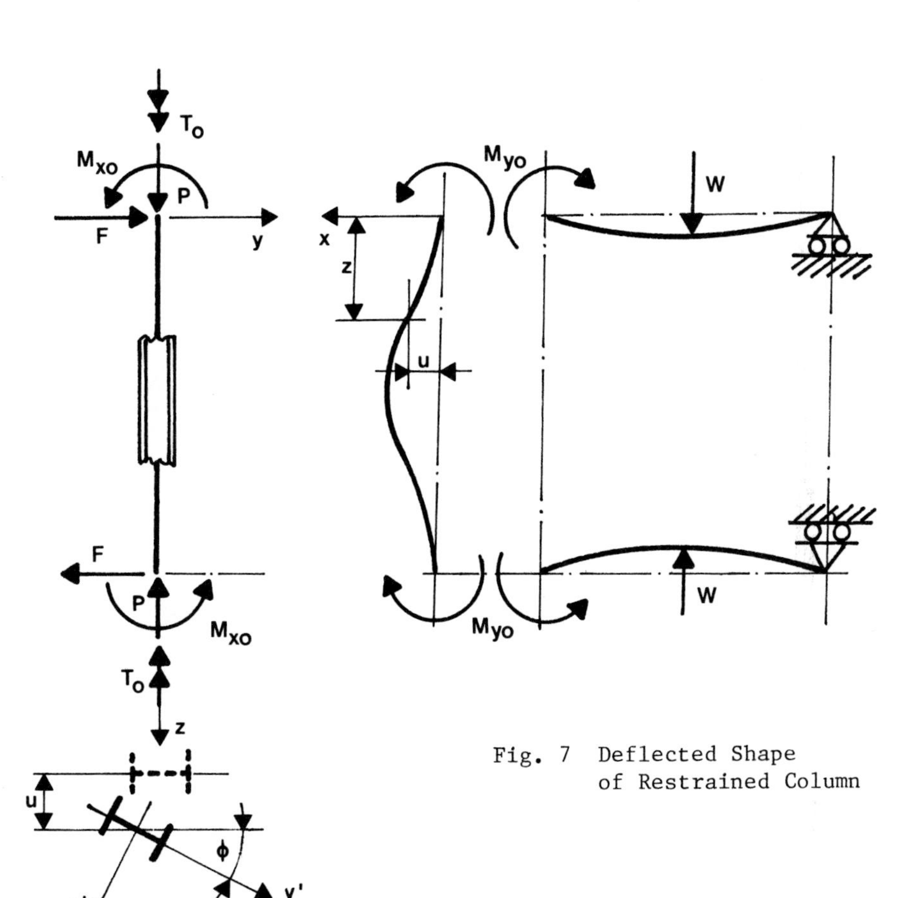

Fig. 7 Deflected Shape of Restrained Column

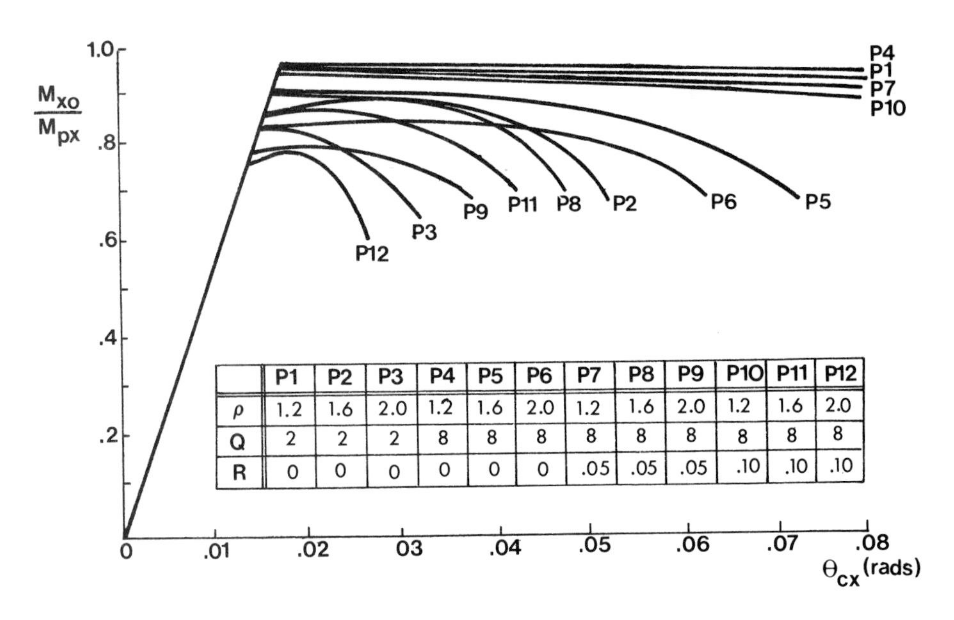

	P1	P2	P3	P4	P5	P6	P7	P8	P9	P10	P11	P12
ρ	1.2	1.6	2.0	1.2	1.6	2.0	1.2	1.6	2.0	1.2	1.6	2.0
Q	2	2	2	8	8	8	8	8	8	8	8	8
R	0	0	0	0	0	0	.05	.05	.05	.10	.10	.10

Fig. 8 Theoretical Moment-Rotation Curves

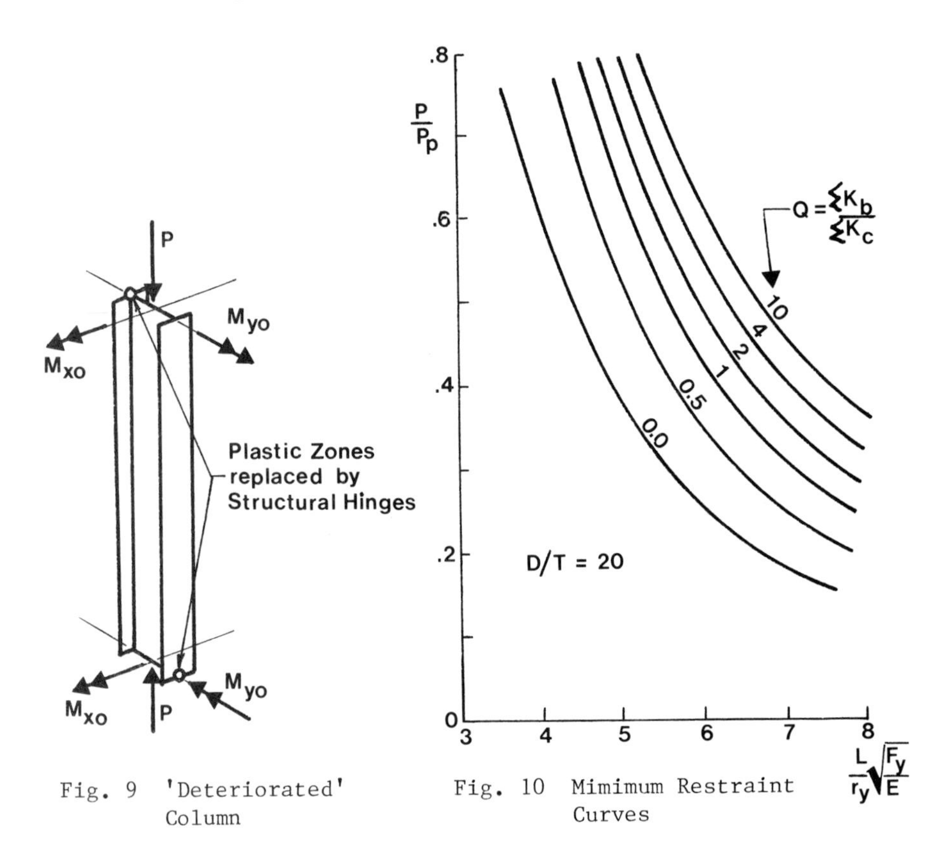

Fig. 9 'Deteriorated' Column

Fig. 10 Mimimum Restraint Curves

THE COLLAPSE BEHAVIOUR OF RESTRAINED COLUMNS

J M Rotter

School of Civil and Mining Engineering, University of Sydney

Synopsis

In this paper a simple analytical model is presented for the collapse states of restrained columns in uniaxial bending, including inelastic flexure. It shows that restrained columns can be separated into two distinct classes (heavily and lightly restrained), and the boundary between these classes is precisely defined. The effects of cross-section shape factor on collapse conditions is demonstrated, and the insensitivity of some collapse states to these factors is shown.

Notation

EI	elastic flexural rigidity of column
k_r	rotational restraint stiffness of beam
L	column length (centre to centre)
M	moment at column centre
M_J	applied moment at joint at collapse
M_U	ultimate moment of cross-section at axial load P
m_j	M_J / M_U
P	axial load on column
P_E	Euler load of column
P_S	squash load of column
p	P / P_S
R	$k_r L / EI$
R_T	tangent value of R, increased by effect of axial load on EI
α	$(\pi/2)\ \sqrt{\rho}$
θ	inelastic rotation at column centre
μ	calibration parameter for inelastic rotations
ρ	P / P_E

Introduction

Many experimental and theoretical studies of the in-plane behaviour of restrained columns have been conducted in the past fifty years. Many of these have dealt with specific classes of problem: elastic buckling (1), plastic collapse of heavily restrained columns (2,3), moment redistribution in columns at high axial load (4), lightly restrained columns (5,6,7). Few attempts have been made to relate these different studies and provide a comprehensive picture which allows the different regimes of behaviour to be defined. Moreover, very few attempts have been made to provide a general elastic-plastic description of restrained columns which can be applied equally to steel, reinforced concrete and composite columns.

In this paper a simple analytical model is presented to describe the collapse behaviour of restrained inelastic ductile columns in uniaxial symmetrical single curvature. The aim is to provide a framework within which different classes of column behaviour may be recognised, and which can be applied with equal validity to steel, reinforced concrete and composite columns. The ideas described here have been extended to columns with cross-sections of limited ductility, to conditions of asymmetric loading, to the effects of hinge formation in restraining beams and to biaxial bending (8). A thorough comparison has also been made with the simple models previously proposed by Merchant (9) and Wood (10), and some of the shortcomings of these descriptions revealed.

In the development of the present model, it is assumed that the basic information about the column is available: the interaction diagram for the cross-section ultimate strength is taken as known and is represented by the cross-section ultimate moment, M_U, at a given axial load, P. The column curve for axially-loaded pin-ended members is also taken as known. The rotational stiffness provided by the restraining beams is taken as k_r, and the collapse description attempts to relate the maximum moment which can be applied at the column end joints, M_J, to the cross-section strength, M_U, for a given ratio of restraint stiffness to column stiffness, $R = k_r L/EI$.

1. COLLAPSE MODES

The in-plane behaviour of restrained columns in symmetrical single curvature at collapse may be divided into two modes. In the first mode, the deformed shape is similar to that for isolated columns, with the central zone of the column experiencing inelastic deformations due to flexure over some part of the cross-section (Fig. 1). The second mode is more closely related to plastic mechanisms (Fig. 2). At collapse, the centre section of the column has reached its ultimate moment capacity, and in columns of strain-softening section, may have passed it. Overall collapse is triggered by inelastic flexural deformation at the ends, which effectively reduces the stiffness of the restraint. For ductile cross-sections, finite hinge rotations have occurred at the centre of the column before collapse. These two modes also occur in asymmetrically loaded columns (8).

2. DEVELOPMENT OF THE MODEL

The model describes the column as linear elastic throughout its length, and accounts for inelastic flexure near the column centre by treating the integrated inelastic curvatures as a concentrated rotation. Inelastic flexure may then be represented as a non-linear spring (Fig 3). The non-linear spring is chosen to be infinitely stiff for small bending moments, giving correct solutions for the elastic critical loads of restrained columns. It is also chosen to have a maximum strength equal to the section ultimate moment, M_U, and the intermediate response is made variable to allow a match with real behaviour.

In the restrained column model described here, the non-linear spring was chosen to display an indefinitely ductile response as the bending moment approaches the ultimate moment. As this model is but one of several (8), it was given the name DUC to distinguish its critical feature. The spring characteristic was chosen to be

$$M^2 + \left(\mu\frac{EI}{L}\right)\theta M - \left(\mu\frac{EI}{L}\right)\theta M_U = 0 \tag{1}$$

If this spring-centred elastic column is loaded as a beam under uniform moment, the moment-end rotation response is comparable with those of steel sections. By suitable choice of μ, it is possible to describe the behaviour of any beam made of ductile material. The curve for typical hot-rolled sections (6) lies between the curves for $\mu=100$ and $\mu=200$, whilst for a solid rectangular section, $\mu=20$ provides a close match.

The response of the column is found by writing slope deflection equations using stability functions (11) for one half of the column (Fig. 3) and enforcing equilibrium and compatibility with the restraining beam and with the inelastic spring. The condition that M_J is at its maximum value then leads to a general equation for all restrained column collapse states in the mode of Fig. 1:

$$M_j = A\cos\alpha\,[1 + 2\chi\,\{1-(1+\frac{1}{\chi})^{\frac{1}{2}}\}] \tag{2}$$

in which $A = 1 + \frac{R}{2}\,\frac{\tan\alpha}{\alpha}$; $\alpha = \frac{\pi}{2}\sqrt{\rho}$

$$\chi = \frac{\alpha\tan\alpha - R/2}{\mu A}\ ; \qquad m_j = \frac{M_J}{M_U}$$

This equation is valid until a hinge forms at the column centre.

For a perfectly elastic restrained column, if the joint moments are distributed between the beams and the column allowing for the loss of column stiffness by using stability functions, the applied joint moment can be related to the column central moment. If the column is then treated as having an ideal elastic-plastic section, the collapse joint moment for this ideal column is obtained:

$$m_j = \frac{M_J}{M_U} = \left(1 + \frac{R}{2}\,\frac{\tan\alpha}{\alpha}\right)\cos\alpha = A\cos\alpha \tag{3}$$

This is essentially the secant formula for a restrained column. If it is compared with the model DUC, the inelastic collapse moment is found to fall below the ideal value by the factor

$$[1 + 2\chi\,(1 - (1 + \frac{1}{\chi})^{\frac{1}{2}})]$$

which is dependent on ρ, R and μ. The unique dependence on the single parameter, χ, is very significant (8). Non-linear springs with different characteristics may be chosen, but the collapse equation always remains a function of χ. Empirical functions of χ may thus be used to improve the match with real behaviour.

The collapse equation is expressed in terms of ρ, the ratio of axial load to Euler critical load. If the buckling condition for the column is to be valid, then ρ must be defined on a tangent modulus basis as ρ_T (Fig. 4). If it is assumed that the reduction in flexural rigidity due to yield under purely axial loads is only dependent on the ratio of axial load to squash load, then ρ_T can be alternatively expressed as $\left(\frac{L}{L_{cc}}\right)^2$ where L_{cc} is the length of a pin-ended axially loaded column of the same section which would buckle at the load P.

By plotting collapse conditions non-dimensionally as ρ_T against m_j, it is possible to compare the results of studies on isolated columns with those of the present model. The predictions of DUC are shown in Fig. 5. Comparisons with the numerical results of Young (12) and Galambos and Ketter (13) for hot rolled I sections bent about the strong axis are shown in Fig. 6. A comparison for composite columns (14,15) is shown in Fig. 7. It is evident that a value of μ of the order of 50 or 100 provides a satisfactory description for all these columns.

3. *RESTRAINED COLUMN BEHAVIOUR*

Following the calibration of the model against known solutions for isolated columns, it can be used to describe the collapse conditions of restrained columns. When the non-dimensional restraint stiffness offered to a column by the beams at its ends is evaluated, it is vital that this should be defined on a tangent modulus basis, as was the axial load to Euler load ratio, ρ. The tangent flexural rigidity of the column has declined from EI to τEI, giving a greater effective restraint

$$R_T = \frac{k_r}{\tau(\frac{EI}{L})} = \frac{\pi^2 k_r}{PL_{cc}} \sqrt{\rho_T} \qquad (4)$$

Choosing a value for μ of 50, Fig. 9 shows the variation of joint moment (m_j) with restraint stiffness (R_T) in the collapse condition, for various values of ρ_T. The curves for different values of μ lie parallel throughout most of their length, differences developing quickly as the restraint stiffness rises from zero to 1.0. The difference between the curves for $\mu = 50$ and $\mu = 100$ is greatest at large values of ρ_T, but is never more than $\Delta m_j = 0.1$. An error in assessing the relevant value of μ is therefore not critical. The curves are terminated by the formation of a plastic hinge at the centre of the column.

At restraint stiffnesses larger than the value corresponding to the central hinge in the column, the model DUC gives no solution. It should then be replaced by a similar description, using finite hinge rotations at the central hinge, and inelastic zones at the ends (Fig. 2). Because these end inelastic zones develop very rapidly, it is not necessary to give them the partially yielded characteristics used in the model DUC at the centre of the column. Instead, an ideal elastic-plastic cross-section can be used. This set of conditions was used by Roderick (2). His equation is, therefore, a suitable counterpart to the model DUC for restraint stiffnesses greater than the value at central hinge formation. In the present notation, Roderick's equation becomes

$$m_j = \frac{R}{2} \frac{\cot \alpha/2}{\alpha} - 1 \tag{5}$$

The point at which the central hinge forms in the model DUC is independent of the shape factor parameter, μ. This is to be expected, since indefinitely large rotations of the central spring occur. The axial load at which the central hinge forms is identical to the critical load of the reduced structure of Fig. 8 which has previously been studied (13) and for which the buckling condition is

$$R_T = 2\,\alpha \tan \alpha \tag{6}$$

with $m_j = \sec \alpha$

This structure gives an approximate bound between those states of collapse which involve partial inelastic behaviour and those which involve finite hinge rotations at the centre with significant inelastic action at the ends. Corroborative evidence to show that a plastic hinge does indeed form at this restraint stiffness has been found (8) derived from the work of Lay (7).

Several descriptions of column collapse are compared in Fig. 10 for $\rho_T = 0.6$. The effect of extrapolating isolated column strengths ($R_T = 0$) to restrained columns can be seen. Partial yielding reduces the isolated column strength below the ideal secant formula value (Eqn 3), but this reduction does not increase as the restraint stiffness increases. The transformation of a restrained column into an equivalent isolated column is therefore very conservative except for lightly restrained columns.

The results of exact analysis (3) are compared with the predictions of DUC in Fig. 11. The latter has been calibrated to match the isolated column strength ($\mu = 63.25$). For a more complete set of comparisons, it is evident (Figs 5 and 6) that the value of μ should change for different levels of axial load. In Fig. 12, a uniform conservative value of $\mu = 20$ has therefore been chosen. The comparison is both close and conservative. More extensive comparisons with steel, reinforced concrete and composite columns are given elsewhere (8). Nevertheless, the value of this model lies not in the accuracy of its predictions (which can be adjusted), but in the description it provides of the modes of column collapse.

4. CONCLUSIONS

A simple model has been presented which describes the collapse states of restrained inelastic columns bent in symmetrical single curvature. It can be applied to steel, reinforced concrete of composite columns if the basic information concerning cross-section strength and the column curve are provided. It can also be extended to examine many other aspects of column behaviour (8). It can be calibrated to match real behaviour and clearly defines the parameters which control the loads at collapse.

References

1. Timoshenko, P.S. and Gere, J.M. 'Theory of Elastic Stability', McGraw-Hill, 1961.
2. Roderick, J.W. 'The Behaviour of Stanchions Bent in Single Curvature', BWRA Report FE1-5/24, 1945.
3. Baker, J.F., Horne, M.R. and Heyman, J. 'The Steel Skeleton - Vol. II Plastic Behaviour and Design', CUP 1956.
4. Gent, A.R. 'Elastic-Plastic Column Stability and the Design of No-Sway Frames', Proc. ICE Vol. 34, June 1966.
5. Bijlaard, P.P., Fisher, G.P. and Winter, G. 'Eccentrically Loaded End-Restrained Columns', Trans., ASCE, Vol. 120, 1955.
6. Driscoll et al, 'Plastic Design of Multi-Story Frames', Fritz Engg Lab Report 273.20, Lehigh University, 1965.
7. Lay, M.G. 'The Static Load-Deformation Behaviour of Planar Steel Structures', PhD Dissertation, Lehigh University, 1964.
8. Rotter, J.M. 'The Behaviour of Continuous Composite Columns', PhD Thesis, University of Sydney, Dec. 1977.
9. Merchant, W., 'Frame Instability in the Plastic Range', Brit. Welding Jnl, Vol. 3, No. 8, Aug. 1956.
10. Wood, R.H. 'Column Design - A New Approach', HMSO, London, Dec. 1973.
11. Horne, M.R. and Merchant, W. 'The Stability of Frames', Pergamon, 1965.
12. Young, B.W., 'The In-Plane Failure of Steel Beam-Columns', The Structural Engineer, Vol. 51, No. 1, Jan. 1973.
13. Galambos, T.V. and Ketter, R.L. 'Columns under Combined Bending and Thrust', Jnl Engg. Mech. Div., ASCE, Proc. Vol. 85, No.EM2, April 1959.
14. Basu, A.K. and Hill, W.F., 'A more exact computation of failure loads of composite columns', Proc. ICE, Vol. 37, May 1968.
15. Basu, A.K. and Sommerville, W., 'Derivation of formulae for the design of rectangular composite columns', Proc. ICE, Supp. Paper 7206S, 1969.

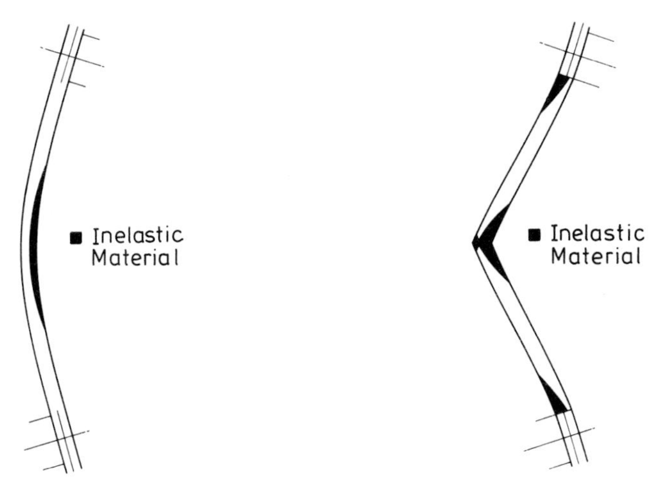

Fig.1 Restrained Collapse Mode 1 Fig.2 Restrained Collapse Mode 2

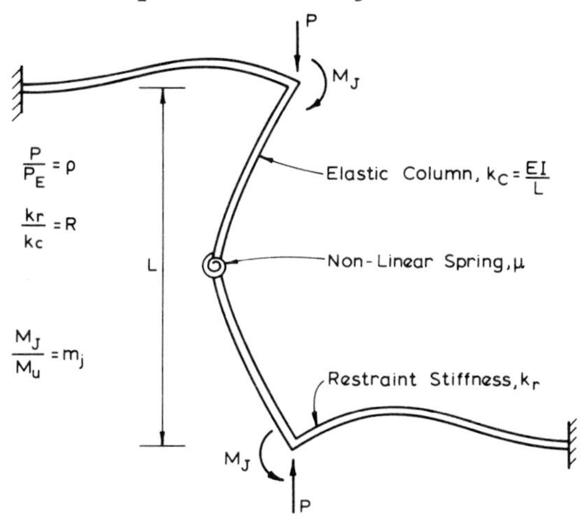

Fig.3 Restrained Column-Spring Model at Collapse

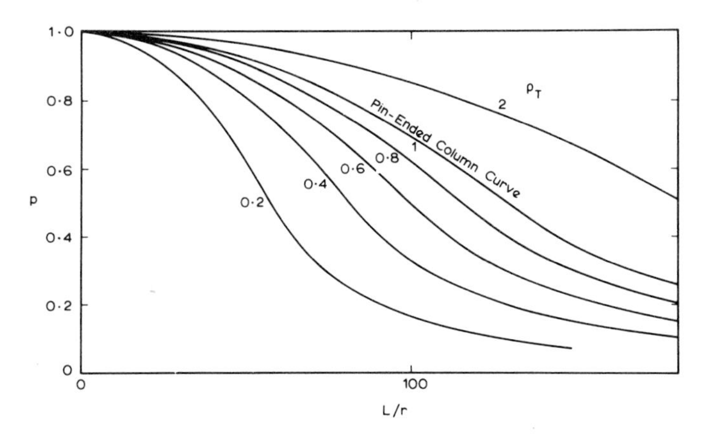

Fig.4 Curves of Constant ρ_T

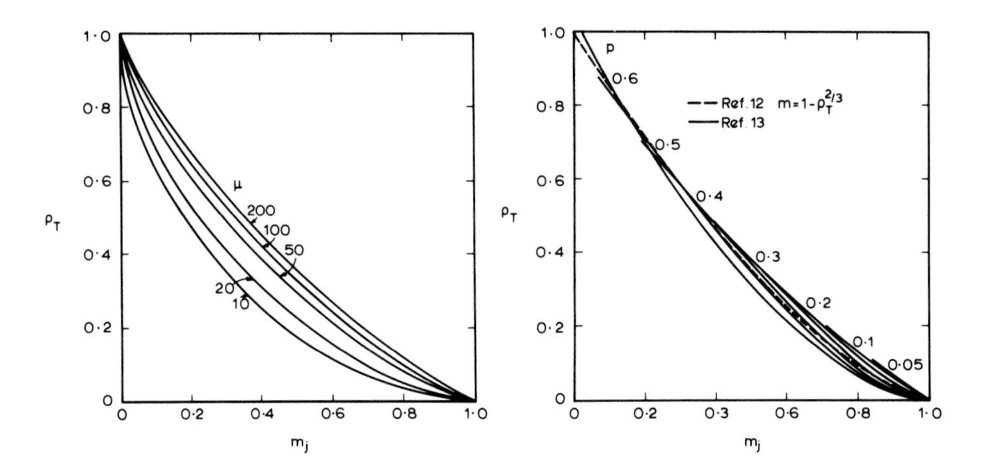

Fig.5 Isolated Column Collapse Loads for Model DUC

Fig.6 Steel Column Collapse Conditions (Refs.12 & 13)

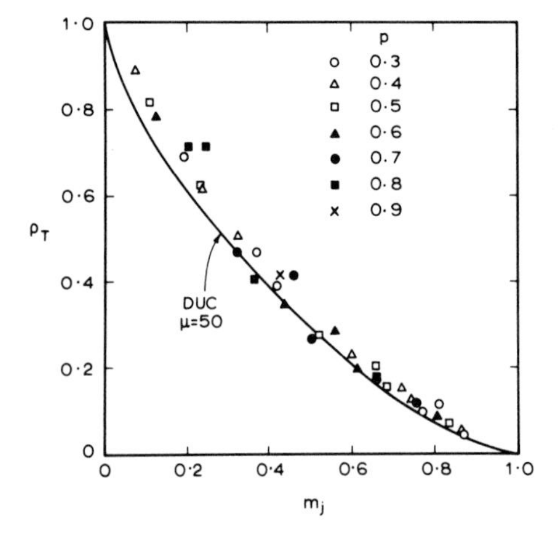

Fig.7 Composite Column Collapse Conditions (Refs.14 & 15)

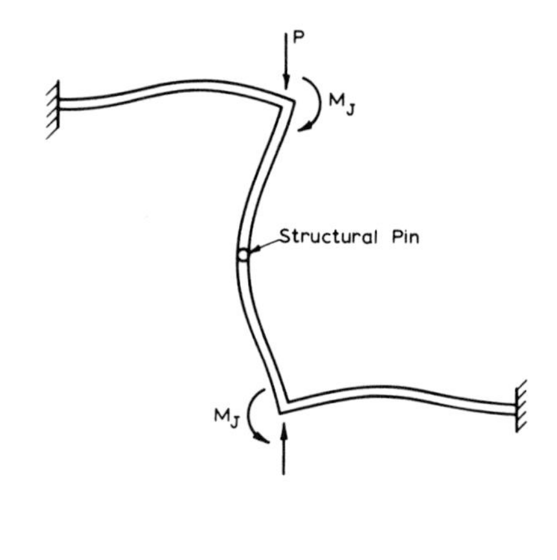

Fig.8 Reduced Structure Due to Merchant

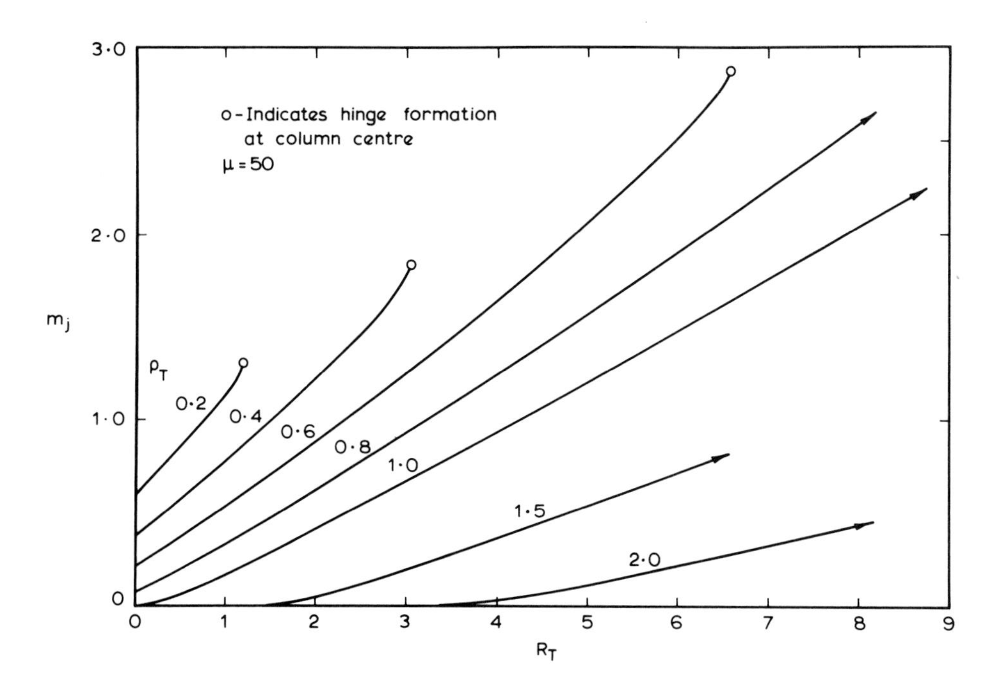

Fig.9 Joint Moment vs Restraint at Collapse for Model DUC

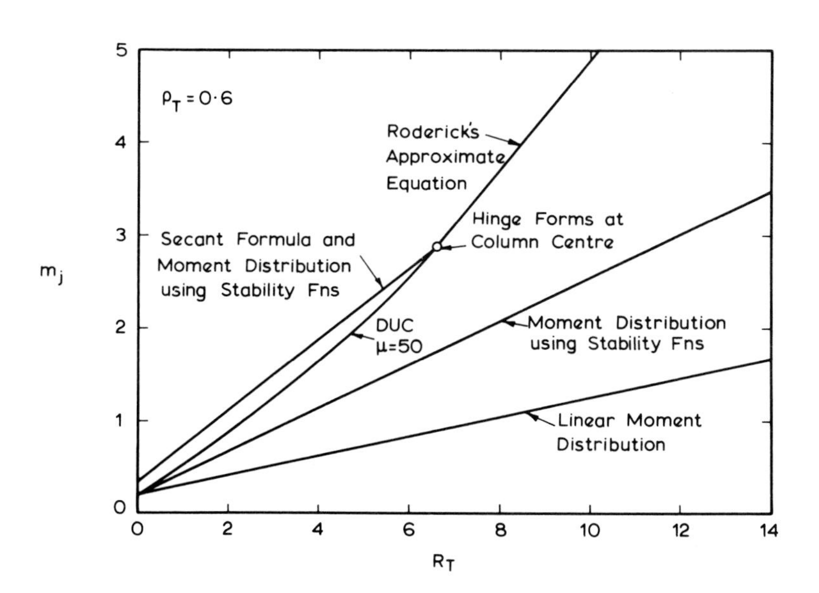

Fig.10 Restrained Column Collapse Predictions Compared

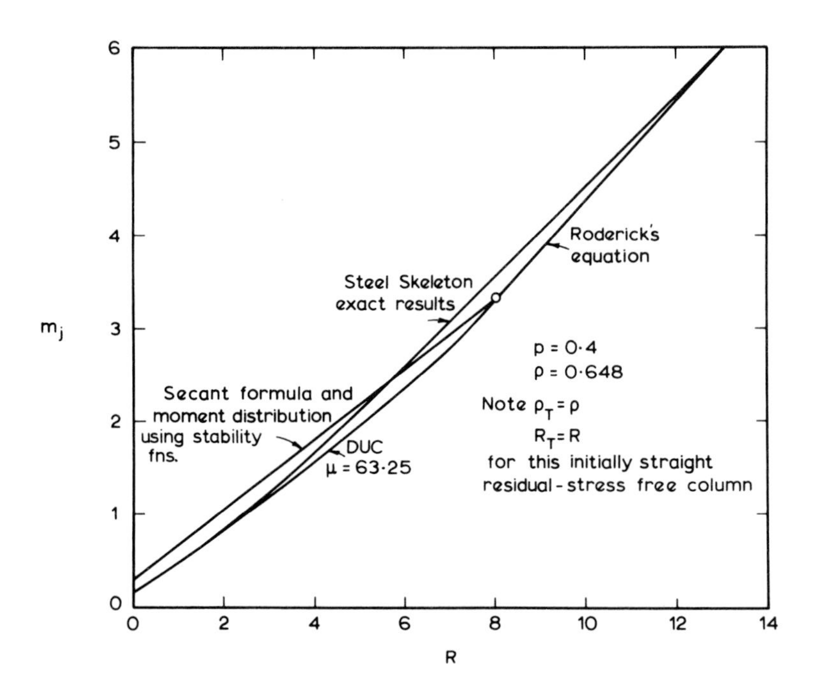

Fig.11 Comparisons with Exact Analysis (Ref.3)

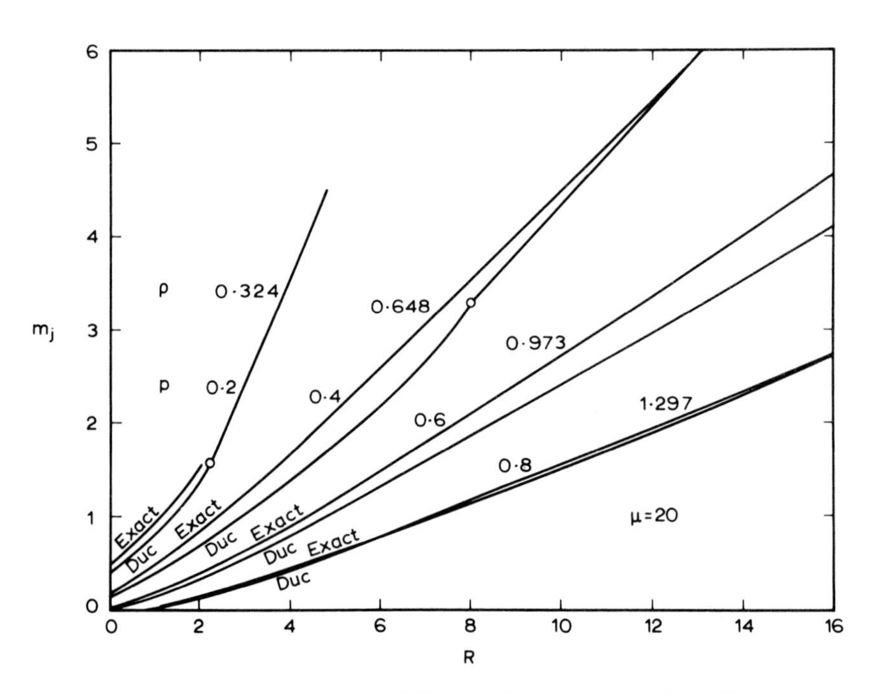

Fig.12 Comparisons with Exact Analysis (Ref.3)

INELASTIC INSTABILITY RESEARCH AT LEHIGH UNIVERSITY

L. W. Lu
Department of Civil Engineering, Lehigh University

Z. Y. Shen
Department of Structures, Tong-Ji University

X. R. Hu
Research Institute of Structural Theory, Tong-Ji University

Synopsis

Results of research on inelastic instability of initially crooked columns and beam-columns are presented. The problems studied are: flexural instability of wide-flange columns and beam-columns, flexural-torsional instability of concentrically and eccentrically loaded single-angle columns, and flexural-torsional instability of H columns subjected to axial force and major-axis bending. Solutions to these problems have been obtained using two separate computer programs, one for flexural instability and the other for flexural-torsional instability. The latter is based on a new finite element procedure. For each problem, typical results are given and the influence of the variables involved is discussed.

Notation

A	cross-sectional area
E	modulus of elasticity
e	load eccentricity
K	effective length factor
L	length of column
M	bending moment
M_p	plastic moment
P	axial load
P_y	axial yield load
R	rotational stiffness of end restraint
r	radius of gyration
S	section modulus
u_o	initial crookedness in x direction
v_o	initial crookedness in y direction
β	end moment ratio
ε	non-dimensional eccentricity

λ non-dimensional slenderness ratio

σ_y yield stress

Introduction

Researchers of steel structures at Lehigh University have enjoyed more than thirty years of continuous cooperation and exchange of information with Professor Michael R. Horne, the man being honored by this conference. The relationship began in the late 1940's when Professor Horne was at Cambridge University working on plastic design research with Lord John F. Baker. Much of the Lehigh research on plastic design and structural stability, including those of the writers have benefitted greatly from the association with Professor Horne, and this paper is prepared as an expression of appreciation to him.

The paper deals with inelastic instability of columns and beam-columns, a subject to which Professor Horne has made distinguished contributions. Among the topics selected for presentation, two are flexural instability columns:

1. axially loaded columns with initial crookedness and end restraint
2. initially crooked beam-columns

and three are spatial instability problems involving lateral and torsional deformations:

3. axially loaded single-angle columns with initial crookedness
4. initially crooked single-angle columns loaded through gusset plates
5. beam-columns with minor-axis crookedness and subjected to major-axis bending.

The results have been obtained from two computer programs developed recently at Lehigh for instability analysis of thin-walled structural members. One of the programs is for members with planar deformation only. The other program considers the effect of spatial deformations and can take into account the secondary bending and torsional moments caused by these deformations. The method of analysis developed for the planar problem is discussed first.

1. PLANAR STABILITY ANALYSIS

A general method of analysis, which can include almost all the known factors affecting the behavior of beam-columns, has been developed. It is similar to the method used previously in analyzing laterally loaded columns by Lu and Kamalvand (1). A description of the present method as applied to axially loaded columns has been presented by Shen and Lu (2). The specific factors that have been included in the development are: 1. initial crookedness, 2. end restraint, 3. eccentricities of axial load, 4. end-moment ratio, 5. residual stresses, 6. variation in mechanical properties of material over cross section, 7. stress-strain characteristics of material, and 8. loading, unloading and reloading of yielded fibers. For the case of a beam-column, it is possible to perform precise analysis for three possible loading paths: proportionally increasing axial load and bending moment, constant axial load and increasing bending moment, and constant bending moment and increasing axial load. The method makes no assumption with regard to the shape of the initial crookedness or of the deflected column under load. The analysis gives both the ascending and descending branches of the load-deflection curve.

2. FLEXURAL INSTABILITY OF COLUMNS

Much of the research conducted on columns in the 1950's and early 1960's placed emphasis on residual stresses and their effect on the tangent modulus load of perfectly straight columns. The tangent modulus load, which is only slightly less than the maximum or ultimate load, was subsequently accepted as a suitable basis for evaluating column strength. Since most of practical columns are initially crooked, the phenomenon of buckling or bifurcation at the tangent modulus load could not occur and failure is usually due to instability. Work on the effect of initial crookedness has been carried out, and the results show that for some columns the reduction in strength is too significant to be ignored in a rational design procedure.

The earlier research on column strength also shows that the magnitude of residual stresses in rolled shapes is approximately the same for steels of different yield stresses. For light column shapes made of structural carbon steel (σ_y = 36 ksi), the maximum compressive residual stress occurs at the flange tips and is about 0.3 σ_y (10.8 ksi). This value of residual stress has also been found in shapes of high strength steels.

Figure 1 shows the ultimate strength of the W8 x 31 columns made of steels with yield stresses of 36, 50 and 100 ksi and bent about its minor axis. The flange tip residual stress has been kept constant at 10.8 ksi. The non-dimensional slenderness ratio λ is defined by

$$\lambda = \frac{1}{\pi}\sqrt{\frac{\sigma_y}{E}}\ \frac{L}{r} \qquad (1)$$

The 100 ksi column is least affected by residual stresses and is the strongest. The initial crookedness for all the cases is assumed to follow a sine curve with a maximum value of $u_{om} = 0.001L$, which is the maximum acceptable sweep of the U.S. mill practice. It is seen that the initial crookedness has its maximum effect for λ between 1.05 and 1.2. For the 36 ksi column the maximum reduction is 31%.

Another factor important in column design is end restraint, which tends to increase the carrying capacity of columns. The exact amount of restraint that exist at the ends of a column in a practical structure is difficult to evaluate, but limited information is available for columns in building frames utilizing "simple" connections. Design recommendations for such columns have been presented by Lui and Chen (3). Figure 2 shows the effect of a small end restraint on the strength of the columns given in Fig. 1. The rotational stiffness R of the restraint is equal to $0.2EI/L$ at both ends. There is a noticeable increase in strength, even for such a small end restraint. It has been found by Shen and Lu (2) that the restraint required to produce an increase in strength which is exactly equal to the reduction caused by a given crookedness varies with λ. For the 36 ksi column, the maximum required R is $2.45EI/L$ for $u_{om} = 0.001L$ occurring at $\lambda = 1.2$.

3. FLEXURAL INSTABILITY OF BEAM-COLUMNS

Analytical studies have also been made on initially crooked beam-columns which fail by in-plane flexural instability. Figure 3 gives the axial load vs. end moment (P-M) interaction curves for the case of symmetrical single curvature bending (end moment ratio $\beta = 1.0$). The solid curves are for crooked W8 x 31 columns ($\sigma_y = 36$ ksi) with $u_{om} = 0.001L$. The dashed curves are for initially straight columns and are the basis of some currently available design formulas. The difference between the solid and dashed curves for a given slenderness ratio becomes significant when the axial load is high. Figure 4 shows comparisons between the P-M interaction curves of the straight W8 x 31 columns and those of the equivalent H columns welded from milled plates. The current formulas are likely to give very unconservative estimate of the strength of welded columns. A detailed discussion of the results obtained on crooked beam-columns can be found in a forthcoming report by Shen and Lu (4).

4. SPATIAL INSTABILITY ANALYSIS

A finite element method has been developed for instability analysis of initially crooked members with spatial deformations. Such a member would usually fail by flexural-torsional instability. The method and the computer program developed are very general and can be used to analyze axially loaded columns, beams and beam-columns with any type of rotational and torsional restraints. Applications have already been made to such problems as single angle columns with concentric and eccentric axial load, segmented beams subjected to transverse load, columns under combined axial load and torsion, H section beam-columns subjected to major axis bending, and biaxially loaded beam-columns. A description of the method has been given by Hu, Shen and Lu (5). The results of some of these studies are presented below.

5. FLEXURAL-TORSIONAL INSTABILITY OF SINGLE-ANGLE COLUMNS

A single-angle column with initial crookedness (or camber) and subjected to a concentric axial load would fail by flexural-torsional instability. The ultimate strength of such a member depends on the crookedness, residual stress, end restraint, and geometrical properties of cross section. The U.S. mill practice permits a maximum camber of L/480 for structural size angles and L/240 for bar size angles.

Numerical calculations have been carried out on four selected angle columns made of 36 ksi material. Two of these are structural size angles, ∠ 125 x 125 x 10mm and ∠ 125 x 80 x 10mm, and the other two are bar size angles, ∠ 2 x 2 x 1/4 and ∠ 3 x 2 x 1/4 (this section normally qualifies as structural size). The columns are simply supported in both bending and torsion. Figure 5 shows the ultimate strength of the columns as a function of the slenderness ratio λ (note: r_{min} is used in determining λ). The camber u_0 is assumed to be in the negative direction of the x axis. For unequal leg angles this represents the most critical situation. The effect of residual stresses has been taken into account in the calculations. The results show that the amount of camber has a significant influence on the strength of the angles when λ is less than about 1.5. Also, the unequal leg angles are weaker than the equal leg angles with the same camber.

6. SINGLE-ANGLE COLUMNS LOADED THROUGH GUSSET PLATES

Angles are often used as secondary members in main structures and connection is usually made through gusset plates. The load transmitted by the gusset plate acts eccentrically on the angle and causes biaxial bending. The strength of an eccentrically loaded angle can be determined by a flexural-torsional instability analysis. For an unequal leg angle, it is generally believed that the member can carry more load if the connection is made through the short leg. This problem has been examined recently as part of a general study on angle columns, and some of the results are shown in Figs. 6 and 7. The shape selected is ∠ 3 x 2 x 1/4 and the thickness of the gusset is 0.424" (this is to simulate the test specimens used in previous studies). The angle is so oriented that the x axis is parallel to the gusset. The camber is assumed to be in the positive direction of the y axis and has a maximum value of L/240 at the mid-height. The direction of the camber is again selected to correspond to the most critical situation. The angles are rotationally restrained at the ends in both the x and y directions. The stiffnesses of the end restraints have been so selected that, if the angles were perfectly straight and concentrically loaded, the effective buckling length would be equal to 0.8L, 0.9L and 1.0L for the three cases studied.

The results of Fig. 6 are for the case where the long leg is the outstanding leg, while in Fig. 7 the short leg is the outstanding leg. A comparison of the results shows that the "long leg outstanding" arrangement is not always the more favorable arrangement. In fact, higher ultimate loads may be obtained with the "short leg outstanding" arrangement for relatively short members ($\lambda < 1.2$ for $K = 0.8$). The results also indicate that end restraint can significantly increase the strength of an eccentrically loaded single-angle column.

7. FLEXURAL-TORSIONAL INSTABILITY OF BEAM-COLUMNS

When an initially straight H column is subjected to simultaneously applied axial load and major-axis bending moment, flexural-torsional buckling may take place at a critical combination of the axial load and bending moment. Further increase of the applied load is possible until the member finally fails by instability. A post-buckling instability analysis is therefore required to determine the ultimate strength of such a member. However, if the member is initially crooked about the minor axis, spatial deformations usually take place as soon as the load is applied and failure occurs when the instability limit is reached. A study of the effect of initial crookedness on the strength of beam-columns subjected to major-axis bending has recently been completed, and Fig. 8 shows some of the results. The load P is applied eccentrically with an eccentricity e. The column is a built-up member with a flange width of 400mm and a web depth of 800mm. The bending moment Pe is applied about the x axis. For the purpose of comparison of similar results obtained for other columns, a non-dimensional parameter $\varepsilon = eA/S$ is used to specify the load eccentricity, where A is the cross sectional area and S is the section modulus for bending about the x axis. The initial crookedness at the midheight is $u_{om} = 0.001L$.

The results show that the initial crookedness has a very substantial effect on the strength of the columns, especially for the case of small ε. This fact is not recognized in most of the current design procedures for laterally unbraced beam-columns.

The results presented in Fig. 8 are for the so-called "warping free" condition at the ends. The strength of the columns may be increased somewhat, if warping deformation is assumed to be fully prevented, as illustrated in Fig. 9.

8. *SUMMARY AND CONCLUSIONS*

Five studies of inelastic instability failure of columns and beam-columns have been described. The results presented indicate that

1. Initial crookedness and end restraint are two important factors affecting the strength of axially load columns failing by flexural instability and should be rationally dealt with in column design
2. Initial crookedness also affects the strength of beam-columns and, for the same crookedness, the strength difference between rolled and welded members is very significant
3. Unequal leg angles are not as strong as equal leg angles when they are subjected to concentric axial compression
4. When an unequal leg angle is loaded through gusset plates, the "long leg outstanding" arrangement does not always give higher ultimate load
5. When a laterally unbraced H column is subjected to combined axial load and major-axis bending, its ultimate strength is substantially reduced by initial crookedness about the minor axis.

References

1. Lu, L.W. and Kamalvand H. 'Ultimate strength of laterally loaded columns.' Journal of Structural Division, ASCE, Vol. 94, No. ST6, 1968.

2. Shen, Z.Y. and Lu, L.W. 'Analysis of initially crooked, end restrained steel columns.' Journal of Constructional Steel Research, Vol. 3, No. 1, 1983.

3. Lui, E.M. and Chen, W.F. 'End restraint and column design using LRFD.' Engineering Journal, AISC, Vol. 20, No. 1, 1983.

4. Shen, Z.Y. and Lu, L.W. 'Strength of initially crooked beam-columns.' Fritz Laboratory Report 471.7, Lehigh University (in preparation).

5. Hu, X.R., Shen, Z.Y. and Lu, L.W. 'Inelastic stability analysis of biaxially loaded beam-columns by the finite element method.' Vol. 2, Proceedings of the International Conference on Finite Element Methods, Shanghai, China, 1982.

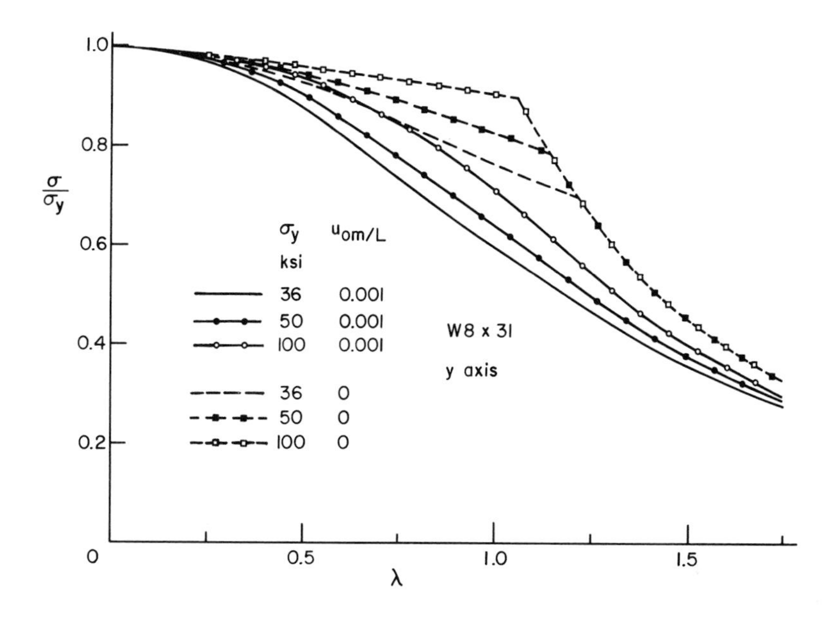

Fig. 1 Ultimate strength of straight and crooked columns

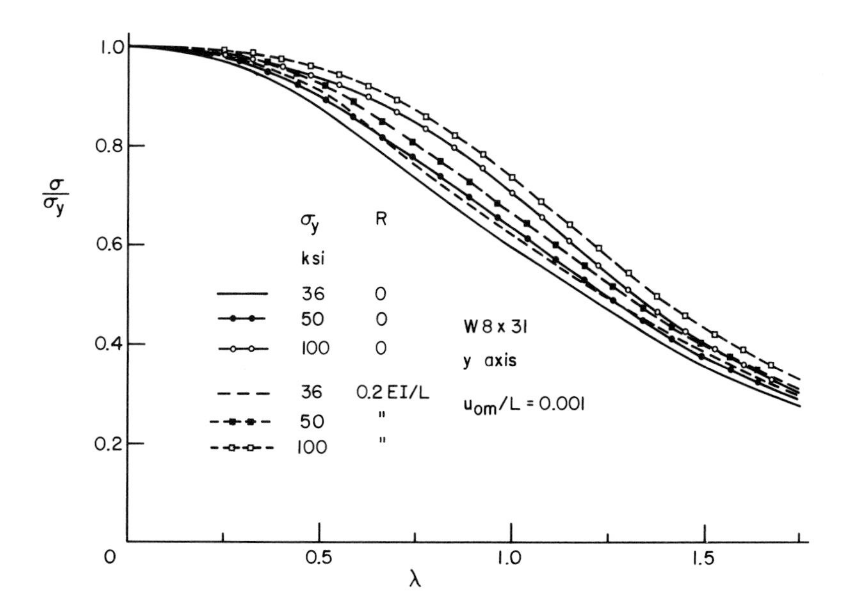

Fig. 2 Effect of end restraint on strength of crooked columns

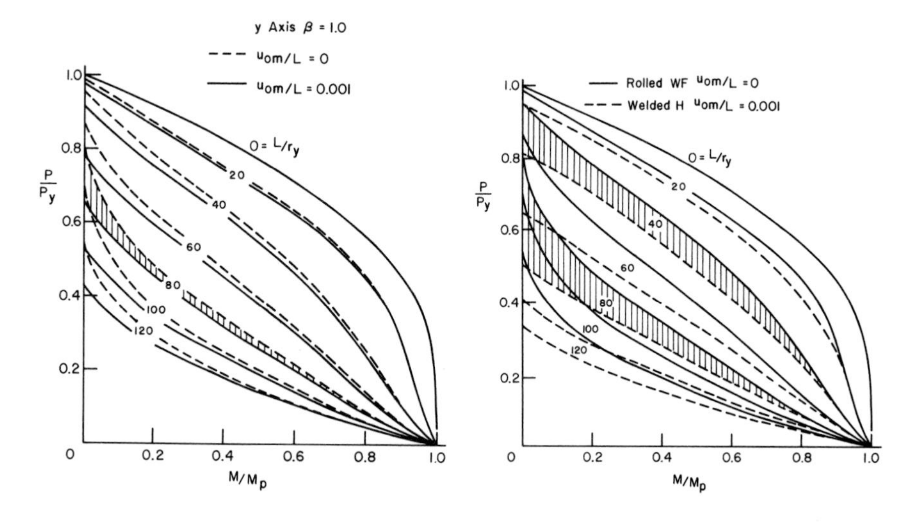

Fig. 3 Interaction curves for straight and crooked beam-columns

Fig. 4 Interaction curves for rolled and welded beam-columns

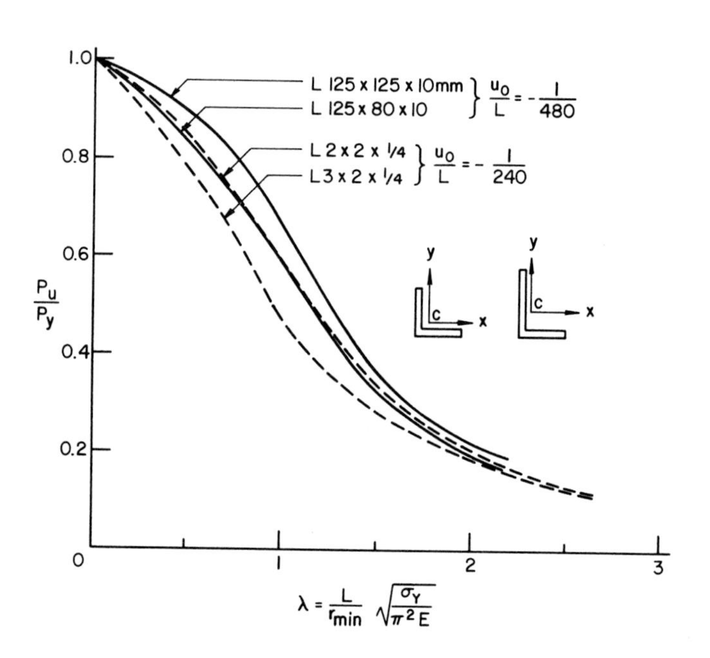

Fig. 5 Ultimate strength of concentrically loaded single-angle columns

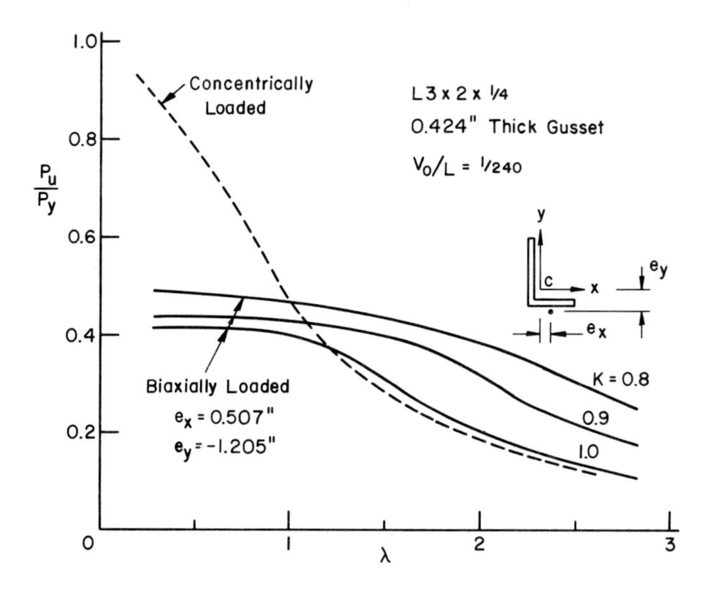

Fig. 6 Ultimate strength of single-angle columns loaded through gusset plates, long let outstanding

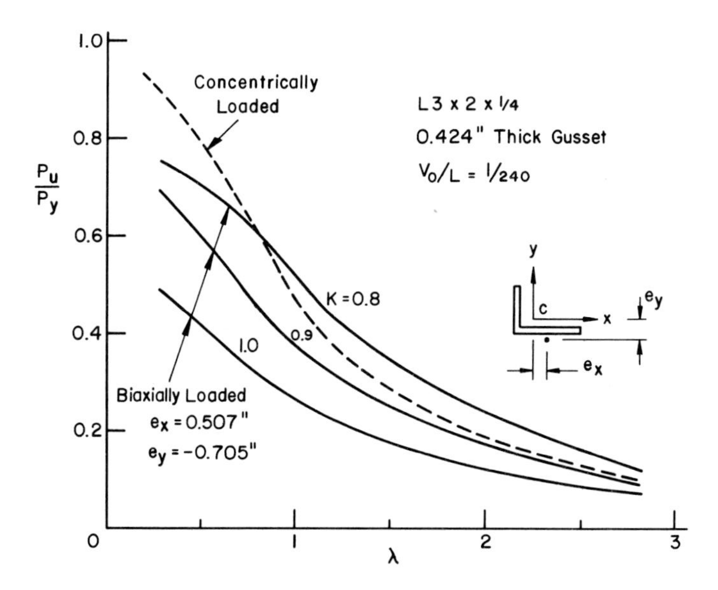

Fig. 7 Ultimate strength of single-angle columns loaded through gusset plates, short leg outstanding

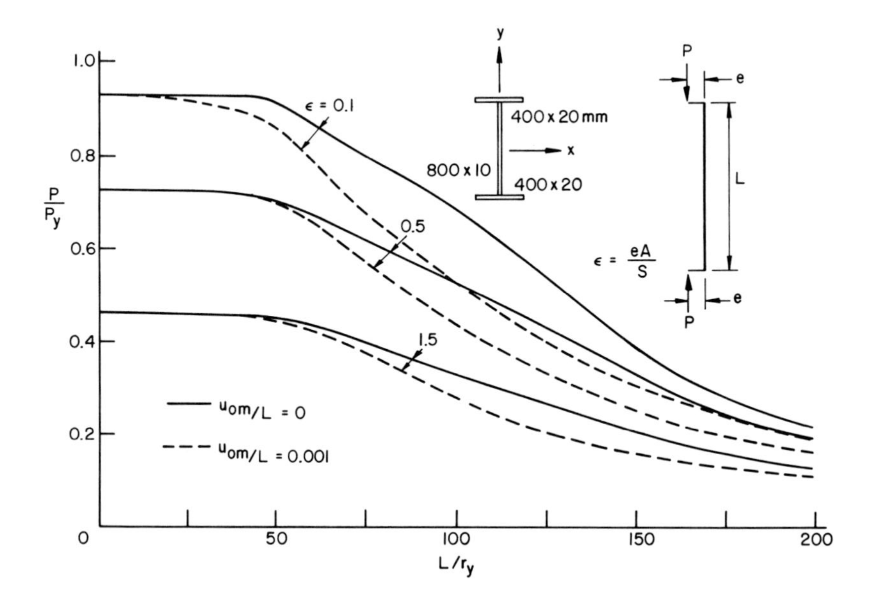

Fig. 8 Ultimate strength of beam-columns with and without minor-axis crookedness

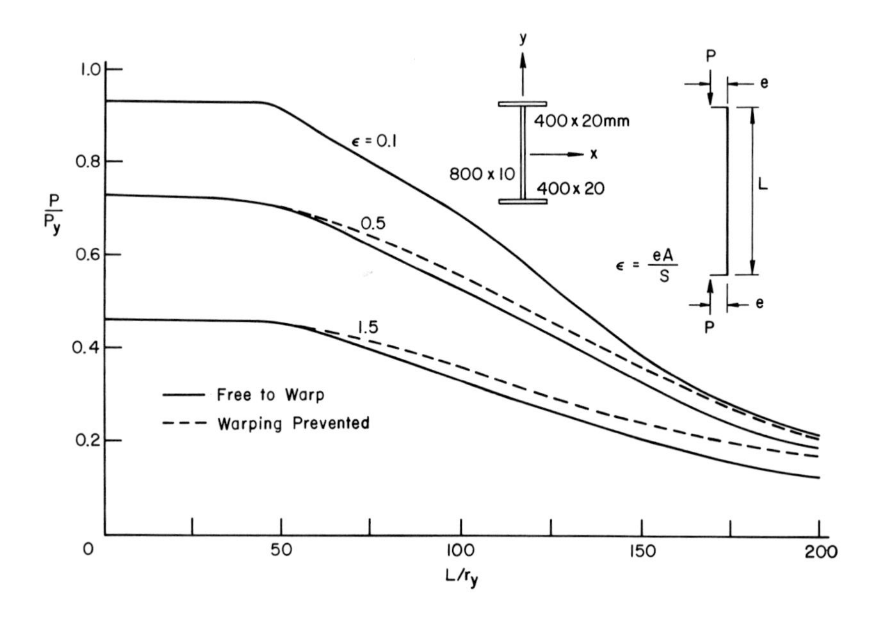

Fig. 9 Effect of warping restraint on strength of beam-columns

THE ANALYSIS OF RECTANGULAR PLATES

A H Chilver

Cranfield Institute of Technology

Synopsis

The paper describes the development of our present understanding of the structural behaviour of rectangular flat plates. It emphasises the importance in structures of metallic flat plates and the behaviour of such plates under in-plane forces. The general conclusion is reached that continuing progess in the analysis of plates is dependent on careful testing to confirm analytical ideas.

1 THE FLAT RECTANGULAR PLATE AS A STRUCTURAL COMPONENT

Most studies of rectangular plates are concerned with detailed analyses of one form or another of plate behaviour. Few studies, if any, look at the flat plate as a structural component. This is surprising, because flat plates are used widely in engineering construction, as for example in the beams and columns of buildings, in aircraft frames and in ship structures. Why is the flat plate so widely used, and what are its main attractions as a structural component?

From a purely structural point of view, the flat plate has a number of characteristics. Under lateral forces, flat plates are relatively flexible, compared for example with slightly-curved or corrugated plates. They are relatively stiff under the action of in-plane forces which do not lead to instability; for example, a flat sheet under combined in-plane tension in mutually perpendicular directions. For in-plane forces which lead to instability, as for example under compressive and shearing stresses, flat plates may buckle initially at lower stresses than curved or corrugated sheets; but flat plates are capable of carrying loads beyond initial buckling, and this behaviour differentiates them from curved or corrugated sheets.

Flat plates are geometrically simple, and there is geometrical symmetry about a plane through the middle surface of the plate. The flat external surfaces of a flat plate make it easy to attach other structural components.

Flat plates are used in modern structural engineering largely because plates, with isotropic and homogeneous properties, can be made relatively easily and cheaply in metallic materials. In particular, steel plates have led the way in the structural use of flat plate components. Other metallic materials, such as light alloys, have also been exploited extensively in structural engineering. But there has been relatively less use of the thin, flat, plate form in other structural materials, such as concrete and structural timber. Thick plates or slabs are widely used in concrete structures, but the use of thin concrete plates is limited by the problems of casting such plates with adequate reinforcement.

It would appear that the main attractions of the thin, flat plate are:-

(1) its geometric simplicity
(2) the relative ease of manufacture in modern metallic materials
(3) its resistance to in-plane forces, which do not lead to instability.
(4) its ability to carry increasing in-plane forces, although these may initially create buckling.

This gives the flat plate a special role in structural engineering. The importance of resistance to in-plane forces has led many analysts to concentrate their interests on the effects of such forces. For this reason, studies of instability of flat plates under in-plane forces have been extensive.

In-plane forces in flat plates can arise in a number of structural uses of flat plates. Plates in tension are most likely to occur on the tension sides of beams, girders and large structures, such as ships. Plates in compression will occur similarly on the compression sides of such structures; but, at the same time, there are numbers of structural columns which involve flat plates in compression. Perhaps the commonest flat plate component - carrying in-plane forces - is the web of a beam or girder, which supports heavy shearing forces, generally combined with overall bending.

In practice, therefore, the flat plate under in-plane forces which involve tension, compression and shear is of major importance. This is reflected in the most modern research, which has tackled in considerable depth and with considerable skill and success the problem of flat plates under in-plane shearing and direct stresses.

2 PLATES UNDER LATERAL FORCES

Having identified the importance of in-plane forces on flat plates, we should not though forget the very considerable knowledge that has been built up over the years of the behaviour of flat plates under lateral loading.

The linear-elastic theory of bending of flat plates is highly developed. With the help of modern finite-element analysis, all lateral loading conditions can be analysed.

The linear-elastic bending of plates is valid for only small lateral deflections of the middle surface of the plate. By 'small' is generally meant lateral deflections which are less than the thickness of the plate. When the lateral deflections approach the thickness of the plate, deviations from linearity increase by about the cube of the deflection, and non-linear effects dominate.

Elastic-plastic type behaviour has been extensively researched. To generalise in this field is especially difficult, because in general the spread of plasticity is complex. Numerical methods, using finite-element analysis, are very powerful tools for the study of specific problems.

Whereas in frame analysis plastic hinges can frequently be regarded as highly localised, for plates the hinges become complex axes or zones of plasticity within and around the middle surface of the plate.

In many practical cases of plates under lateral loading, there is no structural engineering interest in plastic conditions under working loads. Numerical methods of analysis are though very useful to determine the initial onset of plasticity and the ultimate plastic load-carrying conditions.

Limit analysis has been developed extensively to study the collapse loads of flat plates under lateral loading.

3 THE INITIAL BUCKLING OF THIN PLATES

The greatest structural problems of flat plates stem from the tendency of flat plates to buckle under certain types of in-plane forces at relatively low critical stresses. The thin sheet under in-plane forces is very sensitive to out-of-plane deformations. This is particularly so for states of in-plane stresses which are compressive or shear type.

Timoshenko (7) provided one of the earliest reviews of initial buckling, and other researchers have added much to this over the years, while Bulson (2) has given a comprehensive history of the development of knowledge of linear buckling stresses.

Walker (8,3) and his associates have studied extensively the buckling of plates under a wide range of loading conditions, and their work brings together many of the earlier studies.

The result of all this is a wide-ranging knowledge of the initial elastic buckling of rectangular plates under various conditions of loading and with various types of edge support. The field has been greatly strengthened by the powerful analytical tool of finite-element analysis, which opens the way to the assessment of the critical value of any loading condition. The analytical work has dealt with cases not only of continuous support but also of intermittent longitudinal supports and inter-rivet buckling. Solutions have been developed for concentrated edge compressive loads, for plates with holes, thermal buckling, and stiffened plates - with longitudinal and transverse stiffeners.

The initial buckling of very thin flat plates is elastic in form. In the intermediate ranges of thinness between very thin and thick plates buckling is elastic-plastic in form. Our knowledge of elastic-plastic buckling of plates is considerably less extensive than that of elastic buckling. We assume an idealised compressive stress-strain curve of the form shown in Fig 1, the onset of yielding occurring at a compressive yield stress, σ_y. Plastic yielding is simplified to a constant plastic tangent modulus, E_t. In the case of columns, it is now generally accepted that elastic-plastic buckling is determined by the tangent modulus, as expounded by Shanley. For a perfectly-straight, pin-ended column, there are three 'zones' of column buckling, Fig 2. For long columns, with high slenderness ratios (L/r), buckling is elastic at the Euler critical stress. Elastic buckling is often described as 'neutral', but in fact is mildly-stable. For low slenderness ratios, $\sigma_{cr} = \pi^2 E_t/(L/r)^2$, and buckling is again stable. In the intermediate zone, $\sigma_{cr} = \sigma_y$ and buckling is stable for lower slenderness ratios and unstable for higher slenderness ratios. In the case of perfectly-flat plates, initial plastic buckling probably follows similar forms. But, in the case of plates, buckling generally is more stable in form than in columns. For very thin plates, initial elastic buckling is generally stable. For very thick plates, initial plastic buckling is also probably stable. In the intermediate zone, buckling may well occur in a relatively stable state.

The detailed analysis of initial elastic-plastic buckling of thin plates may be of limited practical value, because in design we may be more concerned with the collapse of flat plates. Collapse may occur at very large deformations, and this makes the whole problem of elastic-plastic behaviour a very complex one. Our actual knowledge of elastic-plastic plate buckling is limited largely to test results on a wide range of plates in various materials. But, here again, tests generally give collapse loads and not initial buckling loads.

Because buckled forms of plastic plates may be stable, there has been a tendency for researchers to explain the collapse loads of plates in terms of large-deformation plastic theory rather than incremental theory. It is probable though that large-deformation plastic theory is in effect describing the early stages of post-buckling, and that initial buckling occurs at tangent modulus stresses.

4 THE IMPORTANCE OF ELASTIC POST-BUCKLING OF THIN PLATES

Unlike columns, plates exhibit relatively stable elastic buckling characteristics. For plates with supported edges, the buckled middle-surface of the plate can be sustained in stable equilibrium with large lateral deflections. The post-buckling of thin elastic plates is, in general, a highly non-linear problem. In general, there are no entirely successful analytical solutions, and there is much dependence in engineering design on test results.

The nature of initial elastic buckling of a thin plate can be judged from the initial in-plane stiffness of buckled plates, under various loading conditions.

Consider as an example a long, supported, rectangular, flat plate, which is loaded in compression through rigid end platens, Fig 3. The average compressive stress on the ends is σ_{av}, and the end compressive load is $P = \sigma_{av} bt$, where b is the breadth and t the thickness of the plate.

We shall consider a number of conditions on the longitudinal edges. Suppose the plate contracts longitudinally by an amount εL, so that ε is the overall longitudinal strain of the length L of the plate. In the pre-buckled state, the plate contracts longitudinally by the elastic deformations as given by usual linear-elastic theory. Buckling takes place at a critical stress σ_{cr} and critical longitudinal strain, ε_{cr}, Fig 4. Beyond the critical stress, a post-buckled form is possible, at a 'reduced' contractional stiffness.

Plates under end compression give post-buckling stiffnesses of the forms shown in Fig 5. The post-buckling stiffnesses indicated correspond to the following longitudinal edge conditions:

(a) simply-supported longitudinal sides free to move in the plane of the plate, but longitudinal sides constrained to remain straight
(b) longitudinal sides built-in, but free to move laterally,
(c) one longitudinal edge simply-supported and the other wholly free,
(d) longitudinal sides free to move laterally, and not constrained to remain straight.

From this, we note that initial in-plane stiffness after buckling is of the order of 50% of pre-buckling in-plane stiffness.

In the case of shearing stresses, the initial post-buckling stiffness are closer to pre-buckling stiffness, and attain values of the order of 90% of the pre-buckled stiffness, as shown in Fig 6. In (a), all sides are simply-supported and, in (b), all sides are built-in.

Thus, elastic plates, at least in the early stages of buckling, have considerable in-plane stiffnesses. This is an important structural characteristic which distinguishes flat plates from other structural components. In columns, for example, initial elastic post-buckling is generally fairly neutral. In many situations in which shells are used, initial elastic buckling is highly unstable.

The stable nature of elastic buckling makes it possible to design plates to operate in buckled states. One of the earliest forms of this was the tension-field girder, in which the web of a beam was assumed to have buckled under shearing stresses but still be capable of carrying a shear load.

The stable nature of the buckling of flat plates has made it possible to develop non-linear analysis of buckled plates to a high degree of sophistication. Finite-element analysis can be used to study the buckling forms of elastic plates, and much useful work has been done in this area. As pointed out by Walker (8), the finite-element analysis can be extended to deal with elastic-plastic behaviour. Such analyses have been used powerfully in developing design methods for box-girder bridges. Horne and Narayanan (4) have played an important role in showing how analytical methods correlate with test results of stiffened plates.

But the usefulness of highly-sophisticated methods of numerical analysis is limited by the tendency of buckled plates to change the forms of buckling as buckling develops. The nature of the problem can be illustrated in the following way. Suppose a flat plate of a given 'aspect ratio' is loaded to its initial elastic buckling stress $(\sigma_{cr})_{min}$, and suppose this occurs in a wave form m. The buckling stresses of other wave forms $(m+1), (m-1)$, etc., may be quite close to $(\sigma_{cr})_{min}$ in wave form m. This means the elastic buckling system is one in which there is possibly a cluster of buckling stresses. This clustering of buckling stresses has two important effects:

(i) the initial imperfections in a plate can lead to complicated post-buckling, in which the plate snaps from lower to higher buckling modes;

(ii) two distinct initial buckled forms can coalesce to generate quite new buckled forms.

The importance of (i) is that, unless the initial imperfections are known precisely, no numerical analysis can indicate the precise form of post-buckling behaviour. The importance of (ii) is that new forms of buckling can be generated by coalescence of buckling stresses, and these are difficult to analyse *ab initio*. Such problems are discussed fully by Supple (5) and Thompson (6) and his associates.

For this reason, the careful testing of flat plates is essential to the progress of knowledge of flat plate buckling, if we are concerned in applying this knowledge to real structural engineering.

5 THE ULTIMATE COLLAPSE OF FLAT PLATES

Designers are frequently concerned in estimating the collapse loads of compressed and sheared flat plates.

Where initial imperfections are known precisely, the post-buckling of flat plates can be analysed very successfully, using finite-element methods. Graves-Smith, for example, [see (8)] showed that for compressed flat plates with longitudinal edges free to move in the central plane of the plate, the collapse load is reached when the maximum membrane stress at the edge of the plate reaches the yield stress of the material.

In general, though, we must always turn to test results to confirm the indications of highly-sophisticated analysis. In general, the indications of test results are that collapse stresses, σ_{max}, are functions of initial buckling stresses, σ_{cr}, and the yield stress, σ_y, of a ductile material.

6 *THE PLATE AS A COMPONENT OF A STRUCTURAL MEMBER*

As we have seen, in the initial elastic buckling of a flat plate, the plate is in stable equilibrium, at in-plane stiffnesses of the order of 50% of pre-buckled stiffnesses. This makes the buckled flat plate in general a useful structural component, which can be depended on to maintain considerable in-plane elastic stiffness.

Where flat plates are used in larger structures, consisting of a number of components, the buckled plate can though lead to instability of the structural system as a whole. Conclusions about the overall stability of structures comprised of flat plates should be approached with considerable caution. In a square tubular column, for example, initial local buckling of the column may occur before overall column buckling. On buckling in the compression fibres at the centre of the column, the column as a whole behaves like a Shanley column with a stiffness in the compression fibres of the order of 50% of E.

Although the concept of unstable initial buckling has been known for sometime, the problem of the combined local and overall buckling of columns was first studied in detail by Van der Neut [see(1)]. There are parallels, in this case, with plastic buckling of a column. In such cases, there is an intermediate range of slenderness ratios of the column over which column buckling is unstable.

We learn from this, that, although buckled plate components may take increasing in-plane loads, the stability of the structure as a whole may be unstable. The stability of a structural system as a whole must be studied by looking at the complete system, and not just component parts.

7 *CONCLUSIONS*

The analysis of rectangular plates has attracted much research over the past 100 years or more. The analysis has extended across the whole gamut of problems from linear-elastic bending to stable post-buckling. The progress made has been very considerable. The small linear-elastic deformations of flat plates are well covered by adequate theories.

The initial elastic buckling of flat plates is also well covered by numerous analyses, and more recently by finite-element analysis. The field of post-buckling of flat plates has been well researched. There are limits on the usefulness of theories, because of the problems of initial geometric and loading imperfections. Progress in the field has therefore depended initially on careful testing of buckled plates. As Walker (8) has observed, progress is possible only through a sensible balance of theory and interpretation of test results. The work of Horne and Narayanan (4) on stiffened plates is a very valuable example of this.

In general, the theory of flat plates has flourished because of the use of ductile metals in engineering structures. As other materials come into use, new problems of plates will become evident. The tools of numerical analysis will be very useful in solving structural design problems. Techniques of testing will be combined with these to further our knowledge of the analysis of rectangular plates.

References

1. Budiansky, B. (Editor). 'Buckling of Structures.' IUTAM Symposium 1974, Springer-Verlag, Berlin, 1976.

2. Bulson, P.S. 'The Stability of Flat Plates.' Chatto and Windus, London, 1970.

3. Croll, J.G.A. and Walker, A.C. 'Elements of Structural Stability, Macmillan, London, 1972.

4. Horne, M.R. and Narayanan, R. 'Ultimate Strength of Stiffened Panels under Uniaxial Compression.' Crosby Lockwood Staples, London, 1977.

5. Supple, W.J. (Editor). 'Structural Instability.' IPC Science and Technology Press, London, 1973.

6. Thompson, J.M.T. and Hunt, G.W. 'A General Theory of Elastic Stability.' Wiley, London, 1973.

7. Timoshenko, S. 'Theory of Plates and Shells.' McGraw-Hill, New York, 1940.

8. Walker, A.C. 'A brief review of plate buckling research.' Conference on Thin-Walled Structures, Applied Science, London, 1983.

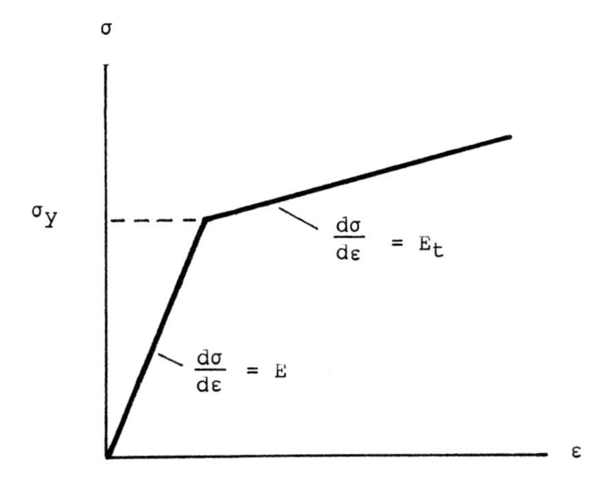

Fig. 1 Idealised compressive stress-strain curve showing plastic deformations at constant tangent modulus

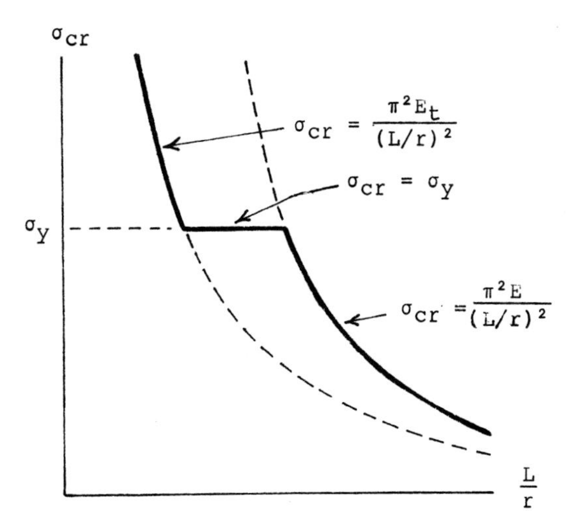

Fig. 2 Elastic-plastic buckling of pin-ended columns

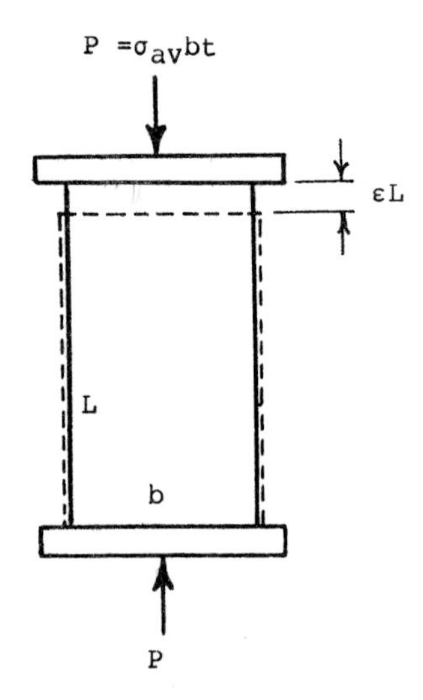

Fig. 3 Post-buckling contraction of a compressed plate

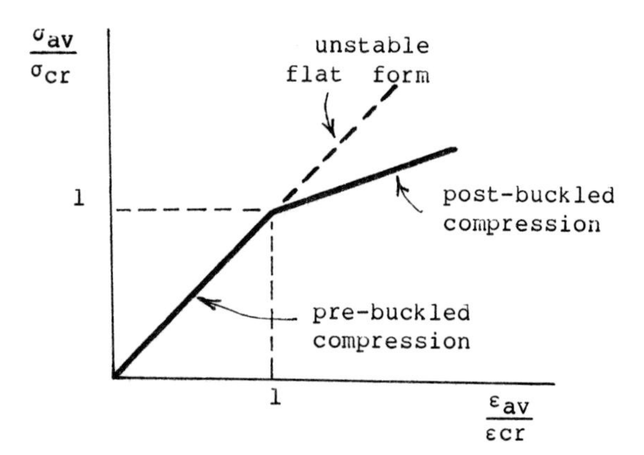

Fig. 4 Contraction of a compressed plate in the range of post-buckling

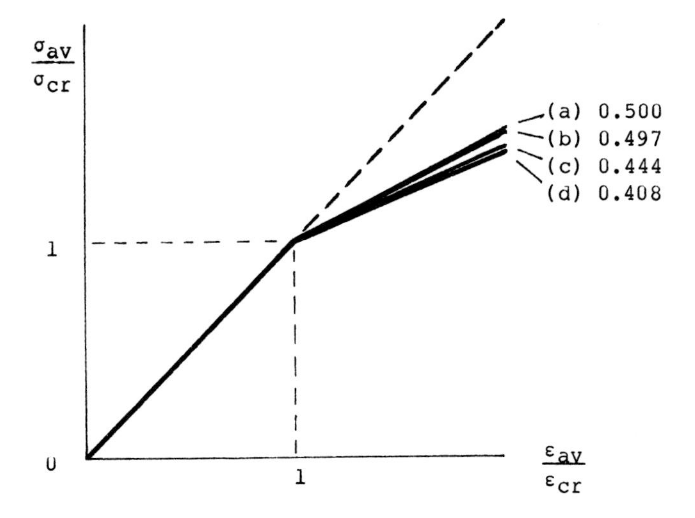

Fig. 5 Post-buckled stiffnesses of flat plates under end compression

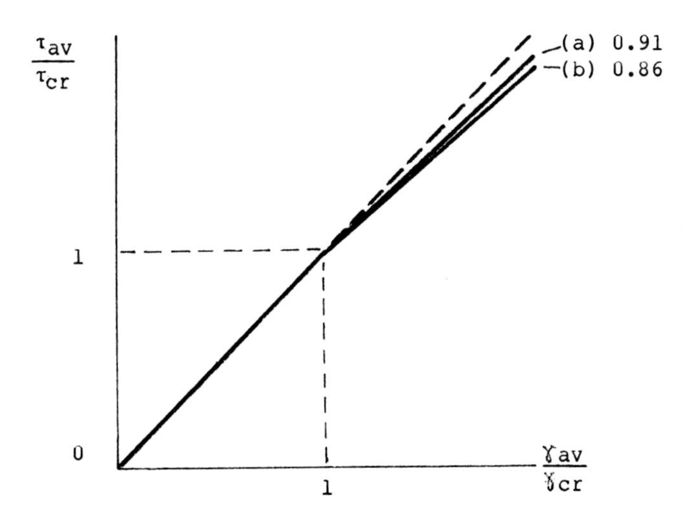

Fig. 6 Post-buckled stiffnesses of flat plates under shear

THE ELASTIC BUCKLING OF RESTRAINED PLATES

I.C. Medland and C.M. Segedin

Department of Theoretical and Applied Mechanics, University of Auckland.

Synopsis

A concise analytical procedure is proposed for the determination of the elastic buckling loads of uniaxially compressed plates supported at regular intervals by Winkler elastic springs across the width. Both linear and rotational supports may be included. A technique for evaluating the strength requirements of the bracing is also outlined.

Nondimensional charts are presented to summarise the brace stiffness versus buckling load relationships.

Notation

A_n	numerical coefficients of shape functions F and G
a	length of an interbrace panel
B_n, B	numerical coefficients of shape functions F and G
b	width of plate
D	flexural rigidity/unit width of plate
f, f_{ocr}	applied compressive force/unit width, critical value of f for simply supported a by b plate,
F	a form of function of x and y or of ξ and η
G	a form of function of x and y or of ξ and η
i	the number of longitudinal half sine waves in a chosen mode of initial distortion
j	the number of longitudinal half sine waves in a chosen mode of buckling
K_L, K_R	nondimensional linear and rotational Winkler spring stiffnesses
k_l, k_r	actual linear and rotational Winkler spring stiffnesses
m	an angle related to the force f

n	the identifier of the general panel and node
N	the number of nodes (Winkler spring attachments) within the length
p	an angle related to the force f
Q_n	numerical coefficients of shape functions F and G
r	aspect ratio a/b of a panel
R_n, R	numerical coefficients of shape functions F and G
w	out of plane displacement
x	longitudinal coordinate in a panel
y	lateral coordinate in a panel
η	nondimensional local lateral coordinate in a panel = y/a
θ	an angle defining a given node in a buckling mode
λ	a nondimensional factor related to the compression f
μ	a factor related to λ
ψ	a factor related to λ
ξ	nondimensional local longitudinal coordinate in a panel = x/a
ω, ω_I	nondimensional out-of-plane displacement of plate.

Introduction

The authors have previously examined the elastic stability of truss chords, columns and beams which are supported by braces at regular intervals (1-4). In Segedin and Medland (2), Medland (3,4) and Taylor (5) a recurrence technique was applied and resulted in a concise formulation of the problems thereby allowing efficient computation of specific results and assembly of design charts.

The associated problem of assessing the brace strength requirements has been addressed by Medland and Segedin (6). The technique involves selecting an assumed initial out-of-straightness of the main member or structure. The destabilising forces in that member or structure magnify the initial deformations and cause the braces to strain from the onset of loading.

The present paper records an application of the method extended to the stiffened plate problem in which Winkler type springs are used to model the support offered by the stiffeners.

1. *THE COMPRESSED PLATE APPLICATION*

Figure 1 illustrates the general form of structure studied in this paper. The plate is b units wide and $(N+1)a$ units long, the length being divided into equal panels by N internal node lines along which are attached Winkler springs. Both linear springs, resisting the out-of-plane displacement w, and rotational springs, resisting the rotation $\partial w/\partial x$, are attached on the nodal lines. The uniform load/

unit width f is applied in the x direction (longitudinal) in the plane of the plate. The longitudinal edges of the plate are simply supported. The end edges at nodes 0 and $N+1$ are simply supported but have rotational Winkler springs of stiffness half that of the internal nodes superimposed. The stiffnesses of each type of spring may be varied independently from zero upwards.

2. DISPLACEMENT FORMULATION

The transverse displacement, w within the general panel n must satisfy the governing differential equation

$$D \nabla^4 w + f \frac{\partial^2 w}{\partial x^2} = 0 \qquad (1)$$

and with the assumption that w takes the form

$$w = W(x) \sin \frac{\pi y}{b} \qquad (2)$$

it follows that W(x) must satisfy the ordinary differential equation

$$\frac{d^4 W}{dx^4} - \frac{2\pi^2}{b^2} \frac{d^2 W}{dx^2} + \frac{\pi^4}{b^4} W + \frac{f}{D} \frac{d^2 W}{dx^2} = 0 \qquad (3)$$

The trial solutions

$$W = \begin{matrix} \sin \\ \cos \end{matrix} \left(\frac{\lambda \pi x}{b}\right) \qquad (4)$$

will satisfy equation (3) provided

$$\frac{\lambda^4 \pi^4}{b^4} + \frac{2\pi^4 \lambda^2}{b^4} + \frac{\pi^4}{b^4} - \frac{f}{D} \frac{\lambda^2 \pi^2}{b^2} = 0 \qquad (5)$$

which leads to the condition

$$f = \pi^2 D(\lambda + \frac{1}{\lambda})^2 / b^2. \qquad (6)$$

When this form is compared with the expression for the critical f value of a single panel plate with an aspect ratio r $(=a/b)$ i.e.

$$f_{ocr} = \pi^2 D(\frac{1}{r} + r)^2 / b^2, \qquad (7)$$

it can be seen that the nondimensional parameter λ in equation (6) represents an aspect ratio. In the cases when f is raised to the critical (buckling) level for the spring supported structures, the λ value represents the aspect ratio of an equivalent simply supported panel - a nondimensional "effective length".

The form

$$W = \begin{matrix} \sin \\ \cos \end{matrix} \left(\frac{\pi x}{\lambda b}\right) \qquad (8)$$

also satisfies equation (3). With the use of the nondimensional parameters

$$\xi = x/a, \quad \eta = y/a, \quad r = a/b, \quad \omega = w/a \qquad (9)$$

the trial solutions of equations (4) and (8) can be written in the form

$$\begin{matrix} \sin \\ \cos \end{matrix} (m\xi) \, , \quad \begin{matrix} \sin \\ \cos \end{matrix} (p\xi)$$

where $\quad m = \lambda \pi a/b \, , \; p = \pi a/\lambda b \qquad (10)$

These forms may be linearly combined to express w in the convenient nondimensional form

$$\omega = [A_n F(\xi) + B_n F(1-\xi) + Q_n G(\xi) + R_n G(1-\xi)] \sin \pi \eta r \qquad (11)$$

2.1 Forms of F(ξ) and G(ξ)

The Functions F and G are designed so that

$$F(0) = \frac{\partial F}{\partial \xi}(1) = \frac{\partial F}{\partial \xi}(0) = 0 \text{ and } F(1) = 1 \qquad (12)$$

$$G(0) = G(1) = \frac{\partial G}{\partial \xi}(0) = 0 \text{ and } \frac{\partial G}{\partial \xi}(1) = 1 \qquad (13)$$

and their forms are

$$F(\xi) = \frac{\dfrac{\frac{1}{m}\sin m\xi - \frac{1}{p}\sin p\xi}{\cos m - \cos p} + \dfrac{\cos m\xi - \cos p\xi}{m \sin m - p \sin p}}{\dfrac{\frac{1}{m}\sin m - \frac{1}{p}\sin p}{\cos m - \cos p} + \dfrac{\cos m - \cos p}{m \sin m - p \sin p}} \qquad (14)$$

$$G(\xi) = \frac{\dfrac{\frac{1}{m}\sin m\xi - \frac{1}{p}\sin p\xi}{\frac{1}{m}\sin m - \frac{1}{p}\sin p} - \dfrac{\cos m\xi - \cos p\xi}{\cos m - \cos p}}{\dfrac{\cos m - \cos p}{\frac{1}{m}\sin m - \frac{1}{p}\sin p} + \dfrac{m \sin m - p \sin p}{\cos m - \cos p}} \qquad (15)$$

Throughout the following the shorthand ω', ω'', F' etc. will be used to symbolise partial derivatives with respect to ξ.

3. *CONTINUITY CONDITIONS*

Displacement and slope are continuous across each internal node line. Moment and shear/unit width are discontinuous by the amount of moment or shear provided by the Winkler springs.

3.1 Displacement Continuity

$$\omega_{n-1}(1,\eta) = \omega_n(0,\eta) \qquad (16)$$

which, upon substitution of the specific end values (summarised in equations (12) and (13)) into equation (11), results in the relationship

$$A_{n-1} = B_n. \qquad (17)$$

3.2 Slope Continuity

$$\omega'_{n-1}(1,\eta) = \omega'_n(0,\eta) \qquad (18)$$

and hence from equation (11)

$$Q_{n-1} = - R_n. \qquad (19)$$

3.3 *Moment Continuity*

The moments/unit width on each side of node n differ by the amount of the rotational Winkler spring stiffness multiplied by ω', now represented in magnitude by R_n. Thus

$$-(B_{n+1} - B_{n-1})\,F''(0) + (R_{n+1} + R_{n-1})\,G''(0)$$

$$-2R_n\,G''(1) - K_R R_n = 0 \qquad (20)$$

in which the nondimensional spring stiffness K_R is defined as

$$K_R = ak_R/D \qquad (21)$$

3.4 *Shear Continuity*

The shears/unit width differ by the amount of the linear spring stiffness multiplied by ω. The continuity equation is

$$2B_nF''(1) - (B_{n+1} + B_{n-1})\,F'''(0) + (R_{n+1} - R_{n-1})G'''(0)$$

$$- K_L B_n = 0 \qquad (22)$$

where $$K_L = a^3k_L/D \qquad (23)$$

4. RECURRENCE FORMULATION

The symbol B_n represents the displacement at node n and R_n the slope at that node. Let B_n and R_n be defined

$$B_n = B\,sin\,n\,\theta\,,\quad R_n = R\,cos\,n\,\theta \qquad (24)$$

where $$\theta = j\pi/(N+1),\quad j = 1,2,.....N. \qquad (25)$$

Substitution of equation (24) into equations (22) and (20) results in the simultaneous pair of homogeneous equations shown in matrix form in equation (26). Obviously the determinant of the coefficient matrix in equation (26) must be zero for nontrivial values of B and R to be obtained.

$$\begin{bmatrix} F'''(0)cos\theta - F'''(1) + \tfrac{1}{2}K_L & G'''(0)sin\theta \\ F''(0)sin\,\theta & G''(1) - G''(0)cos\theta + \tfrac{1}{2}K_R \end{bmatrix} \begin{bmatrix} B \\ R \end{bmatrix} = \underset{\sim}{0} \qquad (26)$$

The factors $F''(0)$ etc. are functions of m and p and are easily derived from equations (14) and (15).

5. BUCKLING LOADS AND MODES

In selecting a value for θ, i.e. a value of j in eqn (25), one is assuming a distribution of the displacement amplitudes of the nodal lines within the length - a buckling mode. Each such assumed longitudinal mode will be associated with a specific critical load. Normally the lowest such load is sought. When j equals $N+1$ no out-of-plane displacement occurs on the nodal lines but the presence of rotational springs will raise the critical load value above that of the unstiffened plate.

The results of selected analyses are shown in Figs 2,3. The buckling values of f are represented by the nondimensional factor f^* which is defined as f/f_{ocr}, where f_{ocr} is the critical load for a simply supported panel of aspect ratio $(N+1)r/j)$. In this way each has a base value of 1.0 when no extra elastic supports are attached.

Figure 2 shows clearly that the lower aspect ratio plates gain more in proportion from the rotational springs than those with larger a/b.

Figure 3 illustrates the effects of the addition of a line of linear springs at mid length for two different aspect ratio values. The scale of the nondimensional stiffnesses K_L and K_R is obvious. K_L, of course has an a^3 multiply and K_R an a, when converted to real stiffnesses/unit length.

6. *INITIALLY DEFORMED PLATE*

For a plate which is not perfectly planar the application of an in-plane compression magnifies the initial imperfections causing bending within the structure. Any braces attached to the structure are strained during this distortion. Whereas for the pure elastic buckling case previously discussed only the brace stiffness was of interest, the strength of the brace becomes important in the study of deformed panels. The following outline indicates how the brace forces could be estimated in the type of plate structures previously analysed.

6.1 *The form of initial shape*

The initial displacement shape, within the general panel is the result of rolling, cooling and/or minor out of plane loading, is assumed to take the form

$$\omega_I(\xi,\eta) = Z \sin i \left(\frac{\eta + \xi}{N + 1}\right)\pi \sin \pi\eta r \qquad (27)$$

where Z is a chosen amplitude for the longitudinal sine shape which contains a chosen number (i) of half waves in the full length. The lateral shape is selected to be that most in sympathy with the buckling mode and hence the one which will be most magnified.

The differential equation to be satisfied is

$$\nabla^2(\omega - \omega_I) + (m + p)^2 \frac{\partial^2\omega}{\partial\xi^2} = 0 \qquad (28)$$

The effect of the axial compressive force in such a case is to magnify the original displacements ω_I by dividing the original amplitude by the factor

$$\left(1 - \frac{f}{f_{cr_i}}\right) \qquad (29)$$

where f_{cr_i} is the critical load/unit width of a simply supported plate which has an aspect ratio (length:width) of

$$\frac{(N + 1)a}{i} : b = \frac{(N + 1)}{i} : r \qquad (30)$$

As with columns, any initial shape which can be expressed in terms of the buckling modes of the plate concerned will deform under compression in a manner where the amplitude of each individual component is magnified in terms of its own buckling load.

7. *CONCLUSIONS*

The technique described in the paper allows very efficient computation of the longitudinal load/unit width which will cause buckling of a flat elastic plate which is elastically braced at regular intervals by lateral linear and rotational braces. A similarly efficient means of determining brace strength requirements has been outlined.

References

1. Medland, I.C. 'A basis for the design of column bracing.' The Structural Engineer, vol. 55, No. 7, 1977.
2. Segedin, C.M. and Medland, I.C. 'The buckling of interbraced columns.' International Journal of Solids and Structures, vol. 14, 1978.
3. Medland, I.C. 'Flexural-torsional buckling of interbraced columns.' Engineering Structures, vol. 1, 1979.
4. Medland, I.C. 'The buckling of interbraced beam systems.' Engineering Structures, vol. 2, 1980.
5. Taylor, J.A. 'Buckling of a periodically constrained plate.' Internal research report, Dept. of Theoretical and Applied Mechanics, University of Auckland, 1977.
6. Medland, I.C. and Segedin, C.M. 'Brace forces in interbraced column structures.' Proceedings ASCE, Journal of the Structural Division, vol. 105, 1979.
7. Timoshenko, S.P. and Gere, J.M. 'Theory of Elastic Stability.' McGraw-Hill Book Co. Ltd., 1961.

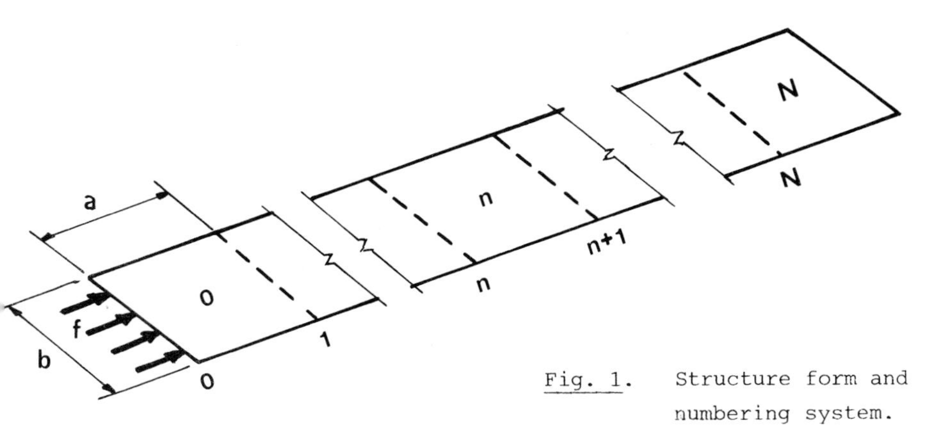

Fig. 1. Structure form and numbering system.

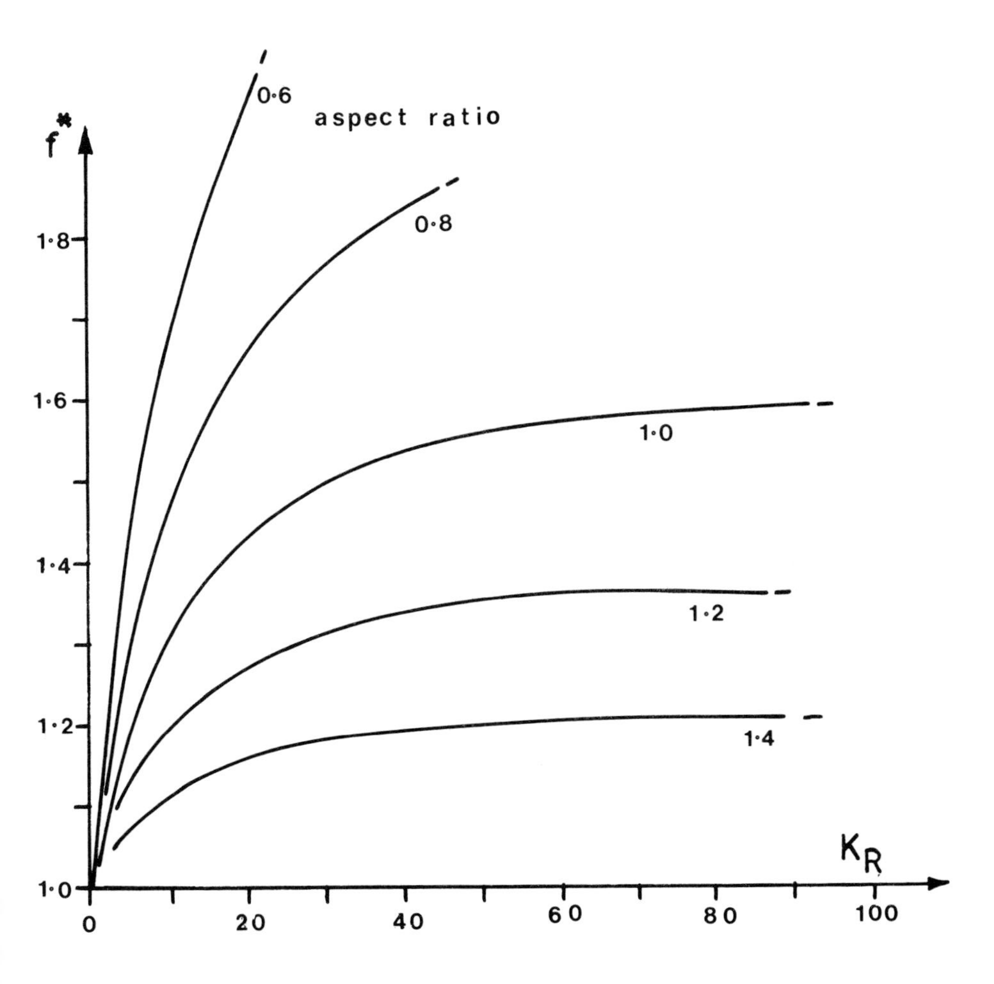

Fig. 2. The effect of rotational springs on buckling load with different aspect ratios.

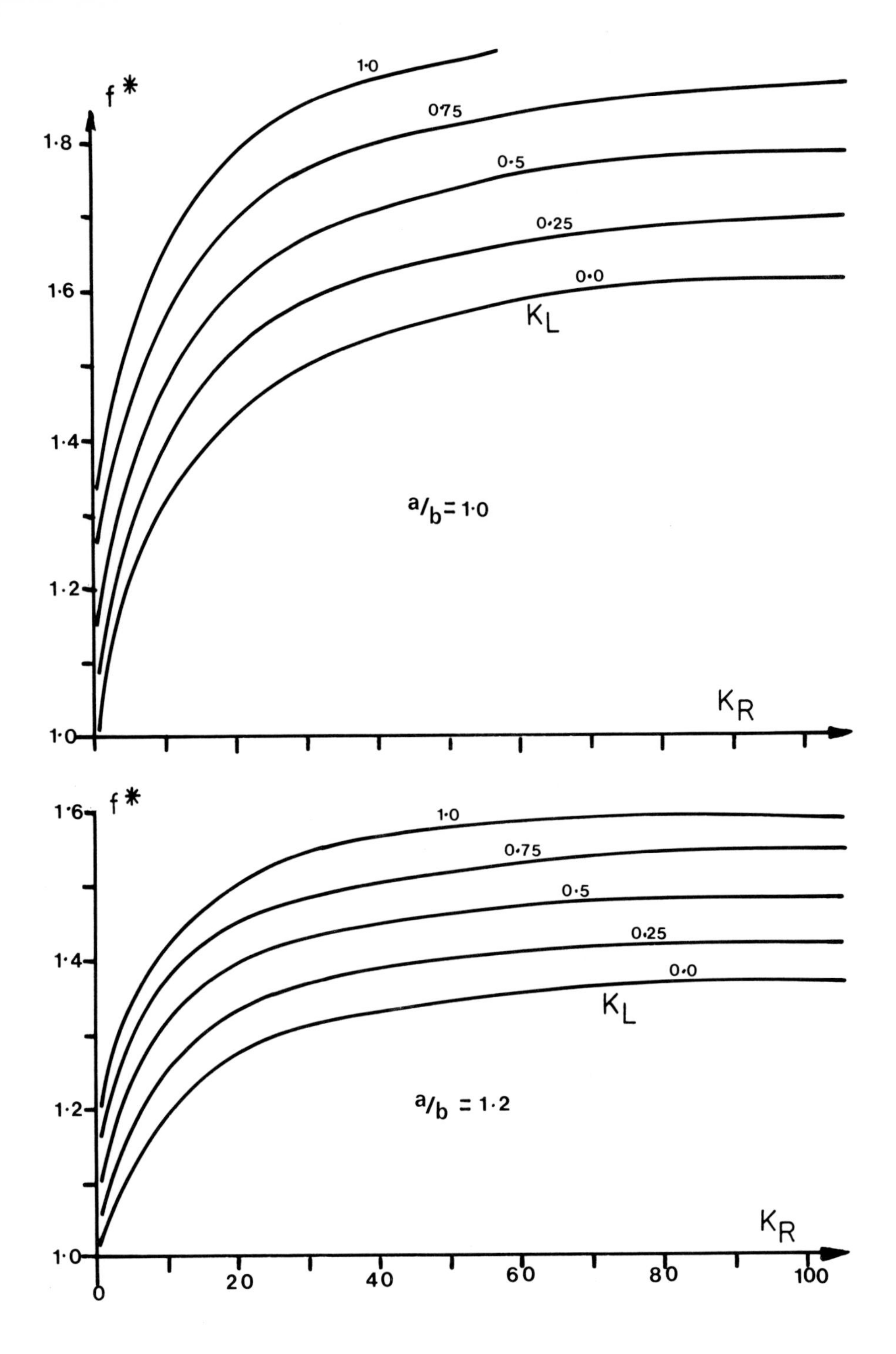

Fig. 3. The effect of linear springs superimposed upon rotational springs.

EFFECT OF BOUNDARY CONDITIONS ON POST-BUCKLING BEHAVIOUR OF ORTHOTROPIC RECTANGULAR PLATES*

C. Y. Chia

Department of Civil Engineering, University of Calgary

Synopsis

This paper is analytically concerned with the effect of boundary conditions on postbuckling behaviour of a rectangular orthotropic plate. A series solution to the von Kármán-type nonlinear equations of the plate under inplane edge compression and edge shear is presented for each of the edges elastically and nonuniformly restrained against rotation. In the formulation the edge moments are replaced by an equivalent transverse normal pressure near the edges. By virtue of the orthogonal functions the governing equations are reduced to an infinite set of algebraoic equations for coefficients in these series. Convergent solutions can be determined to any desired degree of accuracy by successive truncation of this set of equations. Based on the present solution a wide class of boundary conditions can be considered.

Introduction

The geometrically nonlinear behaviour of elastic plates have received considerable attention in recent years. A comprehensive review of the literature may be found in reference (1) or elsewhere. Most of the existing solutions, however, are restricted to simple boundary conditions. In reality a structural plate is generally restrained elastically against rotation and non-uniformly supported along its edge or edges. Using Berger's hypothesis and the Ritz method Ramachandran (2) has discussed the nonlinear flexural vibration of a circular plate with nonuniform edge constraints. In the analysis the strain energy due to stretching of the middle surface has been erroneously formulated and the radial displacement assumed to be independent of the polar angle. This problem has also been considered by Nowinski (3) using the same hypothesis and a corresponding linear solution. The plate edge has been taken to be prevented from the inplane motions. A static analysis for nonlinear bending of a uniformly loaded infinite strip simply supported except for two symmetrically situated built-in segments has been included. Using

*The results presented in this paper were obtained in the course of research sponsored by the Natural Sciences and Engineering Research Council of Canada.

Berger's approximation and an existing linear solution Banerjee (4) has studied the large deflection of a symmetrically loaded circular plate supported at several points along the boundary. In addition nonlinear non-axisymmetric bending of annular plates uniformly clamped along the boundary and bent by prescribing a rotation about a diameter, which is of a similar nature to mixed boundary conditions have been examined by Alzheimer and Davis (5) using an iteration procedure equivalent to the perturbation technique and by Tielking (6) using the Ritz method associated with the von Kármán nonlinear theory of plates.

In the cases of buckling and vibration of circular and rectangular plates with mixed nonuniform boundary conditions a number of research results has appeared in the literature, especially in recent years. A review of these linear problems is referred to several recent works by Kerr et al (7-10).

1. GOVERNING EQUATIONS

Consider a rectilinearly orthotropic, elastic, rectangular plate of length a in the x-direction, width b in the y-direction, and thickness h in the z-direction. The midsurface of the undeformed plate contains the x,y-axes. The von Kármán-type equations for large deflections of the plate under laterally distributed load $q(x,y)$ per unit area may be expressed in terms of the transverse deflection w and a force function ψ. The corresponding nondimensional equations given by the author (1) may be written as

$$\frac{\partial^4 W}{\partial X^4} + 2C_1\lambda^2 \frac{\partial^4 W}{\partial X^2 \partial Y^2} + C_2\lambda^4 \frac{\partial^4 W}{\partial Y^4}$$
$$= C_3\lambda^2(\lambda^2 Q + \frac{\partial^2 W}{\partial X^2}\frac{\partial^2 F}{\partial Y^2} + \frac{\partial^2 W}{\partial Y^2}\frac{\partial^2 F}{\partial X^2} - 2\frac{\partial^2 W}{\partial X\partial Y}\frac{\partial^2 F}{\partial X\partial Y}) \tag{1}$$

$$\frac{\partial^4 F}{\partial X^4} + 2C_4\lambda^2 \frac{\partial F^2}{\partial X^2 \partial Y^2} + C_2\lambda^4 \frac{\partial^4 F}{\partial Y^4} = \lambda^2[(\frac{\partial^2 W}{\partial X\partial Y})^2 - \frac{\partial^2 W}{\partial X^2}\frac{\partial^2 W}{\partial Y^2}] \tag{2}$$

where

$$W = \frac{w}{h} , \quad F = \frac{\psi}{E_2 h^3} , \quad Q = \frac{qb^4}{E_2 h^4}$$

$$X = \frac{x}{a} , \quad Y = \frac{y}{b} , \quad \lambda = \frac{a}{b}$$

$$C_0 = 12(1-\nu_{12}\nu_{21}) , \quad C_1 = \nu_{12}\frac{E_2}{E_1} + \frac{C_0 G_{12}}{6E_1} \tag{3}$$

$$C_2 = \frac{E_2}{E_1} , \quad C_3 = C_0 C_2 , \quad C_4 = \frac{1}{2}(\frac{E_2}{G_{12}} - 2\nu_{12}C_2)$$

In these equations E_1 and E_2 are major moduli of elasticity, ν_{12} and ν_{21} are Poisson's ratios, and G_{12} is the shear modulus.

The nondimensional membrane forces N_1, N_2, N_{12} are related to the nondimensional force function F by

$$N_1 = \frac{\partial^2 F}{\partial Y^2} , \quad N_2 = \frac{1}{\lambda^2}\frac{\partial^2 F}{\partial X^2} , \quad N_{12} = -\frac{1}{\lambda}\frac{\partial^2 F}{\partial X\partial Y} \tag{4}$$

where

$$(N_1, N_2, N_{12}) = \frac{b^2}{E_2 h^3}(N_x, N_y, N_{xy}) \tag{5}$$

with N_x, N_y and N_{xy} being the membrane forces per unit length.

The plate under consideration is subjected to inplane edge compression and edge shear forces per unit length, denoted by $\bar{N}_x$, $\bar{N}_y$ and $\bar{N}_{xy}$ respectively. Furthermore the plate is assumed to be rigidly supported against the transverse deflection and elastically restrained against rotation along its four edges. The rotational edge-restraint coefficient is allowed to vary along each edge. These boundary conditions may be written in the nondimensional form as

$$\begin{aligned}
&W = 0 \quad , \quad \frac{\partial^2 F}{\partial X \partial Y} = -\lambda \bar{N}_{12} \quad \text{at} \quad X = 0,1 \quad \text{and} \quad Y = 0,1 \\
&M_1 = \xi_1(Y)\frac{\partial W}{\partial X} \quad , \quad \frac{\partial^2 F}{\partial Y^2} = -\bar{N}_1 \quad \text{at} \quad X = 0 \\
&M_1 = -\xi_2(Y)\frac{\partial W}{\partial X} \quad , \quad \frac{\partial^2 F}{\partial Y^2} = -\bar{N}_1 \quad \text{at} \quad X = 1 \\
&M_2 = \xi_3(X)\frac{\partial W}{\partial Y} \quad , \quad \frac{\partial^2 F}{\partial X^2} = -\lambda^2 \rho \bar{N}_1 \quad \text{at} \quad Y = 0 \\
&M_2 = -\xi_4(X)\frac{\partial W}{\partial Y} \quad , \quad \frac{\partial^2 F}{\partial X^2} = -\lambda^2 \rho \bar{N}_1 \quad \text{at} \quad Y = 1
\end{aligned} \tag{6}$$

in which ξ_i are the rotational edge-restraint coefficients, functions of X or Y and in which

$$\begin{aligned}
&(M_1, M_2) = \frac{b^2}{E_2 h^4}(M_x, M_y) \\
&(\bar{N}_1, \bar{N}_{12}) = \frac{b^2}{E_2 h^3}(\bar{N}_x, \bar{N}_{xy}) \quad , \quad \rho = \frac{\bar{N}_y}{\bar{N}_x}
\end{aligned} \tag{7}$$

In these expressions M_x and M_y are the bending moments per unit length and $\bar{N}_x$, $\bar{N}_y$ and $\bar{N}_{xy}$ are the prescribed inplane edge forces per unit length.

The system of eqns (1) and (2) thus are to be solved in conjunction with boundary conditions given by eqn (6).

2. *METHOD OF SOLUTION*

A solution of eqns (1) and (2) is sought in the form of double series

$$W = \sum_{m=1}^{\infty} \sum_{n=1}^{\infty} W_{mn} \sin m\pi X \sin n\pi Y \tag{8}$$

$$F = -\frac{\bar{N}_1}{2}(\rho\lambda^2 X^2 + Y^2) - \lambda \bar{N}_{12} XY + \sum_{m=1}^{\infty} \sum_{n=1}^{\infty} F_{mn} R_m(X) S_n(Y) \tag{9}$$

In these expressions each term of the series for W satisfies the boundary condition for zero deflection and R_m and S_n are beam eigenfunctions defined by

$$R_m(X) = \cosh \alpha_m X - \cos \alpha_m X - \gamma_m(\sinh \alpha_m X - \sin \alpha_m X)$$
$$S_n(Y) = \cosh \alpha_m Y - \cos \alpha_m Y - \gamma_n(\sinh \alpha_m Y - \sin \alpha_n Y) \tag{10}$$

These functions possess the following orthogonality properties:

$$\int_0^1 R_i(X)R_j(X)\, dX = \begin{cases} 0 , & i \neq j \\ 1 , & i = j \end{cases}$$
$$\int_0^1 S_i(Y)S_j(Y)\, dY = \begin{cases} 0 , & i \neq j \\ 1 , & i = j \end{cases} \tag{11}$$

The inplane boundary conditions in eqn (6) are satisfied if we take

$$\gamma_k = \frac{\cosh \alpha_k - \cos \alpha_k}{\sinh \alpha_k - \sin \alpha_k} \quad , \quad 1 - \cos \alpha_k \cosh \alpha_k = 0 \tag{12}$$

The roots of the last equation and the corresponding values of γ_k are given by the author and Prabhakara (11).

To satisfy the boundary conditions for edges elastically restrained against rotation we follow Levy (12) and Soper (13) to replace the nondimensional bending moments along the four edges by an equivalent lateral pressure near these edges, denoted by Q_e as shown in Fig 1. This pressure is represented by a Fourier sine series. Taking the limit as the value of d approaches zero we obtain

$$Q_e = 2\pi \sum_{m=1}^{\infty} m\{[M_1(0,Y) - (-1)^m M_1(1,Y)]\sin m\pi X$$
$$+ M_2(X,0) - (-1)^m M_2(X,1)]\sin m\pi Y\} \tag{13}$$

The edge moments in this expression are also expanded into sine series as

$$M_1(0,Y) = \sum_{n=1}^{\infty} a_n \sin n\pi Y \quad , \quad M\ (1,Y) = \sum_{n=1}^{\infty} b_n \sin n\pi Y$$
$$M_2(X,0) = \sum_{m=1}^{\infty} c_m \sin m\pi X \quad , \quad M_2(X,1) = \sum_{m=1}^{\infty} d_m \sin m\pi X \tag{14}$$

in which Fourier coefficients a_n, b_n, c_m and d_m will be determined later. Upon substitution eqn (13) becomes

$$Q_e = \sum_{m=1}^{\infty} \sum_{n=1}^{\infty} r_{mn} \sin m\pi X \sin n\pi Y \tag{15}$$

where

$$r_{mn} = 2m\pi [a_n - (-1)^m b_n] + 2n\pi [c_m - (-1)^n d_m] \tag{16}$$

In this study the lateral load is taken to be zero (q=0) and hence the nondimensional load parameter Q in eqn (1) is to be replaced by the equivalent lateral load Q_e.

The rotational edge-restraint coefficients ξ_i in eqn (6) are expanded into Fourier cosine series as

$$\xi_1(Y) = \frac{A_o}{2} + \sum_{p=1}^{\infty} A_p \cos p\pi Y \quad , \quad \xi_2(Y) = \frac{B_o}{2} + \sum_{p=1}^{\infty} B_p \cos p\pi Y$$

$$\xi_3(X) = \frac{D_o}{2} + \sum_{p=1}^{\infty} D_p \cos p\pi X \quad , \quad \xi_4(X) = \frac{K_o}{2} + \sum_{p=1}^{\infty} K_p \cos p\pi X \tag{17}$$

where

$$A_p = 2 \int_0^1 \xi_1(Y) \cos p\pi Y \, dY \tag{18}$$

Now eqns (8), (14) and (17) are substituted into eqn (6) for edges elastically restrained against rotation. Using the following identity:

$$\sin m\pi\theta \cos n\pi\theta = \frac{1}{2} [\sin(m-n)\pi\theta + \sin(m+n)\pi\theta]$$

we obtain

$$\sum_{k=1}^{\infty} a_k \sin k\pi Y = A_o\pi \sum_{m=1}^{\infty} \sum_{n=1}^{\infty} mW_{mn} \sin n\pi Y$$

$$+ \frac{\pi}{2} \sum_{m=1}^{\infty} \sum_{n=1}^{\infty} \sum_{p=1}^{\infty} mW_{mn} A_p [\sin(n+p)\pi Y + \sin(n-p)\pi Y] \tag{19a}$$

$$\sum_{k=1}^{\infty} b_x \sin k\pi Y = -B_o \sum_{m=1}^{\infty} \sum_{n=1}^{\infty} m(-1)^m W_{mn} \sin n\pi Y$$

$$- \frac{\pi}{2} \sum_{m=1}^{\infty} \sum_{n=1}^{\infty} \sum_{p=1}^{\infty} m(-l)^m W_{mn} B_p [\sin(n+p)\pi Y + \sin(n-p)\pi Y] \tag{19b}$$

$$\sum_{k=1}^{\infty} c_k \sin k\pi X = D_o\pi \sum_{m=1}^{\infty} \sum_{n=1}^{\infty} nW_{mn} \sin m\pi X$$

$$+ \frac{\pi}{2} \sum_{m=1}^{\infty} \sum_{n=1}^{\infty} \sum_{p=1}^{\infty} nW_{mn} D_p [\sin(m+p)\pi X + \sin(m-p)\pi X] \tag{19c}$$

$$\sum_{k=1}^{\infty} d_k \sin k\pi X = -K_o\pi \sum_{m=1}^{\infty} \sum_{n=1}^{\infty} n(-1)^n W_{mn} \sin m\pi X$$

$$- \frac{\pi}{2} \sum_{m=1}^{\infty} \sum_{n=1}^{\infty} \sum_{p=1}^{\infty} n(-1)^n W_{mn} K_p [\sin(m+p)\pi X + \sin(m-p)\pi X] \tag{19d}$$

Equating coefficients of like sine terms in each of these equations, Fourier coefficients a_k, b_k, c_k and d_k can be expressed in terms of the deflection coefficients W_{mn}. It is observed that all the out-of-plane boundary conditions in eqn (6) are satisfied if the deflection function W is given by eqn (8) and conditions described by eq (19) are fulfilled.

Equations (8) for W, (9) for F and (15) for the equivalent lateral pressure are substituted into eqns (1) and (2). Multiplying the first resulting equation by $\sin i\pi X \sin j\pi Y$ and the second by $R_i(X)\ S_j(Y)$, integrating from 0 to 1 with respect to X and Y and using the orthogonality conditions given by eqn (11) we obtain the following two infinite sets of algebraic equations:

$$W_{ij}\pi^4(i^4 + 2C_1\lambda^2 i^2 j^2 + C_2\lambda^4 j^4)$$

$$= C_3\lambda^2[\lambda^2 r_{ij} + \bar{N}_1\pi^2 W_{ij}(i^2 + \rho\lambda^2 j^2)$$

$$+ 4\bar{N}_{12}\pi^2\lambda \sum_{m=1}^{\infty}\sum_{n=1}^{\infty} W_{mn}mn(\frac{\delta_1^{im}}{i+m} + \frac{\delta_2^{im}}{i-m})(\frac{\delta_1^{jn}}{j+n} + \frac{\delta_2^{jn}}{j-n})$$

$$-4\pi^2 \sum_{m=1}^{\infty}\sum_{n=1}^{\infty}\sum_{p=1}^{\infty}\sum_{q=1}^{\infty} W_{mn}F_{pq}(m^2 K_1^{pim} L_2^{qjn} + n^2 K_2^{pim} L_1^{qjn}$$

$$+ 2mnK_3^{pim} L_3^{qjn})] \qquad i,j = 1,2,3,\ldots \tag{20}$$

$$F_{ij}(\alpha_i^4 + C_2\lambda^4\alpha_j^4) + 2C_4\lambda^2 \sum_{m=1}^{\infty}\sum_{n=1}^{\infty} F_{mn}K_4^{im}L_4^{jn}$$

$$= \lambda^2\pi^4 \sum_{m=1}^{\infty}\sum_{n=1}^{\infty}\sum_{p=1}^{\infty}\sum_{q=1}^{\infty} W_{mn}W_{pq}mq(npK_5^{imp}L_5^{jnq}$$

$$- mqK_1^{imp}L_1^{jnq}) \qquad i,j = 1,2,3,\ldots \tag{21}$$

In these equations K_1 to K_5 are constants defined by the following definite integrals:

$$K_1^{pim} = \int_0^1 R_p(X)\ \sin i\pi X\ \sin m\pi X\ dX$$

$$K_2^{pim} = \int_0^1 R_p''(X)\ \sin i\pi X\ \sin m\pi X\ dX$$

$$K_3^{pim} = \int_0^1 R_p'(X)\ \sin i\pi X\ \cos m\pi X\ dX \tag{22}$$

$$K_4^{im} = \int_0^1 R_i(X)\ R_m'(X)\ dX$$

$$K_5^{imp} = \int_0^1 R_i(X)\ \cos m\pi X\ \cos p\ X\pi dX$$

and δ_1 and δ_2 are given by

$$\delta_1^{mn} = \begin{cases} 0\ , & m+n = even \\ 2\ , & m+n = odd \end{cases} \qquad \delta_2^{mn} = \begin{cases} 0\ , & |m-n| = even \\ 2\ , & |m-n| = odd \end{cases} \tag{23}$$

where the primes denote differentiation with reference to the corresponding coordinate. Constants L_1 to L_5 in eqns (20) and (21) are obtained by replacing K, X, i, m and p in eqn (22) by L, Y, j,

n and q, respectively.

Equations (20) and (21) are to be truncated to obtain a solution for coefficients in eqns (8) and (9) for W and F. For various values of i and j these equations give the same number of equations for W_{ij} and F_{ij} as the number of terms taken in these series. Therefore, a solution with any desired degree of accuracy can be obtained for a given set of values in aspect ratio, elastic properties, edge compression and edge shear.

2.1 *Critical load*

The nondimensional critical buckling loads due to edge compression and shear are respectively defined by

$$(N^{+}_{cr}, N^{-}_{cr}) = \frac{b^2}{E_2 h^3}(p^{+}_{cr}, p^{-}_{cr}) \tag{24}$$

where p^{+}_{cr} and p^{-}_{cr} are the applied edge compression and shear, $\bar{N}_x$ and $\bar{N}_{xy}$, at which the plate buckles respectively. The value of N^{+}_{cr} is obtained by dropping all the nonlinear terms and the term containing $\bar{N}_{12}$ in eqn (20) and solving the resulting homogeneous, linear equations for $\bar{N}_1$. It is observed from eqns (16) and (19) that r_{ij} in eqn (20) is expressed in terms of W_{mn}. The lowest eigenvalue of $\bar{N}_1$ found from the system of equations corresponds to the critical load N^{+}_{cr}. Similarly the critical load N^{-}_{cr} can be obtained.

3. *CONCLUSIONS*

A series solution to von Kármán-type nonlinear equations for post-buckling of an orthotropic plate under inplane edge compression and edge shear is formulated for a wide class of boundary conditions, including a plate edge partially clamped, partially simply-supported and partially elastically-restrained against rotation. As expected the effect of boundary conditions is quite significant on the post-buckling behaviour of the plate.

References

1. Chia, C.Y. 'Nonlinear Analysis of Plates'. McGraw-Hill, New York, 1980.
2. Ramachandran, J. 'Large amplitude vibration of circular plates with mixed boundary conditions.' vol. 4, 1974.
3. Nowinski, J.L. 'Some static and dynamic problems concerning nonlinear behavior of plates and shallow shells with discontinuous boundary conditions.' International Journal of Non-Linear Mechanics, vol. 10, 1975.
4. Banerjee, M.M. 'Note on the large deflection of circular plates supported at several points along the boundary.' Bull. Cal. Math. Soc. vol. 68, 1976.
5. Alzheimer, W.E. and Davis, R.T. 'Nonlinear unsymmetrical bending of an annular plate.' ASME Journal of Applied Mechanics, vol. 35, 1968.
6. Tielking, J.T. 'Asymmetric bending of annular plates.' International Journal of Solids and Structures, vol. 16, 1980.

7. Keer, L.M. and Stahl, B. 'Eigenvalue problems of rectangular plates with mixed edge conditions.' ASME Journal of Applied Mechanics, vol. 39, 1972.

8. Leissa, A.W., Laura, P.A.A. and Gutierrez, R.H. 'Vibrations of rectangular plates with nonuniform elastic edge supports.' ASME Journal of Applied Mechanics, vol. 27, 1980.

9. Narita, Y. 'Applications of a series-type method to vibration of orthotropic rectangular plates with mixed boundary conditions.' Journal of Sound and Vibration, vol. 77, 1981.

10. Narita, Y. and Leisaa, A.W. 'Flexural vibrations of free circular plates elastically constrained along parts of the edge.' International Journal of Solids and Structures, vol. 17, 1981.

11. Chia, C.Y. and Prabhakara, M.K. 'Postbuckling of unsymmetrically layered anisotropic rectangular plates.' ASME Journal of Applied Mechanics, vol. 41, 1974.

12. Levy, S. 'Square plate with clamped edges under normal pressure producing large deflections.' NACA R 740 (Washington, D.C.) 1942.

13. Soper, W.G. 'Large deflections of stiffened plates.' ASME Journal of Applied Mechanics, vol. 25, 1958.

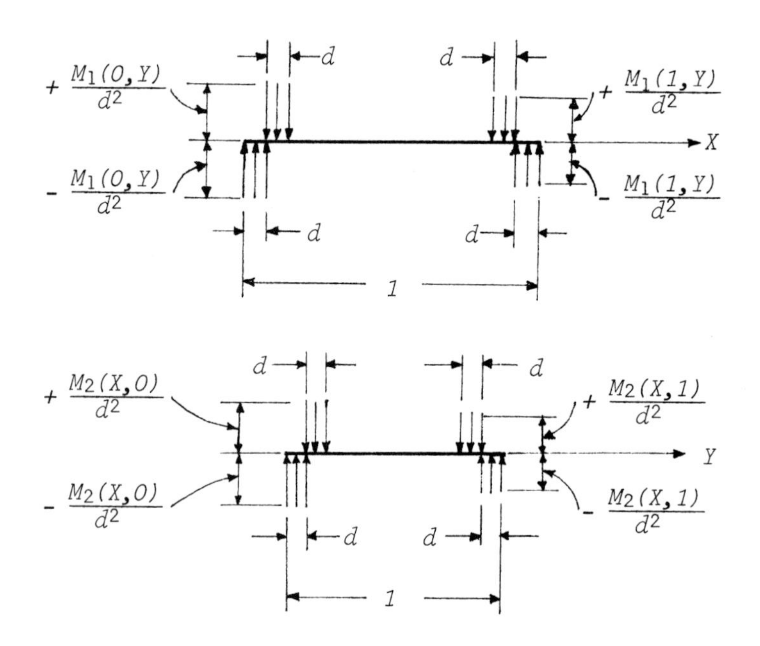

Fig. 1 Equivalent pressure distribution ($d \to 0$) for edge moments

ELASTIC BUCKLING OF FLAT PANELS CONTAINING CIRCULAR AND SQUARE HOLES

A B Sabir and F Y Chow

Department of Civil and Structural Engineering, University College, Cardiff

Synopsis

The finite element method of analysis is employed to determine the elastic critical buckling loads of flat square panels containing circular and square holes. The inplane loadings considered are uniaxial, biaxial or shear distributed uniformly along the straight edges of the plates.

The elements used for calculating the inplane stresses prior to buckling were first developed for general plane elasticity problems, their convergence performance were extensively tested and shown to be superior to other existing elements. They are based on generalised strain rather than displacement assumptions and satisfy the requirement of zero strain due to rigid body displacements. Furthermore these elements were designed to include an additional degree of freedom, namely the inplane rotation. In this way they are made suitable for cases where the inplane rotation as well as other degrees of freedom may be restrained and also when the plate is combined with other structural components having this rotation as an essential external degree of freedom.

The elastic critical buckling loads are obtained for plates with centrally located square and circular holes when subjected to the above mentioned loading cases and having several types of boundary conditions at the straight edges.

Notation

a	length of side of square plate
b	length of side of square hole
B	strain matrix
d	diameter of circular hole
D	flexural rigidity or rigidity matrix
e	eccentricity of centre of hole from centre of plate
E	Young's modulus
k_b, k_s, k_u	buckling coefficients for biaxial, shear and uniaxial loading

K	global stiffness matrix
K_g	global geometric matrix
N_x, N_y, N_{xy}	applied in-plane direct and shearing loads per unit length
t	plate thickness
U	strain energy due to bending
V	potential energy of applied loads
w	normal displacement
x, y, z	co-ordinate axes
ν	Poisson's ratio
λ	critical load parameter
δ	generalised displacement vector
$\sigma_x, \sigma_y, \gamma_{xy}$	in-plane direct and shearing stresses
σ_{cr}	critical stress
π	total potential energy

Introduction

The finite element was first applied to the elastic buckling problems of columns by Gallagher and Padlog (1). This work was followed by an extensive investigation into buckling of plates by Kapur and Hartz (2). They based their computations on the use of a rectangular non-conforming bending element having three degrees of freedom at each corner node. The efficiency of this element whose shape function is attributed to Melosh (3) was demonstrated by calculating the buckling loads for plates under distributed edge loading conditions and agreement with the well-known analytical solutions was obtained. To improve the convergence of the results Pifko and Isakson (4) used the conforming rectangular plate bending elements first suggested by Bogner, Fox and Schmit (5). This element has four degrees of freedom at each corner node and satisfies compatibility of displacements as well as normal slope across interrelement boundaries. This element was also used by Carson and Newton (6) and Sabir (7) who reported identical results for the elastic stability of various plate problems. In the same paper Sabir gave the necessary stiffness matrices explicitly and extended the investigation into the buckling of plates on elastic foundations. The progress into the development of higher order conforming bending elements was continued especially for triangular bending elements. Holland and Moan (8) reported faster convergence to correct buckling loads by using an eighteen degree of freedom triangular element but as the case with the previous conforming rectangular elements the additional unnecessary internal degrees of freedom make these elements difficult to be combined with other structural components not having these internal degrees of freedom.

In all the above mentioned work the buckling results depended on the performance of the bending element and its associated shape function for the lateral deflection from which the geometric matrix is obtained. This is the case for plates subjected to distributed edge loads where the inplane stresses prior and at the instant of buckling are known and need not be estimated. For plates subjected to inplane patch loads and also for plates with cut-outs the inplane stresses vary rapidly with distance from the point load and the edge of the hole respectively, an accurate calculation of the distribution of the inplane stresses within such panels is essential and has an influence on the critical buckling load as obtained by the finite element method.

The finite element method was first employed by Rockey, Anderson and Cheung (9) to examine the buckling of a square plate having a central circular hole when subjected to edge shear. Pennington (10) considered the uniaxial loading case and Shanmugam and Narayanan (11) extended the work to include square and circular central holes under uniaxial, biaxial and shear loadings. In all the above three investigations the inplane stresses were calculated using inplane stiffness matrices based on prescribed displacement functions. Sabir (12) has shown that such inplane finite elements when used to analyse simple elasticity problems give results which converge slowly thus requiring the structure to be divided into large numbers of elements. Shortcomings highlighted in this way were found to be largely removed if assumed strain, rather than displacement functions were used. This work lead to the development of a new class of simple and efficient finite elements for general plane elasticity problems. All these elements satisfy the requirement of zero strains due to rigid body displacements and the stiffness matrices used in the analysis were made to have the inplane rotation as an additional degree of freedom. The performance of these elements for convergence was tested extensively to show that there is no need for using an inordinately large number of elements even for cases where the stresses vary rapidly with distance from the edges of the holes. The use of these inplane elements made it also possible to examine the effect on the buckling loads of more rigorously applied boundary conditions to distinguish between constant strain and constant stress loading cases.

1. *THEORETICAL CONSIDERATIONS*

The elastic buckling load of a structure which is associated with a neutral equilibrium state can be determined by consideration of the total potential energy (π) of the system where

$$\pi = U + V \tag{1}$$

U is the gain in strain energy at the instant of buckling and V is the loss of potential energy of the applied loads. For a plate with Cartesian coordinates x and y, the strain energy stored at buckling is due to bending and is given in terms of the out-of-plane deflection w as

$$U = \frac{D}{2} \iint \left| \left(\frac{\partial^2 w}{\partial x^2} + \frac{\partial^2 w}{\partial y^2}\right)^2 - 2(1-\nu)\left\{\frac{\partial^2 w}{\partial x^2}\frac{\partial^2 w}{\partial y^2} - \left(\frac{\partial^2 w}{\partial x \partial y}\right)^2\right\} \right| dxdy \tag{2}$$

The loss in potential energy of the applied loads can be expressed as

$$V = \frac{1}{2} \iint \left| N_x \left(\frac{\partial w}{\partial x}\right)^2 + N_y \left(\frac{\partial w}{\partial y}\right)^2 + 2\, N_{xy} \left(\frac{\partial w}{\partial x}\frac{\partial w}{\partial y}\right) \right| dxdy \tag{3}$$

The first variation of the total potential energy gives the equilibrium condition and its second variation which is associated with the instability at a neutral equilibrium state can be shown to lead to the matrix form,

$$|K + \lambda K_g|\{\delta\} = 0 \qquad (4)$$

where K is the stiffness matrix, K_g is the so-called geometric matrix and λ is a function of the applied inplane load at buckling.

For an element of the plate of a finite size it can be shown that the stiffness matrix K^e is given by

$$K^e = |C^{-1}|^T\{\int\int B^T \; D \; B \; dxdy\}|C^{-1}| \qquad (5)$$

and the geometric matrix K_g^e as

$$K_g^e = |C^{-1}|^T\{\int\int P^T \; N \; P \; dxdy\}|C^{-1}| \qquad (6)$$

where C is the transformation matrix, D is the rigidity matrix, B is the strain matrix and N is the function of the internal in-plane stresses in the element and is given by

$$N = \begin{vmatrix} \sigma_x & \gamma_{xy} \\ \gamma_{xy} & \sigma_y \end{vmatrix} \qquad (7) \qquad \text{and} \qquad P = \begin{vmatrix} \frac{\partial w}{\partial x} \\ \frac{\partial w}{\partial y} \end{vmatrix} \qquad (8)$$

Equation (4) can be assembled in the usual way from its constituents K^e and K_g^e and the non trivial solutions of (4) are obtained from the criterion for neutral equilibrium i.e.

$$det|K + \lambda K_g| = 0 \qquad (9)$$

The above mentioned formulation shows that once the elements used in the analysis have prescribed displacement function for w and provided that the internal stress in (7) are related to the applied in-plane loads, the buckling load can be computed. For the results given in the present paper we took w for the rectangular elements to be

$$w=a_1+a_2x+a_3y+a_4xy+a_5x^2+a_6y^2+a_7x^3+a_8x^2y+a_9xy^2+a_{10}y^3+a_{11}x^3y + a_{12}xy^3$$

and for the triangular elements from reference (13) as

$$w=a_1+a_2x+a_3y+a_4xy+a_5x^2+a_6y^2+a_7x^3+a_8(x^2y+xy^3)+a_9y^3$$

The internal stresses are calculated using the inplane stiffness matrices described earlier. Exerpience with these strain based elements has shown that they give satisfactory mid element stresses and hence in formulating K_g^e, the inplane stresses through the element are assumed to be equal to their values at the centroids.

2. *FINITE ELEMENT IDEALIZATION AND CONVERGENCE PERFORMANCE*

The performance of the bending elements for the determination of the elastic buckling loads for plate has been thoroughly tested(references 2,4,6,7 and 8) and it is seen that provided the plate is divided into a mesh of 8x8 elements or even less in some cases, the critical buckling loads obtained converge with sufficient accuracy to the analytical values. However this is not the case when displacement based elements are used to obtain the inplane stress distribution which are required in the subsequent calculation of the buckling loads In reference (11), the authors state that convergence studies aimed at obtaining a sufficiently accurate value of critical loads up to 182 elements were used in a quarter of the plate. We noticed that even such an excessive number of elements are used, their results do not converge to the buckling loads for plates with no holes as the size of the hole is made very small. We therefore undertook to examine the performance of the strain based inplane elements for several plane stress problems. The conclusion we arrived at was that for plates with square and circular holes the meshes shown in figures (1) and (2) produce sufficient accurate results for stresses throughout the plate. We see that we require 75 and 85 elements within quarter of the plate with square and circular holes respectively. Furthermore, we have shown (reference 12) that these strain based elements can be used for long and narrow elements of aspect ratio of more than 16:1 without loss of accuracy. The maximum value of this ratio used in the idealizations shown in figures (1) and (2) is about 8:1. In all the cases considered in the present paper the results converge to the classical values for plates with no holes as the sizes of the holes are reduced. For example, a critical buckling coefficient of 9.3395 was obtained for a plate with a very small square hole. This value compares favourably with the analytical buckling coefficient of 9.4 for the same plate without the hole. Throughout this study the NODAL solution routine (14) was utilized for the assembly of the overall stiffness and geometric matrices and for the calculation of the eigen values.

3. CONSTANT STRAIN AND CONSTANT STRESS LOADING CASES

In all the problems analysed in the present paper, two types of loading cases were considered namely, constant stress and constant strain loading systems. The constant stress loading case is simply obtained by applying uniform nodal loads. The case of constant strain is achieved by adding beams of much greater stiffness than the plate to the loaded edges. The stiffness of the beam is established so that a uniform edge displacement correct to 6 significant figures is obtained. It is worth pointing out that both the plate and the beam elements have the same nodal degrees of freedom namely, an inplane rotation as well as the two inplane displacements, and that the in-plane rotation was restrained at all the nodes along the edges to obtain the uniform edge displacement.

4. FINITE ELEMENTS RESULTS FOR BUCKLING LOADS OF SQUARE PLATES WITH CENTRAL SQUARE AND CIRCULAR HOLES

Analytical solutions to buckling of square plates show that the critical buckling stress can be related to the buckling coefficient k by

$$\sigma_{cr} = \frac{k\,\pi^2 E}{12(1-\nu^2)} \left(\frac{t}{a}\right)^2 \qquad (12)$$

In the absence of holes, the values of k for a square plate simply supported along its four edges and subjected to a uniaxial, biaxial or shear loading are 4, 2 and 9.34. These values become 10, 5.3 and 14.7 when the plate is clamped along its four edges. Using the finite element method, we shall show how these coefficients vary as an increasing size of a central circular or square hole is introduced.

(a) square plate subjected to uniaxial edge loads

Curves I, II and III of fig. 3 show the variation of k_u with the ratio of size of hole to width of plate when the square plate is subjected to a uniaxial uniform edge load and when the four edges are clamped. Curves I and II are for a plate with square holes under constant strain loading case and constant stress loading case respectively while curve III is for a plate with circular holes and subjected to a constant stress loading case. We notice that for all these curves the changeover from the first symmetrical to the second asymmetrical mode of buckling takes place at a ratio of b/a or d/a of about one third and that the results for the plate with circular holes lie consistently below those for the plate with square holes. The plate with square holes is insensitive to the loading cases up to the minimum value of k_u for the first mode of buckling which takes place at a value of b/a of about ($^1/5$). This minimum value is about 87% of that for a plate without a hole. For the case of circular holes the minimum k_u is 82% of that for the plate without a hole and takes place at a hole size of ($^1/4$) of the plate.

Figure 4 shows the results for a square plate simply supported along its four edges and having square holes. The variation in the buckling coefficient is more pronounced in the case of constant strain loading (curve I), this curve shows that k_u becomes gradually less up to b/a of 0.3 with a minimum value of 84% of that for a plate without a hole and then rises to become 35% higher than that for a plate without a hole at b/a = 0.7. Curve II for the constant stress loading case appears to reach a minimum value of 74% at about b/a = 0.5 and then rises very slowly. These two sets of results lie consistently above the results given by Hull (15) and Yang (16) which show a continuously decreasing value of k_u for all the hole sizes considered.

The authors results for a simply supported square plate with central circular holes are consistently below those given in references (10, 17 and 18) except for the case of constant stress loading for d/a between 0.4 and 0.5 in which case, they lie between those given in references (17) and (18).

(b) square plates subjected to biaxial edge loading

Figure 5 shows the results for the clamped plate with circular and square holes given by broken and full lines respectively. These curves show a similar pattern to those observed in fig. 3 except that no results were obtained for the second mode of buckling for the case of plates with circular holes. The least values of the buckling coefficients k_b do not appear to change appreciably from those for plates with no holes for holes of less than one fifth of the size of the plate. The results given in reference (11) are also shown by crosses and circles on this figure. The crosses are to be compared with curve II and the circles with curve IV. Satisfactory agreement is indicated but the results of reference (11) were obtained by idealizing quarter of the plate into 182 elements while those for curve II and IV are for idealizations using 75 and 85 elements respectively.

Results for the simply supported plates are also obtained but because of lack of space they are not given.

(c) square plates subjected to uniform shear edge loading

This loading condition is of practical importance for it represents the loading of webs of plate girders. The results for plates with square holes are given in Fig. (6). The upper group of curves are for clamped plates of which curve I represents the constant strain loading case. It is seen that the buckling coefficient k_s for this case decreases continuously with size of the square hole. Curve II represents a constant stress loading case where the inplane rotation at the edges of the plate are restrained while for the results shown in curve III, this degree of freedom is left unrestrained. Again the buckling coefficient continuously decreases to reach a zero value when the entire plate is replaced by the hole. The results obtained in reference (11) for this case are shown by broken lines for the constant stress loading case. These results indicate that there is no convergence to the value of the buckling coefficient for a plate without a hole and that the buckling coefficient would reach a zero value before the square plate is entirely covered by the hole.

A similar set of results for the plate with circular holes are given in fig. (7). They show a similar pattern as for the plates with square holes. The curves represented by the broken lines are from reference (9) which are based on the use of inplane elements with prescribed displacement function. Again convergence to the results for plates with no holes are obtained by the authors results but not with those given in reference (9). It is also noticed that the buckling coefficients for the plate with circular holes are consistently higher than plates with square holes where b = d.

5. *CONCLUSIONS*

The results presented in this paper show

(1) The use of the strain based inplane elements requires less number of elements into which the plate is to be divided into in order that sufficiently accurate converged results are obtained.

(2) The ability of these elements to produce the classical critical loads for plates with no holes in the limit as the size of the hole is reduced. This is not the case with inplane elements based on prescribed displacement functions unless an inordinately large number of elements are used.

(3) Any unusual rapidly changing configuration can be dealt with since no deterioration of results is encountered as the aspect ratio of the elements are varied.

(4) The use of long rectangular elements makes it possible to assemble the overall structural matrices in a simple systematic way and enables the use of the less efficient triangular elements only in regions where they are absolutely needed.

References

1. Gallagher, R.H. and Padlog, J. 'Discrete element approach to structural instability analysis'. AIAA Journal, Vol.1, No. 6, pp. 1437-1439, 1963.

2. Kapur, K.K. and Hartz, B.J. 'Stability of plates using finite element method'. Proceedings of American Society of Civil Engineering, Journal of Engineering Mechanical Division, vol. 92, No. EM2, pp. 177-195, 1966.

3. Melosh, R.J. 'Basic for derivation of matrices for the direct stiffness method'. AIAA Journal, Vol.1,No.7,pp.1631-1637, 1963.

4. Pifko, A.B. and Isakson,G. 'A finite element method for the plastic buckling analysis of plates'. AIAA Journal, 1969.

5. Bogner, F.K.,Fox,R.L. and Schmidt,L.A.'The generation of inter-relement compatible stiffness and mass matrices by the use of interpolation formulae'. Proceedings of the Conference on Matrix Methods in Structural Mechanics. 1966.

6. Carson,W.G. and Newton, R.E. 'Plate buckling analysis using a fully compatible finite element'. AIAA Journal, Vol.7,No.3, pp. 527-529, 1969.

7. Sabir, A.B. 'The application of the finite element method to the buckling of rectangular plates and plates on elastic foundation'. Stavebnicky Casopsis Sav XXI,10,Bratislava, pp.689-712, 1973.

8. Holand, I. and Moan, T. 'The finite element method in plate buckling'. Finite Element Methods in stress analysis. Edited by Holand, I. and Bell, K. The Technical University of Norway, Trondheim, Norway, 1972.

9. Rockey, K.C.,Anderson, R.G. and Cheun, Y.K. 'The behaviour of square shear webs having a circular hole'. Proceedings of the International Conference on Thin Walled Structures, Swansea, pp. 148-169, 1967.

10. Pennington-Wann, W. 'Compressive buckling of perforated plate elements'. Proceedings of the First Speciality Conference on Cold-formed Structures, pp. 52-64, 1971.

11. Shanmugam, N.E. and Narayanan, R. 'Elastic buckling of perforated square plates for various loading and edge conditions'.Int.Conf. on finite element method, Shanghai, 1982.

12. Sabir, A.B. 'A new class of strain based finite elements for plane elasticity problems'. Computational Aspect of finite elements, SMiRT-7, Chicago, 1983.

13. Nath, B. 'Fundamentals of finite elements for Engineers'.Athlow Press. 1974.

14. Sabir, A.B. 'The nodal solution routine for the large number of linear simultaneous equations in the finite element analysis of plates and shells'. Finite Elements for Thin Shells and Curved Members, John Wiley and Sons Ltd. 1976.

15. Hull, R.E. 'Buckling of plates with unknown in-plane stress using the finite element method'. Master Thesis, University of Washington, August, 1970.

16. Yang,H.T.Y. 'A finite element formulation for stability analysis of doubly curved thin shell structures'. Ph.D. Dissertation, Cornell University, January, 1969.

17. Kawai, T. and Ohtsubo, H. 'A method of solution for the complicated buckling problems of elastic plates with combined use of Rayleigh-Ritz's procedure for the finite element method'. Proceedings of the Second Air Force Conference on Matrix Methods in Structural Mechanics.Wright Patterson Air Force Base,Oct.1968.

18. Ritchie, D. and Rhodes, J. 'Buckling and post-buckling behaviour of plates with holes'. Aeronautical Quarterly, pp.281-296, 1975.

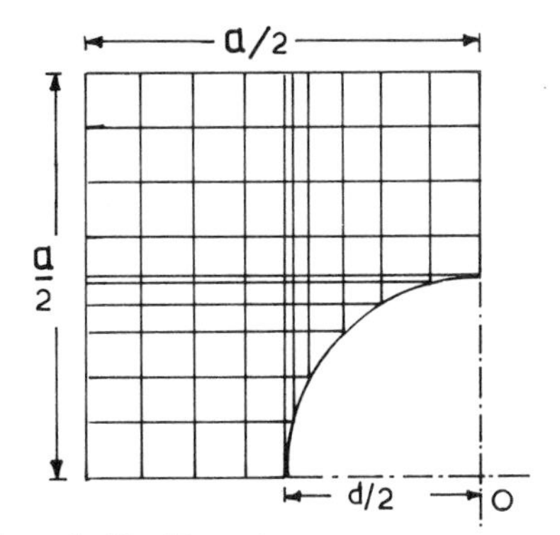

Fig. 1 Finite element idealization for plate with a circular hole

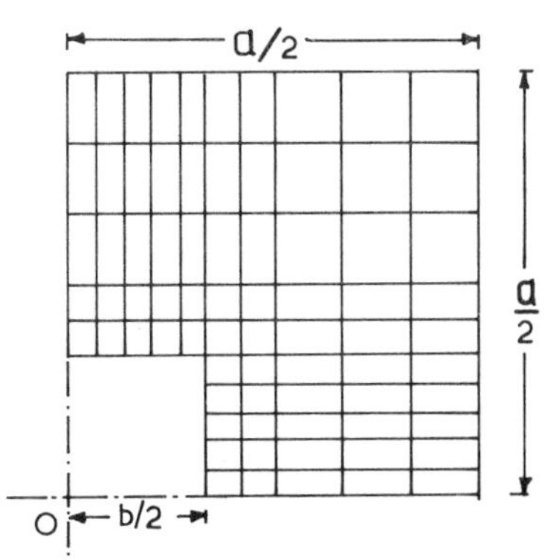

Fig. 2 Finite element idealization for plate with a square hole

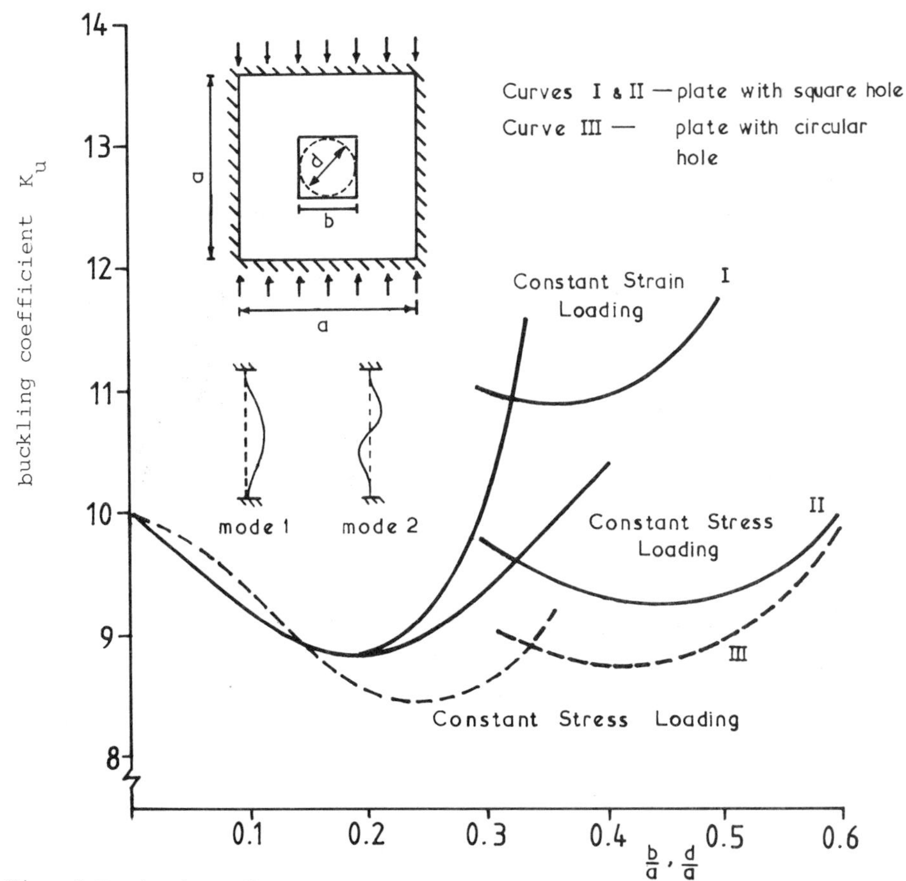

Fig. 3 Variation of buckling coefficient for a clamped, uniaxially loaded square plate with square and circular holes

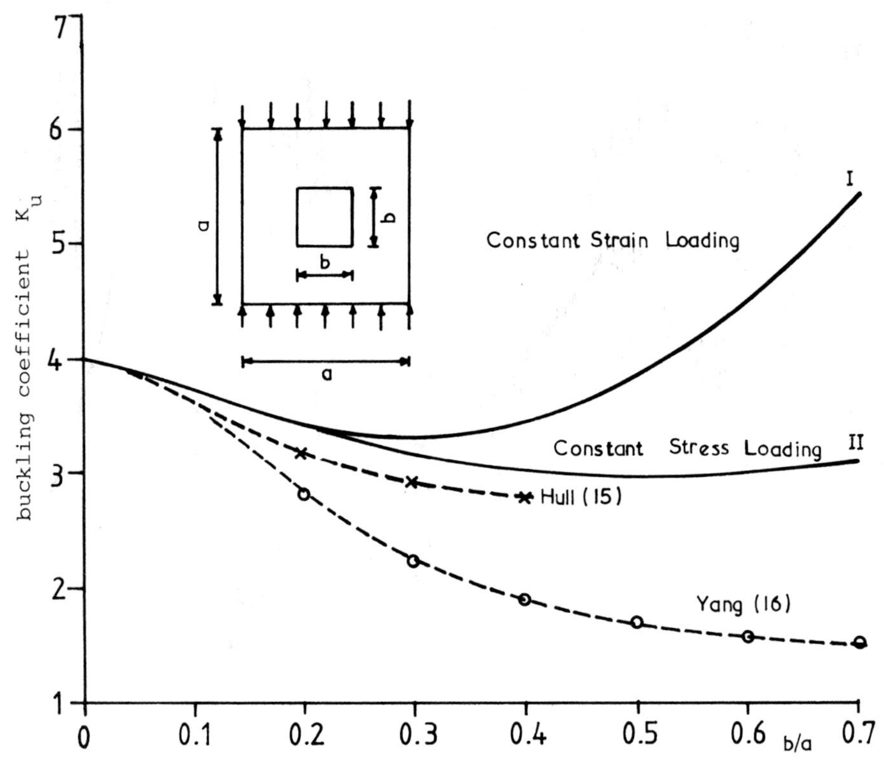

Fig. 4 Variation of buckling coefficient for a simply supported, uniaxially loaded square plate with square holes

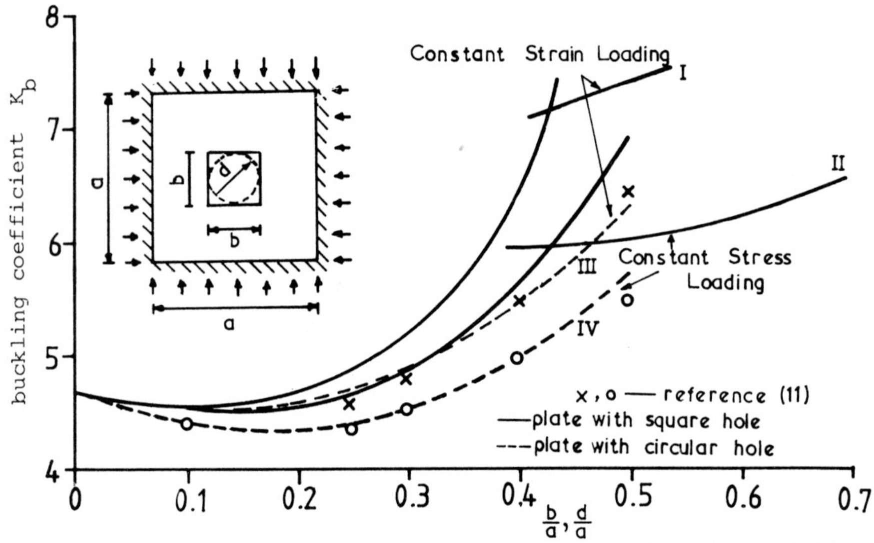

Fig. 5 Variation of buckling coefficient for a clamped biaxially loaded square plate with square or circular holes

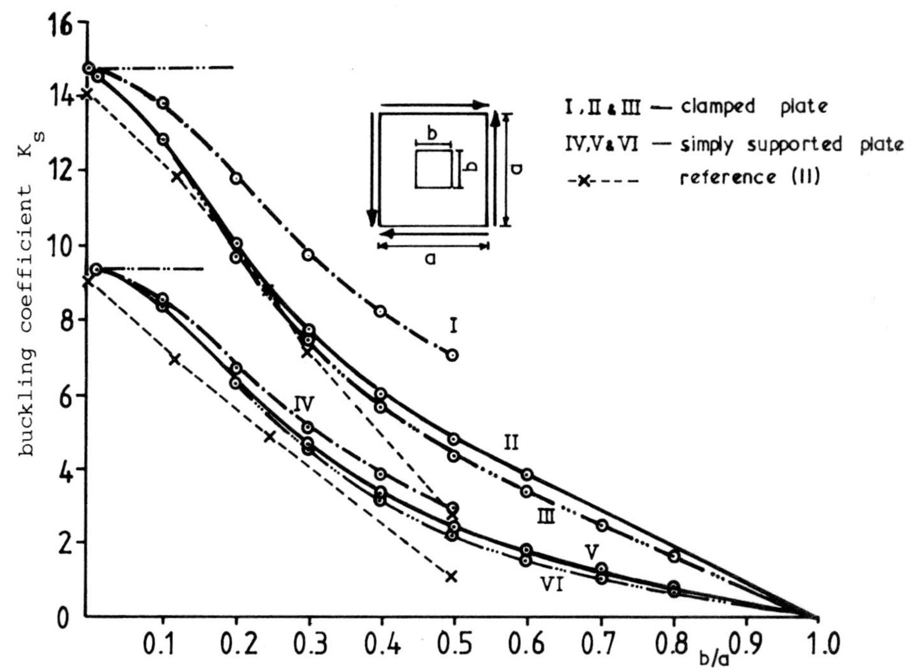

Fig. 6 Variation of buckling coefficient for shear loaded plates with square holes

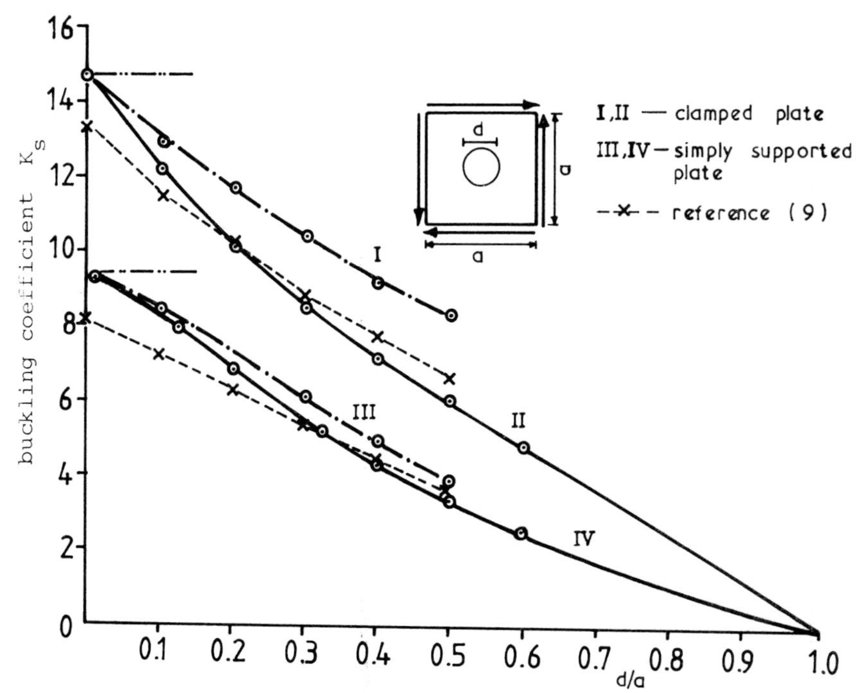

Fig. 7 Variation of buckling coefficient for shear loaded plates with circular holes

BUCKLING AND ELASTO-PLASTIC COLLAPSE OF PERFORATED PLATES

Z G Azizian and T M Roberts

Department of Civil and Structural Engineering, University College, Cardiff

Synopsis

The buckling and geometrically nonlinear elasto-plastic analysis of perforated plates by the finite element method is described. Triangular elements are used to model the plates and a number of solution refinements are discussed. The elasto plastic stress strain relationships are based on Ilyushin's approximate area yield function. Solutions are presented for axially compressed square plates with central square and circular holes.

Notation

a,b	dimensions of plate
d	dimension of hole
E	Young's modulus
f	yield function
K	buckling coefficient
M_x, M_y, M_{xy}	bending stress resultants
N_x, N_y, N_{xy}	membrane stress resultants
P_i	external forces
q_i	displacements
t	thickness of plate
u,v,w	displacements in x,y,z directions
V, V_o, V_{op}	potential energy
w_o, w_{oc}	initial imperfections
x,y,z	coordinate directions
$\gamma_{xym}, \gamma_{xyb}$	membrane and bending shear strains
δ, δ^2	first and second variations
ε_i	strain
$\bar{\varepsilon}_m$	average compressive membrane strain

ε_o	yield strain
$\varepsilon_{xb}, \varepsilon_{yb}$ $\varepsilon_{xm}, \varepsilon_{ym}$	bending and membrane strains
λ	scalar
μ	load factor
ν	Poisson's ratio
σ_{cr}	critical stress
σ_i	stress
$\bar{\sigma}_m$	average compressive membrane stress
σ_o	yield stress
$\bar{\sigma}_u$	average ultimate membrane stress

Introduction

It is often necessary to provide openings in thin plated structures such as cold formed steel members, aeroplane fuselages, plate and box girders and ship structures for access and services. The presence of holes in such structures results in a redistribution of the membrane stresses accompanied by a change in the buckling and strength characteristics.

The buckling of perforated plates subjected to pure shear and uniaxial and biaxial compression has been investigated by Rockey et al (1), Pennington-Wann (2) and Shanmugam and Narayanan (3) using the finite element method. For pure shear, the buckling load decreases continuously with increasing size of the hole. However, for uniaxial and biaxial compression, the buckling load may increase with increasing size of the hole due to the redistribution of the membrane stresses towards the edges of the plate.

Even though the buckling load may increase with increasing size of the hole, it is not to be expected that there will be a corresponding increase in the ultimate or collapse load. This paper describes the buckling and geometrically nonlinear elasto-plastic analysis of perforated plates by the finite element method. Results are presented for uniaxially compressed square plates with central circular and square holes.

Finite element formulations for buckling and geometrically nonlinear elasto-plastic analysis have been presented elsewhere and only a brief outline of the theory is presented herein: Zienkiewicz (4), Kapur and Hartz (5), Roberts and Ashwell (6) and Crisfield (7).

1. ENERGY PRINCIPLES

The total potential energy V of a structural system can be defined by the equation

$$V = V_o - \int P_i \, dq_i + \int \{ \int \sigma_i \, d\varepsilon_i \} dvol \tag{1}$$

in which P_i and q_i represent the external forces and corresponding displacements and σ_i and ε_i represent the internal stresses and corresponding strains. V_o is the potential energy of the system prior to application of external forces.

Along any equilibrium path, V is constant, and hence the first and second variations of V along the equilibrium path, denoted by δV and $\delta^2 V$, are zero. From eqn. (1)

$$\delta V = -P_i\,\delta q_i + \int \sigma_i\,\delta\varepsilon_i\,dvol = 0 \tag{2}$$

$$\delta^2 V = -P_i\delta^2 q_i - \frac{1}{2}\,\delta P_i\,\delta q_i + \int(\sigma_i\delta^2\varepsilon_i + \frac{1}{2}\,\delta\sigma_i\delta\varepsilon_i)dvol = 0 \tag{3}$$

Rearranging eqn. (3) gives

$$\delta P_i\delta q_i = -2P_i\delta^2 q_i + \int(2\sigma_i\delta^2\varepsilon_1 + \delta\sigma_i\delta\varepsilon_i)dvol \tag{4}$$

Equation (4) provides a basis for incremental analysis of nonlinear problems. The right hand side of eqn. (4) is $2\delta^2 V_p$, $\delta^2 Vp$ being the second variation of V for stationary values of the external forces i.e. assuming $\delta P_i = 0$. When $\delta^2 V_p = 0$, eqn. (4) is indeterminate and hence the vanishing of $\delta^2 V_p$ indicates critical conditions on an equilibrium path. Critical conditions occur therefore when

$$\delta^2 V_p = -P_i\delta^2 q_i + \int(\sigma_i\delta^2\varepsilon_i + \frac{1}{2}\,\delta\sigma_i\,\delta\varepsilon_i)dvol = 0 \tag{5}$$

2. *NONLINEAR STRAINS*

An element of a thin plate of thickness t, Young's modulus E and Poisson's ratio ν is shown in Fig. 1. The displacements in the x,y and z directions are denoted by u,v and w and the plate is assumed to have a small initial imperfection w_o. The nonlinear expressions for the membrane and bending strains are : Timoshenko (8)

$$\{\varepsilon_m\} = \begin{bmatrix}\varepsilon_{xm}\\ \varepsilon_{ym}\\ \gamma_{xym}\end{bmatrix} = \begin{bmatrix}u_x + 0.5(w_x^2 - w_{ox}^2)\\ v_y + 0.5(w_y^2 - w_{oy}^2)\\ u_y + v_x + w_x w_y - w_{ox}w_{oy}\end{bmatrix} \tag{6}$$

$$\{\varepsilon_b\} \quad \begin{bmatrix}\varepsilon_{xb}\\ \varepsilon_{yb}\\ \gamma_{xyb}\end{bmatrix} \quad \begin{bmatrix}-z(w_{xx}-w_{oxx})\\ -z(w_{yy}-w_{oyy})\\ -2z(w_{xy}-w_{oxy})\end{bmatrix} \quad \begin{bmatrix}-z(\chi_x-\chi_{ox})\\ -z(\chi_y-\chi_{oy})\\ -2(\chi_{xy}-\chi_{oxy})\end{bmatrix} = -z\{\chi-\chi_o\} \tag{7}$$

in which suffices x,y etc. associated with u,v and w denote differentiation.

3. *STRESS STRAIN RELATIONSHIPS*

For elastic material the membrane forces per unit length, N_x etc. and bending stress resultatns shown in Fig. 1 are related to the strains by the equations

$$\{N\} = [N_x, N_y, N_{xy}]^T = t\,[E]\{\varepsilon_m\} \tag{8}$$

$$\{M\} = [M_x, M_y, M_{xy}]^T = t^3/12\,[E]\{\chi-\chi_o\} \tag{9}$$

in which

$$[E] = \frac{E}{(1-\nu^2)}\begin{bmatrix}1 & \nu & 0\\ \nu & 1 & 0\\ 0 & 0 & (1-\nu)/2\end{bmatrix} \tag{10}$$

The derivation of the elasto-plastic stress strain relationships, using an approximate area yield function proposed by Ilyushin, was presented by Crisfield (7). The approximate yield function proposed by Ilyushin is of the form

$$f = \frac{\bar{N}}{t^2\sigma_o^2} + \frac{4s\ \overline{MN}}{\sqrt{3}\ t^3\sigma_o^2} + \frac{16\ \bar{M}}{t^4\sigma_o^2} \leq 1 \tag{11}$$

in which σ_o is the uniaxial yield stress, $s = \overline{MN}/|\overline{MN}|$ and

$$\bar{N} = N_x^2 + N_y^2 - N_xN_y + 3N_{xy}^2$$

$$\bar{M} = M_x^2 + M_y^2 - M_xM_y + 3M_{xy}^2$$

$$\overline{MN} = M_xN_x + M_yN_y - 0.5(M_xN_y + M_yN_x) + 3M_{xy}\ N_{xy} \tag{12}$$

During plastic flow, the state of stress remains on the yield surface and hence

$$\delta f = \{\partial f/\partial N\}^T\{\delta N\} + \{\partial f/\partial M\}^T\{\delta M\} = 0 \tag{13}$$

Assuming that eqn. (11) serves also as a plastic potential function, the plastic strain increments can be expressed

$$\{\delta\varepsilon_m^p\} = \lambda\ \{\partial f/\partial N\} \quad ; \qquad \{\chi^p\} = \lambda\ \{\partial f/\partial M\} \tag{14}$$

in which λ is a positive scalar which defines the absolute magnitude of the plastic strain increments.

The total strain increments are the sum of elastic and plastic components i.e. $\{\delta\varepsilon_m\} = \{\delta\varepsilon_m^e\} + \{\delta\varepsilon_m^p\}$ etc. Hence

$$\{\delta N\} = t\left[E\right]\{\delta\varepsilon_m^e\} = t\left[E\right]\{\{\delta\varepsilon_m\} - \{\delta\varepsilon_m^p\}\}$$

$$\{\delta M\} = t^3/12\ \left[E\right]\{\delta\chi^e\} = t^3/12\ \left[E\right]\{\{\delta\chi\}-\{\delta\chi^p\}\} \tag{15}$$

Solving eqns. (11) to (15) gives the elasto-plastic stress strain relationships in the form

$$\{\delta N\} = \left[CC\right]\{\delta\varepsilon_m\} + \left[CD\right]\{\delta\chi\}$$

$$\{\delta M\} = \left[CD\right]^T\{\delta\varepsilon_m\} + \left[DD\right]\{\delta\chi\} \tag{16}$$

4. *FINITE ELEMENT ANALYSIS*

Following well known finite element techniques and assuming the material remains elastic, eqn. (5) can be reduced to the form

$$\delta^2 V_p = \frac{1}{2}\{\delta q\}^T\ \left[\left[KL\right] + \mu\left[KG\right]\right]\{\delta q\} = 0 \tag{17}$$

in which $\left[KL\right]$ is the linear stiffness matrix, $\left[KG\right]$ the geometric stiffness matrix which depends on the state of stress prior to buckling and μ is a scalar load factor. $\{\delta q\}$ are the nodal displacement variables, which are linear functions, and hence the term $P_i\delta^2 q_i$ vanishes. Critical conditions occur when $DET\left[\left[KL\right]+\mu\left[KG\right]\right]=0$, the lowest eigenvalue μ defining the critical load factor and the corresponding eigenvector the buckled shape.

Using the elastic or elasto plastic stress strain relationships as appropriate, eqn. (4) can be reduced to the form

$$\{\delta P\} = \left[[KEP] + [KGEP]\right]\{\delta q\} \tag{18}$$

in which $[KEP]$ depends on the material elastic constants and current state of stress and $[KGEP]$ depends also on the current geometry.

Solutions of nonlinear problems were obtained by incrementing displacements and using a mid increment stiffness technique discussed by Roberts and Ashwell (6). Current states of stress were reduced proportionately back to the yield surface to prevent a build up of errors.

The plates analysed were modelled using triangular elements. Bending action was represented by nine degrees of freedom elements, the displacement function being defined in terms of area coordinates : Zienkiewicz (4). Membrane action was represented by six degrees of freedom elements, linear polynomials being used to define both u and v. In deriving the geometric matrices $[KG]$ and $[KGEP]$ in eqns. (17) and (18), linear polynomials were used for u,v and w to ensure constant membrane strains throughout an element, which is advantageous for convergence.

5. *RESULTS*

The incremental finite element program was tested by solving the problem of a simply supported rectangular plate subjected to uniaxial compression, the loaded edges only being constrained to remain straight. The plate was assumed to have an initial imperfection of the form

$$w_o = w_{oc} \sin \pi x/a \sin \pi y/b \tag{19}$$

The number of elements used to model the plate was 288 (12 x 12 rectangular mesh) and results were obtained by incrementing the in-plane displacement of the load. The results and other details of the problem are shown in Fig. 2 and there is excellent agreement with existing analytical and finite element solutions : Moxham (9) and Crisfield (7). $\bar{\sigma}_m$ and $\bar{\varepsilon}_m$ are the average compressive membrane stress and strain respectively and ε_o is the yield strain.

The main problem studied was that of a square plate with sides length b, containing a central circular or square hole of diameter or side length d, and subjected to uniaxial compression. The loaded edges only were constrained to remain straight in the plane of the plate. For displacements normal to the plane of the plate, simply supported and clamped boundary conditions were considered. Values of E, ν and σ_o were taken as 205000 N/mm^2, 0.3 and 245 N/mm^2 respectively. For simply supported plates, w_o was assumed as given by eqn. (19) with $a=b$ while for clamped plates w_o was assumed to be of the form

$$w_o = w_{oc} (1-\cos 2\pi x/b)(1-\cos 2\pi y/b)/4 \tag{20}$$

The initial central deflection w_{oc} was taken as

$$w_{oc} = 0.145\, b\, (\sigma_o/E)^{0.5}$$

which has been recommended in the proposed British Code of Practice for the Design of Steel Bridges.

A number of radial and rectangular meshes were used to test convergence for the eigenvalue solution. When convergence was satisfactory, the same mesh was used for the corresponding nonlinear problem.

The critical compressive membrane stress σ_{cr} for initially flat square plates can be expressed : Timoshenko (8).

$$\sigma_{cr} = K \pi^2 t^2 E/12(1-\nu^2)b^2 \tag{21}$$

in which K is a dimensionless buckling coefficient. K values for perforated plates, obtained from the present analysis are shown in Fig. 3 and there is general agreement with existing results : Pennington-Wann (2) and Shanmugam and Narayanan (3).

The corresponding results for the elasto-plastic analysis are shown in Fig. 4 in which $\bar{\sigma}_u$ is the average compressive membrane stress at failure. There was no significant difference in the results for square and circular holes. To allow for variations in E and σ_o the results can be normalised in accordance with Von Karman's formula for the ultimate strength of uniaxially compressed plates,

$$\bar{\sigma}_u = (\sigma_{cr}\ \sigma_o)^{0.5} \tag{22}$$

by replacing b/t by $(b/t)^*$ where

$$(b/t)^* = (b/t)(205000\ \sigma_o/245\ E)^{0.5} \tag{23}$$

6. *CONCLUSIONS*

The buckling load of a uniaxially compressed plate with a centrally placed hole is almost independent of the hole size up to half the width of the plate and may even increase for larger hole sizes.

The ultimate load of a uniaxially compressed plate with a centrally placed hole is influenced significantly by the size of the hole. The reduction in the ultimate load is most pronounced for low b/t values.

References

1. Rockey, K.C.,Anderson,R.G. and Cheung, Y.K. "The behaviour of square shear webs having a circular hole". Proc.Swansea Conf. Thin Walled Structures, Crosby Lockwood, London, pp. 148-169, 1967.
2. Pennington-Wann, W. "Compressive buckling of perforated plate elements". Proc. 1st Speciality Conf. Cold Formed Structures, University of Missouri-Rolla, pp. 58-64, April, 1973.
3. Shanmugam, N.E. and Narayanan, R. "Elastic buckling of perforated square plates for various loading and edge conditions". Proc.Int. Conf. F.E.M., Paper No.103,Shanghai, August, 1982 .
4. Zienkiewicz, O.C. The Finite Element Method in Engineering Science, McGraw-Hill, London, 1971.
5. Kapur, W.W. and Hartz, B.J. "Stability of plates using the finite element method", Proc.ASCE, Vol. 92, EM2, pp. 177-195, April, 1966.
6. Roberts, T.M. and Ashwell, D.J. "The use of finite element mid-increment stiffness matrices in the post-buckling analysis of imperfect structures". Int.J.Sol.Struct., Vol.7, pp. 805-823, 1971.

7. Crisfield, M.A. "Large deflection elasto-plastic buckling analysis of plates using finite elements". TRRL Report LR 593, 1973.

8. Timoshenko, S.P. and Gere, J.M. Theory of Elastic Stability 2nd. Ed., McGraw-Hill, 1961.

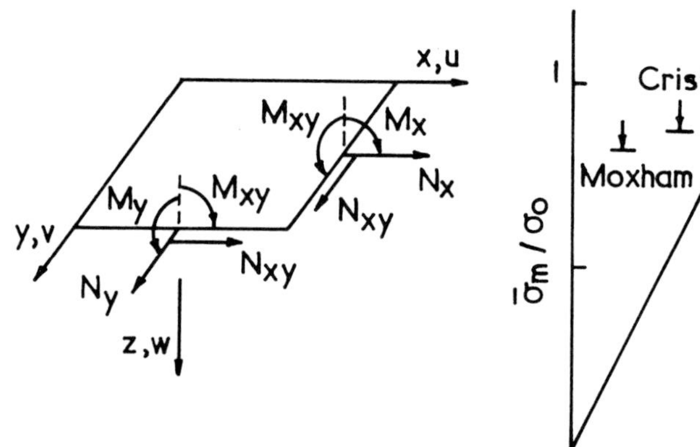

Fig 1 Element of a plate

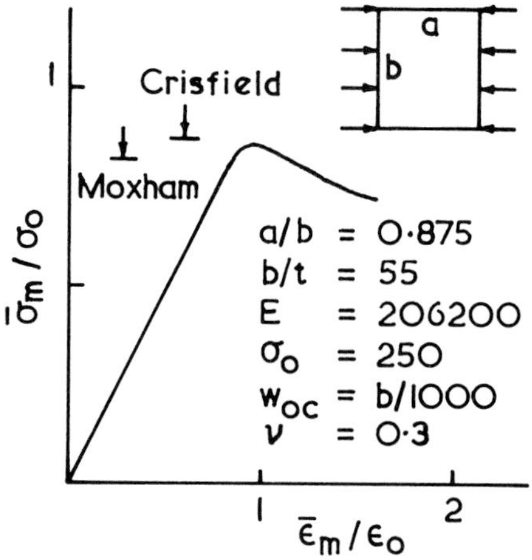

Fig 2 Collapse of axially compressed plate

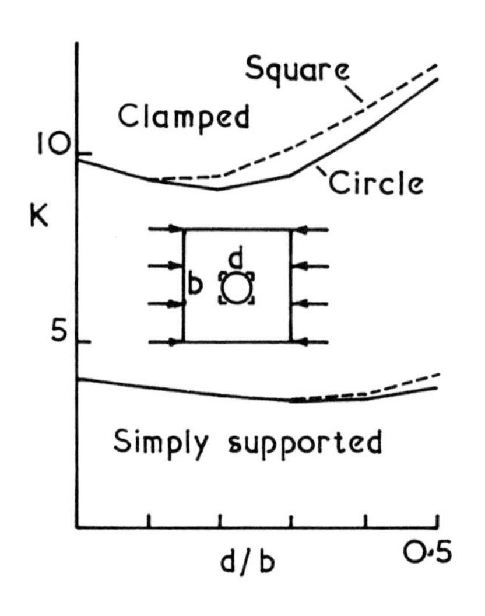

Fig 3 Buckling of perforated plates

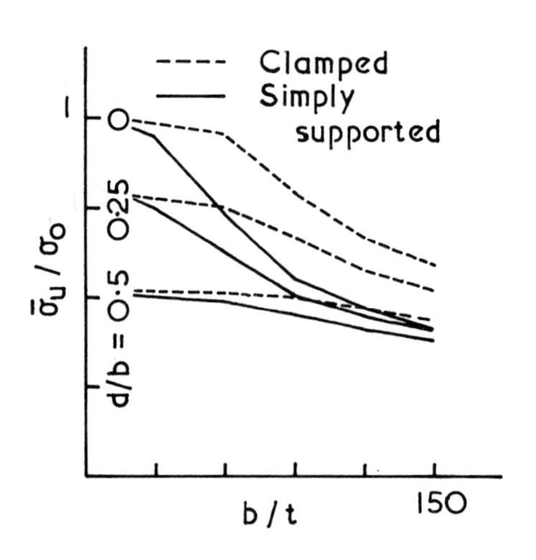

Fig 4 Failure of perforated plates

PARAMETRIC FINITE ELEMENT STUDY OF TRANSVERSE STIFFENERS FOR WEBS IN SHEAR

M.R.Horne

Simon Engineering Laboratories, University of Manchester.

W.R. Grayson

John Cornell and Associates, Melbourne, Australia, formerly Research Associate, Simon Engineering Laboratories, University of Manchester.

Synopsis

An elastic-plastic, large deflection finite element program is used to derive the ultimate strength of transversely stiffened web panels loaded in shear. The effects of alternative initial displacement patterns and of residual stresses are investigated. Both concentric and eccentric stiffeners are considered and the ultimate capacities of the stiffened webs are compared with those of corresponding fully supported individual panels. Stiffener rigidities are derived (the "knuckle" values) such that further increases of rigidity cause only slight increases in the ultimate capacity of a stiffened web. An empirical expression is proposed for the knuckle value which, it is suggested, provides a simple and appropriate basis for the design of transverse stiffeners. Comparisons are made with test results and with corresponding design proposals presented by Rockey et al, and also with corresponding design stiffener values obtained from the British Bridge Code BS 5400.

Introduction

The behaviour of transverse stiffeners in webs subjected to preponderantly shear loading has been the subject of much research, resulting in many alternative design procedures. Much depends on the criteria used in designing the web.

If design is to be based on the critical elastic buckling load of the plate panels in shear, a rational theoretical criterion for the design of the stiffeners is provided by adopting a minimum rigidity (defined non-dimensionally in terms of parameter $\gamma = EI/aD$, see Fig 1) such that the small-deflection buckling load of the *in situ* plate panel is the same as that of a panel of similar dimensions simply supported at the transverse (vertical) edges. However, it is well known that the post-buckling capacity of slender webs may be much in excess of the theoretical initial buckling load, and experimental results (1,2) have led to the proposal that the design rigidity should be three times the above value. Design codes based on elastic behaviour have tended to approximate to this requirement.

The introduction of ultimate load theory for plate and box girders has emphasised the role of post-buckling, post-yield behaviour in webs, and within this context, it is necessary to reconsider the fundamental principles governing the design of web stiffeners. An attempt has been made to do so in the derivation of the requirements of the new British Code of Practice for the Design of Steel Bridges (3). While the resulting stiffener sizes may be regarded as safe, they are highly conservative as compared with previous design criteria, particularly for eccentric stiffeners.

Rockey et al (4) have proposed a design procedure based on stiffener forces calculated from the ultimate load theory of Rockey et al (5) for webs. This attempt at a rational design procedure is extremely valuable and gives stiffness sizes considerably less than those required according to BS 5400. Unfortunately, it suffers from two major disadvantages. Firstly, it is extremely complicated in application, depending as it does on an elaborate estimation of web forces and the plastic design of the stiffener itself with explicit allowance for destabilising forces. The resulting design method is somewhat more involved than that specified in BS 5400. Secondly, it contains an empirical effective width of web plate assumed to act as part of the stiffener, justified by test results for only one particular value of panel aspect ratio ϕ (= a/b) allied with one particular plate slenderness, see Fig 1.

In view of the large number of parameters involved, there is advantage in pursuing an accurate numerical parametric study as a means of establishing a design procedure, although test results will always be required as spot checks on any design method. A large deflection elastic-plastic finite element program has been used by the authors in such a study (6) with the aim of deriving simple design rules for transverse stiffeners in webs subjected to any combinations of shear, longitudinal compression and longitudinal bending stresses. It has been possible to define a design procedure based solely on a rigidity requirement for both concentric and eccentric stiffeners. By far the predominating criterion has been shown to be that for shear loading, and the present paper is restricted to presenting results from this part of the parametric study.

1. *FINITE ELEMENT ANALYSIS*

The finite element program used in the analysis was developed by Grayson (7) from an earlier program by Milner and Grayson (8). For the plates, a rectangular element with 4 corner and 4 mid-side nodes was used. In-plane behaviour was modelled by a standard 8 node quadratic membrane element, while for out-of-plane behaviour, use was made of the four corner node element only. The stress state was monitored at a number of layers through the plate thickness, plasticity being introduced via the von Mises yield criterion and the associated Prandtl-Reuss flow rules.

The stiffeners, with the cross-sections shown in Fig 1, were modelled by linear elements with three nodes and were again divided into a number of elements down the depth to allow for the effects of plasticity. Torsional and lateral bending resistance of the stiffeners was neglected. The finite element geometries used in the analysis are shown in Fig 2.

The finite element analysis followed a step-by-step procedure with continous updating of the stiffness matrix to allow for plasticity. The boundaries were deformation controlled except where the upper and lower boundaries (adjacent to the flanges of the supposed plate or box girder) were assumed to be stress-free for forces normal to the boundary. These latter boundaries were however under strain control in relation to deformation parallel to the boundary. All boundaries were assumed simply supported with respect to out-of-plane forces.

Initial out-of-plane displacement patterns were assumed in double sinusoidal shapes of magnitudes equal to the maximum permitted under BS 5400:Part 6 (9), given for a panel $a \times b$ with a yield stress of σ_y N/mm^2 by

$$\Delta_x = \frac{G}{165} \frac{\sigma_y}{355} \tag{1}$$

with $G = a$ when $a \leqslant b$, $G = 2b$ when $a \geqslant 2b$. Various initial displacement patterns have been assumed in the analysis is shown in Fig 3. For stiffened panels, an additional sinusoidal displacement was superimposed on stiffeners and plate panels to give a maximum stiffener imperfection of

$$\Delta_x = \frac{G}{750} \tag{2}$$

where G = length of stiffener.

In a few cases, residual stress patterns were introduced with compressive stresses in both directions in the plate panels amounting to either 10% or 20% of yield stress. Tensile stresses in the vicinity of the plate/stiffener joints and at the tops of the stiffeners were introduced to establish equilibrium, according to a distribution proposed by Little (10).

2. *PANELS WITH SINGLE TRANSVERSE CONCENTRIC STIFFENERS*

The variables to be considered are (see Fig 1) the panel aspect ratio ϕ, the plate slenderness $\lambda = \frac{b}{t} \sqrt{\frac{\sigma_y}{355}}$ and the rigidity parameter for the stiffener $\gamma = EI/aD$. The second moments of area I of the stiffeners have been calculated without any "effective" plate width associated with them, the view being that such inclusions are empirical and that the simpler parameter is to be preferred. The values of I for concentric stiffeners have been taken about the mid-plane of the plate panel while those of the eccentric stiffeners have been taken about an axis at the surface of the plate on the same side as the stiffener.

The relationship between the imposed normalised shear strain γ' and normalised mean shear stress τ' when the individual plate panels have ϕ = 1.0 and γ = 3, 69 and 100 are shown in Fig 4. The normalised shear strain is the ratio of shear strain to shear strain at yield while the normalised shear stress is the ratio of shear stress to the yield stress in shear. Two alternative initial plate panel displacement patterns are assumed - "cusp" (P3) and "chequerboard" (P5), see Fig 3. Also shown in Fig 4 is the strength of an individual plate panel with the same values of ϕ and λ.

It should be noted that:-

(a) The maximum load capacity occurs at a shear strain between one and two times the elastic shear strain at yield. This is found to be the case for almost all stiffened panels of practical proportions and slenderness values.

(b) The pattern of initial displacement has a discernible although small effect on the ultimate capacity in shear.

(c) While a change in stiffener rigidity from $\gamma = 3$ to $\gamma = 69$ increases the ultimate capacity of the stiffened panel by nearly 30%, a further increase to $\gamma = 100$ has negligible effect.

The effect of a wider range of initial displacement patterns and of variation of stiffener rigidity on the maximum capacity in shear of a stiffened panel, again with $\phi = 1.0$ and with $\lambda = 160$, is shown in Fig 5. Because of the extremely heavy call on computer time made by the finite element program, only five values of γ were investigated ($\gamma = 0$, 3, 30, 69 and 100) and the graph is drawn with straight lines between these points. The true relationship would of course be smooth curves. There is some variation as between the initial displacement patterns in the stiffener rigidity (conveniently referred to as the "knuckle value") at which there is a levelling off of ultimate capacity, but it is seen that a stiffener rigidity of $\gamma = 60$ would be sufficient to ensure the attainment of the individual panel capacity for all initial displacement patterns.

The out-of-plane displacements in the plate panels at a normalised shear strain of $\gamma' = 2.0$ are shown for the chequerboard initial displacement pattern (P5 in Fig 3) for various values of stiffener rigidity in Fig 6. When $\gamma = 30$, the main plate buckle goes clearly through the stiffener and is only slightly reduced in magnitude by its presence. However, with a stiffener with a rigidity of $\gamma = 100$, which exceeds the knuckle value of about 60, the buckle is clearly interrupted by the stiffener, although the direction of the main buckle in the separate plate panels is still nearly across the diagonal of the unstiffened panel.

The areas over the plate panels in which yielding has ocurred on either or both surfaces at the stage when $\gamma' = 0.8$ (see Fig 4) are shown in Fig 7. The stiffener with rigidity $\gamma = 30$ fails to prevent a yielding pattern corresponding to a tension field stretching along a diagonal band of the complete panel, whereas when $\gamma = 100$, the yielding pattern reveals the presence of tension fields stretching along diagonal bands in the separate plate panels. Figures 6 and 7 fill out the significance of the deduction made from Fig 5, that a "knuckle value" of the stiffener rigidity exists between $\gamma = 30$ and $\gamma = 100$.

The behaviour of the stiffeners themselves is illustrated in Fig 8, which refers to a stiffened panel with a concentric stiffener in which the plate panels either side of the stiffener have $\phi = 0.5$ and $\lambda = 240$. The knuckle value for the stiffener panel is found to be at $\gamma = 150$ and the initial plate panel deflections are in the chequerboard pattern. The graph shows the bending moment and axial force distributions in the stiffener (down to the mid-point) when

$\gamma' = 2.0$. The axial force is shown as the mean stress divided by the yield stress and the bending moment as the bending moment in the stiffener divided by the plastic moment. It will be seen that a weak stiffener with $\gamma = 10$ develops plastic hinges at mid-height and quarter points, bending the stiffener into a "zig-zag" reflecting the influence of the initial displacement in the plate panels. Even the stiffener which has a rigidity in the region of the knuckle value develops a plastic hinge at mid-height, reducing the axial force in the stiffener itself to zero or about zero. The axial force in the stiffener therefore has to be transmitted by the plate panel in the vicinity of mid-height, a tendency which is in fact seen to be developing at an earlier stage of shear strain for the stiffened panels dealt with in Fig 7.

The above results illustrate only a small part of the complete parametric study (6). The results of some other important aspects of the investigation are summarised below.

3. PANELS WITH SINGLE TRANSVERSE ECCENTRIC STIFFENERS

The collapse loads of panels with a single eccentric stiffener of either rectangular or T section (see Fig 1) were calculated for a variety of aspect ratios and plate slenderness values. It was found that, with the stiffener I values calculated about an axis at the surface of the plate, the ultimate loads for panels with eccentric stiffeners were identical (within about 3% with a standard deviation of about 1%) with the ultimate loads of panels with concentric stiffeners for the same value of γ. This result enables a great simplification to be made in the design criteria for eccentric stiffeners - and provides an escape from the controversy over the "effective width" of web plate to be included with the stiffener section.

4. INFLUENCE OF RESIDUAL STRESSES

While residual stresses may reduce the ultimate capacity of plate panels in compression they have negligible effects on ultimate capacity in shear (see Harding et al (11)). The presence of residual stresses can also reduce the carrying capacity of stiffeners under axial compression, but it was concluded from the parametric study that the effect of residual stresses on required stiffener rigidity for all forms of stiffened panel loading could be neglected.

5. PLATES WITH MULTIPLE STIFFENERS

Because of the enormous demands on computer time, most of the parametric study was carried out on panels with a single stiffener. Checks were however made on panels with three stiffeners (Fig 2, finite element pattern M10), using the plate panel initial displacement pattern shown as P15, Fig 3. In addition, all stiffeners had the same initial displacement in the same direction as each other. It was found that the ultimate capacity for a given stiffener rigidity and plate panel dimensions was virtually the same for one and three stiffeners when the stiffener rigidity was in the region of the knuckle value for the single stiffener.

6. *EFFECT OF TOP AND BOTTOM BOUNDARY RESTRAINT*

The main part of the parametric study was conducted with the top and bottom edges unrestrained with respect to vertical displacement. This gives a lower bound to the ultimate capacity of any individual plate panel. An enhanced capacity in shear is gained when the top and bottom edges are constrained to remain straight - flanges will in practice give a condition intermediate between the two extremes. An investigation was conducted to determine the knuckle value of stiffener rigidity when the top and bottom edges were constrained to remain straight between the vertical edges of the plate panels, with the stiffeners supporting their share of any tension field loads to which the rigid boundaries were subjected. It was found that keeping the boundaries straight had virtually no effect on the value of stiffener required to achieve single plate panel capacity, and that the knuckle values derived for unrestrained top and bottom boundaries may be used for boundaries held straight - and therefore also for boundaries formed by flanges of any stiffness in the plane of the web.

7. *EMPIRICAL FORMULA FOR THE DESIGN OF TRANSVERSE STIFFENERS IN WEBS IN SHEAR*

The results of the parametric study have led to the following empirical formula for the design of stiffeners which will ensure an ultimate capacity in the web not less than that of the individual plate panel with vertical edges simply supported with respect to out-of-plane deformation and fully restrained against in-plane deformation. The design value of the rigidity γ is $\gamma*$ where

$$\gamma* = \frac{0.6(\lambda - 50)}{\phi} \qquad (3)$$

$$\text{with } \gamma* \quad 0.8(\lambda - 50) \qquad (4)$$

Condition (4) rules when $\phi \leqslant 0.75$.

The design formula is conservative for webs not stressed to the ultimate capacity of single panels in shear. However, parametric studies carried out for applied loading conditions other that pure shear (6) have shown that the design formula may be used as a safe estimate of the required stiffener rigidity for all possible combinations of web loading, so that considerable simplification in design is achieved by adopting the formula as given.

8. *COMPARISONS OF PROPOSED DESIGN RULE WITH TEST RESULTS AND BS 5400*

Rockey et al (4) report the results of tests on eleven model steel girders, four panels long, tested under central point loads, differing only in the cross-section of the transverse stiffeners. The panels had ϕ = 1.0 and λ = 231 and 244 for those with concentric and eccentric stiffeners respectively. Four of the girders had concentric (double sided) rectangular section stiffeners while seven had eccentric (single sided) rectangular stiffeners. For the concentric stiffeners, γ varied from 0 to 62, while for the eccentric stiffeners, the variation was over the range 0 to 142. The variation of test capacity with stiffener rigidity is shown for the concentric stiffeners in Fig 9. and for the eccentric stiffeners in Fig 10. Girders which Rockey et al considered to have shown unsatisfactorily post-buckling behaviour with failure occurring in the stiffeners are indicated by solid symbols.

Finite element analyses have been carried out, using the program described in section 2, of the girders tested by Rockey et al with concentric stiffeners and the results are given in Fig 9. The finite element and experimental results indicated very similar girder behaviour. The finite element analysis assumed unrestrained top and bottom boundaries and the most severe plate imperfections allowed under BS 5400. Allowing for such influences, there is an acceptable correlation between the test results and the finite element analysis.

Rockey et al (4) propose a design method for transverse stiffeners based on the model proposed by Rockey et al (5) for ultimate failure of webs in shear associated with an ultimate (plastic) design criteria for the stiffeners. It is interesting to note that Rockey's assumption of complete plasticity in the stiffeners is borne out in principle by the finite element analysis, see Fig 8. However, his values of required stiffener rigidity are below the values given by the present proposals, being $\gamma = 31$ and 62 for the concentric and eccentric stiffeners respectively in his test girders, compared with values of 108 and 116 respectively given by eqn (3). It is arguable that Rockey's lower values arise because of the sensitivity of the knuckle value to initial imperfection patterns, see Fig 5. Rockey's method was chosen to fit his two experimental results, which are unlikely to reflect the most unfavourable effects of initial plate panel imperfections. On the other hand, it is interesting to note that the BS 5400 design rigidities for the stiffeners are 225 and 800 for concentric and eccentric stiffeners respectively. By way of contrast, the BS 153 and BS 449 requirements for stiffener rigidity give $\gamma = 16.4$ for both the concentric and eccentric stiffeners - a remarkably lower value, reflecting the fact that those code requirements do not allow for the attainment of ultimate strength in the web panels with tension field action.

The above comparisons reveal clearly the difficulties of establishing a consistent treatment for web stiffener design. It would be too impossibly laborious and expensive to attempt to establish empirical rules from test results because of the large number of tests that would have to be performed to cover all the variables of aspect ratio, top and bottom boundary stiffness, type and slenderness of stiffeners, plate slenderness, magnitude and pattern of initial imperfections and combinations of loading conditions. Finite element procedures are now capable of dealing with the large deflection, post-yield behaviour involved, and are the obvious means of deriving sensible design rules.

The parametric studies here described are most promising in their apparent indication of a very simple minimum rigidity criteria for transverse stiffeners. The Rockey et al treatment (4) has been shown to be on the right lines, although somewhat optimistic because of its dependence on only two test results. The BS 5400 treatment for concentric stiffeners appears to be somewhat conservative while its treatment for eccentric stiffeners appears to be quite unreasonably conservative. Finally both the Rockey et al design method and that required in BS 5400 are extremely tedious to apply compared with the simple design method now proposed.

Acknowledgements

The work reported was carried out in the Simon Engineering Laboratories of the University of Manchester. The financial support provided by the Department of Transport is gratefully acknowledged.

References

1. Massonnet Ch. 'Experimental researches on the buckling strength of the webs of solid girders'. Preliminary publication, 4th Congress IABSE, London, 1952, 539-555.

2. Massonnet Ch. 'Stability considerations in the design of steel plate girders'. Am. Soc. Civ. Engrs., J. Struct. Div., 86, Jan 1960, paper 2350.

3. British Standards Institution 'Concrete and Composite Bridges - Code of Practice for Design of Steel Bridges'. BS5 400:Part 3, 1982.

4. Rockey, K.C., Valtinat, G. and Tang, K.H. 'The design of transverse stiffeners on webs loaded in shear - an ultimate load approach'. Proc.Inst. Civ. Engrs. Part 2, Dec 1981, 1069-1099.

5. Rockey, K.C. et al. 'A design method for predicting the collapse behaviour of plate girders'. Proc. Instn Civ. Engrs, Part 2, 65, Mar 1978, 85-112.

6. Horne, M.R. and Grayson, W.R. 'A theoretical study of the behaviour and design of stiffened web panels'. Report, Simon Engineering Laboratories, University of Manchester, Aug 1981.

7. Grayson, W.R. 'Computer Program User Manual for NASP (Mark 2) and NASPLOT'. Simon Engineering Laboratories, University of Manchester, Aug 1981.

8. Milner, H.R. and Grayson W.R. 'Computer Program User Manual for NASP (Non-linear Analysis of Stiffened Plates)'. Simon Engineering Laboratories, University of Manchester, Nov 1977.

9. British Standards Institution 'Concrete and Composite Bridges - Specification for Materials and Workmanship'. BS 5400:Part 6 1981.

10. Little, G.H. 'Stiffened Steel Compression Panels - Theoretical Failure Analysis'. The Struct. Eng., Vol 54, Dec 1976, 489-500.

11. Harding, J.E., Hobbs, R.E. and Neal, B.G. 'Ultimate Load Behaviour of Plates under Combined Direct and Shear In-Plane Loading'. Inter. Conf. on Steel Plated Structures, Ed. Dowling, P.J., Harding, J.E. and Frieze, P.A., Crosby Lockwood Staples, 1977.

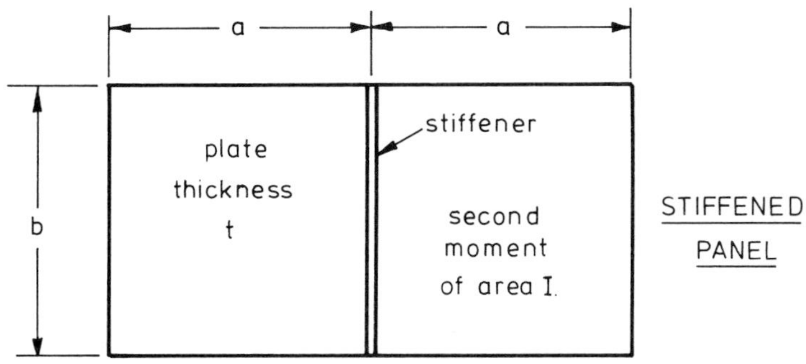

E = Elastic modulus ν = Poisson's ratio σ_y = yield stress (N/mm²)

$\frac{a}{b} = \phi$, $\lambda = \frac{b}{t}\sqrt{\frac{\sigma_y}{355}}$, $D = \frac{Et^3}{12(1-\nu^2)}$, $\gamma = \frac{EI}{aD}$

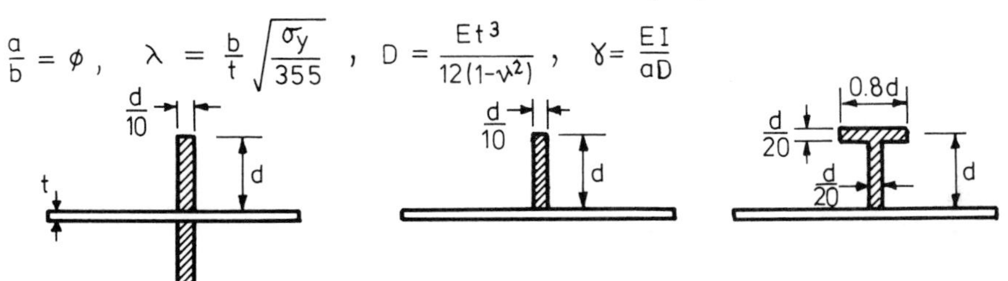

Fig 1 Notation

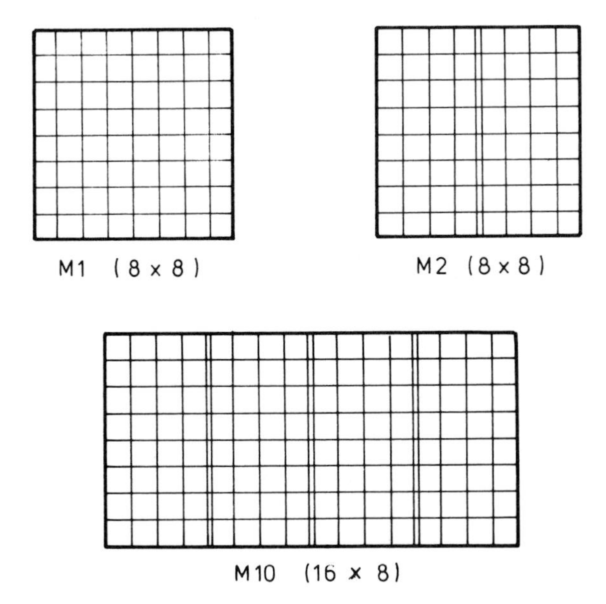

Fig 2 Finite element meshes

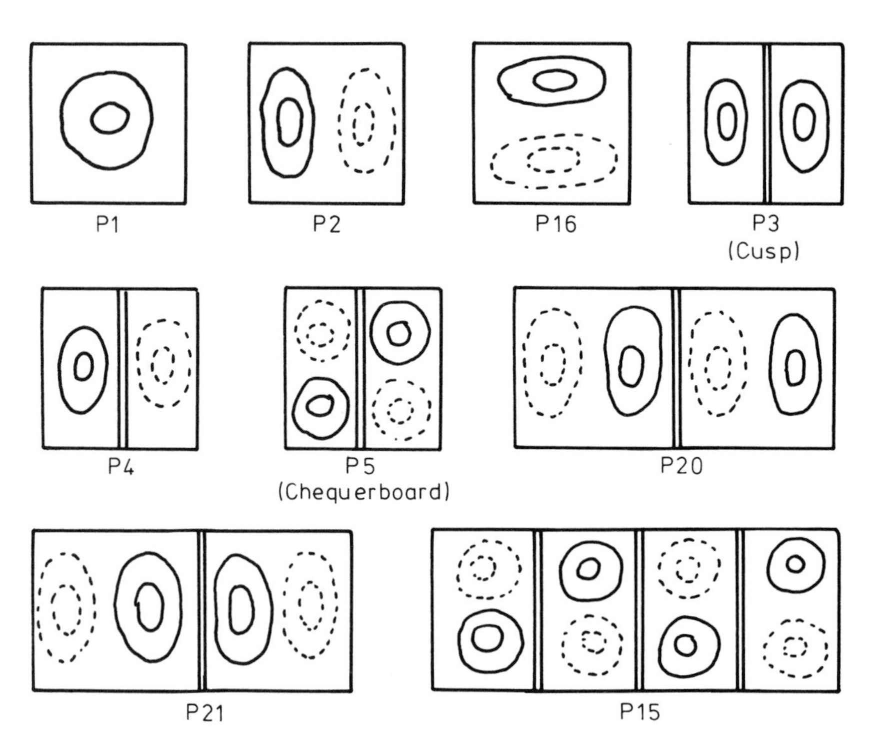

Fig 3 Initial plate panel displacements

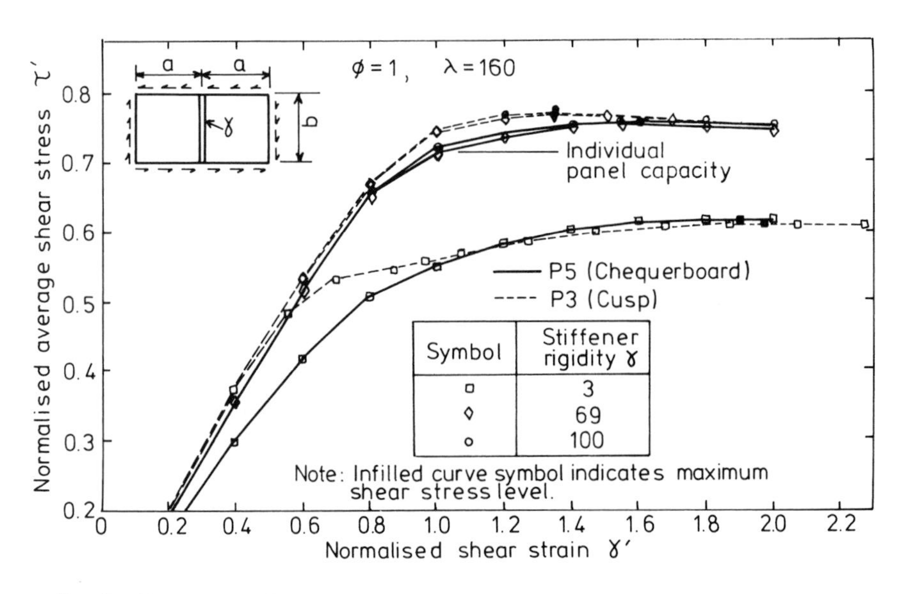

Fig 4 Transversely stiffened plate - shear stress versus strain $\phi = 1$, $\lambda = 160$

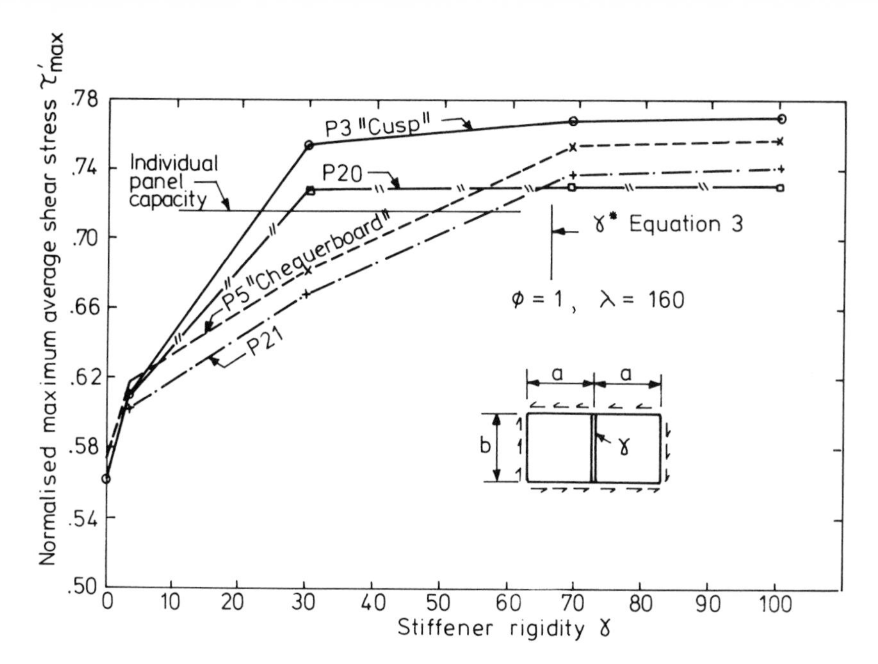

Fig 5 Shear capacity of transversely stiffened plates - influence of initial displacement pattern

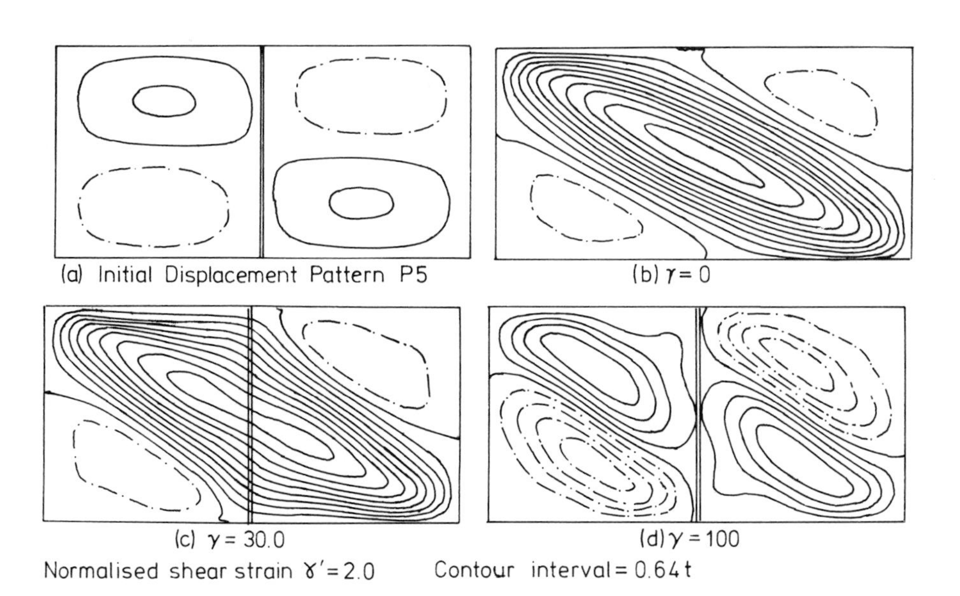

Fig 6 Variation in out-of-plane displacement pattern - chequerboard initial displacements, $\phi = 1$, $\lambda = 160$

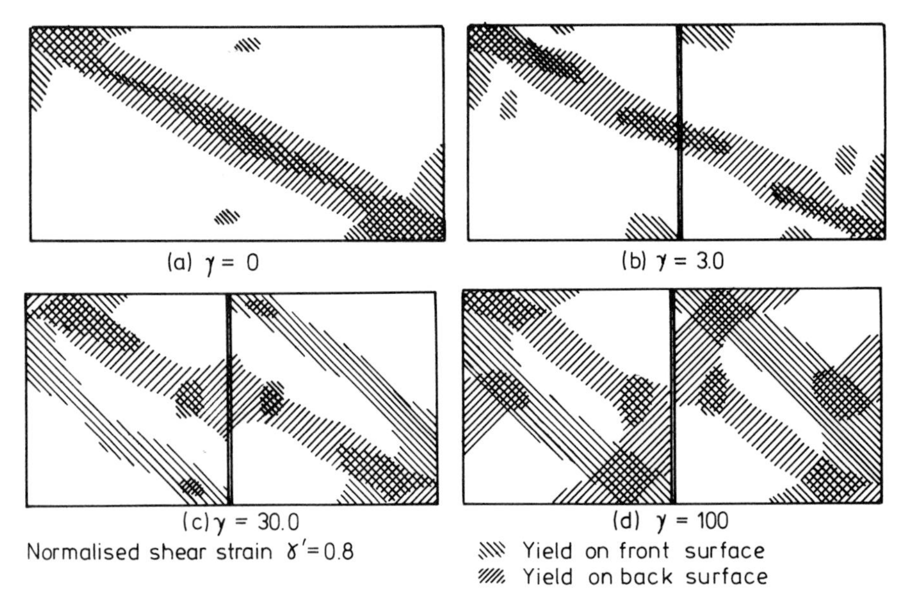

Fig 7 Variation in plastic surface regions - chequerboard initial displacements, ϕ = 1, λ = 160

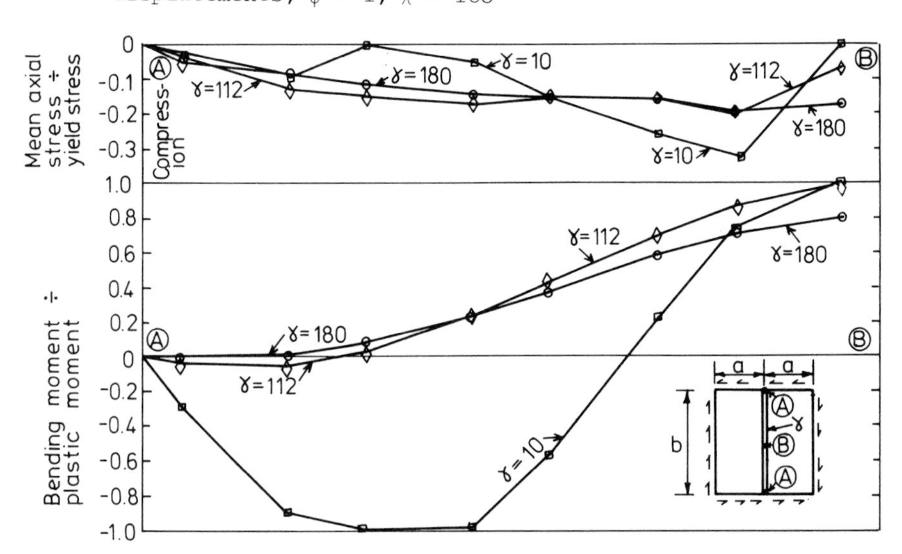

Fig 8 Stiffener axial direct stress and bending moment distributions at normalised shear strain γ' = 2.0 ϕ = 0.5, λ = 240

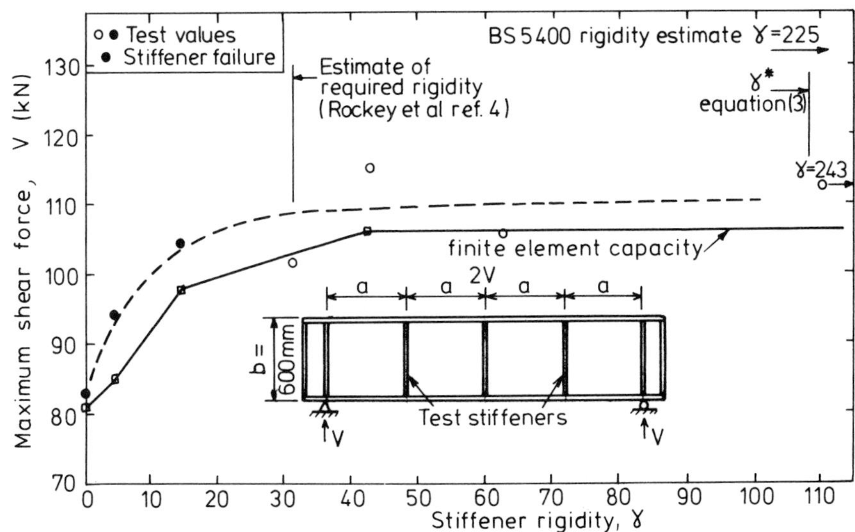

Fig 9 Tests by Rockey et al - comparison of results with concentric stiffeners

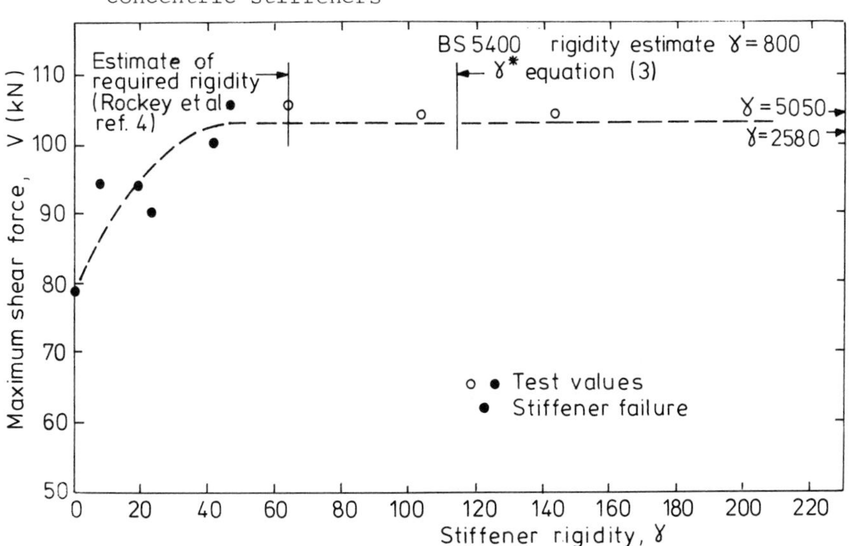

Fig 10 Tests by Rockey et al - comparison of results with eccentric stiffeners

SHEAR STRENGTH OF THIN-WALLED GIRDERS WITH STIFFENED FLANGES

F. Benussi, R. Puhali

Department of Civil Engineering, University of Trieste, Italy

M. Mele

Department of Civil Engineering, University of Rome, Italy

Synopsis

Thin-walled plate girders with stiffened flanges loaded in shear are investigated by numerical analysis and experimental tests. A good agreement has been found between the theoretical approach and tests results, as both point out the significant role of flange stiffness on carrying capacity.

Notation

b	clear width of web panel
b_f	flange width
b_r	rib width
d	clear depth of web panel
f_y	yield stress
l	clear span of girder
P	applied load
P_{cr}	critical load
P_u	ultimate load
s_l	longitudinal rib spacing in the flange
s_t	transverse rib spacing in the flange
t	web thickness
t_f	flange thickness
t_r	rib thickness
v	deflection
α	aspect ratio b/d
θ	inclination of membrane stress field

Introduction

The shear strength of stiffened web plates is usually determined by taking into account the contribution of diagonal tension fields, which arise beyond the critical load in the web panels located between two adjacent stiffeners. These tension fields are anchored both to vertical stiffeners and to flanges, so the bending stiffness of the latter influences the collapse load of the web.

The problem is satisfactorily solved for plate girders with unstiffened flanges; on the contrary, when the flanges are made up of stiffened plates their contribution to web shear strength is unknown. This is the case of steel box girders for which, for example, the AASHTO Design Rules suggest the "true Basler solution", thus entirely neglecting the contribution of the flanges.

The purpose of the present study is to enquire into this matter both theoretically and experimentally.

In particular, a theoretical model is defined in which the tension field action is provided by sloped tendons anchored both to the flanges and to the vertical stiffeners. The structure is studied by a non-linear finite elements programme, following stress and strain development right up to collapse. At the same time, an experimental programme has been planned to test eight girders. The models differ in the position and dimensions of the ribs in the stiffened flange, while the web slenderness and the web panel aspect ratio are held constant.

This paper reports both the theoretical approach and the experimental results obtained up till now.

1. THE THEORETICAL APPROACH

The analysis of the problem, in the post-critical range, has been performed by finite elements which model the real structure properly. The numerical model consists only of beam and truss elements (Fig. 1). The former elements simulate the action of flanges and transverse stiffeners, while the latter have been used to allow for the diagonal tension field action in the webs. For each beam element, section properties have been calculated taking into due account the effective width of the plate (flange or web) acting with it; the diagonal truss elements inclination is assumed to be 45° in any case. This assumption is justified by the value of the web panel aspect ratio of the models (α=0.67).

The analysis is a non-linear one and is performed step by step.

This rather simple model enables one to highlight the fundamental parameters governing the behaviour of the structure: stiffener rigidity and location, end-post stiffness, width of diagonal tension fields, location of plastic hinges.

2. EXPERIMENTAL INVESTIGATION

The experimental programme is performed by testing eight girders, with a 2.4 m span and 0.9 m web depth. Webs are 2 mm thick, while the compressive flanges are made up of 900x2 mm plates, longitudinally and transversally stiffened, as reported in Table 1. Two models (7 and 8) have wine glass flanges. The web slenderness ratio (thickness/height) is in any case equal to 1/450. The value of the aspect ratio was chosen as α=0.67, in order to model the transverse stiffening pattern of large span box girders properly. No longitudinal stiffening was provided so as to achieve low values of the critical load, as well as a wide gap between the collapse load and the critical one; this make it possible to have a large field available for experimental investigation in order to check the reliability of the theoretical approach, thus obtaining better evidence of the influence of the stiffened flange on the collapse load.

Material tests were carried out on the web and flange steel; the average value obtained for the yield stress was 380 MPa.

The deflections of the girder and of the top orthotropic flange were measured by centesimal deflectometers; the strains in the web and flanges were recorded by linear and "rosetta" electrical strain-gauges. A general view of the test rig is reported in Fig. 2. The tests carried out to date are the ones marked with an asterisk in Table 1. The complete geometry of girder 1 is drawn in Figs.3 and 4.

The load on the specimens was held constant at convenient increments throughout the test so that the various data readings could be taken after the specimen had stabilized itself at that load level.

3. *THEORETICAL AND TEST RESULTS*

The behaviour of the plate girder is well described by diagrams reporting the applied load P vs. the mid-span deflection v. In Figs. 5 and 6, P-v curves are plotted for girders 1 and 4; full lines relate to the analytical solution, while the dotted lines interpolate the experimental results. Each curve is characterized by two quasi-linear branches representing the elastic behaviour and the plastic one. The two branches are connected by a "knee" which relates to the development of yielding in the web tension field and in the flanges. Analytical and experimental curves seem to be in good agreement, in spite of the simplicity of the proposed mathematical model when compared with the ones used by other authors (7); the main differences, in the knee region, are essentially due to the assumption of idealized plastic hinges.

The collapse mechanism develops in webs with evident diagonal tension fields anchored both to vertical stiffeners and flanges (Figs.7,8,9).

The influence of the stiffened flange seems to be not negligible, as is made clear by the comparison between the obtained values of collapse load and the ones predicted by the "true Basler solution" (Figs. 5 and 6).

In the girders tested, the loads acted directly on the web; at collapse the flanges were mainly in the elastic range, thus indicating a greater influence of flange rigidity than of resistance on the ultimate load.

The collapse in girder 4 was significantly influenced by the failure of the end-posts (Fig. 10); in Fig. 6 the full line A represents the theoretical behaviour in the case of very stiff end-posts, while full line B describes the case of the actual end-posts.

For the test of girder 1, the webs of the end-posts were therefore properly reinforced to avoid their premature failure.

4. CONCLUSIONS

The following conclusions can be drawn on the basis of the investigation carried out:

- the proposed analytical model seems able to predict satisfactorily the behaviour of plate girders with stiffened flanges loaded in shear;
- the collapse mechanism is characterized by diagonal web strips of tensile stresses anchored both to vertical stiffeners and to flanges, with the strip width anchored to the flanges depending on flange stiffness;
- the ultimate loads are considerably higher than the values calculated by the "true Basler solution", clearly demonstrating the improvement in carrying capacity provided by the stiffened flanges;
- particular care must be paid to the design of the end-posts in order to avoid their premature failure, particularly in the case of not-very-stiff flanges.

The next steps in the research project will be concerned with:

- further confirmations of the results already obtained;
- the behaviour of girders with wine glass flanges;
- the influence of the load position on the carrying capacity of the web, when directly acting on the orthotropic plate in steel bridge decks;
- the proposal of simple design formulas to avoid using the finite element method in practice.

Acknowledgements

The authors are very grateful to A.Rizzo and his technical staff of the Istituto di Scienza delle Costruzioni of the University of Trieste for their skilful work.

References

1. Basler,K. 'Strength of plate girders loaded in shear.' ASCE, Proc. of Struct. Division, n. 2967, St. 7,1961, 151/180.
2. Fujii, T. 'On an improved theory for Dr. Basler's theory.' Proc. 8th Congress IABSE New York, 1968, 477/487.
3. Rockey,K.C., Skaloud, M. 'The ultimate load behaviour of plate girders loaded in shear.' The Structural Engineer, Vol.50, 1972, 29/47.
4. Calladine,C.R. 'A plastic theory for collapse of plate girders under combined shearing force and bending moment.' The Structural Engineer, Vol.51, 1973, 147/154.
5. 'The design of steel bridges.' Proceedings of the Conference, Cardiff 1980. (Editors K.C.Rockey-H.R.Evans) Granada Publishing.
6. Wolchuk, R. 'Design rules for steel box girder bridges.' IABSE Proceedings, P. 41/81.
7. Cescotto,S.,Maquoi,R. and Massonnet, Ch. 'Simulation sur ordi-

nateur du comportament à la ruine des poutres à âme pleine cisal-
lées ou fléchies.' Construction Métallique, 1981, n.2, 27/40.

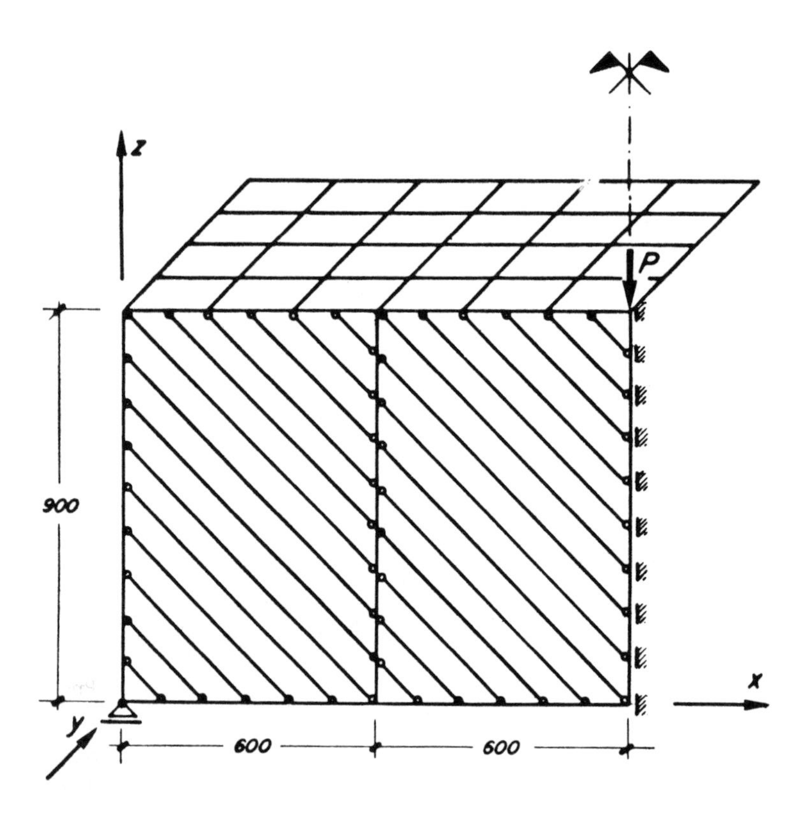

Fig 1 Mathematical model considered by the authors.

GIRDER NUMBER	PARAMETERS OF STIFFENED FLANGE RIBS (FOR EACH WEB PANEL)							
	LONGITUDINAL				TRANSVERSE			
	N°	s_l [mm]	b_r [mm]	t_r [mm]	N°	s_t [mm]	b_r [mm]	t_r [mm]
①*	3+3	100	20	2	2	200	40	4
②	3+3	100	20	2	1	300	40	4
③	2+2	133	30	3	2	200	50	5
④*	2+2	133	30	3	1	300	50	5
⑤ (2)	3+3	100	20	2	2	200	40	4
⑥ (2)	2+2	133	30	3	2	200	50	5
⑦ (1)	3+3	100	20	2	2	200	40	4
⑧ (1)	2+2	133	30	3	2	200	50	5

(1) GIRDERS ⑦ AND ⑧ HAVE A "WINE GLASS" FLANGE (100, 100)

(2) GIRDERS ⑤ AND ⑥ HAVE THE BOTTOM FLANGE STIFFENED

Table 1 Geometrical flange properties.

Fig 2 The test rig.

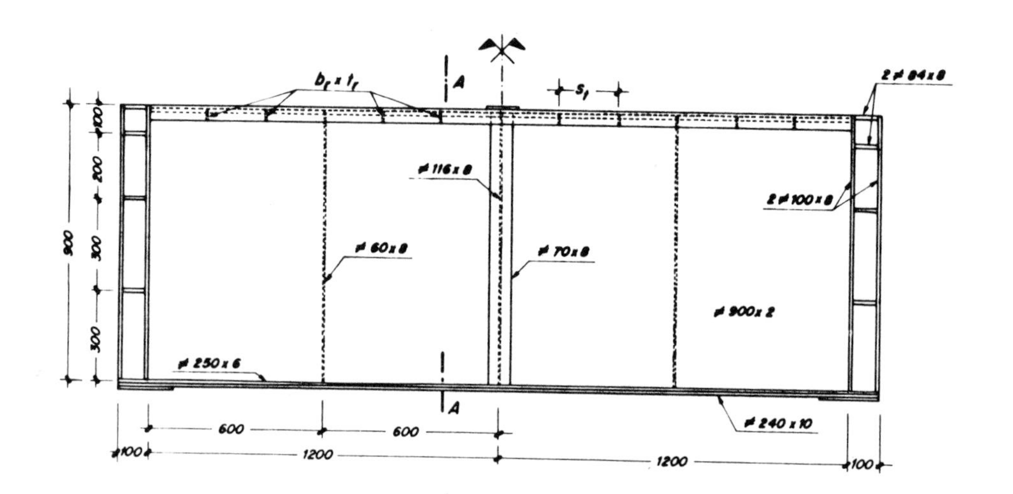

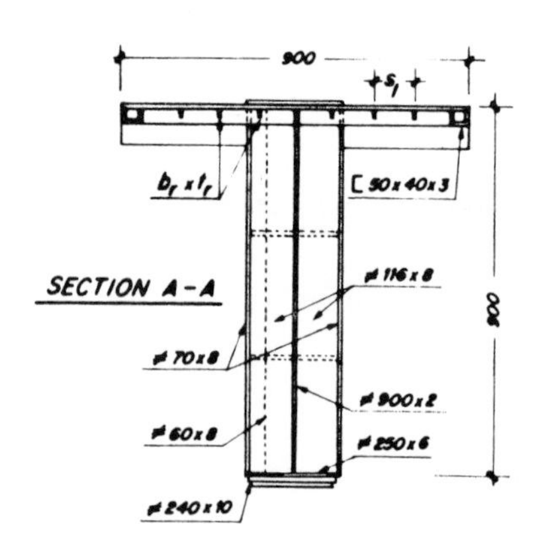

Figs 3 and 4   Details of the model.

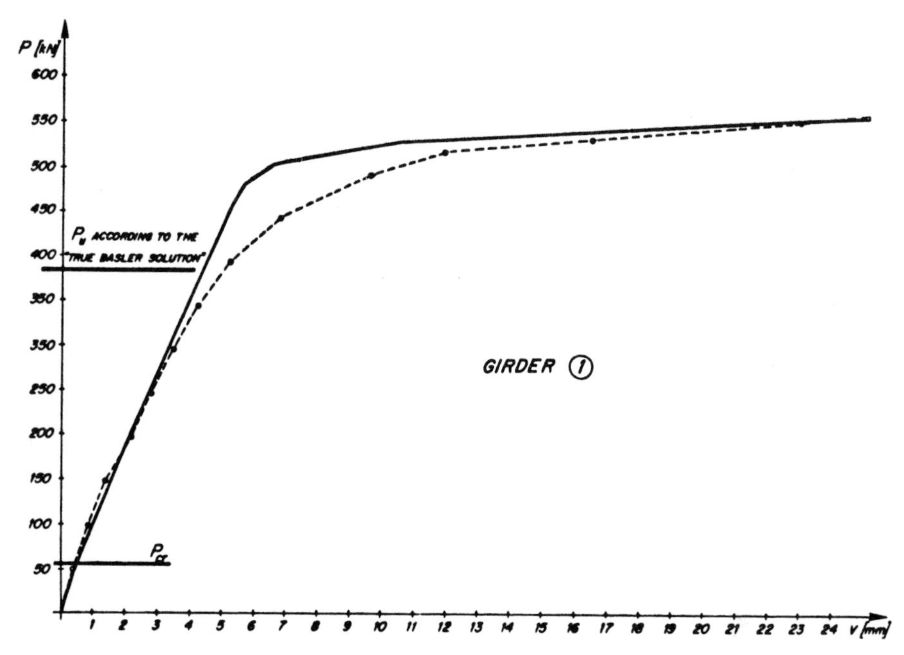

Fig 5 Applied load P vs. midspan deflection v for girder 1.

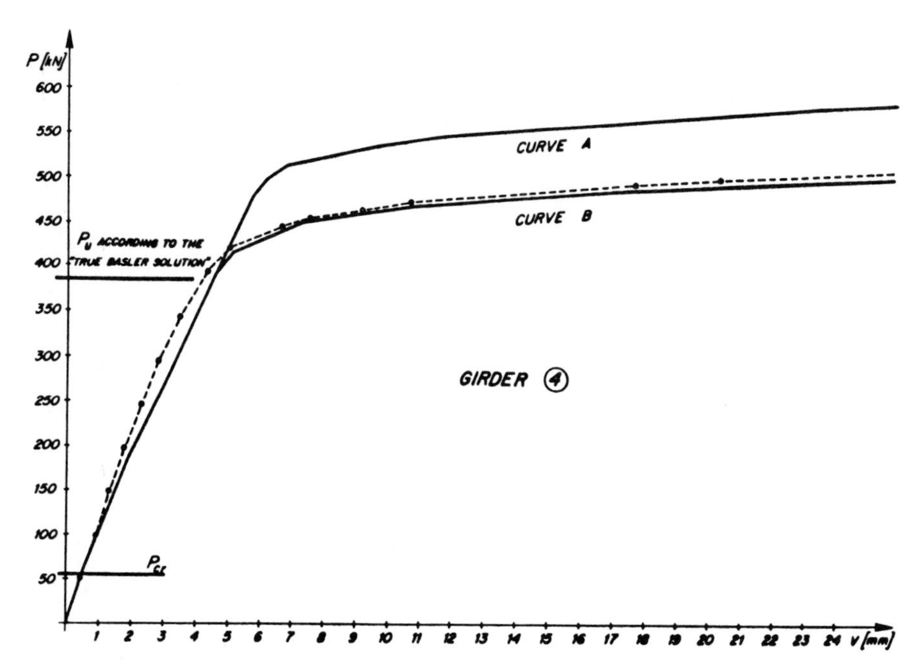

Fig 6 Applied load P vs. midspan deflection v for girder 4.

Fig 7 Girder 1 after test to failure.

Fig 8 Detail of girder 4 after test to failure. Diagonal tension field anchored to the flange and to the vertical stiffener.

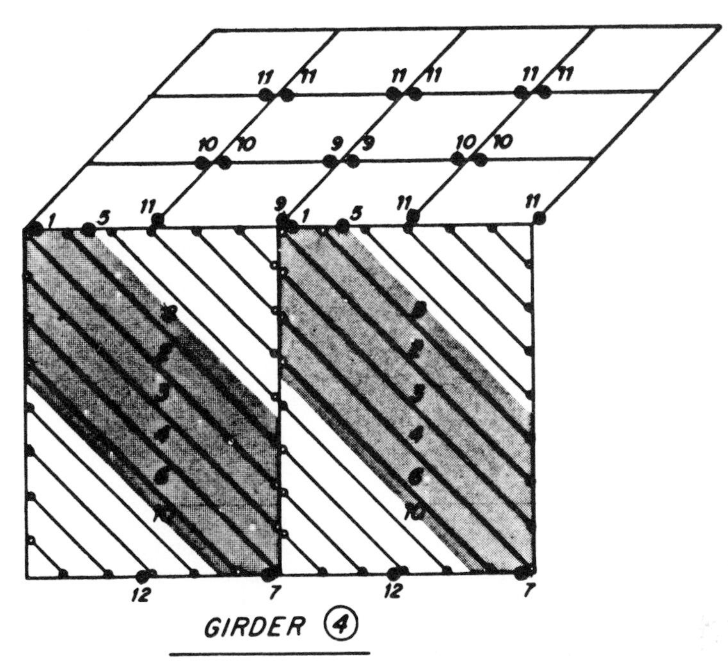

Fig 9 Development of collapse mechanism.
Numbers relate to progressive yielding.

Fig 10 Girder 4 after test to failure.
Well defined end-post mechanism.

INFLUENCE OF CROSS-SECTION DISTORTION ON THE BEHAVIOUR OF SLENDER PLATE GIRDERS

B Coric

Department of Civil Engineering, University of Belgrade, Yugoslavia

T M Roberts

Department of Civil and Structural Engineering, University College, Cardiff

Synopsis

The paper describes an experimental study of the behaviour of slender plate girders which are susceptible to combined local and lateral (or distortional) instability. The girders tested were simply supported and subjected to a central localised load acting on the top flange. The load was applied through a mechanism which permitted lateral displacement and rotation of the loaded flange. It is shown that the classical solutions for lateral and local buckling are not adequate to predict the behaviour of such girders, which failed elastically in a distortional mode. A geometrically nonlinear finite element analysis gives satisfactory agreement with experimental results.

Notation

b	span of girder
b_f	width of flange
C_w	warping constant
d	depth of web
E	Young's modulus
G	shear modulus
I_z	second moment of area about minor principal axis
J	St. Venant torsion constant
K	nondimensional buckling coefficient
M_p	plastic moment
P	load
P_{ex}	experimental failure load
P_{lat}	critical load for lateral buckling
P_{loc}	critical load for local buckling of web
P_p	load corresponding to M_p
r	height of load application above shear centre

t_f thickness of flange

t_w thickness of web

v lateral displacement

v_c lateral displacement of top flange

w vertical displacement of jack

σ_{cr} critical stress for torsional buckling of flange

σ_f yield stress of flange

σ_w yield stress of web

Introduction

Theoretical and experimental research into the buckling of structural members has generally concentrated on either plate buckling of components (local buckling) or overall member buckling. Many studies have been made of local and lateral buckling separately:Timoshenko (1) and Bulson (2). However, few studies have been made of combined local and lateral or distortional buckling:Hancock (3) and Hancock et al (4)

This paper describes tests on three slender plate girders which failed elastically in a distortional mode.

1. *DESCRIPTION OF TESTS*

The girders tested were simply supported and subjected to a central localised load acting on the top flange, as shown in Fig. 1, in which the girder dimensions are also defined. The ends of the beam were constrained against rotation. Details of the dimensions and material properties of the girders are given in Table 1, in which σ_w and σ_f are the static yield stresses of the web and flange respectively.

The load, which was applied by a hydraulic jack, was transmitted to the girders via a mechanism which permitted lateral displacement and rotation of the top flange. The mechanism incorporated a knife edge to permit rotation and a layer of needle bearings to permit lateral displacement. The mechanism is shown diagrammatically in Fig. 2.

The tests were performed using Losenhausen equipment to control the vertical deflection of the jack, which was measured by a transducer. The deflection of the jack was increased at a constant rate. At various load increments, the deflection was held constant for observation of the girder and measurement of the lateral displacements of the vertical centreline of the girder. The lateral displacements, relative to a fixed reference frame, were measured by a transducer mounted so that it could slide along a straight alloy bar. The readings were displayed on an x-y recorder.

2. *RESULTS*

The load P versus vertical displacement of the jack w for girder D5-10 is shown in Fig. 3(a). The corresponding lateral displacement v of the vertical centreline of the girder is shown in Fig. 3(b).

After initial bedding down of the girder, the P-w curve is approximately linear up to a load of 60 kN. Beyond this load, the P-w curve becomes nonlinear with a corresponding gradual increase in the lateral displacement of the girder. At a load of approximately 90 kN there was a rapid increase in the lateral displacement with the girder failing in a distortional mode. The rapid increase in w at failure was due to distortion of the cross section.

Similar results for girders D3-6 and D2-3 are shown in Figs. 4 and 5.

3. *COMPARISON WITH THEORETICAL SOLUTIONS*

The three girders tested failed elastically in a distortional mode which was characterised by pronounced cross section distortion and a rapid increase in the lateral displacement. The experimental failure loads, P_{ex}, are given in Table 2.

The load P_p corresponding to the full plastic moment M_p at the centre of the beam is given by

$$P_p = \frac{4M_p}{b} \tag{1}$$

The elastic critical load for lateral buckling, P_{lat}, neglecting cross section distortion, but allowing for the effective height of load application i.e. at the knife edge, is given in Timoshenko and Gere (1) or by the equation : Azizian (5)

$$P_{lat} = \left[-B + \{B^2 + 294(1 + K^2)\}^{\frac{1}{2}}\right]A \tag{2}$$

in which

$$B = 29.85\mu \quad , \quad \mu = \frac{r}{b}\left\{\frac{EI_z}{GJ}\right\}^{\frac{1}{2}}$$

$$K = \left\{\frac{EC_w\,\pi^2}{GJ\,b^2}\right\}^{\frac{1}{2}} \quad , \quad A = \left\{\frac{EI_z GJ}{b^2}\right\}^{\frac{1}{2}}$$

E, G, r, I_z, J and C_w are defined in the notation.

The elastic critical load, P_{loc}, for local buckling of a simply supported web panel subjected to a concentrated edge load is given approximately by:Kahn and Walker (6)

$$P_{loc} = \frac{K\,E\,t_w^3}{d} \tag{3}$$

in which K is a nondimensional buckling coefficient which is approximately equal to 1·8 for the girders tested.

The critical compressive stress σ_{cr} for local (torsional) buckling of the compression flange, assuming a uniform compression equal to the maximum value at the centre of the girder, is given approximately by: Timoshenko (1)

$$\sigma_{cr} = 1.8E\left\{\frac{t_f}{b_f}\right\}^2 \tag{4}$$

For all the test girders, σ_{cr} is greater than the yield stress of the flange and the corresponding critical load is therefore greater than P_p.

Values of P_p, P_{lat} and P_{loc} are alsogiven in Table 2. The experimental failure loads are significantly less than both P_p and P_{lat}. There is closer agreement between values of P_{ex} and P_{loc} but the correlation is not good.

A geometrically nonlinear, elastic, finite element analysis incorporating an incremental procedure discussed by Desai and Abel (7) has been used by Coric (8) to analyse the test results. The girders were idealised as three dimensional assemblages of plate elements having both bending and membrane stiffness as shown in Fig. 6.

The finite element solution for girder D5-10 is also shown in Fig. 6. The girder was assumed to have an initial sinusoidal lateral imperfection varying linearly from 1mm at the top flange to zero at the bottom flange at the centre of the girder. The results indicate a rapid increase in the lateral deflection of the top flange as the load approaches the experimental failure load.

4. *CONCLUSIONS*

Unrestrained slender plate girders may fail elastically in a distortional mode at loads significantly below the classical value for lateral buckling.

Correlation of experimental failure loads with classical solutions for local buckling is not entirely satisfactory.

Geometrically nonlinear finite element analysis appears able to predict the actual behaviour of such girders.

References

1. Timoshenko, S.P. and Gere, J.H. Theory of elastic stability. McGraw-Hill, N.Y., 1961.
2. Bulson, P.S. The stability of flat plates. Chatto and Windus, 1970.
3. Hancock, G.J. "Local distortional and lateral buckling of I-beams", Proc. ASCE, Vol. 104, ST11, pp. 1787-98, 1978.
4. Hancock, G.J., Bradford, M.A. and Trahair, N.S. "Web distortion and flexural torsional buckling", Proc. ASCE, Vol. 106, ST7, pp. 1557-71, 1980.
5. Azizian, Z.G. Instability and nonlinear analysis of thin walled structures. Ph.D. Thesis, Dept. Civil and Struct. Eng., University College, Cardiff, 1983.
6. Khan, M.Z. and Walker, A.C. "Buckling of plates subjected to localised edge loading". The Structural Engineer, Vol. 50, No. 6, pp. 225-32, 1972.
7. Desai, C.S. and Abel, J.F. Introduction to the finite element method. Van Nostrand, N.Y. 1972.
8. Coric, B. Theoretical and experimental analysis of local and lateral buckling of I-beams with distortional cross section. Ph.D. Thesis, University of Belgrade, Yugoslavia, 1982.

Girder	t_w mm	b_f mm	t_f mm	σ_w N/mm^2	σ_f N/mm^2
D2-3	1.96	80	3.05	178	272
D3-6	3.0	80	6.25	245	298
D5-10	4.94	100	10.0	292	305

For all girders b = 2300 mm and d = 380 mm

TABLE 1

Girder	P_{ex} kN	P_p kN	P_{lat} kN	P_{loc} kN
D2-3	13	65	23	7
D3-6	36	144	51	26
D5-10	94	292	172	117

TABLE 2

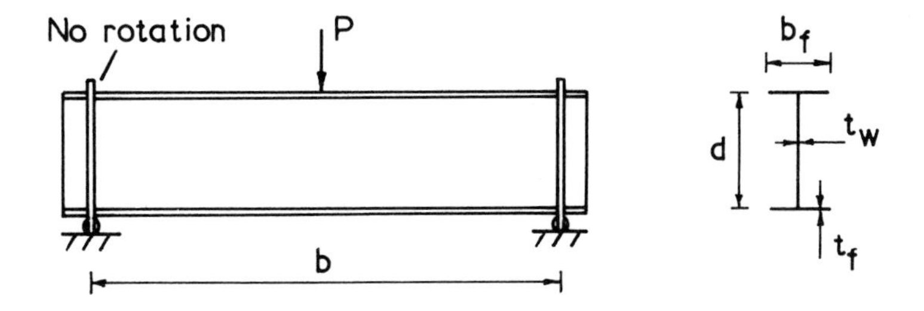

Fig 1 Details of loading and dimensions

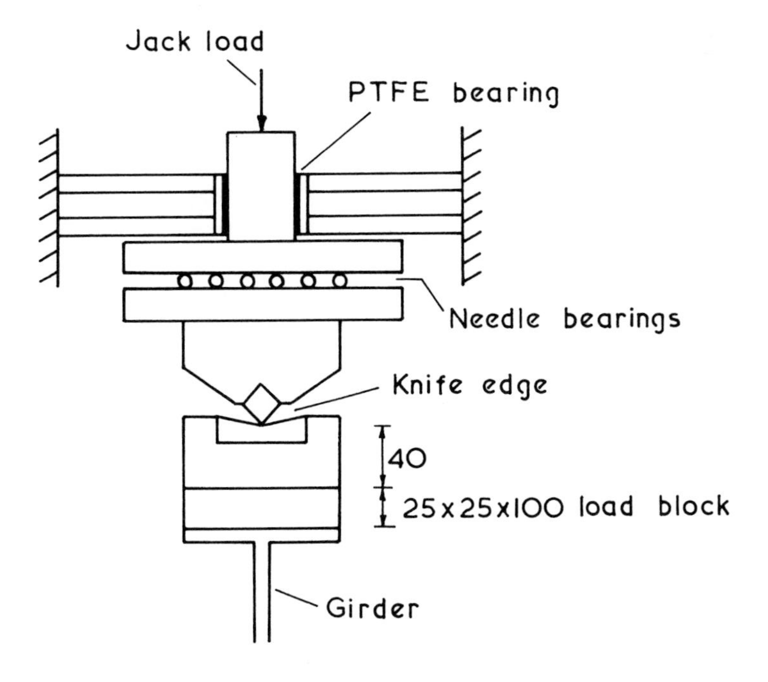

Fig 2 Details of loading members

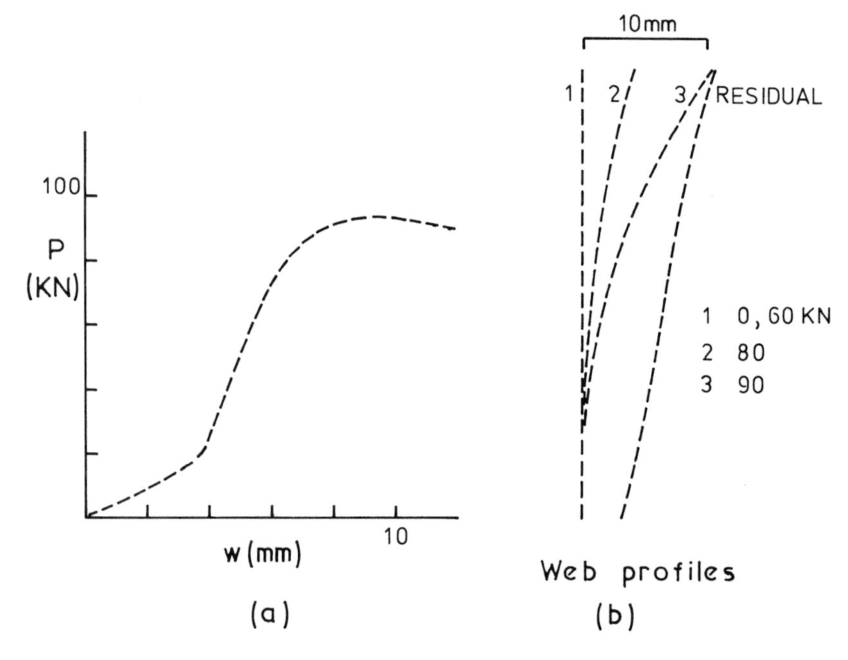

Fig 3 Results for girder D5-10

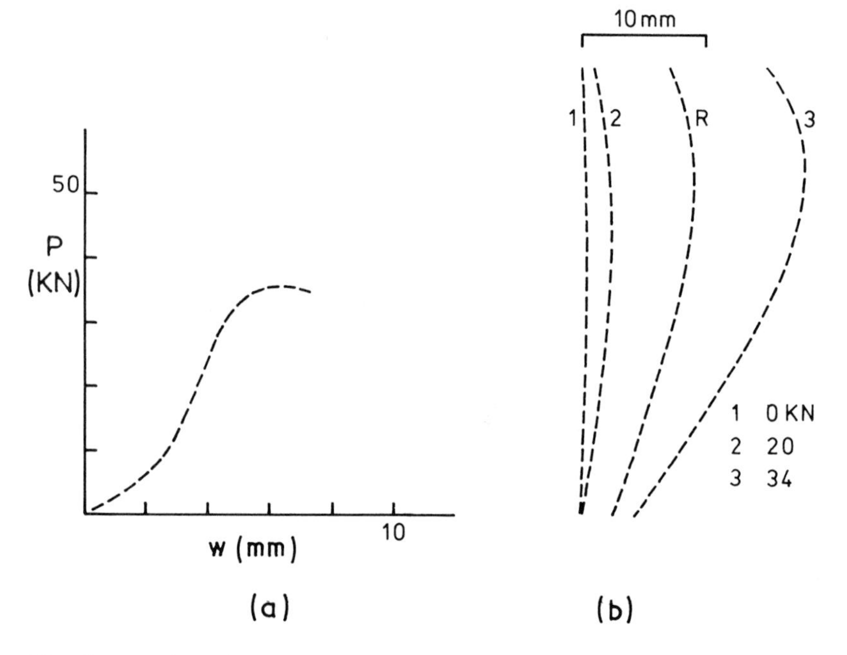

Fig 4 Results for girder D3-6

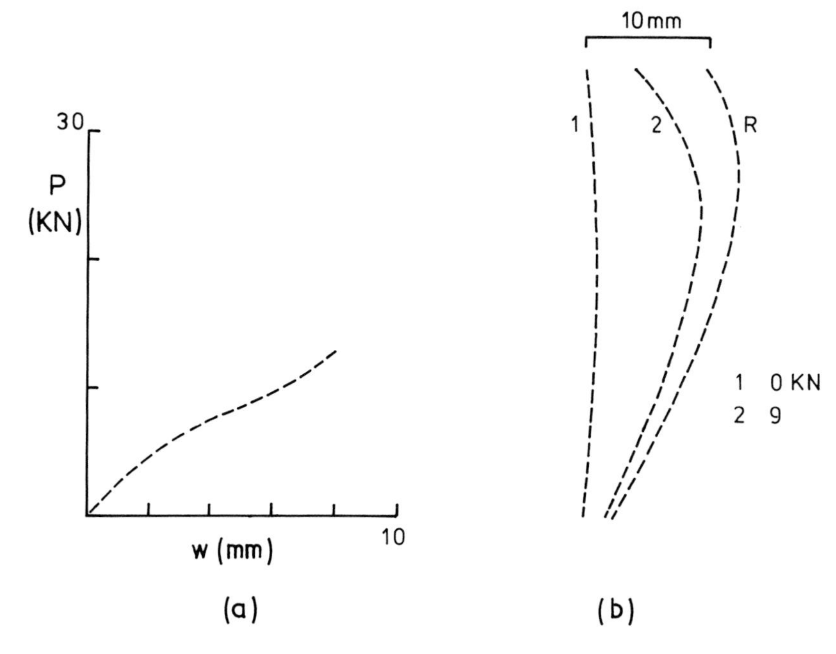

Fig 5 Results for girder D2-3

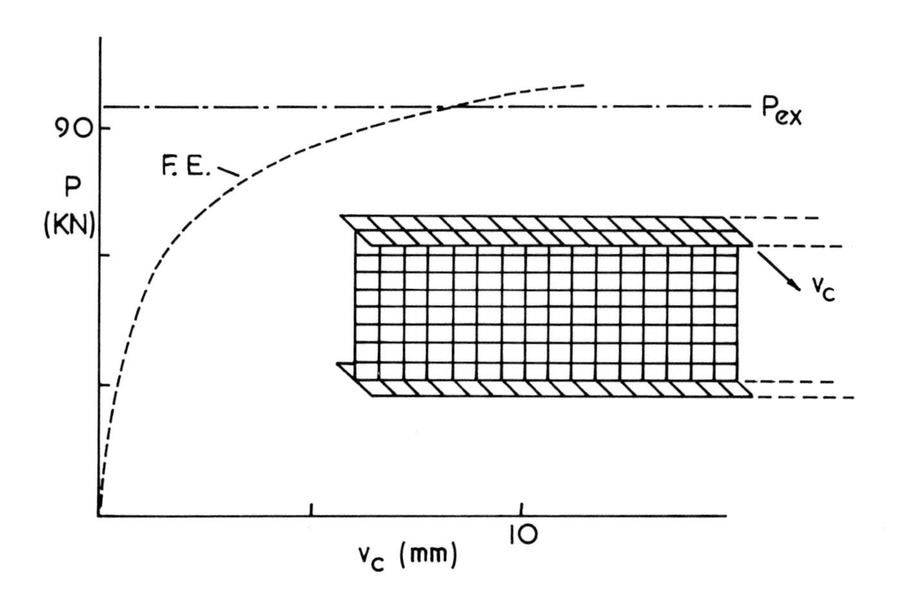

Fig 6 Finite element solution for girder D5-10 [Coric (8)]

ULTIMATE STRENGTH OF PLATE GIRDERS HAVING REINFORCED CUT-OUTS IN WEBS

R Narayanan and N G V Der Avanessian

Department of Civil and Structural Engineering, University College, Cardiff

Synopsis

A theoretical method is proposed for calculating the ultimate capacity of plate girders with reinforced web holes as the sum of four contributions, viz. (i) the elastic critical load, (ii) the load carried by the membrane tension in the web in the post-critical stage, (iii) the load carried by the flange and (iv) the load carried by the reinforcement. Approximate formulae based on a finite element analysis are suggested for calculating the elastic critical loads in shear. The membrane stresses are calculated by using the Von Mises yield criterion; the contributions of the flanges and of the reinforcement are obtained from their plastic moments. Ultimate load tests on plate girder models are presented and compared with the theoretical predictions using the analytical model described in the paper.

Notation

b	clear width of web plate between stiffeners
b_e	effective width of the tension band across the hole
b_f	width of flange plate
b_o	width of rectangular cut-out
c	distance between the flange hinges
c_r	distance between the hinges on the reinforcement
d	diameter of cut-out
d_o	depth of rectangular cut-out
E	Young's modulus
h	clear depth of web plate between flanges
L	length of the reinforcement strips
ℓ	overhanging length of the reinforcement strips
M_p	elastic moment of resistance of flange plate
M_{pr}	elastic moment of resistance of reinforcement strip
t	thickness of web plate

t_f thickness of flange plate

t_r thickness of reinforcement (steel strips)

w_r width of the reinforcement strips

V_{ult} ultimate shear load for perforated web

α angle of inclination of the diagonal of the rectangular cut-out

κ non-dimensional shear buckling coefficient for perforated web

κ_o non-dimensional shear buckling coefficient for unperforated web

ν Poisson's ratio

σ_t^y tensile membrane stress in the web at post critical stage

σ_{yf} yield stress of flange member

σ_{yw} yield stress of web plate

σ_{yr} yield stress of reinforcement

τ_{cr} shear buckling stress in the web

θ angle of inclination of the tensile membrane stress σ_t^y

θ_d angle of inclination of the panel diagonal

Introduction

In webs of plate girders, box girders and ship grillages, cut-outs are provided to facilitate inspection and maintenance. The presence of web holes in these structures will, obviously result in a reduction of strength and alter the buckling characteristics of the plates. The strength lost by providing the cut-outs could, however, be restored by providing suitable reinforcement around the holes.

The behaviour of "thick" walled webs, having typical web slenderness (h/t) values of 60 to 80 and containing circular, elliptical and rectangular holes has been studied extensively but little or no attention has been paid so far to the behaviour of "thin" webs with reinforced holes.

In this paper, analytical models for evaluating the strength of plate girders having reinforced circular and rectangular web holes are presented. Ultimate load tests on plate girder models confirm the validity of the proposed theoretical formulation for a wide range of openings.

1. PLATE GIRDERS WITH REINFORCED CIRCULAR HOLES

A method of assessing the ultimate shear capacity of plate girders with reinforced circular holes can be developed in a manner similar to the one suggested by Porter et al (1). Let us consider an increasing value of the applied shear loading on such a girder. The first stage to be examined is when the load equals the elastic critical load of the web (Fig. 1); its exact value is difficult to compute due to the presence of the hole and the reinforcement. To obtain good estimates of the critical loads, a finite element formulation has been developed, making use of the non-conforming triangular element employed by Rockey, Anderson and Cheung (2). The formulation of the element and the solution procedure are described in reference (3). The elastic critical stress for a perforated web (τ_{cr}), can be expressed in terms of the appropriate buckling coefficient, κ as follows :

$$(\tau_{cr}) = \kappa \frac{\pi^2 E}{12(1-\nu^2)} \left(\frac{t}{h}\right)^2 \tag{1}$$

Based on these finite element studies, the following empirical formula for computing the buckling coefficient, κ, for a clamped plate containing central circular cut-out and provided with a reinforcement ring around the opening is suggested. (Experimental evidence suggests that the behaviour of a perforated plate is similar to the clamped condition rather than a simply supported one):

$$\kappa = \kappa_o \left(1 - \frac{1.5d}{\sqrt{h^2+b^2}}\right) \left[1 + 6\left(\frac{t_r}{t}\right)^2 \cdot \left(\frac{w_r}{h}\right) \sqrt{\frac{d}{\sqrt{h^2+b^2}}}\right] \tag{2}$$

The above formula is subject to the upper limit :

$$\kappa \not> \kappa_o \left(1 - \frac{1.5d}{\sqrt{h^2+d^2}}\right) \left[1 + \frac{17d^2}{h^2+b^2} \left(\frac{w_r}{h}\right)^{\frac{1}{4}}\right]^3 \tag{2a}$$

The second stage in the incremental loading of the plate girder is the post buckled behaviour of the web plate. Once the critical shear stress has been equalled, the web can not sustain any increase in the compressive stress and it buckles. Any additional load would thereafter be supported by a tensile membrane stress developed in the web, thereby causing the flanges to be pulled inwards (see Fig. 1(b)). The extent and inclination of the membrane stress is influenced by the rigidity of the flanges and of the reinforcement provided for the cut-outs. If the reinforcement is slender, it will buckle and the tensile membrane stress will not be uniform; as a result, the tensile membrane stresses will start shedding towards the unbuckled regions of the web (see Fig. 2(a)). For purposes of analysis, the tensile membrane stresses across the cut-out may be replaced by an equivalent width, b_e, of the tension field as shown in Fig. 2(b)). (This concept is similar to the Von Karman's "effective width" of a plate under uniaxial compression).

On further loading, the tensile membrane stress and the buckling stress produces the yielding of the web. The final stage is reached when hinges are formed in the flanges and in the reinforcement as shown in Fig. 3. M_{pr}, can be evaluated from the equilibrium of forces acting across a quarter of a ring (between plastic hinges):

$$M_{pr} = \frac{\sigma_t^y \, b_e^2 \, t}{16} \tag{3}$$

M^*_{pr} is defined as the moment of resistance of the ring, which is just capable of producing a uniform membrane stress σ_t^y in the web, as though there was no cut-out (i.e. when b_e=d).

$$M^*_{pr} = \frac{\sigma^y_t\, d^2 . t}{16} \quad hence \quad \frac{b_e}{d} = \sqrt{\frac{M_{pr}}{M^*_{pr}}} \leqslant 1 \tag{4}$$

The effective width, b_e, can therefore be evaluated if the plastic moment of a reinforcing ring is known :

$$b_e = \sqrt{\frac{16\, M_{pr}}{\sigma^y_t \,.\, t}} \tag{5}$$

From a consideration of virtual work, the ultimate shear can be evaluated in a manner similar to the one derived by Porter et al (1) :

$$V_{ult} = 2c.\sigma^y_t.t \sin^2\Theta + \sigma^y_t t.h(\cot\Theta - \cot\Theta_d)\sin^2\Theta$$
$$- \sigma^y_t.t.d(1 - \frac{b_e}{d})\sin\Theta + (\tau_{cr}).h.t \tag{6}$$

The value of c is obtained from $\frac{2}{\sin\Theta}\sqrt{\frac{M_p}{\sigma^y_t . t}}$ and σ^y_t is evaluated by applying the Von Mises criterion :

$$\sigma^y_t = -\frac{3}{2}(\tau_{cr})\sin\Theta + \sqrt{\sigma^2_{yw} + (\tau_{cr})^2\left[(\frac{3}{2}\sin 2\Theta)^2 - 3\right]} \tag{7}$$

The values of V_{ult} obtained from (6) are dependent on the chosen Θ; the maximum value of V_{ult} is obtained by trial and error, varying Θ.

2. *PLATE GIRDERS WITH REINFORCED RECTANGULAR HOLES*

It is usual to reinforce rectangular holes by welding flat plates to the web, above and below the cut-out (Fig. 4). The reinforcement should project beyond the edges of the hole for a length sufficient to allow the full development of the tension field; this anchorage length (ℓ) will, therefore, depend on the dimensions of the cut-out, a web having a deeper cut-out requiring more anchorage length. Tests conducted at Cardiff suggest that the minimum length of the reinforcement should be $1.5\ b_o$ or $\sqrt{b_o^2 + d_o^2}$ whichever is larger in order that the strength lost by cutting the hole can be fully recovered; in addition, the reinforcement should have sufficient stiffness to resist the membrane tension.

Based on extensive finite element studies, the following minimum reinforcement requirement was found to be essential in order that the buckling coefficient for the perforated web would be at least equal to that of an unperforated web:

$$\left(\frac{t_r}{t}\right)^2 \cdot \left(\frac{w_r}{h}\right) \geq 2.76\sqrt{\frac{b_o \,.\, d_o}{b.h}} \tag{8}$$

The second and third stages are similar to the case of circular cut-outs. Four hinges are formed on the reinforcement as shown in Fig. 5. By the method of Virtual Work, the following equation for the ultimate shear is obtained :

$$V_{ult} = 2c\,\sigma_t^y.t\,\sin^2\Theta + \sigma_t^y.t.h\,(\cot\Theta - \cot\Theta_d)\sin^2\Theta$$
$$- \sigma_t^y.t\sqrt{b_o^2+d_o^2}\,\sin(\alpha+\Theta)\sin\Theta + 2c_r.\sigma_t^y.t.\sin^2\Theta + (\tau_{cr}).h.t \qquad (9)$$

In the above equation c, σ_t^y are calculated as previously indicated for circular cut-outs. If the reinforcement had adequate end fixity, then the hinge distance c_r is obtained from

$$c_r = \frac{2}{\sin\Theta}\sqrt{\frac{M_{pr}}{\sigma_t^y.t}} \qquad (10)$$

3. *EXPERIMENTAL INVESTIGATION*

Recent research has shown (4) that in spite of the inherent problems associated with scale effects and welding stresses, small scale models can be employed satisfactorily to assess the behaviour of plated structures. Model girders made of 1mm webs were therefore used to carry out tests on plate girders containing reinforced web holes. Series 1 consisted of 8 shear panels containing circular holes and Series 2, of 12 shear panels with rectangular holes. The parameters varied were : (a) web aspect ratio (b/h) (b) web slenderness (h/t) (c) sizes of openings and (d) the size of reinforcement.

Figure 6 gives the design details of Series 1 girders; figure 7 gives similar particulars of Series 2 girders.

An Avery - Denison machine capable of applying transverse loads of 600kN was suitably modified to accommodate the model test girders (Fig. 8). The lateral stability of the girders was maintained by purpose-built restrainers, fixed at the supports and at the centre. The machine was set to "deflection control" using an extremely slow rate of deflection-increment, to enable the ultimate behaviour and the post-peak drop-off of load to be determined. Load versus central deflection relationships were obtained by an automatic x-y recorder.

Each of the test girders consisted of two test panels; only one panel was tested at a time. The panel not under test was temporarily stiffened by clamping two stiffeners to its web in order to prevent any distortion occurring on it, while the other panel was being tested. This procedure was repeated with the second panel.

A ripple scanner described in reference (5) was used to determine the flange and web profiles, as well as the deflected shape of the reinforcement after the test. These measurements helped to obtain the locations of the hinges after the conclusion of the test.

Tension tests were carried out on coupons cut from flanges, webs and reinforcing strips. The mean yield stresses are used in computing the theoretically predicted values given in Table 1. At the initial stage of loading, the load-deflection diagrams were linear; at around 80% to 85% of the peak load, the girders exhibited very large deflections before ultimate collapse.

Table 1 gives the measured loads from these tests and compares them with the predicted values of ultimate shear, using the theory outlined previously. The collapse loads obtained using reinforced cut-outs are generally greater than those obtained with unperforated webs.

The loads predicted are seen to be consistently safe and satisfactory. The mean value of $\frac{\text{Predicted Ultimate Load}}{\text{Observed Ultimate Load}}$ for the 6 tests in Series 1 (with circular cut-outs) was 0.895 with a standard deviation of 0.037; the corresponding mean value for 10 tests in Series 2 with rectangular cut-outs was 0.931 with a standard deviation of 0.055.

4. *CONCLUSIONS*

The paper presents methods of predicting the ultimate shear of plate girders containing reinforced rectangular and circular openings. The theory accounts for the buckling load of the web, the post-buckling membrane stress developed in the web and the load carried by the development of plastic hinges in the flanges and in the reinforcement. Test on model girders confirm the validity of the theory.

References

1. Porter, D.M.,Rockey,K.C. and Evans,H.R. 'The collapse behaviour of plate girders loaded in shear'. The Structural Engineer, Vol. 53, No. 8, August 1975, pp. 313-325.

2. Rockey, K.C.,Anderson,R.G. and Cheung,Y.K. 'The behaviour of square shear webs having a circular hole'. Proceedings of a symposium on Thin Walled Structures, Crosby-Lockwood & Sons,1967, pp. 148-169.

3. Zienkiewicz, O.C. and Cheung,Y.K. 'The Finite Element Method for the analysis of elastic isotropic and orthotropic slabs'. Proceedings of the Institution of Civil Engineers, Vol.28, Aug. 1964, pp. 471-488.

4. Narayanan, R. and Adorisio, D. 'Model studies on plate girders'. Accepted for publication in the Journal of Strain Analysis. Institution of Mechanical Engineers, London.

5. Narayanan, R. and Rockey, K.C. 'Ultimate capacity of plate girders with webs containing circular cut-outs'. Proceedings of the Institution of Civil Engineers, London,Part 2,Vol.72, Sept. 1981, pp. 845-862.

TABLE 1

EXPERIMENTAL VALUES OF ULTIMATE LOAD COMPARED WITH PREDICTED VALUES

PANEL DESIGNATION	PREDICTED ULTIMATE LOAD (kN)	OBSERVED ULTIMATE LOAD (kN)	$\frac{\text{PREDICTED LOAD}}{\text{OBSERVED LOAD}}$
SERIES 1:			
TCP1A	31.8	33.1	***
TCP1B	33.5	36.2	0.925
TCP2A	39.1	43.5	0.899
TCP2B	46.6	49.6	0.939
TCP3A	43.1	53.0	***
TCP3B	46.0	55.8	0.824
TCP4A	54.4	61.8	0.880
TCP4B	68.7	75.9	0.905
Mean value of $\frac{\text{predicted load}}{\text{observed load}}$ for 6 tests = 0.895			
standard deviation = 0.037			
SERIES 2:			
GP1A	43.2	46.3	***
GP1B	43.3	46.0	0.941
GP2A	48.3	55.5	0.869
GP2B	48.8	55.0	0.888
GP3A	46.6	46.3	1.007
GP3B	44.2	46.4	0.953
GP4A	44.3	48.0	0.902
GP4B	43.1	45.5	0.947
TCP5A	34.3	36.1	***
TCP5B	39.5	38.0	1.040
TCP6A	33.5	38.6	0.868
TCP6B	33.8	37.8	0.895
Mean value of $\frac{\text{predicted load}}{\text{observed load}}$ for 10 tests = 0.931			
standard deviation = 0.055			

*** These are control tests carried out on panels without openings. It will be seen that the collapse loads of models having reinforced cutouts are at least equal to those with unperforated webs.

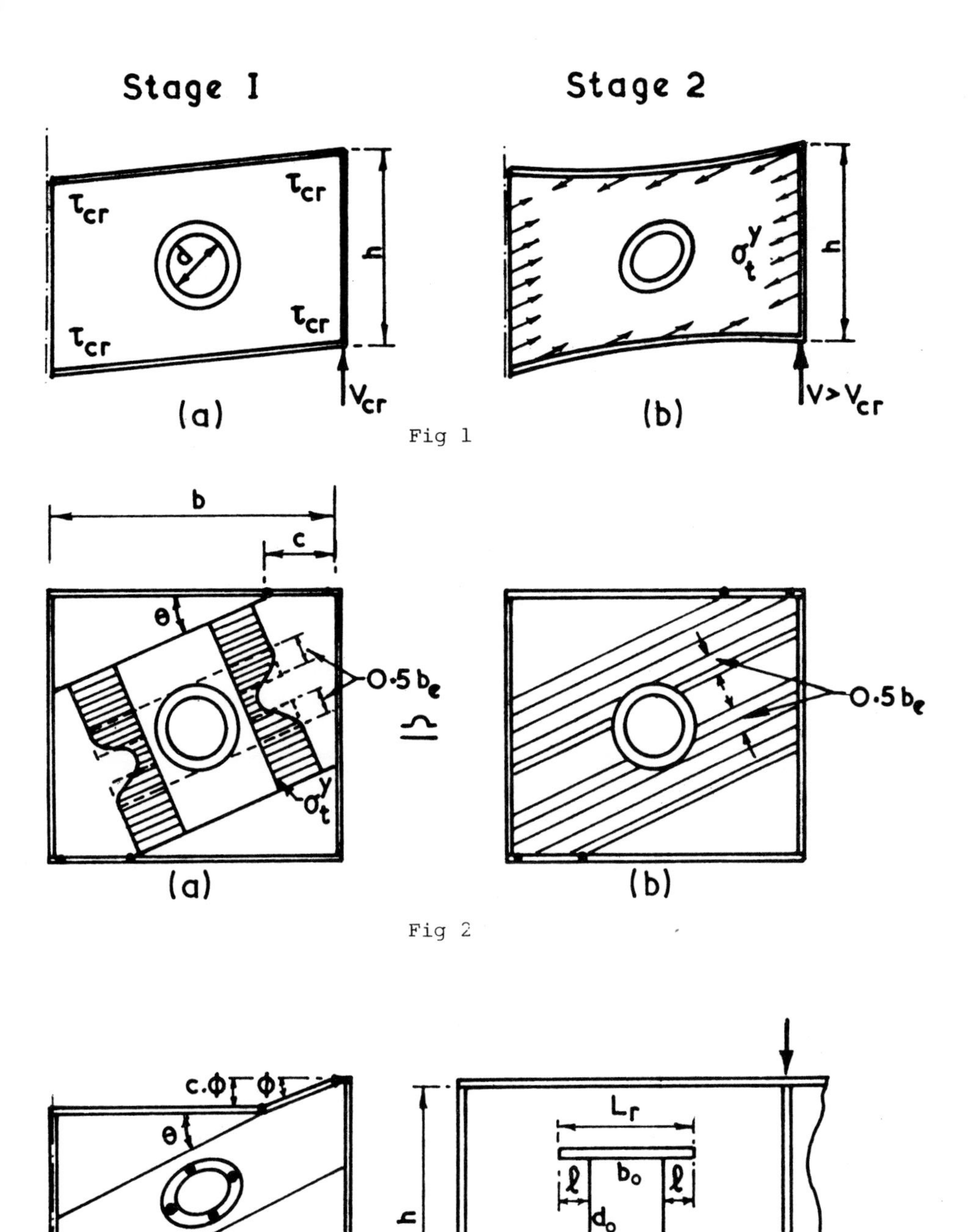

Fig 1

Fig 2

Fig 3 Collapse state

Fig 4 Flat plate reinforcement

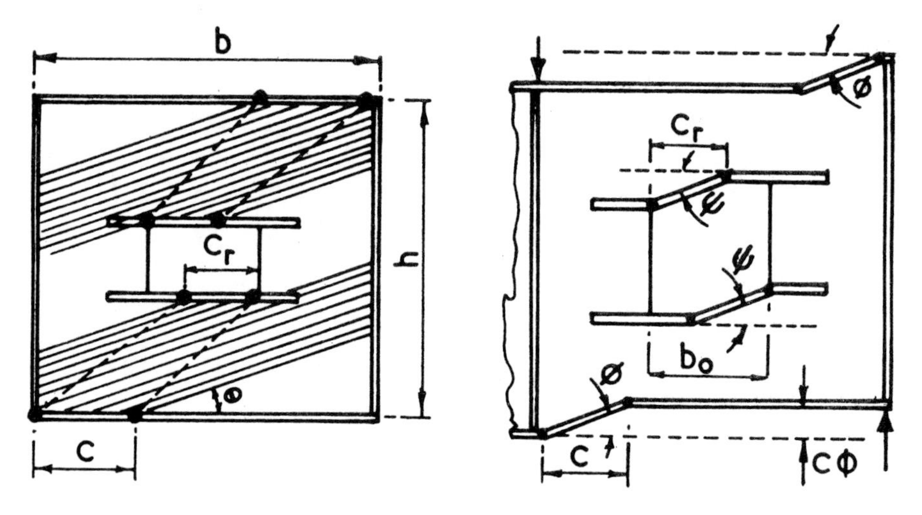

Fig 5

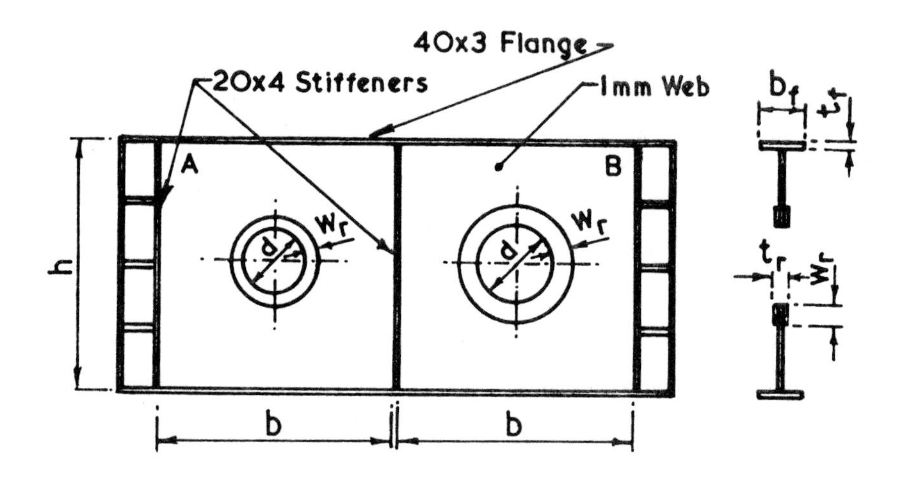

Fig 6 Design details of test specimens - Series 1

Girder No.	Flange $b_f \times t_f$	Web b x h	CUTOUT IN PANEL A		CUTOUT IN PANEL B	
			Diameter	Reinforcement	Diameter	Reinforcement
TCP 1	40 x 3	375x250	0	-	55	11 x 5
TCP 2	40 x 3	375x250	85	17 x 5	125	25 x 5
TCP 3	40 x 3	360x360	0	-	80	16 x 5
TCP 4	40 x 3	360x360	122	25 x 5	180	37 x 5

Notes: All dimensions are in mm. All panels are made of Grade 43 steel. All webs are cut from single sheet of rolling. All flanges are cut from a single batch of rolling.

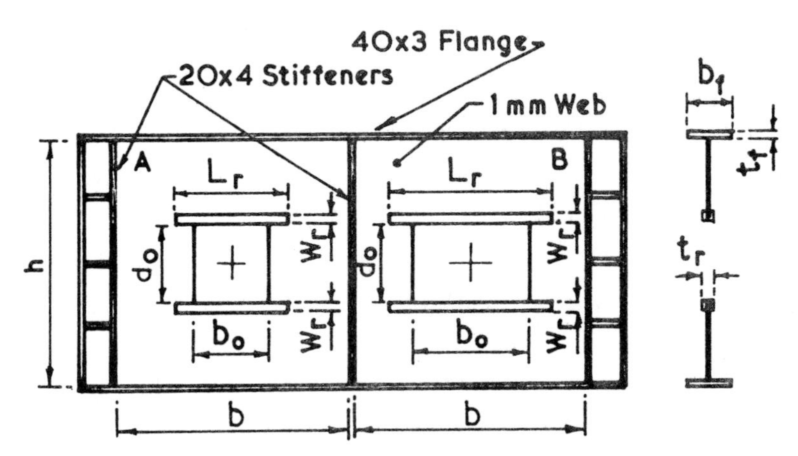

Fig 7 Design details of test specimens - Series 2

Girder No.	Flange $b_f \times t_f$	Web b x h	CUTOUT IN PANEL A		CUTOUT IN PANEL B	
			Size	Reinforcement	Size	Reinforcement
GP 1	40 x 3	300x300	0	=	100x50	18 x 150 x 5
GP 2	40 x 3	300x300	100x150	35 x 260 x 5	100x175	40 x 300 x 5
GP 3	40 x 3	300x300	50x100	21 x 174 x 5	100x100	27 x 174 x 5
GP 4	40 x 3	300x300	150x100	31 x 225 x 5	175x100	34 x 263 x 5
TCP5	40 x 3	450x300	0	=	100x100	26 x 245 x 5
TCP6	40 x 3	450x300	175x100	32 x 260 x 2	250x100	38 x 375 x 5

Notes: As for table under Figure 6.

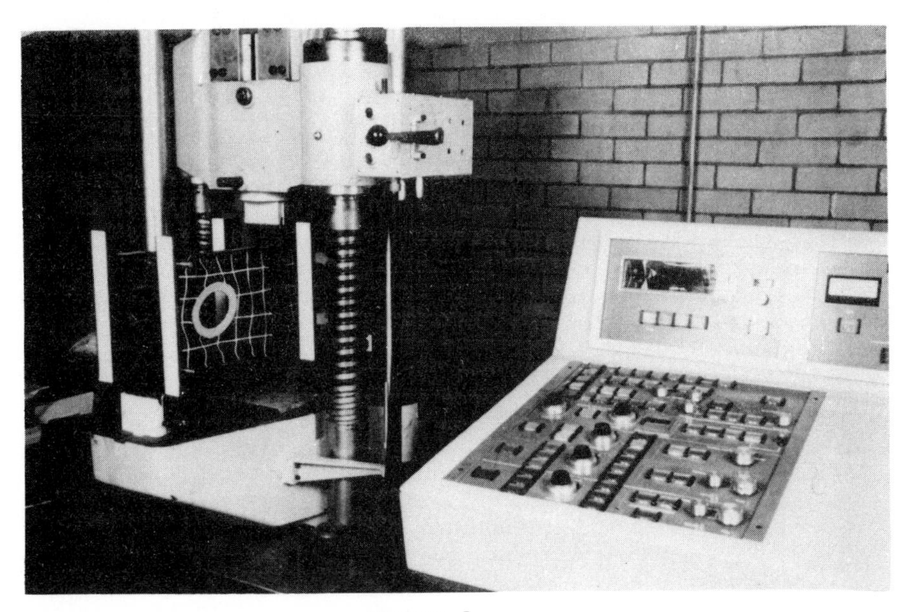

FIGURE 8

AN EXPERIMENTAL STUDY OF THE COLLAPSE BEHAVIOUR OF A PLATE GIRDER WITH CLOSELY-SPACED TRANSVERSE WEB STIFFENERS

H R Evans and K H Tang

Department of Civil and Structural Engineering, University College, Cardiff

Synopsis

The paper describes five tests that were carried out on a large-scale plate girder model to study the mode of failure of the girder and the behaviour of the transverse web stiffeners. The stiffeners were closely spaced on the web and the results show that the tension field mechanism approach can predict the ultimate loads of such girders with good accuracy. The tests also show that methods currently proposed for the design of transverse web stiffeners are safe but conservative.

Introduction

The use of limit state design methods requires the accurate prediction of the collapse loads of plate girders. For the particular case of shear webs, a plastic shear collapse mechanism approach, in which failure is assumed to occur when a yield zone forms in the web under the action of post-buckling tension field stresses and when four plastic hinges form in the flanges, has been proposed (1). This method has been adopted in the new Code of Practice for Design of Steel Bridges - BS5400 Part 3 (2) for the design of transversely stiffened girders.

The method has been validated by comparison (1) with many test results for girders with transversely stiffened webs; invariably, the stiffener spacing exceeded the web depth in these experimental girders. Very few tests have been reported where the web panel aspect ratio, i.e. stiffener spacing/web depth, was less than 1 although a girder with closely spaced transverse stiffeners can provide an efficient design in a high shear situation. The first objective of the test series described herein was to assess the accuracy of the tension field approach for such girders.

Before post-buckling action can develop within a web plate, the boundary elements must be able to support the forces imposed upon them by the tension field. In the case of transverse stiffeners, the imposed loading is complex and, until recently, little was known about the stiffener behaviour when the web was operating in the post-buckled range. It is interesting to note that the importance of transverse stiffening was appreciated by Fairbairn (3) in 1849 in his report on experiments for the Britannia and Conway tubular bridges. "Further experiments illustrated the importance of the pillars in the sides, for, with a small addition of metal to the weight of the tube,...... the ultimate strength was increased considerably". However, as recently as 1980,Bjorhovde (4), in reporting a survey entitled "Research Needs in Stability of Metal Structures" to the American Society of Civil Engineers observed that "Present design methods for vertical stiffeners appear to be conservative. Research should be conducted to produce rational design rules for such elements....".

The second objective of the present test series was to study the behaviour of the transverse stiffeners. The web of the girder was designed to be very slender to ensure the development of extensive post-buckling action during which the effects of the tension field upon the stiffeners could be observed.

1. *TRANSVERSE STIFFENER DESIGN*

The minimum rigidity (γ^*) that a transverse stiffener must possess to ensure that it remains straight and restricts buckling to the individual sub-panels of the web can be determined from linear buckling theory. Tests have shown that the γ^* value must be increased significantly to ensure that the stiffener remains effective in the post-buckling range. Empirical multiplication factors for γ^* were included in several design codes, but limit state design requires the stiffener design to be based on the same theoretical model as the determination of web shear capacity.

Such a consistent approach is adopted in BS 5400 (2), where a transverse stiffener is designed as a strut to carry the axial compression arising from the tension field action. The destabilising influence of the web is taken into account as a notional equivalent axial load. The effective cross-section of the stiffener-strut is assumed to include a width of web plate equal to 32 times the web thickness, thus forming a Tee-section.

The method of stiffener design proposed recently by Rockey, Valtinat and Tang (5) employs a similar Tee-section strut idealisation, although, in this case, an effective web width of 40 times its thickness is assumed. In this method, the loads imposed upon the stiffener by tension field action are calculated more accurately than in BS 5400 by a careful consideration of the geometry of the failure mechanism; this increased accuracy is gained at the cost of considerably more complex calculations. A notional axial load is again imposed upon the stiffener-strut to represent the destabilising influence of the web.

The loads imposed upon the stiffener-strut by the web are applied at the mid-thickness of the web plate and are thus displaced from the centroid of the stiffener cross-section. This eccentricity of loading causes bending of the strut and both BS 5400 and the method of Ref. (5) employ interaction expressions defining the co-existent values of axial compression and bending moments that the stiffener can carry safely. The expression employed in BS 5400 is more conservative than that adopted by Rockey, Valtinat and Tang; in the case of no axial loading, the Code restricts the moment carrying capacity of the strut to that required to produce first yield whereas Rockey, Valtinat and Tang allow the full plastic moment capacity of the stiffener section to be developed.

2. EXPERIMENTAL PROGRAMME

The overall dimensions of the test girder are shown in Fig. 1; the web thickness was approximately *1mm* throughout and all stiffeners were attached to one side of the web only. The girder was fabricated from Grade 43 steel and the yield stresses of the web, flange and stiffener material were measured as *216*, *206* and *299* N/mm^2, respectively.

Five tests were carried out as indicated, the girder being simply supported and subjected to a central load, so that the critical web panel in each successive test was subjected to a different combination of bending moment and shearing force. The moment/shear interaction diagrams predicted by the mechanism approach (1) are also shown in Fig. 1; the superimposed experimental points show all failures to have occurred in a shear mode.

The values of the 3 non-dimensional parameters normally used to define girder proportions are listed in Table 1 for the relevant panel in each test. The table shows the low values of web aspect ratio (b/d) reaching a minimum of *0.62* and the high values of web slenderness (d/t) of around *834*; the values of both these parameters were chosen in accordance with the two main objectives of the study, as discussed earlier. The flange proportions (as represented by the flange strength parameter M_p^*) were representative of normal practice.

In each test, the behaviour of the transverse stiffener immediately adjacent to the failing web panel was carefully observed. The relevant stiffener in each case is indicated in Fig. 1 and its dimensions (width b_s x thickness t_s) are listed in Table 1. The actual stiffener dimensions are compared to the dimensions of optimum stiffeners designed by the method proposed by Rockey, Valtinat and Tang (5). The comparison is made in terms of the cross-sectional area (A_s) and rigidity (γ_s) of the stiffener with the properties of the experimental and optimum stiffeners being indicated, respectively, by the subscripts "EXP" and "OPT". In the first test, the stiffener dimensions were identical to the optimum so that the stiffener was not expected to have any reserve of strength. In tests 2,3 and 4, the stiffeners had varying reserves of strength but in test 5, the stiffener was expected to prove inadequate.

Figures 2a and b show the interaction diagrams employed in Ref.(5) to define the co-existent axial load (P) and bending moment (M) that could be sustained by the stiffeners in the first and fourth tests. These quantities are plotted as ratios of the axial squash load (P_s) and plastic moment capacity (M_p) of the stiffener section respectively. The superimposed design values, showing the axial loads and moments expected to be carried by the stiffeners in the tests, show stiffener SA1 to lie virtually on the curve, and thus have no reserve of strength, whilst stiffener SA2 lies some way within the curve, showing a greater reserve.

The more conservative design curves obtained from BS 5400 (with $\gamma_m = \gamma_{f3} = 1$) are also plotted in Figs. 2a and b. Both stiffeners lie well outside these curves and would not thus satisfy the code requirements. It should also be noted that the stiffener width/thickness ratio (b_s/t_s) exceeded the value of 11 allowed by the Code in each test; such wide stiffeners had to be used to allow adequate strain gauging.

3. *DISCUSSION OF RESULTS*

3.1 *Overall Girder Behaviour*

The displacements of the stiffeners perpendicular to the plane of the web were measured in all the tests, together with the residual deformations of the web and flange panels after collapse. In the first and fourth tests, the axial strains in the critical stiffeners (SA1 and SB1) were measured, 21 gauges being placed on each face of each stiffener.

In all the tests, the transverse stiffeners proved fully adequate right up to collapse and were effective in limiting the buckling to adjacent sub-panels of the web. Figures 3a and b show the two weakest stiffeners (SA1 and SB2) after collapse of the girder in Tests 1 and 5, stiffener SB2 had been expected to fail. Measured displacements of the stiffeners perpendicular to the plane of the web were small in all cases.

Figures 3a and b also illustrate the mode of failure observed in each test. A well-defined shear sway mechanism developed in each case and the load/displacement curve showed a distinct plastic failure plateau.

3.2 *Prediction of Collapse Loads*

The failure load (V_{EXP}) measured in each of the five tests is listed in Table 2, together with the load predicted by the mechanism solution (V_{PRED}). The agreement is excellent and the column headed V_{PRED}/V_{EXPT} shows the maximum error to be an overestimation of 2% for Test 1. The agreement is further illustrated by the superposition of the experimental points on the predicted moment/shear interaction diagrams in Fig. 1.

Further confirmatjon of the validity of the mechanism solution is obtained by comparing the predicted and measured positions of the plastic hinges in the flanges. The column headed C_{PRED}/C_{EXPT} shows good agreement and a maximum difference of 10%; the high moments in the central panels had a considerable influence upon the failure mode in Tests 1 and 2.

The predicted failure loads obtained from BS 5400 are also listed in the table (V_{5400}). These values have been calculated assuming the partial safety factors on loading and material properties (γ_{f3} and γ_m) to be equal to one. Moment/shear interaction diagrams given by the Code are included in Fig. 1 for comparison.

In the Code, the shear capacity is calculated according to tension field theory, but for webs having a slenderness ration (d/t) above *200* (for the test girders d/t was around *833*), a reduction factor of *1.15/1.35* is introduced. The tabulated values show the ratio V_{5400}/V_{EXP} to vary between *0.87* and *0.91* indicating the Code values (with $\gamma_{f3} = \gamma_m = 1$) to be slightly, but not unreasonably, conservative.

3.3 *Stiffener Behaviour*

The measured values showed that the axial strains developed in a stiffener varied linearly across the stiffener width, from a large compressive value at the stiffener/web junction to a small tensile value at the free edge. This is illustrated in Fig. 4 where the strain variation at the mid-depth of the stiffener in Test 1 is plotted for different levels of applied loading.

By extrapolating from the 3 measured values, it is apparent that the compressive strains developed at the stiffener/web junction were in excess of yield (shown as ε_{ys} on the diagram) before failure occurred. The measured yield zone within the stiffener is shown in Fig. 5a and is seen to be localised close to the web and to extend over most of the depth. Figure 5a shows the situation just after the failure plateau was reached; the yield zone spread gradually as the girder moved along the failure plateau and Fig. 5b shows the situation after gross shear deformations of the girder had occurred. Since the load applied to the girder remained constant during this phase, the spread of yield within the stiffener suggests that the amount of web effectively acting with the stiffener must have been diminishing.

The compressive axial loads developed within the stiffener at various levels of applied loading are plotted in Figs. 6a and b for Tests 1 and 4, respectively. These stiffener loads were calculated by multiplying the average axial stresses (obtained from strain plots as in Fig. 4) by the cross-sectional area of the stiffener only. At the collapse load (V_{EXP}) the additional load carried by an effective width ($40t$) of the web was also taken into consideration to enable a direct comparison to be made with the load predicted by the Rockey, Valtinat and Tang (5) method. For stiffener SA1, the agreement between the predicted and measured loading is excellent, but there is a slight underestimation of the measured load in stiffener SB1. The stiffener loading calculated from BS 5400 is also shown in Fig. 6. The Code values are conservative, but again not unreasonably so, although it should be appreciated that these stiffener load values correspond to a lower estimation (i.e. V_{5400}) of ultimate load capacity.

The experimentally measured values of co-existent axial forces and bending moments in the stiffeners at failure are superimposed on Fig. 2 and are seen to lie reasonably close to the design value from Ref. (5). These experimental points lie well outside the design curves given by BS 5400 and the adequate performance of these stiffeners indicates the Code to be conservative in its estimation of stiffener capacity.

4. *CONCLUSIONS*

The tension field mechanism approach enables the ultimate load capacity and failure mode of a plate girder with closely spaced stiffeners to be predicted accurately. BS 5400, with unit values of the partial safety factors, gives a slightly conservative estimate of capacity.

The method of Ref. (5) gives an accurate evaluation of the loading imposed upon the transverse stiffeners. The much simpler approach given in BS 5400, which does not require the determination of the geometry of the failure mechanism, gives a slightly conservative, but satisfactory, estimate of stiffener loading.

The stiffener design criteria proposed in Ref. (5) ensure adequate stiffener performance whilst the criteria specified in BS 5400 are conservative.

ACKNOWLEDGEMENTS

The authors wish to thank the Department of Transport for sponsoring this investigation and Professor K.I. Majid for allowing the facilities of the Department of Civil and Structural Engineering, University College, Cardiff, to be used.

References

1. Rockey, K.C., Evans, H.R. and Porter, D.M. A design method for predicting the collapse behaviour of plate girders. Proc.Instn. Civ.Engrs., Part 2, 1978, 65, Mar., pp. 85-112.
2. British Standards Institution-BS 5400 : Part 3 : 1982; Code of Practice for Design of Steel Bridges.
3. Fairbairn, W. An account of the Construction of the Britannia and Conway Tubular Bridges. London, 1849.
4. Bjorhovde, R. Research needs in stability of metal structures. J.Struct.Div., ASCE, Dec. 1980. pp. 2425-2441.
5. Rockey, K.C., Valtinat, G. and Tang, K.H. The design of transverse stiffeners on webs loaded in shear - an ultimate load approach. Proc. Instn.Civ.Engrs., Part 2, 1981, 71, Dec. pp. 1069-1099.

TEST	PANEL PARAMETERS			STIFFENER	STIFFENER DIMENSIONS			
	b/d	d/t	M_p^*		b_s (mm)	t_s (mm)	$\frac{A_{s,EXP}}{A_{s,OPT}}$	$\frac{\gamma_{s,EPT}}{\gamma_{s,OPT}}$
1	0.75	832	.0136	SA1	69.9	3.67	1.00	1.00
2	0.75	834	.0136	SA2	95.5	3.67	1.36	2.54
3	0.74	834	.0136	SA3	95.5	6.14	2.12	9.56
4	0.62	833	.0136	SB1	75.0	3.66	1.10	1.34
5	0.62	834	.0136	SB2	49.9	3.67	0.72	0.37

TABLE 1 PANEL AND STIFFENER PROPERTIES

TEST	V_{EXP}	V_{PRED}	V_{5400}	$\frac{V_{PRED}}{V_{EXPT}}$	$\frac{V_{5400}}{V_{EXPT}}$	$\frac{C_{PRED}}{C_{EXPT}}$	
						Comp.	Tension
1	81.0	82.8	73.8	1.02	0.91	1.00	0.97
2	83.5	83.9	73.6	1.00	0.88	1.02	1.06
3	85.0	85.7	74.0	1.01	0.87	1.00	0.98
4	90.0	89.1	80.7	0.99	0.90	1.10	1.09
5	91.0	91.0	80.7	1.00	0.89	1.02	1.05

TABLE 2 FAILURE LOADS (kN) AND COMPARISONS

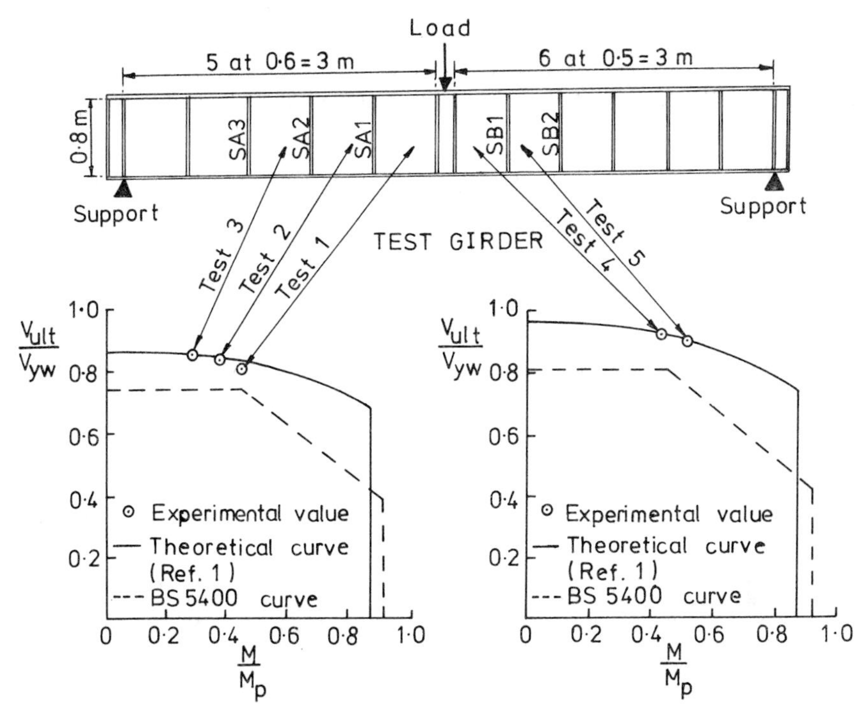

Fig. 1 Details of Test Girder and moment/shear interaction diagrams

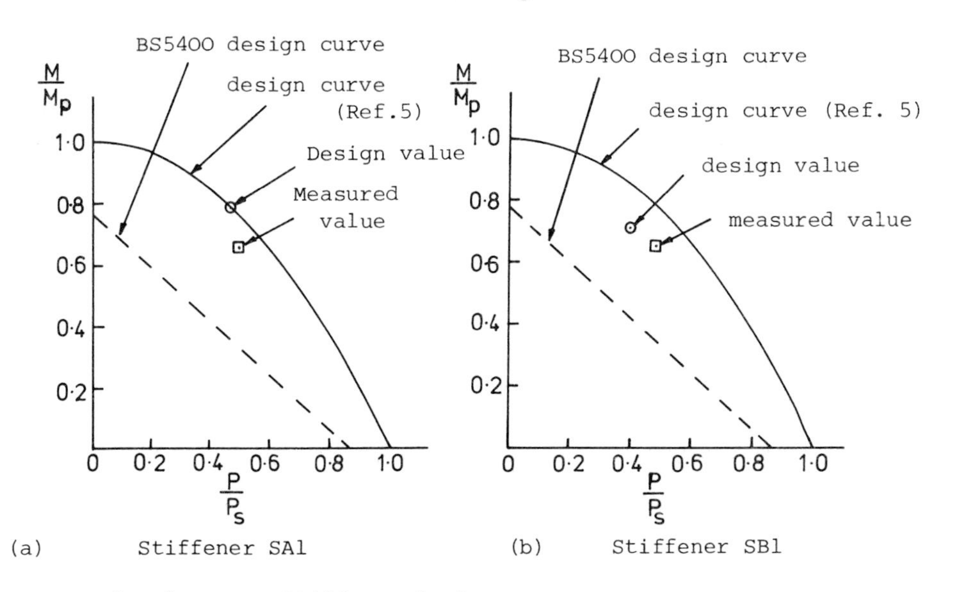

(a) Stiffener SA1 (b) Stiffener SB1

Fig. 2 Stiffener design curves

Fig. 3a View from rear of Girder after Test 1

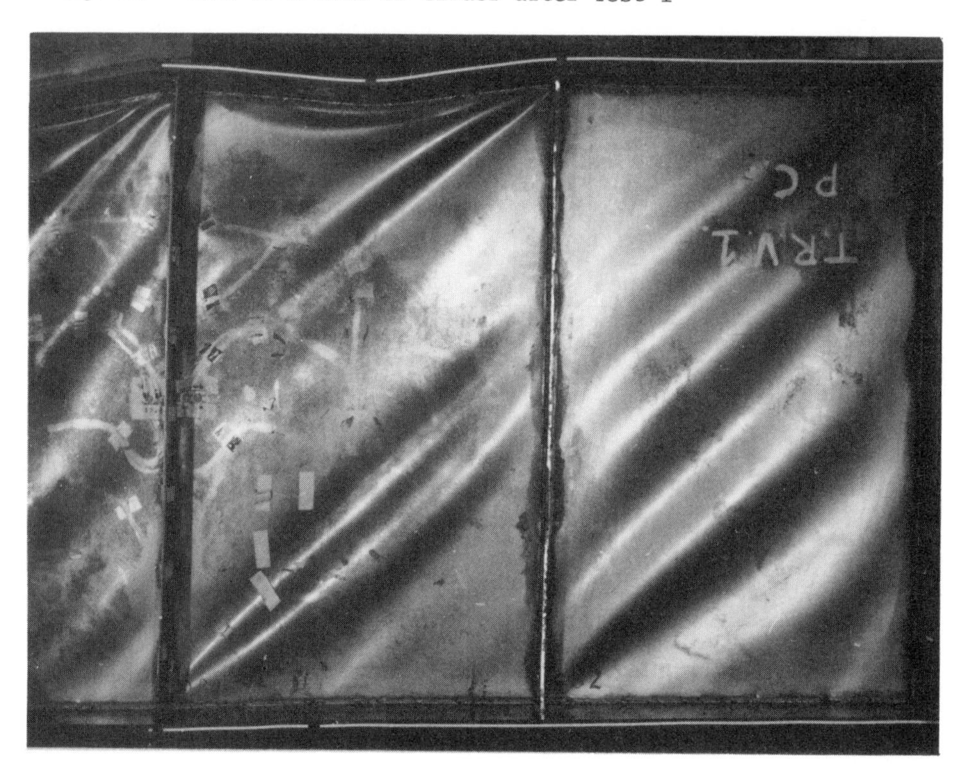

Fig. 3b View of panels PB2 and PB3 and stiffener SB2 after Test 5

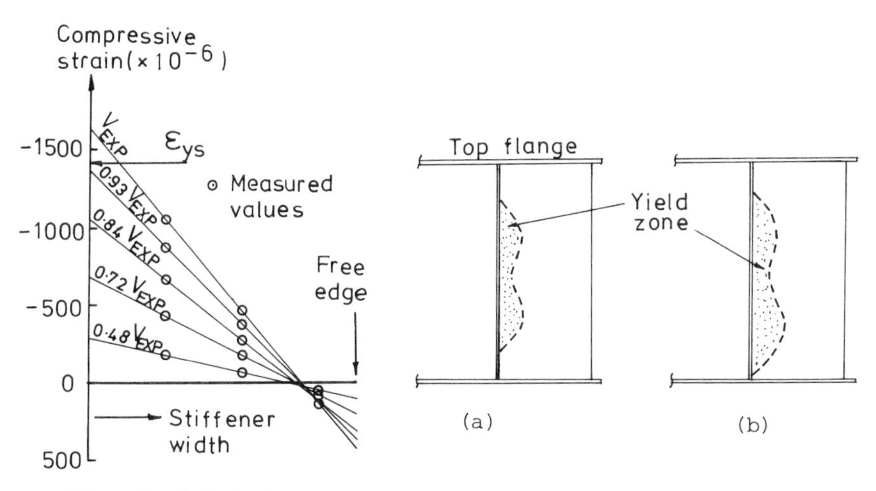

Fig. 4 Axial strain variation at mid-depth of stiffener SA1

Fig. 5 Yield zones in stiffener SA1

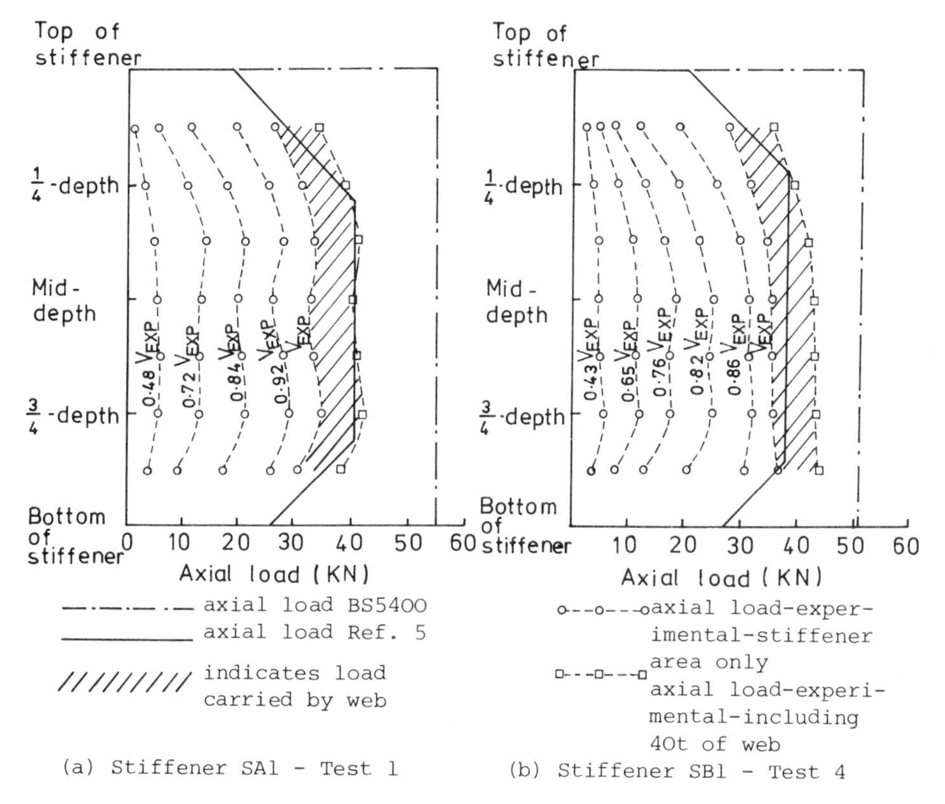

(a) Stiffener SA1 - Test 1

(b) Stiffener SB1 - Test 4

Fig. 6 Axial loads in stiffeners

COLLAPSE OF BOX GIRDER STIFFENED WEBS

P J Dowling, J E Harding and N Agelidis

Imperial College of Science and Technology

Synopsis

The collapse behaviour of longitudinally stiffened webs in four box girder models is described in this paper. A new finite element analysis is used to predict the structural response up to peak load accounting for geometric and material non-linearities, initial imperfections and the complex interaction between webs and flanges and stiffeners and plates. It is shown that the analysis can accurately describe the observed behaviour up to collapse of discretely stiffened webs including both plate panel and stiffener modes.

A comparison is drawn between the test results and the web strengths predicted using the new British bridge code, and it is concluded that using the strengths indicated by the code for the weakest web sub-panel is conservative compared with the collapse of the entire web. The finite element analysis could be used to predict more accurately the failure of the web as a whole including the stiffener behaviour.

Introduction

In 1973 the results of a series of ultimate load tests on eight large models of steel box girders were reported (1). That work formed a major part of the Merrison programme of research which was carried out in the aftermath of the box girder bridge failures at Milford Haven and the Lower Yarra Crossing. The results provided a basis for the Merrison Rules (2) and subsequently the new British bridge code, BS 5400: Part 3(3), which covers the design of box girders.

In the concluding remarks of Ref 1 it was stated that "efforts should be made to develop theories based on the large deflexion elastic-plastic behaviour of initially imperfect stiffened plates under complex loading". Since then tremendous strides have been taken in that direction and many papers (4-7) dealing with different numerical solutions have been published. Among the problems tackled have been plates under complex loading and symmetrically stiffened plates under relatively simple loading (8). However, it is only now, ten years later, that the appropriate analytical tools exist which can be used to analyse accurately boxes containing the irregularly stiffened plates under complex loading represented by the stiffened webs which collapsed in four of the boxes reported in Ref.1.

As the geometrical complexity of the problems which were being studied analytically increased the decision was taken to switch from developing finite difference based numerical solutions to the development of more general finite element programs. Such a program has been developed by Trueb (9) using elements produced by Bates(10) and other research workers at Imperial College. It is used within this paper to study web behaviour by modelling large sections of the box girder and by including the measured initial web distortions.

The experimental and analytical results are also compared with those predicted by the relevant clauses of the new bridge code (3) the bases of which are given in papers by Horne (11) and the first two authors (12). The aims of the paper are:

(1) To show that numerical analyses can now reproduce the response of box girder stiffened webs up to collapse which a decade ago could only be studied with confidence by testing expensive large scale models.

(2) To check the new design methods in BS 5400: Part 3 which are used to estimate the strength of box girder orthogonally stiffened webs failing by inter-stiffener inelastic panel buckling, and

(3) to demonstrate that the analysis could be used to improve the design rules for web stiffeners currently incorporated within BS 5400: Part 3.

1. *DESCRIPTION OF MODEL TESTS*

All the tests relevant to this paper were of boxes loaded by a central point load with ends simply supported. Such tests simulate at the centre the conditions of combined high moments and shears which occur near the internal support regions of a continually supported girder. The four model boxes concerned Models 1, 5, 6 and 7, are described by the numbers ascribed to them in Ref 1. Model 7 was the only one which was subjected to torsion as well as bending.

The materials and fabrication procedures used were representative of those used in full scale box girder bridges although the scaled up weld sizes were somewhat larger than would normally be used. The material was similar to structural steel, grade 43A, although the results of the tensile test specimens showed considerable scatter. Measured yield strengths were used in the finite element and design rule calculations.

The dimensions and material properties of all the models are given in detail in Ref 1 but Table 1 summarises the parameters relevant to this study. Three of the boxes, Models 1, 5 and 7, were of similar construction and had webs with two longitudinal stiffeners in the areas where failures occurred. Model 6 had webs with seven equally spaced longitudinal stiffeners. In addition the heavy tension flange was designed to draw the neutral axis towards it so that compressive bending stresses were induced over much of the web depth.

1.1 Failure of Model 1

The earliest signs of buckling were observed in Model 1 in the plate panels between the longitudinal stiffeners of the compression flange. Before this failure could progress, however, the large web panels in the end web bays buckled in shear at a jack load of 300 KN corresponding to an average web shear stress of 0.62 τ_0. With a small increase in load the centre panel at this section also began to buckle and testing was terminated at a jack load of 316 KN corresponding to an average web shear stress of 0.65 τ_0.

The buckling of the web at the position of least bending moment was due to the absence of the stabilising tensile forces present in the panels nearer the central support point. The positions of the buckles can be seen in Fig 1. Figure 2 gives the overall load deflection response of the model. Also shown on this figure is the predicted behaviour from the finite element correlation. This is discussed later.

1.2 Failure of Model 5

Model 5 was similar to Model 1 except it had thicker flanges and an additional longitudinal web stiffener in the two web panels away from the central diaphragm. This prevented the web failure seen for model 1 and also the simultaneous collapse of the compression flange panels.

Buckling initially occurred in the large web panel adjacent to the centre diaphragm at a jack load of 260 KN corresponding to a web shear stress of 0.67 τ_0. At a slightly increased load level a second buckle occurred in the middle panel which spread through the longitudinal stiffener as shown in Figure 3. The ultimate load reached was 280 KN corresponding to a web shear stress of 0.72 τ_0. The load deflection curve for this model is shown in Fig 4. In this instance no significant unloading was observed, presumably reflecting the relative stability of the panel loaded in shear and tension.

1.3 Failure of Model 6

Model 6 was significantly different to any of the other models tested in that the web had multiple longitudinal stiffeners providing stocky sub-panels, mainly loaded in combined compression and shear because of the presence of an extremely thick tension flange plate.

Failure was initiated in the web panels of the centre bay mid-way between the centre section and the jacking point. At a jack load of 575 KN (τ=0.77 τ_0) a buckle appeared in panel 4 of the web (see Fig 6) about half way between the neutral axis and the compression flange. This is rather surprising as the direct stress is appreciably lower than that in the panel adjacent to the compression flange, and the in-plane restraint available for tension field action should be considerably greater.

The maximum value of jack load reached was some 630 KN corresponding to a web shear stress of 0.85 τ_0 and by this stage additional buckles had appeared in all the panels of the intermediate bay. Two of the buckles twisted and crossed longitudinal stiffeners as shown in Fig 5. Figure 6 shows the load-central deflection response of this box.

1.4 Failure of Model 7

Model 7 was nominally identical to Model 5. The model was tested under combined flexure and torsion so that the webs were subjected to unequal shear and bending stresses. The load was applied by one jack at each end acting at diagonally opposite corners. Buckles first appeared in the large web panels near the centre section on the sides directly loaded by the jacks at a jack load of 300 KN.

Using simple beam and torsion theory the torsional component of the web shear stress is equal to 0.056 times the jack load and the flexural component equals 0.112 times the jack load. The combined shear stress on the heavily loaded web for this load level, assuming no global redistribution of flexural stresses is 0.57 τ_0. At a load of 380 KN ($\tau = 0.72\ \tau_0$) buckles appeared in the adjacent panels of the same bays and at the peak load of 410 KN ($\tau = 0.78\ \tau_0$) further buckles developed in the intermediate bays and spread through the neutral axis stiffener. Figure 7 shows this buckle formation while Fig 8 shows the load deflection response.

Comparing the ultimate stress of this box with that of Model 5 ($\tau = 0.72\ \tau_0$) indicates that a small amount of redistribution may have occurred but initial imperfection differences could also have been responsible. The imperfections of Model 7 are, however, larger than those of Model 5 and in general the failures were initiated in an imperfection insensitive zone so it seems likely that redistribution did occur.

2. FINITE ELEMENT MODELLING

The finite element analyses were carried out using a combination of shell and stiffener elements. The shell element was an eight noded iso-parametric element with six degrees of freedom at each node while the stiffener element is a three noded space beam capable of representing any thin walled open section. The kinematic models for the beam and shell element are fully compatible. Geometric non-linearity is accurately predicted even under a combination of large rotations and translations and plasticity in both elements is based on the von Mises yield function (9,10).

The meshes can be seen in the views of collapse deflections. In general three elements were taken between each vertical stiffener and two between each longitudinal stiffener. In the local areas where the longitudinal stiffener was discontinued the mesh was maintained resulting in a four by three element subdivision for the tension panels adjacent to the centre diaphragms.

All web stiffeners were accurately modelled by elements, and cross frames and diaphragms were also modelled as closely as possible. In the case of flange stiffeners, an equivalent plate thickness was used for the entire flange width giving the same total area of steel except in the case of Box 6 where a reduced number of heavy stiffeners was used to prevent a premature flange failure. It is not thought that this would introduce significant changes in web response although the local bending restraint to inplane movement of the webs, that is the tension field restraint provided to the exterior web panels, will have been slightly over estimated.

In the case of the models subjected to bending and shear, quarter symmetry was assumed so that one web, half the compression flange and half the tension flange were modelled to one side of the central reaction. In the case of the box subjected to torsion half symmetry was assumed and one end of the box was modelled. Because of the size of the problem a coarser mesh was used in areas away from the expected failure zone in this instance.

Geometric imperfections were taken from the actual experimental data for three models and were input as a series of nodal points. In all cases the imperfections for the half web corresponding to the area of initiation of experimental failure were taken. A typical contour plot of the imperfection data, that of Model 5, is shown in Fig 9. In the case of Model 6 experimental data were not readily available and imperfections were incorporated which had values corresponding to BS 5400 design rule tolerances. Panel, and vertical and horizontal stiffener imperfections were superimposed to give as realistic a representation as possible.

Loading was applied as a point load at the position of the jack in the experiments. In the case of Model 6 the loading was spread up the end diaphragm over a distance of two panels to overcome local yielding problems associated with a slight simplification of the end bearing stiffener.

All finite element runs were very large reaching close to both machine time and storage limits. This neccessitated very careful choice of increment size and loading control in order to reduce solution time to a minimum while obtaining sufficient accuracy from the solutions.

In some cases control of loading was exerted by monitoring box displacement. This enabled a more precisely defined peak load to be obtained but also tended to be more time consuming and less numerically stable. Where problems arose, control was exerted by incrementing the load and the peak was obtained as non convergence of the solution.

Although the analysis can in principle incorporate accurately the effect of residual stresses, limitations on the mesh size used in these cases prevented the measured residual stresses being defined precisely. To gauge the effect such stresses might have on behaviour cases were re-run using a yield stress for the plate material reduced by the average value of the compressive residual stresses measured in the failure zones. This conservative assumption provided lower bound analyses results for correlation.

3 RESULTS FROM ANALYSES

The results of the analyses are presented in the form of overall load-deflection responses superimposed on the experimental cuves, and contour and three dimensional views of panel and stiffener deflections at peak load that can be compared directly with the buckle development described in Section 1. The later plots are not precisely comparable as the experimental versions are visual interpretations of buckle development. Nevertheless, they can be used to judge similarities in modes of failure and the loads at which noticable deformations occur.

3.1 Failure of Model 1

Figure 2 shows the load-deflection response obtained from the finite element analysis (the dotted curve) compared with the experimental response. In this instance the curve was obtained by applying load increments and no residual stresses were input. The latter were ommitted because the failure was dominated by shear and previous work (12) had indicated the low sensitivity of shear collapse to residual stresses, expecially for relatively stocky panels. It can be seen that the stiffness taken from the analytical results is very close to that measured experimentally. The final collapse load from the analysis is only marginally in excess of the true failure load and the inclusion of residual stresses would have reduced the failure load slightly.

The deflected shape at peak load shown in Fig 10 is similar to that obtained experimentally. The largest deflections occur in the end web panel adjacent to the tension flange and no significant stiffener movement is apparent in this instance. The three dimensional views also show smaller but significant deflections in the other panels adjacent to the tension flange but nearer to the centre diaphragm. In the analysis, however, the panel buckled inwards while for the experiment the buckle is indicated as outwards. It is not certain, however, that the experimental deflection depicted in the figure is for the corresponding panel and there is no indication in the experimental report of the mode of failure of the other panels.

3.2 Failure of Model 5

Figure 4 shows finite element curves both with and without residual stresses, the latter being modelled by a reduction in yield stress. It can be seen that residual stresses have a large effect on failure in this instance, probably because of the panels concerned. The analysis with residual stress predicts a failure load very slightly below the experimental value. This might be expected as the use of a reduced yield stress in conservative. This is because it does not model the effectivly stronger boundary strips nor the possible stress redistribution capability as it omits the tensile yield zones. Nonetheless the agreement is excellent.

Figure 11 shows that the experimental deflected shape is modelled precisely by the analysis. Both the panal buckles in panels 2 and 3 and the progression of one of the panel 2 buckles across the longitudinal stiffener has been accurately predicted. The relative magnitudes of the buckles is also broadly correct.

3.3 *Failure of Model 6*

Figure 6 shows the load-deflection responses for Model 6 with and without residual stresses. While thee experimental result again falls between the two analysis predictions, the ultimate load predicted by the analysis without residual stresses is significantly closer to the experimental value. The deflections, shown in Fig 12, are larger in the compression panels near the centre diaphragm which is reasonable but does not reflect the larger experimental deflections in the panel 4 region. This is possibly because equal panel imperfections were used in the analysis rather than experimentally measured values. Deflections at peak load were all very small being a fraction of the plate thickness. In the experiments large deflections including the stiffeners were only obtained well after peak load.

3.4 *Failure of Model 7*

Figure 8 shows that the difference between the overall box response with and without residual stresses is similar to that of Model 5. Again, the experimental result falls between these two analytical extremes although in this case the analysis with residual stress is lower, in relative terms,compared with the experimental results, than for Model 5.

Figure 13 shows that the deflected shape is again modelled quite well with the correct panel and stiffener buckles being predicted. In this instance the analysis indicates that the buckle which crosses the longitudinal stiffener was initiated in panel 3 while the experimental observation tends to indicate that there was a higher level of buckling involvement in panel 2. Nevertheless the modes are very similar.

4. *COMPARISON WITH NEW BRITISH BRIDGE CODE, BS 5400*

The new steel bridge design code treats the design of multi-stiffened webs as essentially the design of individual sub-panels, each of which has a certain predicted failure strength, bounded by stiffeners which are intended to remain in position, thus providing the necessary boundary restraint to ensure development of the panel capacities. Panels are designated as restrained or unrestrained depending on the degree of in-plane transverse restraint available at the edges. In general terms this leads to the specification that edge panels, that is panels abutting a flange, are unrestrained and hence weaker, while internal panels are restrained.

The code therefore, in general, does not allow for the possible increases in tension field forces that may occur as the web buckles rotate and traverse the longitudinal stiffeners. (Such increases may be small in any case and are associated with large overall shear strains). Code predictions of shear strength, therefore, would be expected to be on the conservative side.

The sub-panel strength curves within the code have themselves been derived from large deflection elasto-plastic numerical analyses, carried out on unstiffened plate panels with idealised boundary conditions. The results of the analyses are described in reference 5 while the method of incorporation in the code, via an interaction equation, is described in Reference 12.

The idealisation of boundary conditions also means that in the case of edge panels allowance is not made for the flange bending restraint providing even a modest tension field restraint to the web. This, again, would tend to result in the code predictions being on the conservative side.

The models have been analysed using the code for each of the sub-panels between adjacent vertical stiffeners in the area of each box where failure was initiated. Model 7 predictions are directly comparable to those of Model 5. In the case of Model 6 the code would have predicted failure in the web compression panel adjacent to the flange near the centre diaphragm rather than at the section which actually failed and for which the code calculations were performed.

The results of the code calculations are given in Tables 2, 3, 4 and 5 together with the experimental failure loads and failure zones.

The code is only slightly conservative in the case of Model 1 and predicts failure in the correct sub-panel. In the case of Model 5 the code is conservative by some 20%. Similar collapse loads are predicted for the failure loads of panels 1 and 3 while no buckling is apparent in panel 1 in the experiment. The compressive capacity of this panel will certainly be underestimated by the code which is based on the response of a square panel. Panels with higher aspect ratios and single half wave imperfections give higher strengths and steep unloading characteristics. The predictions for Model 7 are more conservative but this is not surprising as the code cannot allow for the global redistribution encountered. No simple formulation would be able to predict this as the phenomenon is very dependent on the distortional stiffness of the box cross section.

The prediction for Model 6 is also conservative by about 20%. The design rules predict failure adjacent to the compression flange while the box actually first buckles in panel 4. Certainly the explanation for the low strength for panel 1 is likely again to be the high strength for the 5:1 aspect ratio panel in compression compared with that for the square panel assumed in the rules. It may also be that for such a high aspect ratio the compression and shear buckle modes are very different and this raises the strength above that of panels with lower levels of compression. The relevance of imperfections has been discussed previously.

While there are obvious discrepancies in the results the slightly conservative predictions seem very reasonable in such a complex interaction area, particularly bearing in mind that some direct stress shedding to flange elements is also possible, especially in the case of Model 6.

There is a general indication that the unrestrained condition is slightly conservative for edge panels but it would be difficult to allow for different degress of practical edge restraint.

It also seems apparent from the experimental results and the correlations that after buckling in an individual sub-panel there is a relatively small amount of vertical redistribution. That is to say that, as long as panels have been designed in a reasonably balanced way, the sub-panel strength gives a suitably conservative indication of overall web strength. Code predictions are reasonable in this instance in spite of the fact that the web sub-panel strengths are not exactly equivalent confirming this small amount of redistribution.

5. *DISCUSSION*

The paper has demonstrated that analysis methods have now reached the stage where assemblages of plates and stiffeners can be accurately modelled in detail and good correlation obtained with experimental inelastic buckling behaviour. In particular, the results have shown that failure modes of complex structures can be predicted even when the failure involves interaction between both plate and stiffener elements.

The failure loads of the models were also predicted with reasonable accuracy but the coarseness of the mesh used meant that a detailed reproduction of residual stresses was not possible. The use of a finer mesh and hence the inclusion of the true compressive and tensile residual stress blocks should have resulted in a closer peak load correlation. The analyses do, however, show a correct bounding of the experimental results and a generally close agreement between theory and test.

The closeness of the results of Model 7 loaded in torsion and flexure indicates that the redistribution of shear stresses between webs has been modelled sufficiently accurately.

The new British Steel Bridge Code, BS 5400, has been shown to be conservative but not unduly so, bearing in mind that the experiments did not represent accurately balanced designs in which all web panels had comparable strengths. Because of this the code predictions for the weakest sub-panel tended to be lower than the overall web collapse strengths.

The Code appears to be conservative in relation to strength predictions for long panels in compression by basing their design on square panel behaviour, and also in the use of completely unrestrained boundaries for external panels, thereby ignoring the presence of any available flange restraint. The former is due to the fact that panels of high aspect ratio loaded in compression along their short sides gain strength at the expense of post buckling stability, because of the presence of imperfection modes which are not affine to the preferred buckling modes. The use of square panel data does not allow for this nor indeed was it intended to as it would be questionable to allow for such an effect in design. This is because of the possibility, however remote, of fully sympathatic modes occurring in an actual fabrication. Analyses such as those presented could certainly be used to investigate the second conservatism by reproducing various flange thicknesses and observing their effect on edge panel strength.

The biggest benefit to be gained from analyses of the type presented is undoubtably the possibility, for the first time, of examining stiffener behaviour and evaluating realistic criteria for their design. The existing design rules for web stiffeners are approximate and of necessity conservative because of the lack of data concerning their real response.

6 *CONCLUSIONS*

The paper has demonstrated the following :-

1. Large deflection elasto plastic methods of analysis now exist which can be used to accurately predict the behaviour including peak load of complex assemblages of plate panels and stiffeners such as box girder structures.

2. It is necessary to incorporate both geometrical imperfections and residual stresses to obtain accurate representations of real behaviour.

3. The new British Steel Bridge Code, BS 5400, provides safe lower bounds to the strength of longitudinally stiffened webs failing by collapse of the web panels. In some cases these appear to be rather too conservative but this is partly because of assumptions relating to imperfections which need to be incorporated in a code formulation.

4. It is now possible, using analyses of this type to perform studies which could result in significant improvements in the design rules, particularly those related to the design of web stiffeners. The effect of destabilising stresses in the web plate on stiffener behaviour can now be studied with confidence.

7 *ACKNOWLEDGEMENT*

The finite element analysis used in this paper (FINAS) was developed at Imperial College by Messrs Trueb and Bates who were sponsored by the SERC and the authors acknowledge gratefully their advice in its use.

References

1. Dowling, P.J. et al 'Experimental investigations on boxes in Steel Box Girder Bridges',Proc of Inf.Mtg. on Steel Box Girder Bridges, Instn.of Civ. Engrs, 28/41, 1973.

2. Inquiry into the Basis of Design and Method of Erecting Steel Box Girder Bridges. Interim Design and Workmanship Rules. Parts 1 to 4. HMSO, London, 1973.

3. BS 5400, Part 3, Code of Practice for Design of Steel Bridges. British Standards Institution, 1982.

4. Frieze, P.A., Dowling, P.J. and Hobbs, R.E. 'Ultimate load behaviour of plates in Compression',Steel Plated Structures, Dowling, P.J et.al.eds, Crosby Lockwood Staples, London, 51/88, 1977.

5. Harding, J.E, Hobbs, R.E. and Neal, B.G. 'Ultimate load behaviour of plates under combined direct and shear in-plane

loading', Steel Plated Structures, Crosby Lockwood Staples, London, 369/403, 1977.

6. Crisfield, M.A. and Puthli, R. 'A finite element method applied to the collapse analysis of stiffened box girder diaphragms', Steel Plated Structures, Crosby Lockwood Staples, London, 311/337, 1977

7. Dier, A.F. and Dowling, P.J. 'The strength of plates subjected to biaxial forces'. Proc.of Conf. on Thin Walled Structures, University of Strathclyde, April, 1983.

8. Dowling, P.J. 'Practical plate collapse problems', Proc. IUTAM Conf. on Collapse, University College London, Sept, 1982 to be published by Cambridge University Press, 1983 .

9. Trueb, U. 'Stability problems of elastic plastic plates and shells by finite elements'. PhD Thesis, (to be submitted), University of London, 1983.

10. Bates, D. 'Nonlinear finite element analysis of curved beams and shells'. PhD Thesis (to be submitted), University of London, 1983.

11. Horne, M.R. 'Basic concepts in design of webs', The Design of Steel Bridges, Rockey K.C. and Evans, H.R. eds, Granada, London 161/188, 1981.

12. Harding, J.E. and Dowling, P.J. 'The basis of the proposed new design rules for the strength of web plates and other panels subject to complex edge loading'. Stability Problems in Engineering Structures and Components, Inst.of Physics, 355/376, 1979.

'INSTABILITY AND PLASTIC COLLAPSE OF STEEL STRUCTURES'

ERRATA

		as printed	correction
Contents	Paper 5.6	Tong-Ji University, Japan	Tong-Ji University, Shanghai, China
Paper 2.5			
p 77	Eqn (6)	$P_{cb} = \frac{4\ ^{2}EI}{h^{2}}$	$P_{cb} = \frac{^{2}EI}{4h^{2}}$
p78	between eqns (10) and (11)	$\therefore\ top = \int_{o}^{H} \frac{wx}{kh}\,dx = \frac{wH^{2}}{2kh} = \frac{wH}{2kh}$	$\therefore\ top = \int_{o}^{H} \frac{wx}{kh}\,dx = \frac{wH^{2}}{2kh} = \frac{WH}{2kh}$
	4th line from bottom	$E = 28\ N/mm^{2}$	$E = 28\ kN/mm^{2}$
	3rd line from bottom	*Frame Beams* : $I = 3.375\ E{-}4,\ m^{4}$	*Frame Beams* : $I = 3.375\ E{-}3\ m^{4}$
Paper 3.1			
90	line 10	Fig 1	Fig 2
91	section 2.3	T_{k}	V_{k}
		$\lvert T_{1}, \ldots, T_{n_b} \rvert$	$\lvert V_{1}, \ldots, V_{nb} \rvert$
92	line 1	mechanics	mechanisms
	line 17	Condition (22)	Condition (15)
	line 27	decomposition (14)	decomposition (16)
93	line 6	T_{k}	V_{k}
	line 9	T	V
94	line 15	rewrite (32),(34),(35) and (36)	rewrite (25),(27),(28) and (29)
	line 19	member	number
	last line	eqn. (4),(7)	eqn. (6),(7)
98	line 10	(49). Problem (56) gives the moments in all hinges	(43). Dual problem $\phi = b^{t}y_{max}$ gives the moments in all hinges
	line 32	(European Construction ...	(European Convention ...
Paper 3.3			
113	line 3	assumption both	assumption of both
115	line 25	$\Delta\underline{\lambda} = \Delta\alpha\underline{\lambda}^{(2)}$	$\Delta\underline{\lambda} = \Delta\alpha\dot{\underline{\lambda}}^{(2)}$
	line 31	$\underset{\sim}{e}_{o}^{1}$	$\dot{\underset{\sim}{e}}_{o}^{1}$
	line 32	$\underset{\sim}{s}_{o}^{1}$	$\dot{\underset{\sim}{s}}_{o}^{1}$
116	line 1	$\underset{\sim}{s}_{o}^{1}$ ..... $\underline{s}$	$\dot{\underset{\sim}{s}}_{o}^{1}$..... $\dot{\underline{s}}$
117	eqn (4)	$(N,\ M,\ h_{u},\ h_{1}) = 0$	$f(N,\ M,\ h_{u},\ h_{1}) = 0$
	line 40	7%	$7^{o}/oo$

		as printed	correction
Paper 3.3 continued			
p118	line 11	$\underset{\sim}{e}_o^1$	$\dot{e}_o^1$
	line 16	to step	go to step
	line 41	factorization is updated	factorization. During the iteratic the vector of known quantities is updated
Paper 3.6			
p145	part eqn (7)	$-[r']\{W'\} =$	$-[r']^T\{W'\} =$
	part eqn (8)	$\frac{P_j \overline{P}_{ji} \alpha_j}{E} +$	$\frac{P_j \overline{P}_{ji} \alpha_j \ell_j}{E} +$
	following line	where, 1_j is	where, ℓ_j is
Paper 7.1			
p331	eqn 2	$\frac{\sigma_y}{355}$	$\sqrt{\frac{\sigma_y}{355}}$
	next line	$a \leqslant b,$	$a \leqslant 2b,$
	line 29	$\frac{b}{t} \; \frac{\sigma_y}{355}$	$\frac{b}{t}\sqrt{\frac{\sigma_y}{355}}$
p334	eqn 4	$\gamma^* \quad 0.8(\lambda - 50)$	$\gamma^* \ngtr 0.8(\lambda - 50)$
Paper 10.5			
p548	penultimate line	7%.	3%.
	Fig 2	Dimension of an actual profile	Dimension of an actual profile (represents most extreme case measured)

Model	t,mm	σ_y (N/mm^2)
1	CF 4.95 TF 4.95 W 3.38 LS 15.88 x 4.76 L TS 50.80 x 6.35 L	247.1 247.1 273.4 328.9 313.5
5	CF 8.12 TF 8.12 W 3.15 LS 38.10 x 6.35 and 25.4 x 6.35 flats TS 101.60 x 63.50 x 6.35 L	264.1 264.1 233.2 296.5 -
6	CF 4.83 TF 6.25 (37.59 cover plate) W 3.35 LS 38.1 x 4.76 flats TS 101.60 x 63.50 x 6.35 L	271.8 - 315.1 312.0 302.7
7	CF 7.80 TF 7.95 W 3.15 LS 38.10 x 6.35 flats and 25.4 x 6.35 flats TS 101.6 x 63.5 x 6.35 L	273.4 - 236.3 296.5 287.3
	Thicknesses of plates and sections are conversions from original imperial sizes. In the case of stiffeners these are nominal dimensions. CF compression flange TF Tension flange W web LS Longitudinal web stiffener TS Transverse web stiffener.	

Table 1 Dimensions and yield strengths of models.

Experiment	BS 5400 Unrestrained	BS 5400 Restrained
Panel 1	0.74*	0.80
Panel 2 +	0.68	0.83*
Panel 3 + (0.62)	0.59*	0.94
Failure 0.65		

All values are τ/τ_o

* Relevant Code Conditions

+ Failure indications

Numbers in brackets relate to first buckling of particular panels

Table 2 Failure predictions for Model 1

Experiment	BS 5400	
	Unrestrained	Restrained
Panel 1	0.59*	0.65
Panel 2 +	0.41	0.73*
Panel 3 + (0.67)	0.61*	0.98
Failure 0.72		

Table 3 Failure predictions for Model 5

Experiment	BS 5400	
	Unrestrained	Restrained
Panel 1 +	0.65*	0.72
Panel 2 +	0.69	0.76*
Panel 3 +	0.73	0.81*
Panel 4 + (0.77)	0.78	0.85*
Panel 5 +	0.82	0.90*
Panel 6 +	0.88	0.96*
Panel 7 +	0.93	1.01*
Panel 8 +	0.97*	1.06
Failure 0.85		

Table 4 Failure predictions for Model 6

Experiment	BS 5400	
	Unrestrained	Restrained
Panel 1	0.59*	0.65
Panel 2 + (0.72)	0.41	0.73*
Panel 3 + (0.57)	0.61*	0.98
Failure 0.78		

Table 5 Failure predictions for Model 7

All values are τ/τ_o

* Relevant code conditions

+ Failure indications

Numbers in brackets relate to first buckling of particular panels.

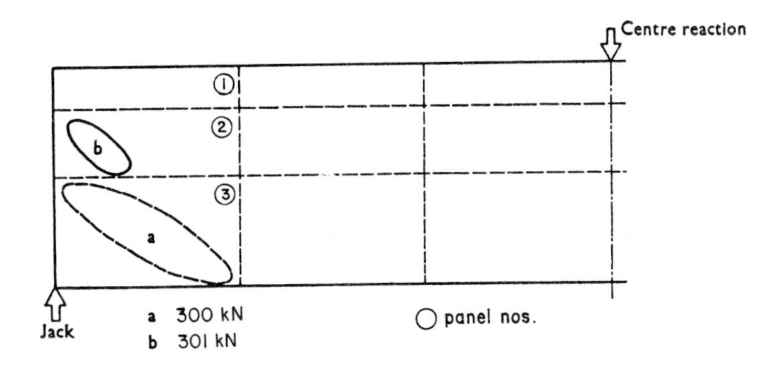

Fig 1   Experimental failure of model 1

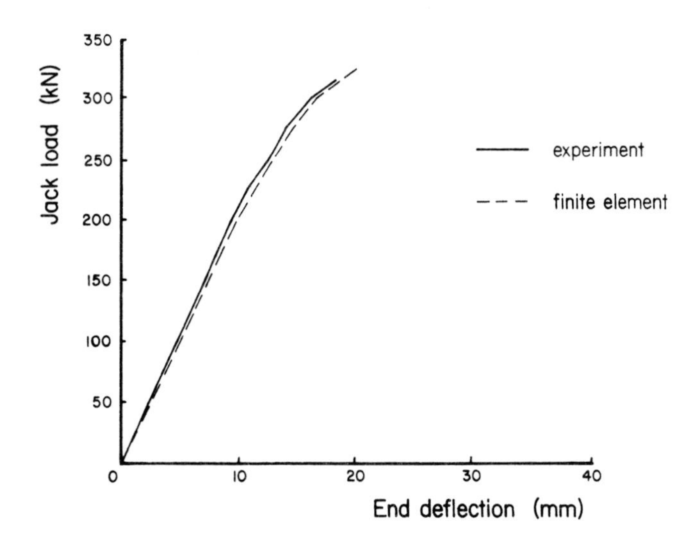

Fig 2   Load-deflection response - model 1

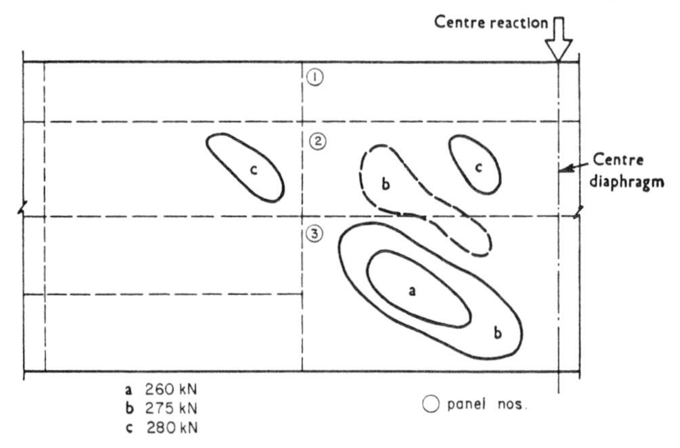

(a) Development of buckles

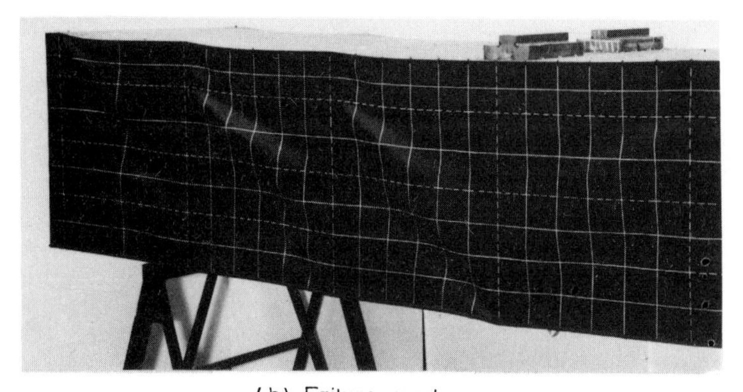

(b) Failure mode

Fig 3 Experimental failure of model 5

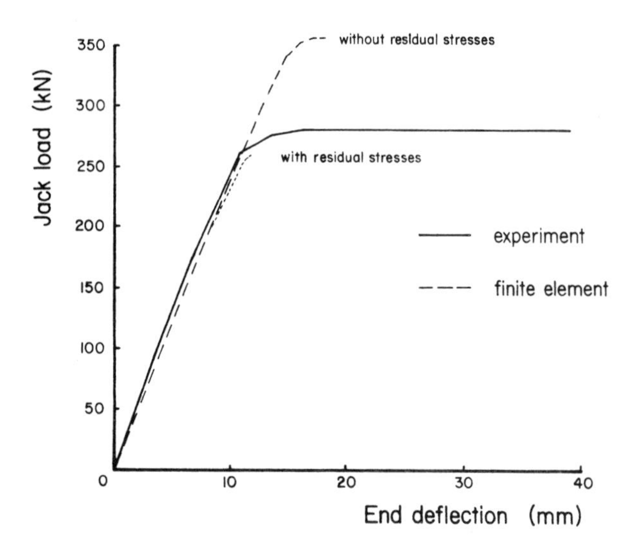

Fig 4 Load-deflection response - model 5

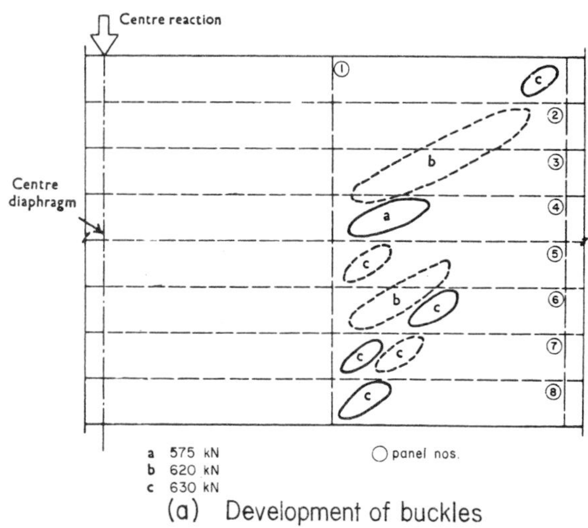

(a) Development of buckles

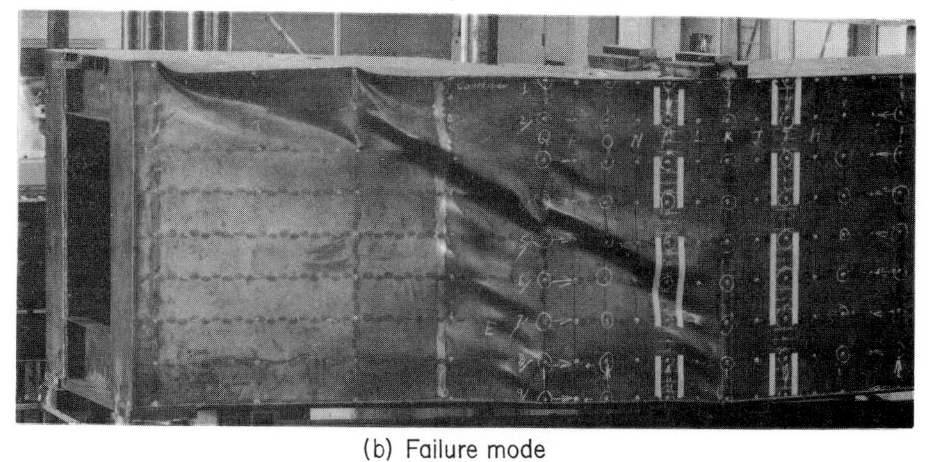

(b) Failure mode

Fig 5 Experimental failure of model 6

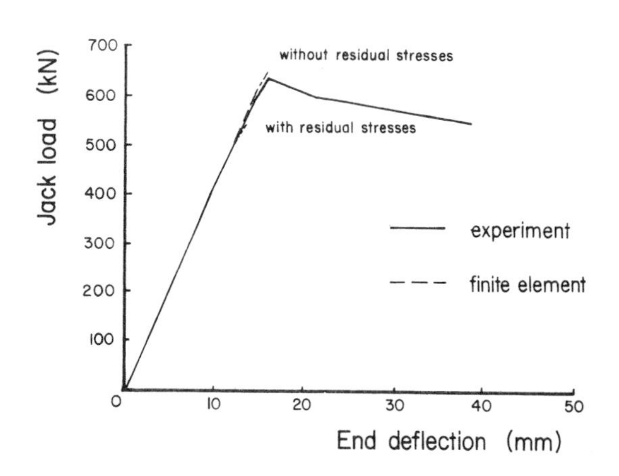

Fig 6 Load-deflection response - model 6

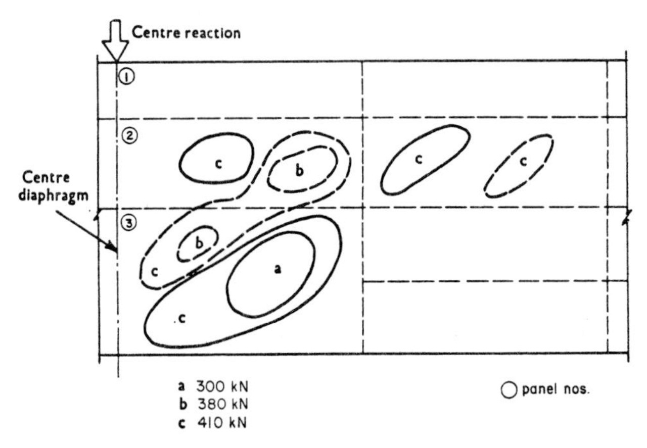

Fig 7 Experimental failure of model 7

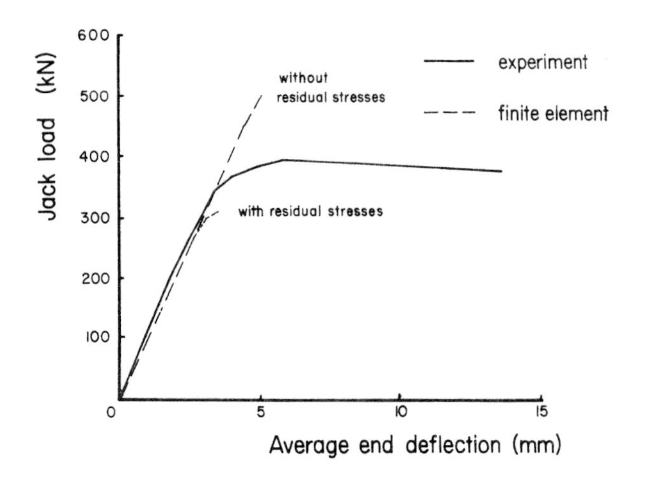

Fig 8 Load-deflection response - model 7

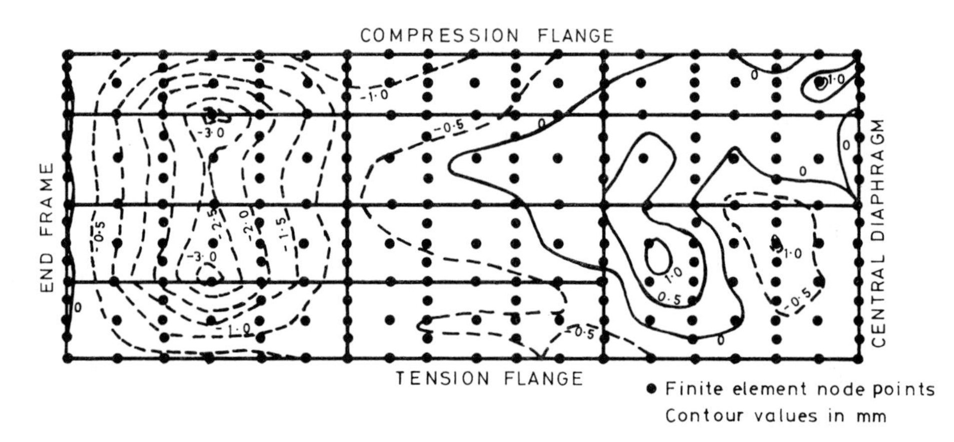

Fig 9 Initial imperfections of model 5

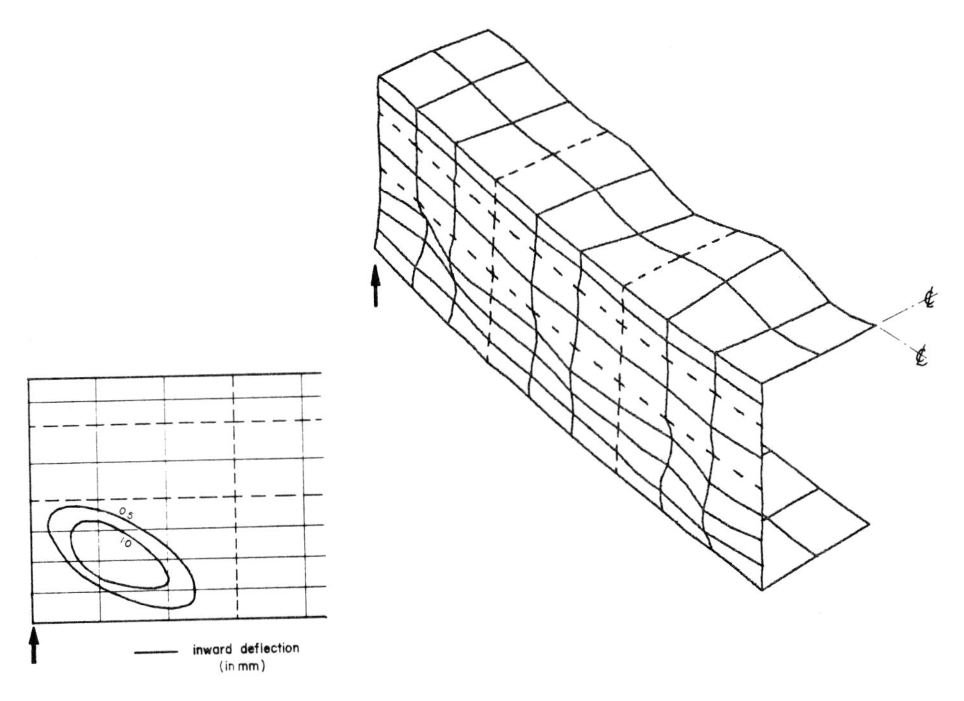

Fig 10 Buckled shape from FE analysis - model 1

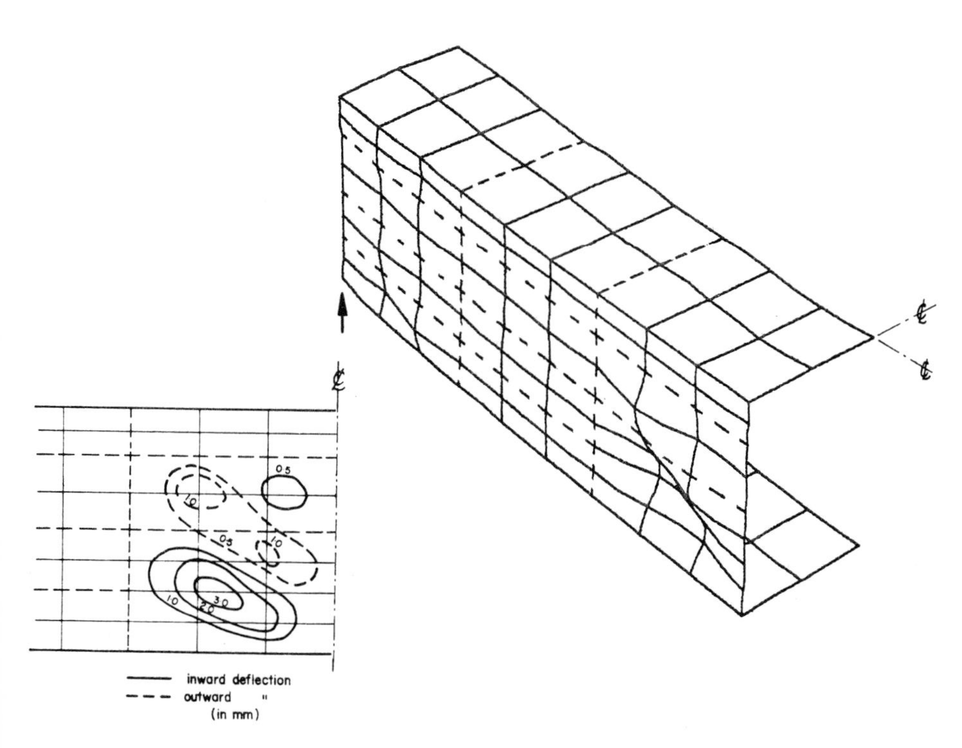

Fig 11 Buckled shape from FE analysis - model 5

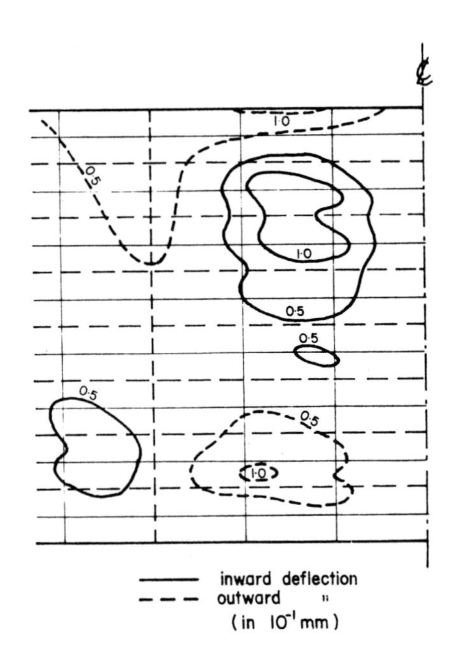

Fig 12 Buckled shape from FE analysis - model 6

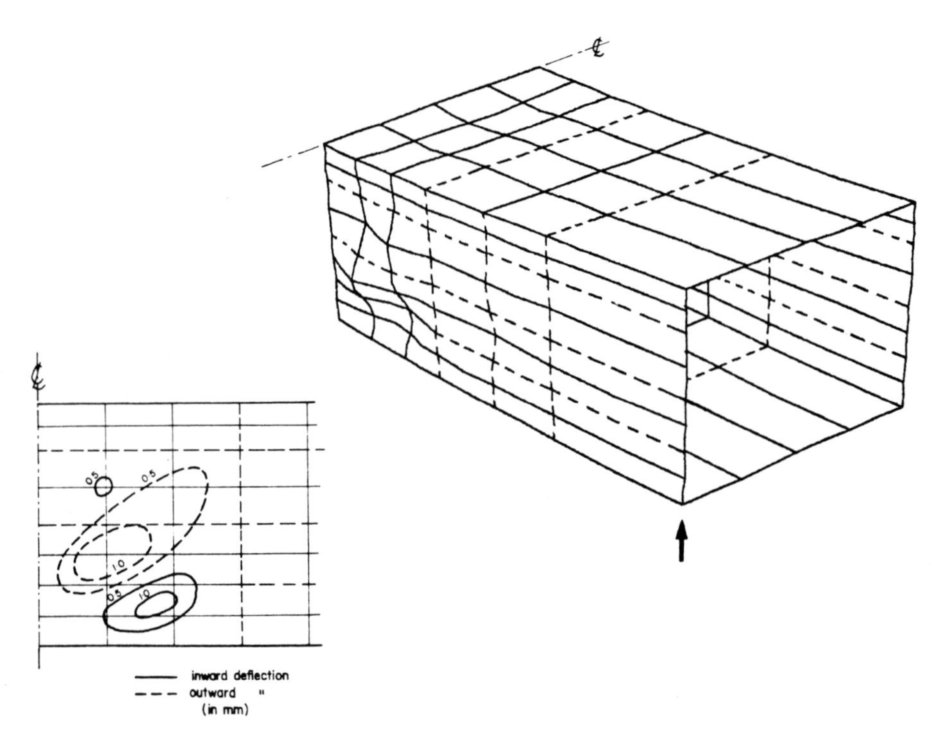

Fig 13 Buckled shape from FE analysis - model 7

A STUDY OF THE ELASTIC-PLASTIC INSTABILITY OF STIFFENED PANELS USING THE FINITE STRIP METHOD

R.J. Plank

Department of Civil and Structural Engineering, University of Sheffield

Synopsis

A brief description is given of a modified finite strip method which allows for non-linear material behaviour in the buckling analysis of any structure which can be considered as an assembly of long, thin, flat plates rigidly connected together at their longitudinal edges. Some indication of the accuracy of the method is given. Typical results are presented demonstrating the capability of the method and indicating the influence of various parameters on the elastic-plastic buckling behaviour of stiffened panels. The accuracy of more simplified approaches is discussed.

Notation

b	width of plating between stiffeners
c	constant defining non-linear stress-strain relationship
E	Young's modulus
E_{sec}	secant modulus
E_t	tangent modulus
h_s	stiffener height
t	thickness of plating
t_s	thickness of stiffeners
β	slenderness of plating $\left[= \frac{b}{\pi t} \sqrt{3\sigma_y (1-\nu^2)/E}\right]$
δ	ratio of area of stiffener to area of plating
ε	longitudinal strain
λ	half wavelength of buckling
μ	σ/σ_y
ν	effective Poisson's ratio
ν_e	elastic Poisson's ratio
ν_p	plastic Poisson's ratio
σ	longitudinal stress

σ_{cr} critical stress

σ_y yield stress

Introduction

Despite the wealth of work which has been published on elastic buckling of stiffened panels, relatively little has been done on studying the interaction between plastic behaviour and instability other than experimentally. This interaction may be important for moderately stocky structures, or where residual stresses are a significant factor and the finite strip method, which is well established as a computationally efficient means of investigating the elastic buckling behaviour of prismatic structures provides a suitable basis for such a study. A brief description is given of the theory, with particular emphasis on the means of incorporating the plastic behaviour of the section into the stability analysis, and the accuracy of the method is discussed. Some typical results obtained using the method are presented, demonstrating the influence of various parameters on the plastic instability of stiffened panels. These include the slenderness of the plating between stiffeners, the size and shape of stiffeners, and the level of residual stress.

Although the finite strip method enables all modes of buckling to be considered and allows for full interaction between different modes, simplified methods for studying particular buckling modes in isolation are popular. Some comparisons are made between results obtained using such methods and those given by the finite strip approach.

1. THE FINITE STRIP METHOD

The basis of the finite strip method described by Cheung (1) is to treat a prismatic structure as an assembly of flat plates, each of which is divided into a number of longitudinal strips. The pattern of buckling displacements in each of these strips is assumed to vary according to a sinusoidal function with a half wavelength λ in the longitudinal direction and polynomial functions across the width. Using classical plate theory and the Principle of Virtual Work, stiffness matrices relating the buckling displacements to the corresponding perturbation forces can be established for individual strips. The derivation of the matrices has been presented by Plank and Wittrick (2) for elastic buckling behaviour, and in considering the elastic-plastic buckling, the work done by the basic applied forces and perturbation edge forces is unchanged. However in establishing the internal virtual work, the simple stress-strain and moment-curvature relationship based on linear eleastic behaviour must be modified to allow for the actual stress-strain curve for steel. This is described in detail by Mahmoud (3). An idealised representation of this is shown Fig.1 and can be represented mathematically in the form suggested by Ylinen (4)

$$\varepsilon = \frac{\sigma\ (1-c\mu)}{E(1-\mu)} \qquad \text{.....(1)}$$

where $\mu = \sigma/\sigma_y$ and c is a constant. A value of c = 0.997 has been found to give a good representation for steel.

The tangent modulus, E_t, secant modulus, E_{sec}, and effective Possion's ratio, ν, can then be expressed as

$$E_t = \frac{d\sigma}{d\varepsilon} = \frac{E\,(1-\mu)^2}{(1-2c\mu + c\mu^2)} \;;\; E_{sec} = \frac{\sigma}{\varepsilon} \frac{E\,(1-\mu)}{(1-c\mu)} \quad \ldots\ldots(2)$$

$$\nu = (\nu_p - \nu_e)\, E_{sec}/E$$

Where ν_p and ν_e are the plastic and elastic Poisson's ratio respectively.

Using these non-linear elastic properties to write the internal stress-strain and moment-curvature relationships enables the internal virtual work to be evaluated. Where the stress is uniform over the strip this can be done by direct integration. However, where the stress varies across the width, for instance when residual stresses are included, it is more convenient to use numerical integration, dividing the strip into a number of substrips over which the stress can be considered constant.

These calculations lead to coefficients in the stiffness matrix which are non-linearly dependent on the level of applied stress. The solution for the critical stress, represented by the vanishing of the determinant of the overall stiffness matrix, cannot therefore be obtained using standard eigenvalue procedures, and instead an algorithm developed by Wittrick and Williams (5) is used.

1.1 Accuracy of the Method

To obtain an indication of the accuracy of the method, results were obtained for an individual simply supported plate in compression with a slenderness, β, equal to one and with the plate divided into two, three and four strips. The values of the critical stress σ_{cr} were found to be 0.874, 0.873 and 0.873 of the yield stress σ_y respectively compared with a value of 0.856 σ_y obtained from ref (6). Similar comparisons have been made with a number of published theoretical and experimental results for plates, stiffened panels and other cross-sectional shapes such as box columns and I-sections, some of which included residual stresses, and these all demonstrated good agreement except where other factors - notably initial imperfections - were significant.

On the basis of these comparisons the component plates of the panels considered in this paper have all been divided into four strips, and where residual stresses are included each strip has been divided into ten substrips for numerical intergration.

2. *SOME ILLUSTRATIVE RESULTS FOR STIFFENED PANELS*

The inelastic buckling behaviour of a number of simply supported panels with four longitudinal stiffeners subject to uniform compression has been studied using the finite strip method. Because of the large number of parameters this is not an exhaustive study but an attempt to highlight general patterns of behaviour and demonstrate the capability of the current approach. Typical panel details together with the assumed pattern of residual stress are shown in Fig. 2.

2.1 Slenderness of Plating between Stiffeners

One parameter which is clearly of importance in the buckling behaviour of stiffened panels is the slenderness of plating between stiffeners, β. The variation of critical stress with β is shown in Fig. 3 for two panels, each with flat stiffeners. The stiffener proportions - $h_s/t_s = 10$ - were selected such that local buckling of the stiffeners would not be significant.

The two curves, corresponding to different values of δ - the ratio of stiffener area to plate area - exhibit three regions. For slender plating (β greater than 1.4) the failure mode is essentially one of elastic instability, whilst for stocky plating (β less than 1.0) the dominant condition is material yielding. At intermediate values of β there is a transition zone in which failure involves significant interaction between yielding and instability.

In deriving these curves both long and short wavelength buckling have been considered (λ/b = 5.0 and 1.0 respectively). It is interesting to note that short wavelength buckling was always critical for the panel with larger stiffeners (δ = 1.0) and that for β less than 1 the critical stress was equal to the yield stress. However for the panel with smaller stiffeners (δ = 0.50) long wavelength buckling was critical for all but the most slender plating, and consequently even for small values of β the critical stress is less than the yield stress.

As expected the critical stresses for the panel with δ = 0.50 are less than for the case of δ = 1.0, the difference being greatest for intermediate values of β.

2.2 The Influence of the Ratio of Stiffener Height to Thickness

Figure 4 shows the variation of critical stress with stiffener size. For small stiffeners, long wavelength buckling is critical, and δ increases as h_s/t increases. Beyond a certain limit short wave-length buckling predominates and increasing stiffener size has little further effect. Note that for the more slender panel, the behaviour is elastic throughout the range considered. The maximum value of $h_s/t = 16$ is such that local buckling of the stiffener will not be important.

2.3 The Effect of Residual Stress

This study has been extended by including various levels of residual stress in the panel, and Fig. 5 shows the variation of σ_{cr}/σ_y against h_s/t for β = 0.88 and various levels of residual stress σ_r. Whilst the residual stress has little influence on long wavelength buckling, it is clearly significant for short wavelength buckling, causing a reduction in critical stress of between 0.9 σ_r and 1.0 σ_r. This pattern is repeated for more slender panels, but as β increases so the effect of residual stresses is reduced. For a panel with β = 1.414 for instance, short wavelength buckling stresses are reduced by between 0.75 σ_r and 0.9 σ_r.

For the purpose of design it may be of interest to know the value of h_s/t above which there is negligible increase in failure stress. This optimum stiffener size can be taken as the value at which short wave-length buckling becomes predominant and is clearly dependent on the slenderness of the plating and the level of residual stress. These parameters are plotted in Fig. 6 and it is interesting to note that the relationships are approximately linear.

2.4 *The Effect of Stiffener Shape*

Panels similar to those described above but with angle or tee stiffeners have been studied and demonstrate a similar pattern of behaviour. In order to compare the effectiveness of the stiffener shape, Fig. 7 shows a plot of σ_{cr}/σ_y against δ for a panel with $\beta = 0.88$ and different stiffener types. In all cases the thickness of the stiffener is the same as that of the plate and for a given value of δ the cross-sectional areas of the stiffeners are identical. The results confirm that for small stiffener sizes long wavelength buckling is critical and that in this mode flat stiffeners are more effective. For larger stiffener sizes, short wavelength buckling predominates and angle or tee stiffeners are marginally more effective in this region. A similar pattern of behaviour has been observed for panels with more slender plating.

3. *SIMPLIFIED METHODS OF LOCAL AND OVERALL BUCKLING ANALYSIS*

It is often assumed that in a local buckling mode the line junctions between stiffeners and plating remain straight and that in an overall buckling mode the stiffeners can be "smoothed out", thereby creating a flat plate with orthotropic properties. (In fact in the latter case it would be more correct to model the panel as an equivalent unsymmetrically layered plate. Unfortunately this introduces coupling between in-plane and out-of-plane effects and is therefore normally ignored although a finite strip method using a general anisotropic plate theory has been formulated by Plank (7) which enables the coupling effect to be considered). These simplified approaches are undoubtedly attractive, and as an indication of the effect of such simplifications on the theoretical buckling behaviour of a panel, Fig. 8 shows a plot of critical stress against wavelength comparing the results obtained using the finite strip method modelling the panel (a) as the full cross-section; (b) as an equivalent orthotropic plate (without coupling); (c) as an equivalent orthotropic plate (including coupling effects); (d) with line junctions between stiffeners and plating held straight. These demonstrate that in the overall buckling region, treating the panel as an equivalent orthotropic plate results in an overestimate of buckling stress of about 13% when coupling is included and 22% when coupling is ignored. In the local buckling region, restraining (in position) the line junctions between plating and stiffeners has negligible effect on the critical load provided the rotational stiffness of the stiffeners is included. Simpler idealisations demonstrate far less accuracy.

The curves shown are entirely within the elastic region, but where conditions are such that plastic behaviour is significant the discrepancies between the approximate solutions and the results of the finite strip method for the full panel are reduced.

4. *CONCLUSIONS*

A brief description has been given of a modified finite strip method which allows non linear material behaviour to be included in the stability analysis of prismatic structures. A number of illustrative results demonstrate the capability of the method and give an indication of the influence of certain parameters on the buckling behaviour of stiffened panels.

References

1. Cheung, Y.K. "The Finite Strip Method in Structural Analysis" Pergammon Press Ltd., London, 1976.
2. Plank, R.J. and Wittrick, W.H. "Buckling under Combined Loading of Thin Flat-Walled Structures by a Complex Finite Strip Method", International Journal for Numerical Methods in Engineering, Vol.8, No.2, pp 323-339, 1974.
3. Mahmoud, N.S. "Inelastic Stability of Plate Structures using the Finite Strip Method", thesis presented to the University of Sheffield in 1981 in partial fulfilment of the requirements for the degree of Doctor of Philosophy.
4. Ylinen, A. "Lateral Buckling of an I-beam in Pure Bending beyond the Limit of Proportionality", Proceedings of the Second Conference on Dimensioning and Strength Calculations, Hungarian Acadamy of Sciences, Budapest, pp 157-167, 1965.
5. Wittrick, W.H. and Williams, F.W. "An Algorithm for Computing Critical Buckling Loads of Elastic Structures", Journal of Structural Mechanics, Vol.1, No.4, pp 497-518, 1973.
6. Scheer, J. and Nolke, H. "The Background to the Future German Plate Buckling Design Rules", An International Symposium on Steel Plated Structures, Imperial College, London, 1976.
7. Plank, R.J. "The Initial Buckling of Thin Walled Structures under Combined Loadings", thesis presented to the University of Birmingham in 1973 in partial fulfilment of the requirements for the degree of Doctor of Philosophy.

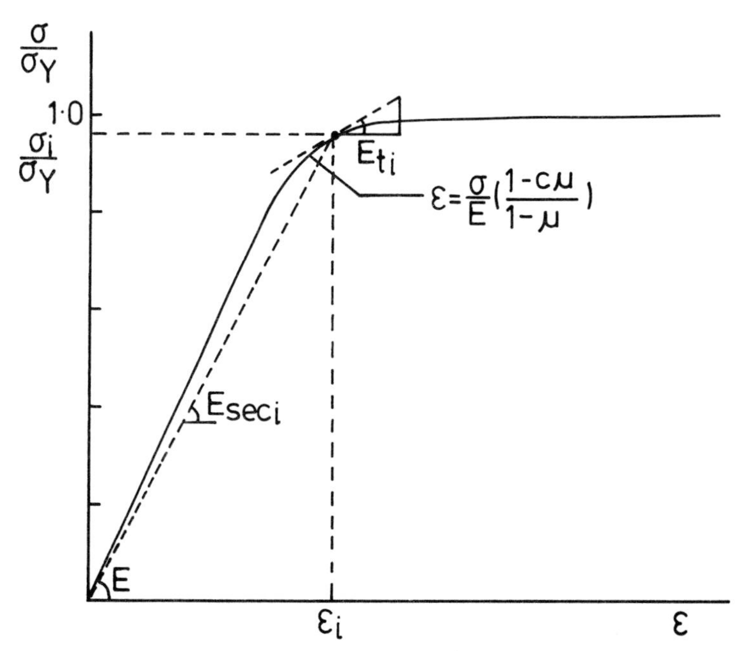

Fig 1 Idealised stress-strain curve

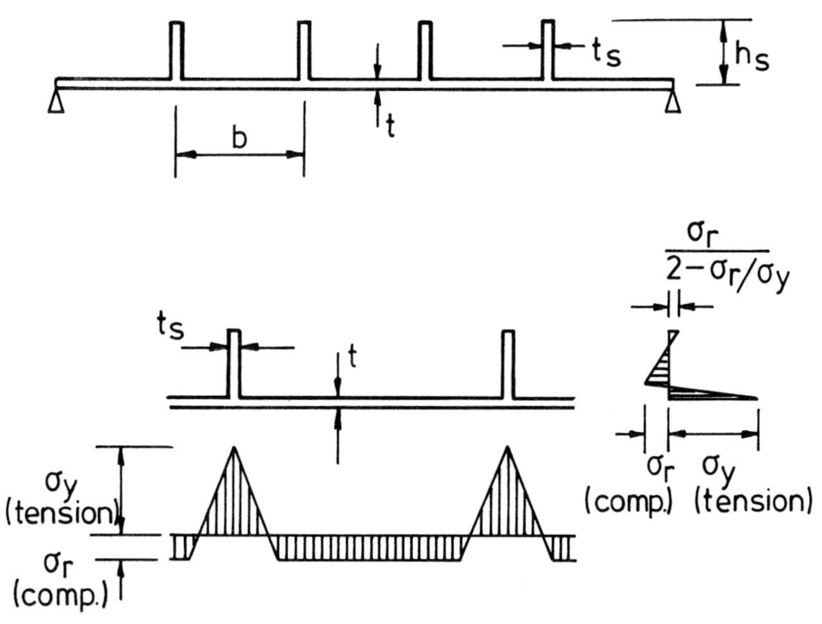

Fig 2 Typical panel details including assumed pattern of residual stress

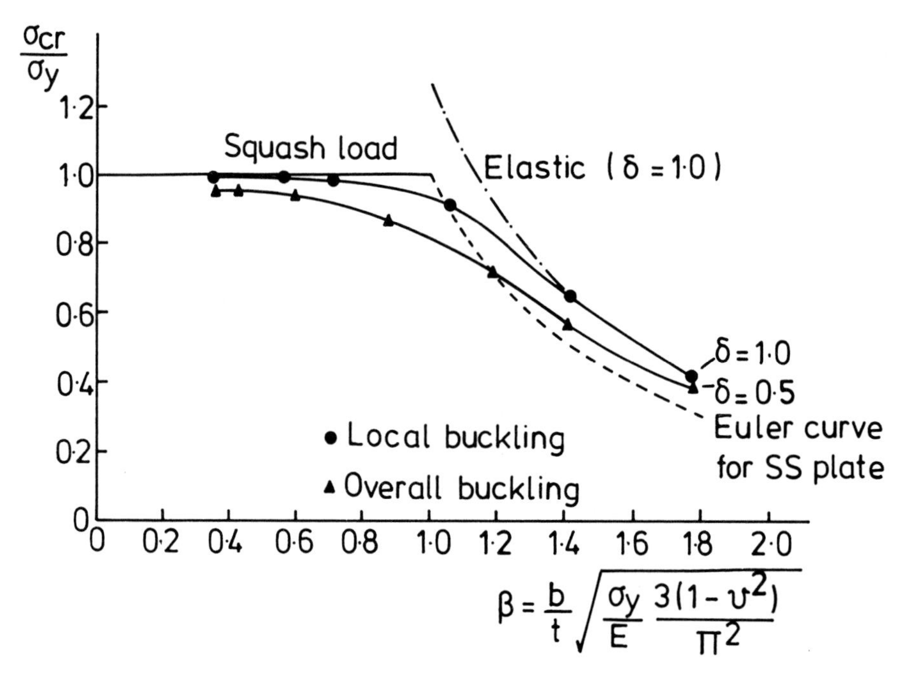

Fig 3 Variation of critical stress with slenderness of plating

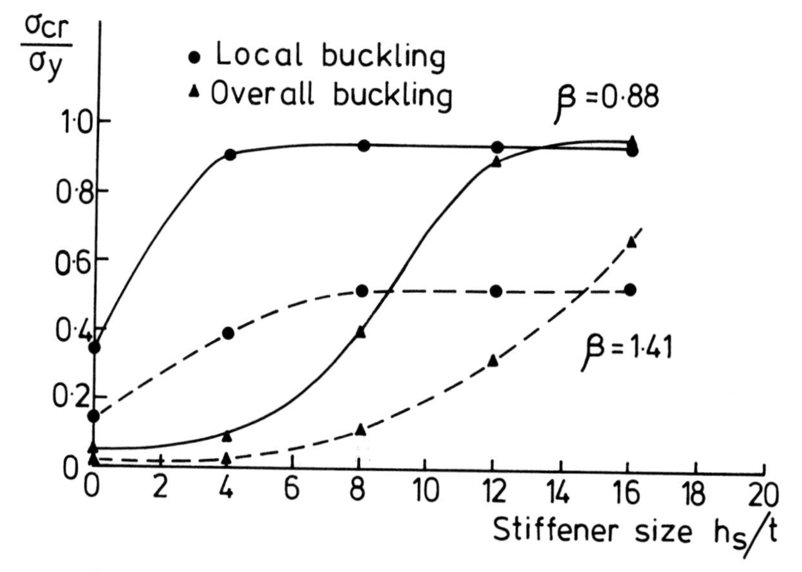

Fig 4 Variation of critical stress with stiffener size

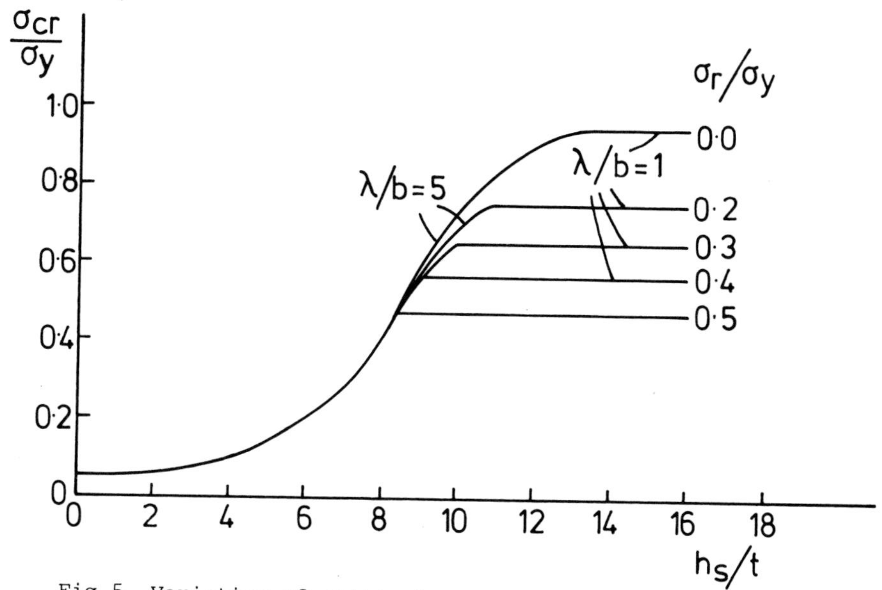

Fig 5 Variation of critical stress with stiffener size for different levels of residual stress (β = 0.88)

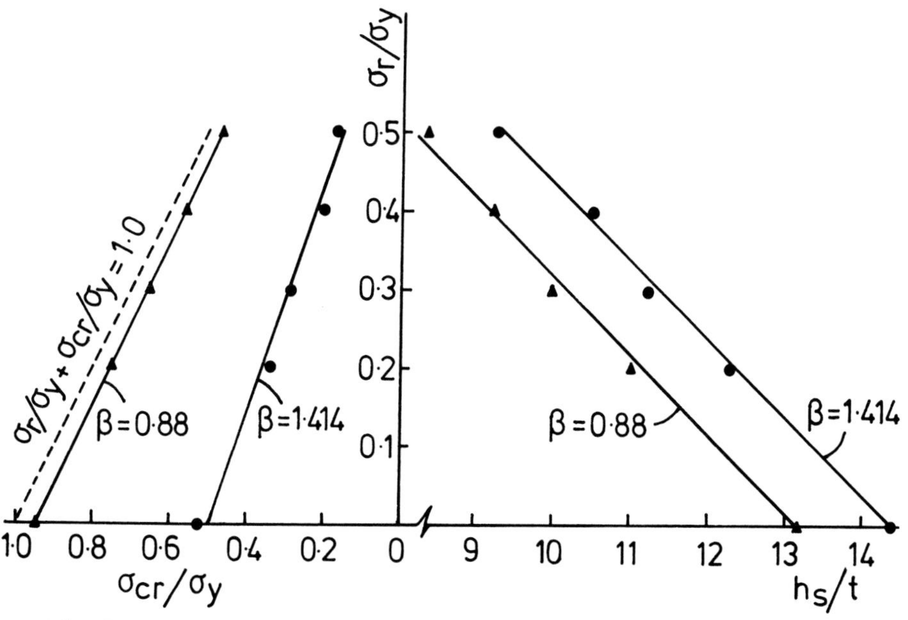

Fig 6 Relationship between optimum stiffener size, residual stress and critical stress

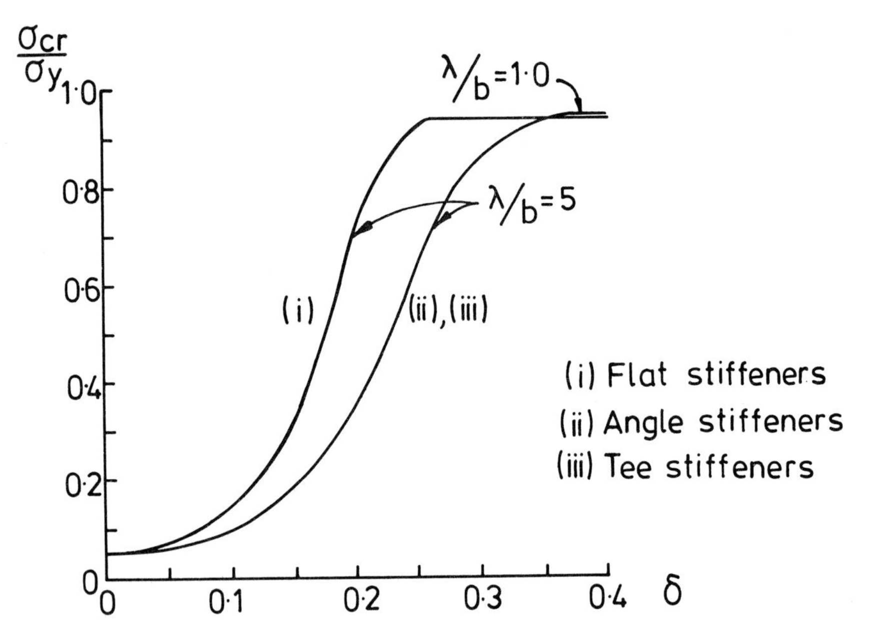

Fig 7 Variation of critical stress with stiffener size for panels with flat, angle and tee stiffeners

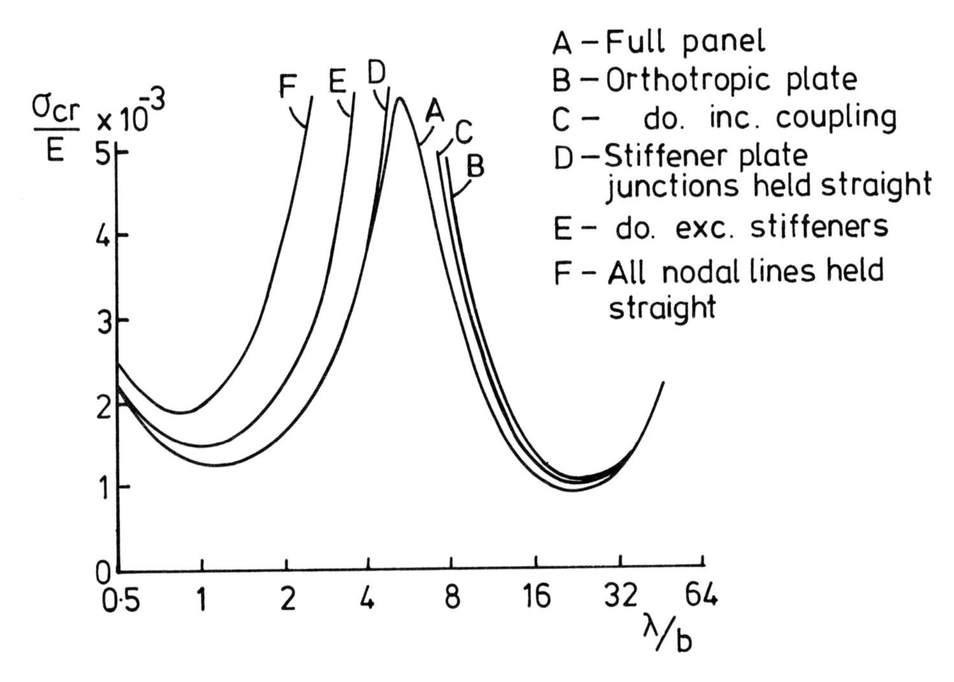

Fig 8 Variation of critical stress with wavelength for different panel idealisations
(h_s/t = 25; b/t = 50; t_s = t)

INTERACTION BETWEEN SHEAR LAG AND STIFFENER-INDUCED BUCKLING IN STEEL BOX GIRDERS

A R G Lamas

Department of Civil Engineering, Technical University of Lisbon, Portugal

P A Frieze

Department of Naval Architecture and Ocean Engineering, University of Glasgow

P J Dowling

Department of Civil Engineering, Imperial College, London

Synopsis

Details of a point load test on a wide stiffened steel box girder failing by stiffener-induced buckling of the compression flange are presented. The model's geometry, material properties, initial imperfections and welding residual strains are reported together with a detailed account of the complex behaviour under load.

The qualitative results of a detailed finite element analysis of the model are also described. They lend support to the interpretation of behaviour derived from the experimental results and earlier numerical studies.

Redistribution in the model is compared with that occurring in a similar model failing by plate-induced stiffened flange buckling. An appropriate note of caution is given concerning allowance for redistribution in the limit state design of girders liable to fail in the stiffener-induced mode.

Introduction

The effect of shear lag on the performance of stiffened compression flanges of box and multi-web girders has been previously examined for behaviour in the elastic and the elasto-plastic range. The latter demonstrates considerable complexity not only because it involves interaction between local and interframe-buckling and shear lag but also because redistribution of stresses takes place due both to yielding and to load-shedding arising from buckling. To date, most studies in this area have concentrated on interframe-buckling precipitated by local plate instability. The relatively gentle unloading that accompanies plate-induced failure has allowed advantage to be taken of redistribution of shear lag effects in ultimate limit state design. However, stiffener-induced interframe-buckling is far more rapid and thus may not suffer such an opportunity. This paper is concerned with a study of stiffener-induced failure of a wide compression flange under pronounced shear lag conditions as a first step to examine how redistribution may occur in such circumstances.

1. BACKGROUND

As part of a major investigation into box girder behaviour, a series of tests was conducted at Imperial College between 1971 and 1976 which involved models of similar geometry being tested under point load and pure moment conditions (1). Apart from enabling comparisons to be made between the behaviour of stiffened panels under uniform stress and longitudinally varying stress conditions, these pairs of tests also provided the opportunity to compare the behaviour of similar panels in the presence and absence of shear lag and therefore to possibly quantify the extent to which redistribution occurs in point loaded models.

Such comparisons are straightforward if the failure mode of each pair of tests is the same. This occurred in two of the companion tests while in the third, flange buckling occurred in opposite directions. In particular, Model 9* a three-bay girder that was tested under point load conditions failed by plate-induced interframe-buckling while its companion, Model 10, loaded in uniform bending suffered stiffener-induced interframe-buckling (2). Both models were stiffened by flat-bar elements.

In order to extend the study of shear lag effects to the case of stiffener-induced failure under point load conditions, a further test, Model 12, was purposely designed (3). It was of similar overall proportions to Models 9 and 10, but in order to ensure the required collapse mode would occur the following modifications to the compression flange were introduced:

(a) the slenderness of the flat-bar stiffeners was increased from 9 to 24 thereby increasing considerably the likelihood of local torsional buckling occurring,

(b) the slenderness of the plate panels was reduced from 48 to 24 to inhibit plate buckling, and

(c) an initial imperfection was built into the centre stiffened panel in the direction opposite the outstands and of a magnitude greater than the effective eccentricity of the stress gradient acting over the height of the stiffeners arising from overall bending of the girder.

In the event, failure occurred in the required mode but in an entirely unexpected sequence. Flexural-torsional buckling of the outstands initiated failure but not in the most highly-stressed region. It occurred first in the stiffeners located at the quarter-width at mid-span, then shortly after in the stiffeners at the ends of the centre panel adjacent to the webs: in this case the corresponding movement of the panel was inwards. Overall buckling of the centre panel occurred shortly after with the full complement of stiffeners deflecting in the direction of the initial buckling: the edge stiffeners thus had to undergo a mode change.

Interpretation of the test data suggested a complex interaction between shear lag and buckling occurred, and in an effort to provide insight into this, a sophisticated numerical procedure was developed (4) which

---

*The actual test numbering sequence is being retained in order to avoid confusion with the original reports.

modelled all aspects of the problem except local buckling of the stiffeners. The procedure enabled interaction between shear lag and particularly plate-induced buckling of the compression flange to be studied in detail (5) and to explain the process of redistribution which occurred in the point loaded models (4). The importance of web shear strength as a factor possibly limiting strength was also highlighted by this work.

The procedure was used to examine Model 12 and the results lent support to the interpretation drawn on the basis of the experimental evidence (4) in spite of the inability to model the full stiffener complement. Since then a finite element system has become available which can examine this problem in greater depth. It has been used for this purpose, and the results of the analysis along with details of Model 12, its testing and behaviour under load, and the interpretation of its complex response are presented in this paper.

2. MODEL DETAILS AND TEST PROCEDURE

2.1 Geometry and Material Properties

The model was 4.875 m long with an effective span of 4.725 m subdivided into three equal length bays by two internal ring-stiffeners and two end cross-frames. The nominal cross-sectional dimensions are summarised in Fig. 1. The measured dimensions of the main elements are listed in Table 1. The shear lag effective width calculated in accordance with (6) is 0.405.

The webs were proportioned so as to prevent their buckling. They were also stiffened at mid-span by tapered stiffeners for connection via high yield pins to restraining units bolted to the laboratory floor (see Fig. 1). These loading stiffeners were connected by an angle welded to the tension flange. However, no transverse reinforcement of the compression flange was used at mid-span to allow unrestricted buckling of the central panel.

The material specified was BS4360, Grade 43A. Tensile tests were conducted on coupons from all elements. The results are summarised in Table 1 from which it can be seen that the yield stress of all components is in excess of the guaranteed minimum with the exception of that of the stiffener material.

2.2 Initial Imperfections

The shape of the compression flange was carefully measured using a bank of deflection transducers the datum for which was an accurately surveyed beam. Deflections were measured along each stiffener and at the nineteen transverse locations. The longitudinal profiles are presented in Fig. 2. They demonstrate clearly the initial outward distortion built into the centre bay of the compression flange. The average outward bow of all the centre bay stiffeners excluding the extreme ones (B and T) was 2.3 mm. Deducting the deflection equivalent to the overall bending stress gradient ($\simeq$ 1.3 mm inwards) the average effective initial distortion at the mid-span cross-section was approximately equal to the panel length ÷ 1590.

Measurement of the stiffener tip profiles in the unloaded, and loaded, state was attempted using photogrammetry but, unfortunately, no useful quantitative interpretation of the recorded data could be derived.

2.3 Residual Strains

Welding strains in the plate and stiffeners of the compression flange were recorded at four stages during fabrication using a demountable strain gauge. Those relevant to the present discussion were the strains recorded over the depth of the stiffeners. The two outer stiffeners on both sides of the flange were found to be in a state of initial tension probably due to the final welding-up of the webs and flange, while the remainder were generally in compression: details can be found in (3).

2.4 Loading and Test Procedure

The model was loaded from underneath by jacks at one end, resisted at the centre by the floor mounted restraining units, and supported at the other end by rocker-roller arrangements. Four jacks were used, one under each web of 1000 kN capacity and two of 500 kN capacity positioned under the end cross-girder, the load being transferred to the webs by angle diagonal members. These end-braces failed during the final test and had to be reinforced. A photograph of the model after testing (Fig. 3) shows the loading arrangements, as well as the post-buckled shape of the compression flange.

Five tests involving sequences of loading and unloading were conducted and always by incrementing the load very slowly. A preliminary one, up to about 10% of the expected failure load, was performed in order to test the equipment. This was followed by a two-stage 'elastic' test (Tests 2 and 3) and then the 'failure' test (Test 4) which had to be interrupted for reinforcement of the end-braces. The second stage of the failure test (Test 5) finally brought about collapse.

3. BEHAVIOUR UNDER LOAD

The overall response is summarised by the load-deflection plots presented in Fig. 4. In the figure both peak and sustained load paths are indicated: important loading stages are identified by letters. The peak path is that which the model follows immediately upon the application of load. The sustained loads reflect the capacity of the model after the effects of yielding have fully developed. Some discontinuities in the load-deflection plots represent external interference, for example, a fault in the servo-mechanism controlling the jacks is indicated by a-a' and buckling of the end-braces occurred at d. Complete details of the tests were presented in (3) so only those aspects relevant to the mode of collapse of the compression flange are described here.

The interest in Tests 2 and 3 relates to the excellent correlation obtained between the experimental results for overall deflection and transverse shear lag stress distribution and those calculated using the rules and formulae proposed in (6): see 'elastic slope', Fig. 4, and (4) respectively. A small permanent set was recorded after Test 2 probably due to relaxation of the strains in the welds and heat-affected zones situated in the tension region of the girder. In Test 3 yielding occurred in the compression flange in the most highly-stressed locations, viz. the plating and the stiffeners adjacent to the webs near mid-span. The strains in the plate generally continued to follow the shear lag distribution. The recorded flange deflections showed the stiffeners near the webs deflecting inwards in the central bay and stiffeners F,G,H and I and, to a lesser extent, stiffeners M, N,O and P deforming in the opposite direction (see Fig. 2 for stiffener identification).

The first phase of Test 4 retraced the unloading path of Test 3. Beyond the previous maximum, yielding continued until at a central point peak load of 4426 kN (a, Fig. 4), stiffeners F,G and H situated to one side of the box at about the quarter-width buckled in a local flexural-torsional mode. This was immediately followed by the failure of the jack servo-control mechanism as signalled by the drop a-a' in the load-deflection curve. After reloading to 4504 kN a similar failure of stiffeners N,O and P located at the other quarter-width occurred (b, Fig. 4).

These failures occurred in the context of outward deflections of the stiffeners located at the quarter-widths while those near the webs showed an increased tendency to deflect inwards following the curvature of the webs (Fig. 5). With further loading local buckles were detected in stiffeners B and T but at the ends of the central bay near the connections to the transverse girders. This was followed by similar failures of stiffeners C,D,Q and S. The load at which this occurred was not accurately recorded but in photographs taken at 4585 kN (c, Fig. 4) this form of failure in stiffeners B and T had already occurred. By now stiffeners E and I on one side of the longitudinal central line and M on the other had also failed by tripping near mid-span. Soon afterwards, at approximately 4800 kN, the end-bracing where the jacking was applied failed and the testing was stopped. At that stage the pattern of buckles in the stiffener outstands had not altered and stiffeners J,K and L appeared to remain straight although R showed a pronounced overall lateral deflection. After unloading, permanent deformations were visible in all the stiffeners that had buckled at mid-span and in the stiffeners adjacent to the webs (B and T).

Final collapse was achieved in the next loading cycle (Test 5) after appropriately reinforcing the end-bracing frames. Ultimate load was marked by an overall outward buckling of the centre bay which involved a reversal of mode of the stiffeners located near the webs. Figure 3, a photograph of the model after testing clearly shows the mode of failure. Very large strains near the webs at mid-span eventually produced localised buckling in those regions.

4. *INTERPRETATION OF BEHAVIOUR*

4.1 Sequence of Buckling

The particular aspect of behaviour of this model under load for which an explanation is sought is the initiation of compression flange buckling by stiffeners located at the quarter-width rather than those positioned in the most highly-stressed region. This complex pattern appears to have involved at least four different structural phenomena, viz.

(a) shear lag,

(b) the column P-Δ effect,

(c) a 'bending' lag, and

(d) anticlastic curvature.

Although a rather obvious contributor, shear lag needs to be mentioned because it involves two separate actions. Firstly, there is the well known variation in average stress over the cross-section. Secondly, there is the bending stress distribution through the depth of the stiffeners arising from overall girder curvature that demonstrates a variation in magnitude across the flange which parallels that of the shear lag direct stress distribution. For any particular girder there will be a fixed relationship between the bending stress level and the direct stress level irrespective of the actual magnitude of the stress.

In the context of the present girder with outward distortions in the centre panel, and inward ones in the end panels, the P-Δ effect at mid-span should have produced tendencies for most of the stiffeners to deflect outwards especially since all of them except one edge stiffener had initial bows in excess of that equivalent to the opposing bending stress gradient. However, because of the column slenderness and of the smallness of the effective initial imperfections involved, this secondary stress would not necessarily become significant until the average stress was close to that corresponding to interframe failure.

Both the above phenomena had been expected to play significant roles in the response of this girder. The involvement of a 'bending' lag and anticlastic curvature had not been anticipated.

Bending lag is here used as an analogy to shear lag implying that the tendency for the stiffened panel to follow the vertical bending of the web gradually diminishes as the transverse distance from the web increases.

Anticlastic curvature arose because the longitudinal curvature imposed on the edge of the flange had the effect, possibly enhanced by the thickness of the web, of causing the adjacent flange plating to deflect further in the same direction. This was initially dominant enough to oppose the preferred direction of movement of the edge stiffeners arising from the P-Δ effect acting on the initial deformations.

The net result of these actions was the combination at the quarter-widths of an imposed inwards bending action which reduced away from the webs and a compressive strain (albeit also reduced compared to the edge due to shear lag) which was sufficient to destabilise the stiffeners locally and thus initiate collapse of the flange in the direction in which it was predisposed by virtue of the imposed outward initial geometric imperfections.

4.2 Redistribution

Studies on isolated compressed stiffened panels have shown stiffener-induced buckling occurs far more rapidly than plate-induced buckling (7). The overall load-deflection response of this girder, however, was not dissimilar to that of Model 9 which failed in the latter mode suggesting possibly some form of redistribution may have taken place. In Model 9 local buckling occurred at a load equal to 72% of its ultimate load. In Model 12, it took place at 89% of maximum load which suggests that if redistribution does occur it is possibly of little benefit for the stiffener-induced mode of failure.

Clearly further work is required before this is clarified. In the meantime it would seem prudent to be cautious in using the results of stiffener-induced compression flange failures of girders under shear lag conditions as the basis for limit state design.

5. *NUMERICAL SIMULATION*

Numerical reproduction of a box girder with characteristics generally similar to those of Model 12 was attempted using a finite difference based technique (4). Although at the time the program only allowed the modelling of a few stiffeners on the compression flange, it was possible to simulate the behaviour of a flange similar to that of Model 12, that is, where the most highly stressed stiffener tips at mid-span were located away from the webs and the centre.

Recently (8) the possibility of accurately modelling the behaviour of this model was explored using a finite element program developed at Imperial College (9). For this study the plates were modelled using 48 d.o.f. isoparametric elements suitable for smooth/discontinuous shell surfaces, and for the stiffeners 21 d.o.f. curved 3-noded stiffener/stand alone elements were employed. Assuming symmetry it was only necessary to analyse one quarter of the girder. The finite element mesh used is indicated in Fig. 6. On the compression flange, the transverse subdivision coincided with the stiffener positions. A more dense distribution of elements was used longitudinally in the central panel than in the end panels. On the web and tension flanges only two elements were considered transversely. The cross-frames and loading stiffeners were modelled considering their real geometry.

The assumed initial shape of the compression flange approximately followed the measured transverse and longitudinal profiles (Fig. 2). Their amplitudes, however, were slightly enhanced to ensure that the anticlastic curvature effect of the web was effectively counteracted by the P-Δ effect.

The material properties where reproduced but in order to account for the reduction in stiffener axial strength due to residual strains, the yield stress of the stiffener material was lowered in proportion to the average measured initial strains, i.e. approximately 200×10^{-6}E. Step-by-step loading was applied uniformly over the web nodes located at mid-span until maximum resistance was reached corresponding to lack of equilibrium in the solution process.

The overall load-deflection response compared remarkably well with the Test 4 curve shown in Fig. 4.

In Fig. 7 the deflected shape of the compression flange at maximum load is illustrated. The close similarity between this and that obtained experimentally as shown in Fig. 5 is clear. The numerical evolution of the distorted shape differed slightly from that exhibited by the model where the flange and stiffeners adjacent to the webs were forced to deflect vertically further than the webs: this was due to the local anticlastic curvature effect mentioned above. In the numerical simulation the P-Δ effect was enhanced by considering larger imperfection amplitudes so that this local effect was not reproduced as in the real model.

The objective of the numerical simulation was to create the conditions where local compression failure of the stiffeners could occur at positions remote from both the webs and the centre due to the combination of the effects described in section 4.1. This appeared to be completely achieved thereby providing full support for the interpretation of the complex behaviour of this model derived on the basis of the experimental and earlier numerical results, as described above.

The numerical formulation of stiffeners as beam elements rotating laterally with respect to their connection with the flange plate did not incorporate localised tripping as a possible failure mode. However, the shape followed by the flange was a sufficient indication that the compressive strain at the stiffener tips would have attained its critical value first at the stiffeners located at the quarter-widths.

6. *CONCLUSIONS*

Details of a three-bay box girder purposely built to precipitate flange stiffener-induced interframe-buckling under pronounced shear lag conditions are reported. The centre compression flange panel was pre-dished to ensure the desired buckling mode would occur: the effective magnitude of the bow was panel length ÷ 1590. The compressive residual strains in the stiffeners resulting from welding and the pre-dishing averaged some 200 μstrain: the edge stiffeners were initially in tension.

Failure of the girder occurred in the desired mode but was complicated by anticlastic behaviour of the flange adjacent to the webs and a form of bending lag of the stiffened flange relative to the webs. The effects of these actions were superimposed upon those arising from shear lag and compression acting on the initial centre panel bow, i.e. the P-Δ effect, and precipitated local torsional buckling of the stiffeners located at the mid-span quarter-widths of the compression flange.

This form of local buckling occurred at a much greater proportion of the ultimate load than that observed in a similar box girder failing by plate-induced stiffened flange buckling. It was concluded that in the present case redistribution could not necessarily be relied upon to provide the safeguards required in design that seems possible when failure occurs in the alternative mode.

Qualitative results of a detailed finite element analysis of the girder are presented. They confirm that local buckling of the stiffeners would have initiated at the flange quarter-widths. The close correlation shown between the experimental and numerical results indicates further numerical work could provide the insight necessary to enable redistribution in these girders to be quantified and therefore their ultimate limit state design.

Acknowledgements

The authors wish to thank Mr U Troebe and Mr D Bates for their help in the use of the elements and the system adopted for the reported numerical simulation. Thanks are also due to Mr N Argelides for assisting in the preparation of data for this simulation, and to Mr V Ferreira for handling the graphical presentation of the results.

References

1. Dowling, et al. 'Experimental and predicted collapse behaviour of rectangular steel box girders'. Proc. Int. Conf. on Steel Box Girders, Instn Civ. Engrs, London, 1973.

2. Dowling, et al. 'The effect of shear lag on the ultimate strength of box girders', in Steel Plated Structures, ed. Dowling, et al, Crosby Lockwood Staples, London, 108-141, 1977.

3. Frieze, P.A. and Dowling, P.J. 'Steel box girders - Model 12 - Progress report 2, Testing of a wide girder with slender compression flange stiffeners under pronounced shear lag conditions'. Eng. Struct. Labs, Dept of Civ. Eng., Imperial College, London. CESLIC Report No. BG49, Dec. 1979.

4. Lamas, A.R.G. 'Influence of shear lag on the collapse of wide-flange girders'. PhD Thesis, University of London, 1979.

5. Lamas, A.R.G. and Dowling, P.J. 'Effect of shear lag on the inelastic buckling behaviour of thin-walled structures', in Thin-Walled Structures, Applied Science Pub., London, 1979.

6. Moffatt, K.R. and Dowling, P.J. 'Shear lag in steel box girder bridges'. The Structural Engineer, vol. 53, no. 10, Oct. 1975.

7. Horne, M.R. and Narayanan, R. 'Ultimate capacity of longitudinally stiffened plates used in box girders'. Proc. Instn Civ. Engrs, Part 2, vol. 61, 253-280, June 1976.

8. Troebe, U. 'Elasto-plastic critical loads of shell structures'. PhD Thesis, University of London (to be presented).

9. Bates, D.N. 'Non-linear elasto-plastic analysis of stiffened plates and shells by finite elements'. PhD Thesis, University of London (to be presented).

Table 1

Component Dimensions and Material Properties

Component	Nominal Size mm	Measured mm	Yield Stress N/mm^2	Elastic Modulus N/mm^2
CF plate	5	4.902	396.3	202500
stiffener	80 × 3.5 Fl	80.64 × 3.384 Fl	238.9	202050
TF plate	6	6.147	339.8	211590
W plate	12.5	12.014	328.8	201570

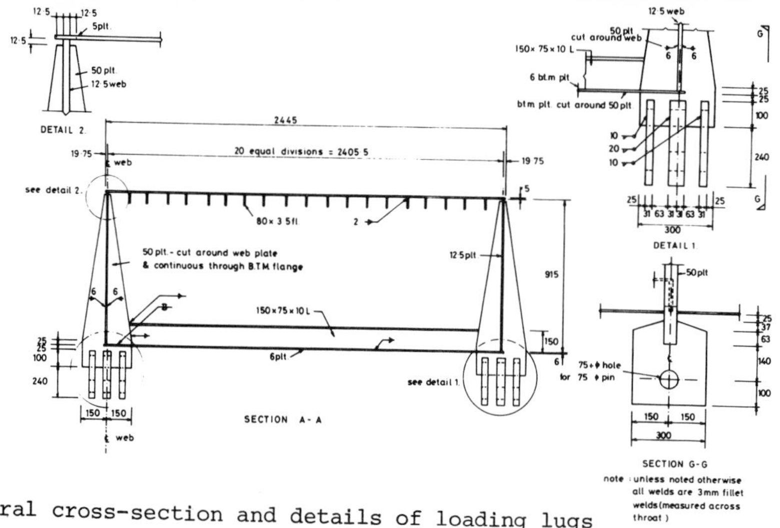

Fig 1 Central cross-section and details of loading lugs

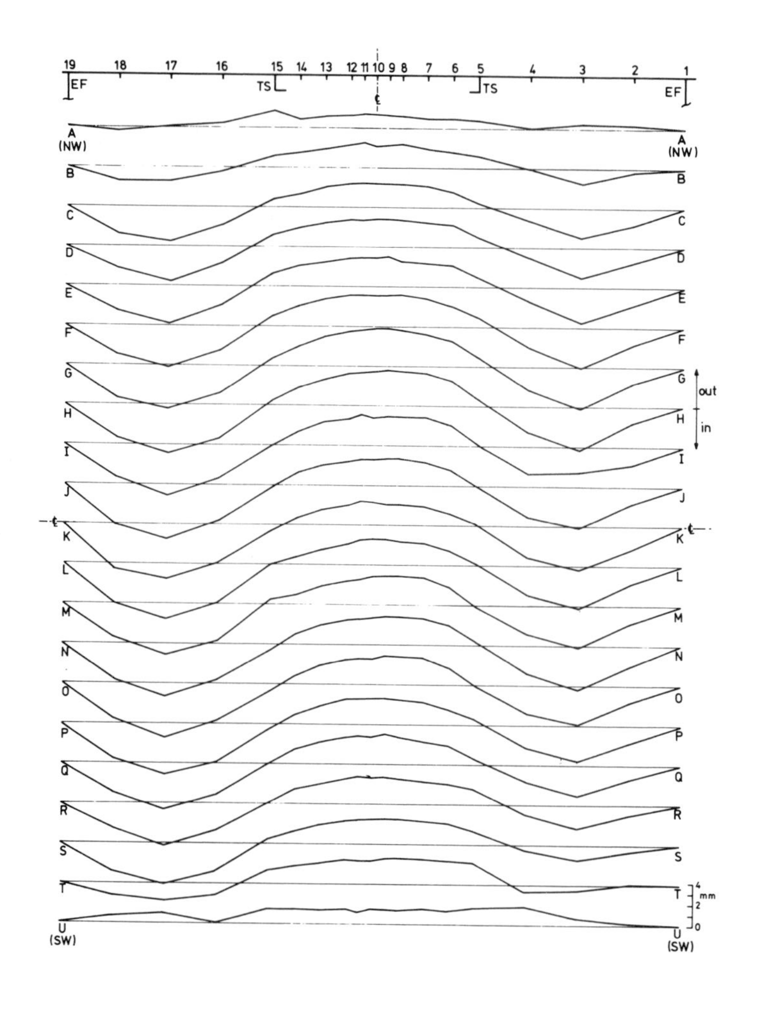

Fig 2 Compression flange initial deformations: Profiles along webs and longitudinal stiffeners

Fig 3 Model in rig after testing

The transducer frame can be seen on top of the model with the jacks and floor mounted restraining units below

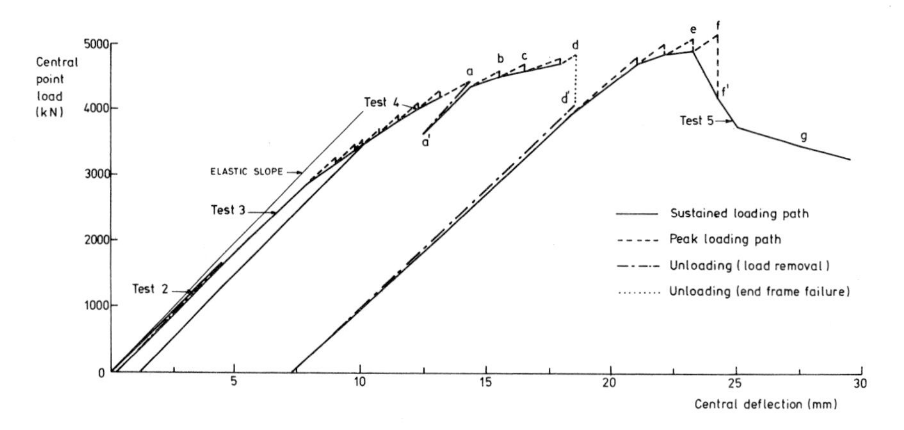

Fig 4 Tests 2-5 Overall load-deflection curve

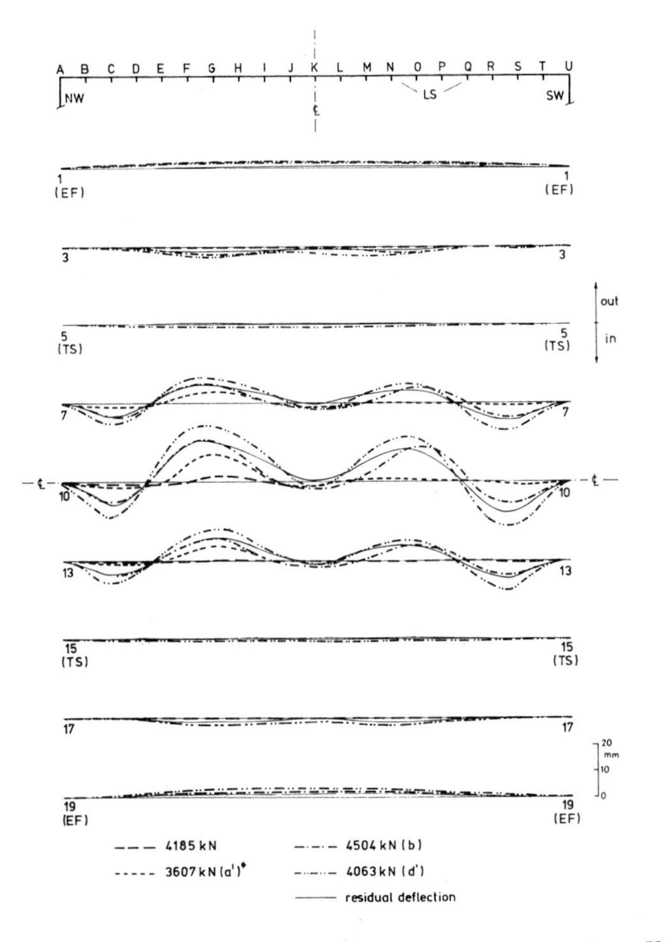

Fig 5 Test 4 Transverse profiles of compression flange

Where not shown, residual deflections are negligible

Initial and previous residual deflections are not included

* see fig. 4 for corresponding points on load-deflecti

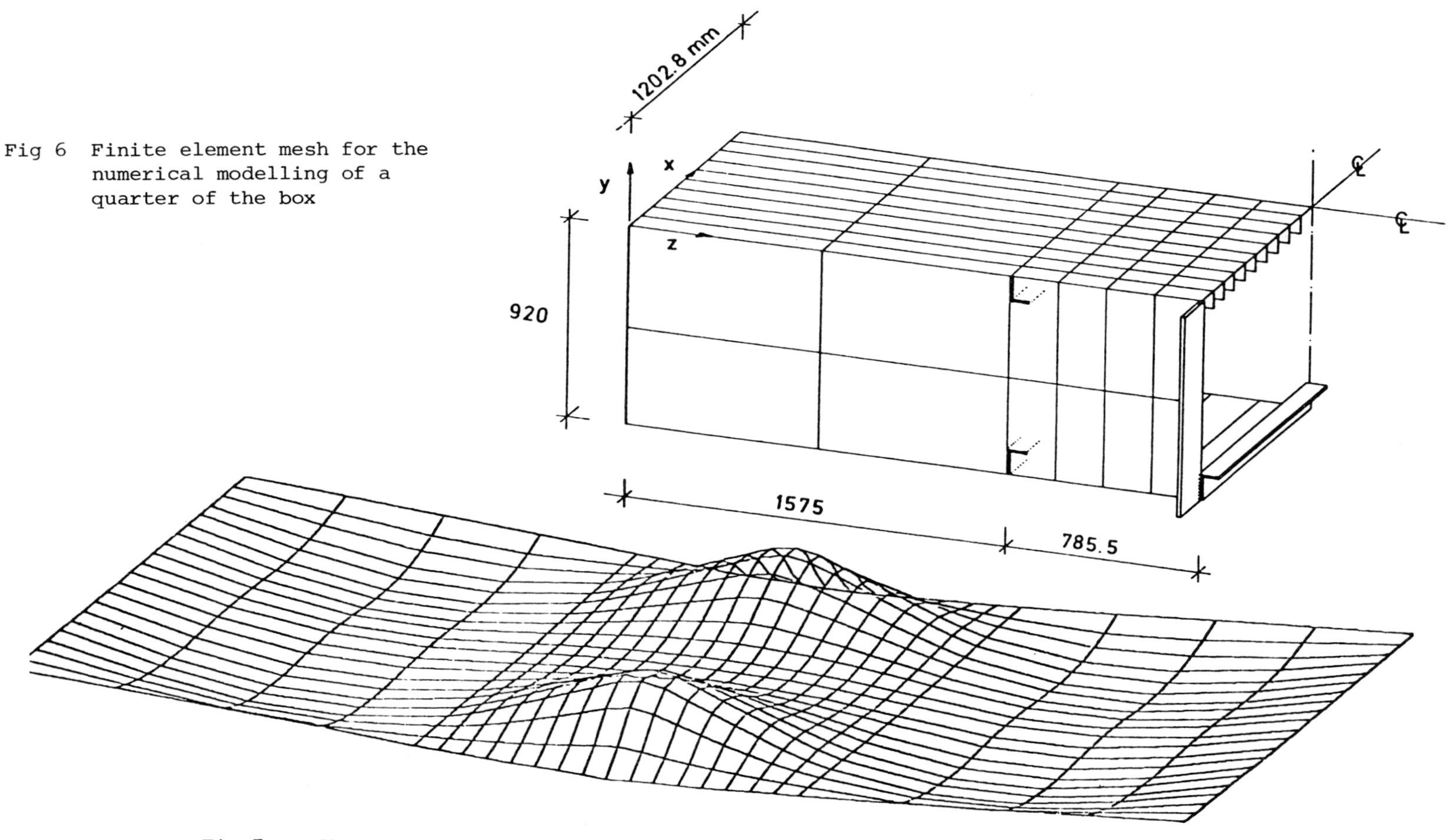

Fig 6 Finite element mesh for the numerical modelling of a quarter of the box

Fig 7 Deflected shape of compression flange at ultimate load obtained from the finite element analysis

INELASTIC POST-BUCKLING BEHAVIOUR OF IMPERFECT LONGITUDINALLY STIFFENED PANELS UNDER AXIAL LOAD

R.S. Puthli

Institute TNO for Building Materials and Building Structures (IBBC) of the Netherlands Organisation for Applied Scientific Research (TNO)

Synopsis

The two classes of instability phenomena, namely bifurcation and snap-through buckling, that occur with elastic and inelastic stiffened steel structures under axial load, are statically analysed for specific cases with the finite element method. The parameters chosen are such as to exhibit the extremely non-linear explosive buckling behaviour that can occur with stiffened plates. Difficulties encountered during the solution of these problems and methods of overcoming them are discussed. The solution of both classes of problems include geometrical as well as material non-linearity, implying a step-wise incremental non-linear analysis for both cases. The methods employed for these severe problems could be used in general for inelastic post-buckling analysis of stiffened plated structures.

Notation

A	cross sectional area
E	Young's modulus
g	out of balance force vector
L	beam length between supports
N_i	desired no. of iterations in prescribed step i
N_{i-1}	actual no. of iterations in previous prescribed step $i-1$
P	total axial load in structure
P_y	squash load = cross-sectional area x yield stress
P_{cr}	elastic critical buckling load of total structure
Δp_i	incremental deflection vector
Δp_i^T	transpose of Δp_i
R	total external force vector = prescribed total loads + reactions

r radius of gyration of beam (major axis)

u axial deformation

w_1, w_2 bending deformations at centre of spans (major axis)

ν Poisson's ratio

σ mean stress in structure

σ_y yield stress

σ_{cr} elastic critical buckling stress

$\sigma_{cr(n.l)}$ non-linear bifurcation stress

$\Delta \ell_i$ arc length of solution path at increment i

Introduction

A limit point type of singularity is usually determined when analysing for collapse or ultimate load in a structure, where the load attains a maximum in the load-deflection relationship. A non-linear analysis normally achieves this load with a monotonically increasing deformation on an initially imperfect structure, so that the buckling and/or yield characteristics of the structure influence the collapse load. Two other classes of singularity, or instability phenomena that can also occur in axially loaded stiffened plates, as with some other slender structures, are: the loss of stability (singularity) at the bifurcation point where two or more load deflection paths intersect, giving a non-unique solution; and the loss of stability due to snap-through buckling. For the latter type of instability, once the structure reaches a limit load (prescribed load) or limit displacement (prescribed displacement), snapping occurs as a dynamic process until a new stable equilibrium configuration is achieved. It is then possible to obtain further limit points.

Stiffened plates are often designed with their squash and buckling loads close to each other. Such structures, together with the difficulties discusses in the choice of initial imperfections, display highly non-linear behaviour, often within confined steps, incorporating explosive snap-through situations, both experimentally and numerically. Explosive or violent instability here refers to cases where snap behaviour (rapid unloading) is also obtained with the prescribed displacement.

In numerical studies, the imperfection patterns that are chosen can either be a combination of initial imperfections observed in the elements of a structure together with deformations due to the welding process, or in the form of a buckling mode (eigenvectors) of the lowest eigenvalue obtained from a linear elastic critical load analysis. The latter imperfection may not be in sympathy with the final collapse mode if the lowest eigenvalues lie close to each other. Also, earlier plastic flow in some elements may give a collapse mode quite different from the elastic critical buckling mode. In both these cases, where the post-buckling (or collapse) mode is different from the imperfection mode, the imperfections, which initially grow with the applied load, will begin to snap through into the preferred shape before the first limit point load or collapse load is reached. Although these two levels are often the same, this is not necessarily the case, especially for very slender structures. With numerical studies, when commencing with certain symmetrical imperfections of exactly the same magnitude and form, bifurcation may occur, inducing snap-through of one of the equally increasing deflections, further increments occurring along one of the post-bifurcation equilibrium paths.

The text discusses the theory used and solves some problems where this behaviour is encountered.

1. *FINITE ELEMENT IDEALISATION (1) AND SOLUTION PROCEDURE*

Plating is idealised by thin 4 noded rectangular elements with bilinear in-plane displacement functions and the restricted quartic non-conforming (2) out-of-plane displacement functions. Stiffeners are idealised as 2 noded elements with linear in-plane functions and cubic out-of-plane functions that match the plate element functions along their edges. Ideal elastic-perfectly plastic stress-strain behaviour is assumed (ignoring strain hardening) and a total Lagrange (original undeformed coordinate) formulation used for the non-linear strains. Geometric and material non-linearities are ignored for torsional effects in the stiffener, so that torsional instability (tripping) is not modelled.

Plasticity in the plate elements is treated using Ilyushin's yield criterion (3), which is a direct function of the six stress resultants in the plate and is derived on the assumption that the equivalent stress (as defined by Von Mises criterion) is at yield throughout the thickness. Intermediate plasticity is ignored, but the approximate treatment gives acceptable accuracy (1). Plasticity in the stiffener elements uses a layered approach (usually 6 layers, each of depth as desired for the problem) based on uniaxial yield in each layer. The flow rules are used to generate elasto-plastic modular matrices for stiffener and plate elements that relate incremental stresses with incremental strains and curvatures. Unloading from the yield surface is modelled.

The solution procedure incorporates improvements to computer program CASPA (1), where the Riks-Wempner arc length constraint (4) as modified by Crisfield (5) and Ramm (6) is used:

$$\Delta p_i^T \, \Delta p_i = \Delta \ell^2 \quad \text{------------------------} \quad (1)$$

It provides an 'arc length' to the solution path as a constraint in load-displacement space rather than the prescribed load or displacement variable, which not only allows limit points to be passed, but is also an efficient convergence procedure. Iterations are performed at each prescribed step so that the Euclidean norm of the out-of-balance forces after each iteration, $(g^T g)^{\frac{1}{2}}$, is less than a specified percentage of the Euclidean norm of the external forces representing prescribed total loads and reactions up to the last iteration, $(R^T R)^{\frac{1}{2}}$. The arc length is updated from one step to the other as suggested by Ramm (6):

$$\Delta \ell_i = \Delta \ell_{i-i} \left(\frac{N_i}{N_{i-1}}\right)^{\frac{1}{2}} \qquad (2)$$

Because the steps occur in load-displacement space, it no longer matters whether load or displacement steps are prescribed in an analysis, and will not specifically be mentioned. However, almost all cases were carried out with prescribed displacement, a few analyses in section 2 also using prescribed loads. Two further refinements have been necessary to the solution process in connection with convergence of the analyses to be discussed.

a) Faster convergence is obtained on highly non-linear paths if the tangent stiffness matrix is updated after the first iteration of a prescribed step. It is also observed that if non-linearity, represented one-dimensionally as a single monotonically changing curvature, gives oscillatory convergence using standard modified Newton Raphson (M.N-R) iterations, on updating the stiffness matrix it changes to the inherently more satisfactory monotonic convergence from outside the step length. In some cases, fully converged solutions are only obtained by updating as described. This strategy is only required where large changes to the stiffness terms occur due to buckling and/or plasticity effects, the latter especially so during unloading in extensive areas.

b) When extensive plasticity in parts of a structure interact with buckling effects in other parts, alternative loading and unloading of the plastic regions can occur, which in turn makes the steps oscillate from positive to negative, giving very slow accumulation of the prescribed step. In such instances, which imply stresses lying about the yield surface during these steps, it is necessary to formulate an 'approximate' tangent stiffness matrix, assuming the whole structure to remain elastic only while formulating the modular matrices (see section 1). Many more iterations are then necessary, but the solution proceeds more smoothly without oscillatory step control.

2. *BEAM COLUMN BUCKLING MODE OF STIFFENED PLATES*

The analysis of the beam-column buckling mode of a stiffened panel with closely spaced stiffeners can be simplified by idealising a section of it as a T-beam. The parameters chosen are of one structure analysed by Bergan et.al (7) that exhibited the most sensitive (violent) instability. See Figs 1 and 2 for details. The structure is divided into 10 stiffener elements for each span. The first analysis involves imperfection A with equal, symmetrical half sinusoidal imperfections (amplitudes = $L/750$) but σ_y = very high, so that elastic bifurcation behaviour could be observed. Very close to bifurcation, the step lengths are automatically made small using eqn (2) along with automatic restart and halving of step size when wide divergence of iterations is observed. It is noted that provided the step lengths are small, no special treatment is necessary at bifurcation σ_{cr} *(n.1)*, because the residuals in the iterative process automatically give small imbalances in the two spans, which magnify themselves at bifurcation. It is necessary to update the stiffness matrix (see section 1(a)) after the first iteration after bifurcation in order to get fully converged iterations for all steps and avoid drift from equilibrium. Some oscillation is observed (See Fig 2) at bifurcation for two steps, before deciding on the correct equilibrium path, to give the non-linear buckling stress ($\sigma_{cr}(n.1)$) of 0.8445 σ_y. The second analysis has the same imperfection as above, but σ_y = 240 N/mm^2, so that inelastic bifurcation could be observed. Updating the stiffness matrix (section 1(a)) at and immediately after bifurcation (0.785σ_y) gives less oscillation of increments near the bifurcation point than without update. Also, without update, iterations close to bifurcation and just before the vertical and horizontal tangents (curve A, Fig 1) are carried out without equilibrating iterations, giving a small drift from equilibrium. Fully converged iterations are obtained with updated stiffness terms, which also gives faster solution near the singular points and strongly non-linear paths. The descent of the new equilibirum path, almost coincident with the ascent of the old path near 0.785 σ_y (Fig 1) indicates violent instability. The effect of bifurcation on the out-of-plane deflections can be observed in Fig 2, where for both the elastic and inelastic cases (A), imperfections w_1 and w_2 increase identically followed by sudden bifurcation of these displacements. It was attempted to analyse the inelastic bifurcation case using the approximate tangent stiffness matrix discussed in section 1(b). Due to the severity of the instability, no solution could be found beyond the maximum load.

Case B involves symmetrical sinusoidal imperfections on both spans of unequal magnitudes (amplitudes = $L/750$, $L/1500$) and σ_y = 240 N/mm^2. Snap-through buckling is involved, with both imperfections growing up to σ = 0.25 σ_y (see Fig 2) and w_2 decreasing with further increase of load. Complete snap-through occurs at 0.673 σ_y, with violent unstable behaviour, again passing a vertical tangent (Fig 1). Use of the approximate tangent stiffness of section 1(b) gives identical results requiring many more iterations. Case C has equal, antisymmetrical imperfection (amplitudes = $L/750$, $-L/750$) and σ_y = 240 N/mm^2. This does not exhibit either bifurcation or snap-through instability, and limit point type of singularity is achieved as usually encountered in analyses, at 0.57 σ_y.

3. *SIMULATION OF THE RAPIDLY UNLOADING POST BUCKLING TEST AT MANCHESTER UNIVERSITY* [8]

The details of the specimen (D21 in reference 8), hinged at the loaded edges along the elastic centroid of the stiffened plate and free along the other two edges, are given in Fig 3. For the analysis, the nominal overall cylindrical imperfection of the panel is represented in one single half sinusoided wave with central amplitude of 1.4 mm. The superimposed dished sinusoidal imperfections of amplitude 5.4 mm are represented transversely with half sine waves of length of 457 mm and longitudinally in half sine waves of length = 1/4 span. The maximum measured imperfection of 5.7 mm is reduced to 5.4 mm to allow an average representation. Two analyses are carried out, first without and then with residual stresses. Squash load of the panel (P_y) is 4684 kN, while elastic critical buckling load of the panel (P_{cr}) is 5189 kN ($\sigma_{cr} = P_{cr}/A$ = 272 N/mm^2) . The plate buckling stress (363.5 n/mm^2) is higher than yield stress (243 N/mm^2) and σ_{cr} (272 N/mm^2), so that overall buckling and/or plastic flow in the plate are predominent.

Horne and Narayanan (8) provide plots of load against three displacements: axial shortening (Fig 4), lateral displacement in the centre of the stiffened plate (Fig 5) and maximum lateral displacement (Fig 6). As anticipated, analysis 1 (residual stresses ignored) gives reasonable agreement up to about half the collapse load, when the effect of residual stresses on the real structure become appreciable (see Figs 4, 5 and 6). The collapse load of 2905 kN is about 8% above the experimental value. Post collapse behaviour is also plotted showing some snap-through behaviour, but this is not totally reliable, since no iterations could be carried out, giving drift from equilibrium. The analysis was carried out before improvements as discussed in sections 1(a) and 1(b) were made. Analysis was therefore abandoned after a while. In analysis 2, the residual stresses in the plate and stiffener are idealised to be individually in equilibrium for an undeformed structure. Iterations are carried out for the unloaded structure with the initial deformations and these stresses to get the new equilibrium state. The prescribed steps were carried out without improvements discussed in sections 1(a) and 1(b) up to a load of 2170 kN (81%) of experimental collapse) in 8 steps. Beyond this load, extensive plasticity in the plate causes sequential positive and negative steps due to loading and unloading of the stresses in the plate from the yield surface. Fully converged iterations are only possible with updated stiffness terms after the first iteration (section 1(a)). A positive step length causes plastic flow in plate, reducing the tangent modular matrix and causing instability. Large out-of-plane displacement (see Figs 5 and 6) causes relaxation of membrane forces and thus unloading from the yield surface in the plate. This gives an elastic modular matrix causing a stable path. Judicious choice of step lengths with larger positive values than negative values increases the total load to 2388 kN (89% of experimental collapse) at the 36th increment, the situation worsening with increasingly non-linear behaviour. Beyond this load, the approximate tangent stiffness matrix discussed in section 1(b) is used, where the modular matrix remains elastic. More iterations are necessary than with standard M.N-R and updated stifnesses particularly beyond collapse, when violent instability occurs (rapid unloading, shown for analysis 2 in Fig 4). A very weak snap-through behaviour is obtained just prior to collapse which is not observed in the test, all increments plotted being fully converged. The rapid unloading, occurring with a near vertical plot (fig. 4) is plotted up to 2315 kN at increment no. 113. The maximum (collapse) load of 2802 kN (at increment no. 59) is 4% higher than the experimental value (2686 kN). Figs. 5 and 6 give satisfactory correlation in lateral displacements between analysis and experiment.

4. *CONCLUSIONS*

It has been shown that severe bifurcation and snap-through problems can be analysed using improved methods with added modifications. Bifurcation problems can only be reliably analysed using the arc-length method in conjunction with an updated tangent stiffness matrix after the first iteration. Snap-through problems involving large plastic zones sometimes require use of the elastic modular matrix for the whole structure when formulating the (approximate) tangent stiffness matrix. Although needing more iterations, growth of incrementation is guaranteed with fully converged increments along the correct equilibrium path.

Acknowledgement

This work forms the analytical basis for parameter studies under way on eccentrically stiffened simply supported plates, within the Marine Technological Research Programme sponsored by the Dutch government for buckling and collapse load determination of offshore steel structures. Ir. F.S.K. Bijlaard, Ir. H.G.A. Stol and the author are members of IBBC-TNO working on this project.

References

1. Puthli, R.S., HERON, Vol. 25, 1980, no. 2.

2. Zienkiewicz, O.C. 'The Finite Element Method in Engineering Science', 3rd edn, Mc. Graw-Hill, London, 1977.

3. Ilyushin, A.A. 'Plasticite' Editions Eyrolles, Paris, 1956.

4. Riks, E. International Journal of Solids and Structures, Vol. 15, 529/551, 1979.

5. Crisfield, M.A. Computers and Structures, Vol. 13, 55/62, 1981.

6. Ramm, E. Proc. Europe-U.S. Workshop on Non-Linear Finite Element Analysis in Structural Mechanics, Bochum, 1980.

7. Bergan, P.G. Holland, I. and Söreide, T.H. in 'Energy Methods in Finite Element Analysis', Eds. Glowinski, R., Rodin, Zienkiewicz, O.C.; John Wiley and Sons, 1979.

8. Horne, M.R. and Narayanan, R. 'Further tests on the ultimate load capacity of longitudinally stiffened panels'. Simon Engineering Laboratories, University of Manchester, July 1974.

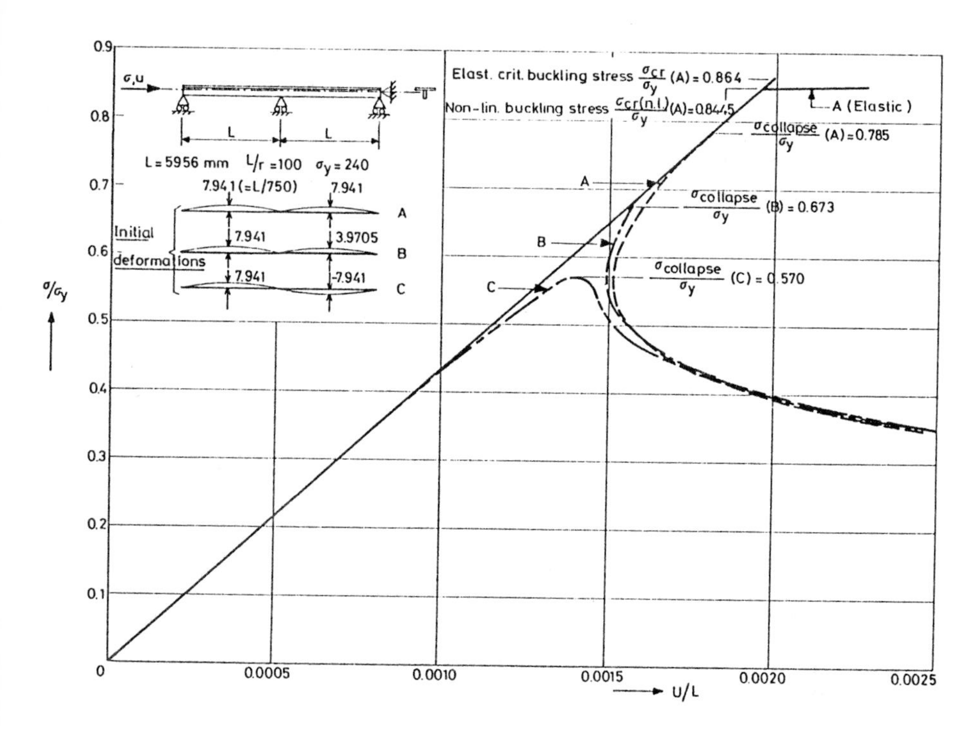

Fig 1 Load-shortening curves for two span beam-column

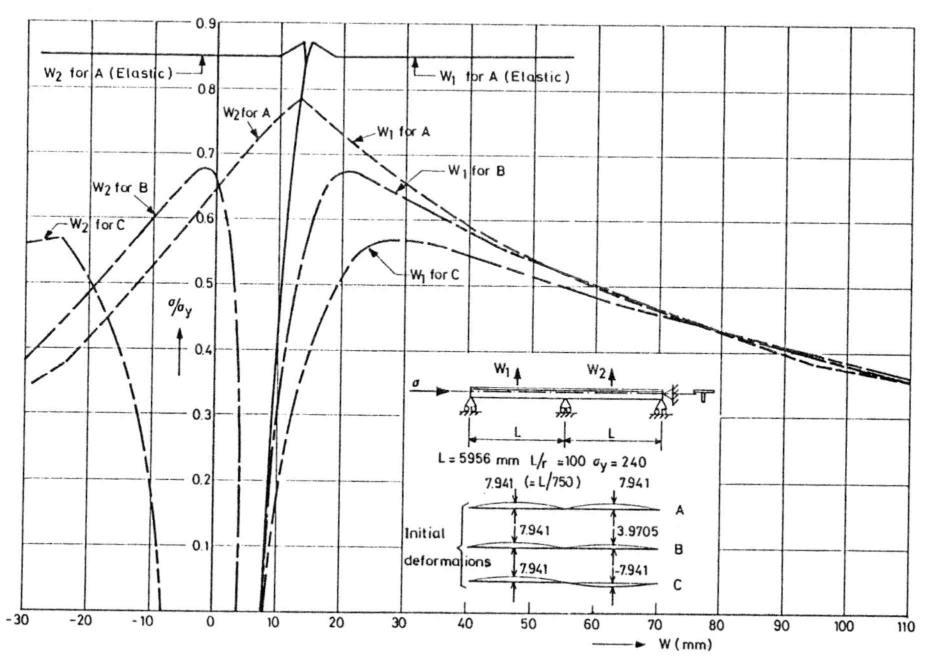

Fig 2 Comparison of out-of-plane deflections of two span beam-column

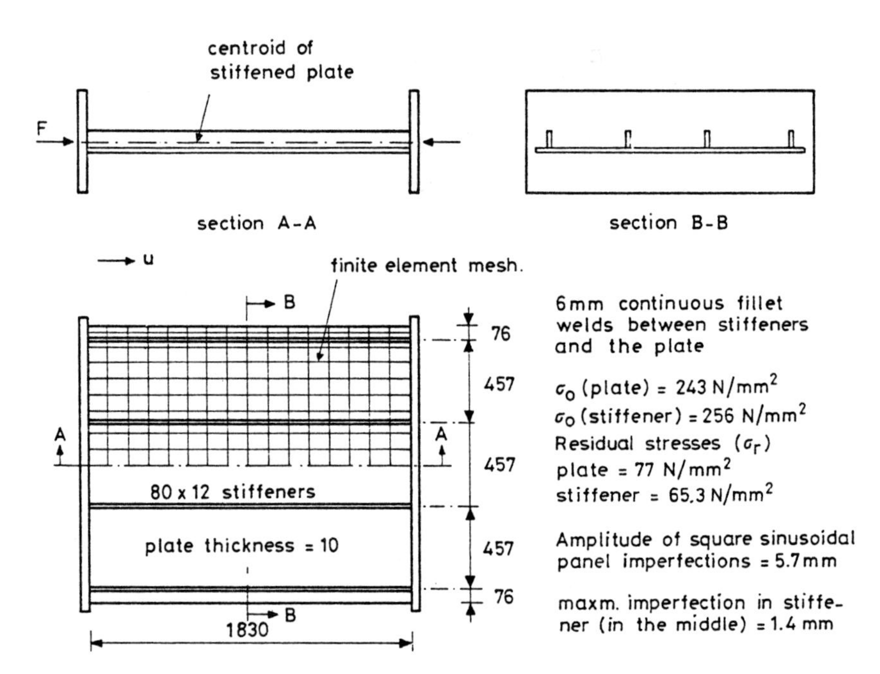

Fig 3 Pin ended stiffened plate 1524 mm wide

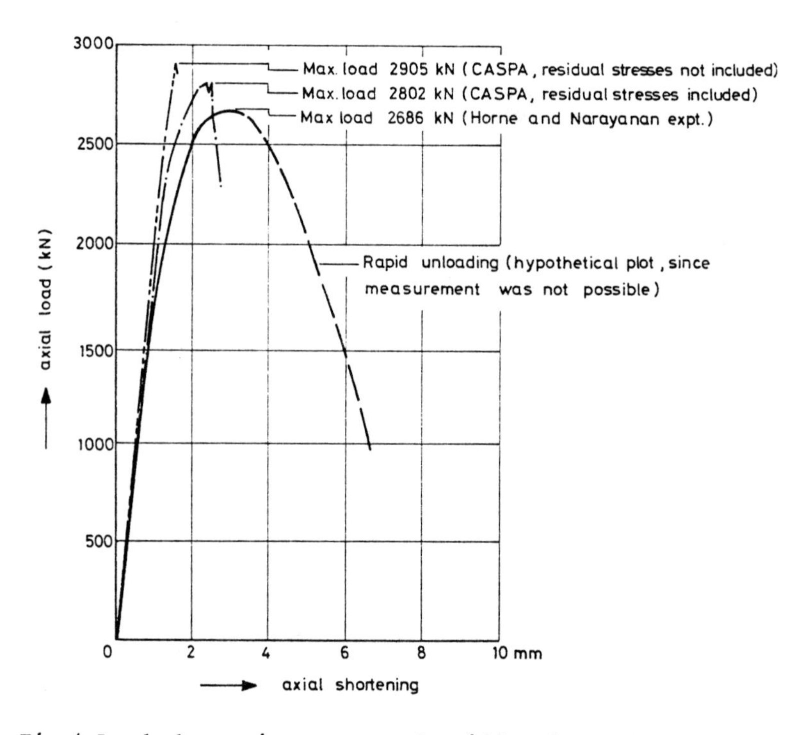

Fig 4 Load-shortening curves of stiffened panel

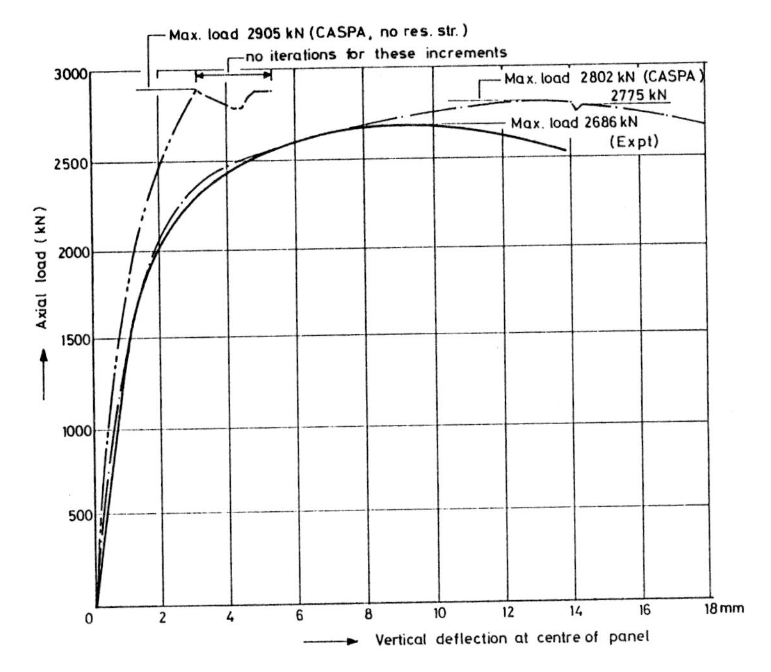

Fig 5 Load vs. out-of-plane deflection curves at centre of stiffened panel

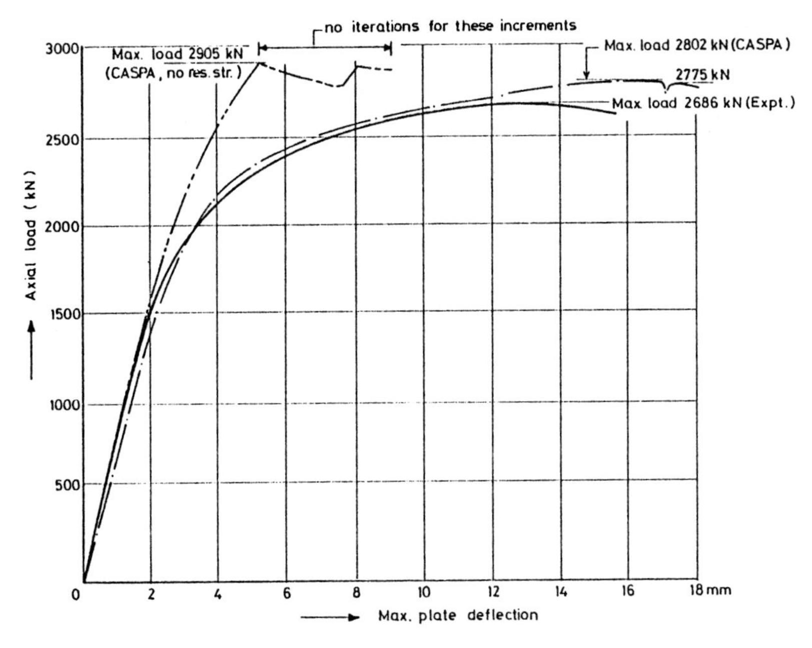

Fig 6 Load vs. maximum out-of-plane plate deflection curves for stiffened panel

THE COLLAPSE BEHAVIOUR OF STIFFENED STEEL PLATES WHICH HAVE HIGH ASPECT RATIOS UNDER AXIAL COMPRESSION

N.W. Murray and W. Katzer

Department of Civil Engineering, Monash University.

Summary

Most methods for evaluating the collapse load of a stiffened plate panel are based on a column model. While these methods are satisfactory for panels without side support they will underestimate the collapse load by a large margin if the aspect ratio (L/B) is large. Experiments demonstrate this and a simple criterion is proposed to define the range of validity of column-based theories.

Introduction

A stiffened panel of overall width B consists of a flat plate with longitudinal stiffeners, the length L of the panel being defined as the distance between the transverse stiffeners. The longitudinal stiffeners are of thin rectangular shape welded to one side of the plate and their spacing b is uniform. The loading is an axial compressive stress whose average value is σ_{av}. When the plate element of width b and thickness t is modelled as a simply-supported plate its theoretical elastic behaviour is summarised in Fig. 1 and 2. Theoretical studies of perfect stiffened plates now enable their critical stresses to be evaluated easily (1). Bilstein's (2) elastic analysis of stiffened plates with initial imperfections is an important step forward.

Although these elastic analyses can be thought of as "exact" the same cannot always be said of collapse analyses, many of which make recourse to empirical rules. The work of Moxham (3,4), Little (5) and Harding et al. (6) on isolated plates are exceptions but for stiffened plates it appears that similar studies do not yet exist.

Most investigators (7-12) model a stiffened panel as a column but some (13,14) treat it as an orthotropic plate. In each case the criterion of failure depends upon the direction of collapse. If the panel buckles towards the stiffener the failure criterion is that the compressive membrane stress in the middle plane of the plate reaches yield (in Ref. 14 the factor 1.065 is used to improve agreement between theory and experiment) and if towards the plate failure is said to occur when the free edge of the stiffeners yield.

Dwight and Little (7) consider that local plate buckling (Fig 1) in a stiffened panel is equivalent to a reduction in yield stress and hence call their method an effective stress method. In other methods (8-10) an effective width b_e of plate is used instead of the full width b in calculating the section properties of the column and the shift e in the position of the neutral axis in the region where local buckling occurs. Generally this local buckling occurs only at the centre of the column so b_e and e are not uniform along its length but for simplicity this is ignored and the panel is treated as an eccentrically loaded column.

Other methods for estimating the failure loads of stiffened panels use empirical formulae of the Rankine type. With these methods the following characteristic loads are first calculated, P_y the squash load (= σ_y x area of cross-section), P_E the Euler buckling load (= $\pi^2 EI/L^2$) and P_{cr} the load at which local buckling occurs (found from reference 1.) With Allen's method (11,12) the failure load P_f is given by

$$P_f^{-2} = P_{cr}^{-2} + P_E^{-2} + P_y^{-2} \tag{1}$$

An analysis of this formula shows that P_f cannot exceed P_{cr}, P_E or P_y whereas it is well-known that P_f can exceed P_{cr}. Because of this Herzog (15) introduced new coefficients and modified the formula to the following form for the mean value of P_f

$$P_f^{-2} = 0.22P_{cr}^{-2} + P_E^{-2} + P_y^{-2} \tag{2}$$

with a lower bound given by

$$P_f^{-1} = 0.47P_{cr}^{-1} + P_E^{-1} + P_y^{-1} \tag{3}$$

From 72 published test results the average ratio of measured strength to that predicted by eqn (2) was 0.928 (7.2% on the unsafe side) and the standard deviation was 0.155. While these methods are easy to use they are liable to be greatly in error in isolated cases. One serious shortcoming of these methods is that they make no attempt to account for the effect of the direction of collapse of a panel on its failure load. Tests indicate that a panel failing towards the stiffeners can generally carry a higher load than one failing towards the plate.

An interesting comparison of theoretical and experimental results has been made by Mikami, Dogaki and Yonezawa (16) who carried out ultimate load tests on four large box-girders. They compared test results from 16 ultimate strength theories and showed that all except Herzog's theory gave good agreement. It should be noted that the b/t ratio (= 17) are lower than average for box girders.

1. *THEORETICAL CONCEPTS*

In this section various ways of treating the ultimate load analysis of a stiffened plate panel are briefly outlined.

(a) The simplest panel to analyse is one without side supports and low b/t values ($b/t < 35$). Such a panel does not exhibit local buckling and behaves as a wide pin-ended column. Let us suppose that the initial deflection δ_{co} at the centre of the panel favours failure towards the stiffeners. As the stress σ_{av} is increased δ_{co} is magnified by the factor $(1 - \sigma_{av}/\sigma_E)^{-1}$ and eventually the middle plane of the plate will start to yield. This value of σ_{av} is defined as the failure stress σ_f and it is given by the Perry-Robertson formula as follows.

$$\sigma_f = \frac{1}{2}[\sigma_y + \sigma_E(1+\eta)] - \frac{1}{2}[\{\sigma_y + \sigma_E(1+\eta)\}^2 - 4\sigma_y\sigma_E]^{\frac{1}{2}} \qquad (4)$$

where $\eta = \dfrac{A\,\delta_{co}\,y_1}{I}$

y_1 = distance from neutral axis to the middle plane of the plate
A = area of cross-section of the panel
I = second moment of area of the cross-section.

When δ_{co} favours failure towards the plate a similar analysis is made. The only difference is that y_1 is replaced by y_2 where y_2 = distance from neutral axis to the free edge of the stiffener.

(b) When the panel described above in (a) is simply-supported along each side it can be analysed as an orthotropic plate by solving the equations of Rostovtsev (17). For axial loading of p_z per unit width the critical load per unit width is

$$p_{z_{cr}} = \frac{\pi^2}{B^2}[D_x\frac{L^2}{B^2} + 2H + D_z\frac{B^2}{L^2}] \qquad (5)$$

where D_x and D_z = average flexural rigidity per unit width for bending in the x and y directions, respectively ($D_x = \dfrac{Et^3}{12(1-\nu^2)}$)

and $H(\simeq Gt^3/12)$ = average torsional rigidity per unit width. [This term is usually very small for panels with stiffeners of open profile.]

The natural buckle length for a panel which is infinitely long is

$$L_{cr} = B[\frac{D_z}{D_x}]^{\frac{1}{4}} \qquad (6)$$

When most of the stiffening is in the z direction, i.e. $D_z >> D_x$ and H, eqn (5) gives $P_{cr} = \pi^2 EI_z/L^2$ = Euler load. Thus the effect of the side supports may be ignored and the panel can be treated as an Euler column. Its failure stress is given by eqn (4) above. This approach is confirmed by the deflected shape of the cross-section of the panel [Fig 3(c)] as it buckles. However, there is another condition which may invalidate this simplification. Examination of eqn (5) shows that if the aspect ratio L/B of the panel is large enough the first term inside of the square brackets may be significant. By ignoring the influence of H and by assuming that the side-support effect must be considered if the first term is 10% or more of the third term it is easily shown that the column models referred to earlier become inaccurate if the aspect ratio exceeds

$$[\frac{L}{B}]_{max} = [\frac{D_z}{10D_x}]^{\frac{1}{4}} \qquad (7)$$

As a typical example consider a cross-section with $B = 2000$, $b = 333$, $t = 15$, stiffener depth $d = 100$ and stiffener thickness, $t_s = 15$. $D_z = 3478 \times 10^6$ Nmm and $D_x = 63.67 \times 10^6$ Nmm and from eqn (7) $(L/B)_{max} = 1.529$ or $L_{max} = 3058$ mm. If L for this panel were infinitely long the natural buckle lengths would be [eqn (6)] 5437 mm and p_{cr} would be [eqn (5)] 5054.7 Nmm^{-1}. An Euler column with this same cross-section but of length $L = 5437$ mm $(L/B = 2.717)$ buckles at 1055 Nmm^{-1}. This indicates that by ignoring the effect of side supports, i.e. by using the column models proposed by various authors, designs can be very conservative if the aspect ratio exceeds that given in eqn (7). One model which considers this effect is the orthotropic plate model of Massonnet and Maquoi (14).

(c) When the panel is free of side support but has a higher b/t ratio and buckles towards the stiffeners it behaves like a pin-ended column. However, local buckling may result in some loss of effective width, i.e. b is reduced to b_e, and in an increase in the eccentricity of the axial loading (Section 2). This can be considered to result in a lower second moment of area for the cross-section and in some effective eccentricity of the axial loading which is applied at each end. The column models, e.g. those of Dwight and Little (7), Horne and Narayanan (8) and Murray (9) take these effects into account. Provided $(L/B)_{max}$ given in eqn (7) is not exceeded these models appear to give accurate results and they are easy to apply.

(d) A panel with side supports and a high b/t ratio [Fig 3(a),(b) and 5] is the most complicated of the four cases considered here. This type of panel is also liable to undergo local buckling resulting in a reduction in the effective cross-section and eccentricity of axial loading. It seems reasonable to assume that the value of D_z should be calculated from the effective cross-section and then $(L/B)_{max}$ obtained from eqn (7). When the aspect ratio L/B is lower than $(L/B)_{max}$ the column model described in (c) above may be used but when it exceeds $(L/B)_{max}$ the orthotropic plate model of Massonnet and Maquoi appears to be the most satisfactory method available.

2. *EXPERIMENTAL METHOD AND RESULTS*

Altogether 24 test panels are considered here. Full details of the fabrication and testing method as well as of the residual stress patterns are given elsewhere (18-20). Each panel was loaded axially in a stiff deflection-controlled testing machine. The sides were constrained by knife edges to approximate to the simply-supported condition. To avoid local buckling at each end it was found necessary to use packing pieces around the profile and this means that the ends of each specimen were much closer to a built-in condition than to simple supports. However, one series of specimens with additional panels at each end (i.e. the specimens were three panels long) was tested to simulate the effect of having a pin-ended panel (to which the central panel approximated). This was to see whether the nearby built-in condition at the ends of the main series of tests resulted in significantly higher failure loads. Within the accuracy of the tests on specimens with high aspect ratios $(L/B = 930 \quad 350 = 2.71)$ there was no measurable difference in either the failure loads or buckling loads of similar panels with built-in and pinned end conditions. The built-in condition appeared to have the advantage that the results were more reproducible because local buckling at the ends was eliminated.

One purpose of the test series was to use it to check the calculated elastic buckling loads and modes obtained from the finite strip method (1,19). The tests were designed to show how either local or global buckling can be controlled by varying the number and size of stiffeners and the spacing of the transverse stiffeners and by using side-supports (19,20). The study involved the use of buckling plots (1), i.e. graphs of buckling stress against length of buckles. Figure 4 shows two typical buckling plots. The first minimum represents a local buckle of short wavelength L_1 which is much shorter than L. The stress at which it occurs is denoted as σ_{cr}. The stress σ_g at which the panel can develop a single global buckle is read from the buckling plot as indicated. In some cases σ_g is less than σ_{cr} and in other cases it exceeds σ_{cr}. In the former case global buckling may be expected first and in the latter local buckling is expected to appear before global buckling. However, this simplistic approach can be modified by the nature of the imperfections.

Details of the panels and test results are contained in Table 1. It is seen that the program has four series described briefly as follows:

Series A Six panels in nominally duplicated pairs with 1, 2 and 3 stiffeners each with $d \simeq 30$ and $t \simeq 1.5$ (i.e. $d/t \simeq 20$, $b/t \simeq 110$, 86 or 64 depending upon the number of stiffeners).

Series B Six panels in nominally duplicated pairs with 1, 2, and 3 stiffeners each with $d \simeq 15$ and $t \simeq 1.5$ (i.e. $d/t \simeq 10$, $b/t \simeq 110$, 86 or 64 depending upon the number of stiffeners).

Series C Six panels in nominally duplicated pairs with 1, 2 and 3 stiffeners each with $d \simeq 40$ and $t \simeq 2$ (i.e. $d/t \simeq 20$, $b/t \simeq 90$, 64 or 40 depending upon the number of stiffeners).

Series T Six panels in nominally duplicated pairs with 1, 2 and 3 stiffeners each with $d \simeq 30$ and $t \simeq 1.5$ (i.e. $d/t \simeq 20$, $b/t \simeq 110$, 86 or 64 depending upon the number of stiffeners). In this series a torsionally weak beam was located across the mid-span of each specimen in order to impose a nodal line. Thus the results of this series should be compared with those of Series A to measure the effect of this constraint which is to halve the length of the panels in Series A.

In Table 1 σ_{cr} and σ_g are the critical stresses estimated from buckling plots as described earlier. The last column describes the mode of collapse. S indicates that the stiffener failed by lateral-torsional buckling and P indicates that collapse was due to the development of a plastic mechanism in the plate. This information is required because the method to be used for calculating the failure load depends upon the direction of failure.

In Table 2 the estimated failure loads from seven published theories are compared with experimental results. Theories 1, 2 and 3 are column models, Theories 4, 5 and 6 are largely empirical and Theory 7 is an orthotropic plate theory. In Table 1 the appropriate form of eqn (4) et seq. has been used for those theories which distinguish between collapse caused by stiffener failure and that caused by plate failure.

3. *CONCLUSIONS*

The following is a brief summary of some of the observations which can be made from these results.

a. It is seen that for Series A and C although the theoretical results for Theories 1 to 6 improve as the number of stiffeners increases, i.e. D_z increases relative to D_x, agreement between theory and experiment is generally poor. In the case of Series B agreement in all cases is very bad. All column theories are conservative in their estimates of the failure stress.

b. Comparison of $(L/B)_{max}$ from eqn (7) with the actual L/B value of *2.71* for Series A, B and C shows that in all cases the column models are outside of their range of validity. However, generally speaking better agreement between theory and experiment is obtained when $(L/B)_{max}$ and the actual L/B ratio are not greatly different. It appears that eqn (7) gives a good indication as to whether a column model can be used.

c. When the aspect ratio is halved by introducing a torsionally-weak stiffener at the horizontal centreline (Series T compared with Series A) $(L/B)_{max}$ is not exceeded and the column models give much more satisfactory results.

d. The results from Massonnet and Maquoi's orthotropic plate theory (Theory 7) show that it gives a much more consistent estimate of the failure stress. However, it does generally overestimate σ_f and it involves a more complicated calculation.

e. A study of experimental results in Table 1 shows that

(i) σ_f increases with n in each series, the greatest increases being in Series C and T.

(ii) The strength of plates A1-8 and A1-9 which had a high b/t ratio was not increased by adding transverse stiffeners midway between the ends (compare with T1-39 and T1-45).

(iii) Increasing the thickness (or stockiness) of the plating and stiffeners in a panel appears to result in a more efficient panel than by increasing the number of longitudinal or transverse stiffeners beyond a certain point.

(iv) Although there was some benefit in increasing the depth d of the stiffeners from 15 to 30 (Series B and A, respectively) further increases in d are probably not worthwhile because of the reduction in σ_{cr} with higher d/t ratios. This point is also illustrated in Reference 1.

References

1. Murray, N.W. and Thierauf, G. 'Tables for the design and analysis of stiffened steel plates.' Vieweg (Braunschweig/ Wiesbaden), 1981.

2. Bilstein, W. 'Anwendung der nichtlinear Beultheorie auf vorverformte, mit diskreten Langsteifen verstärkte Rechhteckplatten under Längsbelastung.' Pub. Inst. für Statik und Stahlbau T.H. Darmstadt, 1974.

3. Moxham, K.E. 'Theoretical prediction of the strength of welded steel plates in compression.' Cambridge University Eng. Dept., Report No. CUED/C-Struct/Tr.2, 1971.

4. Moxham, K.E. 'Buckling tests on individual welded steel plates in compression.' Cambridge University, Dept. Civ. Eng., Report No. CUED/C-Struct/TR.3, 1971.

5. LITTLE, G.H. 'The collapse of rectangular steel plates under uniaxial compression.' The Struct. Eng., vol. 58B, No. 3, Sept. 1980, pp. 45-61.

6. Harding, J.E., Hobbs, R.E. and Neal, B.G. 'The elasto-plastic analysis of imperfect square plates under in-plane loading.' Proc. ICE, Part 2, 63, March 1977, pp. 137-158.

7. Dwight, J.B. and Little, G.H. 'Stiffened steel compression flanges - a simpler apporach.' The Struct. Eng., vol. 54, No. 12, Dec. 1976, pp. 501-509.

8. Horne, M.R. and Narayanan, R. 'An approximate method for the design of stiffened steel compression panels.' Proc. ICE, Part 2, 59, Sept. 1975, pp. 501-514.

9. Murray, N.W. 'Analysis and design of stiffened plates for collapse load.' The Struct. Eng., vol. 53, No. 3, March 1975, pp. 153-158.

10. Yamada, Y. and Watanbe, E. 'On the behaviour and ultimate strength of longitudinally stiffened flanges of steel box girders.' Proc. JSCE, No. 252, Aug. 1976, pp. 127-142.

11. Allen, D. Correspondence on 'Analysis and design of stiffened plates for collapse load' by N.W. Murray. Struct. Eng., vol. 53, No. 9, 1975, pp. 381-382.

12. Allen, D. Discussion for 'An approximate method for the design of stiffened steel compression panels' by M.R. Horne and R.Narayanan. Proc. ICE, Part 2, 61, June 1976, pp. 453-455.

13. Troitsky, M.S. 'Stiffened Plates, Bending Stability and Vibration.' Elsevier, 1976.

14. Massonnet, C. and Maquoi, R. 'New theory and tests on the ultimate strength of stiffened box girders' in 'Steel Box Girder Bridges.' Clowes and Sons, London, 1973, pp. 131-143.

15. Herzog, M. 'Die Traglast dünnwandiger Kastenstützen mit Imperfektionen und Eigenspannungen unter zentrischem Druck nach Versuchen.' VDI-Z., 118, No. 2, Jan. 1976, S.80-85.

16. Mikami, I., Dogaki, M. and Yonezawa, H. 'Ultimate load tests on multi-stiffened steel box girders.' Technology Report of Kansai University, Osaka, Japan, No. 2, March 1980, pp. 157-169.

17. Rostovtsev, G.G. 'Calculation of a thin plate sheeting supported by rods.' Tandy, Leningrad Inst. Inzkenerov, Grazhdanskozo Vasdushnogo Flota, No. 20, 1940.

18. Katzer, W. 'The collapse behaviour of stiffened plate panels of high aspect ratio under uniaxial compression.' M. Eng. Sci. Thesis, Dept. of Civil Eng., Monash University, 1982.

19. Thierauf, G., Katzer, W. and Murray, N.W. 'Application of the finite strip method to the design of stiffened plates for buckling. Proc. Int. Conf. on Finite Element Methods, Shanghai, 1982, pp. 803-810.

20. Katzer, W. and Murray, N.W. 'Elastic buckling of stiffened steel plates of high aspect ratio under uniaxial compression.' To be published. Int. J. Thin-Walled Structures 1983.

TABLE 1 DIMENSIONS AND FAILURE STRESSES OF TEST SPECIMENS

t = thickness of plate and stiffeners   n = no. of stiffeners

All specimens 950 mm long and 350 mm wide.   Series T had central transverse stiffener.

TEST	Cross-Section	Dimensions							Yield & Buckling Stresses			Test Results		Collapse Mode
		n mm	b' mm	b mm	t mm	c mm	b/t –	d/t –	σ_y MPa	σ_{cr} MPa	σ_g MPa	σ_f MPa	σ_f/σ_y –	(See Note)
A1-8		1	175.0	–	1.59	30.0	110	19	205	51.5	96.8	91.8	0.45	S
A1-9		1	175.0	–	1.58	29.5	111	19	215	50.9	93.4	86.7	0.40	S
A2-12		2	111.5	127.0	1.47	31.0	86	21	223	96.7	122.6	107.5	0.48	S
A2-13		2	111.5	127.0	1.47	31.0	86	21	233	96.7	122.6	107.7	0.48	S
A3-11		3	80.0	95.0	1.48	31.0	64	21	227	164.4	141.9	113.3	0.50	S
A3-16		3	80.0	95.0	1.48	30.0	64	20	155	165.5	132.1	106.8	0.69	S
B1-15		1	175.0	–	1.48	15.5	118	10	154	44.1	34.8	72.6	0.47	S
B1-28		1	175.0	–	1.50	15.0	117	10	158	45.3	34.4	66.7	0.42	P
B2-22		2	112.0	126.0	1.47	14.5	86	10	144	97.7	36.8	86.2	0.60	P
B2-29		2	112.0	126.0	1.50	15.0	84	10	158	100.7	37.5	89.7	0.57	P
B3-20		3	80.0	95.0	1.47	15.0	65	10	160	168.4	40.9	96.4	0.60	S
B3-21		3	79.0	96.0	1.48	15.0	65	10	156	168.1	41.1	92.7	0.59	S
C1-17		1	175.0	–	1.95	39.5	90	20	217	77.5	190.9	103.8	0.48	S
C1-23		1	175.0	–	1.92	39.5	91	21	220	79.1	187.0	100.0	0.45	S
C2-24		2	112.0	126.0	1.95	39.0	65	20	204	167.6	222.3	139.9	0.69	P
C2-25		2	113.0	124.0	1.95	39.0	64	20	205	168.6	223.4	135.0	0.66	S
C3-26		3	80.0	95.0	2.00	39.0	40	20	236	273.8	253.0	167.3	0.71	P
C3-27		3	80.0	95.0	1.95	39.0	41	20	226	259.9	251.1	162.4	0.72	P
T1-30		1	175.0	–	1.49	30.0	117	20	181	46.6	228.0	82.2	0.45	SP
T1-45		1	175.0	–	1.50	29.0	117	19	155	45.9	215.7	75.8	0.49	SP
T2-31		2	112.0	126.0	1.49	30.0	85	20	181	103.0	313.0	102.8	0.57	SP
T2-33		2	112.0	126.0	1.50	30.0	84	20	174	101.6	311.8	105.4	0.61	SP
T3-34		3	79.5	95.5	1.50	30.0	64	20	176	174.1	417.4	135.8	0.77	SP
T3-35		3	79.5	95.5	1.50	30.0	64	20	176	174.1	417.4	126.2	0.72	SP

NOTE: S denotes failure caused by stiffener mechanism. P denotes failure caused by plate mechanism. In the T Series the two panels fail in opposite directions, i.e. one S and one P.

TABLE 2. RATIOS OF THEORETICAL TO EXPERIMENTAL FAILURE STRESSES AND TESTS OF RANGE OF VALIDITY FOR THEORIES 1 TO 3 FROM EQUATION (7)

	SPECIMEN	Series A (t = 1.5 d = 30)						Series B (t = 1.5 d = 15)					
		A1-8	A1-9	A2-12	A2-13	A3-11	A3-16	B1-15	B1-28	B2-22	B2-29	B3-20	B3-21
	$(L/B)_{max}$ (eqn (7))	1.84	1.85	2.30	2.30	2.48	2.48	1.16	1.16	1.30	1.30	1.49	1.49
	(L/B) actual	2.71	2.71	2.71	2.71	2.71	2.71	2.71	2.71	2.71	2.71	2.71	2.71
Theory	1 Horne/Narayanan												
	a. Plate mechanism	0.45	0.46	0.55	0.55	0.68	0.68	0.11	0.10	0.10	0.11	0.13	0.13
	b. Stiffener "	0.47	0.48	0.46	0.46	0.50	0.64	0.12	0.11	0.11	0.12	0.14	0.14
	2 Murray (1)												
	a. Plate mechanism	0.67	0.66	0.66	0.66	0.83	0.94	0.11	0.10	0.10	0.10	0.12	0.12
	b. Stiffener "	0.33	0.33	0.53	0.53	0.62	0.53	0.10	0.10	0.11	0.11	0.17	0.17
	3 Dwight/Little (2)	0.59	0.61	0.57	0.58	0.67	0.68	0.28	<u>0.29</u>	<u>0.21</u>	<u>0.21</u>	0.25	0.26
	4 Yamada/Watanabe	0.65	0.68	0.78	0.80	0.98	0.95	0.52	0.54	0.48	0.50	0.58	0.58
	5 Allen	0.43	0.44	0.55	0.55	0.71	0.81	0.13	0.13	0.13	0.13	0.15	0.16
	6 Herzog	0.57	0.57	0.68	0.69	0.81	0.90	0.14	0.13	0.13	0.13	0.16	0.16
	7 Massonnet/Maquoi	1.07	1.13	1.18	1.22	1.38	1.37	0.96	1.02	0.98	1.01	1.13	1.13

Table 2 (cont'd).

	SPECIMEN	Series C (t = 2 d = 40) C1-17	C1-23	C2-24	C2-25	C3-26	C3-27	Series T (t = 1.5 d = 30) T1-30	T1-45	T2-31	T2-33	T3-34	T3-35
	$(L/B)_{max}$ (eqn (7))	2.06	2.06	2.36	2.36	2.55	2.55	1.90	1.90	2.30	2.30	2.48	2.48
	(L/B) actual	2.71	2.71	2.71	2.71	2.71	2.71	1.36	1.36	1.36	.136	1.36	1.36
Theory	1 Horne/Narayanan												
	a. Plate mechanism	0.74	0.75	0.77	0.81	0.88	0.88	1.22	1.15	1.12	1.07	0.99	1.06
	b. Stiffener "	0.57	0.56	0.56	0.51	0.47	0.47	0.99	1.10	0.86	0.85	0.69	0.75
	2 Murray (1)												
	a. Plate mechanism	0.76	0.77	0.78	0.82	0.90	0.89	0.93	0.90	0.96	0.92	0.91	0.98
	b. Stiffener "	0.99	0.99	0.77	0.77	0.81	0.81	0.77	0.77	0.93	0.93	0.83	0.83
	3 Dwight/Little (2)	0.69	0.69	<u>0.59</u>	0.62	<u>0.74</u>	<u>0.73</u>	0.79	0.75	0.82	0.78	0.75	0.81
	4 Yamada/Watanabe	0.71	0.71	0.73	0.76	0.90	0.89	0.70	0.68	0.76	0.73	0.73	0.70
	5 Allen	0.59	0.58	0.66	0.69	0.72	0.73	0.66	0.68	0.83	0.80	0.84	0.90
	6 Herzog	0.81	0.81	0.80	0.84	0.83	0.84	0.98	1.01	1.18	1.13	1.07	1.15
	7 Massonnet/Maquoi	1.09	1.10	1.00	1.05	1.11	1.08	1.17	1.13	1.18	1.12	1.01	1.11

Notes: (1) Theoretical s_f assumes ends are pinned.

(2) Dwight/Little theory only valid for plate mechanism collapse (underlined values).

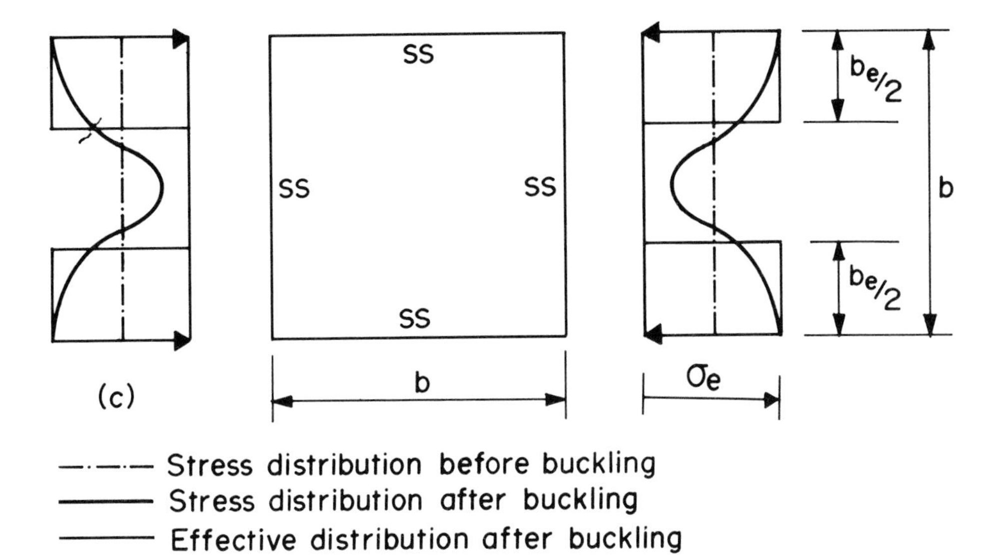

Fig. 1. Stress distribution in a square simply-supported plate with axial stress.

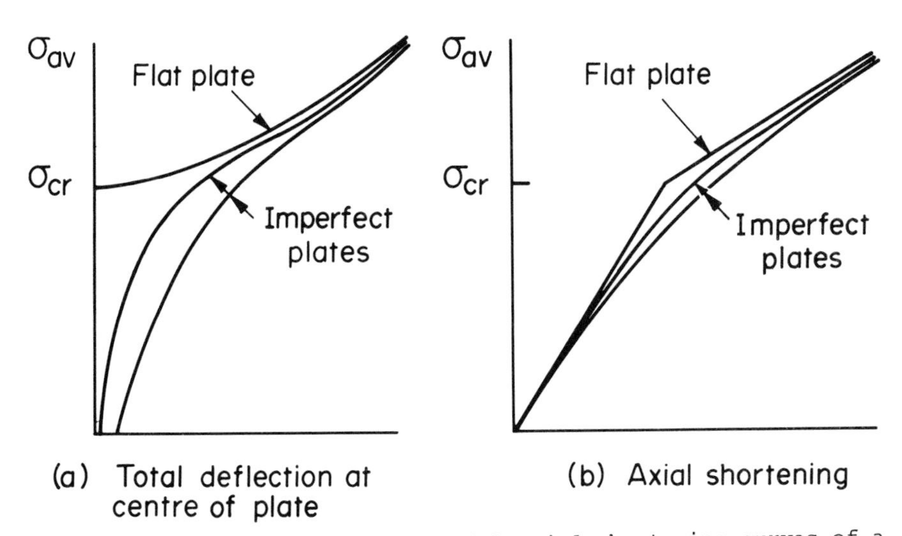

Fig. 2. (a) Lateral deflection and (b) axial shortening curves of a simply-supported plate with axial stress.

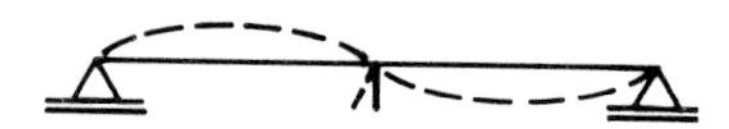

(a) Local buckle

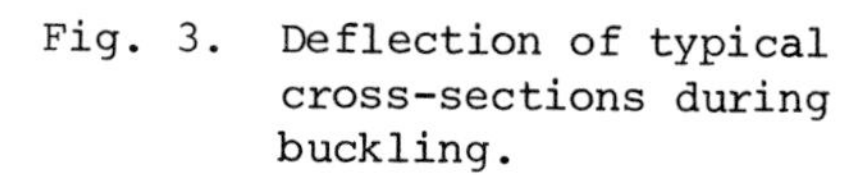

Fig. 3. Deflection of typical cross-sections during buckling.

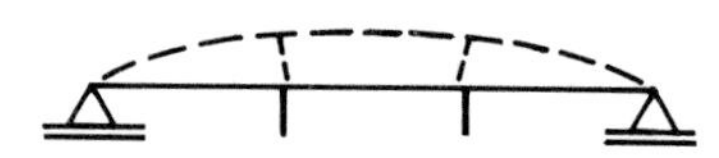

(b) Global buckle
($D_z > D_x$)

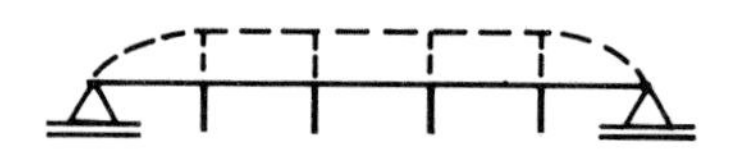

(c) Global buckle
($D_z \gg D_x$)

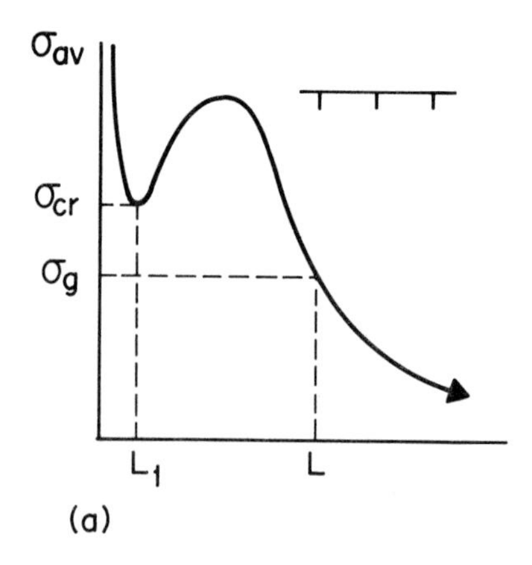

(a)

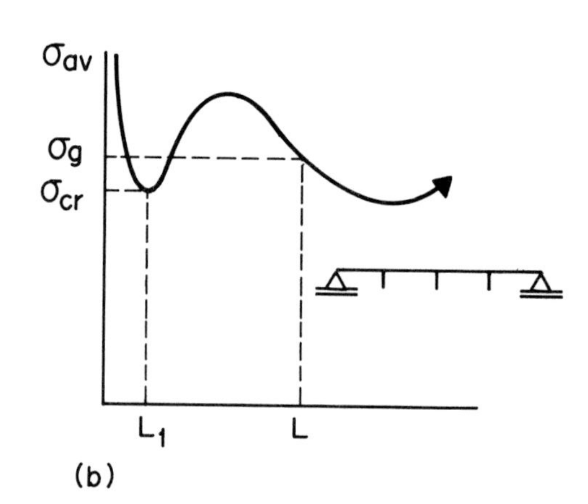

(b)

Fig. 4. Buckling plots show the stress required to maintain a panel of length L in a single buckle.

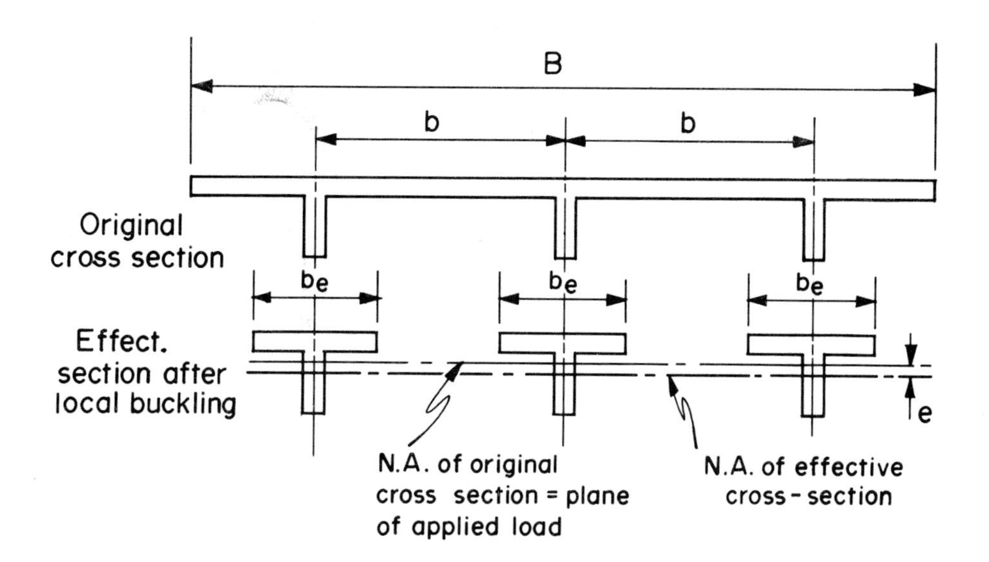

Fig. 5. Panels with high b/t ratio first buckle locally and can then be analysed as an eccentrically loaded column of reduced cross-section.

ULTIMATE STRENGTH OF CURVED FLANGES IN BOX GIRDERS

D Vandepitte and B Verhegghe

Faculty of Applied Science, Ghent University

Synopsis

A method is proposed for the evaluation of the ultimate strength in circumferential tension or compression of curved flanges in box girders, taking into account the loss of strength due to transverse bending stresses. A chart is presented giving directly the effective width of a flange.

Notation

f_y	yield stress
t	thickness of the flange
r	radius of curvature of the middle plane of the flange
$2b$	width of the flange
$2b_{ef}$	effective width of the flange
q	centripetal force per unit area of the flange
m	transverse bending moment per unit length of a longitudinal cross-section of the flange
F	force acting circumferentially in the flange
F_y	$= 2bt\,f_y$
σ	average, over the thickness, of the circumferential direct stress in the flange
$\sigma + \Delta\sigma$, $\sigma - \Delta\sigma$	circumferential direct stress in the upper and lower half, respectively, of the flange in its ultimate state
σ_z	transverse bending stress in the flange
σ_{ef}	local effective stress derived from von Mises's yield criterion
τ	radial shear stress in a longitudinal cross-section of the flange
z	distance between the centre-line of the box and a longitudinal cross-section

ζ	$= 2z/\sqrt{rt}$	ζ_i	$= i\Delta$
ζ^0	$= 2b/\sqrt{rt}$	ω	$= \sigma_z/f_y$
$'$	$= \frac{d}{d\zeta}$	ω_i	$= \omega(i\Delta) = \sigma_z(i\Delta)/f_y$
Δ	increment of ζ	α	$= \frac{3t}{4r}$
i	a positive integer		

Introduction

A box girder is considered. The box is assumed to consist of a single cell of width $2b$. It has a symmetrical cross-section and a lower flange of constant thickness t which is curved in the longitudinal direction of the girder. The flange is assumed to be concave downwards and the radius of curvature of its middle plane is r.

The upper flange of the box girder may be flat or have a curvature of the same sign as the lower flange (Fig 1) or of the opposite sign. The webs are assumed to be thin, so that they do not restrain rotations of the edges of the lower flange in the plane of the cross-section (Fig 2).

$\sigma(z)$ is the average, over the thickness, of the tensile stresses acting in the circumferential direction (perpendicular to Fig 2) in the longitudinal plane defined by the distance z from the axis of symmetry of the cross-section. The tensile force in a circumferential strip of unit width is σt. Due to the curvature of the strip, a centripetal force $q(z) = \sigma t/r$ per unit length tends to pull the strip downwards. The load q produces sagging transverse moments $m(z)$ in the flange. We consider sagging moments to be negative quantities $m(z)$. If σ were constant over the width $2b$, the bending moment $m(o) = -qb^2/2$ would generate elastic transverse stresses

$$\sigma_z(o) = \pm \frac{6m(o)}{t^2} = \pm \frac{3b^2}{rt}\sigma \qquad (1)$$

in the outer fibres along the centre-line of the flange. These stresses may exceed σ considerably.

The assumption that $\sigma(z)$ does not vary with z is an approximation, for the load $q(z)$ causes the flange to deflect. Due to the deflections the circumferential strips are strained less and $q(z)$ and $|m(z)|$ decrease. By the same token the circumferential direct force in the flange decreases and the efficiency and bending strength of the box-section diminish.

Bleich has evolved a theory for the stress distribution in the flange, assuming elastic behaviour. It leads to high values of the maximum effective stress evaluated on the basis of von Mises's yield criterion and to low values of the effective width $2b_{ef}$, b_{ef} being so defined that its ratio to b is equal to the ratio of the maximum usable resultant F of the circumferential stresses in the flange to F_y:

$$\frac{2b_{ef}}{2b} = \frac{F}{F_y} = \frac{F}{2btf_y} \qquad (2)$$

$F_y = 2btf_y$ is the circumferential force which causes the whole cross-section of the flange to yield. f_y is the yield stress.

Massonnet and Save (1) have evaluated the ultimate strength of a curved flange, assuming fully plastic behaviour, by means of a theory based on an approximate version of the plasticity criterion of Tresca-Guest. The effective width resulting from their theory is considerably higher than that found by Bleich. They have also performed six tests on curved I-beams subjected to pure bending and found in every case that the real ultimate strength was somewhat higher than the strength predicted by their theory.

Vandepitte (2) investigated the ultimate strength of curved flanges of I-beams, but based his theory on von Mises's yield criterion, which is not only a little higher than the Tresca-Guest criterion, but which is also in wider use and in better agreement with the experimentally observed behaviour of ductile metals. As might be expected, his theory predicts somewhat higher strength than the theory of Massonnet and Save, and it is thus not at variance with Massonnet's and Save's experimental results.

Later on Vandepitte developed a more complete theory (3) in which the influence of the radial shear stresses in the longitudinal cross-sections of the flange is accounted for. For I-beams a diagram is provided which gives b_{ef}/b directly as a function of the parameter $\zeta^0 = 2b/\sqrt{rt}$.

1. *FUNDAMENTAL EQUATIONS*

The ultimate limit state of a curved flange of a box girder will now be studied. It is assumed that the metal yields *at all points* under the combined effect of the circumferential tension, the transverse bending and the radial shear. Strain hardening is neglected. The transverse bending stress on the upper half of the longitudinal cross-section at distance z from the axis of symmetry is $+\sigma_z$ and that on the lower half is $-\sigma_z$ (Fig 3). A positive σ_z is considered to represent a tensile stress. The actual value of σ_z is negative. The resulting bending moment per unit length of the longitudinal cross-section is $m = t^2\sigma_z/4$. The normal stress on the radial cross-section, in its upper half and in its lower half, is denoted respectively by $\sigma + \Delta\sigma$ and by $\sigma - \Delta\sigma$. Hence σ again is the average circumferential stress in the longitudinal plane defined by z and $q = \sigma t/r$ is the centripetal force per unit area of the flange. The radial shear stress τ produced in the same longitudinal plane by the shear force $\int_0^z q.dz$ is assumed uniform over the thickness t in the ultimate limit state of the flange :

$$\tau = \frac{1}{t}\int_0^z q.dz \qquad (3)$$

This is untrue at the upper and lower edges of the flange, where τ is in fact zero.

The following notation is introduced

$$\omega = \frac{\sigma_z}{f_y} \quad , \quad \alpha = \frac{3t}{4r} \quad , \quad \zeta^0 = \frac{2b}{\sqrt{rt}} \tag{4}$$

$$\zeta = \frac{2z}{\sqrt{rt}} \quad \text{and} \quad ' \equiv \frac{d}{d\zeta}$$

σ , $\Delta\sigma$, σ_z , τ , q and m are functions of ζ. The following relations hold for a tranverse strip of the flange

$$q = \frac{d^2m}{dz^2} \quad \text{or} \quad \frac{\sigma t}{r} = \frac{t^2}{4} \cdot \frac{d^2\sigma_z}{dz^2} \quad \text{or}$$

$$\sigma = \frac{rt}{4} \cdot \frac{d^2\sigma_z}{dz^2} = \sigma_z'' = \omega'' f_y \tag{5}$$

and $$\int_0^z q.dz = \frac{dm}{dz} = \frac{t^2}{4} \cdot \frac{d\sigma_z}{dz} = \frac{t^2}{4} \cdot \frac{2}{\sqrt{rt}} \cdot \sigma_z' \quad \text{or}$$

$$\tau = \frac{1}{2}\sqrt{\frac{t}{r}}\, \omega' f_y \tag{6}$$

The total circumferential tensile force in the flange amounts to

$$F = 2\int_0^b \sigma t.dz = t\sqrt{rt}\, f_y \int_0^{\zeta^0} \omega''.d\zeta = t\sqrt{rt}\, f_y \omega'(\zeta^0) \tag{7}$$

and **eqn** *(2)* now yields

$$\frac{2b_{ef}}{2b} = \frac{F}{2btf_y} = \frac{\omega'(\zeta^0)}{\zeta^0} \tag{8}$$

von Mises's yield condition $\sigma_{ef}^2 = f_y^2$ for points in the upper half of the flange gives

$$(\sigma + \Delta\sigma)^2 + \sigma_z^2 - (\sigma + \Delta\sigma)\sigma_z + 3\tau^2 = f_y^2 \tag{9}$$

and for points in the lower half

$$(\sigma - \Delta\sigma)^2 + \sigma_z^2 + (\sigma - \Delta\sigma)\sigma_z + 3\tau^2 = f_y^2 \tag{10}$$

Subtraction of eqn *(10)* from eqn *(9)* yields

$$2.\Delta\sigma = \sigma_z \tag{11}$$

Substitution of $\sigma_z/2$ for $\Delta\sigma$ in the sum of eqns *(9)* and *(10)*, and use of eqns *(4)*, *(5)* and *(6)* lead to

$$\sigma^2 + 3\tau^2 + \frac{3}{4}\sigma_z^2 = f_y^2 \tag{12}$$

and $$\omega''^2 + \alpha\omega'^2 + \frac{3}{4}\omega^2 = 1 \tag{13}$$

The influence of the radial shear stresses is represented by the term $\alpha\omega'^2$ in eqn *(13)*.

Vandepitte (3) obtained this differential equation. In the present case the appurtenant boundary conditions are

$$\frac{d\sigma_z}{dz} = 0 \quad \text{for} \quad z = 0 \text{ ,} \quad \text{or} \quad \omega'(o) = 0 \tag{14}$$

and $$m(z) = 0 \quad \text{for} \quad z = b \text{ ,} \quad \text{or} \quad \omega(\zeta^0) = 0 \tag{15}$$

Boundary condition *(14)* has already been used implicitly in writing eqn *(3)* and the last expression *(7)*.

2. *SUBSTITUTING DIFFERENCE EQUATIONS FOR DIFFERENTIAL EQUATIONS*

Equations *(13)*, *(14)* and *(15)* determine the function $\omega(\zeta)$. The second-order differential eqn *(13)* is non-linear and eqns *(13)* and *(14)* will therefore be replaced by difference equations and solved numerically, although an analytical treatment is also possible.

ω_{i-1} , ω_i and ω_{i+1} denote the values of ω corresponding to the values $\zeta_{i-1} = (i-1)\Delta$, $\zeta_i = i\Delta$ and $\zeta_{i+1} = (i+1)\Delta$ of the independent variable, i being any integer and Δ a small increment of ζ.

$$\omega_i' = \frac{1}{2\Delta}(\omega_{i+1} - \omega_{i-1}) \tag{16}$$

and $$\omega_i'' = \frac{1}{\Delta^2}(\omega_{i-1} - 2\omega_i + \omega_{i+1}) \tag{17}$$

are inserted into eqn *(13)* and the resulting equation is solved for ω_{i+1} :

$$\omega_{i+1} = \frac{1}{1 + \frac{\alpha\Delta^2}{4}}\left[2\omega_i - \left(1 - \frac{\alpha\Delta^2}{4}\right)\omega_{i-1} \pm \Delta\sqrt{\Delta^2\left(1 + \frac{\alpha\Delta^2}{4}\right)\left(1 - \frac{3}{4}\omega_i^2\right) - \alpha(\omega_i - \omega_{i-1})^2}\right] \tag{18}$$

Boundary condition *(14)* is now written

$$\omega_o' = \frac{1}{2\Delta}(\omega_1 - \omega_{-1}) = 0 \text{ ,} \quad \text{whence} \quad \omega_{-1} = \omega_1 \tag{19}$$

Equation *(13)*, written for $i = 0$, results in

$$\omega_o''^2 = 1 - \frac{3}{4}\omega_o^2 \quad \text{or} \quad \omega_1 = \omega_o \pm \frac{\Delta^2}{2}\sqrt{1 - \frac{3}{4}\omega_o^2} \tag{20}$$

The calculation procedure is as follows. A negative value is selected for ω_o . Equation *(20)* gives ω_1 and eqn *(18)*, written for $i = 1$, gives ω_2 . Repeated application of eqn *(18)* subsequently yields the values of ω_3 , ω_4 ,...., and the numbers ω_1' , ω_2' , ω_3' ,... and ω_o'' , ω_1'' , ω_2'' ,... are then evaluated by means of eqns *(16)* and *(17)*.

3. *NUMERICAL RESULTS FOR HIGH VALUES OF r/t*

It is first assumed that $\alpha = 0$. This is tantamount to ignoring the effect of the shear stresses on plastification when r/t is not great. Equation *(18)* reduces to

$$\omega_{i+1} = 2\omega_i - \omega_{i-1} \pm \Delta^2\sqrt{1 - \frac{3}{4}\omega_i^2} \tag{21}$$

The calculations are performed by means of a small computer program, taking $\Delta = 0.001$ and selecting the plus sign in eqns *(20)* and *(21)*. For every starting value ω_o the computation is stopped when ω_i reaches zero and it produces the variation of $\omega(\zeta)$ with ζ, the value of ζ^0 fulfilling the boundary condition $\omega(\zeta^0) = 0$ and the value of $\omega'(\zeta^0)$. Calculating positive values, up to $\sqrt{4/3}$, for $\omega(\zeta)$ and calculating $\omega(\zeta)$ starting with a positive ω_o not exceeding $\sqrt{4/3}$ would make sense for a flange extending beyond the webs, for hogging bending moments $m(\zeta)$ may then exist in the part of the flange situated between the webs.

The lowest admissible starting value ω_o is $-\sqrt{4/3} = -1.1547$, for eqn *(13)* does not enable real numbers ω'' and ω' to be found when ω_o is lower than $-\sqrt{4/3}$. Figure 4 shows the curves $\omega(\zeta)$ obtained with starting ordinates $\omega_o = -0.05$; -0.2 ; -0.4 ; -0.6 ; -0.8 ; -1.0 ; -1.1 ; -1.15 ; $-\sqrt{4/3}$ and Table 1 contains the corresponding values of ζ^0, $\omega'(\zeta^0)$ and $\omega'(\zeta^0)/\zeta^0$.

It is permissible to switch from the + sign to the - sign and back to the + sign in eqn *(18)* or *(21)* any number of times when calculating an $\omega(\zeta)$ curve. Different curves thus obtained but ending with the same abscissa ζ^0 end with different slopes $\omega'(\zeta^0)$, none of which is higher than the one given in Table 1. The eqns *(13)*, *(18)* and *(21)* are derived from equilibrium considerations and from von Mises's yield criterion. Hence, all stress distributions found by means of those equations are statically admissible and safe in the sense of the theory of plasticity. By virtue of the statical theorem for perfectly elastic-plastic bodies the ultimate circumferential load in the flange is the highest of the loads corresponding to all statically admissible and safe stress distributions. This, together with eqn *(8)*, implies that the numbers in the last column of Table 1 are the correct values of b_{ef}/b.

The lowest curve in Fig 4, when shifted horizontally to the right, still satisfies eqn *(13)*, since the shift does not affect the numerical values of ω, ω' and ω''. After the shift the diagram may be completed by the horizontal straight line $\omega = -\sqrt{4/3}$ extending between the ω-axis and the beginning of the curve ; indeed $\omega' = \omega'' = 0$ along the straight line and thus it satisfies eqn *(13)*. Other $\omega(\zeta)$ curves do not have a steeper right end. Consequently $\omega'(\zeta^0)$ remains constant and amounts to $1.3468 = \sqrt{\pi/\sqrt{3}}$ for all values of ζ^0 exceeding 3.1838. It follows from eqn *(7)* that widening a box girder beyond the width $2b = 3.1838\sqrt{rt}$ does not further increase the effective tensile strength of its curved flange and it follows from eqn *(8)* that

$$2b_{ef} = 1.3468 \times 2b/\zeta^0 = 1.3468\sqrt{rt} \quad \text{if} \quad 2b \geqq 3.1838\sqrt{rt} \qquad (22)$$

The ratio of the effective width $2b_{ef}$ to the actual width $2b$ is represented as a function of ζ^0 by the upper curve in Fig 5. The part of the diagram to the right of the point with abscissa $\zeta^0 = 3.1838$ is the hyperbola defined by eqn *(22)*.

4. *NUMERICAL RESULTS FOR LOW VALUES OF r/t*

The calculation for a sharply bent flange with $r/t = 10$ is the same as in Section 3, except for the use of the complete formula *(18)*, in which we let $\alpha = 3/40$ and $\Delta = 0.001$. The $\omega(\zeta)$ curves have the same general shape as in Fig 4. The diagram giving b_{ef}/b as a function of ζ^0 for a flange with $r/t = 10$ is the lower curve in Fig 5. It differs so little from the curve $r/t = \infty$ obtained in Section 3 that it is not worth-while to calculate curves for other ratios r/t.

The influence of the radial shear stresses on the plastification of a curved flange was shown before to be very slight also for I-beams (3).

5. CONCLUSION

The effective width $2b_{ef}$, for resisting a circumferential direct force, of a curved flange in a box girder in its ultimate limit state can be read directly from the curves in Fig 5. The upper curve is valid for high ratios r/t, the lower one for $r/t = 10$. The trifling difference between the two curves represents the effect of the radial shear stresses on the ultimate strength of the flange. The treatment is limited to flanges that do not extend beyond the two webs of the box girder.

References

1. Massonnet, Ch. and Save, M. 'Résistance limite d'une poutre courbe en caisson soumise à flexion pure.' Amici et Alumni, Hommage à Fernand Campus, 1964.
2. Vandepitte, D. 'Berekening van constructies; Bouwkunde en civiele techniek.' Volume II, Story-Scientia, Gent, Antwerpen, Brussel, Leuven, 1980, 1-709.
3. Vandepitte, D. 'Ultimate strength of curved flanges of I-beams.' Journal of Constructional Steel Research: Vol. 2, No. 3, September 1982.

Table 1

ω_o	ζ^0	$\omega'(\zeta^0)$	$\omega'(\zeta^0)/\zeta^0$
-0.05	0.3163	0.3162	0.9995
-0.2	0.6360	0.6309	0.9919
-0.4	0.9154	0.8853	0.9671
-0.6	1.1587	1.0694	0.9229
-0.8	1.4164	1.2086	0.8533
-1.0	1.7677	1.3074	0.7396
-1.1	2.0945	1.3385	0.6390
-1.15	2.5933	1.3466	0.5192
$-\sqrt{4/3}$	3.1838	1.3468	0.4230

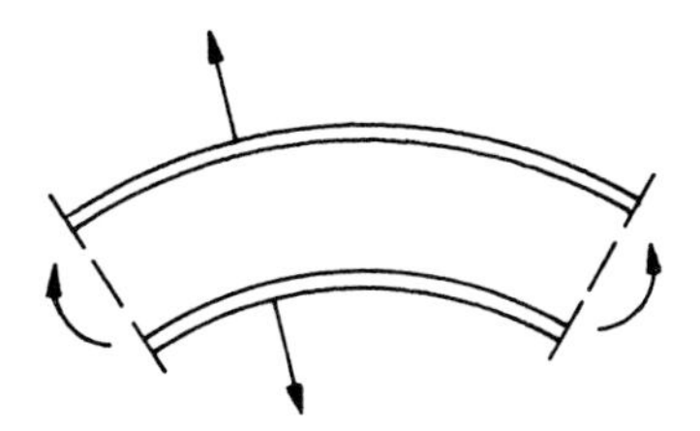

Fig 1

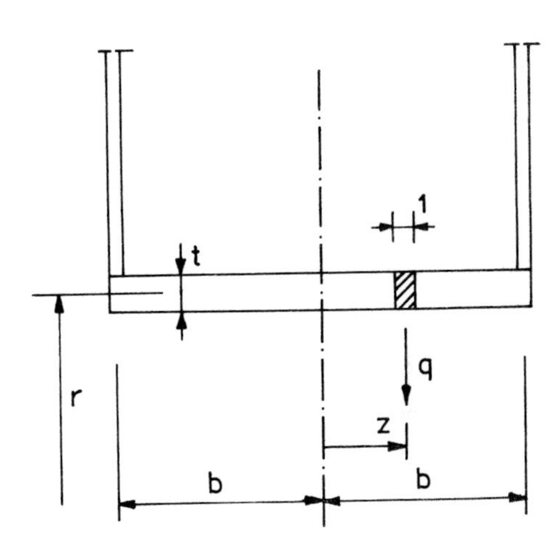

Fig 2

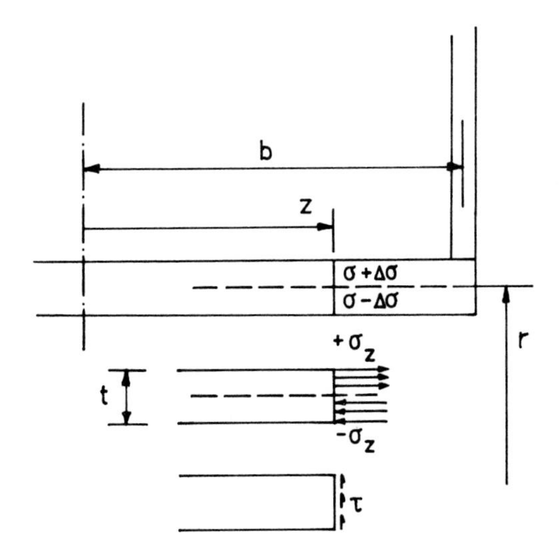

Fig 3

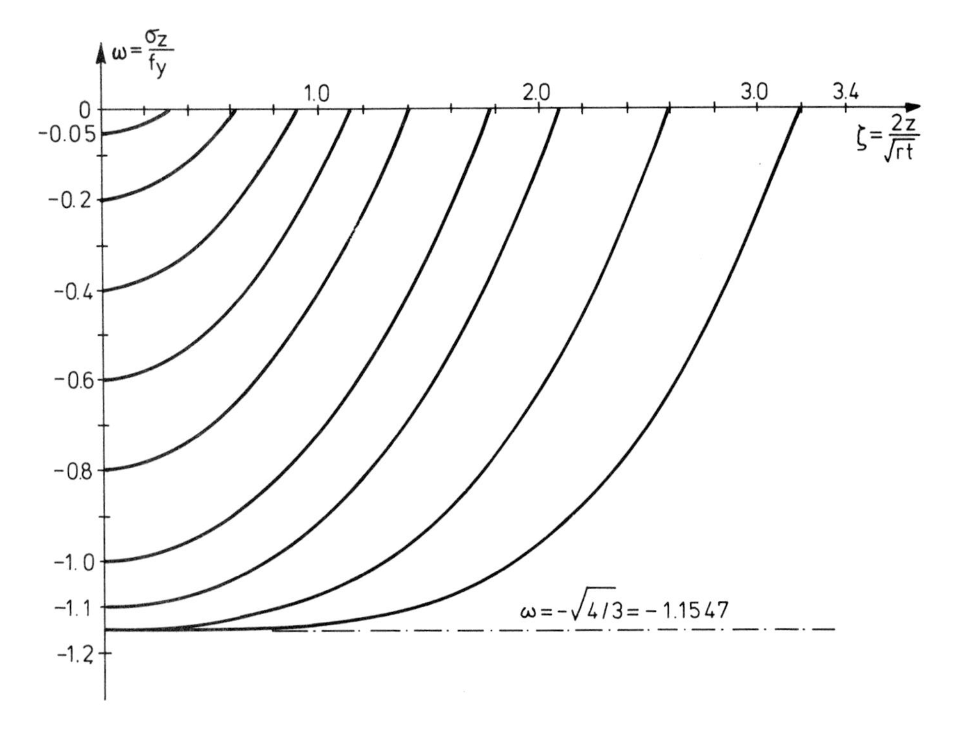

Fig 4 ω-ζ-curves

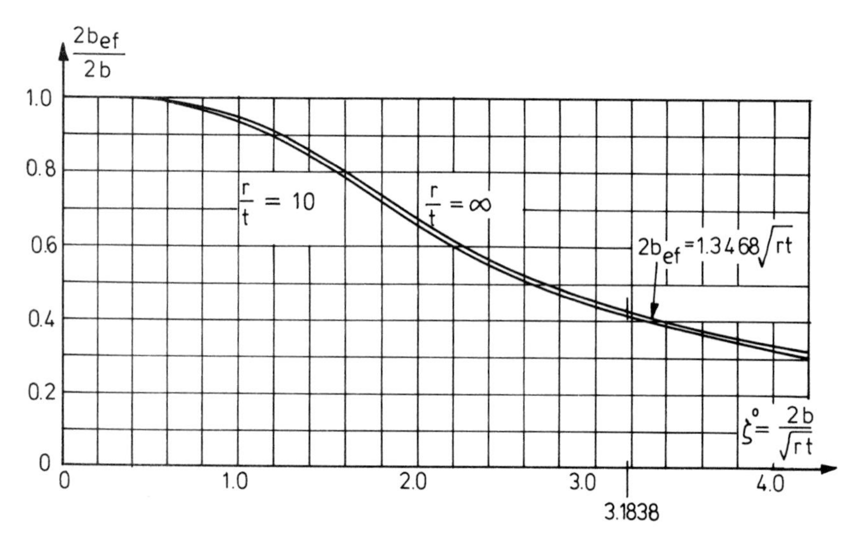

Fig 5 Effective width as a function of ζ^0

COLD-FORMED STEEL STRUCTURAL ELEMENTS: DEVELOPMENTS IN DESIGN AND APPLICATION.

Rolf Baehre

Head of Department of Steel Construction, University of Karlsruhe.

Synopsis

The development of cold-formed steel structural elements is discussed in the light of applications to building construction. Growing experience of the behaviour of thin-walled structures, comprehensive research work, increased material strength and the development of cold forming techniques provide the basis for advanced light-weight building techniques in different areas of the building sector.

1. *HISTORICAL ASPECTS*

Formerly the use of cold-formed thin-walled steel sections was mainly concentrated on products where weight saving was of prime importance, e.g. in the aircraft, wagon and motor industries. On the other hand simple types of profile (mainly corresponding to hot-rolled shapes), as well as profiled sheeting, have been used as non-structural elements in building for about one hundred years.

Structural design based on thin-walled components involves basic research work, experience of the load-bearing behaviour and capacity, critical judgement of safety aspects and acceptance of design rules.

Systematic research work in the field of thin - walled structural elements for use in buildings was commenced at the end of the thirties by George Winter at Cornell University, and as early as 1946, the first edition of the "Specification for the Design of Light Gauge Steel Structural Members", published by the American Iron and Steel Institute, was available. Since then, significant research, product development and specification activities have taken place in many countries. However, there is no doubt that the research efforts at Cornell University have had a strong impact on the world-wide use of thin-walled, cold-formed sheeting and sections together with appropriate research projects in many other parts of the world. The latter is impressively demonstrated by a "Survey on Current Research on Cold-Formed Structures" published by the U.S. Committee on Cold-Formed Members in 1979, which refers to about 70 actual projects in different countries concerned with a wide variety of problems such as torsional-flexural buckling of sections, diaphragm action of sheeting, behaviour of connections, composite action of thin-walled components with other materials and light-weight building techniques in the broadest sense.

However, neither the verified results of scientific work nor useful specifications can alone bring a new technology to success. A market has to be built up and this market has to be served with appropriate products which are favourable in terms of economy and reliability and which meet the market's demand for better quality of the end product. As an example of such market-orientated developments, the use of sheeting in Scandinavia may be quoted. The development of sheeting profiles throughout the past three decades is indicated in general terms in Fig. 1. The progress in profile development from the sinusoidal corrugated sheeting for small spans via trapezoidal sheeting within a span range of 3 to 7 m to high-rise cassette profiles for spans up to about 15 m has been made possible by

- extended knowledge of structural design
- improved cold-forming techniques
- use of high-strength materials

but has also been enforced by the market's demands for

- adjustment to functional requirements
- optimization with respect to building performance
- improved energy conservation
- aesthetic aspects
- competitiveness as against other building methods.

It is a matter of fact that all these factors together have yielded a market share for metal roofs in Sweden of about 90%.

The use of cold-formed structural members offers many advantages since the shape of the section can be optimized to use the material to best advantage. In addition there are ample opportunities for ingenious solutions and the application of these to practical problems has formed a basis for success.

In many countries, cold-formed steel construction is the fastest growing (or only growing) branch of steel structural applications. Sheet steel is used more often in construction today than any other steel mill product.

2. BASIS OF DESIGN

Compared with conventional steel structures, thin-walled structural elements are characterized by

- relatively high "width to thickness" ratios
- unstiffened or incompletely restrained parts of sections
- singly symmetrical or unsymmetrical shapes
- geometrical imperfections of the same order as or exceeding the thickness of the section
- structural imperfections caused by the cold-forming process.

As a consequence, thin-walled structural elements have to be considered with respect to a number of factors, for example:-

- buckling within the range of large deflections
- effects of local buckling on the total stability
- combined torsional and flexural buckling
- shear lag and curling effects
- effects of varying residual stresses over the section.

Under increasing load, thin-walled structrual elements are generally subject to varying non-linear distributions of stress and strain over the cross-section, often in conjunction with substantial out-of-plane deflections. There also exists the possibility of different failure modes, particularly for sections with flat parts in compression that are unstiffened (i.e. elastically restrained along one edge only).

The critical load, i.e. the lowest load at which the structural element assumes a deflection equilibrium position, constitutes an important basis for the estimation of the load-bearing capacity, entailing distinct fundamental modes of failure. Assuming ideal geometrical and structural conditions, the bifurcation load indicates either an upper limit to the load-bearing capacity (unless there is a post-critical region) or the loading level at which the stiffness of the structure decreases rapidly and deflections occur at a faster rate until a new stable equilibrium condition is reached. Provided that geometrical imperfections, or even deflections caused by local buckling, govern the load-deflection behaviour of the structure, a similar increase in deformation can be observed in the region of, or below, the bifurcation load.

As cold-forming techniques allow the geometrical properties of a shape to be readily varied, it is possible to influence the load-bearing behaviour of the structure with respect to strength, stiffness and failure modes by, for example, the introduction of intermediate stiffeners or by ensuring adequate width-to-thickness ratios in adjacent flat parts of the section. These possibilities involve a challenge for the designer.

It is evident from the discussion above that an accurate analyis of the mode of action is usually extremely complicated, especially when imperfections and plasticity have to be taken into consideration. For practical design there is a need for simplified analytical models which allow an approximate but conservative estimate of the failure load and the behaviour of the structure under service load to be made.

A good example is represented by the analysis of the post-critical behaviour of a long compressed plate, where the accurate analytical treatment given by v. Karman's differential equations, has been replaced by the well known "effective width" model. Thus the ultimate load can be determined from a uniform stress distribution within an effective width b_e, which depends on the critical buckling stress (σ_k) (= bifurcation load) and the yield stress (σ_s) of the plate material. The expression for b_e, given by v. Karman, has been subsequently modified by, for example, G. Winter with provision for unintended geometrical imperfections (see Fig. 2). The semi-empirical character of the equation emphasises the importance of experimental investigations and experience, not only in order to verify failure loads based on analytical models, but much more to investigate the behaviour of thin-walled structures and elements during the loading process up to failure. Experience gained through experiments is of the greatest benefit in promoting a feeling for good design and an intuitive understanding of the performance of thin-walled structures.

So far, the discussion has concentrated on basic products, such as cold-formed profiles and profiled sheeting. A structure, however, is built up of both types of components which are connected to each other by means of mechanical fasteners. The question arises as to whether the knowledge of individual parts of the structure is transferable to the behaviour of the full structure. It is necessary to have both a basic knowledge of the characteristic behaviour of different fasteners in connections and of the response of the structure under different loading conditions. In general terms, the use of different types of fasteners can be limited as shown in Fig. 3. Nevertheless, knowledge of the slip and load-bearing capacity of connections is necessary and has to be clarified by tests under conditions which are similar to the behaviour in real structures. In contrast to conventional steel structures, connections in thin-walled structures may cause increased flexibility due to concentrated high load-bearing stresses and slip of the fasteners. Experimental work and experience resulting from full-scale tests on structures again provide a reliable basis for design. Results of scientific work and tests are presented in a great number of papers. A good overall view is provided by references (1) to (5).

3. *HARMONIZATION OF DESIGN RULES*

As mentioned above, international research and development has taken place on a large scale during the past twenty years and, as a consequence, design specifications have been produced in several countries.

About ten years ago, it was decided by the "European Convention for Constructional Steelwork" (ECCS) to set up a Committee on "Cold-formed thin-walled sheet steel in structures" (TC7) charged with the development of European Recommendations for cold-formed sheeting and sections and related activities in design and construction. The Committee is at present chaired by Professor E. R. Bryan from Salford University. The membership includes representatives from Austria, Belgium, Finland, France, Great Britain, Germany (BRD), Italy, Luxembourg, Netherlands, Norway, Sweden, Switzerland and U.S.A. and an associated group of C.I.B. member countries, namely Czechoslovakia, Hungary, Germany (GDR) and Poland. The Committee started its work in 1974 and has so far produced, as a basis for national specifications, Recommendations (E.R.) concerning:

stressed skin design of steel structures
design of profiled sheeting
design of light-gauge steel members
design of connections in thin-walled structural steel elements
testing of profiled metal sheets, sections and connections
good practice in steel cladding and roofing.

The large amount of research and development completed during the past nine years makes it necessary to review the work of Committee TC7, and a revision of the Recommendations is in progress. A new task group has recently started to prepare Recommendations for the design and use of sandwich elements.

The work of Committee TC7 has been widely appreciated and the design rules presented in the Recommendations have been adopted as the basis of many national specifications. This has been possible largely because international harmonization in the above subject areas started before national standards were produced so that progress in developing design rules could be taken into consideration. It should be pointed out that the close co-operation of the ECCS with representatives from the U.S.A has led to a fruitful interchange of experience and a closer harmonization of design procedures.

4. *PRODUCT DEVELOPMENT*

Sucessful product development, as mentioned above, is focussed on the functional and economic demands of the end product - whether it be a structure, a building or a whole technical system - whatever the market strategy of the producer <u>or</u> the market conditions may be.

In many countries, there is a clear trend towards a product refinement in terms of prefabrication, design of multi-functional modes of application, optimization with respect to building performance, and the development of complete structural systems. In such a development, the strength and rigidity requirements become a part of the total functional demands of the end product. This may be defined as the creation of a space which needs to be evaluated in terms of verified qualities, such as durability, climate, sound, fire protection and interior environment (6). The complexity of different functional demands is illustrated in Fig. 4, which demonstrates a quality profile for a flooring system in an office building. The variation of quality demands with respect to different functional requirements is exemplified by six levels, where "level five" means the highest degree of demand. These quality demands are partly made by appropriate building specifications, partly by the user himself, and imply different effects on the product development.

An important conclusion from such quality profiles is that the consideration of space-covering light-weight building systems leads logically to composite structures comprising thin-walled cold-formed products in combination with layers of other materials. In such a system metal panels acting alone or in combination with suitable materials have load-carrying and stablizing functions. These other materials, such as plaster, plywood, board or mineral wool, have been found to have considerable influence on the behaviour of thin-walled steel panels, provided that adequate connection and bonding between the materials are provided.

Trends in product developments, as discussed above, are illustrated in Figs. 5 to 7, showing the evolution from simple and functionally anonymous basic products such as sheeting profiles, sections and suitable composite materials, via building elements designed for special functional purposes, such as beams, columns, walls, roofs and floors, to complex structural building systems.

The trend illustrated above towards building systems implies an exciting challenge for the designer, as these types of products - especially the possibility of creating space-covering building elements - pave the way for new fields of application of steel structures. However, product development involving the different functional requirements discussed in Fig. 4 requires that the designer has an appropriate basic knowledge of a number of scientific areas, such as building physics, material sciences and building economics.

5. *OUTLOOK*

The competitiveness of cold-formed steel structures is documented by applications from all parts of the building sector and from all over the world (7). Extensive experience from theoretical and experimental research work provides a sufficient basis for safe design of building components as individual parts of a structure.

However, when considering complete structural systems, built up of cold-formed sheeting and sections connected to each other by means of mechanical fasteners or spot welding methods, it is apparent that more information about the overall behaviour of the structure is necessary for safe design. This is all the more important if we are seeking novel types of steel structure other than those based on conventional skeletons, girders or frames, using hot-rolled or welded components. This, on the other hand, appears to be the right way to proceed in the development of an appropriate building technology, in which light gauge components are used to the best of their ability in performing functional requirements. This overall view is concluded by a description of the creation of such a system. The basic idea is to develop a roof system (Figs. 8 and 9) for light industrial buildings which meets, as far as possible, the following demands:

(1) Spans from 15 to 30 m approximately shall be covered by only one type of structural element; increased spans shall be possible.

(2) The structural elements shall be produced on line in an industrial process.

(3) The necessary space for storage and transportation shall be minimized.

(4) As the girders are to be placed on the substructure in a single operation, sufficient stability during erection shall be provided by the girder itself.

(5) The roof sheeting shall consist of trapezoidal sheeting, minimized in weight and able to provide diaphragm action, covered with insulating layers and a waterproof outer skin.

(6) The structural girder system shall provide space for functional requirements such as sound insulation and installations for heating, ventilation and lighting.

Based on the considerations discussed above, and with regard to the functional requirements, a V-shaped structural girder system, as illustrated in Figs. 8 and 9, has been investigated. This comprises plane elements built up from thin-walled low-rise trapezoidal sheeting and cold-formed edge profiles which can be manufactured on line in the workshop and transported to the building site while occupying a minimum of space. Using simple V-shaped frames, the two plane beams can be assembled on the ground to form a V-girder by connecting the lower edge profiles and by providing temporary stiffeners or trapezoidal sheeting at the top of the V. The girders can then be placed on the superstructure. By varying the distance between the V-girders, spans between 15 and 30 m can be obtained with the same section of about 1200 mm depth. A particular advantage of the system is that the span of the horizontal trapezoidal sheeting - due to the V-shaped support - is relatively small and this allows the use of light-weight sheeting. For the maximum span, where the top flanges of the V-girders are close together, an orthotropic structural system of beams in one direction and trusses in the other direction is obtained by providing the lower flanges of the beams with another course of profiled sheeting. This two-way girder system makes substantially increased spans possible.

The form of the structure allows provision for lighting and other services to be made within the V-beams. The webs of the V-beams can also be perforated in order to obtain sound insulation or ventilation. For architectural effects, plastic coatings can be applied to webs, profiles and roof parts.

A current research project at the University of Karlsruhe is concerned with the technical development of this form of construction and its practical realisation. It is thought that it may perhaps illustrate an attempt to find a new structural concept using light-gauge building techniques.

References

1. International Specialty Conferences on Cold-Formed Steel Structures, 1971 - 1982, Proceedings, University of Missouri-Rolla.
2. Rhodes, J. and Walker, A.C. 'Thin-walled Structures.' Granada, 1980.
3. Davies, J. M. and Bryan, E.R. 'Manual of Stressed Skin Diaphragm Design.' Granada, 1982.
4. Rhodes, J. and Walker, A.C. 'Developments in Thin-walled Structures.' Applied Science Publishers Ltd., 1982.
5. Baehre, R. 'Entwicklungsmerkmale der Leichtbautechnik: Aussteifungen, Komponenten, Verbund.' Swedish Council for Building Research, Doc. D8:1978, Stockholm.
6. Baehre, R. 'Raumabschließende Bauelemente.' Stahlbau Handbuch, Band 1, Kap. 17, Stahlbau Verlags GmbH, Köln, 1982.
7. Baehre, R. et al. 'Cold-formed Steel Applications Abroad.' Sixth International Specialty Conference on Cold-formed Steel Structures, University of Missouri-Rolla, 1982.

TIME	DEMANDS →	MEASURES FOR DEVELOPMENT →	TYPE OF PROFILES
	Light weight roofing	Cold-forming techniques (transversal corrugations)	Sinusoidal sheeting
	Aesthetics	Press-brake techniques Coating techniques	"Architectural" cladding
50's	Load bearing capacity	Design criteria and calculation methods	Profiled sheeting
	Mass production	Cold-rolling techniques (longitudinal profiling)	Trapezoidal sheeting
60's	Functional requirements	Building performance (e.g. insulation, deflection)	Functional adjustment
	Increased bearing capacity and spans	Preventing of local buckling, high-strength materials high-rise profiles	Profiles with intermediate stiffeners
70's	Increased competition	Utilization of diaphragm action, weight reduction, building systems	Optimized profile programmes
	Increased spans Reduction of work on building site	Cold-rolling techniques (transversal + longitudinal profiling); calculation methods	High-rise casette profiles
80's	Replacement of combustible roof covering; energy conservation	Roof systems with thick insulation layer and outer sheeting (building physics)	Profiles for roof covering, adjusted to climatic requirements
	Aesthetic demands with respect to cladding and roofing	Aesthetic studies concerning profile shapes, depth, colours, types of coating Maintenance	"Architectural" profiles

Fig. 1: Development of profiled sheeting

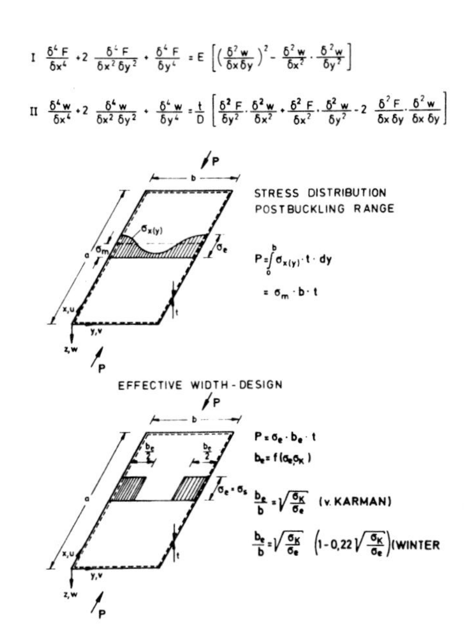

Fig. 2: Non-linear buckling equations and the effective width concept

	FUNCTIONAL REQUIREMENT	SPECIFIC REQUIREMENT
LOADBEARING CAPACITY	BENDING MOMENT	PROVISIONAL MEASURES
	SHEAR FORCES	FOR LATER INCREASE
	SINGLE LOAD	OF LOADING
	DIAPHRAGM ACTION	
	NORMAL FORCES	
STIFFNESS	BENDING STIFFNESS	
	SHEAR STIFFNESS	
	NATURAL FREQUENCY	
	DAMPING	
DURABILITY	CORROSION PROT	
	BIOLOGICAL PROT	
	TIME LIFE	
	FUNCTIONAL DURAB	
FIRE PROT	FIRE RESISTANCE	
	EXPOSED SURFACE	
	SMOKE FORMATION	
SOUND PROT	IMPACT SOUND	SPECIAL WORKING
	AIRBORNE SOUND	CONDITIONS
	FLANK TRANSMISSION	
CLIM PROT	THERMAL INSULATION	
	COLD BRIDGES	
	CONVECTION	
ROOM ENV	SURFACE TREATMENT	
	SOUND ABSORPTION	
	INSTALLATIONS	FLEXIBILITY

QUALITY DEMAND: 0 1 2 3 4 5

MINIMUM REQUIREMENT — ADDITIONAL REQUIREMENT

Fig. 4: Example for a quality profile concerning a flooring system in an office. Classification of functional requirements and quality demands

TYPE OF LOADING	TYPE OF CONNECTION: THIN-TO-THIN	TYPE OF CONNECTION: THIN-TO-THICK	TYPE OF FASTENER	TYPE OF FAILURE: ACCEPTABLE	TYPE OF FAILURE: UNDESIRABLE
SHEAR LOAD	X		BLIND RIVETS		
	X		SELF - DRILLING - SCREWS		
	X	X	SELF - TAPPING - SCREWS		
	X	X	WELDS		
		X	CARTRIDGE FIRED PINS		
		(X)	BOLTS		
	(X)	(X)	FRICTION - GRIP BOLTS		
	(X)	(X)	ADHESIVES		
TENSILE LOAD	X		BLIND RIVETS		
	(X)		SELF - DRILLING - SCREWS		
	X	X	SELF - TAPPING - SCREWS		
		X	CARTRIDGE FIRED PINS		
	(X)	(X)	BOLTS		
	(X)	(X)	FRICTION - GRIP BOLTS		

Fig. 3: Types of fasteners and connections for thin-walled structural elements

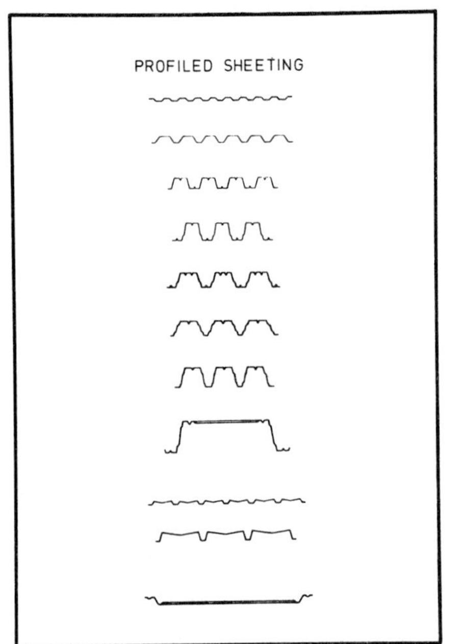

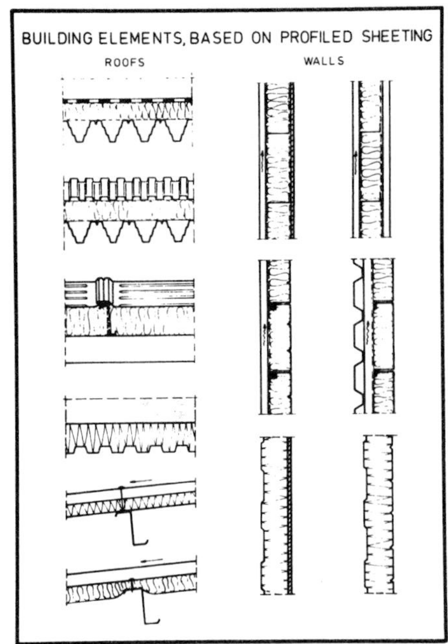

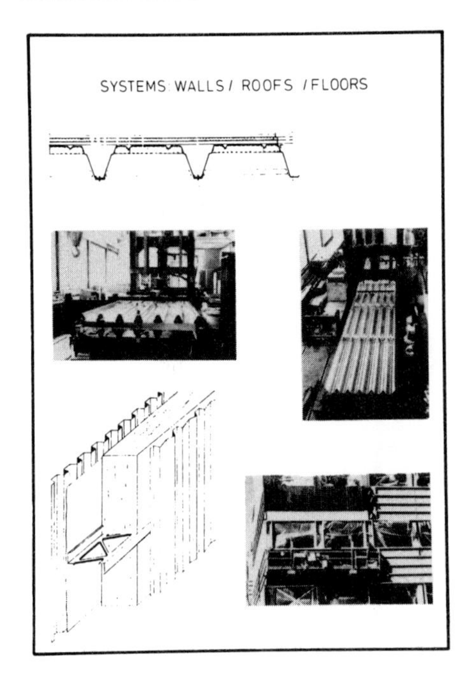

Fig. 5: Trends in product development from simple sheeting towards building elements and systems

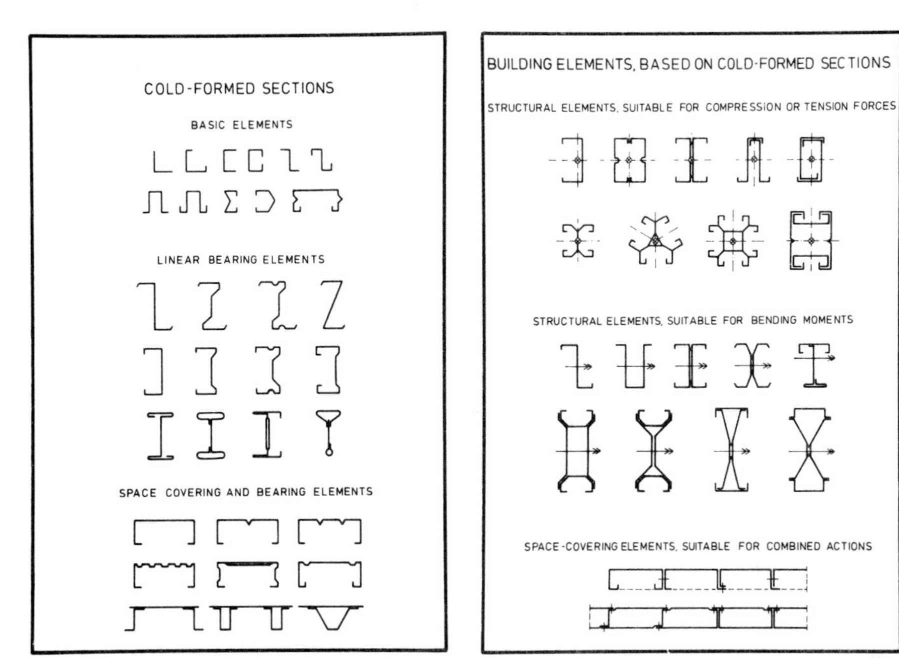

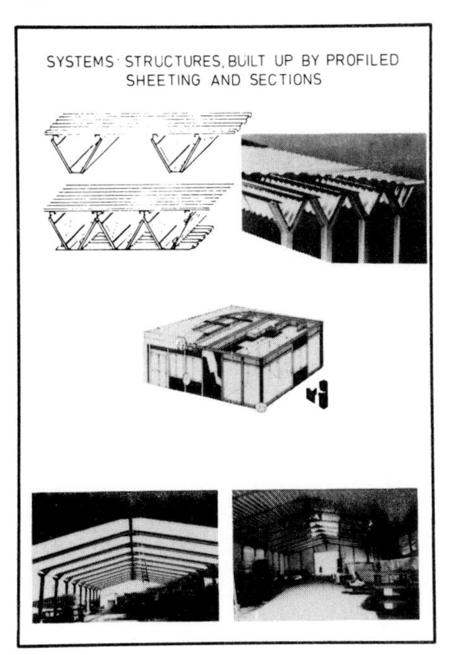

Fig. 6: Trends in product development from simple sections towards building elements and systems

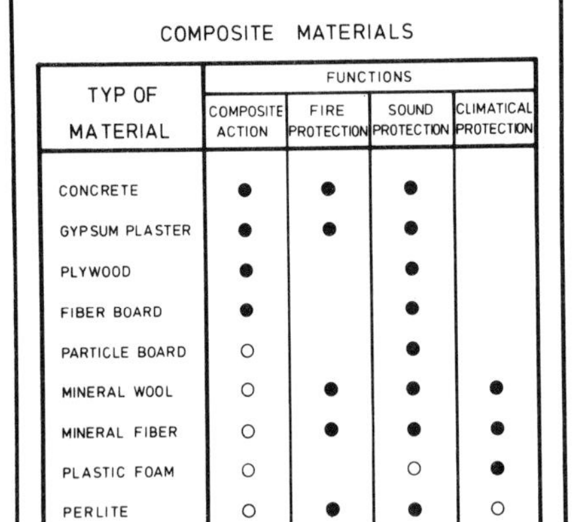

COMPOSITE MATERIALS

TYP OF MATERIAL	FUNCTIONS			
	COMPOSITE ACTION	FIRE PROTECTION	SOUND PROTECTION	CLIMATICAL PROTECTION
CONCRETE	●	●	●	
GYPSUM PLASTER	●	●	●	
PLYWOOD	●		●	
FIBER BOARD	●		●	
PARTICLE BOARD	○		●	
MINERAL WOOL	○	●	●	●
MINERAL FIBER	○	●	●	●
PLASTIC FOAM	○		○	●
PERLITE	○	●	●	○

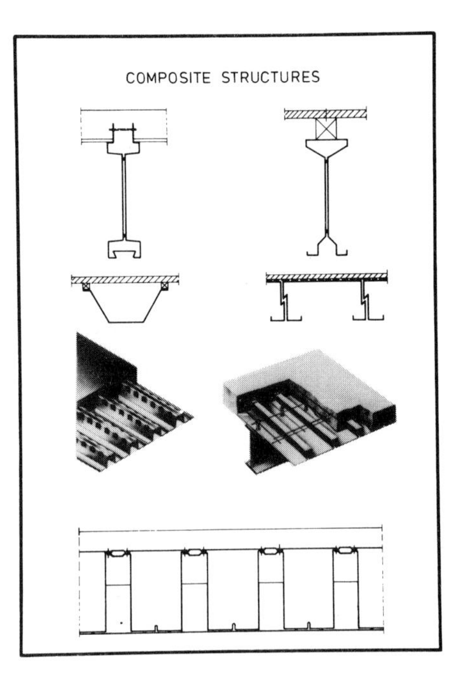

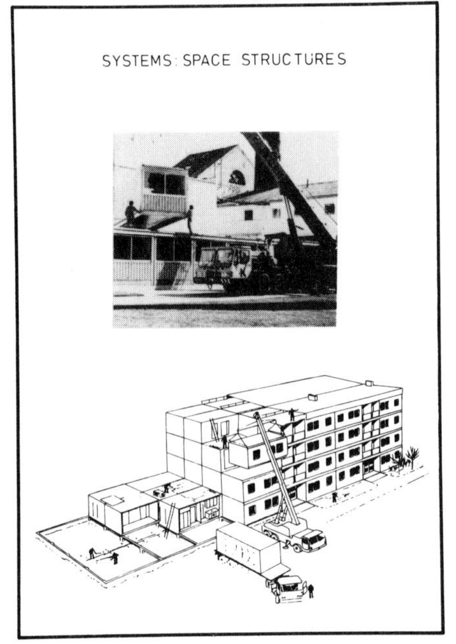

Fig. 7: Trends in product development towards building elements and systems according to quality demands as illustrated in fig. 4

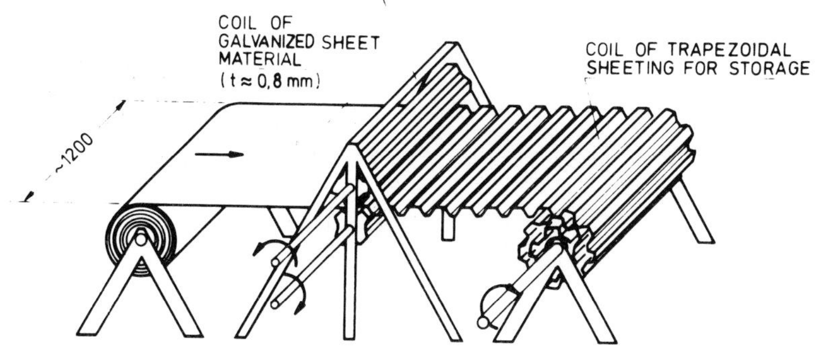

PRINCIPLE OF MANUFACTURING WEBS

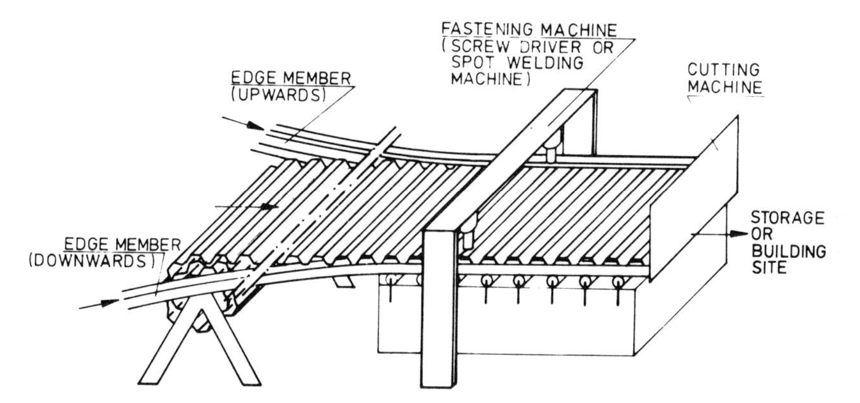

PRINCIPLE OF PRODUCING PLANE BEAM ELEMENTS ONLINE

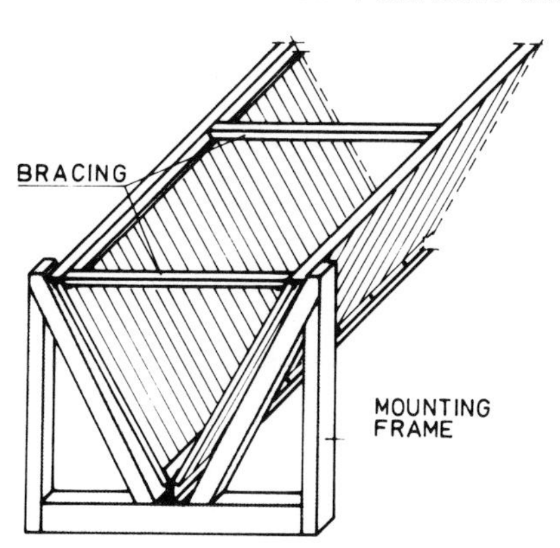

ASSEMBLY OF PLANE BEAM ELEMENTS TO V-BEAMS (AT BUILDING SITE)

Fig. 8: Development of a roof system for light industrial buildings, produced on line in an industrial process

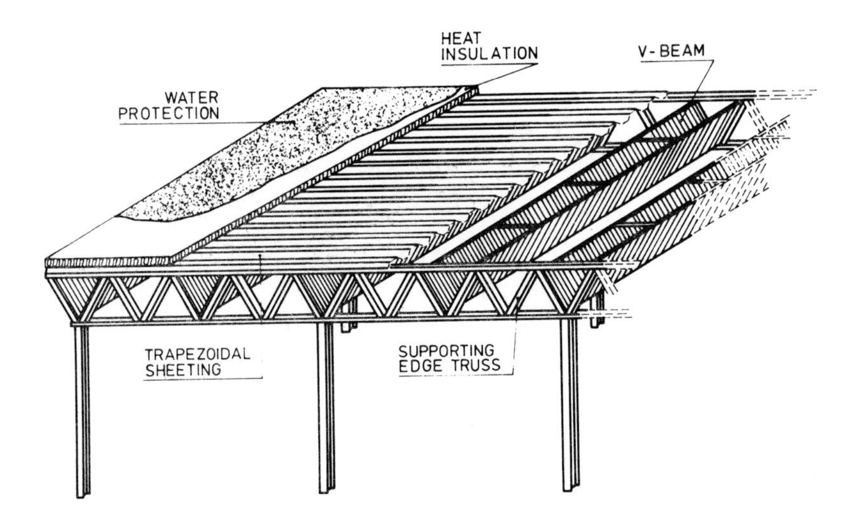

Fig. 9: Performance of the roof system, built-up by sections and trapezoidal sheeting for spans between 15 and 30 m

THE GENERALIZED BEAM THEORY

R. Schardt

University of Darmstadt

Synopsis

The main ideas of the generalized beam theory are presented. A simple basic version is already sufficient for many problems in cold rolled sections and folded plate structures, where distortional effects must be regarded. Some extentions for application in special fields like stability problems are discussed. Examples demonstrate that for exact results small computer capacity is sufficient.

Notation

s	mid-line axis in cross-section element
$\bar{s}$	axis rectangular to x and s
x	longitudinal axis
r	index of node or cross-section element
$'$	derivative in x-direction
$\cdot$	derivative in s-direction
$\sim$	sign for orthogonalized values
$u(s,x)$	displacement in x-direction
$f_L(s,x)$	displacement in s-direction
$f(s,x)$	displacement in $\bar{s}$-direction
σ_x	longitudinal stress
m	transverse bending moment
${}^kV$	deformation-resultant
${}^kW$	stress-resultant, generalized warping moment
k	upper left index for number of differential equation
B	generalized transverse bending stiffness
C	generalized warping-constant
D	generalized torsion-constant

A	cross-section area
$\varkappa$	values for 2nd order theory
ℓ	span
Δx	subelement of span for finite differences
E	Young's modulus of elasticity
G	shear modulus of elasticity

Introduction

The engineers' bending theory describes the deformation and the stress distribution of a beam supposing that the shape of the cross-section is not deformed but only displaced. Then four independent and orthogonal displacement functions and their derivatives are sufficient to express the behaviour of the beam submitted to a load. On the other hand there are different theories for folded plate structures, which do not provide a natural transition to the ordinary beam theory. The generalized beam theory combines both fields in such a way that the basic ideas, expressions and formulas are extended to a more general meaning (Schardt (1), Sedlaceck (2)). So the necessary number and the selection of the functions can optimally be adapted to the complexity of the individual problem. Between the basic and the complete version many special versions can be used as the examples demonstrate.

1. THE BASIC VERSION

1.1 The differential equations

The following assumptions are made:

1. The axial displacements u, the axial stress σ_x and the transverse bending moment m are linear functions between two nodes.

2. The transverse membrane strain ε_s and the membrane shear deformation are neglected.

From 1 and 2 follows that, if the cross-section has n plate-elements, the number of independent deformation functions is $n + 1$, that is the number of the nodes. The independent warping ordinates u_r at the nodes are the elements of the general vector

$$u = \{u_1, \; . \; . \; u_r, \; . \; . \; u_{n+1}\} \qquad (1)$$

by which all other deformation components can be expressed, except they are constant in x - direction. Therefore a type of functions ${}^kV$ is introduced which can be explained as deformation-resultants. With a special set of $n + 1$ warping vectors $\tilde{u}$ the general vector is expressed by the series

$$u(x,s) = \sum_k {}^k\tilde{u}(s)\, {}^kV'(x) \qquad (2)$$

The special vectors ${}^{k}\tilde{u}$ are orthogonal eigenvectors and are found as the solution of an eigenvalue-problem

$$(C - \lambda B)\ u = 0 \qquad (3)$$

in which the elements of matrix C represent the internal virtual work of the axial stress σ_x and B the work of the transverse bending moment m_s. The eigenvalue λ is the relation between the transverse bending stiffness and the warping stiffness. Equation (3) can be solved by the Jacobi-method. Four of the eigenvalues $\lambda_1 \div \lambda_4$ are zero and the corresponding vectors are the warping functions due to the first derivatives of the rigid body displacements for which the transverse bending stiffness is not affected. The displacements f and f_L in the plane of the cross-section are defined by $V' = 1$, the transverse bending moment $m(s)$ by $V = 1$. The displacements f_L are constant between two nodes, the transverse deflection f is a cubic function for $k > 4$. All functions depending on s can be expressed by u. They are steady between two nodes and can be evaluated by integration-constants.

$${}^{ik}C = \int {}^{i}\tilde{u}\,{}^{k}\tilde{u}\ dA \qquad (4)$$

$${}^{ik}B = \int {}^{k}\tilde{m}\,{}^{k}\tilde{f}\ ds \qquad (5)$$

The off-diagonal values C and B are zero for the eigenfunctions. The virtual work of the torsional stresses

$${}^{ik}D = \int {}^{i}\tilde{m} \cdot {}^{k}\tilde{f}\ ds \qquad (6)$$

is not orthogonalized but the off-diagonal values are not important and can be neglected. So a set of $n+1$ ordinary and orthogonal differential equations

$$E\,{}^{k}C\,{}^{k}V'''' - G\,{}^{k}D\,{}^{k}V'' + {}^{k}B\,{}^{k}V = {}^{k}q \qquad (7)$$

remains to describe the beam behaviour as well as the distortional behaviour. The load term $\tilde{q}$ is the virtual work of the transverse load at the basic deformations $\tilde{f}_L$ and $\tilde{f}$. The stress distribution is defined by

$$\sigma_x(s,x) = -\sum_k {}^{k}W(x) \cdot {}^{k}\tilde{u}(s) / {}^{k}C \qquad (8)$$

where the stress-resultant or internal force ${}^{k}W$ is

$${}^{k}W(x) = -E\,{}^{k}C\,{}^{k}V''(x) \qquad (9)$$

Table 1 shows the relations between usual notation in ordinary beam theory and the unified notation in the generalized beam theory.

1.2 Solutions of the differential equations

Formulas for the solution of the differential equations (7) are given by Hetenyi (3) for special cases of load distribution and end-conditions, evaluation for extreme values in (4). For general cases a special version of finite differences is proposed, which uses the deformation V_i and the internal force W_i as unknowns. So the number of subelements Δx can be reduced and concentrated loads F_c are realized correctly. The coefficients of the two equations for the endpoint of a subelement $i-1$ are

	V_{i-1}	V_i	V_{i+1}	W_{i-1}	W_i	W_{i+1}	=	R	
1.	$-a$	$2a$	$-a$	-1	10	-1		$-F_{ci}\cdot\Delta x$	(10)
2.	$-b_1$	b_2	$-b_1$	-1	2	-1		$(F_{ci}+F_{qi})\,\Delta x$	(11)

with the abbreviations

$a = 12\,E\,C/\Delta x$
$b_1 = G\,D - B\,\Delta x\,/12$
$b_2 = 2G\,D + 5\,B\,\Delta x\,/6$
F_{ci} = concentrated load at i
$F_{qi} = \Delta x(q_{i-1} + 10q_i + q_{i+1})/12$

For single support conditions $(V=0,\ W=0)$ no special equations are necessary. Other supports need special end-formulas. Example 1 demonstrates the efficiency and accuracy of this method.

2. *SPECIAL CONDITIONS FOR CROSS-SECTIONS*

2.1 Cross-sections with links

Links at the cross-section, which are continuous in x-direction, change the number of distortional and undistortional deformations. They are regarded in the eigenvalue problem and influence only the basic functions $\tilde{u}$ and the coefficients of the differential equations. Links may occur at internal nodes or at the end nodes e.g. to express symmetric or antimetric conditions. Example 2 shows a cross-section with end-links.

2.2 Closed sections

For closed sections (single-cell) the last element of vector u is identic to the first $(u_1 \equiv u_{n+1})$. It is replaced by the deformation of constant shear-flows. This case and the multi-cell-sections are treated very thoroughly by Sedlacek (2) and Saal, G. (5). A comprehensive review with special accent on box-beams is given by Maisel (8).

2.3 Plate effects

For concentrated loads or local buckling effects between the nodes the number of independent deformation functions can be enlarged by introduction of intermediate nodes, the number of which is only limited by the requested accuracy. The transverse displacements f of these nodes are additional elements in the general vector $\tilde{u}$. In this case it is essential to regard the longitudinal bending stiffness of the plates. Miosga (6) and Möller (7) use this possibility.

3. FURTHER EXTENSIONS

3.1 Stability Problems

In the deformed structure the stresses σ_x cause deviation forces q , which affect the equilibrium in a similar manner as the load q. In the simple case of constant stress σ_x in x-direction for each differential equation a term of

$$q_{II} = \int_A \sigma_x ({}^j f_L {}^k f_L + {}^j f {}^k f) {}^j V'' dA \qquad (12)$$

must be added. After replacing σ_x by W and integration the deviation force is

$$q_{II} = \sum_i {}^i W \; {}^{ijk}\chi \; {}^k V'' \qquad (13)$$

where $${}^{ijk}\chi = \frac{1}{{}^i C} \int {}^i\tilde{u} ({}^j\tilde{f}_L {}^k\tilde{f}_L + {}^j\tilde{f} {}^k\tilde{f}) dA \qquad (14)$$

The coupling values χ describe the effects of 2nd order for simple lateral and torsional buckling as well as for distortional buckling The complete differential equation for varying stresses σ_x in x-direction is $(2 = j \; ; \; k = n+1 \; ; \; 1 = i = n+1)$

$$E \, {}^k C \, {}^k V'''' - G \, {}^k D \, {}^k V'' - \sum_i \sum_j {}^{ijk}\chi ({}^i W \, {}^j V')' + {}^k B \, {}^k V = 0 \qquad (15)$$

Example 2 shows an application of the buckling analysis for a segmental cylindrical shell with longitudinal groove stiffeners. Miosga (6) presents an analysis of stability problems with an extension to postcritical behaviour. For nonlinear behaviour the stress distribution in eqn (12) cannot be expressed by the warping ordinates $\tilde{u}$ only. So another term ν is introduced into the differential equations (15) by Miosga, which takes into account the strain

$$\varepsilon_x = u' + (f_L'^2 + f'^2) \qquad (16)$$

The additional values ν are integrals on a product of four deformation functions. The nonlinear diff. eqns. are solved by iteration. Also direct iteration is possible. Stress redistribution is calculated when the buckling deformation is defined by eqn(15) and an amplitude of the deformation is assumed. Improved χ-values are introduced into eqn(15) and so on. Example 3 shows the results of this procedure.

3.2 Dynamic Problems

Dynamic effects which can be treated in a similar manner as stability effects have been investigated by Saal (5) and Möller (7), using the generalized beam theory.

3.3 Completion by transverse Strain

For some problems e.g. very short prismatic folded plate structures, wall-beams or postcritical behaviour with large deflections, the transverse strain ε_s must also be regarded. For this reason a third type of onedimensional functions is added, which are defined by the independent displacements f_L in s-direction of the intermediate nodes of the plain elements. The results of present investigations are expected in future.

4. *EXAMPLES*

4.1 *Example 1: Stress-distribution for hat-section*

A beam with cold rolled hat-section is loaded by a concentrated force at node *2* and $x = 20$ cm (Fig 1). The mechanical properties of the cross-section are calculated by solving the eigenvalue-problem eqn(3) with 6 unknowns. The numerical results are shown in Table 2 (warping vectors) and Table 3 (stiffness values). To apply the finite differences method the span ℓ is divided in 5 sections that means a set of 8 linear equations has to be solved. A comparative calculation with 20 sections gave the same results.

Example 2: Buckling-stress for stiffened cylindrical shell

A partial cylindrical shell with longitudinal groove stiffeners (Fig 2) is subjected to a constant compression stress at the end sections. All edges of the shell are only displaceable in longitudinal direction. This condition requires, that both end-elements have equal warping displacements for both nodes. So only 15 independent deformations remain, for which the eigenvalue-problem (eqn 3) has to be solved. In the next step the matrix of $\varkappa$-values is calculated for $i = 1$ (eqn 14). The assumed functions

$$^{k}V(x) = {}^{k}a \sin \pi x/\ell \qquad 2 \leq k \leq 15 \tag{17}$$

satisfy the differential eqn (15) and the support conditions at the end sections. Introduced into eqn (15) we get 14 linear and homogeneous equations of type

$$E\,{}^{k}C\,\pi^2/\ell^2 + G\,{}^{k}D + {}^{k}B\,\ell^2/\pi^2 + {}^{1}W_{cr} \sum_j {}^{1jk}\varkappa = 0 \tag{18}$$

The lowest eigenvalue ${}^{1}W_{cr}$, that is the critical normal force, can be determined by vector-iteration. The elements of the corresponding eigenvector are the amplitudes ${}^{k}a$ of the assumed functions ${}^{k}V$. Figure 2 shows the critical stress for various depths of the grooves as function of the slenderness ratio L/R. The capacity of a microcomputer is sufficient for calculation.

4.3 *Example 3: Postcritical behaviour of plate with longitudinal edge-stiffener*

A rectangular plate simply supported at three edges and with a longitudinal edge stiffener is loaded by constant longitudinal stress The load is applied by constant strain mode. No imperfections are regarded. The cross-section is divided in 6 equal sections (5 intermediate nodes, total 8 nodes) and the lip. The basic warping vectors and stiffness-values are determined by an eigenvalue problem of 8 unknowns. Critical stress and buckling deformation are found by eqn (18). Up to the critical stress the stress distribution remains constant. An amplitude is assumed for the deformation and the relating stress-distribution is calculated. Now the $\varkappa$-values have to be recalculated and the procedure is repeated by solving eqns(18) once more until accuracy is sufficient. Then the amplitude can be enlarged. Figure 3 shows the results in deformation and stress-distribution for four different amplitudes.

References

1. Schardt, R. 'Eine Erweiterung der technischen Biegelehre für die Berechnung biegesteifer prismatischer Faltwerke. (An extension of engineers' theory of bending for the analysis of prismatic folded plates with flexural stiffness.)' Der Stahlbau, Vol. 35, No. 6, June 1966, pp. 161-171, No. 12, December 1966, p. 384.

2. Sedlacek, G. 'Systematische Darstellung des Biege- und Verdrehvorganges für prismatische Stäbe mit dünnwandigem Querschnitt unter Berücksichtigung der Profilverformung. (Systematic description of the process of bending and torsion for prismatic beams of thin-walled cross-section, considering distortion of cross-section.)' Fortschritt-Berichte VDI-Zeitschrift, Reihe 4, Nr. 8, September 1968, 110 pp.

3. Hetenyi, M. 'Beams on elastic foundation.' The University of Michigan Press, 1951.

4. Schardt, R. and Okur, H. 'Hilfswerte für die Lösung der Differentialgleichung $a\, y''''(x) - b\, y''(x) + c\, y(x) = p(x)$. (Auxiliary values for the solution of the differential equation.)' Der Stahlbau, Vol. 38, No. 1, January 1971.

5. Saal, G. 'Ein Beitrag zur Schwingungsberechnung von dünnwandigen, prismatischen Schalentragwerken mit unverzweigtem Querschnitt.' Dissertation, Technische Hochschule Darmstadt, 1974.

6. Miosga, G. 'Vorwiegend längsbeanspruchte dünnwandige prismatische Stäbe und Platten mit endlichen elastischen Verformungen. Dissertation, Technische Hochschule Darmstadt, 1976.

7. Möller, R. 'Zur Berechnung prismatischer Strukturen mit beliebigem nicht formtreuem Querschnitt.' Institut für Statik, Technische Hochschule Darmstadt, Bericht Nr. 2, 1982.

8. Maisel, B.I. 'Analysis of concrete box beams using small-computer capacity.' Development Report 5, Cement and Concrete Association, December 1982.

Deformation component	1	2	3	4	k
Deformation function	$\int u \cdot dx$	v	w	Θ	--
	${}^1V$	${}^2V$	${}^3V$	${}^4V$	${}^kV$
Warping function	-1	ζ	η	ω	--
	${}^1u$	${}^2u$	${}^3u$	${}^4u$	${}^ku$
Internal force	N	M_ζ	M_η	W_ω	--
	${}^1W$	${}^2W$	${}^3W$	${}^4W$	${}^kW$
Stiffness acting by V''	A	I_ζ	I_η	C_ω	--
	${}^1C$	${}^2C$	${}^3C$	${}^4C$	${}^kC$
V'	k_x	N	N	I_D	--
				${}^4D$	${}^kD$,
V		k_ζ	k_η	k_ω	--
		${}^2B$	${}^3B$	${}^4B$	${}^kB$
Load transverse		q_ζ	q_η	m_ω	--
		${}^2q$	${}^3q$	${}^4q$	${}^kq$
longitudinal	n_x	m_ζ	m_η	--	--
	${}^1Q$	${}^2Q$	${}^3Q$	${}^4Q$	${}^kQ$

Table 1 Traditional and generalized notation

r \ k	1	2	3	4	5	6
1	-1	6.210	2.473	-7.002	1.000	1.000
2	-1	4.210	2.473	5.407	0.006	-0.400
3	-1	2.500	-2.226	-3.744	-0.576	0.080
4	-1	-2.500	-2.226	3.744	0.576	0.080
5	-1	-4.210	2.473	-5.407	-0.006	-0.400
6	-1	-6.210	2.473	7.002	-1.000	1.000

Table 2 Matrix of warping-vectors for hat-section (example 1)

k	1	2	3	4	5	6
C	1.9	23.53	6.777	15.43	0.299	0.149
D	0	0	0	0.0063	0.00020	0.000116
B	0	0	0	0	0.0871	0.0204

Table 3 Stiffness-values of hat-section (example 1)

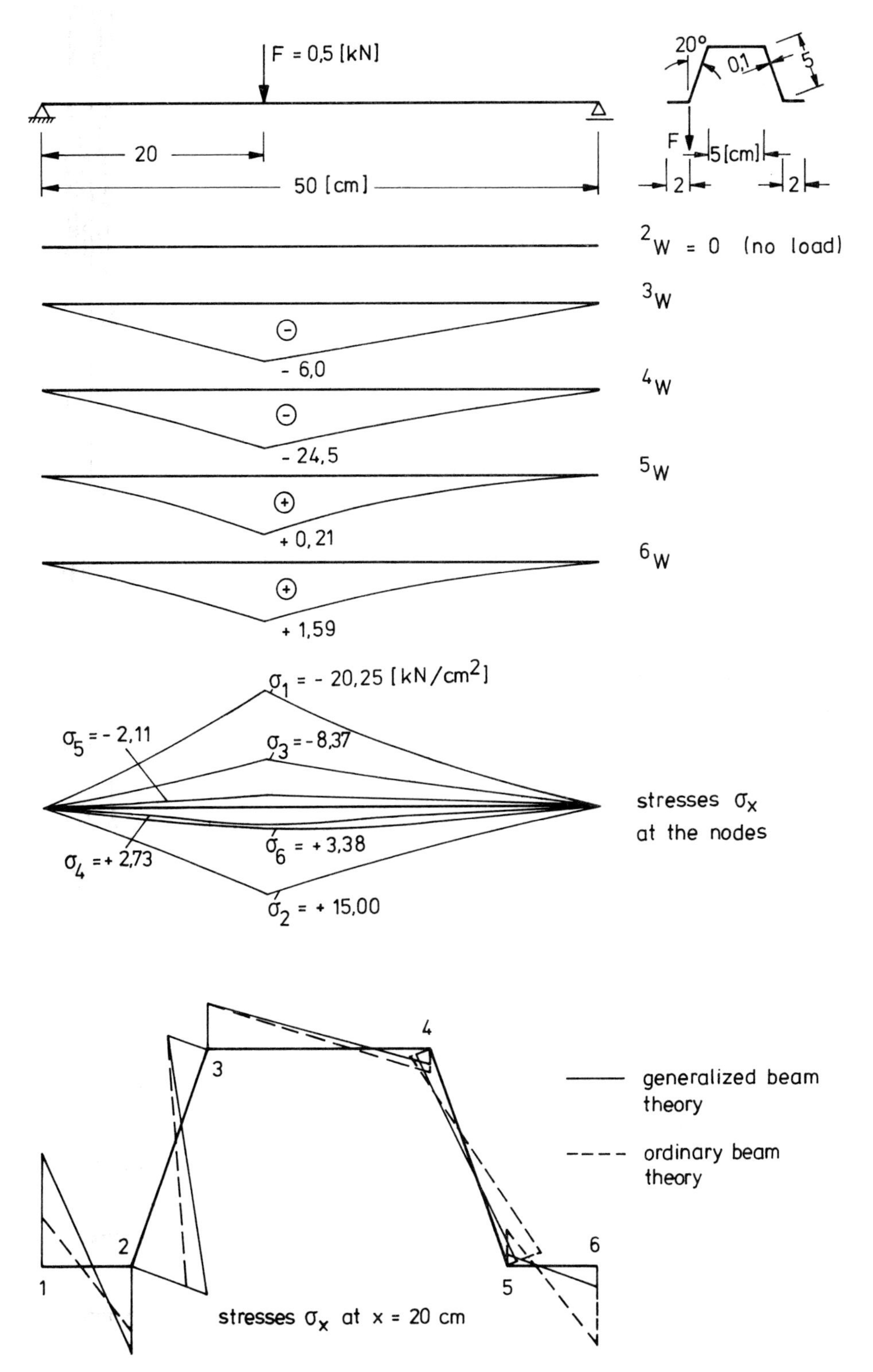

Fig 1 Stress distribution of hat-section

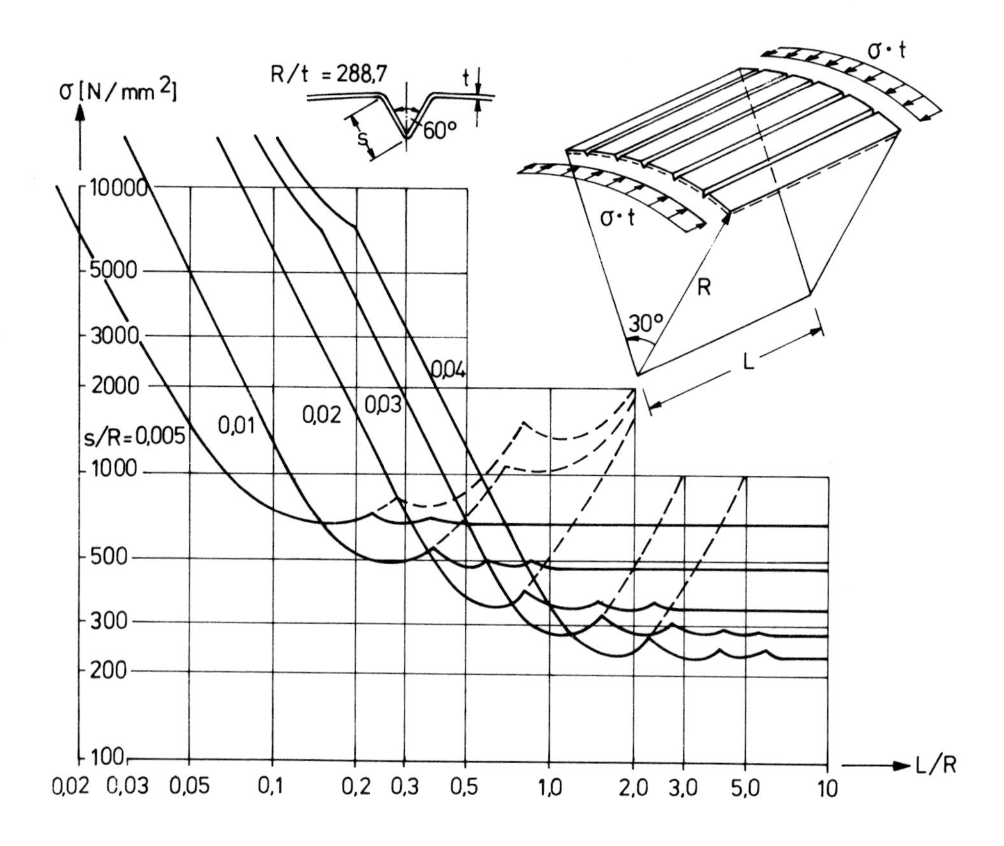

Fig 2 Buckling stress of stiffened cylindrical shell

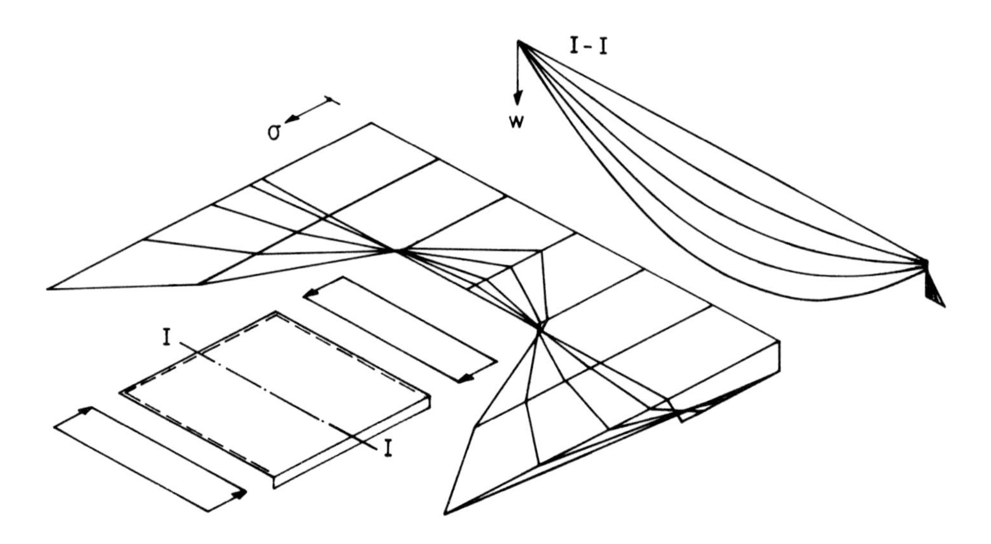

Fig 3 Postcritical behaviour of plate with edge-stiffener

LOCAL AND OVERALL BUCKLING OF LIGHT GAUGE MEMBERS

J. M. DAVIES

Department of Civil Engineering, University of Salford.

P. O. THOMASSON

Division of Steel Construction, Royal Institute of Technology, Stockholm.

Synopsis

The design of cold-formed columns and beams which fail by a combination of local and overall buckling is shown to be amenable to a Perry-Robertson approach. Suitable forms of the Perry coefficient are considered and comparisons with the experimental results of several other authors are given.

Notation

A	gross cross-sectional area
A_{eff}	effective cross-sectional area
b	width of plate element
b_{eff}	effective width of plate element
c	half-width of I section
d	depth of I section
E	Young's Modulus
G	shear modulus
h	depth of RHS section
I_y	second moment of area about minor axis
J	St. Venant torsional constant
K	reduction factor on yield force in columns (= σ/σ_y) or yield moment in beams (= $M/Z\,\sigma_y$)
K_σ	plate buckling coefficient
ℓ	effective length of column (= actual length when nominally pin-ended)
ℓ_a, ℓ_b	lengths of plate elements
L	span of beam
m	factor in expression for K_σ for I-sections
M	uniform bending moment at failure

M_{crit} critical bending moment for elastic lateral buckling

Q strength reduction factor for short lengths (= A_{eff}/A or Z_{eff}/Z)

r radius of gyration

t, t_a, t_b net thickness of plate material

t_w thickness of web of member

Z gross section modulus

Z_{eff} net section modulus taking account of effective widths

α, α' factors in Perry coefficient for columns

η Perry coefficient

η' imperfection parameter

$\bar{\lambda}$ slenderness parameter = $\sqrt{\sigma_y/\sigma_e}$ (columns) or $\sqrt{\sigma_y/\sigma_{crd}}$ (beams)

$\bar{\lambda}_{el}$ slenderness parameter for plate element = $\sqrt{\sigma_y/\sigma_{cr}}$

ρ effective width reduction factor = b_{eff}/b

σ axial stress in column at failure

σ_{cr} elastic critical stress for buckling of plate elements

σ_{crd} elastic critical stress for lateral buckling of beams

σ_{crlb} elastic critical stress for local buckling of flange in compression.

σ_e Euler stress = $\pi^2 E/(\ell/r)^2$

σ_y yield stress

Γ warping constant

INTRODUCTION

Light gauge steel members which fail by a combination of local and global buckling provide structural engineers with one of their more difficult theoretical problems and a number of complex mathematical studies have been undertaken by research workers in this field. However, for practical purposes something much simpler is required and this paper has been prepared in the conviction that suitable forms of the Perry-Robertson equation provide a generally acceptable approach to all such problems. The history of European Column Curves provides a hint that this may be the case and the comparison between proposed design curves and experimental results presented in this paper provides further evidence. Indeed, the authors maintain that the only problem lies in finding suitable expressions for the "Perry coefficient" for the various cases that must be considered.

1. *THE EUROPEAN COLUMN CURVES*

The non-dimensional European Column Curves, which describe the buckling strength of centrally compressed columns, were derived empirically following an extensive experimental and theoretical programme (1,2,3). They were later shown (4) to be capable of representation with remarkable accuracy by the Perry-Robertson equation:

$$(\sigma_E - \sigma)\ (\sigma \quad - \sigma) \quad = \eta\ \sigma_E \sigma \tag{1}$$

In which the Perry coefficient η is given by

$$\eta \quad = \quad \alpha(\frac{\ell}{r} \quad - 0.2 \quad \frac{\pi^2 E}{\sigma_y}\ ) \tag{2}$$

The factor α varies between the four column curves as follows:

= 0.0020 for curve 'a' (e.g. hollow sections)
= 0.0035 for curve 'b' (e.g. rolled I-sections about minor axis)
= 0.0055 for curve 'c' } e.g. welded I-sections about
= 0.0080 for curve 'd' } minor axis

Tables showing the applicability of the four curves to the range of possible sections are given, for example, in the current draft British Standard for the "Structural Use of Steelwork in Buildings"(5).

For computational purposes, eqns. (1) and (2) may be conveniently expressed as follows:

$$\text{Let } \bar{\lambda} = \sqrt{\frac{\sigma_y}{\sigma_E}} = \frac{\ell}{r}\sqrt{\frac{\sigma_y}{\pi^2 E}} = \text{slenderness parameter} \tag{3}$$

$$\text{and} \quad \eta' = \sqrt{\frac{\pi^2 E}{\sigma_y}}\ (\bar{\lambda} - 0.2) = \text{imperfection parameter} \tag{4}$$

Then the reduction factor $K = \sigma/\sigma_y$ on the yield strength is given by:

$$K^2 - K\ [1 + \frac{(1 + \eta')}{\bar{\lambda}^2}\ ] + \frac{1}{\bar{\lambda}^2} = 0 \tag{5}$$

i.e $K = F - \sqrt{F^2 - {}^1/_{\bar{\lambda}^2}}$

$$\text{where} \quad F = \tfrac{1}{2}\left\{1 + \frac{\left[1 + \alpha\sqrt{\frac{\pi^2 E}{\sigma_y}}\ (\bar{\lambda} - 0.2)\right]}{\bar{\lambda}^2}\right\} \tag{6}$$

the equations in this form can be conveniently modified to include an allowance for local buckling.

2. *MODIFICATION OF THE PERRY-ROBERTSON EQUATION TO ALLOW FOR LOCAL BUCKLING OF COLUMNS.*

The study reported herein was carried out as part of the work of European Convention for Constructional Steelwork Committee TC7 who are responsible for drafting European Recommendations for the design of light gauge sections. This Committee had adopted the Winter approach to the effective width of elements subject to local buckling which may be expressed as follows:

$$\text{Let} \quad \bar{\lambda}_{el} = \sqrt{\frac{\sigma_y}{\sigma_{cr}}} = \frac{1.052}{\sqrt{K_\sigma}} \quad \frac{b}{t}\sqrt{\frac{\sigma_y}{E}} \tag{7}$$

where the buckling coefficient K_σ may be conservatively taken as equal to 4.0 for a stiffened element or 0.426 for an unstiffened element. Alternatively, advantage may be taken of continuity within the cross-section in order to use higher values of K_σ as discussed below.
The effective width b_{eff} is given by ρb where

$$\rho = \left[1 - \frac{0.22}{\bar{\lambda}_{el}}\right] \frac{1}{\bar{\lambda}_{el}} \quad ; \quad (\rho \leqslant 1.0) \tag{8}$$

Applications of eqns. (7) and (8) to each plate element within a cross-section results in effective cross-sectional area A_{eff} which is less than or equal to the gross area A.

If
$$Q = \frac{A_{eff}}{A} \tag{9}$$

Q can be seen as a reduction factor on the gross cross-sectional area or alternatively as a reduction factor on the yield stress taken on the gross area. Therefore, replacing σ_y in the Perry-Robertson formula by $Q\,\sigma_y$, we have

$$K = F - \sqrt{F^2 - Q/\bar{\lambda}^2}$$

$$\text{where } F = \tfrac{1}{2}\left\{Q + \left[\frac{1 + \alpha\sqrt{\dfrac{\pi^2 E}{\sigma_y}}\,(\bar{\lambda} - 0.2)}{\bar{\lambda}^2}\right]\right\} \tag{10}$$

As formulated above, equation (10) assumes that local buckling has no influence on the Perry coefficient which proves not to be the case. It has been found empirically that better correlation with test results is found by introducing a factor of $(4-3Q)$ into the Perry coefficient. Introducing at the same time the simplification that $\sqrt{\dfrac{\pi^2 E}{\sigma_y}} \simeq 90$ gives

$$F = \tfrac{1}{2}\left\{Q + \left[\frac{1 + \alpha'(4 - 3Q)\,(\bar{\lambda} - 0.2)}{\bar{\lambda}^2}\right]\right\} \tag{11}$$

where $\alpha' = 0.21,\ 0.34,\ 0.49$ for the first three European column curves respectively. This is the formulation which is proposed for the purposes of design and which is the subject of further investigation in section 4 of this paper.

3. *INFLUENCE OF CONTINUITY WITHIN THE CROSS-SECTION ON K_σ*

Values of the element buckling coefficient K_σ considering interaction with adjacent plate elements have been given by various authors (e.g 6,7). For most common sections (I-sections, channels, zeds, rectangular hollow sections) charts of values are available but these do not, in general, admit of simple explicit design expressions so that it is more convenient to use approximations for the systematic computations which are presented later in comparision with test results.

For the most slender element of a rectangular hollow section, Fig. 1 shows the exact values of K_σ together with a simple approximation suggested by Johnston (6), namely

$$K_\sigma = 7 - \frac{15}{7}\left(\frac{t}{t_w}\right)^2 \left(0.4 + \frac{b}{h}\right) \tag{12}$$

This expression with $t = t_w$ has been used in the present study.

For the flanges of I-sections, Fig. 2 shows exact values of K_σ together with an approximation given by Bleich (7), namely:

$$
\left.
\begin{aligned}
K_\sigma &= \left[0.65 + \frac{2}{3m + 4}\right] \\
\text{where} \quad m &= 2\frac{t^3}{t_w^3}\,\frac{d}{c}\left[\frac{1}{1 - 0.106\,\dfrac{t^2 d^2}{t_w^2 c^2}}\right]
\end{aligned}
\right\} \qquad (13)
$$

It is important to realise that the above values of K_σ are based on the conditions pertaining at initial elastic buckling of the cross-section whereas cold-formed section design is concerned with stress distributions in the post-buckled condition. The stress at which local buckling first occurs is a secondary consideration and it is necessary, therefore, to enquire whether the above values of K_σ are equally applicable to the calculation of effective widths for design purposes at mean stresses in excess of the buckling stress. There appears to be little information available on this point.

Curves giving the 'exact' values of K_σ for the webs of I-sections, channels and zeds and for the longer sides of rectangular hollow sections are also included in the Swedish Code of Practice (8). Corresponding values of K_σ for the other elements of a given cross-section may be obtained by reading off the appropriate curve for element 'a' and factoring this by $\left[\dfrac{t_a\,\ell_b}{t_b\,\ell_a}\right]^2$ to give the value of K_σ for the adjacent element 'b'. The second author confirms that this approach is intended solely to permit the calculation of the initial buckling stress when element 'b' buckles first. However, it is possible to interpret the Swedish Code to require that the above factor should always be applied, whichever element buckles first, and it has been implied in informal discussions that this provides an approach which takes into account the fact that the act of restraining one element may materially weaken the other element concerned. Although this was not the intention of those responsible for the drafting, calculations on the above basis have been included in this study and will be referred to as 'according to the Swedish Code'.

If the values of K_σ given by the Swedish Code for the web buckling of I-sections are appropriately factored to give the corresponding values of K_σ for the flanges, the short broken curves on Fig.2 are obtained. These web values were in some cases significantly less than 4.0. For the purposes of this study, in which flange buckling is much more important than web buckling, the effective widths of flanges were calculated using K_σ given by equation (13). For I-section columns, these values of K_σ were then factored as described above in order to calculate corresponding effective widths for the webs. In other words, the converse of the Swedish approach was used.

The webs of beams are subject to a stress gradient and the considerations become rather different. Because the I-section beams considered later had relatively stocky webs, the compression flanges were calculated with K_σ given by equation (13) and the webs were assumed to be fully effective.

As the influence of values of K_σ other than 4.0 and 0.426 has significant implications for design, the opportunity is taken of examining alternative approaches in comparison with the results of tests carried out by a number of different investigators.

4. COMPARISON WITH THE TEST RESULTS OF OTHER AUTHORS.

The interaction between local and overall buckling in axially loaded columns has been investigated experimentally by a number of authors. The sources considered in this present study are summarised in Table 1. Some typical design curves are given in Figs 3 to 10. The complete set of 25 are available in Reference 14.

Source	Number of different cross-sections	Material	Details	Typical results
CIDECT (9)	9	Steel	Rectangular hollow sections, some with large corner radii, some annealed.	Figs 3 - 6
De Wolf (10)	4	Steel	Rectangular hollow sections fabricated from two cold-formed channel sections screwed and glued together.	Fig.7
Kalayanaraman (11)	5	Steel	I-sections fabricated from two cold-formed channel sections screwed and glued together.	Fig.8
De Wolf (10)	4	Steel	I-sections fabricated from two cold-formed channel sections screwed and glued together.	Fig.9
Bijlaard and Fisher (12)	3	Aluminium	Extruded I-sections	Fig.10

TABLE 1. Summary of test results considered

The CIDECT (9) results are of particular importance because each individual test was repeated a number of times so that it is possible to show not only the individual results but also the mean plus or minus two standard deviations. Typical results for a non-annealed rectangular cross-section are shown in Fig.3 and for an annealed square section in Fig. 4. The theoretical results are based on equations (10) and (11) with $\alpha' = 0.21$ for rectangular hollow sections.

Some of the tests were carried out in the as-fabricated condition and others were annealed before testing so that the influence of residual stresses can be considered. It would appear that annealing results in a significant improvement in performance as shown in Figs 4 and 5.

Another factor revealed by the CIDECT tests is the influence of the corner radii. Fig. 6 shows a typical set of design curves in which the stub-column results are lower than those for columns with a significant length. This has been observed in other tests and is apparently an end effect caused by local stresses adjacent to the loading platten. There is an evident need for caution in interpreting stub-column tests when the sections have significant corner radii.

De Wolf's tests (10) had quite different proportions, the most extreme cross-section being as shown in Fig. 7. This series is significant, not only because of the success of the proposed equations in representing the measured behaviour, but also because of the results given by the alternative treatments of the buckling coefficient K_σ. The 'Swedish' approach falls a long way below the simple treatment and is clearly unrealistically pessimistic.

With I-sections with slender flanges, element continuity plays a significant role because the local buckling coefficient of 0.426 for simply supported flanges is frequently increased to the order of 1.0 as a result of continuity with the web. The results for I-section columns are therefore rather more sensitive in this respect than those for rectangular hollow sections. Typical results obtained by Kalayanaraman and De Wolf are shown in Figs. 8 and 9 respectively. The theoretical curves are with $\alpha' = 0.34$.

The results obtained by Bijlaard and Fisher on extruded aluminium sections are primarily of interest because, being based on the 0.2% proof stress and presumably also because of reduced residual stresses, they perform relatively better than similar cold-formed steel sections. This observation is of importance in the interpretation of some tests on the lateral buckling of beams considered later.

5. *CONCLUSIONS FOR COLUMNS.*

These conclusions are based on the complete investigation, including results not reported in detail here.

(a) The suggested Perry-Robertson approach with $K_\sigma = 4.0$ for stiffened elements and $K_\sigma = 0.426$ for unstiffened elements is consistently safe when compared with test results. This is true over the whole range of parameters considered with the exception of rectangular hollow stub columns with excessively large bending radii. The reason for this has already been discussed.

(b) The most optimistic interpretation of the equations governing local buckling behaviour is to take the enhanced values of K_σ for the element which buckles first and the base value of 4.0 for the restraining element. This neglects the detrimental effect of plate continuity on the post-buckling behaviour of the restraining element and gives the upper curves in Figs 3 to 10. These curves tend to give a rather good account of the test results with hardly any test results falling below them. However, this apparent success may be due in some measure to the conservative nature of the effective width equation (8) and therefore it is probably more prudent to be content with the simple values of 4.0 and 0.426 for K_σ as appropriate.

(c) The pessimistic interpretation of the Swedish Code referred to in section 4 is clearly unwarranted. The influence of plate continuity often appears to cause a reduction in carrying capacity which hardly appears logical and does not accord with the test results.

6. *APPLICATION OF THE PERRY-ROBERTSON EQUATION TO THE LATERAL BUCKLING OF LIGHT GAUGE BEAMS.*

The application of the Perry-Robertson equation to the lateral stability of thin-walled beams follows an identical course. It is merely necessary to define the various quantities in terms of bending moment rather than axial force. For the purpose of this exercise, the beam will be assumed to be subject to uniform bending moment, thus:-

$$M = K\sigma_y Z \tag{14}$$

where

$$K = F - \sqrt{F^2 - Q/\bar{\lambda}^2} \tag{15}$$

and

$$F = \tfrac{1}{2}\left\{Q + \frac{[1 + \eta'(\bar{\lambda} - \bar{\lambda}_o)]}{\bar{\lambda}^2}\right\} \tag{16}$$

Equations (14) to (16) are of course identical in form to (10)and (11). There are, however, two additional difficulties in implementing this approach in practice. In the first place the determination of σ_{crd} may be by no means elementary for practical combinations of support conditions and bending moment distribution. Design aids for this problem allowing approximate solutions for many situations are now available (5) and such aids will be necessary if the above approach is to be used in practice. Secondly, the correct form of the expression $\eta'(\bar{\lambda} - \bar{\lambda}_o)$ which contains two unknowns, is not at all obvious as a consequence of the lack of suitable test results.

7. *FORM OF THE PERRY COEFFICIENT FOR LIGHT GAUGE BEAMS.*

One possible form of the Perry coefficient expression $\eta'(\bar{\lambda} - \bar{\lambda}_o)$ is to adopt the well-researched expression for hot rolled beams (5), namely

$$\eta'(\bar{\lambda} - \bar{\lambda}_o) = 0.007\left[\sqrt{\frac{\pi^2 E}{\sigma_{crd}}} - 0.4\sqrt{\frac{\pi^2 E}{\sigma_y}}\right] \tag{17}$$

An alternative suggested by the second author for possible inclusion in the European Recommendations is:

$$\eta'(\bar{\lambda} - \bar{\lambda}_o) = 0.85\left[1 - \frac{\sigma_{crlb}}{Q\sigma_y}\right](\bar{\lambda} - 0.2) \text{ but } \nless 0.1 \tag{18}$$

In the present state-of-the-art, these must be seen as tentative suggestions which may be superseded in the light of current and future research.

8. *EXPERIMENTAL RESULTS FOR THE INTERACTION BETWEEN LATERAL AND LOCAL BUCKLING OF LIGHT GAUGE BEAMS.*

The authors are only aware of a single set of test results relevant to this problem, namely those obtained by Cherry (15) in 1960. As Cherry's results were obtained using extruded aluminium I and T sections, they are of limited value in deriving suitable parameters for the design of cold-formed steel beams but they are, nevertheless, useful in justifying the basic approach. These results will form the basis of discussion pending the availability of test results from a program currently in progress at the University of Salford.

Cherry completed six series of tests on beams of varying length loaded by pure end couples. The supports were such as to permit the ends of the beams to rotate freely about their principal axes while preventing warping and twist about the longitudinal axis. Beams of varying lengths were tested, consistent dimensions within each series being obtained by reducing the cross-sections in a milling machine.

Comparison between the ultimate loads obtain in four of the test series with those predicted by the alternative versions of the Perry-Robertson formula are given in Figs. 11 to 14. For comparison purposes, the elastic critical buckling curves are also shown. These were obtained using the following expression for the critical moment M_{crit} which was obtained from a Galerkin solution of the governing differential equation.

$$M_{crit} = \frac{2\pi}{\sqrt{3}L}[EI_yJG + \frac{4\pi^2}{L^2} EI_yE\Gamma]^{\frac{1}{2}} \qquad (19)$$

This equation is close enough for all practical purposes to the much more complicated equation given by Cherry.

The comparison of results is quite consistent and the following conclusions may be drawn:

1. The Perry-Robertson approach gives a perfectly satisfactory representation of the transition from local buckling to lateral buckling.
2. Very safe results are given by the use of $K_\sigma = 0.426$ for the buckling coefficient of the flanges.
3. With theoretically more correct values of K_σ, quite accurate values of the failure load are obtained.
4. For the I-section beams the Perry coefficient suggested for the European Recommendations (eqn. (18)) gives a satisfactory lower bound whereas the alternative (eqn. (17)) is slightly less consistent.
5. The T-section beams provide a special problem because the first yield took place in the web in tension. The theoretical results shown in Fig 14 are based on tensile yield and, on this basis, eqn. (17) provides an excellent lower bound whereas eqn. (18) appears to be over-conservative. An alternative basis, resulting in higher theoretical moments of resistance, would be to allow limited yield in tension up to the point where yield is predicted in the compression flange.

9. *CONCLUSIONS.*

The Perry-Robertson equation, suitably modified by an appropriate factor Q to take account of local buckling, provides an entirely satisfactory approach to the design of light gauge steel columns and beams. There remains some doubt regarding the precise form of the

Perry factor for beams. Further work is also required in order to investigate the lateral buckling of sections without a vertical axis of symmetry.

References

1. Sfintesco, D. 'Experimental basis of the European column curves.' Construction Metallique, No. 3, 1970 (in French).
2. Beer, H. and Schultz, G. 'The theoretical basis of the new column curves of the European Convention for Constructional Steelwork.' Construction Metallique, No. 3, 1970 (in French).
3. Young, B.W. 'Axially loaded steel columns.' C.I.R.I.A. Tech. Note No. 33, 1971.
4. Dwight, J.B. 'Use of Perry formula to represent new European strut curves.' Univ. of Cambridge, Dept. of Eng. Report No. CUED/C-struct/TR.30, 1972.
5. 'The structural use of steelwork in buildings.' Draft for Comment of BS 5950 Part I, 1982.
6. Johnson, B.G. (ed) 'Guide to Stability Design Criteria for Metal Structures.' 3rd edition, John Wiley & Sons, New York, 1976.
7. Bleich, F. 'Buckling Strength of Metal Structures.' McGraw Hill, New York, 1952.
8. 'Swedish code for light gauge metal structures.' Swedish Institute of Steel Construction, Publication 76, March 1982, (in English).
9. Braham, M. et al. 'Flambement des profils creux a parois minces, cas des profiles rectangulaires, charges axialement.' Comission des Communautes Europeanes Recherche Technique Acier, Projet de Rapport Final, Convention No. 6210 SA/3/301, 1979 (in French).
10. De Wolf, J.T. 'Local and overall buckling of cold-formed compression members.' Dept. of Structural Engineering, Cornell University, Report No. 354, May 1973.
11. Kalyanaraman, V. 'Performance of unstiffened compression elements.' Dept. of Structural Engineering, Cornell University, Report No. 362, February 1978.
12. Bijlaard, P.P. and Fisher, G.P. 'Column strength of H-sections and square tubes in post-buckling range of component plates.' NACA TN 2994, 1963.
13. Cherry, S. 'The stability of beams with buckled compression flanges.' The Structural Enginer, September 1960.
14. Davies, J.M. 'Further results in connections with the local and overall buckling of light gauge members.' University of Salford, Dept. of Civil Engineering Report Ref. 83/178, February 1983.

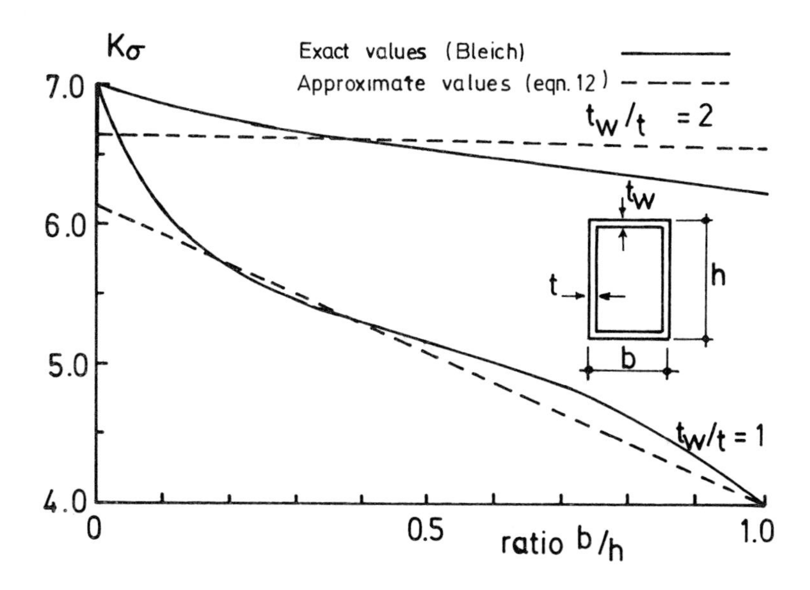

Fig.1. Buckling coefficient K_σ for rectangular hollow sections

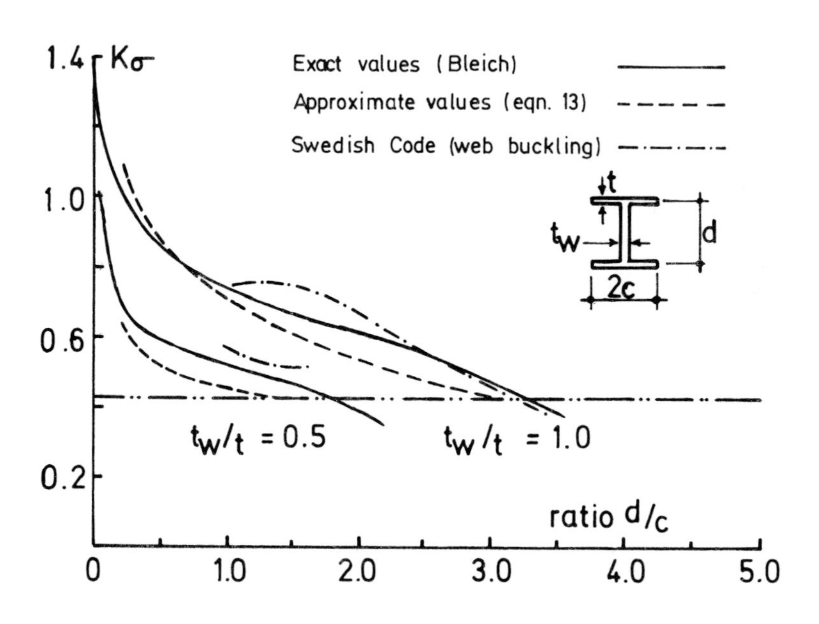

Fig.2. Buckling coefficient K_σ for the flanges of I-sections

TESTS ON RECTANGULAR HOLLOW SECTION COLUMNS

Theoretical curves with $K_\sigma = 4.0$ ———

Theoretical curves with $K_\sigma > 4.0$ on long side – – – –

Theoretical curves with K_σ according to Swedish Code –·–·–·

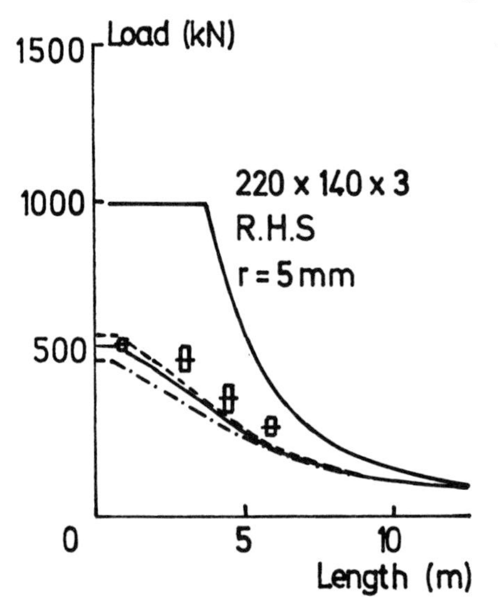

Fig.3. CIDECT Series VII

Fig.4. CIDECT Series VIII

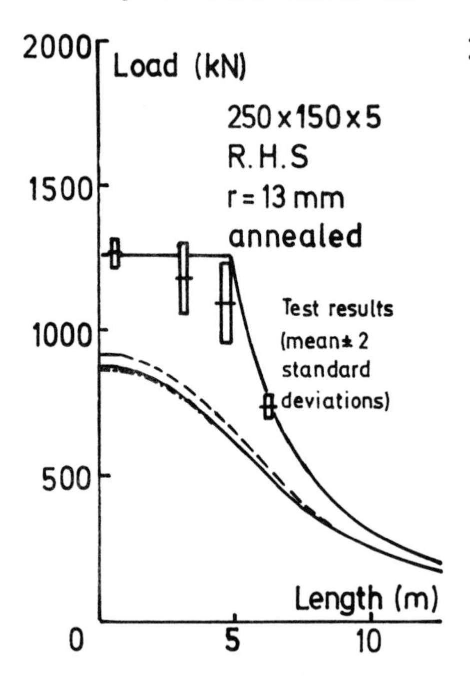

Fig.5. CIDECT Series III

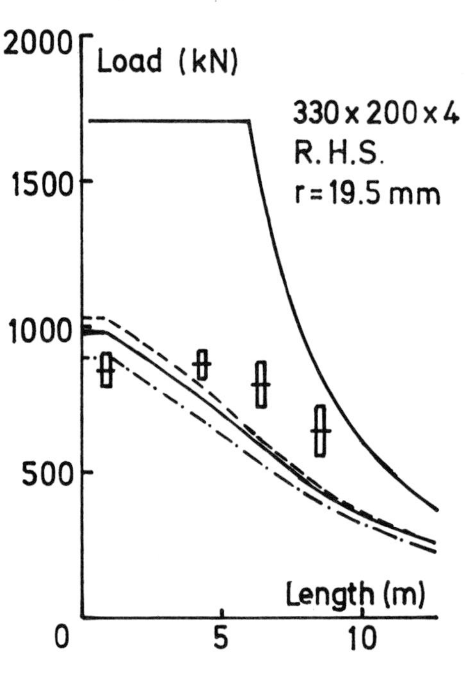

Fig.6. CIDECT Series I.

TESTS ON COLD-FORMED STEEL AND ALUMINIUM COLUMNS

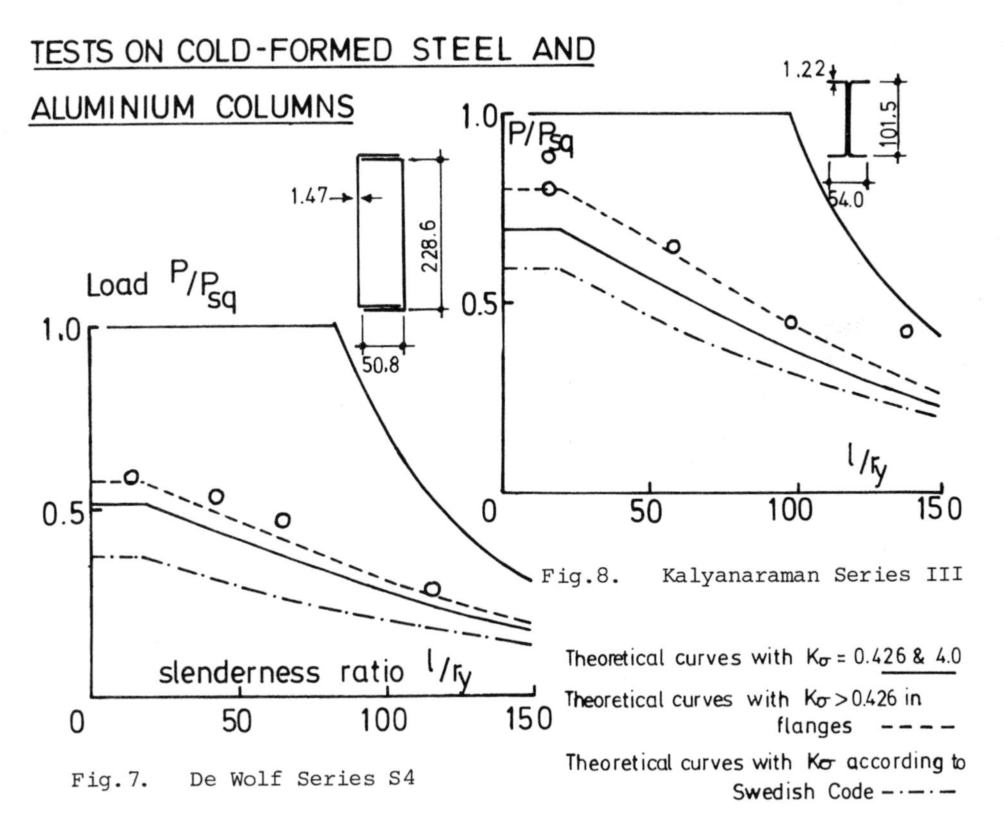

Fig.7. De Wolf Series S4

Fig.8. Kalyanaraman Series III

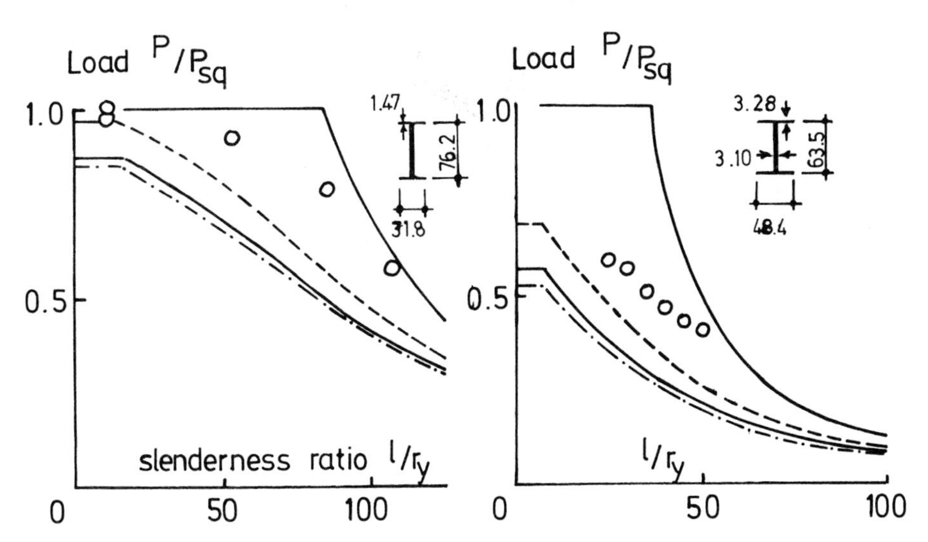

Fig.9. De Wolf Series U2

Fig.10. Bijlaard and Fisher Series K (aluminium)

TESTS ON EXTRUDED ALUMINIUM BEAMS LOADED IN PURE BENDING (CHERRY)

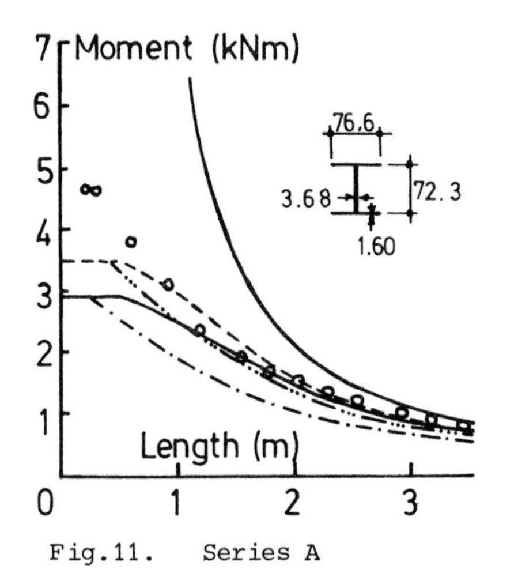

Fig.11. Series A

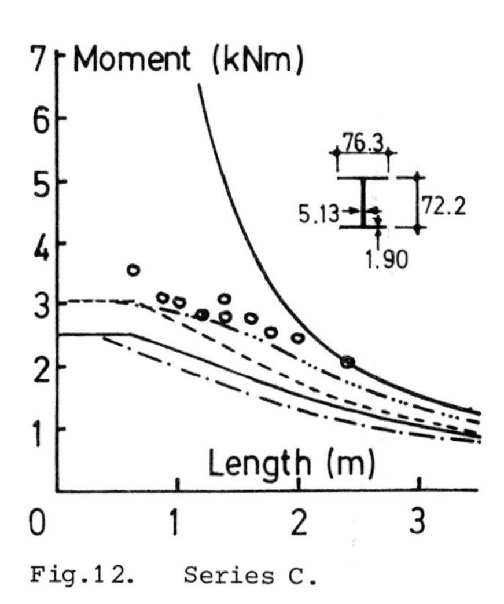

Fig.12. Series C.

Theoretical curves: $K_\sigma = 0.426$ & 4.0, Perry coeff. eqn 17 ————

Perry coeff. eqn 18 — · — · —

$K_\sigma > 0.426$ in flanges, Perry coeff. eqn 17 — — — — —

Perry coeff. eqn 18 —··· —··· —···

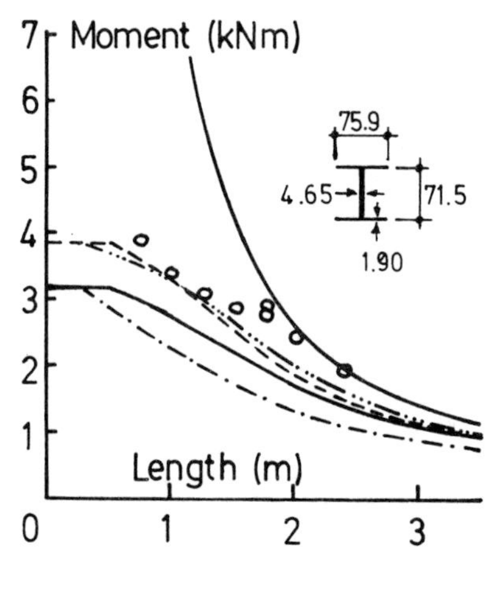

Fig.13. Series D

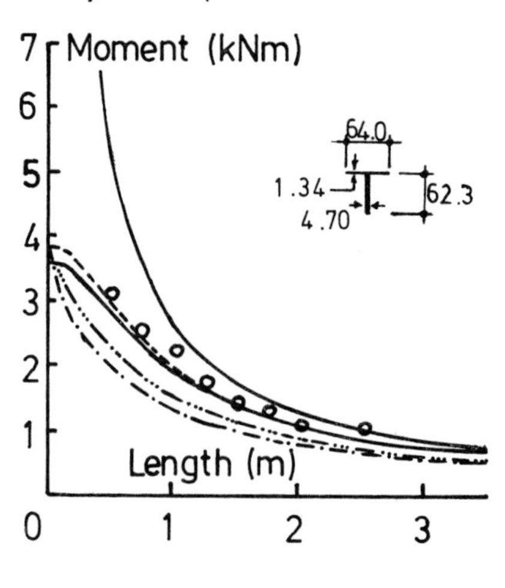

Fig.14. Series T

THREE-DIMENSIONAL FAILURE CRITERIA FOR COLD-FORMED STEEL SECTIONS

Frank Monasa

Department of Civil Engineering, Michigan Technological University, Houghton, USA

Jeffrey Bushie

Sargent and Lundy, Chicago, USA

Synopsis

A general procedure is presented to determine the three-dimensional failure criteria for a thin-walled cold-formed section with a specific geometry; namely, an equal flange zee. The same experimental procedure can be applied to derive the flexural failure envelopes for any other thin-walled section with a specific geometry and, consequently, the failure criteria can be defined. The scope of this paper focuses mainly on cold-formed members as structural elements for vehicular-type structures, such as automobile and bus body structures.

Introduction

Analytical procedures, by Monasa (1) and Kecman and Miles (2), to evaluate the structural integrity of vehicle structures in crash situations require, usually, a three-dimensional quasi-static collapse analysis to determine the ultimate strength capacity of vehicle body structures. Using this approach, the ultimate moment capacity of the structural members of vehicle framework is required. And because the crash could cause loading at any inclination to the member, it is important to determine the ultimate moment capacity of the member for any direction of loading. Therefore, three-dimensional failure criteria for combined flexural and torsional moments for a variety of shapes is essential.

This paper presents a general experimental procedure to determine the two-dimensional flexural failure criteria for thin-walled cold-formed steel cross-sections with specific geometry. These criteria can then be extended to three dimensions by including torsional ultimate moments. Cold-formed steel members of vehicular structures can be made in a variety of shapes. Some of the more common shapes are: tubular, channel, I, Zee, hat, and angle sections. Several investigators, such as Dawson and Walker (3), Reck, Pekoz, and Winter (4), Yener and Pekoz (5), and Yu (6), presented methods for determining the ultimate flexural moment capacity of cold-formed steel sections. These methods are only applicable to limited type of cross sections and to a specified application of the bending moment. They consider that the stiffened and the unstiffened elements of the section are subjected to a uniform compression stress distribution in the elastic pre-buckled state, and therefore, these methods are only applicable for moment determination about an axis parallel to the plane of the compression element. As a result, only one point on the failure envelope can be defined by these methods. At any other angle of loading the stress distribution is no longer uniform and consequently these methods are not applicable. The experimental procedure used in this paper treats each section geometry separately; and therefore, it takes into account the effect of having other than a uniform elastic stress acting on the compression elements, and the effect that the adjoining element has on the unstiffened compression flange.

1. *FAILURE CRITERIA*

The procedure presented in this paper relies on an experimental approach that results in construction of failure envelopes for combined states of stress of the sections tested. These failure envelopes are similar in some degree to the yield envelopes (surfaces) of cross-sections that would be obtained if their stiffened and unstiffened plate elements yielded before local buckling occurred. To further define the properties of the yield and failure envelopes, the following is presented.

1.1 *Yield envelope (surface)*

The yield envelope is an important concept in plastic theory. A point in the plane of the figure represents a certain combination of bending and torsional moments for a given section having known values of M_p and M_{pt}. M_p is the value of full plastic flexural moment of the cross-section in the absence of torsional moment, whereas M_{pt} is the torsional plastic moment in the absence of bending moments. A point on the boundary of the convex yield surface represents a combination of bending and torsional moments that just causes the cross-section to become fully plastic.

In this paper the effect of combined biaxial bending and torsional moments is considered. The axial and transverse shear forces are neglected in the formulation of failure criteria. Figure 1 shows a yield surface for a zee section where the location of the strong and weak plastic principal axes are indicated on the diagram. It is noted that the same general shape of the yield surface will be preserved for any zee section. A general property of the yield envelope is that a normal to its surface defines the direction of the plastic neutral axis, i.e., the normal to the yield surface must coincide with the direction of the zero stress axis. For a general unsymmetrical section, the yield surface will, in general, be skew-symmetric, and the plastic principal axes are not compelled to be orthogonal, as shown in Fig. 1. It is noted that the portions AB and CD of the yield surface in Fig. 1 are made of straight lines parallel to the axis of M_x. This occurs because the zero-stress axis lies entirely within the web of the zee section, and with the approximation made here for thin-walled sections where the centerline dimensions are used, the value of M_x will be indeterminate.

The following assumptions were made in the formulation of this paper to simplify the complexity of the yield criterion:

1. A simplified form of the von Mises yield criterion (strain ener-

gy of distortion, or octahedral shear stress) is used to describe the yield condition of a cross-section under combined stresses. The axial and shear forces are neglected in formulation of the yield function.

2. Warping stresses are neglected.
3. The stress-strain relationship is assumed to be linear elastic, and perfectly plastic, and a shape factor of unity is assumed.
4. Plastic yielding is localized and does not spread lengthwise.

The von Mises yield condition for a plane stress condition in which one plane is acted upon by a normal stress and a shear stress can be written as (7):

$$\sigma^2 + 3\tau^2 = F_y^2 = 3\tau_y^2 \qquad (1)$$

where F_y is the yield stress in pure tension or compression, and τ_y is the yield stress in pure shear. Equation (1) in the normalized form becomes:

$$(\sigma/F_y)^2 + (\tau/\tau_y)^2 = 1 \qquad (2)$$

Equation (2) may be written as:

$$m_r^2 + m_\tau^2 = 1 \qquad (3)$$

where $m_r = \sigma/F_y = M_r/M_p$, M_r = moment resultant = $(M_x^2 + M_y^2)^{\frac{1}{2}}$, where M_x and M_y are the bending moments about the x and y axes, respectively; $m_t = \tau/\tau_y = M_t/M_{pt}$, M_t = torsional moment, M_{pt} torsional plastic moment = $\sum_{i=1}^{n} \frac{1}{2} wt^2\tau_y$ for open sections and is equal to $2\,a\,b\,t\,\tau_y$ for closed sections; w and t are the width and thickness of a thin rectangular plate element, respectively, n is the number of elements, and a and b are the centerline dimensions of the cross-section.

From eqn (3), the yield surface, in two-dimensional space, for flexural biaxial moments can be defined by

$$(M_x/M_p)^2 + (M_y/M_p)^2 \leqslant 1.0 \qquad (4)$$

where $M_p = (M_{px}^2 + M_{py}^2)^{\frac{1}{2}}$, and M_{px}, M_{py} are the plastic flexural moments about the x and y axes, respectively. The yield surface can now be expanded to three-dimensional space by including the torsional moments as follows:

$$(M_x/M_p)^2 + (M_y/M_p)^{\frac{1}{2}} + (M_t + M_{pt})^2 \leqslant 1.0 \qquad (5)$$

1.2 *Failure envelope (surface)*

To develop failure criteria for thin-walled cold-formed sections, an experimental approach is used in this paper. An analytical procedure seems very tedious due to the complexity and nonuniformity of the state of stresses in the plate elements of the cross-section under combined loading conditions. Zee sections were selected to conduct the experimental procedure because it is a point-symmetric section; that is, it has no planes of symmetry and would seem to be a difficult section to analyze. As mentioned earlier the yield envelope for zee sections is skew-symmetric, and the exact inclination of the plastic principal axes is a direct function of the web/flange ratio as shown in Fig. 1. A total of seventy zee sections of various thicknesses and web/flange ratios were tested. The purpose of the tests was to develop a series of two-dimensional flexural failure envelopes that is representative of the practical range of cold-formed structural zee sections used for vehicular-type structures. These failure envelopes allow prediction of the ultimate flexural moment capacity of the cross-section for any axis of loading. The two-dimensional failure criterion can then be extended to three dimensions by including torsional moments.

2. *TEST SETUP AND PROCEDURE*

All specimens were tested as cantilever beams ranging in length from 18 to 36 inches (0.46 to 0.92 m). The reason for variable lengths was to assure that each specimen had an adequate factor of safety against transverse shear buckling.

Although the flange width was specified to be 0.85" (22 mm) for all specimens the thickness was varied from 0.0235" (0.60 mm) to 0.048" (1.22 mm) within the specific web/flange ratio; this allowed for testing of various w/t ratios. Three different web/flange ratios were used; these were *1/1*, *2/1*, and *4/1*. Yield strength of the specimens ranged between 27 to 35 ksi (186-241 Mpa). The limiting w/t ratio for an unstiffened element which yields before it buckles locally is (6)

$$(w/t)_{lim} = 63.3/\sqrt{F_y} \tag{6}$$

The above dimensions were chosen, so that the w/t value for each specimen was larger than w/t_{lim}.

The loads were applied at the free end of the beam. Different ratios of horizontal to vertical load were chosen so as to define the entire failure surface. The loads were applied gradually in about 2% increments before failure, but because of deflection these loads did not remain strictly horizontal and vertical. The loads were adjusted, however, to obtain the true horizontal and vertical components at each load increment. Both horizontal and vertical deflections were measured at the load end through the use of dial indicators and a specially machined plate.

The zee section, as mentioned earlier, is point-symmetric and its principal axes (elastic and plastic) are oblique. It will therefore deflect both vertically and horizontally under the influence of a vertical load passing through the elastic shear center. In the initial stage of loading, the stresses are elastic and there will be only a small torque generated at the fixed end due to the small displacement at the free end of the beam. However, when the section becomes plastic, there is substantial displacement and, consequently, there is appreciable torque transmitted to the fixed end. Also, for a zee section, the shear center will not retain its elastic position when the section goes plastic. In order to construct the failure envelope of a cross-section for biaxial bending moments, it was necessary to eliminate the effect of torsional moments, by preventing the rotation of the free end of the beam. This was accomplished by requiring all loads to be applied through knife-edge supports at the free end which allow for vertical and horizontal displacements, but prevented rotation.

For all of the specimens tested, failure was assumed to occur when the specimen could no longer take an increase in load. This was the failure condition regardless of the amount of deflection. In general, three types of failures occurred. The first was brought about by local out-of-plane elastic buckling of the compression flange. The second mode of failure was of the yielding type. The section did not buckle out of plane but rather continued to sustain load well past the point of full plasticity and up into the strain hardening region. The third mode of failure was intermediate between the two just described. Although local buckling may have occurred, the section was still capable of sustaining an increase in load. This is due to the post-buckling behavior, as described by Winter (8), of the compressive thin-walled elements and the ability of the other stable elements to resist and transfer loads.

3. *CONSTRUCTION OF FAILURE ENVELOPES (9)*

Upon completion of testing, the flexural failure moments were computed and plotted for each section type. Figures 2 to 4 show typical experimental failure envelopes superimposed on the yield envelope for specimens with web/flange ratios of *1/1* and $(w/t)\sqrt{F_y}$ values of 89, 131, and 195, respectively. Figure 5 shows the flexural failure envelope for a specimen with a web/flange ratio of *2/1* and $(w/t)\sqrt{F_y}$ value of 196. Diagrams for specimens with web/flange ratio of *4/1* were also prepared. It is shown that specimens with $(w/t)\sqrt{F_y}$ ratios of about 90 are fully effective (failure envelope is outside of yield envelope) for any moment ratio as shown in Fig. 2. The failure envelopes for the 3 sets of specimens with $(w/t)\sqrt{F_y}$ ratios of about 130 generally fall within the yield envelope in the straight-line portion of the curve as shown in Fig. 3. The reason for this is because in this region the plastic neutral axis falls within the plane of the web, consequently, the entire flange is under compressive stress which may induce local buckling. The failure envelopes for the 3 sets of specimens with $(w/t)\sqrt{F_y}$ ratios of about 195 are shown in Figs. 4 and 5. As might be expected, these sections are not fully effective (failure envelope inside of yield envelope).

Sections that attain fully plastic moment obey always the normality condition. This is not the case, however, for a thin-walled cold-formed section in which local buckling may predominate. Indeed, even if local buckling does not initiate the failure, local distortion of a thin-walled section will always follow the formation of a plastic hinge and any normality condition, therefore, can not be satisfied.

4. *ULTIMATE TORSIONAL MOMENT CAPACITY*

Thin-walled sections may fail due to shearing stresses. Torsion will cause primarily shearing stresses in the cross-section, consequently, the ultimate torsional moment depends on the ultimate shearing strength of each plate element of the cross section. The ultimate torsional moments for thin-walled sections will be determined analytically using the procedure developed by Monasa and Chipman (10). The ultimate shearing stresses, τ_{ult}, are presented by Yu (6) according the w/t ratio of the element. The following ultimate shearing stresses are used:

$$\tau_{ult} = \tau_y = F_y/\sqrt{3} \quad \text{for} \quad w/t < 380/\sqrt{F_y} \tag{7a}$$

$$\tau_{ult} = \tau_{icr} = (0.8\, F_y\, \tau_{cr}/\sqrt{3})^{\frac{1}{2}} \quad \text{for} \quad 380/\sqrt{F_y} < w/t < 547/\sqrt{F_y} \tag{7b}$$

and

$$\tau_{ult} = \tau_{cr} = 5.35\, \pi^2\, E/12(1-\nu^2)(w/t)^2 \quad \text{for} \quad w/t > 547/\sqrt{F_y} \tag{7c}$$

where τ_{icr} = inelastic shear buckling stress, τ_{cr} = initial elastic critical shear buckling stress, E is the Young's modulus, and ν is Poisson's ratio. The ultimate torsional moment, $M_{t,ult}$, for open and closed cross-sections is then determined similar to the torsional plastic moment with τ_{ult} substituted for τ_y.

5. *FAILURE CRITERIA EQUATIONS (9)*

The failure criteria in two dimensions can be defined by substituting, in the yield criteria, M_{ult} for M_p and $M_{t,ult}$ for M_{pt} and it can can be as written as

$$(M_x/M_{ult})^2 + (M_y/M_{ult})^2 \leqslant 1.0 \tag{8}$$

where M_{ult} is the resultant ultimate (failure) flexural moment for a specific value of moment ratio, $R = M_x/M_y$. M_{ult} can be determined as follows:

$$M_{ult} = \left[(M_{x,ult})^2 + (M_{y,ult})^2\right]^{\frac{1}{2}} \tag{9}$$

where $M_{x,ult}$, and $M_{y,ult}$ are ultimate flexural moment about the x and y-axes, respectively. The failure envelope can then be extended to three-dimensional space by including the torsional moments:

$$(M_x/M_{ult})^2 + (M_y/M_{ult})^2 + (M_t/M_{t,ult})^2 \leqslant 1.0 \tag{10}$$

It is noted that while the failure envelope for flexure is exact, the envelope for torsion is approximated by an ellipse.

6. *DETERMINATION OF ULTIMATE FLEXURAL MOMENT FOR ZEE SECTIONS (9)*

In order to develop a general procedure to determine the ultimate moment capacity for any equal flange zee section, the failure envelopes were nondimensionalized with respect to their yield envelopes. To be specific, the value of the experimental failure (ultimate) moment was divided by the plastic moment for that particular moment ratio. Figures 6, 7, and 8 show nondimensionalized plots for web/flange ratios of *1/1*, *2/1*, and *4/1* zee sections, respectively. With the inclusion of the common parameter $(w/t)\sqrt{F_y}$ in the nondimensionalized plots, the ultimate flexural moment capacity for any equal flange zee section can be predicted.

6.1 *Procedure*

The following describes the procedure for predicting the ultimate moment capacity of a general equal flange zee section. It should be noted that the procedure will be greatly simplified for a section with at least one plane of symmetry. The procedure follows these steps.

1. Calculate the moment ratio, $R = M_x/M_y$ assuming that R is positive if the M_x and M_y vectors are in the positive directions of the axes.
2. The yield surface (excluding the straight portion of the curve) is defined by: $M_{px} = F_y\ t\ d\ (b-d/4-2z)$, and $M_{py} = F_y\ t\ (b^2-2z^2)$, where z is the distance measured from the centerline of the web and along the flange.
3. Solve for M_{px} and M_{py} with $z = 0$. This allows one to find the transition point between the straight line portion of the curve and the rest of the curve.
4. If $M_{py}\ (z = 0)/M_{px}\ (z = 0) \geqslant R$, then calculate $M_{px,max}$ and $M_{py,max}$ for the section being analyzed. This can be obtained by summing moments about the x and y-axes, respectively, as follows:

$$M_{px,max} = F_y\ t\ (\ bd + d^2/4), \quad \text{and}$$

$$M_{py,max} = F_y\ t\ b^2.$$

If $M_{py}(z = 0)/M_{px}\ (z = 0) \geqslant R \geqslant M_{px,max}/M_{py,max}$ then R intersects the straight line portion of the yield envelope. M_{py} is then equal to $M_{py,max}$ and M_{px} is simply equal to $M_{py,max}/R$.

5. For any other ratio, R intersects the continuous portion of yield envelope. Then the value of z can be obtained from

$$R = M_y/M_x = F_y\ t\ (b^2 - 2z^2)/F_y\ t\ d\ (b - d/4 - 2\ z).$$

This value of z is used to determine M_{px} and M_{py}, and

$$M_p = \left[(M_{px})^2 + (M_{py})^2\right]^{\frac{1}{2}}.$$

6. Calculate the position of the strong plastic principal axis by using

$$\tan \alpha = M_{py,max}/M_{px,max},$$

and calculate the angle of the axis at which the applied load acts about by using

$$\tan \beta = M_y/M_x.$$

Now, from Figs. 6, 7, and 8, whichever is applicable, rotate an imaginary axis from the strong plastic principal axis an angle equal to $(\alpha - \beta)$ and in the same direction as α to β. This imaginary axis is the axis at which all interpolations will be made between the two figures with the closest web/flange ratio as that of the section being analyzed. Note that this imaginary axis effectively superimposes the strong plastic principal axes for the web/flange ratio considered.

7. Find the values of M_{ult}/M_p along the imaginary axis for the $(w/t)\sqrt{F_y}$ ratio of the specimen being analyzed, using Figs. 6, 7, and 8 as applicable.

8. Find the final value of M_{ult}/M_p for the section being analyzed by interpolating between Figs. 6, 7, and 8 as applicable, for the specific web/flange ratio of the section being analyzed.
9. The ultimate flexural moment can then be found by taking the value of M_{ult}/M_p just found and multiplying it by M_p as found in step No. 5.

The authors will be pleased to provide a numerical example if requested.

References

1. Monasa, F. 'Engineering analysis of the structural integrity of small buses.' Report submitted to the Michigan Department of Transportation, Contract No. 79-2533, Nov. 1981, 334 pages.
2. Kecman, D. and Miles, J. 'Application of the finite element method to the door intrusion and roof crush analysis of a passenger car.' Proceedings of the 3rd International Conference on Vehicle Structural Mechanics, P-83, Society of Automotive Engineers, October 10-12, 1979, pp. 191-197.
3. Dawson, R. G. and Walker, A. C. 'A proposed method for the design of thin-walled beams which buckle locally.' The Structural Engineer, Vol. 50, No. 2, Feb. 1972, pp. 95-105.
4. Reck, H. P., Pekoz, T. and Winter, G. 'Inelastic strength of cold-formed steel beams.' Journal of the Structural Division, ASCE, Vol. 101, No. ST11, Proc. Paper 11713, Nov., 1975, pp. 2193-2203.
5. Yener, M. and Pekoz, T. 'Inelastic load carrying capacity of cold-formed steel beams.' Proceedings of the Fifth International Specialty Conference on Cold-Formed Steel Structures, November 18-19, 1980, University of Missouri-Rolla, pp. 145-174.
6. Yu, Wei-Wen. 'Cold-Formed Structures.' McGraw-Hill, New York, N.Y., 1973, pp. 53-102.
7. Horne, M. R. 'Plastic Theory of Structures.' 2nd ed. Pergamon Press, Oxford, England, 1979, p. 72.
8. Winter, G. 'Commentary on the 1968 edition of the specification for the design of cold-formed steel structural members.' American Iron and Steel Institute, Washington, D.C., 1968.
9. Bushie, J. D. 'Ultimate strength predictions for thin-walled sections.' A report submitted in partial fulfillment of the requirements for the degree of Master of Science in Civil Engineering, Michigan Technological University, 1982, 101 pages.
10. Monasa, F. and Chipman, T. 'Ultimate moment capacity of thin-walled cold-formed members of vehicular structures.' Proceedings of the Sixth International Specialty Conference on Cold-Formed Steel Structures, Nov. 15-17, 1982, University of Missouri-Rolla, pp. 235-257.

Fig. 1 Yield surface for 0.85" x 17" x 24 gage zee section with $F_y = 30$ *ksi*.

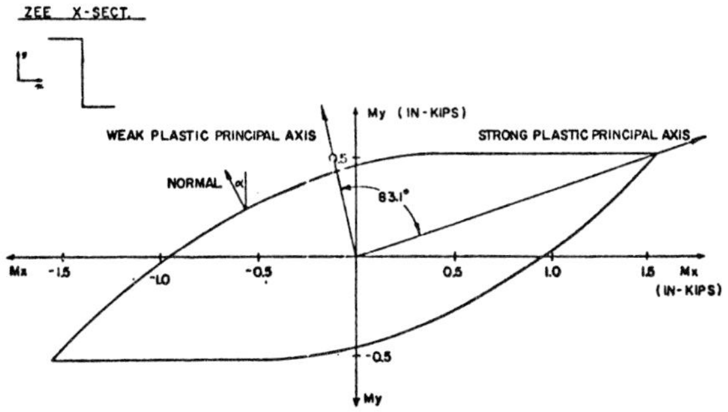

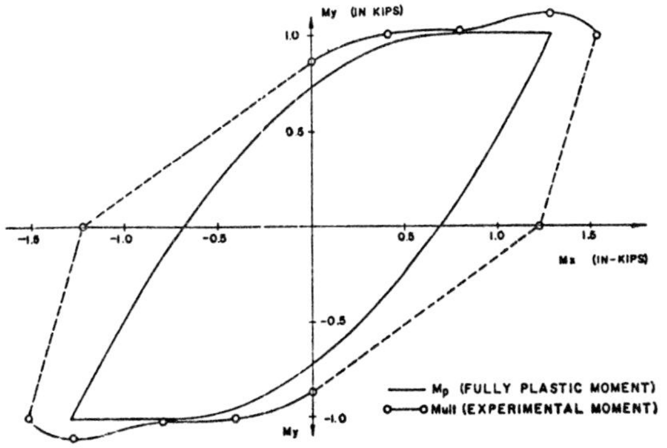

Fig. 2. Experimental failure surface for *1/1* zee with $(w/t)\sqrt{F_y} = 89$.

Fig. 3 Experimental failure surface for *1/1* zee with $(w/t)\sqrt{F_y} = 131$.

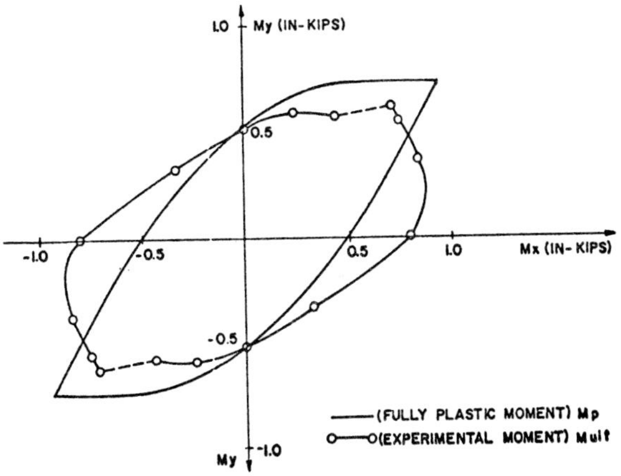

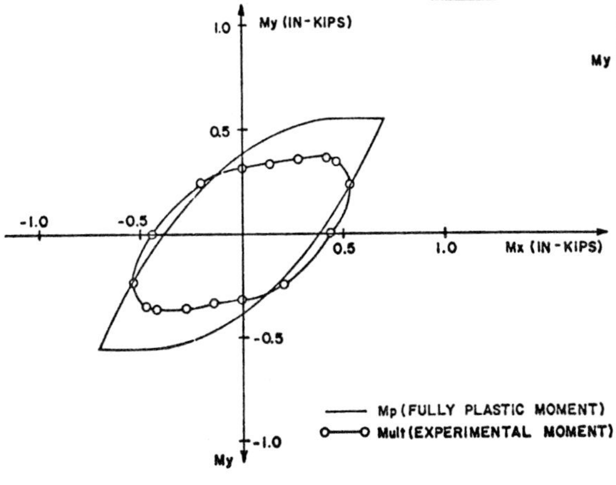

Fig. 4 Experimental failure surface for *1/1* zee with $(w/t)\sqrt{F_y} = 195$.

M_{γ} Value of M for interior segment.

M_x Major axis bending moment.

P End load.

P_o, P_{∞} Values of P for overhanging segments free to warp and prevented from warping.

R Factor reflecting relative importance of the support and torsional stiffnesses.

u Buckling deflection.

w Intensity of uniformly distributed load.

w_o, w_{∞} Values of w for overhanging segments free to warp and prevented from warping.

α_{RZ} Stiffness of torsional restraint.

β $=(\alpha_{RZ}L/GJ)/(1+\alpha_{RZ}L/GJ)$.

γ Segment length ratio $= L_I/L$.

ε Dimensionless load height $=(\bar{a}/L)\surd(EI_y/GJ)$.

ϕ Buckling twist rotation.

1. *INTRODUCTION*

1.1 *Lateral buckling of elastic I-beams*

The study of the lateral buckling of elastic beams began with the investigations of narrow rectangular section beams by Michell (22) and Prandtl (34) in 1899. These works were extended to doubly symmetric I-beams by Timoshenko in 1905, who also investigated the effects of load distribution, load height, and support and restraint conditions. By the time of the publication in 1961 of the second edition of his "Theory of Elastic Stability" (Ref. 39), it could be said that the elastic buckling of simply supported beams and cantilevers was well understood.

Timoshenko and others of that period solved their numerical problems by hand methods. The advent of the digital computer in the 1950's revolutionised the prediction of elastic buckling loads, and nowadays most researchers have access to a finite element computer program such as that developed by Barsoum and Gallagher (2). The power of the computer methods has allowed many investigations to be made, elaborating on the previous studies of the buckling of beams and cantilevers, and extending them to restrained beams, continuous beams, frames, and arches.

Some of the results of these studies are given in an extensive survey made in 1971 by the Column Research Committee of Japan (10), and in a number of more recent textbooks (9, 17, 41), while other surveys will soon be published (26, 31). However, it can be said that there is no complete survey of available knowledge. The most comprehensive survey by the CRC of Japan was markedly incomplete in 1971, and the proliferation of studies since then makes it unlikely that a complete survey will ever be attempted or successful.

1.2 *Related lateral buckling problems*

The effects of unequal flanges on the buckling of I-beams has long been of interest. It was generally recognised that it would be better to have the larger flange of a simply supported beam acting in compression rather than tension, and theoretical analyses were made by extending the Wagner theory (47) used to explain the torsional buckling of compression members. In 1980, Kitipornchai (19) produced simple approximations for monosymmetric beams which permit much more accurate designs to be made.

Anderson (1) considered the buckling of monosymmetric cantilevers, and showed that the tension flange plays an important role, because it generally buckles the further in a cantilever. Recently, Ojalvo (32) has cast some doubt on the basis of the Wagner hypothesis, although it is believed that the results obtained by using it are correct (42).

A study has been made of the strengthening effect of the prebuckling in-plane curvature (44), and this has been extended to explain the reductions in the buckling strengths of arches and other members with convex in-plane curvature (45). These studies led to an investigation of the post-buckling behaviour of beams (49), and it has been shown that a significant reserve of strength exists only in very slender redundant beams (50).

There have been very many studies made of the inelastic lateral buckling of steel beams, and an increasing number of systematic investigations of the strengths of real beams. Some of these are summarised in two recent surveys (30, 43).

The effects of web distortion generally and at supports have received increasing attention (4, 5), and the distortional buckling of composite bridge girders in negative bending has been investigated (15). Studies are also in progress on the non-linear interaction between local and lateral buckling (3), in which local buckling of the compression flange reduces both its in-plane and out-of-plane stiffnesses, and consequently reduces the lateral buckling resistance.

1.3 *Present and future lateral buckling problems*

Some of the problems discussed above will be studied further in the future. There are also a number of other problems which are at present under study or are likely to be. These include the buckling of light gauge roof purlins; the effects of bracing on the buckling of the rafters of portal frames (14, 36); the role of geometrical imperfections on the ultimate strength of beams of intermediate slenderness (46); optimising the resistance of a beam to lateral buckling; non-conservative loading; and the influence of support conditions on the buckling of built-in cantilevers and of overhanging beams.

2. *LATERAL BUCKLING OF BUILT-IN CANTILEVERS*

The lateral buckling of built-in cantilevers has been studied by a number of investigators, most of whom were concerned with the elastic buckling of uniform doubly-symmetric I-section cantilevers under various loading conditions (1, 23, 28, 33, 37-39). Other studies have included the effects of cross-section monosymmetry, (1), of variations of the cross-section along the cantilever (8, 20, 21, 24), of end and intermediate restraints (16, 18, 27), of post-buckling behaviour (48, 49), and of premature yielding caused by residual stresses (25).

Some of the elastic buckling moments predicted for uniform doubly symmetric I-section cantilevers which are built-in at their supports are shown in Figs 1-3, together with other values obtained by using a finite element computer program (12). For cantilevers in uniform bending (Fig. 1), the elastic buckling moment M can be approximated by using

$$M_{\infty}L/\sqrt{(EI_y GJ)} = 1.6 + 0.8K \tag{1}$$

in which $$K = \sqrt{(\pi^2 EI_w/GJL^2)} \tag{2}$$

The linear form of eqn (1) differs markedly from

$$ML/\sqrt{(EI_y GJ)} = \pi\sqrt{(1+K^2)} \tag{3}$$

which is the exact solution (39) for a simply supported beam in uniform bending (Fig. 1).

The elastic buckling moments of cantilevers with end loads P acting at the top flange, centroid, or bottom flange are shown in Fig. 2. These can be approximated by

$$P_{\infty}L^2/\sqrt{(EI_y GJ)} = 11\{1+1.2\varepsilon/\sqrt{(1+1.2^2\varepsilon^2)}\}$$
$$+4(K-2)\{1+1.2(\varepsilon-0.1)/\sqrt{(1+1.2^2(\varepsilon-0.1)^2)}\} \tag{4}$$

in which $$\varepsilon = (\bar{a}/L)\sqrt{(EI_y/GJ)} = (2\bar{a}/h)(K/\pi) \tag{5}$$

since $$I_w = I_y h^2/4 \tag{6}$$

for an I-section member. The approximately linear variation of the cantilever buckling moment with K is again demonstrated by Fig. 2 and eqn (4). A comparison of Figs 1 and 2 shows that the buckling resistance of a cantilever with a centroidal end load is substantially greater than with an end moment. This is so because in the case of end load, the major axis moment M_x is high only near the restrained support. The effect of load height $\bar{a}$ is demonstrated in Fig. 2, and it can be seen that while bottom flange loading significantly increases the buckling resistance, top flange loading substantially reduces it, especially for beams with high values of K. The non-linear effect of load height is suggested by the approximate formulation of eqn (4).

The elastic buckling moments of cantilevers with uniformly distributed load w are shown in Fig. 3. These can be approximated by

$$w_{\infty}L^3/(2\sqrt{(EI_y GJ)})=27\{1+1.4(\varepsilon-0.1)/\sqrt{(1+1.4^2(\varepsilon-0.1)^2)}\}$$
$$+10(K-2)\{1+1.3(\varepsilon-0.1)/\sqrt{(1+1.3^2(\varepsilon-0.1)^2)}\} \tag{7}$$

These buckling moments are even higher than those shown in Fig. 2 for beams with end loads, because the moment distribution caused by distributed load is even more concentrated near the support than is the case for end loads. The effects of load height for distributed loads are similar to those for end loads.

3. *LATERAL BUCKLING OF OVERHANGING BEAMS*

3.1 *General*

An overhanging beam is a beam which is continuous over an end support and cantilevered beyond it (Fig 4a). The elastic resistance to lateral buckling of the overhanging segment is frequently taken as that of a cantilever built-in at its support (Fig. 5a). It is shown in this paper that the built-in cantilever model overestimates the buckling resistance of an overhanging segment.

The reason for this is demonstrated by the buckled shapes shown in Figs 4b and 5b. For the built-in cantilever, twisting and warping are prevented at the support. However, this is not the case for an overhanging beam, as warping of the overhanging segment is at best elastically restrained by continuity with the interior segment of the beam. Further, in many practical cases, the support details of overhanging beams only provide elastic restraints against twisting, as shown in Fig. 6. Although some design codes (6, 7) clearly recognise the difference between built-in cantilevers and overhanging beams, the only published study of the latter appears to be that by Nethercot (23).

In this paper, the effects of the support conditions on the elastic buckling of uniform I-section overhanging segments are investigated for the special case where warping is free at the supports. The circumstances under which an overhanging beam acts as if free to warp are studied, and an approximate method is suggested for the more general case in which warping of an overhanging segment is elastically restrained by its continuity with the interior segment. Finally, the effects of elastic restraints against twisting at the supports are examined.

3.2 *Buckling of overhanging segments free to warp*

The support conditions for a built-in cantilever (Fig. 5) may be specified as being prevented from deflecting and twisting ($u = \phi = 0$), from rotating about the minor axis ($u' = 0$) and from warping ($\phi' = 0$) at $z = 0$. However, these two latter conditions should be modified in the case of an overhanging segment (Fig. 4), since continuity at the support will generally result in $u' \neq 0$ and $\phi' \neq 0$ at $z = 0$. While the actual conditions will vary with the load and geometry of the overhanging beam, there will be one particular arrangement for which there is no buckling interaction between the overhanging and the interior segments. In this particular case, the minor axis end moment and the warping stresses will be zero, so that $u'' = \phi'' = 0$ at $z = 0$. It is this case that is treated in this section.

However, when the support condition $u' = 0$ is replaced by $u'' = 0$ for an overhanging segment, it is found that it buckles at zero load in a rigid body motion by rotating in a horizontal plane about the support. While such a buckling mode satisfies a mathematical definition of instability, it does not correspond to an engineering concept of failure. Indeed, such an overhanging segment is able to support non-zero loads without failure, even though it may rotate in the horizontal plane. The reason for this is that the boundary condition $u' = 0$ ensures the rigid body stability of an overhanging segment, without having any effect on the bending and twisting which take place during buckling. It may therefore be enforced to eliminate mathematically the rigid body buckling mode without changing the flexural-torsional buckling and failure loads. It may also be noted that the condition $u'' = 0$ is satisfied automatically when $\phi = 0$, since this requires the minor axis component of the major axis moment to be zero. Thus the boundary conditions taken for an overhanging segment are $u = \phi = u' = \phi'' = 0$ at $z = 0$.

The elastic buckling of an overhanging segment free to warp has been analysed for a number of geometry and loading conditions by using a finite element computer program (12) with 8 equal length elements in the overhang. Some of the results obtained are shown as dimensionless buckling moments in Figs 1-3, where they are compared with the correponding values for built-in cantilevers.For overhanging segments in uniform bending (Fig. 1), the elastic buckling moment M can be approximated by using

$$M_o L/\sqrt{(EI_y GJ)} = 1.6 + 0.05K \tag{8}$$

This equation is virtually independent of K, because the absence of any warping restraint at the support allows a buckling mode in which the angle of twist ϕ varies almost linearly along the overhang. Thus the overhanging segment is almost in uniform torsion, and there is very little warping torsion $EI_w\phi'''$ generated.

The elastic buckling moments of overhanging segments with end loads P are shown in Fig. 2. These can be approximated by

$$P_o L^2/\sqrt{(EI_y GJ)} = 6\{1+1.5(\varepsilon-0.1)/\sqrt{(1+1.5^2(\varepsilon-0.1)^2)}\}$$

$$+ 1.5(K-2)\{1+3(\varepsilon-0.3)/\sqrt{(1+3^2(\varepsilon-0.3)^2)}\} \tag{9}$$

For centroidal loading $(\varepsilon = 0)$, these values are again virtually independent of K. While they are the same as the values for built-in cantilevers when $K = 0$, they diverge from these as K increases, and are substantially less for high values of K. This is because the high resistances of built-in cantilevers for which K is large, which depend substantially on warping torsion generated through the warping rigidity, are not realised in isolated overhanging segments free to warp, in which only small warping torques are generated during buckling.

This absence of warping torsion is demonstrated in Fig. 7, in which are compared the buckled shapes u (shown dashed) and $h\phi/2$ (shown dotted) of cantilevers and overhanging segments with centroidal end loads. For high values of K, warping restraint at the support of the cantilever causes considerable curvature of the twist shape $h\phi/2$, thereby inducing significant warping torques. However, for the free to warp overhanging segment, $h\phi/2$ varies almost linearly, and there is very little warping torque. For low values of K, warping is relatively unimportant, as can be seen from the twist shapes for $K=0.2$ in Fig. 7, and it makes little difference whether the member is prevented from warping or free to warp.

The elastic buckling moments of isolated overhanging segments with uniformly distributed loads w are shown in Fig. 3. These can be approximated by

$$w_o L^3/(2\sqrt{(EI_y GJ)}) = 15\{1+1.8(\varepsilon-0.3)/\sqrt{(1+1.8^2(\varepsilon-0.3)^2)}\}$$
$$+ 4(K-2)\{1+2.8(\varepsilon-0.4)/\sqrt{(1+2.8^2(\varepsilon-0.4)^2)}\} \quad (10)$$

Again, the values for centroidal loading are virtually independent of K, and the resistances of isolated overhanging segments with high values of K are substantially lower than those of built-in cantilevers.

3.3 Interaction buckling of overhanging beams

In the previous section, it was noted that when there is no buckling interaction between the overhanging and interior segments of an overhanging beam, then each segment may be assumed to be free of warp. In this case, the buckling moment of an overhanging segment with end load can be obtained from eqn (9), and the buckling moment of the interior segment by substituting $L_I = \gamma L$ for L in eqn (2) and (3), so that

$$M_{\gamma o} L/\sqrt{(EI_y GJ)} = (\pi/\gamma)\sqrt{(1+K^2/\gamma^2)} \quad (11)$$

The variations of these moments are shown by the dashed lines in Fig. 8.

More generally, there will be an interaction between adjacent segments during buckling, and the less critically loaded segment will elastically restrain the more critically loaded segment. In these cases, a lower bound estimate of the buckling moment can be obtained from the lesser of the two moments $P_o L$ and $M_{\gamma o}$ calculated for segments free to warp (and shown by the dashed lines in Fig. 8). An upper bound estimate can be obtained from the lesser of the two moments $P_\infty L$ and $M_{\gamma\infty}$ calculated by assuming that the segments are prevented from warping. The moment $P_\infty L$ for the overhanging segment can be obtained from eqn (4), and the moment $M_{\gamma\infty}$ for the interior segment from the approximation (11)

$$M_{\gamma\infty} L/\sqrt{(EI_y GJ)} = (\pi/\gamma)\sqrt{(1+4K^2/\gamma^2)} \quad (12)$$

which incorporates a warping effective length factor of $k_w = 0.5$. These upper bound estimates are shown by the dash-dot lines in Fig. 8. (Note that for beams with $K=0$, warping is unimportant, and the upper and lower bounds coincide.)

Accurate solutions for the interaction buckling moments have been obtained, and these are shown by the solid lines in Fig. 8. The corresponding buckled shapes $(u \pm h\phi/2)$ of the top (T) and bottom (B) flanges are shown in Fig. 9. It can be seen from Fig. 9 that for short interior segments $(\gamma \to 0)$, the overhanging segments receive substantial warping restraint, and from Fig. 8 that the buckling moment PL approaches the warping prevented value $P_\infty L$ given by eqn (4). As the length γL of the interior segment increases, so the warping restraints it provides decrease, and become zero when $\gamma \approx 1.2$ (for $K=2$), so that the segments buckle independently. For higher values of γ, the overhanging segments provide warping restraints to the interior segment. However, the effects of these warping restraints on the interior segment are not generally as significant as are those on the overhanging segments.

Approximate solutions for the interaction buckling moments can also be obtained by extending the method developed in References 29 and 41. Lower bound estimates $P_o L$ and $M_{\gamma o}$ can be obtained as explained above. When $M_{\gamma o} < P_o L$, then Fig. 8 indicates that the lower bound will be sufficiently accurate. When $P_o L < M_{\gamma o}$, then an improved estimate for PL can be obtained by approximating the warping effective length factor k_w of the overhanging segment by

$$k_w = 0.5 + 0.5\sqrt{(P_o L/M_{\gamma o})} \tag{13}$$

and by using this in the approximation

$$P = P_o(2 - 1/k_w) + P_\infty(1/k_w - 1) \tag{14}$$

Equation (13) ensures that k_w is equal to 1 at zero interaction (when $P_o L = M_{\gamma o}$), in which case eqn (14) yields P equal to P_o. As γ approaches zero, eqn (13) allows k_w to approach 0.5, and eqn (14) allows P to approach P_∞.

The approximate buckling moments calculated by this method are shown by the circled points in Fig. 8. It can be seen that they agree favourably with the accurate finite element solutions.

4. *EFFECTS OF ELASTIC TORSIONAL END RESTRAINT*

4.1 *General*

It is usually assumed that flexural members are prevented from twisting at their supports by rigid restraints. However, it is not uncommon in practice that twisting at a support is only elastically restrained, as shown in Fig. 6. A reduction in the stiffness of a support restraint causes a reduction in the resistance of the member to lateral buckling.

The case where support twisting is permitted by local distortions of a slender unstiffened web (Fig. 6b) has been studied recently (4, 5), and improved design approximations have been suggested for simply supported beams. Some studies have also been made of the case where the flexibility of the support itself permits end twisting (Fig. 6a). An unpublished investigation of the buckling of narrow rectangular cantilevers (16) was used in the predecessors of recent design codes (6, 7), while other studies have investigated the buckling of simply supported beams (35, 40), and the restraint stiffnesses provided by some common beam end connections (13). The presence of a flexible torsional end restraint increases the total torsional flexibility of a member from L/GJ to $(L/GJ+R/\alpha_{RZ})$, in which α_{RZ} is the stiffness of the restraint and R is a factor which reflects the relative importance of the support stiffness to the torsional stiffness GJ/L. Thus the effective torsional stiffness

stiffness of the member is decreased to $(GJ/L)/\{1+(GJ/L)/(\alpha_{RZ}/R)\}$

The influence of this decreased torsional stiffness on the buckling resistances of overhanging segments and beams is studied in the following sub-sections by using a modification of a finite element program (12) which allows for concentrated elastic restraints (31).

4.2 *Overhanging segments free to warp*

The buckling end loads of overhanging segments free to warp but prevented from twisting at the support are shown in Fig. 5 and approximated by eqn (9). The reduced non-dimensional buckling loads of overhanging segments with elastic torsional restraints are plotted in Fig. 10 for a range of values of K from 0.05 to 5 and of the dimensionless load height $2\bar{a}/h$ from -1 (top flange) to $+1$ (bottom flange).

For centroidal loading $(2\bar{a}/h = 0)$, the non-dimensional curves are grouped, and can be closely approximated by

$$\frac{P}{P_o} = \sqrt{\{\frac{\alpha_{RZ}L/GJ}{(5+4K^2/(1+K^2)+\alpha_{RZ}L/GJ)}\}}\{1-\frac{2\bar{a}}{h}K\beta^2(1-\beta)\} \tag{15}$$

in which

$$\beta = (\alpha_{RZ}L/GJ)/(1+\alpha_{RZ}L/GJ) \tag{16}$$

This formulation is equivalent to approximating the relative torsional stiffness factor R by

$$R = 5+4K^2/(1+K^2) \tag{17}$$

so that R varies between 5 (for $K=0$) and 9 (for $K=\infty$).

For non-centroidal loading of segments with $K=2$, the approximation of eqn (15) and (16) is accurate for bottom flange loading and conservative for top flange loading while $\beta>0.2$, as can be seen in Fig. 10. It can be expected that its accuracy will not decrease for segments with $K<2$, since the effects of non-centroidal loading decrease with K (Fig. 2).

4.3 *Overhanging beams*

The buckling end loads of overhanging beams with centroidal loading are shown in Fig. 11 for a range of span ratios γ. For span ratios less than 2, these loads may be obtained by replacing $(\alpha_{RZ}L/GJ)$ in eqns (15) and (16) by

$$(\alpha_{RZ}L/GJ)_\gamma = (\alpha_{RZ}L/GJ)\left(\frac{0.35}{1-0.8(\gamma-1.2)}\right) \tag{18}$$

However, for higher span ratios, the interior segment dominates in the buckling action. In this case, it may be modelled as free to warp, and the moment at buckling approximated by using

$$\frac{M}{\pi\sqrt{(EI_y GJ/L^2)}\sqrt{(1+K^2)}} = \sqrt{\{\frac{\alpha_{RZ}L/GJ}{(4.9+4.5K^2)+\alpha_{RZ}L/GJ}\}} \tag{19}$$

while $K\leqslant 5$. This equation is equivalent to approximating the relative torsional stiffness factor R by

$$R = 4.9 + 4.5K^2 \tag{20}$$

The accuracy of eqn (19) is demonstrated in Fig. 12.

5. *CONCLUSIONS*

A cantilever model of an overhanging beam overestimates its lateral buckling resistance, because it assumes that warping is prevented at the support. In the worst case, an overhanging segment acts as if free to warp, and the warping rigidity is not mobilised as it is in a cantilever. Because of this, the buckling resistance of a centroidally loaded overhanging segment is almost independent of the warping rigidity.

Approximate equations have been developed for estimating the buckling resistances of built-in cantilevers and of overhanging segments free to warp. These take markedly different forms to those commonly used for simply supported beams.

In an overhanging beam, there is in general a warping interaction between the interior and overhanging segments. A short interior segment provides substantial warping restraints to the overhanging segments. The effects of these on the buckling resistance of the overhanging segments can be estimated approximately. The restraining effects of short overhanging segments on the buckling resistance of the interior segment are small, and it may be assumed that the latter is free to warp.

Elastic torsional end restraints decrease the effective torsional rigidity of a flexural member, and consequently reduce its buckling resistance. Approximate equations have been developed for estimating the reduced buckling loads of overhanging segments free to warp and of overhanging beams.

References

1. Anderson, J.M., and Trahair, N.S., 'Stability of Monosymmetric Beams and Cantilevers', Journal of the Structural Division, ASCE, Vol. 28, No.ST1, Jan. 1972, pp.269-286.

2. Barsoum, R.S., and Gallagher, R.H., 'Finite Element Analysis of Torsional and Torsional-Flexural Stability Problems', International Journal of Numerical Methods in Engineering, Vol.2, 1970, pp.335-352.

3. Bradford, M.A., and Hancock, G.J., 'Interaction of Local and Lateral Buckling in Beams', Research Report R408, School of Civil and Mining Engineering, University of Sydney, March 1982.

4. Bradford, M.A., and Trahair, N.S., 'Distortional Buckling of I-Beams', Journal of the Structural Division, ASCE, Vol.107, No.ST2, Feb. 1981, pp.355-370.

5. Bradford, M.A., and Trahair, N.S., 'Lateral Stability of Beams on Seats', Research Report No.R423, School of Civil and Mining Engineering, University of Sydney, November 1982.

6. British Standards Institution, 'Draft Specification for the Structural Use of Steelwork in Buildings: Part 1 - Simple Construction and Continuous Construction', BSI, London, 1977.

7. British Standards Institution, 'BS5400: Part 3 Code of Practice for Design of Steel Bridges', BSI, London, 1982.

8. Brown, T.G., 'Lateral-Torsional Buckling of Tapered I-Beams', Journal of the Structural Division, ASCE, Vol.107, No.ST4, April 1981, pp.689-697.

9. Chen, W.F., and Atsuta, T., 'Theory of Beam-Columns. Vol.2. Space Behaviour and Design', McGraw-Hill, New York, 1977.

10. Column Research Committee Japan, 'Handbook of Structural Stability', Corona, Tokyo, 1971.

11. Flint, A.R., 'The Stability and Strength of Stocky Beams', Journal of the Mechanics and Physics of Solids, Vol.1, No.2, Jan. 1953, pp.90-102.

12. Hancock, G.J., and Trahair, N.S., 'Finite Element Lateral Buckling of Continuously Restrained Beam-Columns', Civil Engineering Transactions, Institution of Engineers, Australia, Vol.CE20, No.2, 1978, pp.120-127.

13. Hogan, T.J., and Thomas, I.R., 'Standardized Structural Connections. Part B: Design Methods', Australian Institute of Steel Construction, Sydney, 1978.

14. Horne, M.R., 'Plastic Design of Single Storey Frames', Lecture 9 in 'A Short Course on the New British Standard for Structural Steelwork', Simon Engineering Laboratories, University of Manchester, September 1977.

15. Johnson, R.P., and Bradford, M.A., 'Distortional Lateral Buckling of Unstiffened Composite Bridge Girders', Research Report R432, School of Civil and Mining Engineering, University of Sydney, February 1983.

16. Kerensky, O.A., Flint, A.R., and Brown, W.C., 'The Basis for Design of Beams and Plate Girders in the Revised British Standard 153', Proceedings, Inst. Civ. Engrs, Vol.5, Part III, Aug. 1956, pp.396-461.

17. Kirby, P.A., and Nethercot, D.A., 'Design for Structural Stability', Granada Publishing, St. Albans, 1979.

18. Kitipornchai, S., Dux, P.F., and Richter, N.J., 'Buckling and Bracing of Cantilevers', Research Report No.CE41, Department of Civil Engineering, University of Queensland, March 1983.

19. Kitipornchai, S., and Trahair, N.S., 'Buckling Properties of Monosymmetric I-Beams', Journal of the Structural Division, ASCE, Vol.106, No.ST5, May 1980, pp.941-957.

20. Krefeld, W.J., Butler, D.J., and Anderson, G.B., 'Welded Cantilever Wedge Beams', Welding Journal, Vol.38, No.3, 1959, pp.97s-112s.

21. Massey, P.C., and McGuire, P.J., 'Lateral Stability of Non-Uniform Cantilevers', Journal of the Engineering

Mechanics Division, ASCE, Vol.97, No.EM3, June 1971, pp.673-686.

22. Michell, A.G.M., 'Elastic Stability of Long Beams Under Transverse Forces', Phil. Mag., Vol.48, 1899, pp.298-309.

23. Nethercot, D.A., 'The Effective Lengths of Cantilevers as Governed by Lateral Buckling', The Structural Engineer, Vol.51, 1973, pp.161-168.

24. Nethercot, D.A., 'Lateral Buckling of Tapered Beams', Publications, IABSE, Vol.33, 1973, pp.173-192.

25. Nethercot, D.A., 'Inelastic Buckling of Steel Beams Under Non-Uniform Moment', The Structural Engineer, Vol.53, No.2, Feb. 1975, pp.73-78.

26. Nethercot, D.A. 'Elastic Lateral Buckling of Beams', 'Developments in the Stability and Strength of Structures. Vol.2. Beams and Beam-Columns', Applied Science Publishers, 1983.

27. Nethercot, D.A., and Al-Shankyty, M.A.F., 'Bracing of Slender Cantilever Beams', 'Stability Problems in Engineering Structures and Components', Applied Science Publishers Ltd, London, 1979, pp.89-99.

28. Nethercot, D.A., and Rockey, K.C., 'A Unified Approach to the Elastic Lateral Buckling of Beams', The Structural Engineer, Vol.49, 1971, pp.321-330.

29. Nethercot, D.A., and Trahair, N.S., 'Lateral Buckling Approximations for Elastic Beams', The Structural Engineer, Vol.54, 1976, pp.197-204.

30. Nethercot, D.A., and Trahair, N.S., 'Design of Laterally Unsupported Beams', 'Developments in the Stability and Strength of Structures'. Vol.2. Beams and Beam-Columns, Applied Science Publishers, 1983.

31. Nethercot, D.A., and Trahair, N.S., 'Bracing Requirements in Thin-Walled Structures', Developments in Thin-Walled Structures, Applied Science Publishers, 1983.

32. Ojalvo, M., 'Wagner Hypothesis in Beam and Column Theory', Journal of the Engineering Mechanics Division, ASCE, Vol.107, No.EM4, August 1981, pp.669-677.

33. Poley, S., 'Lateral Buckling of Cantilevered I-Beams Under Uniform Load', Transactions, ASCE, Vol.121, 1956, pp.786-790.

34. Prandtl, 'Kipperscheinungen', Thesis, Munich, 1899.

35. Schmidt, L., 'Restraints Against Elastic Lateral Buckling', Journal of the Engg Mech. Dvn, ASCE, Vol.91, No.EM6, Dec. 1965, pp.1-10.

36. Singh, K.P., 'Ultimate Behaviour of Laterally Supported Beams', Ph.D. Thesis, University of Manchester, 1969.

37. Timoshenko, S.P., 'Einige Stabilitatsprobleme der Elastizitatstheorie', 'The Collected Papers of Stephen P.

Timoshenko', McGraw-Hill, London, 1953, pp.1-50.

38. Timoshenko, S.P. 'Sur La Stabilite des Systemes Elastiques', 'The Collected Papers of Stephen P. Timoshenko', McGraw-Hill, London, 1953, pp.92-224.

39. Timoshenko S.P., and Gere, J.M., 'Theory of Elastic Stability', 2nd ed., McGraw-Hill, New York, 1961.

40. Trahair, N.S., 'Stability of I-Beams with Elastic End Restraints', Journal, Inst. Engrs, Aust., Vol.37, No.6, June 1965, pp.157-168.

41. Trahair, N.S., 'The Behaviour and Design of Steel Structures', Chapman and Hall, London, 1977.

42. Trahair, N.S., Discussion of Ref. 32, Journal of the Engineering Mechanics Division, ASCE, Vol.108, No.EM3, June 1982, pp.575-578.

43. Trahair, N.S., 'Inelastic Lateral Buckling of Beams', 'Developments in the Stability and Strength of Structures. Vol.2. Beams and Beam-Columns', Applied Science Publishers, 1983.

44. Trahair, N.S., and Woolcock, S.T., 'Effect of Major Axis Curvature on I-Beam Stability', Journal of the Engineering Mechanics Division, ASCE, Vol.99, No.EM1, Feb. 1973, pp.85-98.

45. Vacharajittiphan, P., and Trahair, N.S., 'Flexural-Torsional Buckling of Curved Members', Journal of the Structural Division, ASCE, Vol.101, No.ST6, June 1975, pp.1223-1238.

46. Vinnakota, S., 'Finite Difference Method for Plastic Beam-Columns', Chapter 10 in 'Theory of Beam-Columns. Vol.2' by W.F. Chen and T. Atsuta, McGraw-Hill, New York, 1977.

47. Wagner, H., 'Verdrehung und Knickung von Offenen Profilen', 25th Anniversary Publication, Technische Hochschule, Danzig, 1904-1924, Translated in NACA Technical Memorandum No.807, 1936.

48. Warnick, W.L., and Walston, W.H., 'Secondary Deflections and Lateral Stability of Beams', Journal of the Engineering Mechanics Division, ASCE, Vol.106, No.EM6, Dec. 1980, pp.1307-1325.

49. Woolcock, S.T., and Trahair, N.S., 'Post-Buckling Behaviour of Determinate Beams', Journal of the Engineering Mechanics Division, ASCE, Vol.100, No.EM2, April 1974, pp.151-171.

50. Woolcock, S.T., and Trahair, N.S., 'Post-Buckling of Redundant I-Beams', Journal of the Engineering Mechanics Division, ASCE, Vol.102, No.EM2, April 1976, pp.293-312.

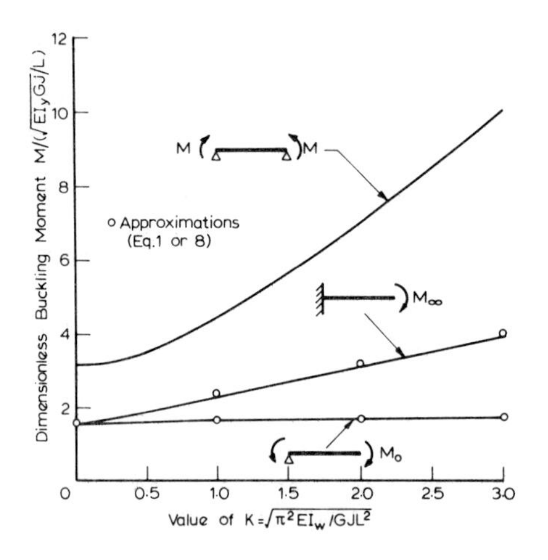

Fig.1 Buckling under end moments

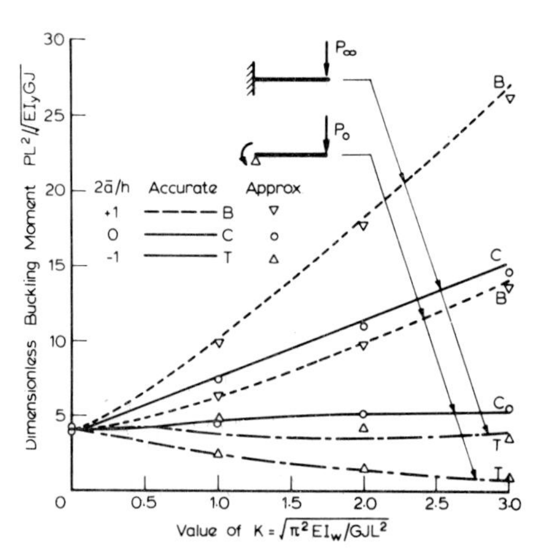

Fig.2 Buckling under end loads

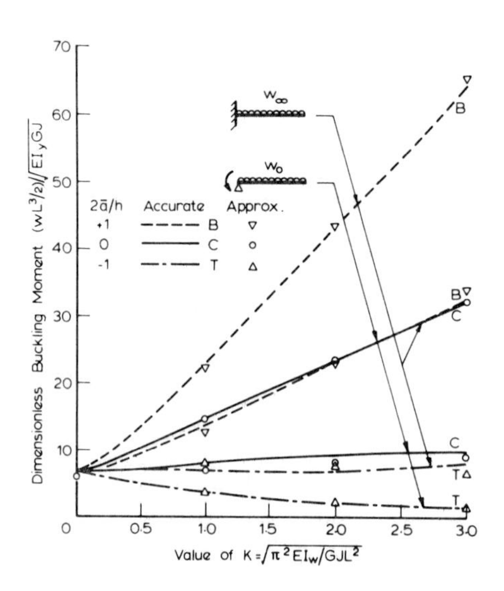

Fig.3 Buckling under distributed loads

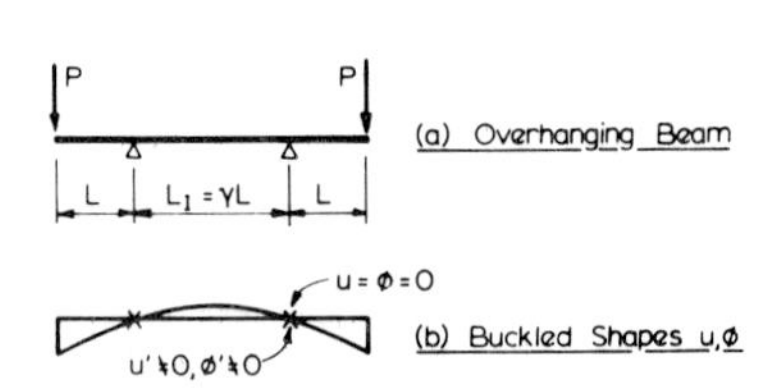

Fig.4 Buckling mode of an overhanging beam

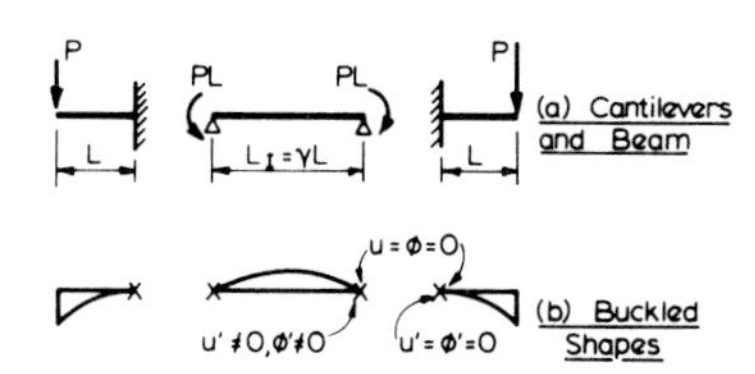

Fig.5 Cantilever model of an overhanging beam

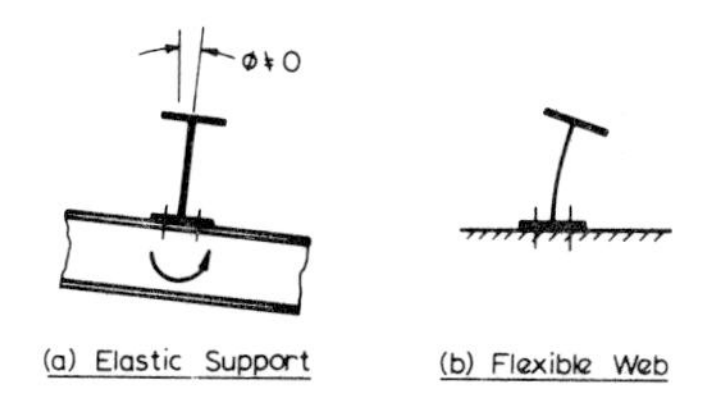

Fig.6 Elastic restraints against twisting

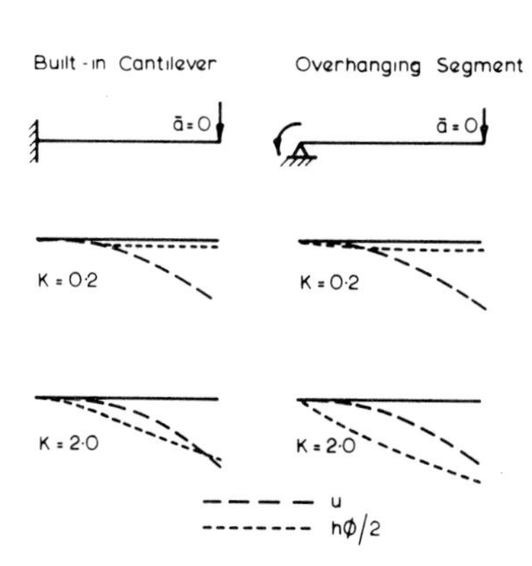

Fig.7 Buckled shapes for cantilevers and overhanging segments

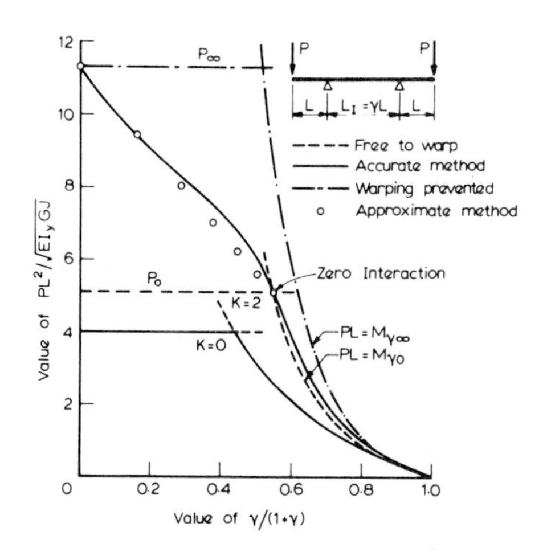

Fig.8 Interaction buckling of overhanging beams

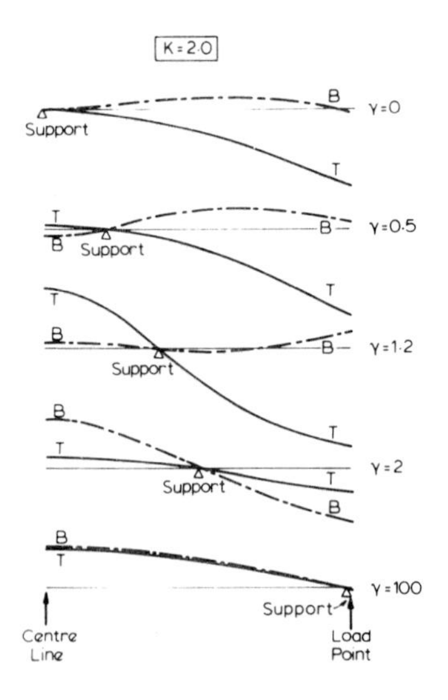

Fig.9 Buckled shapes of overhanging beams

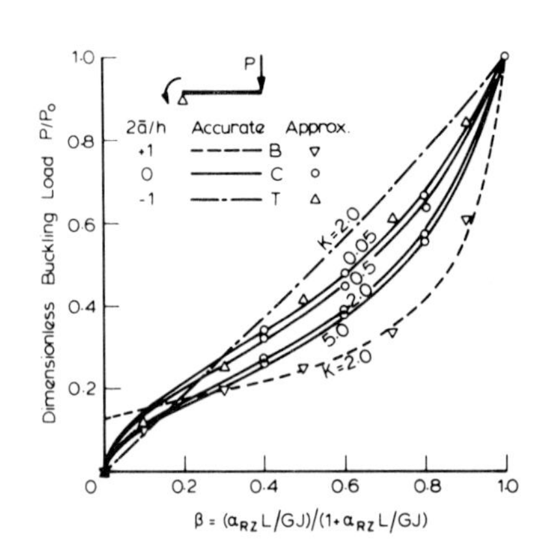

Fig.10 Overhanging segments with torsional restraints

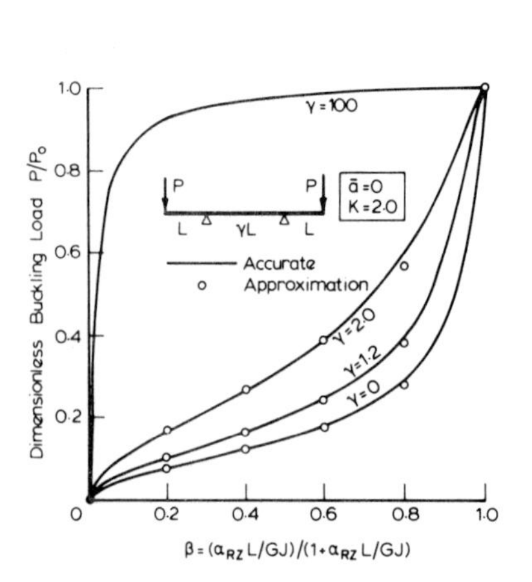

Fig.11 Overhanging beams with torsional restraints

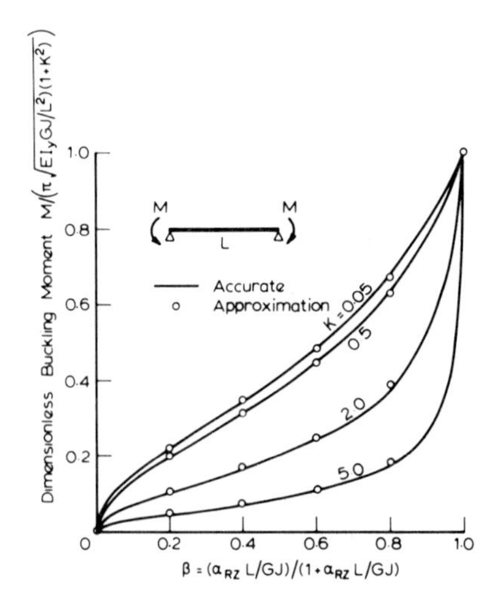

Fig.12 Simply supported beams with torsional restraints

PLASTIC DESIGN AND LOCAL INSTABILITY PHENOMENA IN ROLLED STEEL ANGLES

R Bez and M A Hirt

Swiss Federal Institute of Technology, Lausanne

Synopsis

The development of design charts for angles subject to biaxial bending and axial load is discussed. The theory necessary for establishing these interaction curves in the plastic domain for any angle is compared with the results of a series of tests. The limitations of the application of plastic design due to local instability problems are also examined.

Introduction

The I section is undoubtedly the most common steel section in applications where bending and possibly axial loads have to be carried. The angle is generally reserved for use as a tension or compression element in bracing or lattice systems. Occasionally however its form is convenient for use in built up sections with specific architectural applications and this was the origin of the work to be presented.

The *USM MIDI 1000* beam system (1) was conceived as a building component offering maximum possible horizontal and vertical permeability for services (Fig 1). The beam is a "box" formed by the four flange angles. In span it behaves partly as a Vierendeel girder, due to the pressed plate verticals bolted to the flange angles, and partly as a lattice girder due to the welded diagonals. In consequence the angles forming the flanges are simultaneously subjected to bending and axial loads.

In order to understand the behaviour of these angles in the plastic domain a research project was initiated. Its aims were to:

- determine the ultimate plastic strength of angles under combined bending and axial load,
- consider possible phenomena of instability in such a member.

Apart from a general theoretical treatment a series of tests were conducted on flange assemblies built up from *L 150 • 100 • 10* angles.

This article presents a discussion of the behaviour of an angle up to the plastic limit and an examination of the instability phenomena which may manifest themselves.

1. *DESIGN SELECTION OF ANGLES*

Angles distinguish themselves from other sections commonly used in steel construction by their lack of symmetry. The behaviour of the section is thus intuitively more difficult to grasp as we are accustomed to working essentially with doubly symmetrical members.

The aim of this section is to examine the behaviour of an angle under different loading conditions, so that a design selection might be possible.

1.1 Elastic design

The first condition considered is with the angle subject to bending forces only and staying in the elastic range. Figure 2 shows the stress distribution in the section for the following three types of loading:

- the maximum possible moment M_x in the elastic range (Fig 2a),
- the maximum possible moment M_y in the elastic range (Fig 2b),
- a superposition of these two moments (Fig 2c).

With respect to this stress distribution two remarks should be made.

(a) The neutral axis resulting from horizontal (Fig 2a) or vertical bending (Fig 2b) is neither horizontal nor vertical. This is because the principal axes of inertia of an angle are themselves neither horizontal nor vertical. Hence, in order to have a horizontal or vertical neutral axis, it is necessary that the section should be simultaneously subject to moments M_x and M_y.

(b) In the superposition (Fig 2c) of the two moments a stress condition is obtained whose maximum value may be less than in the two former cases.

Recognising this fact, it becomes interesting to understand the interaction of the two moments M_x and M_y. This may be done by finding all possible positions of the neutral axis such that, at the elastic limit, the equilibrium condition, $\int_A \sigma \, dA = 0$, is satisfied.

If, in addition to the bending loads, an axial force N is introduced then it is possible to establish curves of the interaction between M_x, M_y and N for different states of compression and tension; the value of the normal force being $N = \int_A \sigma \, dA$. The interaction curves thus obtained are presented in Fig 3 where the difference is drawn between a tensile (Fig 3a) and compressive force (Fig 3b).

These curves clearly show the effect that, having reached yield point with bending about a horizontal axis, further bending may be imposed about a vertical axis before the elastic limit is reached. It should also be noted that:

- the inclination of the neutral axis varies rapidly depending on position along the interaction curve,
- the neutral axis is only horizontal or vertical when there is a superposition of M_x and M_y,
- the reduction in carrying capacity due to an axial load is not always proportional to its intensity,

- the interaction curves are not the same according to whether the axial load is one of compression or tension.

1.2 Plastic design

The angle loaded purely in bending will now be reconsidered above the elastic limit by admitting total plastification of the section. The internal distribution of stresses is still of the same type as that obtained for the elastic calculation except that there will be, at all points of the section, a stress equal to the yield stress σ_y of the steel.

By following the same process as for the elastic design it is again possible to establish curves for the interaction between M_x, M_y and N. This has been done by applying the theory presented by Chen and Astuta in Chapter 5 of reference (2) which treats precisely the curves of interaction between bending and axial force, whatever the cross-section. This then permitted the presentation of plastic interaction given in Fig 4.

It may be seen that the plastic interaction curves have generally the same shape as their elastic counterparts and that the remarks made with respect to them are still valid for the plastic curves.

It is worth noting that reference (3) gives examples of plastic interaction curves for various symmetric and asymmetric sections.

1.3 A comparison of elastic and plastic calculation

The interaction curves established clearly show that, for angles, the understanding of interaction between bending and axial load is not intuitive in the first instance. In addition they demonstrate that by calculating plastically a "gain in strength" may be obtained over elastic design which can, depending on the case, exceed *100%* and is clearly far from negligible.

If the evolution of the position of the neutral axis was continuously observed during the passage of the section from the elastic to the plastic limit state it would be seen that it varies sensitively, especially if there is an axial load. This phenomena corresponds with internal redistribution of stresses which naturally take place with a plastification of a section. It is consequently essential that this redistribution should be able to take place effectively in order that a plastic design be applicable.

Thus although the plastic design of angles brings about an appreciable "gain in strength" its application is subject to certain restrictions which are discussed in the following section.

2. INSTABILITY PHENOMENA

It is necessary to check the limits of application by ensuring that possible phenomena of instability cannot intervene as the limits are approached. This will be examined in this section, firstly in a theoretical manner and then on the basis of tests from which it is possible to draw some practical conclusions.

2.1 *Limits imposed by the SIA 161 Standard*

The SIA 161 Standard (4) gives slenderness limits which must not be exceeded. The limit values, summarised in Table 1, depend both on the steel grade and type of design calculation to be made.

Table 1 Slenderness limits prescribed by the Swiss SIA 161 Standard for plane elements in compression

TYPE OF CALCULATION		LEG SLENDERNESS LIMIT *(leg/t)*		
MEMBER FORCES	SECTIONAL RESISTANCE	*Fe 360* (σ_y = 235)	*43 B* (σ_y = 255)	*Fe 510* (σ_y = 355)
ELASTIC	ELASTIC	17	16	14
ELASTIC	PLASTIC	13	12	11
PLASTIC	PLASTIC	11	10	9

On the basis of these limit values and knowing that *L 150 • 100 • 10* angle has a slenderness of *14,5* it may be concluded that the standard permits the use of this section for an elastic strength calculation only.

If, moreover the theories of Winter (5) and von Karman (6) concerning local buckling (used as a basis for the SIA 161 Standard) are applied then one arrives similarly at the conclusion that there may well be a problem in trying to take account of the ultimate plastic strength of the section.

Given that the Standard forbids the use of such angles because of the problems of local instability then it may be interesting to explore the problem more thoroughly by considering only the slenderness of the part of the element actually in compression. With this intention the plastic interaction curves of Fig 4 have been shaded to show the zone where the slenderness of the part of the angle web in compression exceeds *13*. This is the limit given by the standard for an elastic-plastic calculation of a section in *Fe 360*. It shows that, even by adopting less severe conditions than the standard in considering only the slenderness of the compressed part and not the full section, there remains an important region where the phenomena of local instability may still occur.

From the theoretical considerations above it is clearly probable that a phenomena of local instability (web buckling) might appear if a plastic design calculation is carried out for a *L 150 • 100 • 10* angle. With the aim of confirming this experimentally a series of tests were initiated which are described in the following sections.

2.2 *Description of the tests*

The principle of the tests carried out was the following (Fig 5):

- Statical system: a beam on two supports, *1200 mm* apart, with its ends cantilering *520 mm* and loaded there by vertical forces P. The beam is simultaneously compressed axially by a force N (Fig 5a).

- Cross-section: formed by two *L 150 • 100 • 10* angles *380 mm* apart which corresponds closely to the *USM MIDI 1000* system (Fig 5b).

With such a system and the application of the vertical forces P a zone of constant moment is created between the supports which adds itself to the compression introduced by the axial force N. It is thus possible to generate any combination of bending and compression. In addition, because the cross section of the test beam has a vertical axis of symmetry its neutral axis will remain horizontal throughout the application of load. Thus whilst in the elastic domain a curve corresponding to the "horizontal neutral axis" line of Fig 3b will be followed and then, on continuing to a complete plastification of the section the corresponding curve of Fig 4b.

The application of load was carried out in the following way. Firstly an axial compressive force N was applied and maintained constant throughout the test.

The force P in the vertical rams was then increased in steps up to the failure of the beam. At each loading step a recording was made of the magnitude of deformations and displacements at several points of the section, in order to permit a comparison with calculated values.

Seven tests were carried out with *L 150 • 100 • 10* angles, each differing from one another by virtue of the quality of steel and the magnitude of the axial load applied. These differences were (with $N_p = \sigma_y A$):

- 4 tests with beams in *Fe 360* steel and axial loads of $N/N_p = 0 - 0{,}2 - 0{,}4 - 0{,}6$
- 3 tests with beams in *Fe 510* steel and axial loads of $N/N_p = 0 - 0{,}2 - 0{,}4$.

2.3 Results of tests

The relationship between the load P_{fail} observed at the failure of the beam and the theoretical load P_p, which should have been necessary to attain full plastification, is presented in Fig 6 for the seven tests carried out. This diagram shows that the loss of carrying capacity due solely to the appearance of local instability phenomena is far from negligible. (It may be noted that the test carried out with $N = 0{,}6\ N_p$ gave a value of P_{fail} which was too large. This was due to the necessity of applying vertical force P at the beginning of the test to compensate for uplift produced during the application of axial load).

The following conclusions may be drawn.

(a) The failure of the beam occurred before total plastification of the section except for the tests with no axial load where the plastic limit state was reached.

(b) The elastic limit state was always exceeded.

(c) The difference in steel strength is not significant.

(d) In relation to a plastic calculation value a reduction was observed in the failure load P_{fail} of the order of:

- *0%* for $N = 0$
- *25%* for $N = -0{,}2\ N_p$
- *40%* for $N = -0{,}4\ N_p$
- *40%* for $N = -0{,}6\ N_p$

The behaviour is illustrated by Fig 7 which shows the development of stresses in a section as a function of the vertical load P. In comparing the evolution of stresses predicted by the theory with those actually measured during the test it may be seen that there is good correspondence up to the elastic limit state. In all cases, after the beginning of plastification of the section, the measured curve diverges from that predicted and a very rapid failure of the beam is observed before reaching the load corresponding to the plastic limit state. It may thus be seen that the internal redistribution of stresses, necessary for a full plastification of the section, could not take place and that, in consequence, the compressed web of the angle was forced out of its plane by buckling. It is worth noting in this respect that for the tests without axial force the internal redistribution of stresses necessary to attain the plastic limit state is not very important; it was therefore able to take place normally which permitted the plastification of the entire section.

3. *CONCLUSIONS*

The study of rolled steel angles presented in this article has firstly permitted the preparation of a series of elastic and plastic curves of interaction between biaxial bending and axial load (7). They reveal the unusual behaviour of such a section and give the practising engineer a very straightforward means of design. The comparison of the elastic and plastic methods of calculation show that the latter permits an appreciable 'gain in strength' to be obtained which, depending on the case, may exceed *100%*.

The use of a plastic design calculation of the ultimate strength of the sections is however always subject to certain conditions which limit the slenderness of elements in compression. It consequently appears that a large number of unequal angles available commercially exceed this limiting value which means that they may not be used in plastic design.

A series of tests, limited to a single type of angle *(L 150 • 100 • 10)* has demonstrated, without being able to establish a general rule, that in certain cases the phenomenon of local buckling actually appears before total plastification of the section has been achieved.

The SIA 161 Standard thus places itself on the side of safety in imposing an elastic design on certain angles, which are too slender, in order to avoid the appearance of local instability phenomena.

Acknowledgements

Thanks are due to R. Mathys, owner of the consulting engineers at Bienne who placed this research contract with ICOM (Institute of Steel Structures). We similarly acknowledge the help given by the company who supplied the test beams, U. Schärer Söhne AG of Münsingen and the financial support of the Swiss Institute of Steel Construction (SZS). The theoretical work was carried out within the framework of stability research funded by the Swiss National Science Foundation. Finally our thanks go to our colleagues at ICOM for their help in this work and, in particular, B. Kerridge for adaption from the original French text (8).

References

1. USM Bausystem MIDI 1000. Münsingen, U. Schärer Söhne AG, 1980.
2. Chen, W.F. and Atsuta, T. 'Theory of Beam-Columns.' vol. 2: Space Behaviour and Design. McGraw-Hill, New York, 1977.
3. Chen, W.F. and Atsuta, T. 'Interaction curves for biaxial bending of plastic columns.' Lehigh University, Fritz Engineering Laboratory report no. 331.21, 1972.
4. SIA, Standard 161, 1979 edition. 'Steel Structures.' SIA, Zurich, 1979.
5. Winter, G. 'Performance of thin steel compression flanges.' IABSE, Preliminary Report, Third Congress, Liège, 1948.
6. Von Karman, I., Sechler, E.E. and Donnel, L.H. 'The strength of thin plates in compression.' Transactions of the American Society of Mechanical Engineers, vol. 54, 1932.
7. Bez, R. 'Diagrams d'interaction de cornières métalliques.' ICOM report 111, Swiss Federal Institute of Technology, Lausanne.
8. Bez, R. and Hirt, M.A. 'Dimensionnement plastique et phénomènes d'instabilité de cornières métalliques.' Construction Métallique, Puteaux, vol. 19 no. 1, 1982.

Fig 1

Construction using *MIDI 1000* beams.

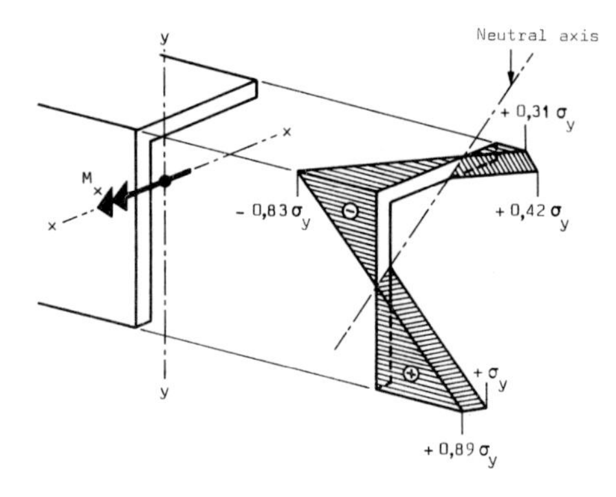

a) Bending moment M_x.

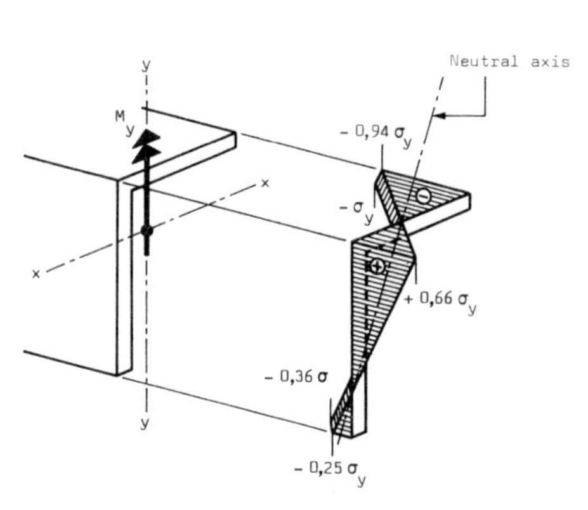

b) Bending moment M_y.

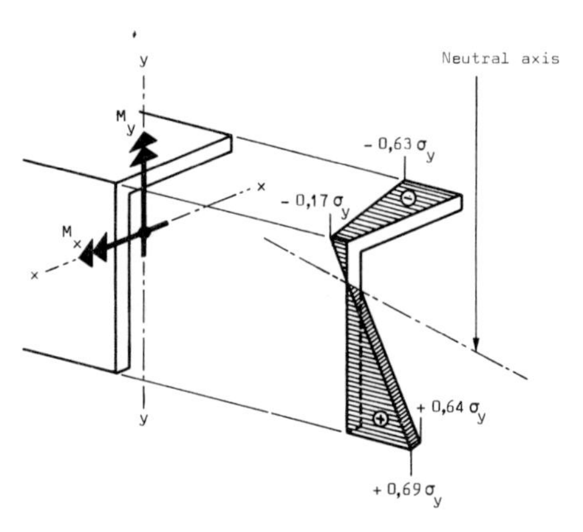

c) Superposition of moments M_x and M_y.

Fig 2

Internal stress distribution in the elastic range for a *L 150 • 100 • 10* in bending.

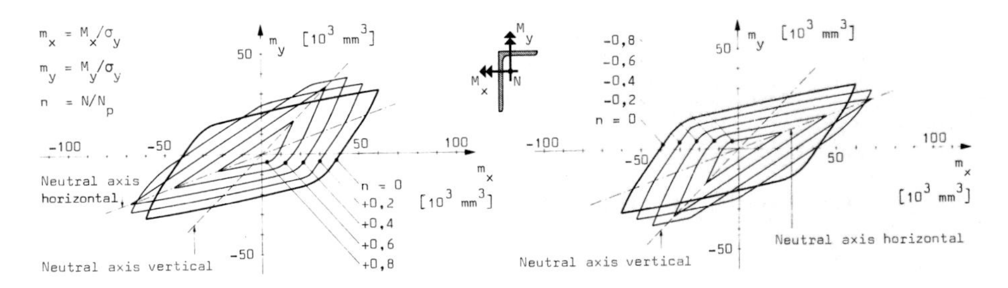

Fig 3

Curves of interaction between M_x, M_y with an axial force N for a *L 150 • 100 • 10* angle in the elastic limit state.

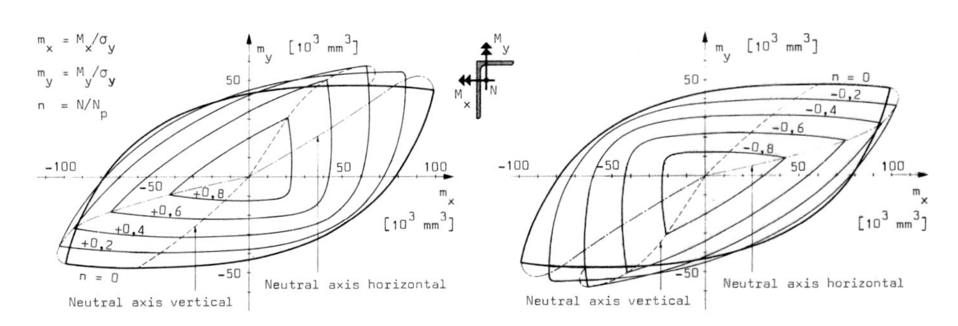

Fig 4

Curves of interaction between M_x, M_y with an axial force N for a *L 150 • 100 • 10* angle in the plastic limit state

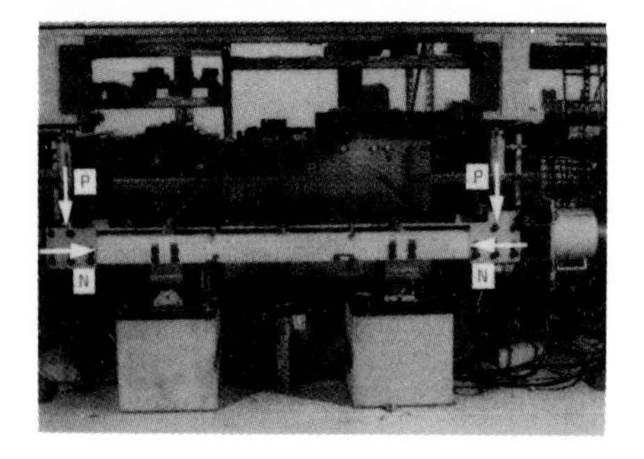

a) Test set-up.

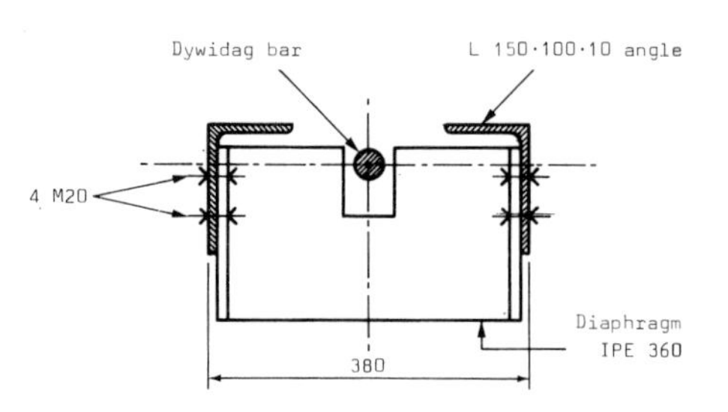

b) Cross section of the test beam.

Fig 5

Description of tests.

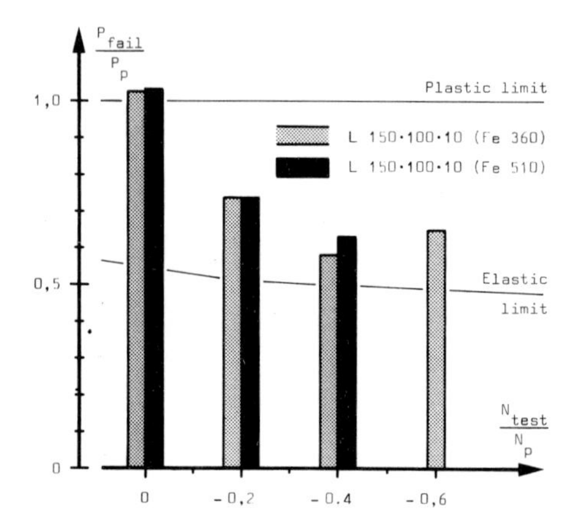

Fig 6

Test results : ratio between the failure load P_{fail} and the load necessary to attain the plastic limit P_p as a function of the axial load.

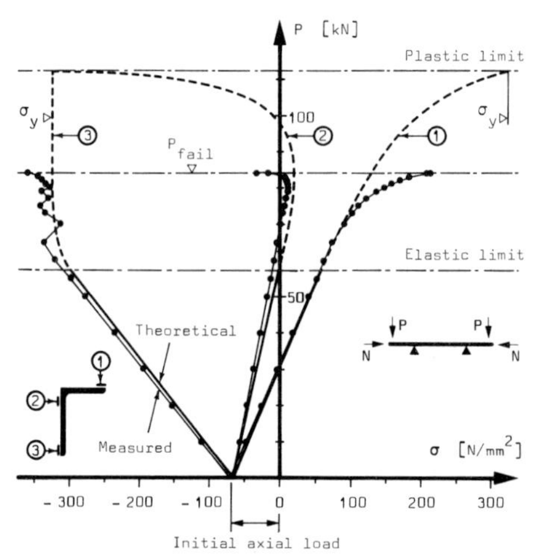

Fig 7

Development of stresses during test.

LOCAL BUCKLING OF I-SECTIONS IN PLASTIC REGIONS OF HIGH-MOMENT GRADIENT

A R Kemp

Professor of Structural Engineering, Department of Civil Engineering, University of the Witwatersrand, Johannesburg

Synopsis

This paper describes the influence of local buckling of I-sections in regions of moment gradient on the flexural ductility as measured by the Rotation Capacity. Different potential modes of failure are described and approximate relationships identified for assessing the rotation capacity. Comparisons are made with observed behaviour in appropriate tests on local buckling.

Notation

b	breadth of flange
E	stress/strain modulus (E = elastic, E_s = strain-hardening)
e	ratio of elastic modulus, E/ strain-hardening modulus, E_s
f	shape factor of section
G_s	modulus of rigidity in strain-hardened region
h_w	clear web depth
L	span of cantilever (Figure 1)
L_b	full wave-length of local buckle
L_p	region of significant yielding ($l = L_p/L$)
M	bending moment (M_m = maximum; M_p = plastic resistance)
R	rotation capacity (R_m at max.moment; R_u at level of M_p)
s	strain at onset of strain-hardening, ε_s /yield strain, ε_y
t	thickness property (t_w = web; t_f = flange)
ε	strain (ε_s = onset of strain-hardening; ε_y= yield; ε_c = critical)
Θ_h	plastic hinge rotation at end of cantilever (Figure 2)
Θ_p	equivalent elastic rotation at end of cantilever at M_p

Introduction

In this paper attention is given to the influence of local buckling of I-sections on flexural ductility in regions of moment gradient. A typical arrangement of the relevant portion of the member is illustrated by the fixed-ended cantilever in Figure 1a and this may be considered representative of the region of a continuous beam adjacent to a support or the portion of a portal rafter adjacent to the eaves. In both cases this includes the plastic hinge in the structure in which the largest rotations will generally be required in plastic analyses. The flexural ductility is assessed by considering the relationship between moment at the fixed end, M , and the rotation, Θ , at the free end of this cantilever.

1 *CODE REQUIREMENTS*

Two levels of flexural ductility are recognised in a number of recent design codes and are illustrated by the two-moment rotation curves illustrated in Figure 2 as follows :-

a) Members possessing "*Compact* Properties" are assumed to be capable of developing their fully-plastic moment-of-resistance, as shown by curve a).

b) Members possessing "*Plastic* Properties" are assumed to be capable of maintaining this resistance moment without loss of strength during the rotations of the plastic hinges which are required to develop a complete mechanism of collapse of the structure, as shown by curve b).

Typical limitations on the flange width-thickness ratio, b/t_f, and clear web depth-thickness ratio, h_w/t_w, for mild steel contained in recent codes are described in Table 1 and are aimed at maintaining ductility despite local buckling.

	Compact Properties	Plastic Properties
Flange Width-Thickness Ratio, b/t_f	*20 - 21*	*17 - 18*
Clear Web Depth-Thickness Ratio, h_w/t_w	*90 - 110*	*66 - 72*

TABLE 1 : LIMITATIONS ON SLENDERNESS FOR LOCAL BUCKLING IN RECENT CODES (Mild Steel in European, British, American, Canadian, Australian and South African Codes)

2 *ROTATION CAPACITY*

The maximum angle, Θ_h , in Figure 2 through which one of the first hinges to form in a structure is required to rotate in order to develop the collapse mechanism, is the subject of some uncertainty. The Rotation Capacity, R , is used to measure this ductility and is defined as the ratio of plastic to elastic hinge rotations, Θ_h/Θ_p, in Figure 2 at the end of the cantilever under uniform moment-gradient as shown in Figure 1. Kemp (1) has evaluated the research on this subject conducted by Driscoll (2,3), Lukey and Adams (4),

Kerfoot (5), Axelrod (6) and Korn and Galambos (7). Considering both portal and beam structures, he has proposed the minimum Rotation Capacities specified in Table 2 for plastic and compact properties, depending on the nature of the assessment and whether the plastic hinge rotation is measured up to the maximum moment (Θ_{hm} in Figure 2) or the point at which the moment falls below the plastic resistance (Θ_h).

METHOD OF EVALUATION	PLASTIC PROPERTIES		COMPACT PROPERTIES	
	$R_m=\Theta_{hm}/\Theta_p$	$R_u=\Theta_h/\Theta_p$	$R_m=\Theta_{hm}/\Theta_p$	$R_u=\Theta_h/\Theta_p$
Tests on Specimens Where Θ_h Includes All Yielding	3	5	1,0	1,7
Theoretical Assessment With Concentrated Plastic Hinges	2	3	0,7	1,0

TABLE 2 : MINIMUM ROTATION CAPACITIES FOR PLASTIC AND COMPACT PROPERTIES

Assuming the distribution of strain in the plastic region shown in Figure 1b, the Rotation Capacity at maximum moment may be assessed as follows:

$$R_m = \left[(s/f)(2l-0,053)-l+e(l-0,05)^2/0,95(1-l) \right] \quad (1)$$

where $l = L_p/L$ and L_p, the length of the region of significant yielding is assumed to correspond to the portion where the moment exceeds $0,95M_p$. Other terms are defined under Notation.

For known material properties the Rotation Capacity is therefore a function of the proportion of the length in which significant yielding has occurred. Local buckling and the plastic region of interaction between local and lateral buckling is assumed to be confined within this portion of the length. If the yielded lengths can be assessed at which these modes of buckling first develop, the Rotation Capacity can be evaluated, as well as the maximum moment in Figures 1 and 2 :

$$\frac{M_m}{M_p} = 0,95/(1-l) \quad (2)$$

3 EXPERIMENTAL RESULTS

For the purposes of studying the different modes of failure influencing local buckling and attempting to predict the Rotation Capacity, the results of 27 tests are summarised in Table 3. They were conducted by Lukey and Adams (4), Adams, Lay and Galambos (8), Climenhaga and Johnson (9), Axelrod (6) and the author (tests L1-L5). In all cases the moment-rotation characteristics illustrated in Figure 2 were documented for I-sections in which local buckling occurred in regions of moment-gradient. It is interesting to note that 20 of the specimens satisfy the average dimensional requirements for Plastic Properties in modern codes (Table 1) for nominal mild steel, but only 12 meet the requirements for Rotation Capacity defined in Table 2. This is partly due to variations in yield strength identified in column (2) although this could not be allowed for in design. A number of other tests on local buckling (10-12) have not been included due to lack of comparable information.

4 MODES OF FAILURE

Based on these tests and approximate theoretical studies using simple equivalent spring models, it is proposed that failure associated with local buckling in regions of moment gradient may be considered in terms of the following possible modes of failure :

i) Stowell (13), considered the post-elastic local-buckling of flanges in a thorough theoretical study and, including Southward's simplifications for web effects (14), the full wave-length, L_b, may be rearranged as follows :

$$\frac{L_{b1}}{b} = \Pi / \left[1{,}5\varepsilon_c (b/t_f)^2 - 6(G_s/E_s) \right]^{\frac{1}{2}} \qquad (3)$$

The critical compressive strain, ε_c, may be assumed to be the strain midway along the length of the strain-hardened region in Figure 1 :

$$\varepsilon_c = \varepsilon_y \left[s+0{,}5ef(l-0{,}05)/(1-l) \right] \qquad (4)$$

The moduli of elasticity, E_s, and rigidity, G_s, in the strain-hardened region have been considered by Haaijer (15) and Lay (16), but Southward (14) and Lukey (4) have shown that their proposals overestimate the wave-length. Adopting a poissen's ratio for plastic strain of $0{,}5$, Southward proposed $G_s = E_s/3$ which is adopted in eqn.3.

The predictions of maximum-moment ratio obtained for this pure torsional mode of local flange buckling from equations 2,3 and 4 are given in column 7 of Table 3.

ii) Kemp (17) has assessed the elastic-plastic shape in which the compression flange will buckle laterally in members under moment-gradient containing plastic hinges. These simplified spring-energy models allow for the different moduli in the elastic and plastic regions, the moment gradient and the possibility of only half the flange width resisting lateral buckling in the plastic region due to coincident local buckling of the flange on one side of the web. Columns 8 and 9 contain the maximum-moment ratios predicted in this way as follows :

a) Column 8: based on the compression flange being fully effective in the plastic region.

b) Column 9: based on only half the flange width resisting lateral buckling.

The span-breadth ratio in column 5 is clearly of prime importance in these results.

iii) Lay (16), Stowell (13) and Southward (14) derived expressions for the flange wave-length in local buckling which indicate increases in this length as the web slenderness increases. This would imply an increased Rotation Capacity, but this is not substantiated by test results. In fact if web buckling occurs a loss of potential energy will result and the wave length will be reduced. This is particularly the case when there is a coincident axial force supported by the web (tests SB2, SB8, HB41, W1, B2 and L2). The results of tests L1 and

L2 which possess almost identical properties apart from the axial force (as reflected by the additional effective depth of the web in column 4), indicate the reduction in rotation capacity due to the increase of the depth of web in compression. On the other hand the incidence of web buckling alone has not been reported as causing a major reduction in ductility for moment-gradient conditions.

These different modes of failure do not often occur in isolation in I-sections under moment-gradient, partly due to the restraint to local buckling provided by membrane action as a result of strain incompatibilities across the width and along the length of the buckled portion of the flange. The following combined modes of failure and predicted Rotation Capacities are therefore proposed, as illustrated by the test results in Table 3 :

Mode a : If the maximum moment for lateral buckling assuming only half the flange width effective in the plastic region (column 9), is greater than the maximum moment for local buckling (column 7), no lateral buckling will occur. This is likely to represent the most ductile form of local buckling and strain imcompatibilities are satisfied either in an extended plastic region or by lateral deflection. The predicted Rotation Capacity (column 11) is obtained by using the buckled length (eqn.3) as the plastic length in eqn.1 (eg. the first 14 tests in Table 3).

Mode b: If the maximum moment for lateral buckling considering the full flange width (column 8) is less than that for local buckling (column 7), lateral buckling will proceed local buckling and a significant loss of ductility will result. The predicted Rotation Capacity is obtained from eqn.1 using the plastic length derived from the maximum moment ratio in Column 8 (eg. most of the tests with the lowest rotation capacities in Table 3).

Mode c: Lateral buckling conditions intermediate between modes a and b, in which case a combined mode of lateral and local buckling will result. The predicted Rotation Capacity is obtained from eqn. 1 using the average value obtained from the plastic lengths corresponding to the three moment ratios (columns 7 to 9) (eg. tests L1 and L2).

Mode d: Combined web and flange buckling in cases where the effective web slenderness including axial force (column 4) exceeds about 60. This will lead to a loss in predicted ductility and explains the low rotation capacities observed in tests W1, B2, L2 and L3.

The influence of the span in these combined modes of failure can be seen by comparing the Rotation Capacities in the tests L1 and L5, as well as tests HT28, HT43 and HT52. In each case the properties were almost identical apart from the span, but the ductility was increased by reducing the span-breadth ratio (column 5). This is due to the influence of span on both the propensity to lateral buckling (columns 8 and 9) and the Rotation Capacity predicted by eqn.1 for a particular yielded length, L_p, but reduced span, L, (Figure 1).

5 *CONCLUSIONS*

The predicted maximum moments and Rotation Capacities based on the different modes of failure described in this paper, compare favourably with the observations in local-buckling tests. All 15 specimens with a Rotation Capacity less then the limit of *3,0* for Plastic Properties in Table 2 were predicted to fall into this category and 8 of the 12 specimens exceeding *3,0* were correctly categorised. This correspondence is despite the simplified nature of the theoretical models and the uncertainties in defining important parameters such as the stress-strain modulus in the strain-hardening region. Further attention should be given to specimens likely to fail in mode b when lateral buckling just preceeds local buckling of the flange and web with properties similar to test L3.

The flexural ductility of I-sections in regions of moment gradient, as measured by the Rotation Capacity, is inhibited by a combined mode of failure involving both local and lateral buckling of the compression flange. The ratio of the length between the sections of maximum and zero moment to the flange breadth, L/b, therefore becomes an important additional parameter in predicting flexural ductility not only because it measures the propensity for elastic-plastic lateral buckling, but also because the Rotation Capacity is reduced as L/b is increased for a particular yielded length, L_p , within which failure occurs. This effect is not allowed for in existing codes and many tests have incorporated considerably shorter L/b ratios than would be found in actual portal frames.

In plastic design of pitched-roof portals it is a common requirement to provide restraints against lateral buckling at the plastic hinge in the vicinity of the eaves and additionally at a short distance on either side of it. On the basis of the discussion in this paper it would appear that the limitations on flexural ductility due to combined local and lateral buckling could be enhanced if the additional lateral restraints were placed closer to the hinge near the end of the plastic region.

References

1. Kemp, A.R. 'Commentary on Chapter 11 (Plastic Analysis) of the S.A. Code : S.A.B.S. 0162:1983 - Draft Code of Practice for the structural use of steel.' S.A.Inst.of Steel Construction,1983.

2. Driscoll, G.C. 'Rotation capacity requirements for single-span frames.' Fritz.Engr. Lab. Report No. 268.5, Lehigh Univ., Sept.1958.

3. Driscoll, G.C. 'Rotation capacity of a 3-span continuous beam.' Fritz. Engr.Lab. Report No. 268.2, Lehigh Univ., June 1957.

4. Lukey, A.F. and Adams, P.F. ' Rotation capacity of wide-flange beams under moment gradient.' Dept. of Civ.Engr., Univ. of Alberta, Canada, Behaviour of High Strength Steel Members Report No.1 May 1967.

5. Kerfoot, R.P. 'Rotation Capacity of beams.' Fritz. Engr.Lab. Report No. 297.14, Lehigh University, March 1965.

6. Axelrod, M.S. 'Local Web buckling and lateral buckling of plastically-designed steel beams and beam-columns.' Thesis presented in partial fulfilment of MSc degree, Univ. of Witwatersrand, Johannesburg, 1982.

7. Korn, A. and Galambos, T.V. 'Behaviour of elastic-plastic frames.' Proc. Am.Soc. of Civ. Engrs., Journ. of Struc. Div., V94 (ST5), May 1968, 1119-1142.

8. Adams, P.F., Lay, M.G. and Galambos, T.V. 'Experiments on high

strength steel members.' Welding Research Council Bulletin No. 110, November 1965, 1-16.

9. Climenhaga, J.J. and Johnson, R.P. 'Local buckling in continuous composite beams.' The Struct. Engr., V50(9), September 1972, 367-374.

10. Augusti, G. 'Experimental rotation capacity of steel beam-columns.' Proc. Am. Soc. of Civ. Engrs., Journ. of Struct. Div., V90(ST6), December 1964, 171-188.

11. Croce, A.D. 'The strength of continuous welded girders with unstiffened webs.' Thesis for MSc degree, Univ. of Texas, Austin, 1970.

12. Sawko, F. 'The effect of strain-hardening on the elastic-plastic behaviour of beams and grillages.' Proc. Inst. of Civ. Engrs., London, V28, August 1964, 489-504.

13. Stowell, E.Z. 'Compressive strength of flanges.' National Adv. Comm. for Aero., Tech. Note No. 2020, 1950.

14. Southward, R.E. 'Local buckling in Universal Sections.' Internal Rep., University of Cambridge, 1969.

15. Haaijer, G. 'Inelastic buckling in steel.' Trans.Am. Soc. of Civ. Engrs., V125, 1960, 308-338.

16. Lay, M.G. 'Flange local-buckling in wide-flange shapes.' Proc.Am. Soc. of Civ. Engrs., Journ. of Struct.Div., V91(ST6), December 1965, 95-116.

17. Kemp, A.R. 'Simple, spring-energy models in elastic-plastic buckling.' Struct. Ductility Res. Rep. No.3, Dept. of Civ. Engr., University of the Witwatersrand, April 1983.

Acknowledgements

The research described in this paper was supported by the South African Institute of Steel Construction, the South African Iron and Steel Corporation and Speedy Welders, Johannesburg. It was partly undertaken during a period of sabbatical leave at the Fritz Engineering Laboratory, Lehigh University with the assistance of Dr Lynne Beedle and Dr Le Wu Lu, and the additional support of the Council for Scientific and Industrial Research and the University of the Witwatersrand.

Test & Ref Nos.	Flange Yield Stress	Slenderness		Span Ratio L/b	Max. Moment Ratio M_m/M_p				Rotn.Cap. R_m	
		Flange b/t_f	Web h_w/t_w		Test	Mode i)	Mode iia)	Mode iib)	Test	Prediction & Mode
(1)	(2)	(3)	(4)	(5)	(6)	(7)	(8)	(9)	(10)	(11)
A1-4	285	18,8	29,9	8,55	1,38	1,12	2,21	1,82	5,9	2,3-a
A2-4		16,3	29,9	8,37	1,41	1,16	2,21	1,82	6,8	3,1-a
B1-4		19,4	42,7	7,57	1,11	1,07	1,78	1,54	1,45	2,3-a
B2-4		14,0	42,7	7,01	1,15	1,15	1,77	1,54	5,2	4,3-a
B3-4	373	16,3	42,7	7,29	1,13	1,11	1,77	1,54	3,35	3,2-a
B4-4		17,8	42,7	7,43	1,05	1,09	1,77	1,54	1,7	2,7-a
B5-4		18,3	42,7	7,48	1,04	1,08	1,77	1,54	1,6	2,6-a
C1-4		19,4	54,5	6,73	1,11	1,10	1,90	1,63	2,1	2,6-a
C2-4		14,0	54,5	6,52	1,26	1,18	1,79	1,54	6,85	4,5-a
C3-4	352	16,3	54,5	6,80	1,16	1,13	1,79	1,54	4,0	3,3-a
C4-4		17,8	54,5	6,93	1,12	1,11	1,80	1,55	2,1	2,8-a
C5-4		17,1	54,5	6,90	1,14	1,12	1,79	1,54	3,25	3,0-a
HT28-8		15,7	32,5	7,37	1,14	1,12	1,77	1,49	5,2	2,8 a
HT43-8	423	15,7	32,5	4,82	1,13	1,21	2,51	2,27	4,9	4,9 a
HT52-8		15,7	32,5	15,2	1,14	1,04	1,07	0,96	3,4	0,9 c
SB2-9		16,5	47 +	14,7	1,08	1,05	1,05*	0,99*	1,1	1,3*b
SB8-9	325	16,5	47 +	10,1	1,03	1,09	1,58*	1,36*	1,4	2,1*a
HB41-9		15,1	51 +	13,2	1,02	1,08	1,06*	0,99*	0,9	1,4*b
W1-6	343	16,9	98 +	22,4	0,88	1,02	1,01*	0,96*	0,1	0,6*b
B2a-6	355	14,6	61 +	21,8	0,99	1,03	1,01*	0,96*	0,3	0,6*b
B2b-6	317	15,3	65 +	21,0	1,04	1,03	1,01*	0,97*	0,5	0,6*b
B2c-6	351	14,7	61 +	21,4	0,95	1,03	1,01*	0,96*	0,2	0,6*b
L1	340	18,5	35	12,3	1,12	1,05	1,18*	1,02*	1,4	1,7*c
L2	340	16,1	57 +	12,6	1,11	1,07	1,18*	1,02*	0,6	1,8*c,d
L3	375	17,4	64	12,6	1,06	1,06	1,05*	0,99*	0,7	1,3*b,d
L4	285	13,7	34	12,7	1,21	1,12	1,32*	1,43*	3,9	2,6*a
L5	340	17,4	34	6,1	1,16	1,17	2,51*	2,11*	7,6	3,6*a

Notes: All tests, except reference 16, conducted on simply-supported beams with central point-load, but measurements refer to half-span illustrated in Figure 1.

\+ Web depth increased due to coincident axial force to represent twice depth of plastic web in compression.

* Approximate values based on assumed strain-hardening modulus of 4000 MPa.

TABLE 3 : TEST RESULTS AND PREDICTED BEHAVIOUR

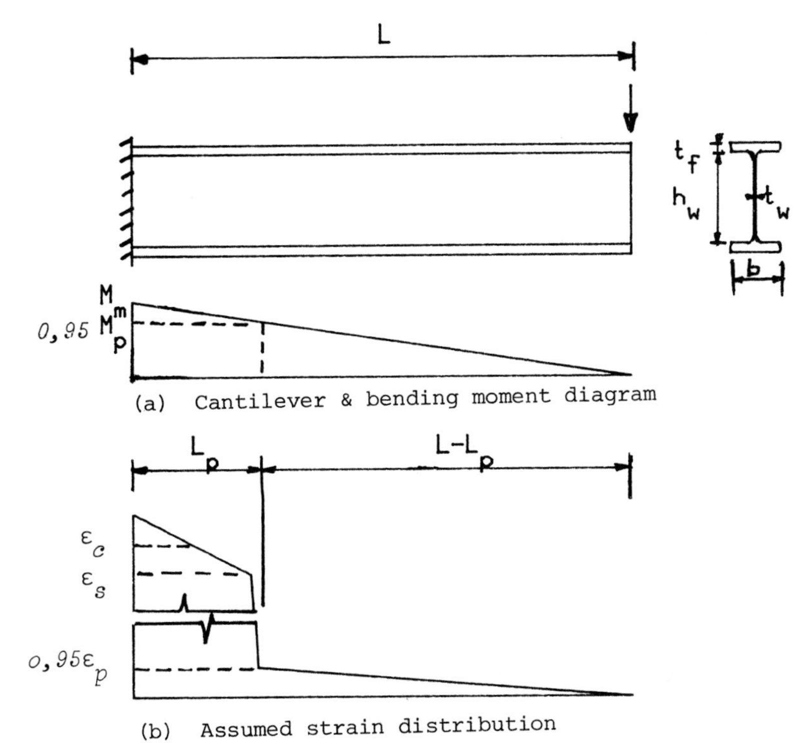

(a) Cantilever & bending moment diagram

(b) Assumed strain distribution

Fig 1 Region of member under moment gradient

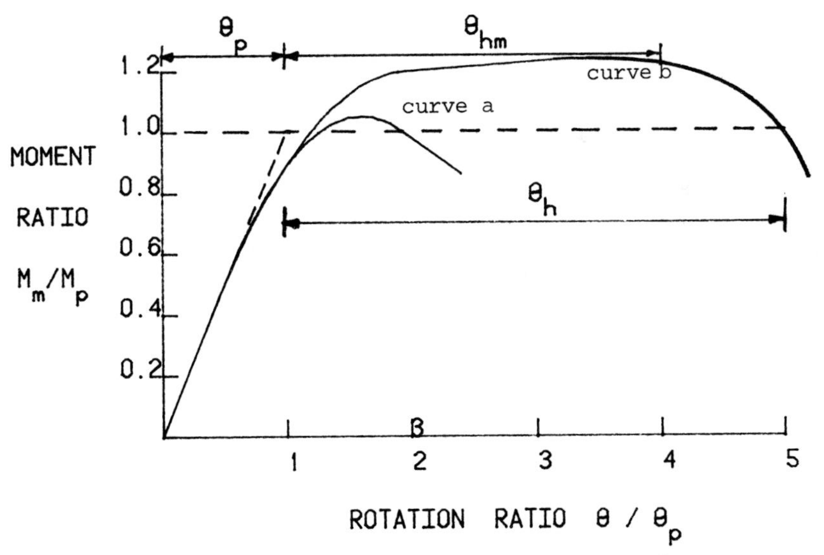

Fig 2 Moment/Rotation curve for member

INFLUENCE OF END-PLATES ON THE ULTIMATE LOAD OF LATERALLY UNSUPPORTED BEAMS

J. Lindner and R. Gietzelt

Fachgebiet Stahlbau, Technische Universität Berlin, W. Germany

Synopsis

End-plate connections effect a stabilizing warping restraint if lateral torsional buckling may occur. A method to calculate this stabilizing effect is proposed for open I-sections. It is based on the determination of the critical load and on the use of the ECCS-lateral torsional buckling curve.

Ultimate load calculations and test results confirm this proposal. Finally an example on how to use the method in practice is given.

Notation

M_E	elastic critical moment capacity including warping restraint
$\overline{M}_E$	elastic critical moment capacity of a simply supported beam with ends free to warp
$M_{pl,y}$	full plastic moment capacity (in-plane bending)
$M_{u,y}$	ultimate moment capacity including lateral-torsional buckling
M_{tr}	ultimate moment-calculation
M_{ex}	ultimate moment-test
E	elastic modulus of material
G	shear modulus
t_p	thickness of the end-plate
t_g	thickness of the flange
h, h_p	depth of section, depth of the end-plate
h'	reduced depth = $h - t_p$
b, b_p	width of flange, width of the end-plate
I_z	moment of inertia (weak axis)
I_T	torsion constant
I_ω	warping constant

β_o	elastic warping restraint factor
ζ	moment pattern factor
ℓ	length of the beam
z_p	level of application of the load
c_ω	elastic warping restraint
χ	torsional factor (to determine ζ) = $EI_\omega / GI_T \ \ell^2$
$\overline{\lambda}_M$	equivalent slenderness for lateral torsional buckling
β_s	yield strength

Introduction

When loaded in its plane of greatest stiffness a laterally unsupported beam will continue to deflect in its plane until the load reaches a certain critical value, whereupon it will buckle sideways in a combined mode involving both lateral deflection and twist. This phenomenon is known as lateral torsional buckling or torsional flexural buckling (1).

Depending on the critical load and on the yield strength of the beam this buckling will be either elastic or inelastic.

The risk of inducing lateral torsional buckling depends on the following parameters (2):

- initial twist and bow
- yield strength
- residual stresses
- type of cross-section
- deformation of the web
- type of loading and level of application of the load
- restraint from adjacent members
- support conditions at the ends of the beam

In calculating the critical load it is usually assumed that the end conditions correspond to simple support in the lateral plane. Thus lateral deflection and twist are prevented but no resistance is provided against either lateral bending or warping. But practical constructional details frequently produce a stabilizing effect on the beam due to the real support conditions. For instance end-plate connections (Fig 1), often used in practice, effect an elastic warping restraint, which depends mainly on the thickness of the welded end-plate.

This paper proposes a method to calculate the lateral torsional buckling load including the stabilizing effects due to welded end-plates.

1. *CRITICAL LOAD*

The elastic critical load is the upper bound for the lateral torsional buckling load. This load is calculated using the energy method. This is done in two different ways (beam-theory and plate-theory to prove the influence of the web deformations and the influence of the deformation of the welded end-plates.

Using the beam theory the end-plates are taken into account as an elastic warping restraint. Using the plate-theory they are considered directly as stiffened rectangular plates (Figs 2,3). A comparison of these results is given in Fig 2. There are only small differences between both methods in the range of $\bar{\lambda}_M \geq 0{,}6$. Web buckling occurs (plate-theory) in the range of $0 \leq \bar{\lambda}_M \leq 0{,}6$ and must therefore be taken into account separately.

Due to these results it is justifiable to calculate the elastic critical lateral torsional buckling load on the base of the beam theory.

For the practical design eqn (1) gives satisfactory results, wherein the parameter β_o affects the elastic warping restraint and the parameter ζ should be chosen corresponding to different moment patterns.

$$M_E = \zeta \cdot \frac{E I_z \pi^2}{\ell^2} \cdot \left(\sqrt{\frac{1}{I_z} \cdot \left(\frac{I_\omega}{\beta_o^2} + \frac{G I_T \cdot \ell^2}{E \pi^2}\right) + \left(\frac{5 z_p}{\pi^2}\right)^2} + \frac{5 z_p}{\pi^2} \right) \qquad (1)$$

The parameter β_o is determined with regard to the properties of the beam and the thickness of the end-plate. Analytical comparisons lead to eqn (2) giving the actual value of β_o, (3).

$$\beta_o = 1 - \frac{0{,}5}{1 + \frac{2 E I_\omega}{c_w \cdot \ell}} \qquad (2)$$

$$c_w = \frac{1}{3} G \, t_p^3 \, h' \, b \qquad (3)$$

From Fig 3 the parameter ζ may be chosen, depending on various moment patterns and depending slightly also on χ.

The ζ-values (Fig 3) are very similar to those given by Roik et al (4) which had been calculated without the influence of an elastic warping restraint.

Furthermore the inelastic material behaviour, initial out-of-straightness and residual stresses should be taken into account. It is checked whether the ECCS lateral torsional buckling curve (Fig 4) may also be used here with regard to the above mentioned warping effect. Therefore ultimate load calculations and a test series have been carried out.

2. *ULTIMATE LOAD CALCULATIONS*

The ultimate loads are calculated with an extended program described in (5). Residual stresses (max. compression stress = $0{,}3\ \beta_S$) as well as an initial out-of-straightness ($v_o = \ell/1000$) and inelastic material behaviour are taken into account. Some results are shown in Fig 5. Here the ultimate moments of an IPE 200 with a uniform moment distribution are seen. Different rates of thickness of the end-plate t_p to thickness of the flange t_g are investigated, wherein $\overline{M}_E$ has been used to calculate $\overline{\lambda}_M$. By this method the increase of the ultimate load can be shown graphically. Using M_E (eqn 1) to calculate $\overline{\lambda}_M$ all curves gather near the ECCS buckling curve, see (3).

Further investigations given by Gietzelt (3) produce the same tendency. The main result of all calculations is a significant increase of the ultimate load depending on the relation t_p/t_g, for instance about 15% when $t_p/t_g \sim 4$.

3. TEST RESULTS

Fig 6 shows a general sketch of the test rig. A detailed description of the test rig, the instrumentation and the test procedure is given in (6), (7). Six test specimens were chosen for the test program, two IPE 200/St 37 and four IPE 160/St 37. The load at midspan is applied at top of the beam ($z_p = - h/2$). The loading system ensures that the applied loads remain vertical during the test procedure.

The boundary conditions are similar to that for a simple support. The cross section may rotate about the major and minor axis but it is fixed against twisting. Warping restraint only should be affected by the welded end-plates.

Stub column test and tension coupon tests were undertaken to determine the material properties. Because the failure in the main tests is similar to the buckling of the compression flange, the results of the stub column test were taken for further evaluations. All relevant test data is given in Table 1.

Table 2 gives all essential data for the evaluation of the tests.

Fig 7 shows the test results, wherein $\overline{M}_E$ is used to calculate $\overline{\lambda}_M$. The test results are significantly above the lateral buckling curve, giving the same tendency as in Fig 5.

A comparison between the ECCS curve and the test results is shown in Fig 8, wherein M_E is calculated as proposed before (eqn 1).

This comparison as well as further investigations and calculations (3) show the ECCS curve may also be used for beams with end-plate connections.

4. EXAMPLE

IPE 200 with end-plate $t_p = 25$ mm

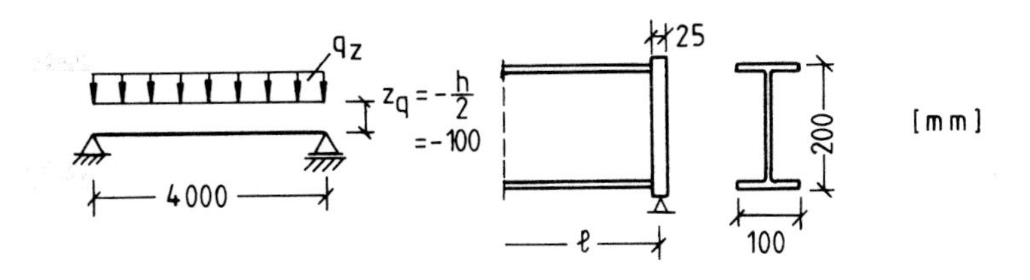

$I_z = 142\ cm^4; \quad I_T = 7,02\ cm^4; \quad I_\omega = 12990\ cm^6$

$M_{pl,y} = 52,6\ kNm$

$$c_w = 8100 \cdot 2,5^3 \cdot 10 \cdot (20,0-0,85)/3 = 8078900\ kNcm^3 = 8,08\ kNm^3 \qquad \text{eqn. (3)}$$

$$\beta_o = 1 - \frac{0,5}{\left(1 + \frac{2 \cdot 2,1 \cdot 1,299}{8,08 \cdot 4,0}\right)} = 0,572 \qquad \text{eqn. (2)}$$

$$\chi = \frac{21000 \cdot 12990}{8100 \cdot 7,02 \cdot 400^2} = 0,030$$

$$\rightarrow \quad \zeta = 1,16$$

$$M_E = 1,16\ \frac{21000 \cdot 142 \cdot \pi^2}{400^2} \left(\sqrt{\frac{1}{142}\left(\frac{12990}{0,572^2} + \frac{8100 \cdot 7,02 \cdot 400^2}{21000}\right) + \left(\frac{5 \cdot 10}{\pi^2}\right)^2} - \frac{5 \cdot 10}{\pi^2} \right) \qquad \text{eqn. (1)}$$

$M_E = 42,1\ kNm$

ECCS lateral torsional buckling curve:

$$\bar{\lambda}_M = \sqrt{52,6/42,1} = 1,12 \quad , \quad \kappa_M = 0,666$$

$$\underline{M_{u,y}} = 52,6 \cdot 0,666 = \underline{\underline{35,0\ kNm}}$$

5. CONCLUSION

It has been shown that welded end-plates effect a significant increase in the ultimate lateral torsional buckling load. A method is proposed to take this effect into account using the ECCS lateral torsional buckling curve.

Additionally tests had been carried out to confirm this method.

Finally an example for a beam used in building structures is given.

References

1. Bleich, F. 'Buckling Strength of Metal Structures.' McGraw Hill Book Co. Inc., New York, 1961.

2. Lindner, J. 'General Report - Lateral Buckling.' Second International Colloquium on Stability of Steel Structures, Liege, 13-15, April 1977, Final Report, pp 103-109.

3. Gietzelt, R. 'Grenztragfähigkeit von querbelasteten Trägern mit I-Querschnitt unter Berücksichtigung wirklichkeitsnaher Lagerungsbedingungen an den Trägerenden.' Dissertation, TU Berlin 1983.

4. Roik, K., Carl, J., Lindner, J. 'Biegetorsionsprobleme gerader dünnwandiger Stäbe.' Verlag Wilhelm Ernst & Sohn, Berlin-München-Düsseldorf, 1972.

5. Lindner, J. 'Einfluß von Eigenspannungen auf die Traglast von I-Trägern.' Der Stahlbau 43(1974), pp 39-45, 86-91.

6. Lindner, J. 'Developments on lateral torsional buckling.' 2nd International Colloquium on Stability, Washington D.C., May 1977, pp 532-539.

7. Lindner, J. and Geitzelt, R. 'Biegedrillknicklasten von Walzprofilen IPE 200 and IPE 160 mit angeschweißten Kopfplatten und baupraktischen Ausklinkungen an den Trägerenden.' Versuchsbericht VR 2042, Institut für Baukonstruktionen und Festigkeit, TU Berlin, 1982.

Table 1 Measured test data

Test No.	cross-section	system, moment	ℓ	Dimension of the end-plate t_p	h_p	b_p	β_s	M_{ex}	M_{pl}	$\kappa_M = M_{ex}/M_{pl}$
·/·	·/·	·/·	m	mm	mm	mm	kN/cm²	kNm	kNm	·/·
W1	IPE 200	P_z, $z_p = -h/2$, ℓ, M_{ex}	1,80	25	220	120	25,7	52,0	54,9	0,947
W2	IPE 200		1,80	25	220	120	25,7	51,2	54,9	0,933
W3	IPE 160		2,20	25	180	120	26,6	30,0	33,0	0,909
W4	IPE 160		2,20	25	180	120	26,6	31,6	33,0	0,958
W5	IPE 160		3,80	25	180	120	26,6	23,7	33,0	0,718
W6	IPE 160		3,80	25	180	120	26,6	24,6	33,0	0,745

Table 2 Evaluation of the tests

Test No.	I_z	I_T	I_ω	Eval. $\bar{M}_E$ M_E	$\bar{\lambda}_M$	Evaluation M_E c_w	β_0	$\sqrt{X}$	$\zeta_{(z_p=-h/2)}$	M_E	$\bar{\lambda}_M$
·/·	cm⁴	cm⁴	cm⁶	kNm	·/·	kNm³	·/·	·/·	·/·	kNm	·/·
W1	132	6,62	12200	92,9	0,769	8,08	0,630	0,384	1,24	135	0,638
W2	132	6,62	12200	92,9	0,769	8,08	0,630	0,384	1,24	135	0,638
W3	68,7	3,67	4018	35,5	0,964	5,28	0,563	0,242	1,27	49,8	0,814
W4	68,7	3,67	4018	35,5	0,964	5,28	0,563	0,242	1,27	49,8	0,814
W5	68,7	3,67	4018	20,0	1,286	5,28	0,539	0,140	1,31	24,2	1,17
W6	68,7	3,67	4018	20,0	1,286	5,28	0,539	0,140	1,31	24,2	1,17

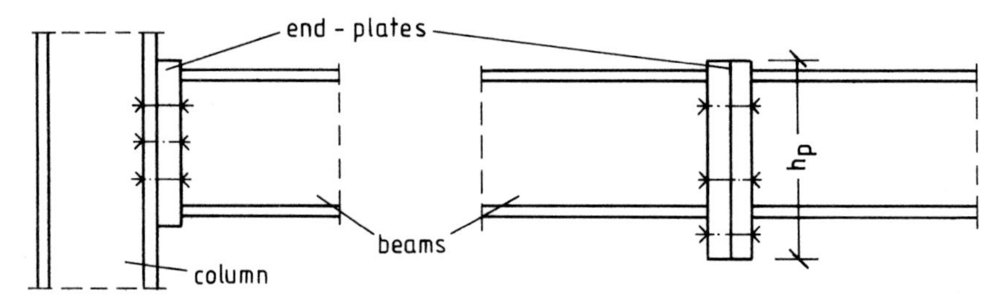

Fig. 1: End-plate connections beams and column

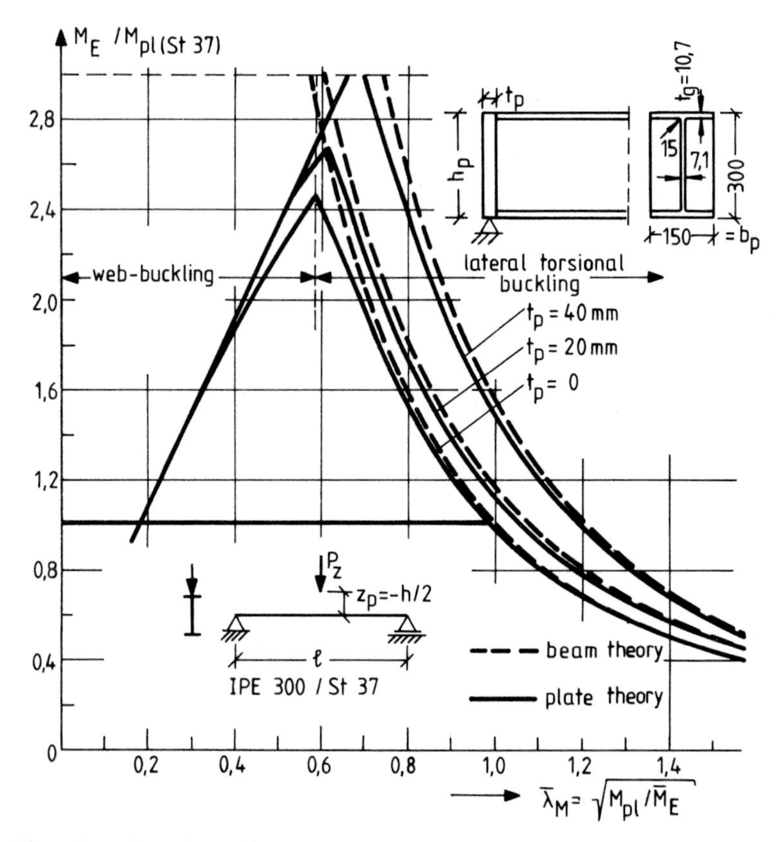

Fig. 2: Elastic critical moment depending on the thickness of the end-plate

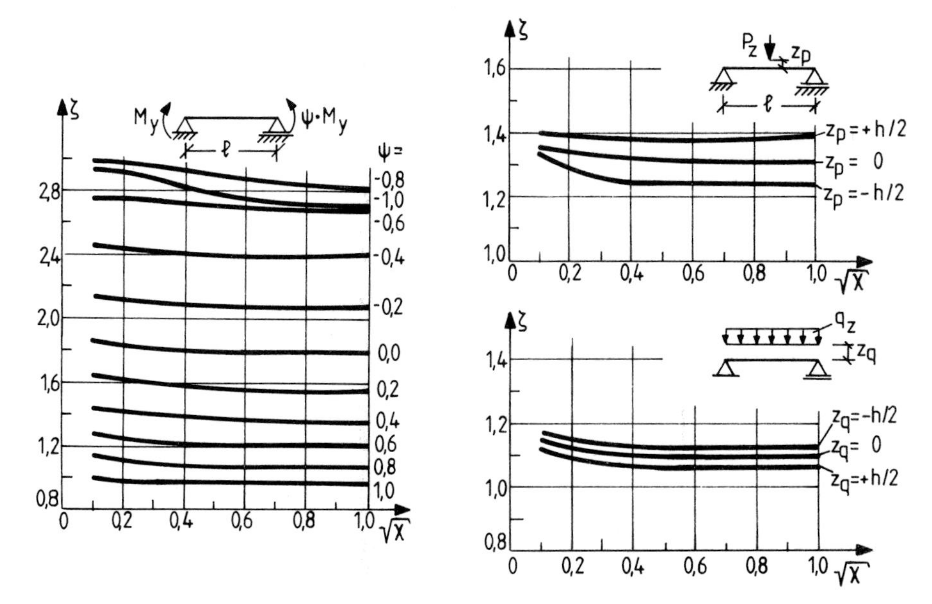

Fig. 3: ζ-values depending on moment pattern

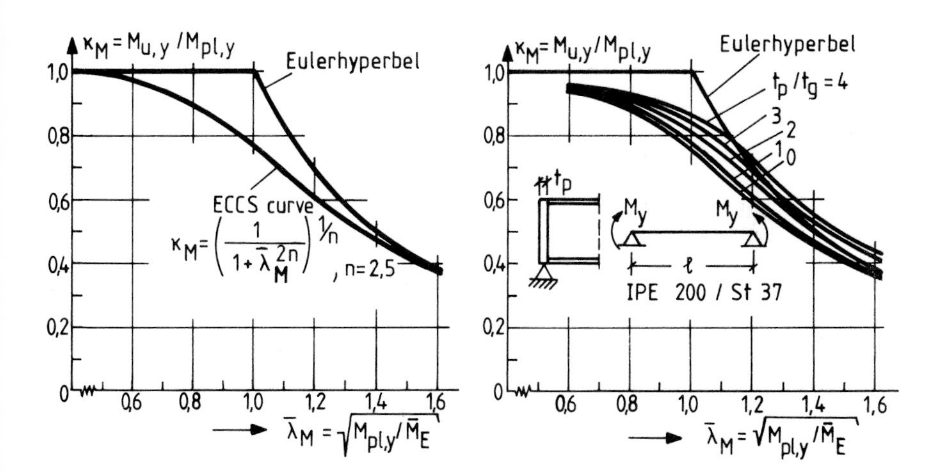

Fig. 4: ECCS buckling curve

Fig. 5: Ultimate loads calculated as (5)

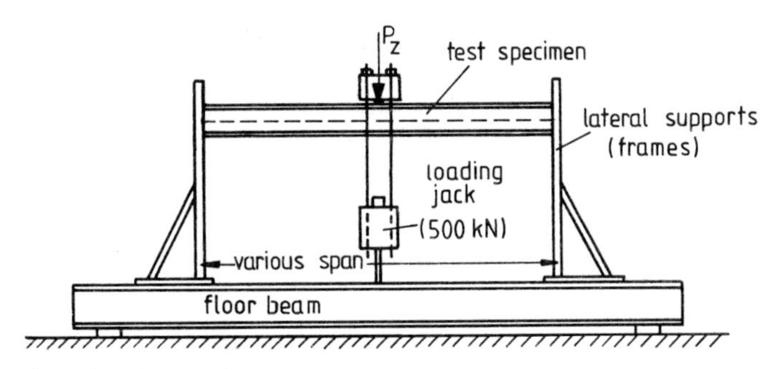

Fig. 6: Test rig

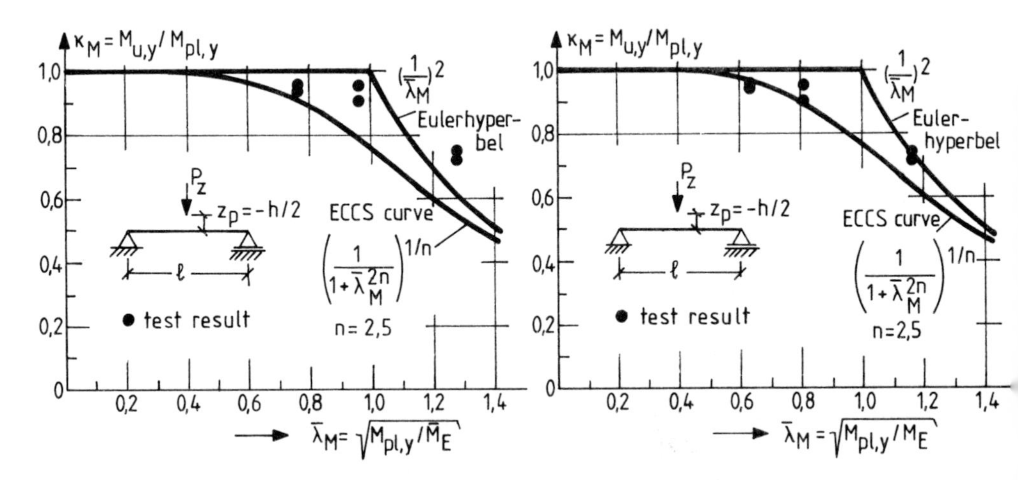

Fig. 7: Test results using $\bar{M}_E$

Fig. 8: Test results using M_E

EXPERIMENTAL BEHAVIOUR OF HAUNCHED MEMBERS

L.J. Morris

Simon Engineering Laboratories, University of Manchester

K. Nakane

University of Canterbury, Christchurch, New Zealand

Synopsis

A series of fifteen tests on haunched members with varying degrees of restraint to both the tension and compression flanges are described. The haunched members are similar to those found in portal frame construction in the United Kingdom. The experimental behaviour of the haunched members is discussed with respect to lateral torsional instability and the ability (or otherwise) of such members to develop a plastic hinge at the haunch/rafter intersection.

Introduction

Previous experimental evidence (1) had indicated that lateral instability was only one of several design criteria which influence the behaviour of a plastically-designed steel portal frame in the region of the eaves. Therefore, it was decided to study the stability of haunched rafters in isolation from the other effects, such as excessive deformations within the connection and the column web panel. Figure 1 illustrates a typical haunched rafter for portal frames constructed in the United Kingdom, with the depth of the haunch being approximately twice that of the basic rafter section. This rule allows a cutting of the same basic section or even a heavier section to be used for the haunch. This structural detail automatically introduces a third or 'middle' flange within the haunched zone. The principal aim of the investigation has been to assess the benefits of this middle flange on the overall ability of the haunched member to remain stable when subjected to large bending moments. An possible economic benefit would be the relaxation of the present spacing requirements of lateral restraints.

Horne et al (2) have investigated the lateral stability of two-flanged, tapered and haunched members, which resulted in design formulae which can only be used to estimate the limiting lengths of such members. Further research is necessary to extend this work to three-flanged sections. In previous tests undertaken at Manchester by Morris and Packer (1) on the type of haunched detail being considered, the design of the specimens was such as to force a plastic hinge to form in the column member immediately below the haunch while the stresses at the haunch/rafter intersection remained elastic. The present series of tests were designed to cause the reverse to happen, i.e. a plastic hinge at the haunch/rafter intersection with the column member remaining elastic. Therefore, the column member was deliberately overdesigned so that it would remain well within the elastic range during the testing to failure of the haunched members. By making the column stiffer than would be found in practice the possiblity of excessive deformation in the column web panel was eliminated. Moreover, this meant that the same column could be used for all tests in the series.

1. *DETAILS OF THE TESTS*

Each haunched member was tested in the arrangement shown in Fig 1. Generally, the end-plates were made thicker than normal practice (30mm) in order to minimise the effect of any end-plate deformation. The exceptions were the 20mm plates used for specimens RS6S and RS6W. Each specimen was attached by preloaded bolts to the stiff column member which in turn was fastened to the laboratory strong floor. When testing the larger haunched members (356x127UB33) the major axis bending of the column was restricted by means of two inclined rods taken back to the strong floor, see Fig 1. The free end of the cantilevered specimens was caused to move vertically downwards between two guides by means of a hydraulic jacking system. The load was applied to the bottom flange of the specimen and its magnitude was controlled by means of a calibrated load cell. Such an arrangement ensured, for all practical purposes, that a specimen was restrained torsionally at both ends of the specimen while allowing the free end to warp freely and rotate about the major and minor axes.

There were basically three types of test. The first group of specimens were tested with no physical restraint between the end restraints to either flange of the haunched member. The second group (denoted by the letter T or W) were restrained by small horizontal members, which were bolted at one end to a plate welded to the tension flange immediately above the haunch/rafter intersection. This restricted any lateral movement, but allowed rotation of the tension flange about the point of attachment. The other end of these restraining members were adjusted so that they remained in a horizontal position throughout the test. The third group of tests (denoted by the letter S) were similar to the second group, but had diagonal braces connecting the compression flange of the haunched member to the horizontal restraints, see Fig 1.

The specimens were fabricated from Grade 43 steel to BS 4360 and tensile coupons were obtained from material adjacent to that used for both the basic rafter member and haunch cuttings. The average yield stresses, based on the measured flange and web yield stresses, are noted in Table 1. It was observed and recorded that there were significant differences, resulting from the rolling process in the flange thicknesses. This is similar to that noted by Lindner (3), except in some instances the web was also offset from the nominal centreline of the profile, see Fig 2. The values shown in Fig 2 are in fact for the worse case measured; most values were of the order of 7%. Apart from RS5P, RS6S and RS6W the specimens had half-depth web stiffeners welded to the member at the haunch/rafter intersection.

Specimen RS5P had no such stiffeners; RS6S had full-depth stiffeners, while RS6W had plates welded to the tips of the flanges at the intersection, i.e. these 'doubler' plates were parallel to the web.

Dial gauges were positioned horizontally against the column flanges, in line with the flanges of the haunch, in order to measure both the effect on the column member of the in-plane moment being applied to the specimen, and the relationship between the torsional rotation of the column and the development of instability in the haunched member. The in-plane end deflection and the lateral (horizontal) movement of the compression flange at the haunch/rafter intersection were measured. Appropriate areas of the specimens were cleaned from loose mill-scale and coated with a thin layer of brittle lacquer (colophany). In the presence of large strains this lacquer cracked and the resulting crack patterns indicated clearly the onset of yielding and the subsequent spread of plasticity.

In the some tests a rocker plate was inserted between the end-plate and the column flange as it was felt that in practice there would be lack of fit present, due to plate distortion resulting from welding. Nevertheless, a certain amount of lateral restraint at the bolted end of the specimen was provided by preloaded bolts. The rocker plate was only used for the smaller specimens (254x102UB22).

The specimens were loaded in 5kN increments until such time (as indicated by a continuous plot of the applied load v. end deflection) when it became prudent to reduce the magnitude of the load increment in order to effect a controlled failure of the specimen. In the later stages of the tests the control moved from one of load to that of deflection. Initial lateral movement was detected easily when the specimen made contact with one of several metal vertical wire grids. These grids were placed adjacent to both sides of the specimen before the test, some 2mm away from the compression flange.

2. *DISCUSSION OF TEST RESULTS*

2.1 *General*

The experimental evidence is discussed with respect to the present investigation. It should be noted that the results from specimen RS5S were suspect due to an inadequate stabilising system. The load-deflection characteristics are examined, together with other observations to see whether or not any particular characteristic was acceptable. That is, was there adequate rotation capacity while sustaining a constant moment, e.g. M_p, so that moment redistribution can take place, being a primary requirement of plastic analysis. However, it can be argued that a slight reduction in the moment capacity of an inelastic member is not as important for the purposes of plastic design as its ability to provide sufficient rotation of the plastic hinge without a *rapid* fall-off in resisting moment. The variation of applied moment v. bending strains in both the tension and compression flanges are discussed, particularly with respect to the onset of assymetrical strains.

In most of the tests the unloading phase was precipitated by local buckling of the compression flange in the basic rafter, immediately adjacent to the haunch/rafter intersection. The initiation of local buckling can be attributed to the accumulation of the compression stresses resulting from,

(a) major axis bending due to the applied load,
(b) minor axis bending due to lateral-torsional instability,
(c) the resultant upward force due to the change in direction of the compression flange at the intersection and
(d) residual stress pattern resulting from the manufacturing and welding processes.

The exceptions to this phenomenon were the specimens RS5P (web buckling at intersection) and RS6S (web buckling near bolted connection).

2.2 *In-plane deflections*

The applied moments (relative to the haunch/rafter intersection) have been plotted in terms of M/M_p, where M_p is the reduced plastic moment due to the effect of the vertical shear, see Figs 3(a)(b) and (c). Most of the moment-deflection curves obtained from the tests displayed a relatively poor characteristic, i.e. there was a significant fall-off in moment capacity as rotation increased. The specimens which exhibited a reasonable response were those which were diagonal braced at the haunch/rafter intersection, particularly RS6S. It is interesting to note that all characteristics became non-linear at about $0.35M_p$. This initial non-linearity reflected the premature yielding caused by the presence of residual stresses. The non-linearity continued to increase until the maximum load was reached. The tests were continued well beyond this point until a definite unloading stage had been attained when the test was terminated.

In order to discuss the various characteristics the moment-deflection curves have been grouped. Figure 3(a) shows the moment-deflection curves for specimens RS1A, RS1T and RS1S, each having the same basic rafter size (254x102UB22). The curves for RS1 and RS2 were almost identical to that of RS1A and have been omitted for clarity. The specimens RS1, RS1A and RS2 attained a maximum value for M/M_p of about 0.76, followed by a significant fall-off in moment capacity as the rotation increased. The prime cause of this reduction was local buckling of the basic section at the haunch/rafter intersection, initiated by the twisting out-of-plane of the specimen. There was no significant effect on the load-deflection behaviour resulting from the presence of the rocker plate (RS1) or by having all the 'tension' bolts within the depth of the haunch (RS2). A comparison of the responses of the specimens RS1A, RS1T and RS1S indicates clearly the influence of increasing restraint on the performance of the haunched member, resulting in higher moment capacity. A similar effect was noted for the specimens which had been fabricated with heavier haunches, i.e. those specimens prefixed RS3 and RS4. Comparisons between the curves in Figs 3(a) and 3(b) illustrate the benefit of using a thicker compression flange in the haunch zone, i.e. the haunched length becomes less susceptible to instability. Horne (5) made this point in a recent discussion on portal frames (4). Nevertheless, though the curves in Fig 3(b) show higher values of M/M_p than those given in Fig 3(a) there was no significant improvement in the moment-rotation performance

The remaining group of tests (prefixed RS5 and RS6) were fabricated from a larger basic rafter size (354x127UB33) and the moment-deflection curves are given in Fig 3(c). The poor characteristic of specimen RS5P illustrates the necessity for a stiffener to be positioned at the haunch/rafter intersection in order to resist the inward thrust from the inclined haunch flange. The better initial stiffness of RS5P can be attributed to negligible inital out-of-plane straightness.

Specimen RS6W showed a marked improvement over RS5T, see Fig 4(b), both being laterally restrained only on the tension flange. This improvement was probably due to the stiffer horizontal restraint system used and the use of doubler plates. In the case of the 'fully' restrained specimen RS6S a good moment-rotation curve was produced but unfortunately the maximum value of M/M_p was only 0.82. This particular specimen became unstable between the bolted connection and the braced position and eventually failed by buckling of the haunch web near the end-plate. Note that the end-plate in this test was more flexible than previous tests, i.e. 20 mm. This flexibility would have caused an increase in the compression strain, resulting from a reduction in the effective moment arm. This test emphasised the point that it is essential that the staying system must be stiff enough to force failure to occur between restraints rather than allowing overall instability of the member.

The failure loads of the last group (RS5P-RS6W) tended to be relatively lower than those of the smaller specimens. This was probably due to a longer rafter length which caused a slightly more severe condition at the haunch/rafter intersection and also the 356x127UB33 section is torsionally weaker than the 254x102UB22.

2.3 Lateral deflections

The load at which the initial lateral movement was detected always corresponded (with hindsight) to the slight deviation from the initial linear elastic response between the applied load and end deflection. Lateral deflection at the haunch/rafter intersection was measured only for the larger specimens (RS5P-RS6W) and Figs 4(a)(b) and (c) depict the relationship of the lateral movement with the in-plane deflections for the tests RS5P, RS5T and RS6S. There was a noticeable change in all lateral deflection curves at approximately 0.8 of the failure load when the twisting of the specimens began to have an effect. Apart from RS5P and RS6S, significant lateral movement had taken place as the failure load was reached. The lateral deflection continued to increase as the specimens began to unload, Fig 4(b) illustrates this point.
In the case of specimen RS5P (Fig 4(a)) the sudden onset of web buckling as indicated by the in-plane deflection is reflected by a similar behaviour in the lateral movement. The lateral deflection of specimen RS6S (Fig 4(c)) proved to be interesting. At about 0.8 of the failure load there was a reversal in direction of the lateral movement as the load approached failure, when a second reversal occurred. The resulting lateral deflection at the haunch/rafter intersection was relatively small, indicating the effectiveness of the restraint provided by the diagonal braces.

Due to the flexibility of the horizontal restraining members the bracing system proved to be ineffective and specimen RS5S failed at a lower load than predicted. This illustrates the point that the flexural out-of-plane rigidity of the roofing system (cold-formed purlins and sheeting) is an important factor in providing adequate restraint via the diagonal braces to the compression flange. A moment-resistant purlin connection is essential in going some way towards supplying the necessary restraint. The minimum amount of in-plane rigidity to be supplied by the roofing system needs to be quantified, especially when the roof is to be supported by very large span portal frames.

2.4 Twisting of the column member

The torsional rotation (of the column) v. applied moment curves fall into two distinct groups. In the first group of specimens (RS1-RS4) the column member was not restrained by the inclined bars, while in the second group (RS5P-RS6W) the column was restrained, see Fig 1. This had the effect of producing different rotational behaviour.

In the first group there was a gradual increase in the torsional rotation during the early stages of loading (< 0.8 of the failure load). This early rotation was always in one direction, irrespective of the direction of the later stages of rotation and suggested a slight out-of-straightness in the column member. However, as soon as lateral instability began to affect the haunched member there was a similar response in the column member, i.e. the column rotations increased significantly, reflecting the direction of the lateral movement in the haunched member. Generally, rotation in the later stage of the tests in this group were less significant when the rafter member was 'partially' or 'fully' restrained or even when the haunch cutting was stronger than the basic section.

In the second group the effect of the stayed column resulted in virtually no rotation in the early stages of loading (< 0.85 of failure load). Nevertheless, as the loading approached the failure load significant rotation occurred, again responding in a like manner to the movement caused by instability in the rafter.

As the haunched member started to go unstable significant torsional moment was induced in the column member, indicating the need for restraint at the inside corner of the column/haunch intersection. It should be remembered that the test column was stiffer than would have been the case in practice, i.e. one would expect the appropriate column size, commensurate with the rafter size, to twist more. In the case of an adequately braced haunched member in which overall lateral-torsional instability is prevented then the such bracing may not be necessary.

2.5 Bending strains

The strains due to bending were measured at the extreme fibres of both the tension and compression flanges - at three sections along the haunched member, i.e. near the end-plate, mid-haunch and in the basic section adjacent to the haunch/rafter intersection. The strains were recorded from pairs of ERS gauges, placed symmetrically at the quarter points across the flange.

The strains in the tension flange near the connection remained linear almost up to the failure load. There was virtually no difference between the tension strains as measured by any one pair of gauges. On the other hand, the strains associated with the compression flange near the end-pate showed that in the 'partial' or 'fully' restrained cases premature yielding occurred at about $0.4M_p$, probably due to residual stresses resulting from welding and the response of the specimen in resisting the initial lateral movement between supports. In these cases, each pair of compression strains continued to increase with load, whereas in the cases of the unrestrained specimens a decrease in strain was noted as the failure load was approached.

At mid-haunch, each pair of tension strains remained linear and were virtually identical until about 0.95 of the failure load. The magnitude of these strains at failure for the 'fully' restrained specimens were approximately 1.5 to 2 times larger than the strains recorded for the specimens with less restraint. That is, there was a tendency to fail between restraints, though only specimen RS6S actually failed in this mode. The corresponding compression strains were linear until instability started, after which each pair of strains tended to diverge as the test continued. Yielding did occur in the 'partial' and 'fully' restrained specimens.

The strains in the tension flange measured within 30mm of the haunch/rafter intersection remained linear to within 0.9 of the failure load, thereafter indicated a reasonable amount of yielding (5000 $\mu\varepsilon$) which is commensurate with a plastic hinge being formed. The associated compression strains indicated gross yielding had taken place following a reasonable initial linear behaviour. In the case of the smaller specimens (RS1-RS4) there was a significant increase (10%) in moment capacity during the yielding stage, while the larger specimens (RS5P-RS6W) exhibited a lower increase ($< 5\%$). It was interesting to note that the specimen with no stiffeners at the haunch/rafter intersection (RS5P) showed very little sign of yielding before failing 'instantly'.

2.6 Load in diagonal braces

One of the grey areas of design has been defining the magnitude of the load induced in the diagonal braces when restraining the compression flange against instability. The typical restraining force in the braces in this series of tests was found to be about 1.5% of the *squash load* of the compression flange being braced. However, when the lateral deflection began to increase rapidly, loads of about 3.5% were recorded, albeit tension, compared with about 2% compression, see Fig 5. The values shown are for RS6S, for which the mode of failure was instability within the haunch length caused by an adequate bracing system. It would appear from this evidence that the present design rule of 2.5% is acceptable. Generally the changes in the restraining load reflected similar changes in the lateral movement of the haunched member.

The design slenderness ratio of the test diagonal braces used during the tests varied between 90-100. These braces showed no signs of distress. However, practical braces, particularly those fabricated from narrow plates, would be more prone to imperfections and therefore would have a lower effective slenderness ratio than the test braces would indicate.

3. *CONCLUSIONS*

This series of tests have been extremely useful in providing experimental evidence with regard to the behaviour of haunched members with differing degrees of restraint at the haunch/rafter intersection, particularly with respect to those parameters that influence the failure of such members. Due to the complex nature of the stresses in the vicinity of the intersection, difficulty was experienced in forming a plastic hinge at the intersection before instability occurred. The results provided a means by which different methods, advocated for checking the adequacy of haunched members against instability, could be assessed.

Acknowledgments

The authors are grateful to Constrado for their financial support of this experimental project and in particular to their project director A.D. Weller.

References

1. Morris, L.J. and Packer, J.A., 'The behaviour and design of steel portal frame knees'. (to be published)

2. Horne, M.R., Shakir-Khalil, H. and Akhtar, S., 'The stability of tapered and haunched beams'. Proc.I.C.E., Vol 67, Sept 1979, 677-694

3. Lindner, J., 'Developments on lateral torsional buckling' 2nd Int Colloquium on Stability, Washington, 1977, 532-539

4. Morris, L.J., 'A commentary on portal frame design' The Structural Engr., Vol 59A, Dec 1981, 394-404

5. Horne, M.R. - a contribution to the discussion on ref 4 The Structural Engr., Vol 61A, June 1983, 181-189

Table 1 Details of specimens

	Basic section	Haunch section	Rocker plate	Bolts within depth of section	Stiffener (half depth)	Restraints	Haunch length (m)	Total length (m)	Failure load (m)	Average yield stress (N/mm^2)
RS1	254x102UB22	254x102UB22	x	x	✓	x	1.50	2.43	64	319
RS2	"	"	✓	x	✓	x	"	"	64	322
RS1A	"	"	✓	✓	✓	x	"	"	73	331
RS1T	"	"	✓	✓	✓	LL	"	"	68	275
RS1S	"	"	✓	✓	✓	LL + B	"	"	78	273
RS3	"	254x102UB25	✓	✓	✓	x	"	"	63	290
RS3T	"	"	✓	✓	✓	LL	"	"	71	278
RS3S	"	"	✓	✓	✓	LL + B	"	"	74	275
RS4	"	254x102UB28	✓	✓	✓	x	"	"	77	293
RS5P	356x127UB33	356x127UB33	x	✓	None	x	1.80	3.20	98	325
RS5	"	"	x	✓	✓	x	"	"	91	322
RS5T	"	"	x	✓	✓	LL	"	"	89	325
RS5S	"	"	x	✓	✓	LL + B	"	"	95	324
RS6S	"	"	x	✓	Full depth	ML + B	"	"	107	318
RS6W	"	"	x	✓	Side stiffener	ML	"	"	96	316

LL light lateral restraint   ML medium lateral restraint   B diagonal braces

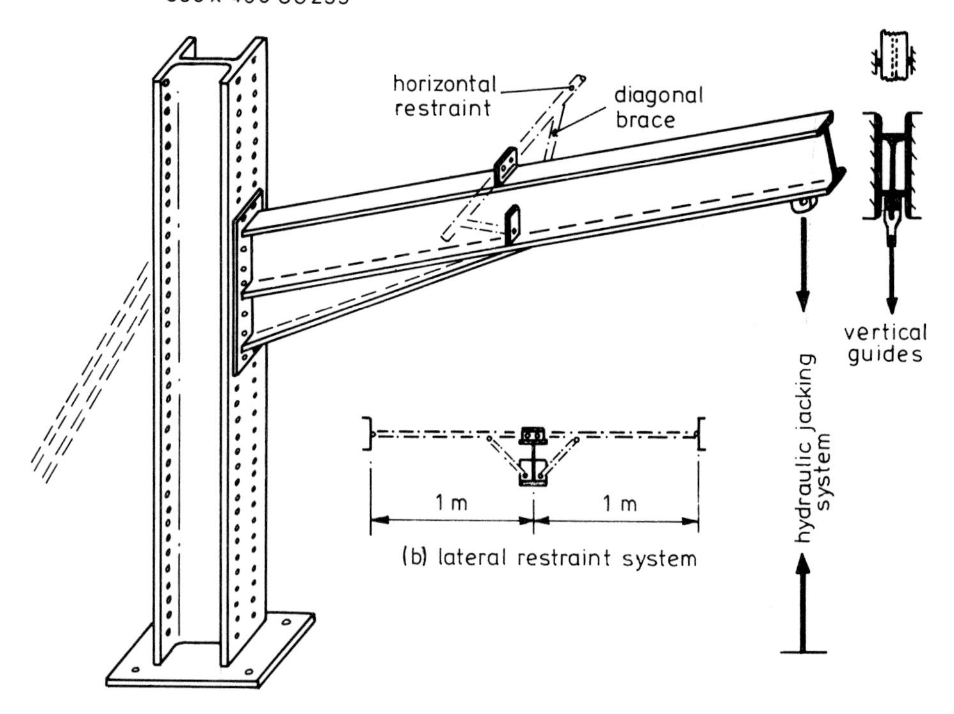

Fig 1 Test arrangement

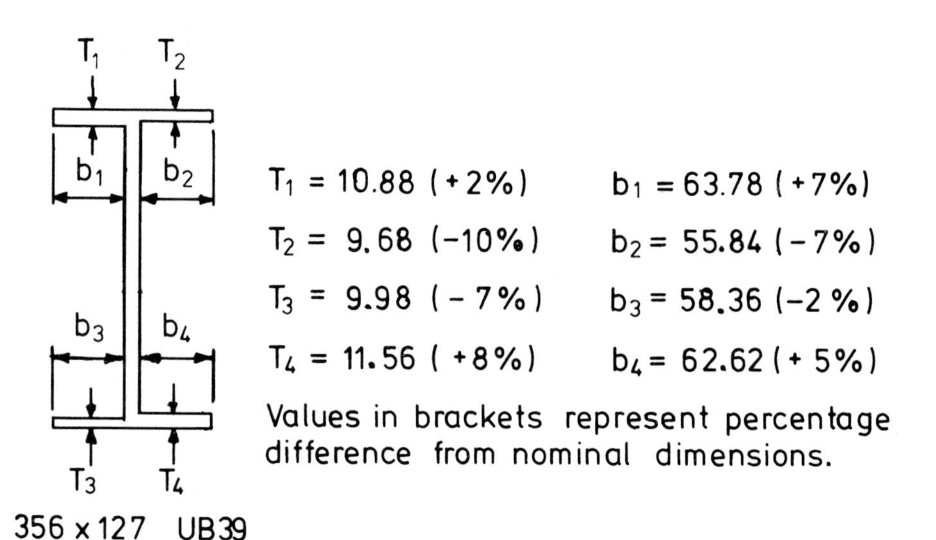

Fig 2 Dimensions of an actual profile

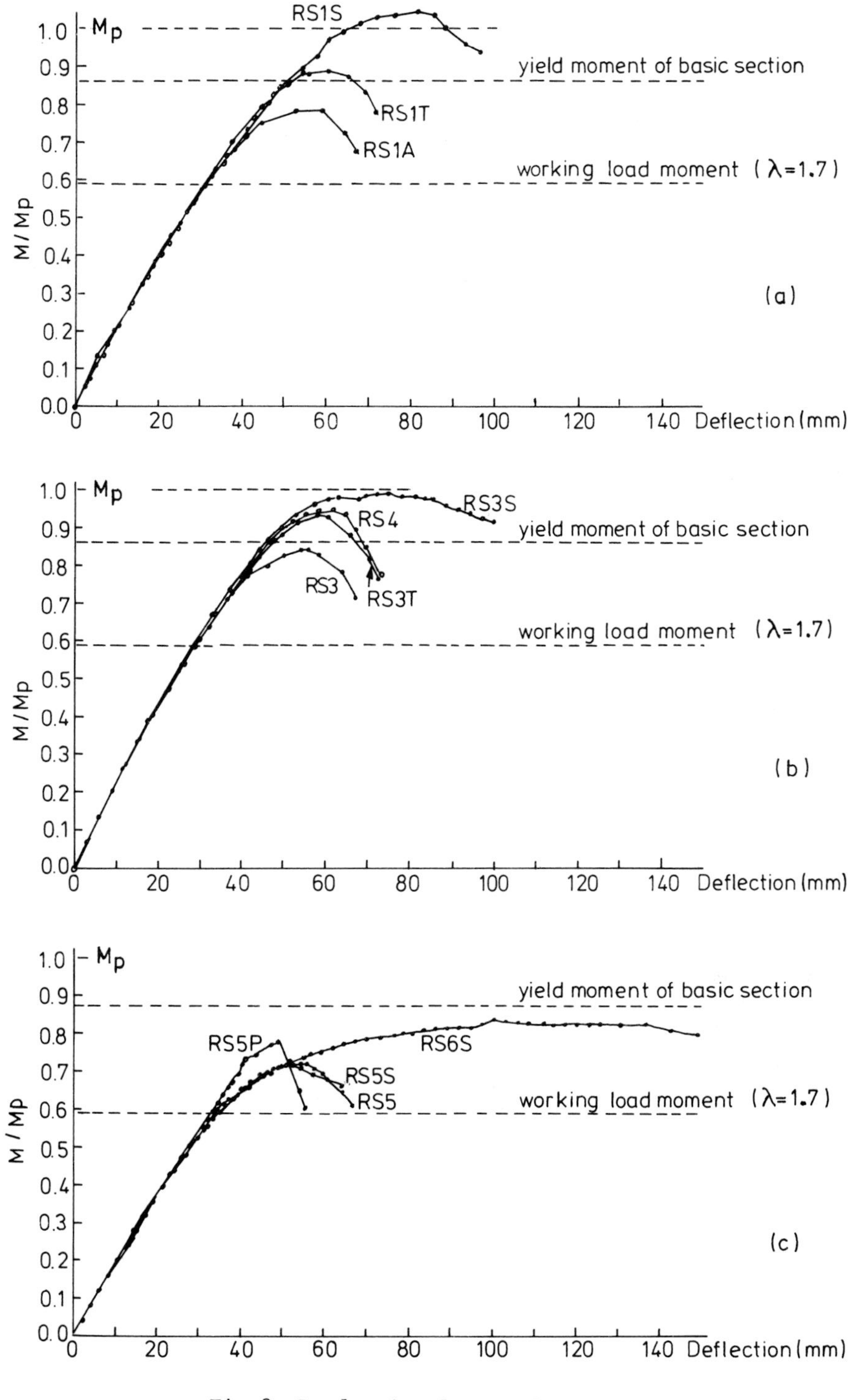

Fig 3 Load v in-plane end-deflection

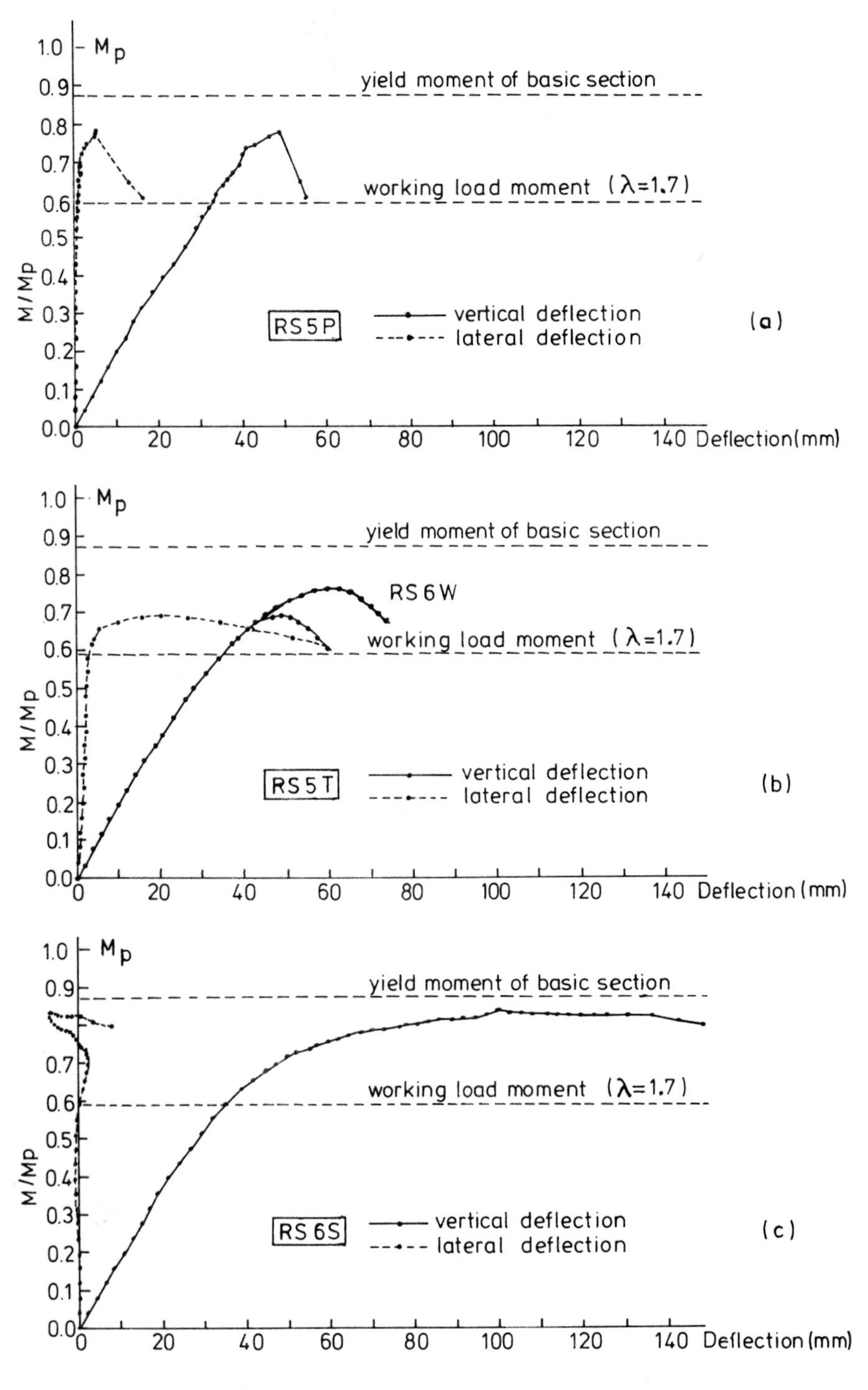

Fig 4 Lateral and in-plane deflections

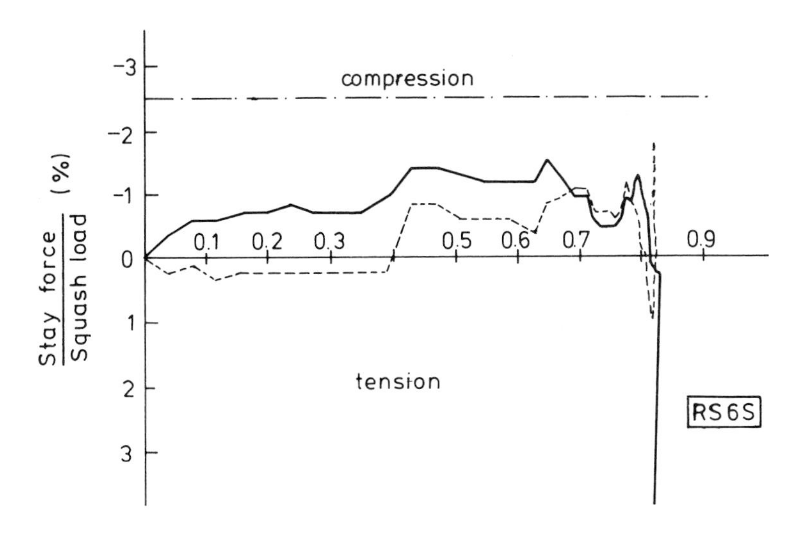

Fig 5 Axial load in diagonal braces

THE DESIGN OF TENSION FLANGE BRACED BEAMS

H R Milner and S N Rao

Department of Civil Engineering, Chisholm Institute of Technology, Australia

Synopsis

It is common practice to brace beams against lateral buckling by attaching braces to the compression flange. However in some circumstances it may not be architecturally desirable to do so and, in such cases, designers may choose to attach braces to the tension flange. This paper describes the theory of tension flange bracing and the design procedures.

Notation

a	distance from beam centroid to brace centroid
B	flange width
d	distance centre to centre of flanges of beam
D	overall depth of beam $\doteqdot d$
E	Young's modulus
F	stress in beam flange
G	shear modulus
I_B	second moment of area of brace
I_y	minor axis second moment of area of beam
J	St. Venant's torsion constant of beam
K_E	effective brace stiffness with respect to twist
K_B	$6(EI/L)_B$ or $3(EI/L)_B$ where all terms refer to the braces
K_J	stiffness of the purlin-beam connection
K_W	local web stiffness
l	overall length of the beam
L	length of brace
m	l/s = number of subspans
M_{cr}	critical moment of beam
M_J	moment developed in purlin-beam connection

n number of half-waves of twist

P axial load on beam

P_E $\pi^2 EI_y/s^2$

P_f force in braced flange

q_z wind pressure

r_y minor axis radius of gyration of beam

s spacing of braces or design effective length

t web thickness

T flange thickness

Introduction

A not uncommon problem facing steel frame designers is that of bracing beams and frames against lateral-torsional buckling by braces attached to only one flange. In a simple supported beam, if that flange is the compression (critical) flange, a series of equally spaced braces, sufficiently stiff and strong to fix the flange in position, enables a designer to proportion the braced member on the basis of its effective length being equal to the brace spacing; see Clause 5.9.5.1 of AS1250 (1). However, if the braces are attached to the tension flange, AS1250, Clause 3.3.4.7, implies the bracing is completely ineffective even though some, although reduced, benefits do accrue. The purpose of this paper is to quantify the effectiveness of tension flange bracing and provide the necessary technical data for routine design in such situations.

Fig 1 shows a tension flange brace acting on a beam, under test, in which it can be seen that twist has been eliminated at the brace attachment point. When the bracing system is designed to eliminate twist at a tension flange brace it acts in an identical manner to a fly brace system implying the effective length of the braced beam may be taken as equal to the brace spacing and designed for relatively high allowable stress. On the other hand, if a brace is attached to the tension flange which fixes the flange in position only (twisting not prevented), the benefits which accrue are very much reduced, although the brace is not completely useless. As the stiffness and strength of the brace increase with respect to resisting twist the effective length of the braced beam reduces, rapidly at first, until a limit is reached in which no further reduction occurs and the beam buckles in the form typified by the beam shown in Fig 1. Designers should be aware that it is the twist resisting character which is important and, if they wish to obtain worthwhile benefits, then the brace and its connections must have this capability.

The basic problem analysed in this paper is the buckling of a simply supported beam under uniform bending moment to which discrete braces are attached at equally spaced intervals to the tension flange; see Fig 2. The braces exert rigid positional restraint and finite torsional restraint about a longitudinal axis lying outside the beam section, assumed to coincide with the centroid of the braces.

There are two basic modes of buckling:

mode 1, which is associated with twisting only about a longitudinal axis through the brace centroid, and

mode 2, which is associated with lateral-torsional buckling between brace points with zero twist and lateral displacements at the brace attachment points.

1 THEORY AND PRINCIPLES

1.1 *Brace stiffness and affective length*

The requirements for complete bracing are determined, Horne and Ajmani (4), by equating the moment for overall buckling to the moment for buckling between brace points; see Fig 2.

For overall buckling between brace points, the critical moment is given by:

$$\left[\frac{2a\,M_{cr}}{GJ}\right]_1 = 1 + \left(1 + 4\frac{a^2}{d^2}\right)\left(\frac{\Pi^2\,EI_y}{4GJ}\right)\left(\frac{d}{s}\right)^2\left(\frac{n}{m}\right)^2 - \frac{P}{P_E}\left(1 + 4\frac{a^2}{d^2}\right)\left(\frac{d}{s}\right)^2 + \frac{1}{\Pi^2}\left(\frac{m}{n}\right)^2\left(\frac{d}{a}\right)^2\left(\frac{s}{d}\right)^2\left(\frac{a^2k}{GJ}\right) \qquad 1(a)$$

and for buckling between brace points by:

$$\left[\frac{2a\,M_{cr}}{GJ}\right]_2 = 4\left(\frac{a}{d}\right)\left(\frac{d}{s}\right)\left[\left(\frac{\Pi^2\,EI_y}{4GJ}\right)\left(1 - \frac{P}{P_E}\right)\left\{1 + \left(\frac{\Pi^2\,EI_y}{4GJ}\right)\left(\frac{d}{s}\right)^2\left(1 - \frac{P}{P_E}\right)\right\}\right]^{\frac{1}{2}} \qquad 1(b)$$

In equation 1(a), n is an integer $(1 \leq n \leq m)$ indicating the number of half waves of twist; it is selected so as to minimise $(2a\,M_{cr}/GJ)1$. In selecting n, note that the end supports prevent twist entirely and are equivalent to rigid braces. With a large value of m, there is a large number of subspans and the effect of the supports on M_{cr} declines. In the limit, where $m = \infty$, the supports have no influence on the buckling loads and values of $(2a\,M_{cr}/GJ)1$ are least for all values of $m>2$. In this case the value of (m/n), which minimises $(2a\,M_{cr}/GJ)1$, is given by:

$$\left(\frac{m}{n}\right)^4 = \Pi^2\left(1 + 4\frac{a^2}{d^2}\right)\frac{\left(\Pi^2\,EI_y\right)}{4GJ}\left(\frac{d}{s}\right)^4\left(\frac{a}{d}\right)^2 \Big/ \left(\frac{a^2k}{GJ}\right) \qquad (2)$$

By equating the moments given by 1(a) and 1(b), taking (m/n) as given by (2), an expression may be obtained which is non-linear in (s/d) for given distributed brace stiffness a^2k/GJ, beam properties $\Pi^2EI/4GJ = (6.2\,I_y/J)$ and brace connection geometry a/d.

For $a/d = 0.75$ values of s/d are given in Table I. To use Table I it is necessary to:

i. compute $6.2\,I_y/J$ and a^2k/GJ,
ii. select the appropriate value of s/d,
iii. design the beam as having an effective length given by the value of s.

For values of $P/P_E = 0.5$ and 1.0 values of s/d are given by Milner (5). It will be noted that (s/d) values for $P/P_E = 0$ lead to the largest (s/d) values and accordingly Table I may be used conservatively for varying P/P_E.

1.2 Strength requirements

The requirements of the bracing system for strength have been investigated by Milner and Rao (6). It was concluded that a modification of the current 2.5 percent rule of Cl.3.3.4.2 of AS1250 shown in Fig 3 should be adopted as the design criterion. The underlying principle is based on the assumption that, if it is sufficient to design position fixing braces attached to the compression flange for a force of $0.025\ P_f$, where P_f is the force in the braced flange, then it is also sufficient to design a brace and its connection to a tension flange for a moment of $0.025\ P_f\ (a + d/2)$ together with the force of $0.025\ P_f$; see Fig 3.

The moment strength required of the brace is given by:

$$M_J = 0.025\ BT\ (a + d/2)\ F \qquad (3)$$

1.3 Brace stiffness and effective brace stiffness

The equations 1(a), 1(b) are based on an analysis in which the braces are fully effective. However, as the beam is loaded about the major axis, the twist is resisted by the braces and significant local web distortions occur in the vicinity of the brace and the connections also distort.

The brace, brace connection and the web may be regarded as a set of rotational springs in series with an effective stiffness:

$$K_E = (K_B^{-1} + K_J^{-1} + K_W^{-1})^{-1} \qquad (4)$$

The term K_B may be computed as $(3EI_B/L_B)$ where L_B is the length of the brace.

The term K_J, the joint stiffness, has been measured by Rao (8) for a wide range of typical beam to purlin bolted connections and generally speaking their initial tangent stiffness is large compared with K_W. It is satisfactory to put K_J^{-1} to zero in equation (4).

Finally, the term K_W is given by:

$$K_W = 0.5\ Et^3 \qquad (5)$$

an expression which has been evaluated theoretically, Milner and Rao (7), using the finite element method and its validity verified by experiment.

2 DESIGN PRINCIPLES

The design procedure is based on the following principles:

(i) Isolation of a beam segment for design purposes

A beam segment must be isolated from a structural framework in such a way that the segment is tension flange braced and the boundary conditions are satisfied in a conservative sense. This means that beam segments adjacent to the segment under consideration are in stable or neutral equilibrium with respect to lateral-torsional buckling.

(ii) Smearing of brace stiffness

Provided the braces are uniformly spaced along the beam, the bracing against twisting may be considered as uniformly distributed and of stiffness $k = K_E/s$.

If the brace spacing is not uniform or the effective brace stiffness varies along the beam, then a conservative computation can be performed by computing the equivalent distributed brace stiffness using $k = (K_{Ei}/s_i)_{min.}$ where the min. subscript implies the selection of the minimum value.

(iii) Limits on effective length

Although the effective length may be obtained from k, it is not appropriate to select an effective length less than $s_{i\ max}$.

(iv) Sharing brace strength requirements

Cl.3.3.4.4 of AS1250 permits brace strength requirements to be shared where multiple braces are used.

(v) Variation in the ratio a/d

Within the range $0.5 \leq a/d \leq 1.25$ it is satisfactory to compute the (s/d) value using the Table I values and the interpolated value $(s/d) = (s/d)_{0.75}\ (a/0.75d)$ where $(s/d)_{0.75} = s/d$ value corresponding to $a/d = 0.75$.

2.1 Capacity of beam-purlin connections

Moment capacities and joint stiffness have been measured, Rao (8) for joints involving light gauge channels, the use of the AISC Standard Connections (3) and both HS and MS bolts. With HS bolts the joint can be designed as a friction grip connection. Similarly MS bolted joints can be designed on the basis of an initial tension of 20 kN.

3 DESIGN EXAMPLE

The structure is assumed to consist of the frames shown in Fig 4 spaced at 6 m intervals, and for simplicity the member sizes are set for wind uplift load condition.

3.1 Bending moments

Concentrating on the member BC the moment under dead load and wind loads to AS1170 (2) varies in accordance with:

$$M = 172.81 - 56.85x + 2.87x^2$$

where x is measured from B.

The section revised to resist the positive moment is a 410 UB 59 member. Relevant section parameters are $z_x = 1.06 \times 10^3\ mm^3$, $I_y = 12 \times 10^6\ mm^4$, $J = 0.329 \times 10^6\ mm^4$, $D = 406$, $T = 12.8$, $t = 7.8$, $d = D - T = 394$, $r_y = 39.7$, $B = 177.8$ (all mm units), $E = 200$ MPa $G = 80$ MPa.

3.2 Torsion resisting brace-design

For an internal frame the stiffness of the brace itself is equal to:

$$K_B = 2(3EI/L)_B = 888.6 \times 10^6 \ Nmm/rad$$

$$K_W = 0.5 \ Et^3 = 47.5 \times 10^6 \ Nmm/rad$$

The joint is assumed to be of a type (8), for which $K_J = 1270 \times 10^6$ Nmm/rad

$$K_E = 43.5 \times 10^6 \ Nmm/rad$$

$$k = 16900 \ N$$

$$a = 325 \ mm$$

$$a/d = 0.8, \ 6.2 \ I_y/J = 226, \ a^2k/GJ = 0.0678$$

$(s/d) = 11 \times (0.8/0.75)$ (See Table I and Design Principle 5)

$$= 11.7$$

$s = 11.7 \times 394 = 4610 \ mm$, less than actual brace spacing.

$$s/r_y = 116, \ D/T = 31.8, \ d/t = 48.8, \ T/t = 1.64.$$

Permissible stress = $122 \ MPa$ (Table 5.4.1 (3) of AS1250).
Moment capacity = $122 \times 106 \times 10^3 \times 10^{-6} = 129 \ kNm$.

There is sufficient moment capacity to support the maximum negative moment acting on beam BC = 108.7 kNm at x = 9.90 m.

The required joint strength is given by:

$$M_J = 0.025 \ BT \ (a + d/2) \ F = 3.08 \ kNm$$

However, this is supplied by more than one brace. Each brace is assumed to provide:

$$3.08 \times \frac{2.57}{4.61} = 1.72 \ kNm$$

Assume M 16 bolts are used in a two hole standard connection shown in Fig 4.

Bolt proof load = 90 kN
Moment capacity = 0.6 x 0.33 x 0.11 x 90 = 1.96 kNm > 1.72 kNm.

4 CONCLUSIONS

The design method presented herein is a straight forward procedure which can be easily incorporated in current codes of practice.

References

1. AS1250-1975. 'Standards Association of Australia Steel Structures Code'.

2. AS1170 Part 2-1973. 'Standards Association of Australia Loading Code', Part 2 - Wind Forces.

3. Hogan, T.J. and Firkins, A. 'Standard Structural Connections Part A : Details and Design Capacities', AISC Publication, Sydney 1978.

4. Horne, M.R. and Ajmani, J.F. 'Stability of Columns Laterally Supported by Side Rails', International Journal of Mechanical Sciences, Vol. 11, 1969, pp. 159.

5. Milner, H.R. 'The Design of Simply Supported Members Braced Against Twisting on the Tension Flange', Civ. Eng. Trans., I.E.Aust., Vol. CE19, No. 1, 1977.

6. Milner, H.R. and Rao, S.N. 'Strength and Stiffness of Moment Resisting Beam-Purlin Connections', Civ. Eng. Trans., I.E.Aust., Vol. CE20, No. 1, 1978.

7. Milner, H.R. and Rao, S.N. 'Twisting Members by a Torsion Moment Applied to One Flange', Sixth Australasian Conference for the Mechanics of Structures and Materials, Christchurch, 1977.

8. Rao, S.N. 'Second Report on an Investigation into Tension Flange Bracing of Steel UB Members', Report to AEBIRA by Department of Civil Engineering, Chisholm Institute of Technology, April 1982.

$P/P_E = 0$ $a/d = 0.75$

a^2k/GJ	$6.2\ I_y/J$		
	100	200	400
0.000	31.48	44.52	62.96
0.002	19.86	24.71	30.20
0.004	17.48	21.35	25.72
0.006	16.08	19.47	23.29
0.010	14.36	17.23	20.49
0.020	12.18	14.49	17.15
0.030	11.02	13.06	15.44
0.040	10.24	12.13	14.34
0.060	9.24	10.92	12.91
0.080	8.58	10.14	11.98
0.100	8.10	9.57	11.31
0.120	7.72	9.13	10.79
0.140	7.42	8.77	10.37
0.160	7.17	8.47	10.02
0.180	6.95	8.22	9.72
0.200	6.76	8.00	9.46
0.220	6.60	7.80	9.23
0.240	6.45	7.63	9.03

TABLE I: Values of s/d for various a^2k/GJ, and $6.2\ I_y/J$.

Fig. 1 Tension flange braced beam under test

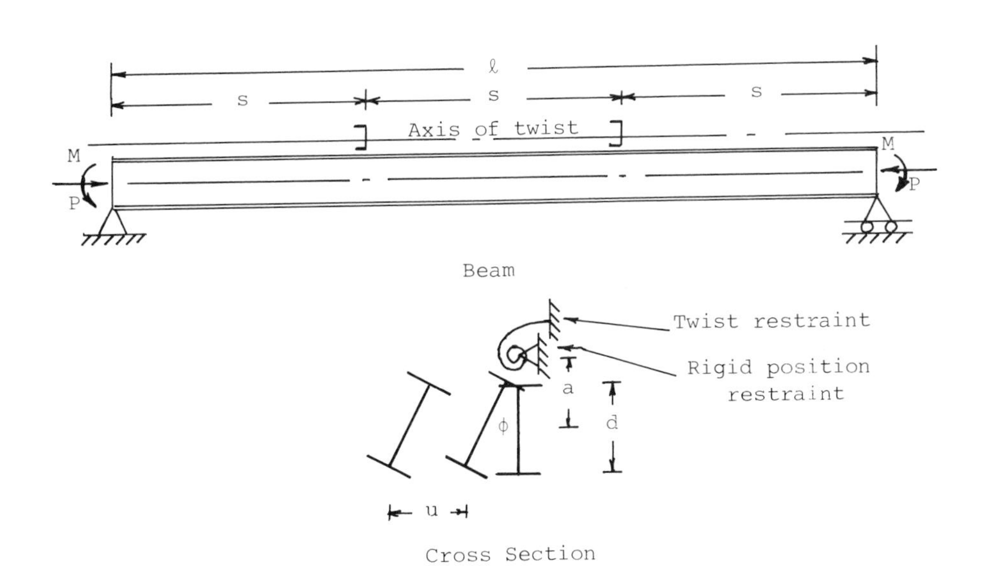

Fig. 2 System analysed

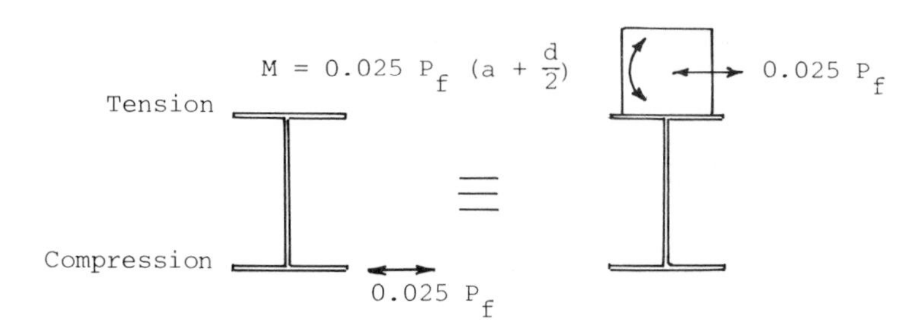

Fig. 3 Strength of bracing system

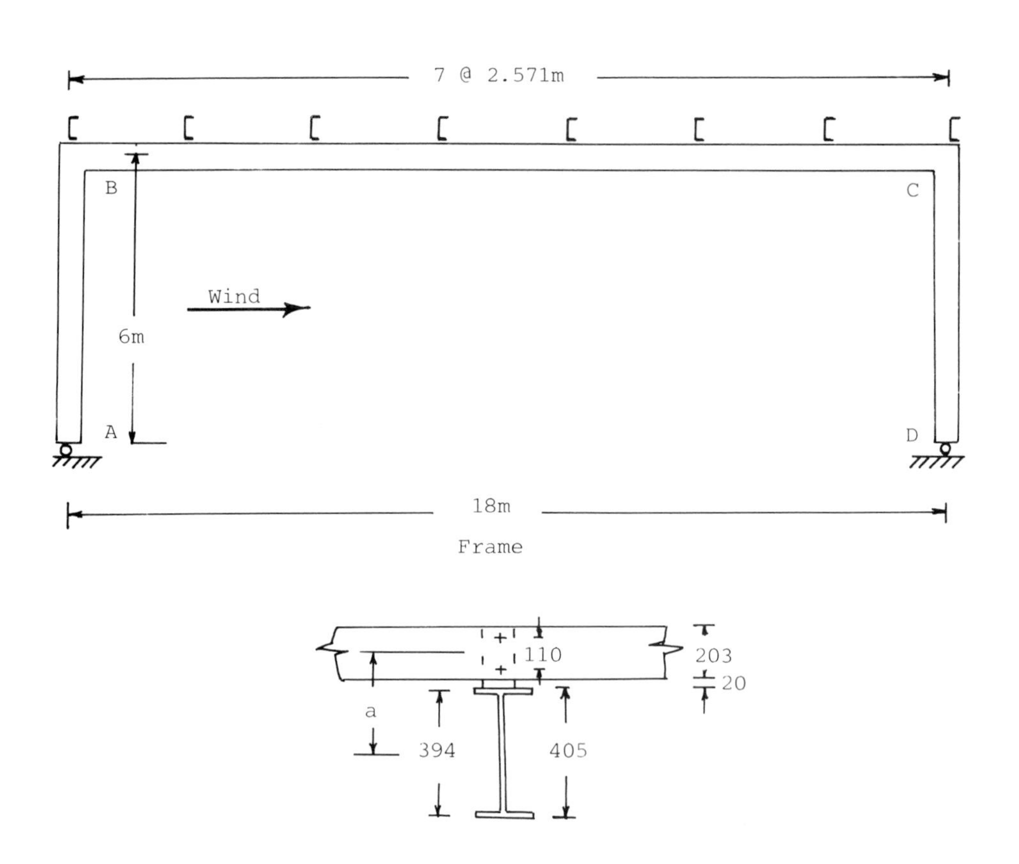

Fig. 4 Details of design frame

DISTORTIONAL LATERAL BUCKLING OF UNSTIFFENED COMPOSITE BRIDGE GIRDERS

R.P. Johnson,

Engineering Department, University of Warwick.

M.A. Bradford,

School of Civil and Mining Engineering, University of Sydney.

Synopsis

A finite element parametric study is reported of elastic distortional lateral buckling in laterally unstiffened fixed-ended composite girders with web depth/thickness ratios ranging from 39 to 100. The work is relevant to the design of highway viaducts with spans up to about 30 m. It shows that there is much room for improvement in existing design methods, that further study is needed of the interaction between lateral buckling and inelastic local buckling, and that the latter is the more likely to govern design.

Notation

B, b, d, T, w dimensions defined in Fig. 4.

$D, M_D, y_t, \beta, \gamma, \delta, \lambda_{LT}, \sigma_{\ell c}, \sigma_{\ell i}$ symbols defined in clause 9 of BS 5400: Part 3.

f_{cr} elastic critical stress in bottom flange

f_u design ultimate bending stress in compression

f_y yield stress of structural steel

L span of beam

M_p hogging plastic moment of resistance of composite section

M_u design ultimate hogging bending moment at support

M_y hogging yield moment of composite section

S shape factor of composite section in hogging bending

η imperfection constant

λ_{cr} lowest eigenvalue for elastic buckling

Introduction

Continuous composite beam-and-slab viaducts have been found to be an economic form of construction. Plate or box girders are used for the longer spans, but the minimum economic span can be reduced to well below 20 m by the use of unplated universal beam (UB) rolled sections.

Design calculations for one such structure are presented in courses on composite bridges run by CONSTRADO. It is a five-span viaduct with internal spans of 23 m. The concrete deck, 12.5 m wide, is composite with four 914 × 419 UB 388 sections of Grade 50 steel, with transverse spacing of 3.5 m. In design to BS 5400:Part 3 (1), transverse bracing is required at bottom flange level, and was provided at cross-sections 6.0 m either side of each internal support. The need for bracing of this type will be studied in this paper.

The work relates to isolated composite T-beams with negligible torsional moments, and assumes that no failure occurs in the deck slab or the shear connection. Lateral buckling of the steel bottom flange influences design only near internal supports. The governing longitudinal hogging moments in bridges of this type occur when the two bogies of an abnormal vehicle straddle a support. The beams then behave almost as if fixed-ended. The moment gradient at the other end of the span is less steep. It has little influence on buckling at the end under the vehicle, which can be studied using a single fixed-ended span under loading symmetrical about mid-span.

The scope of the work is limited to steel rolled sections or plate girders of uniform section, with webs of slenderness (d/w) up to 100. The rolled sections range in depth from 850 to 920 mm. The members are restrained by the concrete deck slab, but other bracing and stiffening are provided only at the supports. Slip at the shear connection and the effects of temperature, shrinkage of concrete, and local wheel loads are neglected.

The classical theory of elastic lateral-torsional buckling is not applicable, for it assumes that each cross-section of the member rotates as a whole, without distortion. In a composite member without web stiffeners, lateral buckling of bottom flanges is restrained by a continuous inverted U-frame consisting of two or more adjacent steel beams and the deck slab to which they are attached (Fig. 6). The design method of BS 5400:Part 3 is based on this U-frame action, but also uses the conservative assumptions that the compressive force in a bottom flange is everywhere equal to its peak value at the internal support, and that torsional stiffness is negligible. It leads to effective lengths that exceed the length of bottom flange in compression, so that its neglect of moment gradient greatly reduces the design compressive stress.

Little useful information on this subject is available from tests, for the cost of a realistic test specimen that models U-frame behaviour in a continuous girder is prohibitive. Scores of hogging moment regions have been tested as isolated T-section cantilevers. The usual mode of failure has been yield of longitudinal reinforcement followed, at large rotation, by inelastic local buckling of the bottom flange. Most specimens were braced against lateral buckling, which has occurred only in a few unbraced specimens (2), and after local buckling.

This paper presents a parametric study of elastic distortional buckling in fixed-ended spans under uniformly-distributed loading. A curve relating design compressive stress to the relevant slenderness is deduced from the critical stress curve by the Perry-Robertson method that is used for lateral-torsional buckling of beams in BS 5400:Part 3. Thus it is assumed, in absence of direct evidence, that the destabilising effects of residual stresses, geometrical imperfections, and (for compact members) yielding of steel, are the same for distortional buckling under moment gradient as in the method of BS 5400.

1. *ELASTIC DISTORTIONAL BUCKLING ANALYSIS*

The overall buckling mode for an I-beam whose tension flange is restrained against lateral displacement and twist is characterised by distortion of the web in bending (3). Computer programs for the study of such buckling are now available. That used here, FEBGEN, was developed by the second author and Trahair (4) at the University of Sydney. It is a finite element analysis of prismatic thin-walled members under the action of shear and longitudinal bending stresses that may vary along the length of the member.

The critical load factor λ_{cr} for a given initial stress distribution is given by the eigenvalue equation

$$\left| [K] - \lambda_{cr}[G] \right| = 0 \qquad (1)$$

in which $[K]$ is the elastic stiffness matrix that expresses the resistance of the beam to lateral-torsional buckling, and $[G]$ is the stability matrix that expresses the tendency of the stress distribution in the beam to cause lateral-torsional deformations.

The beam is partitioned into a number of longitudinal member elements. Each consists of a rectangular web panel with membrane, torsional, and bending flexibilities, to which is connected flange assemblies of general cross-section. They have membrane and torsional flexibilities only, and their cross-sections remain undistorted during buckling. These flanges consist of rectangular strip sub-elements. Here, each member element consists of a web and two flange sub-elements as shown in Fig. 1(a).

The infinitesimal buckling deformations of a member element are described by polynomial displacement fields that are related to eleven degrees of freedom at each member node: four at the bottom of the web, a similar four at the top, and three at mid-depth (Fig. 1(b), in which a prime indicates differentiation with respect to z). The displacements u and ϕ at top and bottom of the web, and also v, are represented by cubic functions of z, and the displacement w by a linear function; thus there are 22 displacement coefficients per member element. In this application eight of these, relating to the top of the web, are set at zero. At the fixed end of the beam, u, ϕ, v, v' and w are also set at zero, but not u' and ϕ', so that rotation of the bottom flange about a vertical axis is allowed. The buckling deformation of the web in a cross-sectional plane is represented by a cubic polynomial.

Expressions for the increase in strain energy and decrease in potential of the loads are obtained from these displacement fields, and the matrices $[K]$ and $[G]$ for the whole beam are assembled in the usual way (5).

From symmetry, it is evident that the critical buckling modes are either symmetrical (mode 1) or anti-symmetrical (mode 2) about mid-span, as shown in Fig. 2. For this reason only half of a typical span need be analysed provided that the type of buckling mode is determined by the boundary conditions defined at mid-span (either $u' = \phi' = 0$ or $u = \phi = 0$).

The end 20% of a typical span was represented by n_1 member elements and the next 30% by n_2 elements, with $n_1 = 3n_2$ so that shorter elements were used in the region where the bottom flange is in compression. Figure 2 shows to scale the location of member nodes for $n_1 = 6$, $n_2 = 2$.

Initial values of bending and shear stress were required at each node, and at the corners of each flange sub-element. It was assumed that over the length of a member element, the longitudinal stresses vary linearly and the shear stresses are constant. The elastic critical stresses are these initial stresses multiplied by the lowest eigenvalue λ_{cr} computed as explained above.

In comparing a design method with theoretical results for elastic buckling of steel members with yield stress f_y, it is convenient to use variables f_u/f_y and $\sqrt{(f_y/f_{cr})}$, as in Figs. 5 and 7, and to focus on the longitudinal compressive stress at a "reference point" in the member. Here, this point is at the lower surface of the bottom flange at the face of the support. f_u is the design ultimate stress, calculated by elastic theory, implied by the design method, and f_{cr} is the computed critical distortional buckling stress at the same point. If elastic critical theory were used as the design method curve AB in Fig. 7 would be obtained, with a cut-off at $f_u = f_y$.

The bending moment distribution for each beam and the stresses at nodes were determined by elastic theory, taking account of cracking of the concrete slab. The value of the modular ratio, taken as 13.6, had little influence on the results. The calculations were done for design end moments M_D given for the unbraced member by the method of BS 5400:Part 3 for the ultimate limit state. The shear stresses in the elements were found from the vertical shear at mid-length of each element, so that the distributed load was modelled as a series of point loads at the nodes.

An increase in the number of elements gives a different value for the lowest eigenvalue, λ_{cr}, for two reasons: the approximations made within the elements become more accurate, and the point loads more closely represent distributed load. Typical results from studies of convergence with both effects present are shown in Fig. 3, in which λ_{cr} is plotted against $1/(n_1 + n_2)$ for $n_1 + n_2 = 4, 8, 12$ and 16. Extrapolation gives $\lambda_{cr} \rightarrow 4.64$ as $n_1 + n_2 \rightarrow \infty$. This result is 1.4% below that for 8 elements, which was the number used in all subsequent calculations.

2. *PARAMETRIC STUDY*

The calculations were done either for specific universal beam sections (Table 1), or for non-standard I-sections (Table 2), treated as three plates each of rectangular section. The dimensions of a typical section are defined in Fig. 4. The main variables were the ratios

L/B (L = span), B/T, and d/w, given in the tables. The yield stresses of the I-sections and of reinforcement were taken as 355 N/mm^2 and 428 N/mm^2. Three sizes of concrete flange were used, S, M, and L in Table 1. Their breadths b were respectively 2.0 m, 3.0 m, and 4.0 m. The reinforcing bars were placed in two identical layers, and the ratios of total bar area to slab area were respectively 0.01, 0.015, and 0.02. All flanges in Table 2 were type M.

Eight pairs of beams were analysed for both symmetrical and anti-symmetrical buckling. The modal shapes are most clearly defined by lateral displacement of the centre of the bottom flange, u. Figure 2 shows the variations of u along a half span for beam B11, for which the two eigenvalues differed by only 0.3%. The symmetrical mode always gave the lower eigenvalue λ_{cr}. These results are given in Tables 1 and 2 and Fig. 5.

3. *DESIGN ULTIMATE STRESS* f_u, *to BS 5400: Part 3*

The relevant safety factor format defined in Part 3 is:

$$\text{(effects of } \gamma_{fL} \times \text{ loading)} \leq \frac{\text{resistance of section or member}}{\gamma_{f3}\,\gamma_m}$$

where the resistance is calculated using an unfactored yield strength for structural steel, f_y, and the appropriate geometric variables. The stress f_u, as used here, is a measure of this unfactored resistance, and excludes the partial safety factors γ_{fL}, γ_{f3}, and γ_m.

The ratio f_u/f_y was calculated for each beam as follows. The slenderness ratio λ_{LT} was found in accordance with clauses 9.6.6, 9.7.2 (with $\eta = 1$) and 9.8. Table 10 gives $\sigma_{\ell i}/f_y$. The method of clause 9.3.7 was used to determine if the member was compact or slender. For a slender section, the design stress $\sigma_{\ell c}$ is given by clause 9.8.3. The design ultimate moment is

$$M_u = M_y(\sigma_{\ell c}/f_y) \tag{2}$$

where M_y is the moment for first yield. For a compact section $\sigma_{\ell c}$ is taken as $\sigma_{\ell i}$ and the design ultimate moment is

$$M_u = M_p(\sigma_{\ell i}/f_y) \tag{3}$$

where M_p is the plastic moment of resistance.

From eqns.(2) and (3), the ultimate stress is $\sigma_{\ell c}$ for a slender section or $S\sigma_{\ell i}$ for a compact section, where S is the shape factor (M_p/M_y) for the section. A few of the sections studied are compact if of mild steel and slender if of Grade 50 steel. For these beams, f_u has been calculated by both methods, giving points on both curves AB and CD in Fig. 5. All of the ultimate stresses are well below yield, but all the critical stresses are above yield, the lowest for a UB section being 2.47 f_y (beam B13), so that distortional buckling, if it occurred, would be inelastic.

3.1 *Influence of parameters on relationship between f_u and f_{cr}.*

Tables 1 and 2 and Fig. 5 show that the influence of L/B is negligible within the range 48 to 90 (compare beams B1,2, and 3; B6 and 7; B8 and 9; and B10 and 11). The influence of the B/T ratio for the compression flange is small (beams P21 and 22; P23 and 24), with an increase in B/T at constant B causing a smaller increase in f_u than in f_{cr} (1% cf. 6% for beams P21 and 22). The principal influence on both f_u and f_{cr} is the web slenderness d/w; all of the results lie close to the same curve, irrespective of the ratios L/B and B/T, for d/w ranging from 39 to 100.

An increase in the size of the concrete slab increases the depth of web in compression and hence reduces the stability of the member, but the reduction in f_u is much greater than in f_{cr} (11% cf. 2.5% for beams B4 and 5), because in BS 5400 increase in lateral-torsional slenderness leads to reduction of design stresses in the whole cross-section, not just in the compression zone.

4. *FLEXIBILITY OF CONCRETE SLAB AND OF SHEAR CONNECTION*

The assumption in the parametric study that the upper edge of the web is fixed in direction as well as position leads to an over-estimate of the elastic critical stress. The curves in Fig. 5 were corrected for this error as follows. The deflection δ of the bottom flange (Fig. 6) due to unit lateral force per unit length of beam is a measure of the flexibility of the inverted U-frame. The formula given for δ in clause 9.6.6.2 of BS 5400: Part 3 has two terms, one for the web, proportional to $(d/w)^3$, and one for the concrete flange. By reducing w slightly, an effective d/w can be found such that the flexibility due to the web alone equals the true flexibility. For a given concrete slab, the correction is greatest at low d/w. For example, for an edge girder with b = 3.5 m and h_c = 0.22 m, d/w is increased from 40 to 41.3 and from 80 to 80.3. As f_{cr} in Fig. 5 depends almost entirely on d/w, the curves can be modified to allow for an increase in this ratio. The revised curve AB is shown as EF; the change in curve CD is insignificant.

The contribution of the shear connection to the total flexibility was also estimated, and found to be at least an order of magnitude less than that due to the concrete slab.

5. *TENTATIVE DESIGN METHOD*

Curves EF and CD on Fig. 5 are re-plotted at smaller scale on Fig. 7. They lie far below the 'yield' line $f_u = f_y$ and the elastic critical curve $f_u = f_{cr}$ because f_u in BS 5400 is related to a critical stress derived for a uniformly compressed strut that is far less stable than a real compression flange under moment gradient.

Design curves for members liable to buckle are usually derived from the relevant elastic critical curve in a way that allows for a level of imperfections and/or residual stresses that is either assumed or deduced from tests. Lay (6) has compared and discussed the methods that have been used. The Perry-Robertson equation is one of the earliest. In the present notation it is given by Trahair (7) as

$$f_u/f_y = \frac{1}{2}\{1+(1+\eta)(f_{cr}/f_y)\} - \frac{1}{2}[\{1+(1+\eta)(f_{cr}/f_y)\}^2 - 4(f_{cr}/f_y)]^{\frac{1}{2}} \quad (4)$$

where η is an imperfection constant and $f_u \not> f_y$.

The relationship between $\sigma_{\ell i}/f_y$ and f_{cr} given in BS 5400: Part 3 for lateral-torsional beam buckling (clause G.7) is in this form. It is obtained by assuming that

$$f_{cr}/f_y = 5700/\beta^2 \quad (5)$$

where $\beta = \lambda_{LT}(f_y/355)^{\frac{1}{2}}$ and λ_{LT} is a slenderness function that is proportional to the slenderness of the beam (ℓ/r) for minor-axis bending. The imperfection constant is given by

$$\eta = 0.005(\beta-45) \quad (6)$$

Elimination of η and β from eqns.(4) to (6) enables the design curve of BS 5400 to be plotted on Fig. 7, as GHJ.

For compact sections, $f_u = S\sigma_{\ell i}$. A typical value of S is 1.3 (Tables 1 and 2), so the design strength f_u, as defined here, is obtained from curve GHJ by multiplying its ordinates by about 1.3 (curve KNP in Fig. 7). For slender sections, $f_u = \sigma_{\ell c}$, and for the sections studied here, $\sigma_{\ell c}$ as given by BS 5400: Part 3 exceeds $\sigma_{\ell i}$ by about 20%. It is doubtful whether this adjustment, intended mainly for asymmetric steel sections, should be used for composite sections. From Fig. 7 it appears to be excessive in relation to a realistic curve for critical stress, and so is omitted from the tentative design method.

The use of GHJ as a design curve requires knowledge of the critical stress f_{cr}. It cannot easily be calculated, but as it depends almost solely on the web slenderness d/w, an approximate method can be derived. A logarithmic plot was made of f_{cr}/f_y against d/w for all the beams analysed. A lower bound to f_{cr} was found to be

$$f_{cr}/f_y = 600(d/w)^{-1.4} \quad (7)$$

Substituting from eqn.(5) gives $\beta = 3.08(d/w)^{0.7}$. This was increased by about 10% to give the proposed design relationship,

$$\beta = 3.4(d/w)^{0.7} \quad (8)$$

5.1 *Summary of design method.*

The method is a modification of the method of BS 5400: Part 3 for beams continuously restrained by a deck not at compression flange level (clause 9.6.6.2). It applies to composite beams with web stiffeners and lateral bracing only at supports, not within the span.

The slenderness function β is calculated from eqn.(8). The stress $\sigma_{\ell i}$ is found from Fig. 10 of Part 3. The method is then as in Part 3, except that for slender sections, $\sigma_{\ell c}$ is taken as the lesser of $(D/2y_t)\sigma_{\ell i}$ and $\sigma_{\ell i}$.

The ratios f_u/f_y have been calculated by this method for all the beams studied here, and are plotted on Fig. 8. The curve GLM through these points (also shown on Fig. 7) gives the design stress f_u for a slender cross-section. For a compact section, its ordinate has to be multiplied by the shape factor. The design method is everywhere conservative in comparison with the "Perry-Robertson" beam curve GHJ, yet it gives moments of resistance between 1.5 and 2.5 times those given by the existing method (curves EF and CD).

This proposed use of much higher design stresses than at present, in conjunction with a plastic section modulus for beams of compact section, raises the question whether inelastic local buckling of the bottom flange near an internal support may interact with lateral buckling in a way not considered in the preceding analysis.

6. *INTERACTION BETWEEN INELASTIC LOCAL AND LATERAL BUCKLING*

In a fixed ended beam of the type studied here, subjected to increasing load, yielding begins in the bottom flange at each support, and spreads along the flange and upwards into the web. The resulting deterioration in stiffness reduces the local buckling load for the bottom flange and the distortional buckling load for the member.

This effect has been studied in wide-flange steel beams under moment gradient by Lay and Galambos (8). They present simplified analyses for the separate influences of lengths of yielded flange on distortional lateral buckling and on local buckling. For lateral buckling the steel is assumed to possess strain-hardening properties within the yielded region (e.g., a Young's modulus of $E_s/33$ for A 36 steel). When web restraint is included, local flange buckling is considered to occur in compact wide-flange sections (9) under moment gradient when the yielded length L_y reaches the wavelength of the local buckle:

$$L_y/B = 1.42(d/B)^{0.25}(T/w)^{0.75} \tag{9}$$

where the notation is as in Fig. 4.

It is shown (8) that throughout the practical range of values, the yielded length for local buckling is significantly less than that for lateral buckling, so that local buckling will occur first. The results of many tests on hogging moment regions of composite beams (2,10,11) are consistent with this conclusion, in that local buckling almost always occurred. Lateral buckling was rare, and never occurred until after severe deformations at the local buckle.

Lay and Galambos suggest that local buckling can reduce the bending stiffness of the flange about a vertical axis by a factor of about eight, which reduces the yielded length for lateral buckling to below that for local buckling, so that lateral buckling follows at once. The authors know of no systematic experimental study relevant to this suggestion. In an imperfect beam under moment gradient, susceptibility to lateral buckling would increase the compressive stress on one side only of a flange. It seems unlikely that this could significantly reduce its local buckling load; but again there is no direct evidence.

The effects of residual stresses, neglected in Reference 8, influence the spread of yielding along the bottom flange, and computer programs now available can provide more accurate analyses of lateral buckling

(12) than the simple lower-bound method of Lay and Galambos. Research therefore continues on the spread of yielding along the bottom flange, its influence on local and lateral buckling, and its interaction with redistribution of moments in a fixed-ended beam.

7. *CONCLUSIONS*

Elastic critical buckling loads have been computed for fixed-ended laterally unbraced composite beams under uniformly distributed load. The critical buckling mode is always symmetrical about mid-span. There is lateral displacement and twist of the bottom flange throughout the span (Fig. 2).

The critical compressive stress in the bottom flange at the face of a support is not influenced by the span, or by the ratio of span to flange breadth in the range 48 to 90. The influences of the size of the concrete top flange and the breadth/thickness ratio of the bottom flange are small. The stress f_{cr} is determined mainly by the depth/thickness ratio of the web, eqn.(7). The influence of the transverse flexibility of the slab and the shear connection is negligible except at low d/w ratios (Fig. 5).

A tentative design method, related to the critical stresses by the Perry-Robertson method for beams, suggests that design moments given by BS 5400: Part 3 for laterally unbraced members could be doubled. Inelastic local buckling could determine the strength of some beams if this were done. Such buckling, and its interaction with redistribution of moments, is being studied further at the University of Warwick, from which a fuller version of this paper is available.

Acknowledgements

The first author is grateful to the School of Civil & Mining Engineering, University of Sydney for the facilities provided during a period of study leave at Sydney. Both authors acknowledge with thanks the contribution made by Professor N.S. Trahair in discussions of this subject.

References

1. BS 5400. Steel, Concrete, and Composite Bridges, British Standards Institution, London, 1978-1983.
2. Climenhaga, J.J. and Johnson, R.P. Local Buckling in Continuous Beams. Structural Engineer, 50, 367-374, Sept. 1972.
3. Hancock, G.J., Bradford, M.A. and Trahair, N.S., Web Distortion and Flexural-Torsional Buckling. Proc. A.S.C.E., 106, No. ST7, 1557-1571, July 1980.
4. Bradford, M.A. and Trahair, N.S., Distortional Buckling of Thin-Web Beam-Columns.Engineering Structures, 4, No.1,2-10, Jan. 1982.
5. Zienkiewicz, O.C., The Finite Element Method in Engineering Science, McGraw Hill, London, 1971.
6. Lay, M.G., Elastic Buckling and Structural Reality. Civ. Eng. Trans., Instn. of Engineers, Australia, CE22,No.3,186-192,Nov.1980.
7. Trahair, N.S., The Behaviour and Design of Steel Structures. Chapman and Hall, London, 1977.
8. Lay, M.G. and Galambos, T.V., Inelastic Beams Under Moment Gradient, Proc. A.S.C.E.,93, No. ST1, 381-399, Feb. 1967.

9. Lay, M.G., Flange Local Buckling in Wide-Flange Shapes, Proc. A.S.C.E., 91, No. ST6, 95-116, Dec. 1965.
10. Johnson, R.P. and Willmington, R.T., Vertical Shear in Continous Composite Beams.Proc. Instn. Civ. Engrs.,53, 189-205, Sept. 1972.
11. Hope-Gill, M. and Johnson, R.P., Tests on Three 3-Span Continuous Composite Beams. Proc. Instn. Civ. Engrs., Part 2, 61, 367-381, June 1976.
12. Baigent, A.H. and Hancock, G.J. The Strength of Portal Frames Composed of Cold-Formed Channels. School of Civil and Mining Engineering, University of Sydney, Research Report R407, Mar.1982.

TABLE 1

Beam No.	Steel mm, mm, kg/mm	Con-crete flange	B/T	d/w	L/B	S	$\left(\frac{f_y}{f_{cr}}\right)^{\frac{1}{2}}$	$\frac{f_u}{f_y}$ compact	$\frac{f_u}{f_y}$ slender
B1		M	11.5	39.0	48.0	1.29	.500	.761	-
B2	920.5	M	11.5	39.0	54.7	1.29	.486	.761	-
B3	x420.5	M	11.5	39.0	65.0	1.29	.481	.761	-
B4	UB 388	S	11.5	39.0	54.7	1.23	.483	.784	-
B5		L	11.5	39.0	54.7	1.31	.489	.701	-
B6	926.6	M	9.6	43.7	65.0	1.35	.558	.564	-
B7	x307.8	M	9.6	43.7	90.0	1.35	.562	.564	-
	UB 289								
B8	918.5	M	10.9	49.5	65.0	1.36	.601	.501	.431
B9	x305.5	M	10.9	49.5	90.0	1.36	.598	.501	.431
	UB 253								
B10	903.0	M	15.0	56.4	65.0	1.40	.635	-	.391
B11	x303.4	M	15.0	56.4	90.0	1.40	.629	-	.391
B12	UB 201	S	15.0	56.4	90.0	1.31	.618	-	.392
B13		L	15.0	56.4	90.0	1.43	.636	-	.388

TABLE 2

Beam No.	B mm	D mm	w mm	B/T	d/w	L/B	S	$\left(\frac{f_y}{f_{cr}}\right)^{\frac{1}{2}}$	$\frac{f_u}{f_y}$ slender
P21	303.4	929.6	19.6	15.0	45.2	90	1.44	.559	.459
P22	303.4	949.4	19.6	9.6	45.2	90	1.36	.575	.454
P23	307.8	880.0	15.2	15.0	55.2	89	1.39	.614	.417
P24	307.8	903.0	15.2	9.6	55.2	89	1.32	.636	.400
P25	303.4	903.0	12.7	15.0	68.0	90	1.37	.699	.337
P26	303.4	903.0	10.8	15.0	80.0	90	1.34	.780	.289
P27	303.4	903.0	8.63	15.0	100	90	1.31	.927	.231

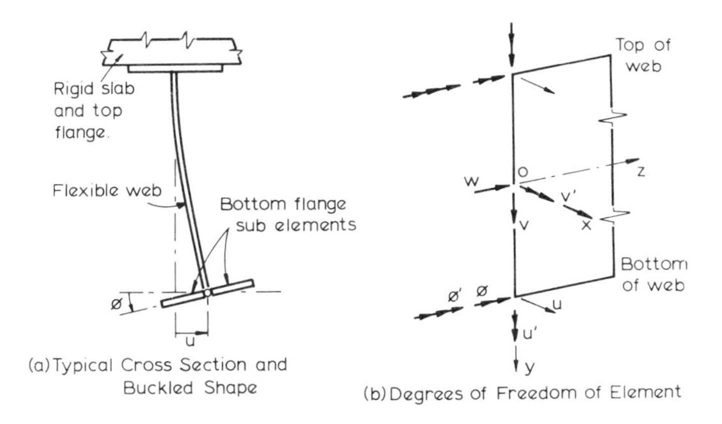

Fig.1 Finite Element Model

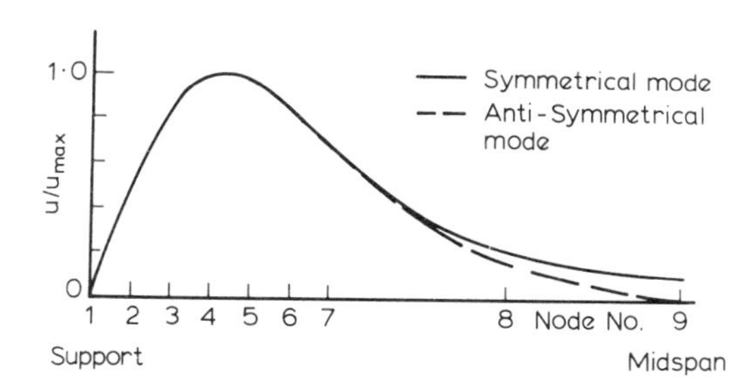

Fig.2 Lateral Displacement of Bottom Flange

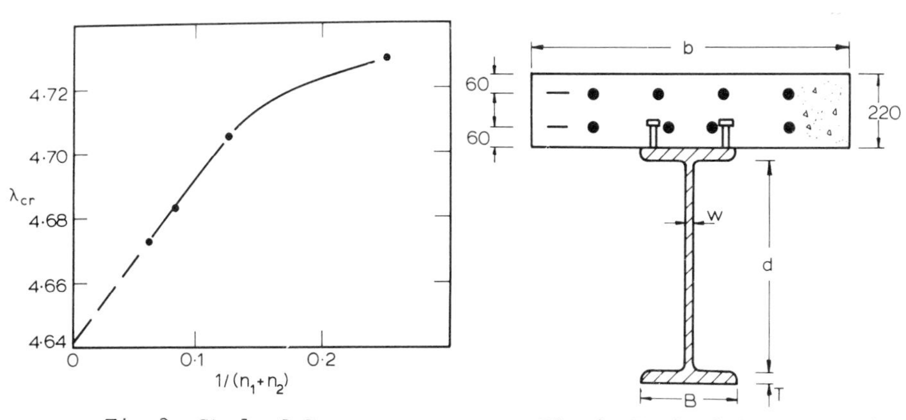

Fig.3 Study of Convergence

Fig.4 Typical Cross-Section

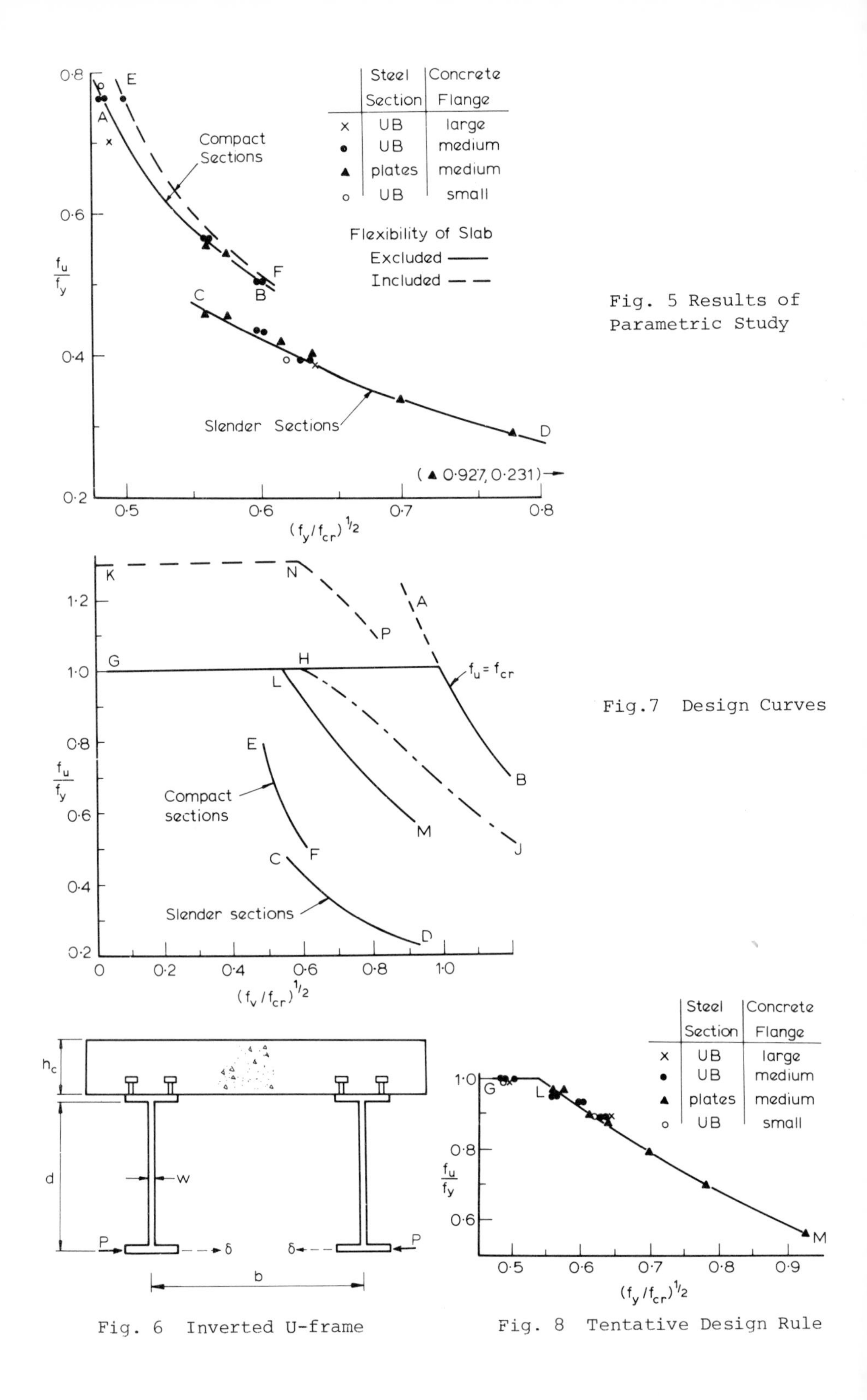

Fig. 5 Results of Parametric Study

Fig.7 Design Curves

Fig. 6 Inverted U-frame

Fig. 8 Tentative Design Rule

THE BEHAVIOUR OF CIRCULAR TUBES WITH LARGE ASYMMETRIC OPENINGS SUBJECTED TO TORSION AND BENDING

Peter Montague

Simon Engineering Laboratories, University of Manchester

Synopsis

An analysis is given for the elastic warping stresses around a large asymmetric opening in a circular tube subjected to torsion. This, together with a non-linear bending analysis, is illustrated by reference to eight experimental specimens subjected to various combinations of torsion and bending. Suggestions are made for the safe limiting values of torque, bending moment and combined torque and bending moment on such tubes.

Notation

C	torsional rigidity = $GJ = \bar{a}t^3G/3$
C_1	warping rigidity = EC_w
C_w	warping constant $= 2tR^3[\theta^3 - 6(\sin\theta - \theta\cos\theta)^2/(\theta - \sin\theta\cos\theta)]/3$
E	Young's modulus
G	shear modulus
L	total length of tube tested
M_y	moment about OY (Fig 1)
$M_{y,exp}$	maximum experimental value of M_y
$M_{y,\ell}$	estimated limiting value of M_y (eqn (11) and Fig 8)
$M_{y,max}$	empirical estimate of maximum M_y (eqn (12))
$M_{y,o}$	full plastic value of M_y (eqn (10))
P	axial load on cylindrical tube
R	mean radius of tube (Fig 1)
S	$(t^4/R^2\ell^2)^{1/3}.10^3$
T	torque
T_{exp}	maximum experimental value of T
T_{max}	empirical estimate of maximum T (eqn (14))

T_o	plastic limit value of T (eqn (13))
a	half-width of cut-out = $R\sin\alpha$
$\bar{a}$	$2R\theta$
c	$(1 - \cosh n\ell)/\sinh n\ell$
e	distance of shear centre from O
	$= 2R(\sin\theta - \theta\cos\theta)/(\theta - \sin\theta\cos\theta)$ (Fig 3)
$f(\theta)$	$[\theta - (e\sin\theta)/R]/[\theta^3 - 6(\sin\theta - \theta\cos\theta)^2/(\theta - \sin\theta\cos\theta)]$
ℓ	length of straight, free edge of cut-out
ℓ_e	equivalent length of straight free edge (eqn (6))
m	magnification of initial free edge radial displacement
w_s	warping function = $\int_o^s (R + e\cos\delta)ds$ (see Fig 3)
$\bar{w}_s$	mean value of $w_s = (\int_o^s w_s.ds)/\bar{a}$
α	half-angle of cut-out
β	see Fig 4
$\bar{\beta}$	$a\{12(1 - \nu^2)^{\frac{1}{4}}\}/(8Rt)^{\frac{1}{2}}$
δ	see Fig 3
$\bar{\delta}$	$(\delta - \alpha)$
δ_o	shortening at cut-out edge due to bending and compression
η_{of}	initial free-edge radial displacement
$\bar{\eta}_f$	loaded free edge radial displacement at $z = \ell/2$
η_{of}	initial free-edge radial displacement at $z = \ell/2$
λ	see Fig 4
ν	Poisson's ratio
ϕ	angle of twist of tube
ψ	half-angle of buckled leaf
w	see Fig 4
ρ	mean bending curvature of tube at cut-out section
σ_y	yield stress
σ_{zb}	longitudinal bending stress
σ_{zw}	longitudinal warping stress
τ_{T1}	$Gtd\phi/dz$ = shear stress associated with T_1 (eqn (1))
τ_{T2}	$-\frac{E}{t}\cdot\frac{d^3\phi}{dz^3}\int_o^s(\bar{w}_s - w_s)ds$ = shear stress associated with T_2 (eqn (1))
θ	$(\pi - \alpha)$

Introduction

The paper describes part of a study examining the applications of axial compression, shear, bending and torsion to lighting column bases which contain large, asymmetrically-placed access holes. A review of previous work on the behaviour of cylindrical tubes with a single hole was included in a previous paper (1), where it was pointed out that theoretical attention had concentrated on the application of shallow shell theory to relatively thin shells ($R/t > 100$), with low values (0 to 4) of the curvature parameter $\bar{\beta}$, and penetrated by a relatively small circular hole rather than a large, elongated one. Rectangular holes were considered by Almroth and Holmes (2), but their shells had R/t equal to 430 and 675. No theoretical study could be found which was relevant to the lighting column base which, typically, has R/t of about 20, β about 10 and a hole occupying some 20% of the circumference with a length to width ratio of about 5.5. Tests have been conducted on column bases by Fontaine (3) and by Little and Baban (4), who produced empirical expressions for bending and torsion capacity.

Montague and Horne (1) presented two analyses of the column base subjected to axial compression and bending in the OX plane (see Fig 1) and compared their predictions with results from 13 model tests. The first analysis examined the outward radial growth of the initial imperfection (lack of straightness) of the long, free edge of the cut-out with consequent retreat of the peak compressive stress away from the edge and the onset of failure when the peak stress reached the yield value of the material. The second analysis dealt with local buckling of the free edge at an applied load level well below the capacity of the cut-out cross-section.

On the limited evidence of the test results available, a tentative distinction was made, in terms of the slenderness parameter $S = (t^4/R^2\ell^2)^{1/3}.10^3$, between tubes, with $S > 3$, in which steady growth of the free edge radial displacement would occur until failure of the entire section and tubes, with $S < 3$, in which inconsequential buckling of the free edge would take place at loads below those required to cause total failure.

The test specimens discussed here have $S = 3.4$ and their behaviour is in accord with the above distinction. They are the closest parametric representation of real column bases (with $S \simeq 5$) which could be conveniently handled in the laboratory.

1. THEORETICAL CONSIDERATIONS OF BENDING AND TORSION

The total spectrum of structural behaviour at the cut-out section will include early small-deformation stress distributions and concentrations, the non-linear growth of free edge displacements, the development of local plasticity, the redistribution of stresses due to these last two effects, the possibility of local instability at the free edge and eventual failure due to strength breakdown or overall instability at the cut-out section. Some of these aspects are examined.

1.1 Small-deformation elastic stresses at the cut-out section

The stress distributions at the cut-out section due to axial load, bending moment and shear are readily calculated. Of more interest are the stresses due to torsion and particularly the longitudinal warping stress, σ_{zw}, arising from the warping restraint provided by the complete cross-section at each end of the cut-out section (see Fig 3).

A torque T applied to the tube consists of two parts, T_1 and T_2, such that

$$T = T_1 + T_2 = C\frac{d\phi}{dz} - C_1\frac{d^3\phi}{dz^3} \qquad (1)$$

and gives rise to three distinct stresses τ_{T1}, τ_{T2} and σ_{zw} (defined in the Notation).

If it is assumed that $d\phi/dz = 0$ at $z = 0, \ell$ (Fig 3) and warping is prevented at these end cross-sections, it follows that

$$\tau_{T1} = \frac{GtT}{C}[1 - \{\cosh nz + c \sinh nz\}] \text{ where } n^2 = \frac{C}{C_1} \qquad (2)$$

$$\tau_{T2} = \frac{R^2T}{C_w}[\cosh nz + c \sinh nz][R\bar{\delta}(\theta + \frac{\bar{\delta}}{2}) + e\ (\cos\delta - \cos\alpha)],$$

$$\alpha \leqslant \delta \leqslant (2\pi - \alpha) \qquad (3)$$

$$\sigma_{zw} = -\frac{3ERT}{2\theta t^3\ GC_w}[R(\pi - \delta) - e \sin\delta][\sinh nz + c \cosh nz],$$

$$\alpha \leqslant \delta \leqslant (2\pi - \alpha) \qquad (4)$$

The greatest magnitude of σ_{zw} will occur at $z = 0$, $\delta = \alpha$, $(2\pi - \alpha)$ and at $z = \ell$, $\delta = \alpha$, $(2\pi - \alpha)$, see Fig 3. By making the approximation that, for small values of n, $c = -n\ell/2$, the maximum value of σ_{zw} reduces to

$$\sigma_{zw\,(z=0,\ \delta=\alpha)} = \frac{T\ell}{4\pi R^3 t} f(\theta) \qquad (5)$$

where $f(\theta)$ is defined in the Notation.

The importance of σ_{zw} is illustrated on Fig 3. A standard 10-20 column base has been subjected to a bending moment of 1 kNm in a plane to cause the maximum compressive bending stress (σ_{zb}) along the cut-out edge AA'. A torque of 0.25 kNm is also applied and, as can be seen, produces warping stresses σ_{zw} in excess of the maximum bending stress.

1.2 Overall torsional buckling at the cut-out section

This mode has been examined (5) under the actions of an applied moment M (with components M_x and M_y about the principal axes of the cut-out section) and an applied compressive axial load P. It was found that, for sections with geometric parameters in the range of lighting column bases, the loads required to cause elastic buckling of this kind are too high to be of any practical interest.

1.3 *Large deformation of the long, free edge under the action of M_y and P*

If it is assumed that a 'leaf', which extends through an angle 2ψ from the free edge (see Fig 8) has, along the edge, an initial shape of the form

$$\eta_{of} = \bar{\eta}_{of} (1 - \cos 2\pi z/\ell_e)/2 \qquad (6)$$

and that the maximum initial lack of straightness, $\bar{\eta}_{of}$, is magnified under load by the factor *m*, then it can be shown (1) that the solution for *m*, at any combination of M_y and *P*, lies in the equation

$$m^3 \alpha_1 \bar{\eta}_{of}^2 + m(\alpha_2 M_y + \alpha_3 P - \alpha_1 \bar{\eta}_{of}^2 - \alpha_4) + \alpha_4 = 0 \qquad (7)$$

where α_1, α_2, α_3, α_4 are functions of ψ and the properties of the tube.

Having found *m*, the longitudinal stress distribution at the mid-length $(z = \ell/2)$ cross-section where the maximum leaf-edge displacement $(\bar{\eta}_f)$ occurs is given (see Fig 4) by

$$\sigma_{zw} = E.\delta_o/\ell_e + R\rho E \sin w, \qquad w \leqslant \lambda \qquad (8)$$

$$\sigma_{zw} = E.\delta_o/\ell_e + R\rho E \sin w - \frac{\pi^2 E}{4\ell_e^2}(m^2 - 1)\{\frac{\sin \beta/2}{\sin \psi}\}^2 \bar{\eta}_{of}^2, \qquad w > \lambda \qquad (9)$$

where δ_o is the shortening of the leaf due to axial compression and bending and ρ is the mean bending curvature of the tube.

2. *THE EXPERIMENTAL BEHAVIOUR OF THE TUBES RELATED TO THEORY*

Figure 1 and Table 1 give details of the eight specimens tested in bending, torsion and combinations of both, using the rig shown in Fig 2, and two full-scale bases tested by Little and Baban (4) in bending. The rig (Fig 2) allowed the application of a constant bending moment M_y over the length of the specimen. Loading was by dead weights and spring balances which gave the advantage of controlled failure. Displacements of the tubes and the free edges were measured by dial gauges and considerable numbers of strain gauges (applied in pairs on opposite sides of the wall) were used in the vicinity of the openings.

2.1 *Tubes subjected to bending only (Tubes 2 and 6)*

The behaviour of the two specimens was similar, so attention is confined to Tube 2. Figure 4 shows the experimental radial outward displacement at the centre of the free edge with increasing M_y; a steady growth with no suggestion of snap-through or sudden increase. The longitudinal elastic stress (σ_z) profiles near the edge of the cut-out (deduced from strain readings), shown on Fig 5, are interesting. They demonstrate a reluctance for σ_z to rise, even to the values derived from simple bending theory, until the middle half of the cut-out length, where there is a sudden rise; a phenomenon noted also in tubes under direct axial compression in Ref (1). This pattern of behaviour is also shown by the mean elastic strain profiles on Fig 6 (lines 1, 2 and 3). Lines 4, 5 and 6 on Fig 6 indicate large distortions occurring only over this middle half of the cut-out edge, i.e. extending to approximately $\ell/4$ each side of the centre-line, giving an effective length for a cosine-type displacement of $\ell_e = \ell/2 = 140$ mm (see the left-hand free edge of Tube 2 on Plate 1).

The use of eqn (7), with $P = 0$, to analyse the behaviour of the cut-out section demands knowledge of 2ψ, $\bar{\eta}_{of}$ and ℓ_e. However, although the value of m (the magnification factor) is sensitive to these variables, the increase in displacement of the leaf i.e. $(m - 1)\bar{\eta}_{of} = (\bar{\eta}_f - \bar{\eta}_{of})$, is insensitive to both 2ψ and $\bar{\eta}_{of}$. This is demonstrated on Fig 4, where theoretical values of $\bar{\eta}_f$ are shown using $\bar{\eta}_{of} = 0.4$ mm; $\psi = 10^o, 20^o$ and $\bar{\eta}_{of} = 0.8$ mm; $\psi = 10^o, 20^o$.

The application of eqn (8) and (9) to Tube 2, using $\ell_e = 140$ mm, $\eta_{of} = 0.4$ mm (the measured value) and ψ equal to 10° and 20°, give the elastic stress (σ_z) distributions across the cut-out cross-section shown on Fig 7a and 7b respectively. The bending of the leaf (which subtends 2ψ) causes a loss of stress at the free edge, with a consequent movement of the peak stress away from the edge. The limits of the elastic solution (maximum $\sigma_z = \sigma_y$) occur at $M_y = 2.6$ kNm and 2.8 kNm for $\psi = 10^o$ and 20° respectively. Tube 2 failed at 3.0 kNm The true value of ψ in practice would be difficult to determine, depending no doubt on the initial imperfection of the tube and possibly changing with M_y. The analysis demonstrated on Fig 7 does, however, provide an insight into the physical behaviour of the tube and is referred to again below.

The full plastic value of M_y viz $M_{y,o}$ is given by

$$M_{y,o} = 4\sigma_y tR^2 \cos\frac{\alpha}{2}\ (1 - \sin\frac{\alpha}{2}) \qquad (10)$$

$M_{y,o}$ will always give a high estimate of the cut-out bending capacity because, apart from anything else, it takes no account of the low stress in the buckled leaf. A modified limit stress distribution, suggested by the stress patterns on Fig 7, is shown on Fig 8, where it is assumed that the free edge stress has reduced to zero and rises linearly with angular distance round the leaf, reaching σ_y at 2ψ from the edge. This limiting moment $M_{y,\ell}$ is defined by

$$M_{y,\ell} = \sigma_y tR^2 [\{\cos(2\psi + \alpha) - \cos\alpha\}/\psi + \{4\sin(\pi - \alpha - \psi)/2\}] \qquad (11)$$

$M_{y,\ell} \rightarrow M_{y,o}$ as $\psi \rightarrow 0$. Values of $M_{y,\ell}$ for $\psi = 20^o$ are listed in Table 2.

A sensible empirical expression for the maximum value of M_y (reflecting the interaction between strength and stability at failure) is suggested as

$$M_{y,max} = M_{y,o}\ [1/\{1 + A(\frac{R\ell}{t^2})(\frac{\sigma_y}{E})\}] \qquad (12)$$

where A is an experimentally determined constant. An appropriate value for A, from the limited evidence of Tubes 2, 6, ACM1 and ACM2, is 0.08 (see Table 2).

2.2 Tubes subjected to torsion only (Tubes 1 and 3)

The twisting of Tube 3 (over its entire length L), and the outward radial displacement of the highly compressed 'knuckle' point A, are shown on Fig 9. It is clear that the buckling displacement of A is a good indicator of the tube's torsional capacity.

Fig 10 shows the experimental longitudinal elastic stress (σ_{zw}) distribution, at $T = 0.4T_{exp}$, in a tensile quadrant of Tube 3 and compares the free edge stress pattern with that predicted by eqn (4). In spite of the high predicted stress at C (Fig 10) it underestimates the actual peak value. The general shapes of the stress profiles across the top of the cut-out (EF) and around the circumference of the tube (CD) are as expected (see Fig 3), with extremely rapid damping away from C. Subsequent measurements at $T > 0.4T_{exp}$ (not shown on Fig 10) demonstrated the continuing local nature of the high warping stress near the knuckle (at C). The phenomenon is further illustrated on Fig 11a and 11b which show the growth of σ_{zw} at the cross-sections LL and MM on Tube 1. (The solid lines on Fig 11a and 11b arise directly from elastic strain measurements and the dashed lines are deduced from equilibrium at the two sections i.e. $P = 0$, M_y = known applied moment). Quadrants 1 and 4 experience compression at the cut-out edge and the phenomenon noted in bending is evident again here viz as the edge buckles, it loses stress and the peak stress retreats. Extensive tensile yielding occurs at the free edge in quadrants 2 and 3 and spreads along LL and MM before the tube reaches T_{exp}. Comparison of the stress patterns at low values of T in quadrants 2 and 3 with the theoretical distribution deduced for the column base in Fig 3, show considerable similarity.

Based upon an ultimate σ_{zw} distribution as shown in Fig 11c, Horne (6) has calculated the plastic value of T as

$$T_o = \frac{R^3 t\sigma_y}{\ell}\left[\frac{10\cos^2\alpha/2}{1+\sqrt{3}\tan\alpha}\right]\left[\frac{1 + 2.15\tan\alpha + 0.85R/\ell}{1 + 2.15\tan\alpha + 0.85\frac{R}{\ell} + 3.8\left(\frac{R}{\ell}\right)^2}\right] \qquad (13)$$

The experimental patterns of Fig 11a and 11b are clearly in accord with an (idealised) final development to that of Fig 11c, the major difference being the inability of the buckled knuckle to sustain σ_y. As can be seen from Table 2, eqn (13) does give an optimistic estimate of the tube's torsional capacity.

On a basis similar to that of eqn (12), an empirical expression for the maximum torque (see Table 2) is suggested as

$$T_{max} = T_o\left[1/\left\{1 + 0.05\left(\frac{R\ell}{t^2}\right)\left(\frac{\sigma_y}{E}\right)\right\}\right] \qquad (14)$$

2.3 Tubes subjected to bending and torsion (Tubes 4,5,7 and 8)

When torsion is combined with M_y, the bending compression on the cut-out side of the tube is supplemented by warping compression in alternate quadrants of the cut-out edge and opposed by warping tension in the other two quadrants. The ratio M_y/T for Tubes 4,5,7 and 8 was approximately 1.75, 3.5, 8.8 and 17.6 respectively through most of the loading range. The precise values of M_y and T at failure in each case are given in Table 2.

On Tube 5, measurements of strain were taken at locations similar to those on Tube 3 (Fig 5). In spite of the presence of M_y, the warping tension remained the dominant stress at point C (see Fig 5). The decay of this tensile stress concentration along CD was not significantly more rapid when $M_y = 3.5T$ than when $M_y = 0$ (as on Fig 10). As expected the dominant influence on the ultimate performance of the combined loading cases was compressive buckling at the knuckle C (see Plate 1).

For design purposes, safe expressions are required for the capacity of the tubes in bending, torsion and combinations of both. Table 2 lists the experimental values of $M_{y,exp}$ and T_{exp} and compares them with $M_{y,o}$, $M_{y,max}$, T_o and T_{max}. Not surprisingly, $M_{y,o}$ and T_o, when each acts alone, always overestimate the capacity of the tube. $M_{y,\ell}$ and $M_{y,max}$ give very similar values for the Tubes 1 to 8 but, probably more significantly because they were full-size specimens, ACM1 and ACM2 are estimated more conservatively by $M_{y,\ell}$. T_{max} must be regarded as tentative because it is based upon very limited experimental evidence.

On Fig 12, the experimental results in Table 2 are compared with three simple interaction curves of the forms,

$$\frac{M_y}{M_{y,o}} + \frac{T}{T_o} = 1 \qquad (15)$$

$$\frac{M_y}{M_{y,\ell}} + \frac{T}{T_{max}} = 1 \qquad (16)$$

$$\frac{M_y}{M_{y,max}} + \frac{T}{T_{max}} = 1 \qquad (17)$$

Eqn (15) is clearly unsafe, but either eqn (16) or (17) would appear to give an acceptable estimate of interactive capacity (see Fig 12b and 12c). The values of $M_{y,\ell}$, $M_{y,max}$ and T_{max} are very simple to calculate and, in any particular case, it would be prudent to use the lower of either $M_{y,\ell}$ or $M_{y,max}$.

3. *CONCLUSIONS*

The experiments carried out on the bending and twisting of circular tubes with large asymmetric cut-outs have provided some insight into their structural behaviour and have led to suggestions for the simple estimation of capacity when subjected to such loading.

References

1. Montague, P. and Horne, M.R. 'The behaviour of circular tubes with large openings subjected to axial compression.' J.Mech.Eng.Sci., vol. 23 no. 5, 1981.

2. Almroth, B.O. and Holmes, A.M.C. 'Buckling of shells with cut-outs, experiment and analysis.' J. Solids Structures, 8, 1057-1071, 1972.

3. Fontaine, B. 'The behaviour of steel tubes with apertures and subjected to bending and torsion.' Acier (Belgium), 31, 347-350, 1966.

4. Little, G.H. and Baban, S.A. 'The bending strength of steel lighting columns.' Red. Report, Dept. of Civil Eng., University of Birmingham, 1978.

5. Horne, M.R. and Montague, P. 'Interim report on the theoretical study of lighting column door openings.' For Ministry of Transport, Simon Engineering Laboratories, University of Manchester, 1977.

6. Montague, P. and Horne, M.R. 'Third report on the study of lighting column door openings.' For Ministry of Transport, Simon Engineering Laboratories, University of Manchester, 1980.

Table 1 The tubes tested

Tube	R mm	α degrees	t mm	a mm	E kN/mm^2	σ_y N/mm^2	$\bar{\eta}_{of}$ mm	test*
1			2.01		195	266	0.48	T
2			2.05		190	215	0.40	M_y
3			2.0		195	220	0.61	T
4			2.03		203	260	0.50	$M_y + T$
5	58.1	37.04	2.04	35.0	200	205	0.35	$M_y + T$
6			2.01		196	272	0.57	M_y
7			2.00		198	281	0.40	$M_y + T$
8			2.01		210	280	0.51	$M_y + T$
ACM1+	107.5	38.55	5.45	67.0	206	390	not known	M_y
ACM2			5.26		208			M_y

*T = subjected to torsion; M_y = subjected to M_y

+full scale lighting column bases (ℓ = 466 mm). Ref (4)

Table 2 Experimental results compared with theory

Tube	Nm							$\frac{M_{y,exp}}{M_{y,o}}$	$\frac{M_{y,exp}}{M_{y,\ell}}$	$\frac{M_{y,exp}}{M_{y,max}}$	$\frac{T_{exp}}{T_o}$	$\frac{T_{exp}}{T_{max}}$
	$M_{y,exp}$	T_{exp}	$M_{y,o}$	$M_{y,\ell}$	$M_{y,max}$	T_o	T_{max}					
1	0	955	4671	3376	3245	1172	919	0	0	0	0.81	1.04
2	3000	0	3851	2783	2852	966	792	0.78	1.08	1.05	0	0
3	0	833	3863	2792	2833	969	790	0	0	0	0.86	1.05
4	1264	702	4611	3333	3283	1157	924	0.27	0.38	0.39	0.61	0.76
5	1682	543	3654	2640	2767	916	763	0.46	0.64	0.61	0.59	0.71
6	3303	0	4776	3452	3301	1198	936	0.69	0.96	1.00	0	0
7	2583	293	4910	3548	3359	1231	955	0.53	0.73	0.77	0.24	0.31
8	2884	166	4917	3553	3440	1233	972	0.59	0.81	0.84	0.13	0.17
ACM1	53300	0	62128	44644	49524	16535	14258	0.86	1.19	1.08	0	0
ACM2	48300	0	59886	43088	47095	15959	13643	0.81	1.12	1.03	0	0

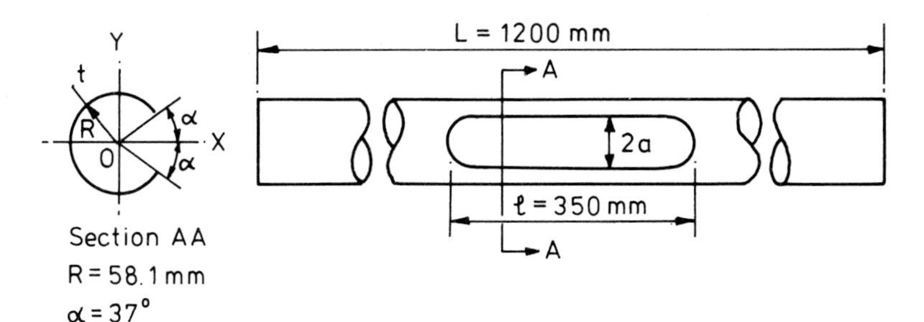

Fig 1 Common dimensions of Tubes 1-8 (Table 1)

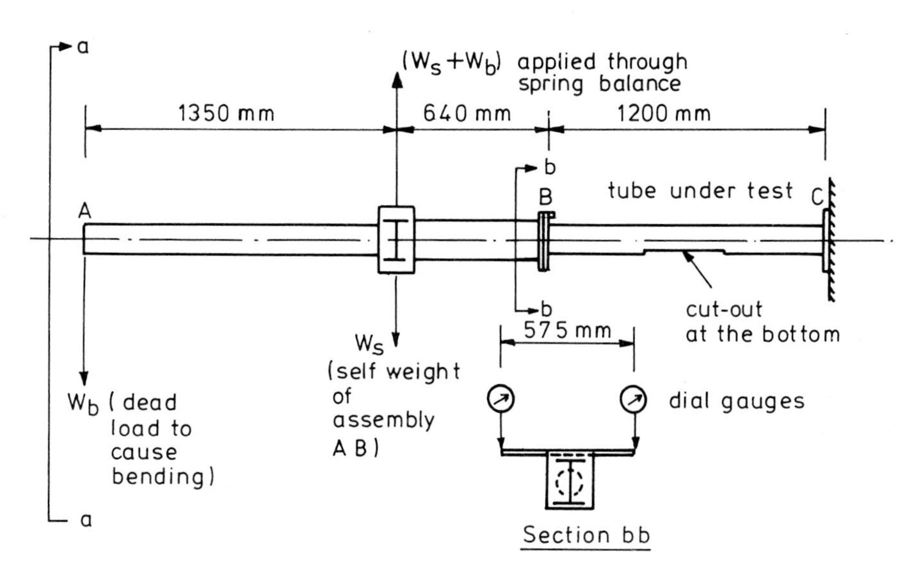

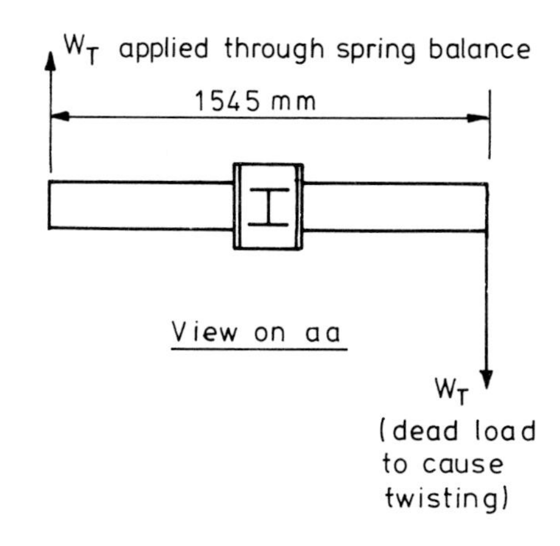

Fig 2 Simple rig for applying bending moment (constant along the length of the tube) and torque to the tube specimens

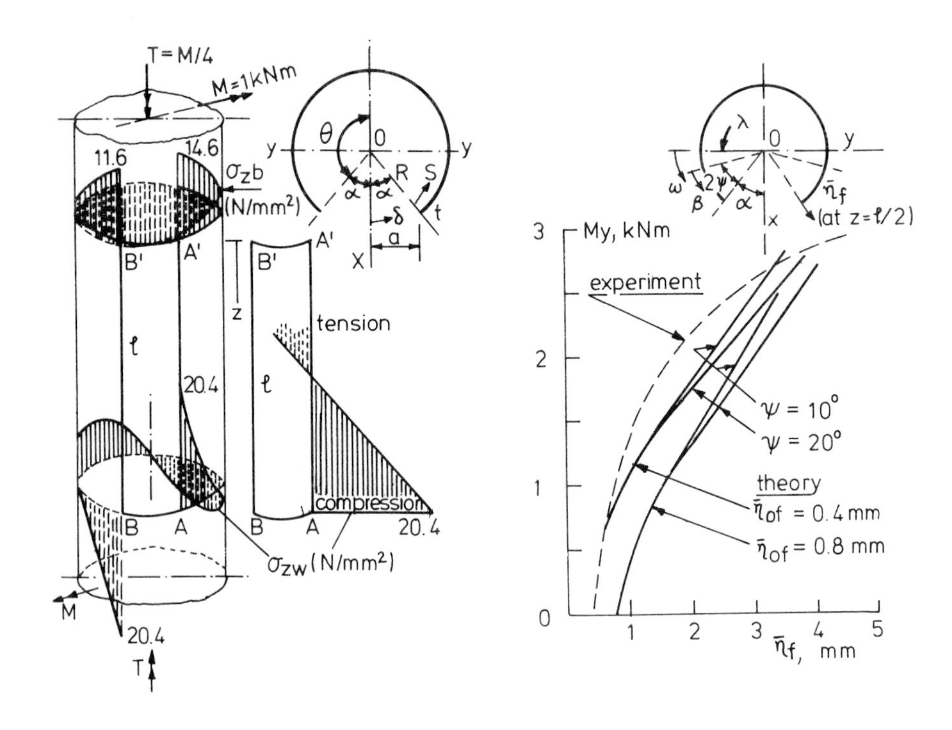

Fig 3 Edge warping stresses and bending stresses on a standard 10-20 base

Fig 4 Tube 2 with ℓ = 140 mm

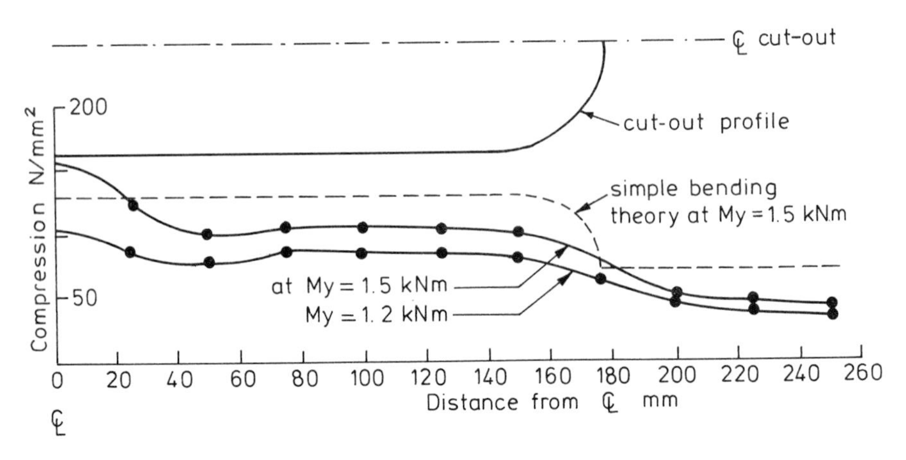

Fig 5 Tube 2 longitudinal elastic stress at 5 mm from cut-out

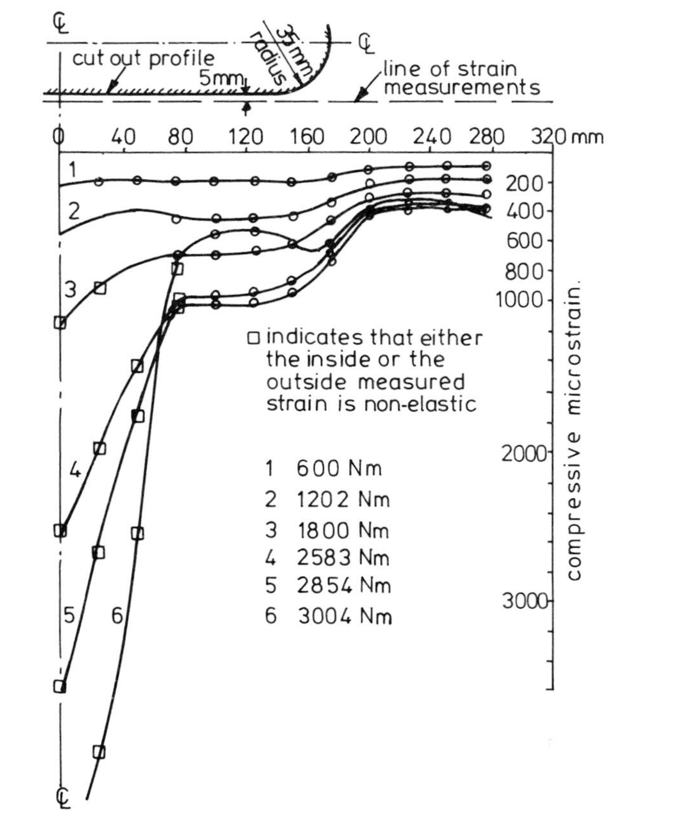

Fig 6 Tube 2. Longitudinal measured mean strains at 5 mm from cut-out

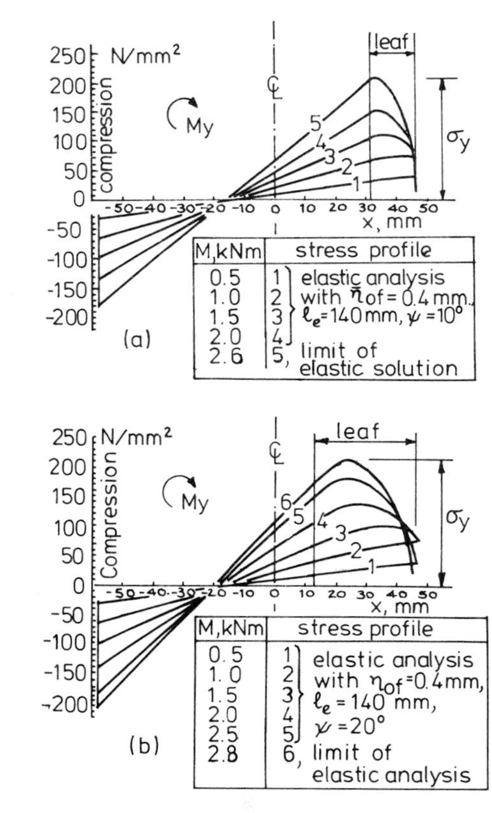

Fig 7 Theoretical longitudinal stress distributions across the central cross-section on Tube 2

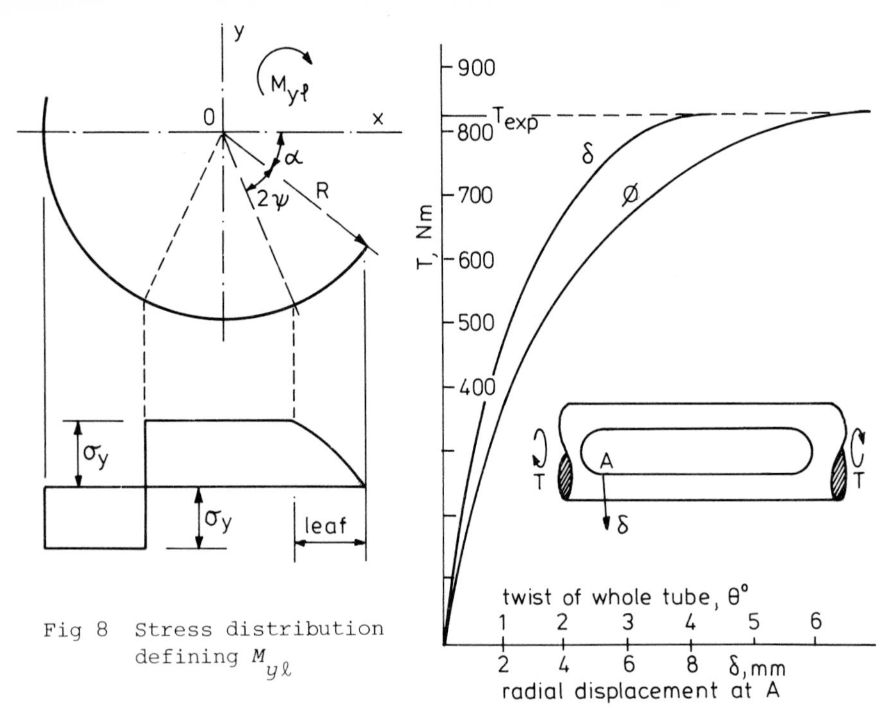

Fig 8 Stress distribution defining $M_{y\ell}$

Fig 9 Response of Tube 3 to torsion

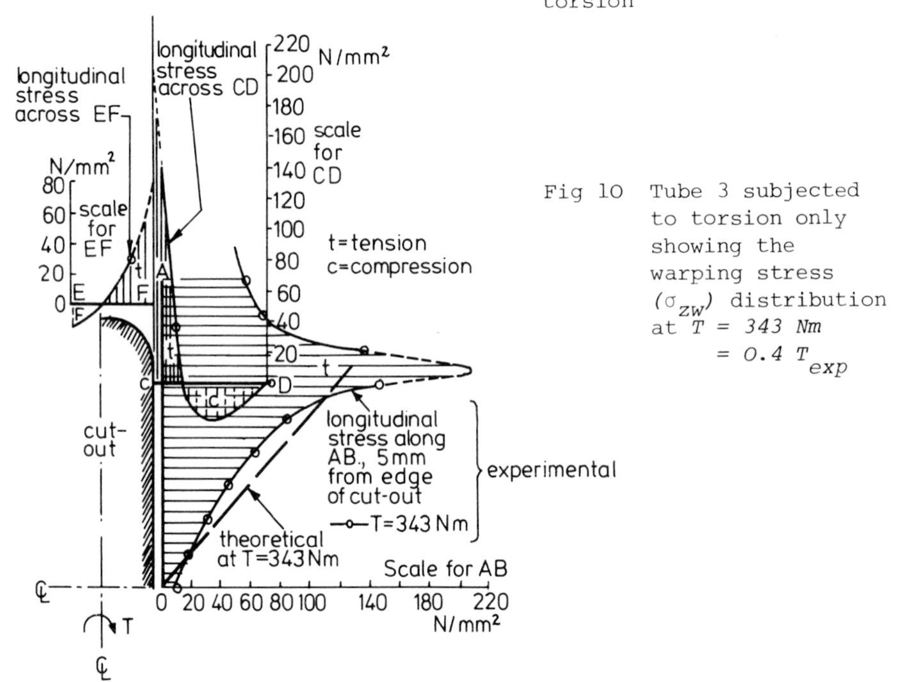

Fig 10 Tube 3 subjected to torsion only showing the warping stress (σ_{zw}) distribution at $T = 343$ Nm $= 0.4\ T_{exp}$

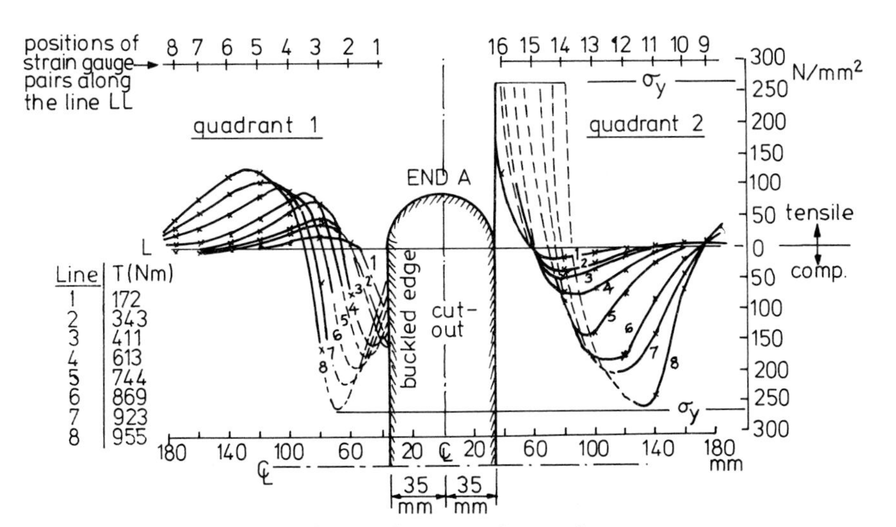

Fig 11a Tube 1: experimental σ_{zw} values along LL

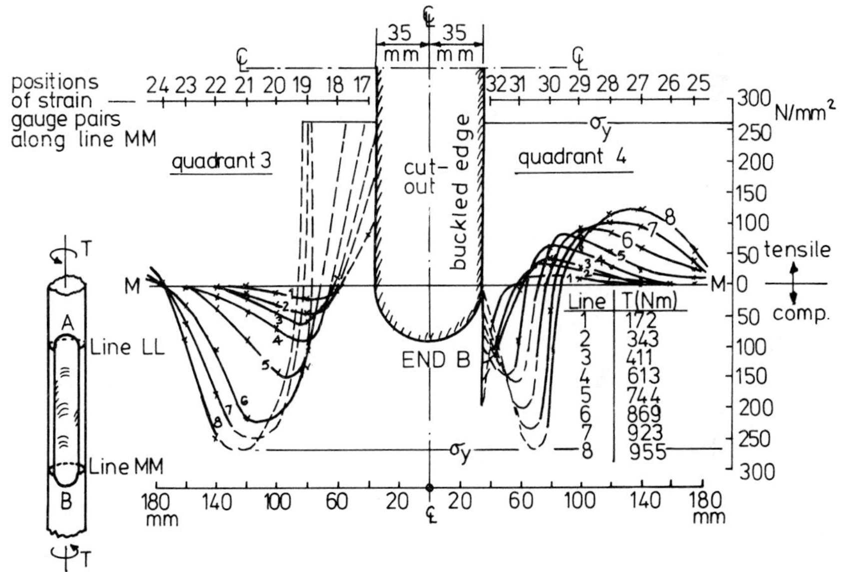

Fig 11b Tube 1: experimental σ_{zw} values along MM

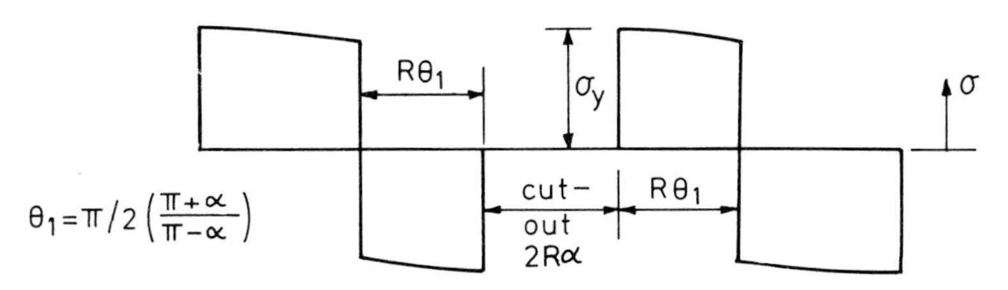

Fig 11c Ultimate stress distribution (σ) at T_O. Ref (6)

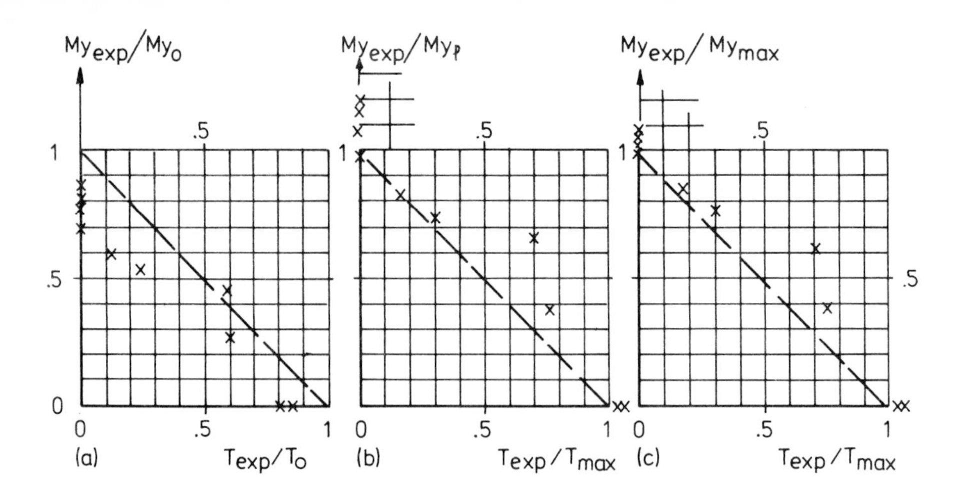

Fig 12 Interaction curves for M_y and T

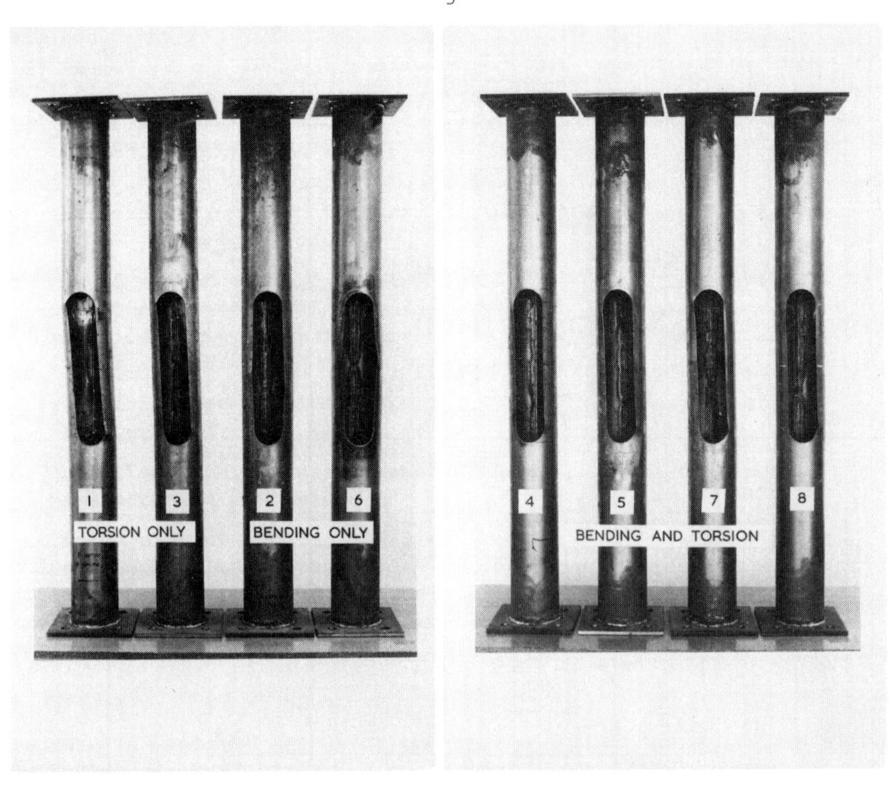

(a) Tubes 1 and 3 (torsion) and 2 and 6 (bending) after testing

(b) Tubes 4,5,7 and 8 (bending plus torsion) after testing

Plate 1

LOCAL BUCKLING IN RECTANGULAR STEEL HOLLOW SECTIONS COMPOSED OF PLATES OF VARYING b/t RATIOS.

M.J. Gardner, A. Stamenkovic

School of Civil Engineering, Kingston Polytechnic

Synopsis

Results of a theoretical investigation into the behaviour of short lengths of hollow steel section under axial compression are presented. The results have been obtained using a large displacement non-linear finite element program with a semi-loof shell element. The solution process involves a total Lagrangian formulation to account for geometric non linearity and a rigorous multi-layer procedure based on Von-Mises yield criterion to model material non linearity. Comparison is made, where possible, with experimental results, alternative theoretical analyses and relevant design proposals.

Notation

b	Plate width
b_f	Flat plate width of flange
b_w	Flat plate width of web
c	Width of full yield tension residual stress block
r_i, r_m, r_o	Inner, mean and outer corner radii of section
t	Plate thickness
δ_o	Initial imperfection in flange and web (+ve outward direction, -ve inward direction)
ε_y	Yield strain
σ_r	Value of mid plate compressive residual stress
σ_y	Uniaxial yield stress
$\bar{\sigma}$	Average stress in plate element

Introduction

The problem of overall buckling of pin ended columns is complicated by local buckling effects in those sections which have sufficiently high plate width/thickness *(b/t)* ratios for them to be unable to sustain full yield. Various attempts to develop design rules to allow for the interaction of local and overall buckling of columns have been made including those of Little (1) at Cambridge and Rondal (2) at the University of Liege. Both of those methods are based on an effective stress approach whereby a reduced yield stress is calculated for a short length of section which reflects the effects of the local buckling component of the problem. This is then used to obtain a column buckling curve for overall buckling whereby the whole section is assumed to be composed of material with this reduced yield stress. Perry Robertson type functions are used in both methods, the Cambridge approach being slightly more involved in so much that a composite of 2 curves is used (one of which governs at low slenderness ratios and reflects more directly, local buckling).

Little's method applies only to square sections and is theoretically based using single plate results obtained by Moxham (3). The Liege method is based on an extensive experimental program and the derived design proposals fit well these results; the resulting design method involves calculating a reduced yield stress for the section based upon the behaviour of the individual plate elements and involves no interaction between adjacent flange and web plates.

The work reported in this paper was initiated as part of a larger research program into the measurement and effects of residual stresses on column buckling behaviour of hot rolled hollow steel rectangular sections (4). The development of a large displacement non linear finite element program LUSAS (5) at Kingston Polytechnic provided an opportunity to examine local buckling in these sections and obtain information hitherto not available, viz:

(a) the stabilizing effect of web elements of varying b/t ratio on a buckling flange element.

(b) the effect of corner geometry on (a) including relatively large corner radii (typical of cold formed sections), smaller corner radii found in hot rolled sections or an abrupt right angle corner (as in welded built up box sections).

(c) geometric imperfections especially when these may be of a different nature in flange and web.

(d) the effect of residual stresses inherent in the different manufacturing processes.

The finite element program LUSAS has been used by Irving (5) to analyse single flat plates loaded by in-plane compression with various initial imperfections and residual stresses, the results agreeing excellently with corresponding solutions obtained by Little (6), Moxham (3), and Crisfield (7).

1. FINITE ELEMENT MODEL

Figures 1(a) and (b) represent the finite element mesh used in all of the analyses. This was based on mesh convergence data obtained by Irving (5). As it is purely local buckling which is being investigated symmetry requires only one eighth of the specimen to be considered. For the box section two 4 x 4 meshes have been used and for the RHS the corners have been divided into a 4 x 2 mesh as shown (i.e. a total of 32 or 40 elements). For the curved corners the radii used are such that maximum curvature conditions for the semi-loof element are satisfied. The results for single flat plates which are used for comparison on some of the succeeding figures are based on either a 4 x 4 or a 6 x 6 mesh depending on the residual stress distribution considered.

Along vertical and horizontal axes of symmetry movement across and rotation about the axis is restrained. Along the loaded top edge (a uniform displacement loading has been used to monitor post collapse behaviour) rotation is allowed. No external boundary constraints are imposed along the vertical junction between flange and web. For ease of data preparation the wall thickness, t, has normally been held constant at 5mm and plate widths varied to give differing b/t ratios. Moxham (3) has shown that the critical aspect ratio is a function of b/t ratio but for the b/t ratios under consideration here collapse loads are not significantly dependent on this. Obviously for differing plate widths the requirement of a fixed length of section leads to differing aspect ratios for flange and web plates. A constant total length of section of 200 mm has been adopted in the parametric study (Section 2) which corresponds to an aspect ratio of 0.889 for a plate width with b/t = 45. Plate imperfections have all been calculated on a double sine wave giving a maximum amplitude at the intersection of vertical and horizontal axes of symmetry. Imperfection and b/t values quoted have been based on the flat plate width. In the parametric study the uniaxial yield stress has been kept constant at 250 N/mm^2, with E = 210 000 N/mm^2. A total of 50 analyses have been carried out using a VAX 11/780 computer, each analysis using approximately 60 minutes of CPU time (generally for a total of 10 separate imposed deformation increments).

The residual stress values and distributions used in the analyses have been chosen to be representative of the sections under consideration. The right angle corner sections correspond to welded made-up box sections and a residual stress pattern relevant to this has been adopted.

The magnitude and distribution of residual stresses in hot rolled RHS have been the subject of extensive examination by the authors (4) and typical values for these types of section have been used in the analyses. These are very much less severe than welded sections with a more gradual variation away from the corners. Figures 2(a) and (b) show the residual stress values adopted for the two cases. Also plate slenderness values used for RHS reflect typical values for standard U.K. sections (a maximum b/t of 45).

2. *PARAMETRIC STUDY*

As mentioned in the Introduction the main object of the work reported here was to obtain information concerning the effect of residual stresses, varying b/t ratios and initial imperfections for flange and web elements, and the way in which the corner geometry influences interaction between flange and web plates. It was decided to undertake a parametric study in which each variable could be examined in turn for a typical range of sections. Space dictates that only a selection of results obtained can be reported here.

2.1. *Initial Imperfections in Flange and Web*

Figures 3(a) and (b) show for a RHS with corner and right angled box section respectively the effect of varying the nature of the initial geometric imperfection in flange and web on the load carrying capacity of the section. As can be seen there is a significant reduction in strength when the imperfections are of an 'unsympathetic' nature, i.e. outwards in the flange and inwards in the web (or vice versa). Figure 3 relates to two plates of b/t = 45 but this reduction is found in all the cases studied and represents the biggest single effect in the parametric study. The effective restraint offered by adjacent plates is such that when the imperfection is sympathetic (both plates initially deformed in the same direction) the section can sustain 98% of the squash load even with b/t = 45, with no unloading. The reduction in strength when an unsympathetic imperfection is considered is greater in the curved corner section than the box section - this can partly be explained by the fact that the overall section is marginally more slender as no attempt has been made here to calculate an "effective" flat plate width to include the curved corner, the b/t value being calculated in all cases on the flat plate width. All subsequent analyses are thus based on the lower bound unsympathetic imperfection values.

2.2. *Effect of Residual Stress Levels*

Figures 4(a) and (b) represent the effect on load-shortening curves of different residual stress levels together with the stress free curve for comparison. As expected the level of compressive residual stress affects the peak load attained. The peak load for the box section is some 3% higher than for the corresponding curved corner section probably for the same reason as stated in Section 2.1. A series of analyses for a box section subjected to full yield residual stresses with b/t values up to 80 have also been obtained but is not reported here.

2.3. *Flange - Web Interaction*

Figures 5(a) and (b) show the effect on load carrying capacity of a flange plate of fixed b/t supported by web plates of varying b/t ratio with the curve for a single simply supported plate of b/t = 45 for comparison. As can be seen, when the ratio of the supporting web exceeds approximately 30 for the RHS section and 40 for the box section the inward buckling of the web encourages increased outward buckling of the flange and reduces the load carrying capacity of the flange below that of the simply supported case. (This is confirmed by out of plane displacement results which are not reported here).

3. *COMPARISON WITH EXPERIMENTAL WORK*

Local buckling tests are being undertaken at Kingston Polytechnic using the most slender of the BSC range of hollow sections but no definitive results are available as yet. However, a major program of experimental work (on cold formed sections) has been completed at the University of Liege (2) and sufficient data has been provided for these results to be used as a comparison with the finite element analysis. A series of tests involving nine different sections were carried out at Liege and results for section geometry, material properties, and residual stresses (for two sections only), reported together with the load-deformation behaviour for local buckling tests.

For the two series of tests where detailed residual stresses were reported, analyses have been carried out using measured values for geometry, yield stresses and residual stresses. Local plate imperfections were not reported and a value of $b/1000$ has been assumed of the 'unsympathetic' type (outwards in the flange and inwards in the web).

Figure 6 presents the load-shortening curves for three analyses conducted on the two sections (the square section has been analysed using two different column lengths). The theoretical peak loads obtained are shown on Figures 7(a) and (b) together with experimental values obtained from Liege. As can be seen, two of the results agree extremely well with experimental values. The analysis related to the 500 mm long 265 x 265 x 4.04 section shows a higher collapse load (by some 16%) but this is almost certainly due to the theoretical analysis converging on a higher order buckling mode (each plate forming three half sine waves along its length instead of one) which in turn is probably due to too coarse a finite element mesh being used along the column length. Figures 8(a) and (b) show the deformed shape of the 265 x 265 x 4.04 section for the two lengths considered and the relevant buckling modes are clearly visible.

Figure 6 represents the behaviour of the overall section and as stated peak loads compare very well with experimental values. However when the theoretical behaviour of the individual plate elements is examined (Figure 6 curves 3(a),(b)) (it is not possible to monitor this experimentally), they differ from their predicted behaviour using the proposed Liege design method. If the 330 x 200 x 4.05 section is taken as an example, the finite element results show a far higher load-carrying capacity for the flange (0.775Psq), whereas the web (although stockier, the residual stresses are compressive) unloads rapidly from a peak load of 0.42 Psq. The corresponding values for the flange and web in the Liege design method are 0.344 Psq and 0.631 Psq respectively (where Psq = 'squash load').

4. *CONCLUSIONS*

A parametric study of hollow steel sections composed of plates of various b/t ratios connected by curved or right angle corners with various combinations of residual stresses and initial imperfections subjected to in-plane compression has been carried out and the following conclusions can be drawn.

(i) the major effect on load-carrying capacity is due to unsympathetic imperfections in the two adjacent plates. Thus buckling initiated in one direction in one of the plates 'encourages' buckling in the other direction in the adjacent plate. This interaction occurs in both types of corner geometry considered.

(ii) residual stresses of varying type and magnitude reduce load-carrying capacity by a similar order to that found by other researchers when considering single plate behaviour.

(iii) the support offered to a buckling flange element is dependent on the b/t value of the web plate and this support is more significant in a right angled corner than a curved corner. Assumption of a simply supported flange plate corresponds approximately, in load-carrying capacity, to a supporting web plate of b/t = 40 and 30 in a right angled corner or curved corner respectively.

Theoretical results agree excellently with experimental results available from Liege, but behaviour of individual plate elements does not correspond with that assumed in their proposed design method.

References

1. Little, G.H. 'The Strength of square steel box columns - design curves and their theoretical basis' The Structural Engineer, Vol.57A, No.2, Feb. 1979

2. Rondal, J., et alia. 'Flambement des profils creux a parois minces cas des profils Rectangulaires charges axialement', CIDECT 2H 79/A9.

3. Moxham, K., Bradfield, C. 'The Strength of welded steel plates under in-plane compression'. University of Cambridge Report CUED/C-Strut/TR.65, 1977.

4. Gardner, M., Stamenkovic, A. 'Effect of Residual Stresses on the Column Behaviour of Hot Formed Steel Structural Hollow Sections'. Summary Report to Tubes Division, BSC, Nov.1982.

5. Irving, D.J. 'Large Deformation Elasto-plastic Finite Element Analysis of Plates, Shells and Tubular Joints using Semi-loof Elements. PhD. Thesis, 1982. Kingston Polytechnic.

6. Little, G.H. 'Rapid Analysis of plate collapse by live-energy minimisation'. Int.J.Mech.Sci. Vol.19. 1977.

7. Crisfield, M.A. 'Full range analysis of steel plates and stiffened plating under uniaxial compression'. Proc.Instn.Civ. Eng., Pt.2, Vol.59, 1975.

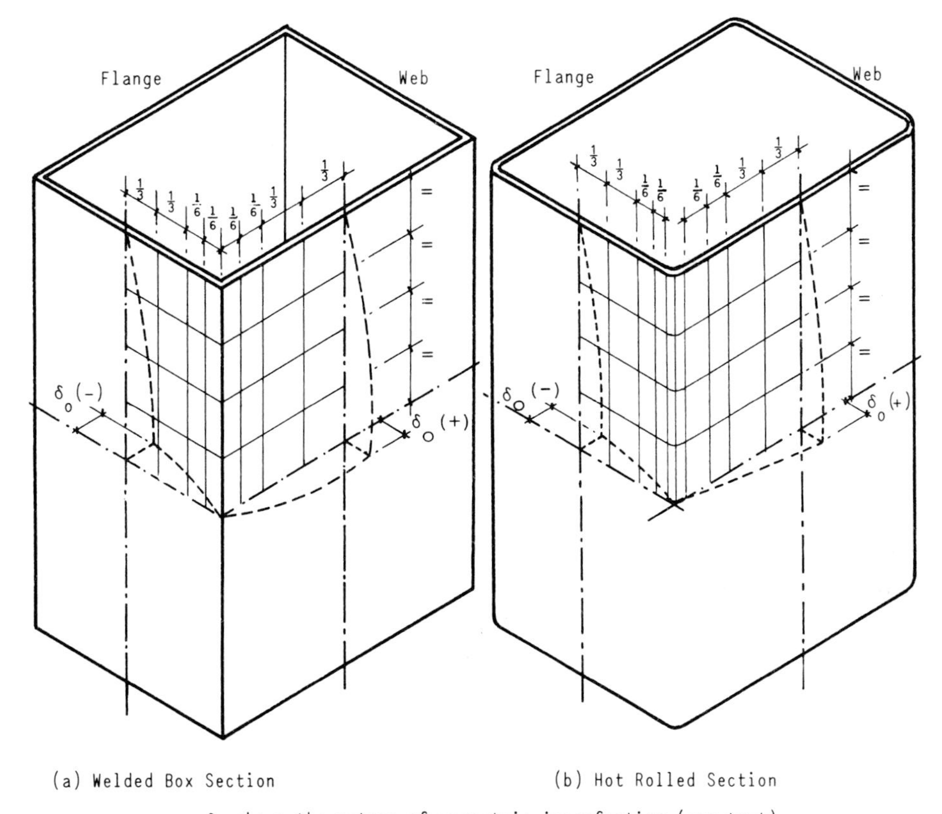

δ_o shows the nature of geometric imperfection (see text)

Fig 1 Finite element mesh used

σ_y (250 N/mm²)

t

σ_r

C

b/2

50 N/mm²

25 N/mm²

b/2

(a) Typical Welded 'Box' Section

(b) Typical Hot Rolled Section

Fig 2 Longitudinal residual stress patterns used in the analysis

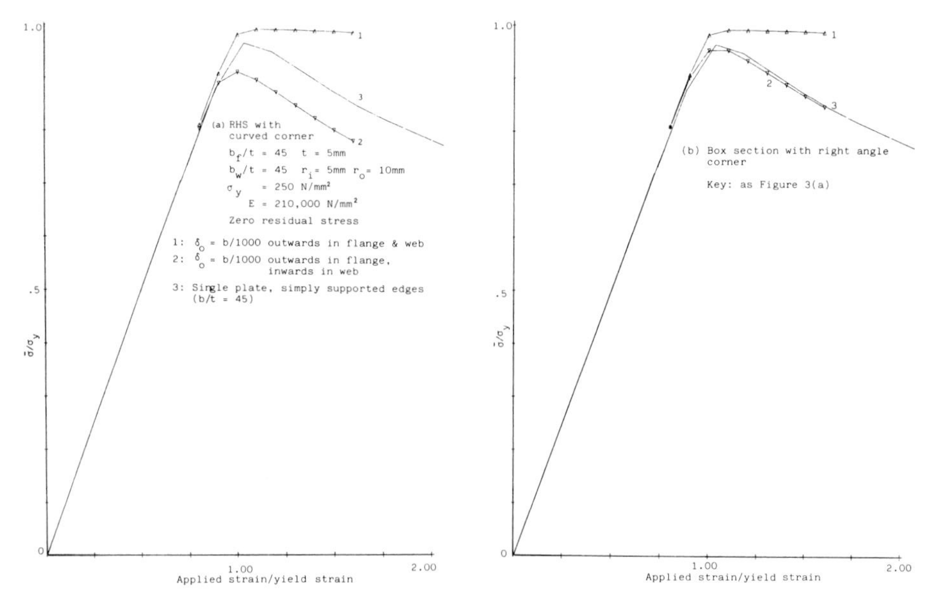

Fig 3 Effect of nature of initial imperfection on load carrying capacity

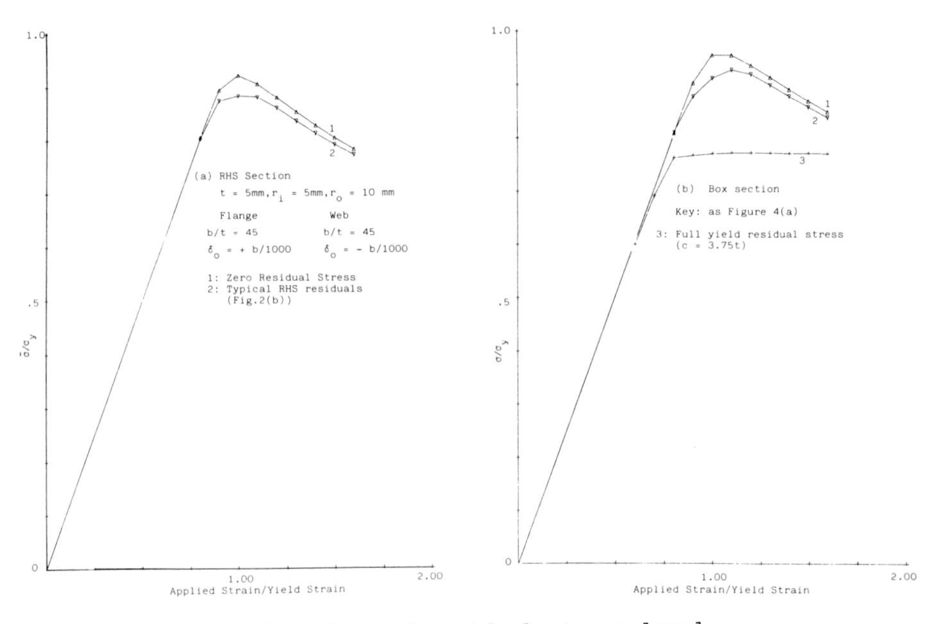

Fig 4 Effect of residual stress level

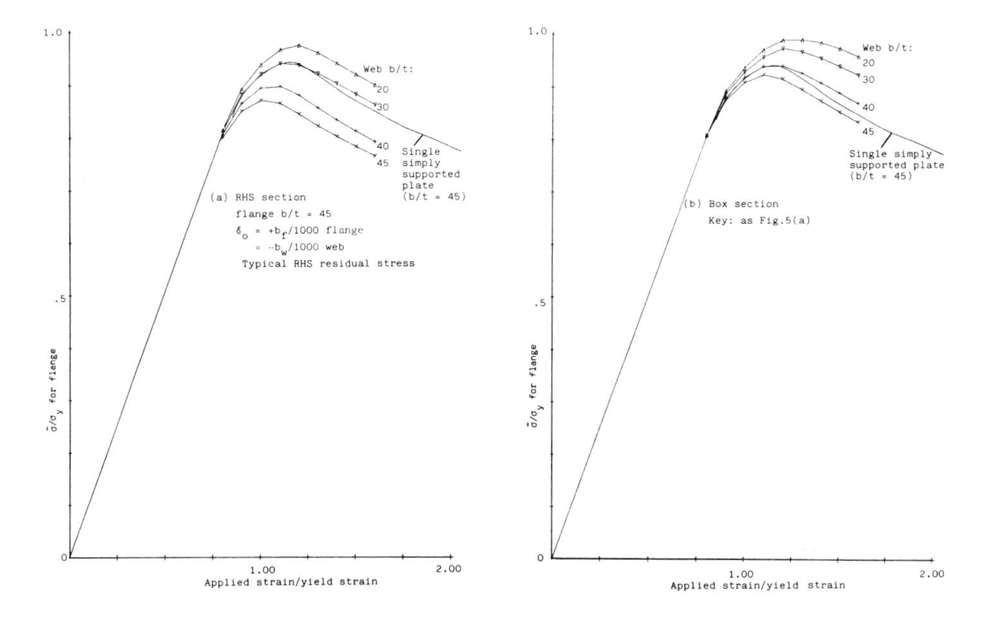

Fig 5 Flange-web interaction

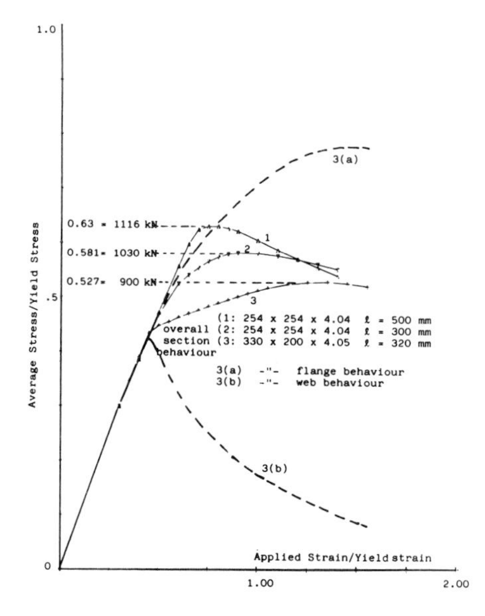

Fig 6 Load/shortening curves for two Liege cold formed sections

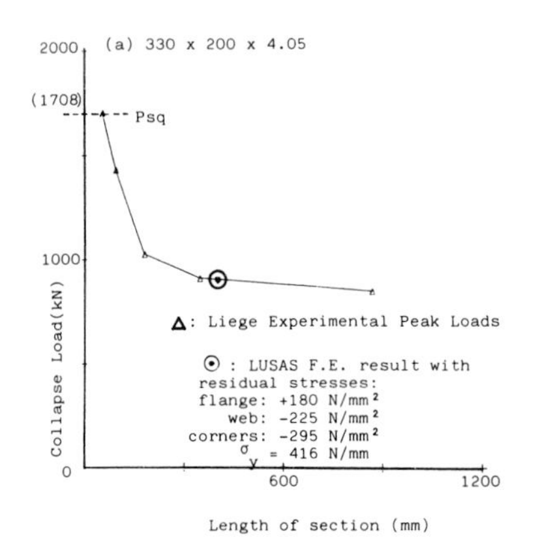

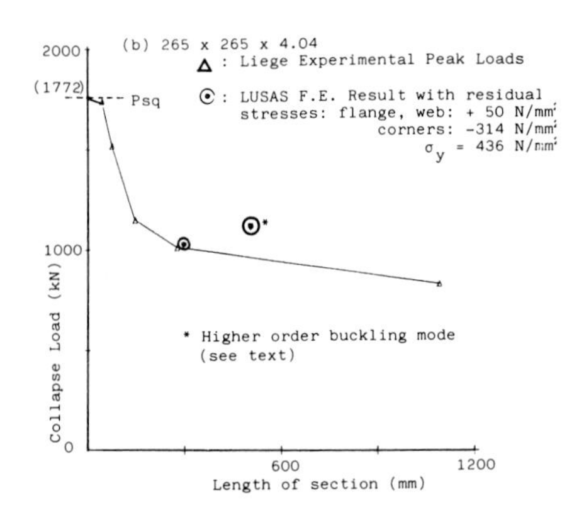

Fig 7 Peak loads obtained in local buckling experiments

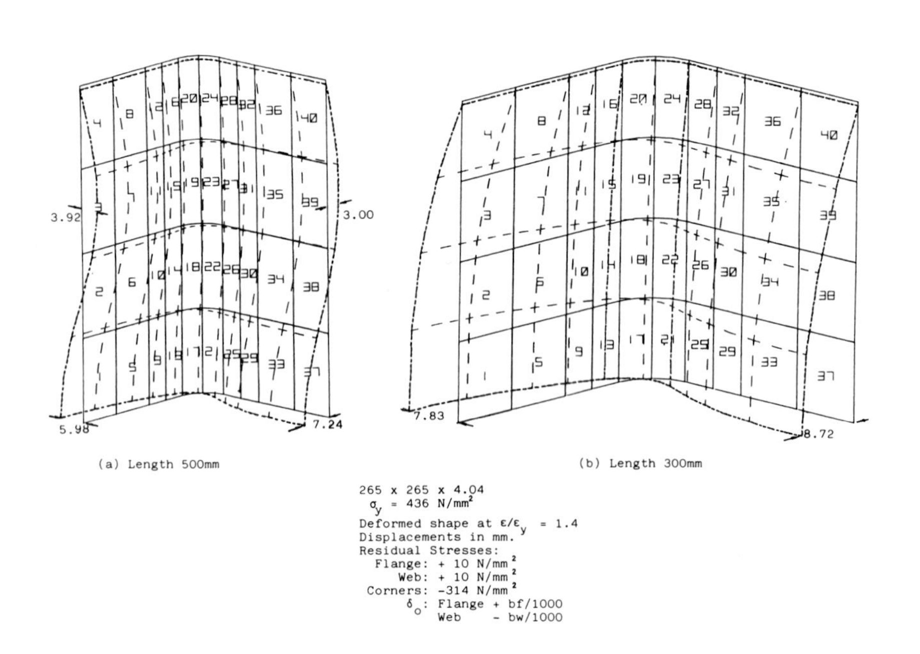

Fig 8 Deformed shapes for 265 x 265 x 4.04 Liege specimen

BEARING AND BUCKLING STRENGTH OF RECTANGULAR HOLLOW SECTION CROSS JOINTS

J.A. Packer

Department of Civil Engineering, University of Toronto, Canada

G. Davies, M.G. Coutie

Department of Civil Engineering, University of Nottingham, U.K.

Synopsis

Joints between Rectangular Hollow Sections (RHS) may fail by web crippling. This paper reviews the various methods which have been proposed for estimating the web crippling strength of such joints when subjected to localised transverse compressive loads. This mode of failure has been studied extensively for I beams over a long period of time and its significance is recognised in most structural design codes, with the same rules often being applied to the RHS situation too. Although design recommendations for the strength of some RHS joints often automatically include for the effect of web bearing and buckling, amongst other factors, there will often be cases where it is necessary to check for web bearing and web buckling resistance separately in RHS connections. These failure modes can be observed in full-width welded connections to RHS chord members, where the branches (in the form of plate, square, or rectangular hollow sections) are loaded in axial compression.

A comparison is made of the various methods of assessing the web crippling strength against test data collected from Great Britain, Canada and the Netherlands. The parameters covered include chord wall slenderness, chord aspect ratio, forming process (hot rolled or cold rolled stress relieved), yield stress, the inclination of the branch member to the chord and chord axial load. Test results are correlated with each of the methods of strength estimation and a suitable approach for use with limit states design is discussed.

Notation

A_o	cross-sectional area of the chord member
b_o, b_1	width of the chord and branch member respectively (90° to plane of joint)
E	Modulus of elasticity (taken here as 210,000 N/mm^2)
h_o, h_1	depth of the chord and branch member respectively (in plane of joint)
N_{1u}	compression force in branch member at joint ultimate strength

t_0, t_1	wall thickness of the chord and branch member respectively
β	width ratio between branch and chord member = b_1/b_0
θ_1	acute angle between the branch and chord member
λ	slenderness of the chord side walls (webs)
σ_{eo}	yield stress of the chord member
σ_k	buckling stress of the chord side walls (webs)
σ_0	axial compressive stress in the chord member

Introduction

A considerable number of studies have now been undertaken on RHS T and Cross joints, principally in Canada, Europe and Japan, to determine the joint strength, stiffness and rotation capacity with branch members welded to a single RHS chord member. These investigations have been both theoretical and experimental, but are generally restricted to joints for which the width ratio between branch and chord members (β) is less than 1.0, and/or the branches are subjected to moment loading (1,2,3,4,5,6,7,8,9,10,11,12,13).

It is well known that for T or Cross joints, with the branches in axial compression, an increase in joint strength and stiffness is obtained with increasing β values, (in the low to medium β range). For low β values joint failure occurs by flexural yielding of the chord connecting face, or a combination of the chord connecting face and side walls. For such joint failures, yield line models have been found to give a good prediction of the limit state of excessive joint local deformations, if a limit of about 1% of the chord width is used for the allowable vertical deflection of the chord face at the service load level (3,14). The yield line model also generally gives a reasonable estimate of the joint ultimate strength, but becomes increasingly conservative at low β values and for high chord slenderness ratios (b_0/t_0). Nevertheless, the local deflection limit state typically governs the joint design at low to medium β values and so yield line theory is always applicable in this range (14). At large β values the mode of joint failure changes from a predominantly chord flange failure to a chord web crippling failure and the application of yield line theory leads to unsafe estimates of the joint ultimate strength (8,11,12,14). Brodka and Szlendak (12) considered yield line failure models for the chord flange to be inapplicable at width ratios (β) greater than 0.8; Kato and Nishiyama (11) at width ratios greater than 0.83; and Davies and Coutie (8) and Wardenier (14) at width ratios greater than 0.85. Although research has generally focussed on Cross joints having width ratios less than 1.0, improved joint performance is obtained with full width branch members. Therefore this important structural connection is considered further in this paper, with a view to establishing a preferred design approach.

1. *REVIEW OF FULL-WIDTH CROSS JOINT RESEARCH*

The most extensive research on this topic has taken place in Poland where over 400 cross joints of varying β have been tested, with the RHS members formed by welding two cold formed channel sections together at the toes - a manufacturing technique apparently typical in Poland (12). Czechowski and Brodka (4) developed a semi-empirical ultimate strength equation for cross joints with $\beta=1$, where the maximum load in the compression branch (N_{1u}) is given by:

$$N_{1u} = [1.06 - 0.021 (h_o/t_o)] \sigma_{eo} A_o \qquad (1)$$

This equation was developed for orthogonal joints but might be modified for inclined branch members to the form:

$$N_{1u} = [1.06 - 0.021 (h_o/t_o)] \sigma_{eo} A_o/Sin\theta_1 \qquad (1a)$$

Further ultimate strength equations were derived by Czechowski and Brodka for less than full width Cross joints and these also included a joint strength reduction factor to take account of the amount of axial compression force in the chord member. Czechowski et al (7) extended the study with an analysis based on a frame analogy to determine the ultimate joint strength for a specified β, and then Brodka and Szlendak (12) developed this into a simplified method applicable to the continuous range of β from 0.3 to 1.0. The resulting joint ultimate strength expression was based upon an effective length for the chord wall of 0.7 h_o but is still far too complex for application to design. Nevertheless, Kato and Nishiyama (11) applied Brodka and Szlendak's predictor equation to their cold formed T joint test results having $\beta > 0.80$ and found that Brodka and Szlendak's prediction erred excessively on the unsafe side in all instances. Kato and Nishiyama attribute this serious discrepancy to the fact that the secondary moment caused by axial force and deflection of the chord web was not considered in the strength evaluation of the chord web. Consequently Brodka and Szlendak's method is not considered further.

The experimental work by Kato and Nishiyama (11) consisted of 50 T joint tests on either cold formed or cold formed stress relieved tubes with β values between 0.28 and 0.89. For T or Cross joints having width ratios greater than 0.83, Kato and Nishiyama simulated the loading on the chord webs by using the analogy of an eccentrically loaded column. Whilst this yields promising results it is not in a form suitable for design.

In Great Britain a number of Cross joint tests have been carried out by Davies and Coutie (8). They observed that a compressive chord axial load reduced the joint strength when $\beta < 1.0$, but had no effect upon the ultimate strength of the Cross joint at $\beta = 1.0$, which is also supported by Wardenier (14) and implicit in the strength equations proposed by Czechowski and Brodka (4).

In the Netherlands, as part of a vast investigation into tubular joints, Wardenier et al (14,5) have undertaken both experimental and theoretical research on RHS T and Cross joints as well as plate to RHS connections, with the branches loaded both in compression and tension. The outcome of this research programme has been implemented in CIDECT* Monograph No. 6 on tubular connections (15) after discussion within CIDECT and the International Institute of Welding (IIW). For Cross joints having a width ratio $\beta = 1.0$ the strength of the joint is determined by the strength of the chord webs, which can experience web crippling in the form of either a bearing or buckling failure mode. For bearing failure of the chord webs a dispersion of the

*Comité International pour le Développement et l'Etude de la Construction Tubulaire.

bearing force at 22° (2.5:1) through the tube wall is assumed, similar to beam-to-column moment connections, and the bearing strength is hence given by:

$$N_{1u} = 2\,\sigma_{eo}\,t_o\,[h_1/Sin\theta_1 + 5\,t_o]/Sin\theta_1 \quad (2)$$

Dutch tests have indicated that bearing rather than buckling of the chord webs is the critical failure mode for h_o/t_o values less than 20-25. For slenderness values greater than this limit, web buckling could be predicted in a similar manner to eqn (2) by the following:

$$N_{1u} = 2\,\sigma_k\,t_o\,[h_1/Sin\theta_1 + 5\,t_o]/Sin\theta_1 \quad (3)$$

where σ_k is based on the ECCS buckling curve "a" (Fig 1). Thus web bearing and buckling could be conveniently combined into one composite equation, provided that one assumed the effective width of the chord web for buckling was the same as the bearing length, which is probably very conservative. The buckling stress σ_k in eqn (3) is obtained from Fig 1 with a column slenderness (λ) of 3.46 $(h_o - 2\,t_o)/t_o$. This assumes a pin-ended strut of length $(h_o - 2\,t_o)$ and a radius of gyration (r) of $t_o/3.46$ (I = effective column width x $t_o^3/12$ = effective column width x t_o x r^2).

Equation (3) has been checked against Dutch and Polish Cross joint tests, all of which were orthogonal, and has been found to be a lower bound (14). Davies et al (16) have shown theoretically that the elastic buckling strength of RHS chord webs in Cross joints increases with increasing inclination of the load, with a conservative estimate being given by:

$$N_{1u(\theta_1 < 90°)} = N_{1u(\theta_1 = 90°)}/Sin\theta_1 \quad (4)$$

or $N_{1u}Sin\theta_1$ is approximately constant. Since for elastic buckling the critical buckling stress is a function of $1/\lambda^2$, Wardenier (14) amended the chord wall slenderness (λ) for inclined branch members to:

$$\lambda = 3.46\,(h_o - 2\,t_o)/(t_o \cdot \sqrt{Sin\theta_1}) \quad (5)$$

Thus, to summarize, Wardenier (14) advocates that the web bearing or buckling strength can be determined by the lower bound eqn (3), with σ_k obtained from Fig 1 using λ specified by eqn (5). In addition, Wardenier applies a further partial safety factor of 0.8 to eqn (3) to obtain a limit states design strength for the chord webs (14). One disadvantage of Wardenier's lower bound approach is that it yields values of $\sigma_k < \sigma_{eo}$ for chord wall slenderness values $h_o/t_o \geq 10$, although web buckling does not govern until at least $h_o/t_o = 20$ is exceeded. A re-examination of Wardenier's orthogonal cross joint data (Fig 5.15, ref.14) indicates that a simplification to the above approach could be obtained if one used an empirical fit:

$$\sigma_k = \sigma_{eo} \quad for\ h_o/t_o \leq 15$$
$$\sigma_k = \sigma_{eo}\,[1 - (h_o/t_o - 15)/62] \quad for\ h_o/t_o > 15 \quad (6)$$

In determining the strength of a Cross joint by any of the above methods it is assumed that joint failure is governed by failure of the chord webs and that premature local buckling of the compression branch does not occur. Wardenier's recommendations are also subject to a restricted range of parameter validity for the chord member:

$$b_o/t_o \text{ and } h_o/t_o \leq 35; \quad 0.5 \leq h_o/b_o \leq 2.0$$

Since it has been found that T joints have very similar web crippling loads to their Cross joint counterparts, expressions for the ultimate strength of full-width T joints could also be considered. Packer and Haleem (17) proposed a means of calculating the mean ultimate strength of T joints, which was later modified again (18). Modifying the effective width term slightly the ultimate strength of a full-width T or Cross joint could be determined by:

$$N_{1u} = \sigma_{eo} b_o^{0.3} t_o^{1.7} [3.8 + 10.75(\frac{b_1 + h_1}{2 b_o})^2]/Sin\theta_1 \qquad (7)$$

Giddings (19) has also proposed a method for determining the ultimate strength of a Cross joint by:

$$N_{1u} = 2.3 \sigma_k t_o [b_e] (b_o/h_o)^{0.5}/Sin\theta_1 \qquad (8)$$

where $b_e = h_o$ if $h_1/Sin\theta_1 < h_o$ or $b_e = h_1/Sin\theta_1$ if $h_1/Sin\theta_1 > h_o$. This is a variation upon Wardenier's eqn (3) with λ calculated from eqn (5).

2. *GENERAL WEB BEARING/BUCKLING CONSIDERATIONS*

Web bearing and buckling in steel I sections have been studied by numerous researchers for as long as 70 years (e.g. 20,21,22) and have received further attention recently (23) as structural codes endeavour to take account of different effective web heights, for web buckling, depending upon the type of flange restraint provided, (e.g. British Standards new Limit States Steelwork Code provisions and proposed revisions to the American Institute of Steel Construction code). The treatment of web bearing failure in I sections has changed very little with some codes having slightly different dispersion angles of the load to the end of the root fillet radius. Equation (2) is a direct application of this principle, and the Australian Institute of Steel Construction (24) has implemented this approach for RHS members with a dispersion angle of 30°. For inclined branch members this would lead to the equation:

$$N_{1u} = 2 \sigma_{eo} t_o (h_1 + 2\sqrt{3}\, t_o)/Sin\theta_1 \qquad (9)$$

Web buckling has traditionally been based, for I section webs, upon the buckling of a strut whose width was the load dispersion width at midheight of the web, using a dispersion angle of 45°. The Australian Institute of Steel Construction (24) again implemented this approach for the web buckling of RHS members. For inclined branch members this would lead to the equation:

$$N_{1u} = 2 \sigma_k t_o (h_1 + h_o)/Sin\theta_1 \qquad (10)$$

where σ_k is based upon a column slenderness (λ) of 1.73 $(h_o - 2\, t_o)/t_o$. This assumes a strut with fixed ends.

An interesting amendment has been proposed to the American Institute of Steel Construction Code by Yura (University of Texas, 1982) for I section web bearing failure, which when applied to RHS members, becomes:

$$N_{1u} = 2\,\sigma_{eo}\,t_o(\sqrt{h_1} + 5\,t_o) \qquad (11)$$

By using the square root of the bearing length the equation is dimensionally unbalanced and so h_1 and t_o must be in inches, σ_{eo} in k.s.i. and N_{1u} is given in kips. Adaptation of this equation to allow for SI units and inclined branch members leads to the form:

$$N_{1u} = 10\,\sigma_{eo}\,t_o\,(\sqrt{h_1} + t_o)/Sin\theta_1 \qquad (11a)$$

where t_o and h_1 are taken in mm, σ_{eo} in N/mm^2 and N_{1u} in Newtons.

Another proposed revision to the A.I.S.C. code by Yura, based upon Swedish work by Bergfelt for I section web buckling failure, when applied to RHS members becomes:

$$N_{1u} = 2560\,t_o^{\,2} \qquad (12)$$

This is again dimensionally unbalanced and t_o must be expressed in inches, with N_{1u} given in kips. Adaptation of this equation to allow for SI units and inclined branch members leads to the form:

$$N_{1u} = 17.65\,t_o^{\,2}/Sin\theta_1 \qquad (12a)$$

where t_o is taken in mm and N_{1u} in kiloNewtons.

All these web bearing or buckling prediction methods reviewed assume that the connection is an "interior" type and that load dispersion takes place to the maximum extent on either side of the connection within the RHS chord.

3. *COMPARISON WITH TEST DATA*

Detailed test data for 58 RHS to RHS and Plate to RHS Cross joints has been collected by the authors from work carried out by:

Barentse, Delft University of Technology (5)	:	19 tests
Davies, Platt, Peksa and Bettison, Univ. of Nottingham	:	21 tests
Dixon and Erdosy, University of Toronto	:	18 tests

A typical joint is shown in Fig 2. The tests covered both hot formed and cold formed stress relieved RHS members. Tests having failure other than by web bearing or buckling of the RHS chord have been excluded. Unfortunately it has not been possible to include the Polish results already referred to, as the individual test details have not been available. Additionally, the specimens were of a type not used outside Poland.

Correlations between the 58 test results and the 6 methods of Cross joint strength prediction discussed earlier are shown in Figs 5 to 11. In computing the joint strength, the dimension h_1 has been taken as the basic branch width and no addition has been made for the weld size. The influence of compressive chord force on ultimate joint strength is shown in Fig 3, for both plate and RHS branch members. A small reduction in joint strength is generally experienced, with the maximum observed being 20%. There is no clear behavioural trend evident with increasing chord force, even though chord forces up to 85% of the squash value have been used. The scatter of test results is no greater than that experienced in the other tests without chord force and it is suggested that no specific chord force parameter need be included in a general joint strength equation. (This graph includes 5 additional results by Bettison for which partial information is available). Fig 4, showing the influence of bracing angle on ultimate joint strength, indicates an increase in strength as angle reduces. However, the scatter is considerable and the strength increase is generally less than that associated with $1/Sin\theta_1$. An influence function such as $1/\sqrt{Sin\theta_1}$ rather than $1/Sin\theta_1$ might better indicate the influence of branch angle, but as will be seen later, the scatter of results does not suggest that such sophistication will be warranted.

Figure 5 (Czechowski and Brodka) shows a large scatter indicated by the high Coefficient of Variation (C.O.V.), giving a prediction that is extremely unsafe for plate to RHS joints yet very conservative for joints of large A_o. Although Wardenier's equation (Fig 6) provides a good lower bound the scatter, like Fig 5, is still very large. This equation is very conservative for plate to RHS joints leading to a mean greater than 2.0. Like eqn (3), eqn (7) due to Packer (Fig 7) provides a lower bound. However, the scatter here is much smaller with all results lying between 1.0 and 1.8, the C.O.V. being only 0.14. Giddings' equation, (Fig 8), provides a good estimate of mean joint strength but is unsafe for plate to RHS joints. The C.O.V. is large at 0.40. The Australian Institute of Steel Construction procedure (Fig 9) works well for RHS to RHS joints, but is very conservative for plate to RHS joints (results above 1.6). All failures are predicted to be by bearing. Yura's proposals are seen in Fig 10 to provide a lower bound and a small C.O.V. of 0.13. Again all failures are predicted to be by bearing. It might be thought that a web slenderness ratio corresponding to pinned ends, rather than fixed ends, might be more appropriate for the buckling of RHS webs, although the fixed end assumption has traditionally been used for I beams. In Fig 11 the results of Fig 9 are re-presented based on the pinned end assumption. This change seems to produce an improvement but it is only slight.

4. *CONCLUSIONS*

The two most satisfactory methods at present available for the estimation of the ultimate strength of RHS to RHS or plate to RHS Cross joints are those due to Packer (eqn (7), Fig 7) and to Yura (eqn (11a), Fig 10). This is valid for both hot formed or cold formed stress relieved tubes, and these ultimate strength equations can be easily adapted for use with limit states design. As Yura's eqn (11a) is dimensionally unbalanced, eqn (7) is perhaps to be preferred. This has the additional merit of being similar in form to the RHS to RHS gapped K joint strength expression used by CIDECT (15). Although an angle function of $1/Sin\theta_1$ is not ideal, further refinement is unwarranted at this stage for the range $90° \geq \theta_1 \geq 45°$. Similarly, no joint strength reduction need be made for chord forces. These results may be applied to the chord slenderness range covered by the tests, namely up to $h_o/t_o = 42$, and also to chord aspect ratios in the range $0.50 \leq h_o/b_o \leq 1.50$.

5. *ACKNOWLEDGEMENTS*

The authors are particularly indebted to the North Atlantic Treaty Organization for support of this work, and to Messrs. Barentse, Platt, Peksa, Bettison, Dixon and Erdosy.

References

1. Redwood, R.G. 'The behaviour of joints between rectangular hollow structural members.' Civil Engineering and Public Works Review, Vol. 60, No. 711, pp. 1463-1469, 1965.

2. Jubb, J.E. and Redwood, R.G. 'Design of joints to box sections.' Institution of Structural Engineers, Conference on Industrial Building and the Structural Engineers, May 1966.

3. Mouty, J. 'Calcul des charges ultimes des assemblages soudés de profils creux carrés et rectangulaires.' Construction Métallique, No. 2, Paris, pp. 37-58, June 1976.

4. Czechowski, A. and Brodka, J. 'Etude de la resistance statique des assemblages soudés en croix de profils creux rectangulaires.' Construction Métallique, No. 3, Paris, pp. 17-25, 1977.

5. Barentse, J. 'Investigation into static strength of welded T-joints made of rectangular hollow sections.' Stevin Reports 6-76-23 and 6-77-7, Delft University of Technology, 1976 and 1977.

6. Korol, R.M., El-Zanaty, M. and Brady, F.J. 'Unequal width connections of square hollow sections in vierendeel trusses.' Canadian Journal of Civil Engineering, Vol. 4, No. 2, pp. 190-201, 1977.

7. Czechowski, A., Szlendak, J. and Zycinski, J. 'Investigation of selected tubular joints in elements of a "Mostostal" integrated system - Load capacity of frame joints having shape of letter T - Stage 2: Theoretical analysis of joints.' COBKM Mostostal Report No. 07/1/13.3.4.N, Warsaw, 1979.

8. Davies, G. and Coutie, M.G. 'The collapse behaviour of steel welded cross joints in rectangular hollow section.' Institution of Structural Engineers/Building Research Establishment Seminar on "The use of physical models in the design of offshore structures", B.R.E., Watford, November 1979.

9. Mang, F., Bucak, O. and Hummel, T. 'Investigations into the behaviour of high tensile steel joints of rectangular hollow sections.' I.I.W. Document No. XV-416-78, May 1978.

10. Korol, R.M. and Mansour, M.H. 'Theoretical analysis of haunch-reinforced T-joints in square hollow sections.' Canadian Journal of Civil Engineering, Vol. 6, No. 4, pp. 601-609, 1979.

11. Kato, B. and Nishiyama, I. 'T-joints made of rectangular tubes.' 5th International Specialty Conference on Cold-Formed Steel Structures, University of Missouri-Rolla, St. Louis, pp. 663-679, November 1980.

12. Brodka, J. and Szlendak, J. 'Strength of cross joints in rectangular hollow sections.' XXVI Scientific Conference of the Civil and Hydraulic Engineering Section of the Polish Academy of Science and of the Science Division of PZITB, 1980.

13. McCarthy, J.R. 'Welded connections of shaped structural steel tubes.' M.Eng. thesis, University of Wisconsin-Milwaukee, 1976.

14. Wardenier, J. 'Hollow section joints.' Delft University Press, 1982.

15. Giddings, T.W. and Wardenier, J. 'The strength and behaviour of statically loaded welded connections in structural hollow sections.' CIDECT Monograph No. 6 (Draft), 1982.

16. Davies, G., Platt, J.C. and Snell, C. 'The buckling of long simply supported rectangular plates under partially distributed skew pinch loads.' Report NUCE/ST/10 - 1982, Department of Civil Engineering, University of Nottingham, 1982.

17. Packer, J.A. and Haleem, A.S. 'Ultimate strength formulae for statically loaded welded HSS joints in lattice girders with RHS chords.' Canadian Society for Civil Engineering Annual Conference, Fredericton, Proceedings Vol. 1, pp. 331-343, May 1981.

18. Packer, J.A. 'Developments in the design of welded HSS truss joints with RHS chords.' Canadian Journal of Civil Engineering, Vol. 10, No. 1, pp. 92-103, 1983.

19. Giddings, T.W. 'Welded joints in tubular construction.' Conference on Joints in Structural Steelwork, Middlesbrough, Proceedings pp. 4.3 - 4.24, April 1981.

20. Moore, H.F. and Wilson, W.M. 'Strength of webs of I-beams and girders.' University of Illinois Engineering Experiment Station, Bulletin No. 86, May 1916.

21. Winter, G. and Pian, R.H.J. 'Crushing strength of thin steel webs.' Cornell University Engineering Experiment Station, Bulletin No. 35, Part 1, April 1946.

22. Graham, J.D., Sherbourne, A.N. and Khabbaz, R.N. 'Welded interior beam-to-column connections.' American Institute of Steel Construction, New York, 1959.

23. Holmes, M., Astill, A.W. and Martin, L.H. 'Web buckling of rolled steel beams.' CIRIA Report ISBN: 086017-153-1, London, November 1980.

24. Australian Institute of Steel Construction. 'Safe load tables for structural steel - Metric units.' 3rd edition, Sydney, 1975.

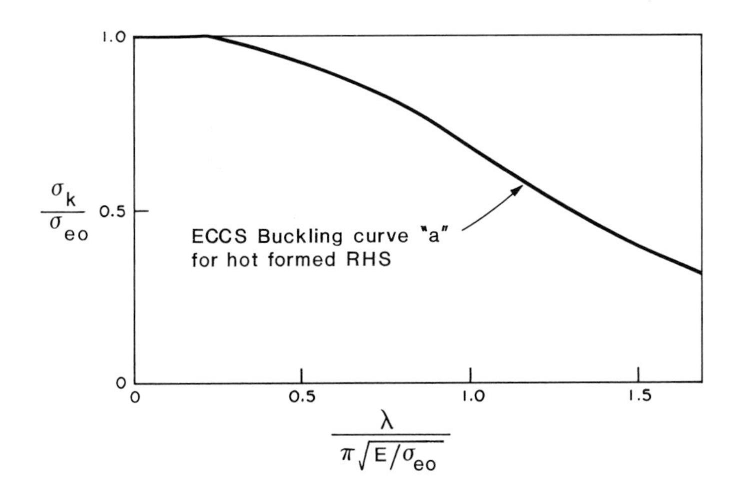

Fig 1 ECCS buckling curve "a" for hot formed RHS.

Fig 2 Typical RHS to RHS Cross joint test with inclined branch members (Dixon, University of Toronto)

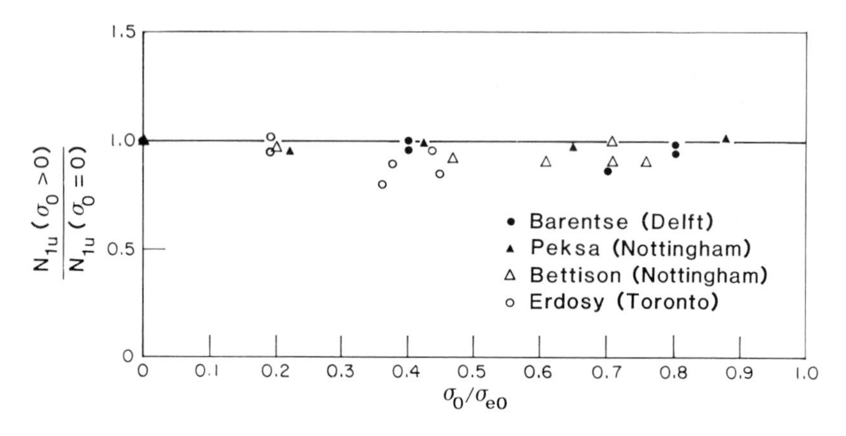

Fig 3 Effect of compression chord force on joint ultimate strength

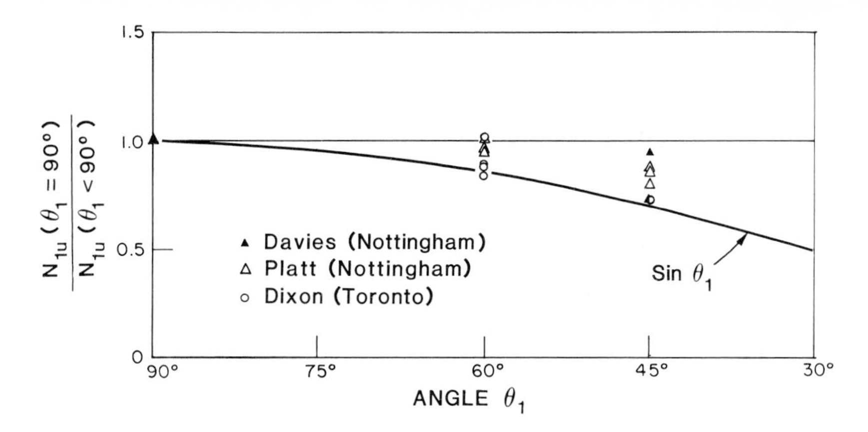

Fig 4 Effect of branch member inclination on joint ultimate strength

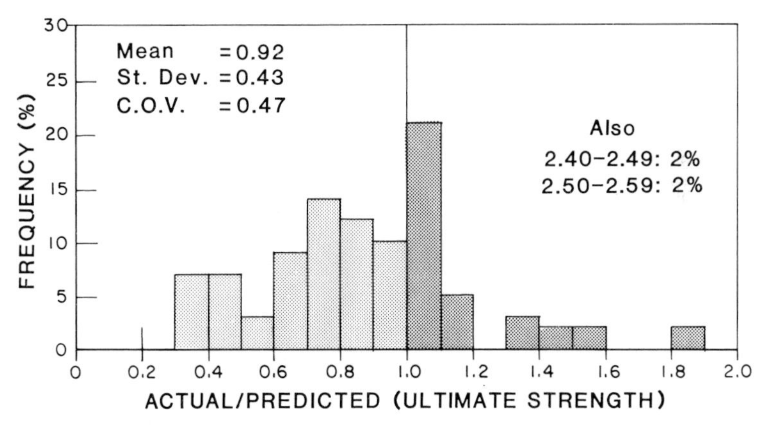

Fig 5 Correlation between test results and equation (1a) - Czechowski and Brodka

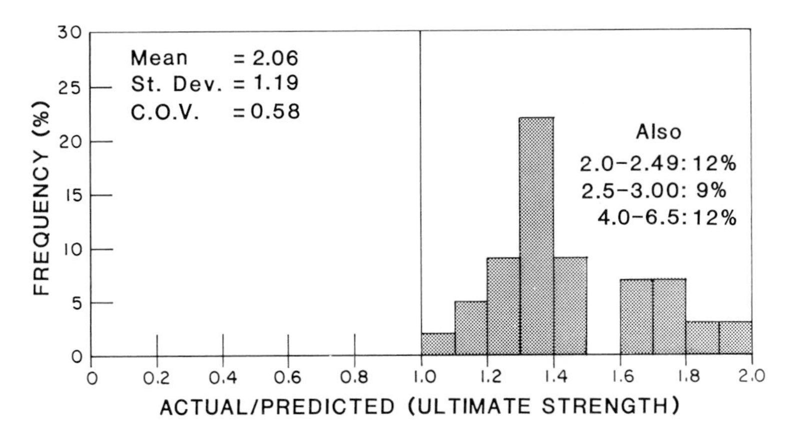

Fig 6 Correlation between test results and equation (3) - Wardenier

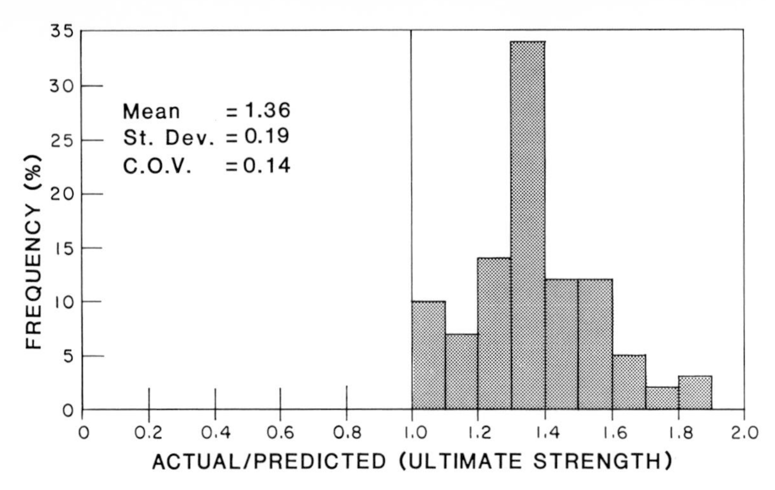

Fig 7 Correlation between test results and equation (7) - Packer

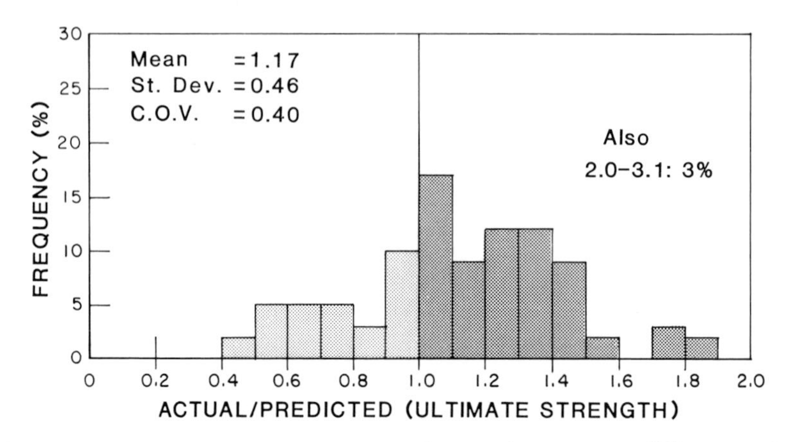

Fig 8 Correlation between test results and equation (8) - Giddings

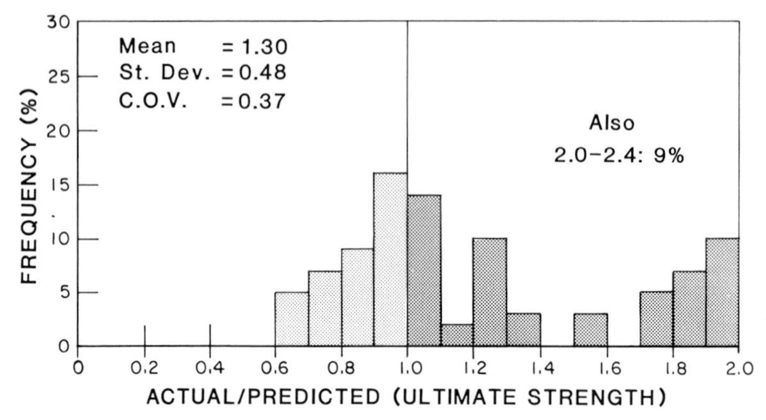

Fig 9 Correlation between test results and equations (9) and (10) - Australian Institute of Steel Construction

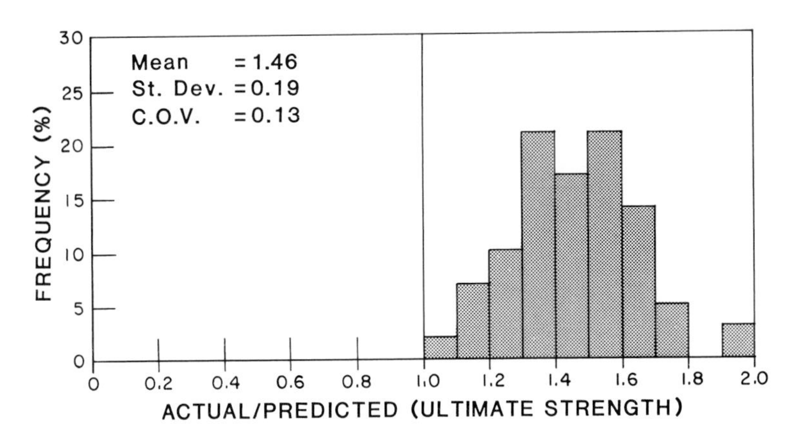

Fig 10 Correlation between test results and equations (11a) and (12a) - Yura

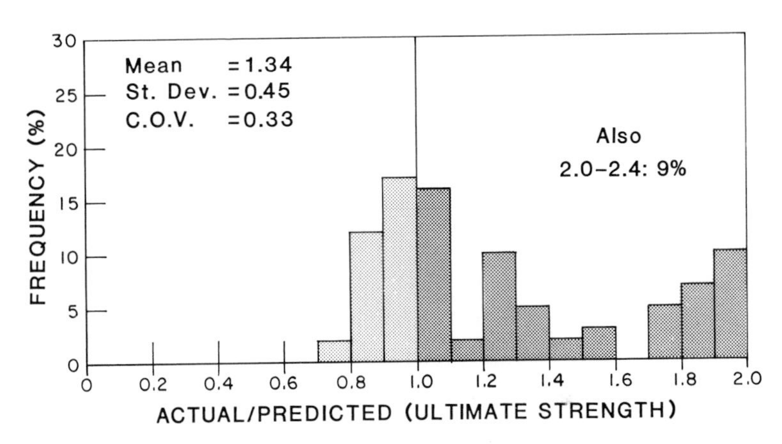

Fig 11 Correlation between test results and equations (9) and (10) with a modified web slenderness ratio.

ULTIMATE LOAD ANALYSIS OF TUBULAR JOINTS

D J Irving, L P R Lyons

Finite Element Analysis Limited, London

A Stamenkovic

School of Civil Engineering, Kingston Polytechnic

Synopsis

This paper presents the results obtained from the nonlinear finite element analysis of a series of tubular T-joints under axial loading. In the work presented the semiloof shell element has been incorporated within the LUSAS (1) finite element system. Geometric nonlinearities are accounted for using a total Lagrangian formulation based on a large displacement, small rotation strain-displacement relationship, while material nonlinearities are accounted for using a multilayer procedure based on the von-Mises yield criterion.

Notation

a	vector of displacements and derivatives at any point
δ	vector of nodal displacements u, v, w
ε	vector of continuum strains ε_x, ε_y, ε_{xy}
ε^r	vector of Green Lagrange strains ε^r_x, ε^r_y, ε^r_{xy}
σ	vector of continuum stresses
σ^r	vector of Piola Kirchhoff shell stress resultants
t	thickness
A	area
d	brace diameter
D	chord diameter
T	chord thickness

Introduction

Since rolled hollow sections were first manufactured they have been widely used in the construction of numerous structures. Their popularity over conventional open sections has primarily been due to economical and functional advantages. Circular hollow sections are particularly useful because they provide little resistance to drag forces and exhibit non-directional section properties. One drawback however is their susceptibility to local damage due to loads applied vertically onto their surface. This problem is particularly apparent at the intersection of connecting members and makes joints the weakest part of any tubular structure.

In this paper the collapse strength of tubular joints is assessed numerically using the nonlinear finite element technique. The joints are modelled using thin doubly curved shell elements. Geometric and material nonlinearities are simultaneously accounted for using a total Lagrangian technique combined with a multilayer plasticity procedure based on the von-Mises yield criterion.

1 THEORY

Full details of the semiloof shell element may be found elsewhere (2). The element is initially formulated with 45 degrees of freedom which are reduced to 32 by combining the central translational freedoms, discarding the inplane components, and applying appropriate shear constraints at discrete points within the element to comply with the thin shell assumptions. The resulting element's nodal configuration is ideal for tubular joint analysis since the freedoms will model a linear stress field in both bending and membrane action and will allow sharp corners and multiple junctions to be modelled without difficulties.

1.1 Geometric nonlinearity

Following standard finite element theory the local displacements and their derivatives at any point (ξ,η) within the shell may be obtained in terms of the nodal freedoms as

$$a = N_s\delta \tag{1}$$

where N_s is the constrained shape function array and

$$a = \left\{u, v, w, \frac{\partial u}{\partial x}, \frac{\partial u}{\partial y}, \frac{\partial v}{\partial x}, \frac{\partial v}{\partial y}, \frac{\partial w}{\partial x}, \frac{\partial w}{\partial y}, \frac{\partial^2 u}{\partial x\partial y}, \frac{\partial^2 u}{\partial y\partial z}, \frac{\partial^2 v}{\partial x\partial z}, \frac{\partial^2 v}{\partial y\partial z}\right\}^T \tag{2}$$

By combining the continuum mechanics theory, introduced by Green and St. Venant, with the thin shell assumptions a set of large displacement, small rotation strain-displacement relations which are independent of geometrical parameters and applicable to both deep and shallow shells can be written for any point on the mid-surface of the shell as

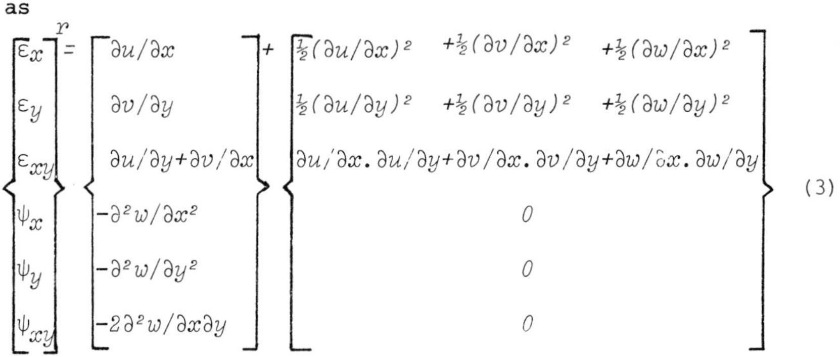

$$\begin{Bmatrix} \varepsilon_x \\ \varepsilon_y \\ \varepsilon_{xy} \\ \psi_x \\ \psi_y \\ \psi_{xy} \end{Bmatrix}^r = \begin{Bmatrix} \partial u/\partial x \\ \partial v/\partial y \\ \partial u/\partial y + \partial v/\partial x \\ -\partial^2 w/\partial x^2 \\ -\partial^2 w/\partial y^2 \\ -2\partial^2 w/\partial x\partial y \end{Bmatrix} + \begin{Bmatrix} \frac{1}{2}(\partial u/\partial x)^2 + \frac{1}{2}(\partial v/\partial x)^2 + \frac{1}{2}(\partial w/\partial x)^2 \\ \frac{1}{2}(\partial u/\partial y)^2 + \frac{1}{2}(\partial v/\partial y)^2 + \frac{1}{2}(\partial w/\partial y)^2 \\ \partial u/\partial x.\partial u/\partial y + \partial v/\partial x.\partial v/\partial y + \partial w/\partial x.\partial w/\partial y \\ 0 \\ 0 \\ 0 \end{Bmatrix} \tag{3}$$

Using the Kirchhoff hypothesis, the Green Lagrange strains at any point within the shell a distance z from the shell mid-surface may be related to the Green Lagrange strains at the shell mid-surface by

$$\begin{Bmatrix} \varepsilon_x \\ \varepsilon_y \\ \varepsilon_{xy} \end{Bmatrix} = \begin{bmatrix} 1 & 0 & 0 & z & 0 & 0 \\ 0 & 1 & 0 & 0 & z & 0 \\ 0 & 0 & 1 & 0 & 0 & z \end{bmatrix} \begin{Bmatrix} \varepsilon_x \\ \varepsilon_y \\ \varepsilon_{xy} \\ \psi_x \\ \psi_y \\ \psi_{xy} \end{Bmatrix}^r \qquad (4)$$

With these large deformation relationships established, other matrices required for the analysis may be derived in the standard manner (3).

From eqn (3) it can be seen that the slope terms $\partial w/\partial x$ and $\partial w/\partial y$ not used previously in linear analysis are required. Examination of these terms, which are of special importance since they are squared in the strain displacement relations, has been carried out successfully on flat rectangular and quadrilateral elements (4).

1.2 *Material Nonlinearity*

Material nonlinearities are generally predominant during the collapse analysis of tubular structures and must thus be dealt with accordingly. In the analyses presented a multilayer procedure which allows progressive growth of plasticity through the thickness has been adopted to satisfy the von-Mises yield criterion and Prandtl-Reuss flow rules at five levels through the shell thickness. To ensure the incremental flow rules are satisfied a novel step selection algorithm has been implemented. This technique automatically selects the number of sub-incremental strain steps required to accurately position the stress on the yield surface.

1.3 *Through Thickness Integration*

In order to satisfy the stress-strain relationship throughout the shell thickness the strain variation at each layer must be computed. By expressing eqn (4) in shortened form as

$$\varepsilon = H\,\varepsilon^r \qquad (5)$$

the continuum mechanics strain increments at any point a distance z from the shell mid-surface are related to the incremental strains by

$$\Delta\varepsilon = H\,\Delta\varepsilon^r \qquad (6)$$

From the constitutive law relating Piola-Kirchhoff stress resultants to Green Lagrange strains the continuum mechanics stress increments at any point in the shell may be computed in incremental form as

$$\Delta\sigma = D_{ep}\,\Delta\varepsilon \qquad (7)$$

where D_{ep} is the elasto plastic modulus matrix relating continuum stress to strain obtained by combining the von-Mises yield criterion and Prandtl Reuss flow rule (3).

The incremental Piola Kirchhoff shell stress resultants may be obtained by integrating the continuum mechanics stress increments through the shell thickness as

$$\Delta\sigma^r = \int_t H^T \Delta\sigma \, dt \tag{8}$$

Combining eqns (6), (7) and (8) the Piola Kirchhoff stress resultants are related to the Green Lagrange strains by

$$\Delta\sigma^r = D^* \Delta\varepsilon^r \tag{9}$$

where

$$D^* = \int_t H^T D_{ep} H \, dt \tag{10}$$

is the elasto-plastic rigidity matrix.

1.4 Formulation of Tangential Stiffness Matrix

For nonlinear analysis which takes account of both material and geometric nonlinearities, an incremental solution scheme is required. The total Lagrangian formulation may be derived from the virtual work equation as

$$K_T \Delta\delta = R-F \tag{11}$$

where $\Delta\delta$ are the displacement increments, R is the nodal force vector, F is the nodal equivalent force vector equivalent to the elements internal stress resultants, and K_T is the tangential stiffness matrix given by

$$K_T = K_\ell + K_\sigma \tag{12}$$

where

$$K_\ell = \int_A B^T D^* B \, dA \tag{13}$$

and

$$K_\sigma = \int_A G^T S G \, dA \tag{14}$$

is the initial stress matrix.

2 VERIFICATION

Verification of the mathematical model has been carried out by examining the geometric and material nonlinearities independently and in a combined form (4). This has been carried out in three stages:-

(i) the results obtained for the geometric nonlinearity have been compared against those available for a simply supported plate under UDL (5), for a snap through problem (6), and for an imperfect plate subjected to in-plane compression (7)

(ii) the nonlinear material model has been verified against hand calculations based on fundamental plasticity theory and (iii) the combined geometric and material nonlinearities have been extensively tested and found to produce reliable results which are in close agreement with solutions presented by Little (8) for imperfect plates subjected to in-plane compression.

3 RESULTS OF TUBULAR JOINT ANALYSIS

Since 1974 a testing programme has been taking place at Kingston Polytechnic in collaboration with the Tubes Division of the British Steel Corporation to study the effects of varying geometric parameters and load types on the ultimate static strength of tubular joints. In this paper the results obtained experimentally from the axial load tests are compared to finite element results obtained using the procedures previously described.

A typical idealisation which considers the double symmetry of both the joint and loading is presented in Fig 1. In the analyses both 'welded' and 'unwelded' joints have been considered. The welds have been idealised by increasing the thickness of the shell elements around the joint intersection.

Loading in all the analyses was applied in 14 increments using displacement control. A Newton Raphson iteration scheme with a simple line search per iteration was adopted to eliminate the out of balance forces and convergence to equilibrium was assumed when the displacement and residual norms were reduced to less than 1% and 5% respectively. From the comparison of experimental (9) and finite element results presented in Fig 3 it can be seen that simulation of the weld by thickening the elements around the joint intersection increases the stiffness of the joint and thus reduces the flexural and membrane strains. This retards the onset of yield and transfers the deformations in the chord away from the chord-brace intersection Fig 2.

One technique currently being used to reduce the local deformations in the chord around the joint intersection in large diameter tubular joints is to add ring stiffeners inside the chord member. In Fig 4 results from the finite element analysis of a stiffened and unstiffened T-joint ($d/D=0.45$, $D/T=72$) are compared to experimental results (10). The deformed shapes at ultimate load exaggerated by a factor of 3 are shown in Figs 5 and 6. Close agreement is obtained with the experimental stiffened T-joint results. For the unstiffened T-joint however agreement is not so good. This was not unexpected since the ultimate strength of joints with high D/T ratios, such as that analysed, would be severely affected by minor imperfections and residual stresses. These effects could not be accounted for in the analysis since no published information was available.

4 CONCLUSIONS

The comparisons carried out between experimental and analytical results show that the nonlinear semiloof shell element presented can be successfully used to produce a reliable prediction of the collapse strength of tubular joints under axial loading.

References

1. Lyons, L.P.R. 'The LUSAS user's manual.' Finite Element Analysis Limited, London, 1980.
2. Irons, B.M. 'The semiloof shell element.' Finite elements for thin shells and curved members. Ed. Ashwell and Gallagher, Chap. 11, Pub. J. Wiley, London, 1976.
3. Zienkiewicz, O.C. 'The finite element method.' 3rd ed., Pub. McGraw-Hill, London, 1977.
4. Irving, D.J. 'Large deformation elasto-plastic finite element analysis of plates shells and tubular joints using semiloof elements.' Ph.D. Thesis, Kingston Polytechnic, 1982.
5. Rushton, K.R. 'Large deflection of plates with initial curvature.' Int. J. Mech. Sci., Vol. 12, pp1037-1051, 1970.
6. Sabir, A.B., Lock, A.C. 'The application of finite elements to the large deflection geometrically nonlinear behaviour of cylindrical shells.' Variational methods in engineering. Ed. Brebbia et al. Pub. Soton. Univ. Press., 1972.
7. Yamaki, N. 'Post buckling behaviour of rectangular plates with small initial curvature loaded in edge compression.' J. Appl. Mech., ASME, Vol. 26, Pt. 3, pp407-414, 1959.
8. Bradfield, C.D., Chladny, E. 'A review of the elastic-plastic analysis of steel plates loaded in in-plane compression.' Univ. Cambridge, Dept. of Eng. Report No. CUED/D-Struct/TR.77, 1979.
9. Sparrow, K.D., Stamenkovic, A. 'Experimental determination of the ultimate static strength of T-joints in circular hollow sections subject to axial load and moment.' Joints in structural steelwork. Ed. Howlett et al., Pub. Pentech Press, Plymouth, 1981.
10. Sawanda, Y., Idogaki, S., Sekita, K. 'Static and fatigue tests on T-joints stiffened by an internal ring.' OTC 3422, 1979.

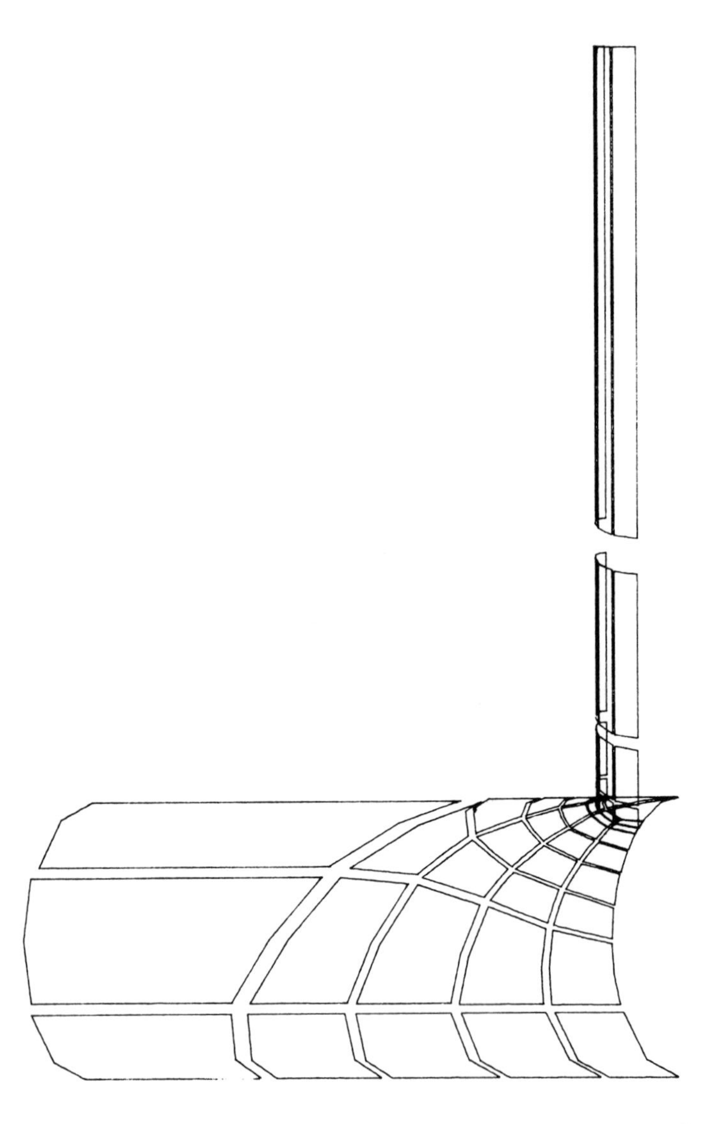

Fig 1 Typical exploded mesh idealisation of T-joint analysis in large deformation elasto-plastic finite element analysis

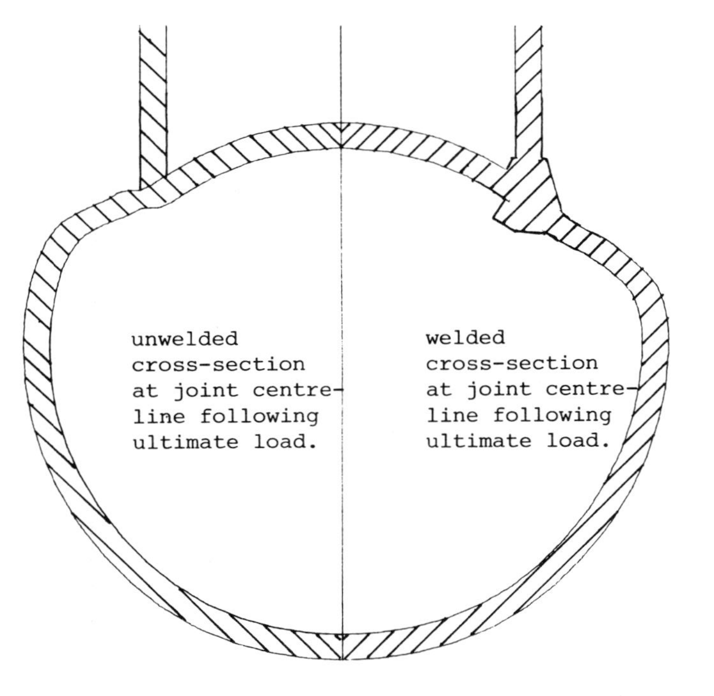

Fig 2 Comparison between analytical and experimental deformed cross-sections following ultimate load

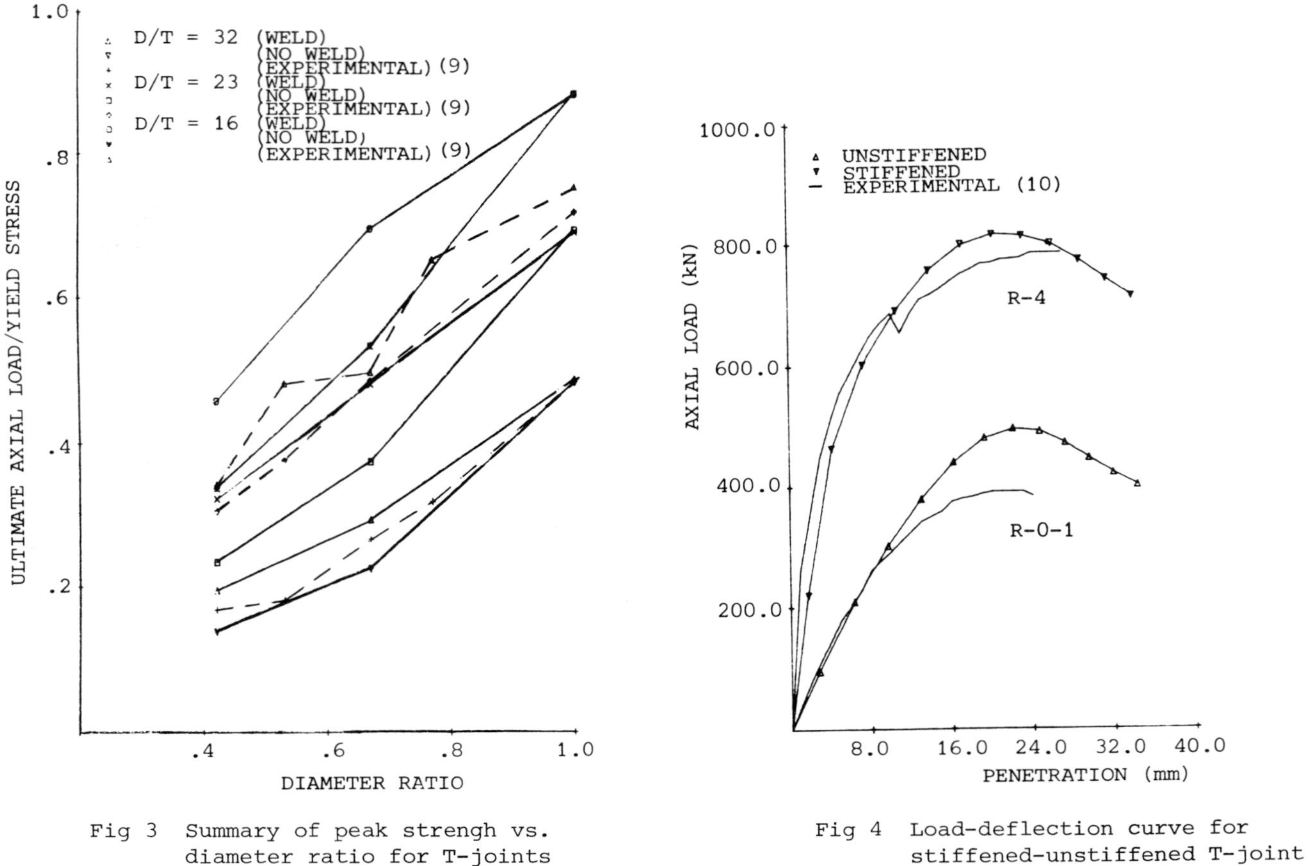

Fig 3 Summary of peak strengh vs. diameter ratio for T-joints subjected to axial loading

Fig 4 Load-deflection curve for stiffened-unstiffened T-joint under axial compressive loading

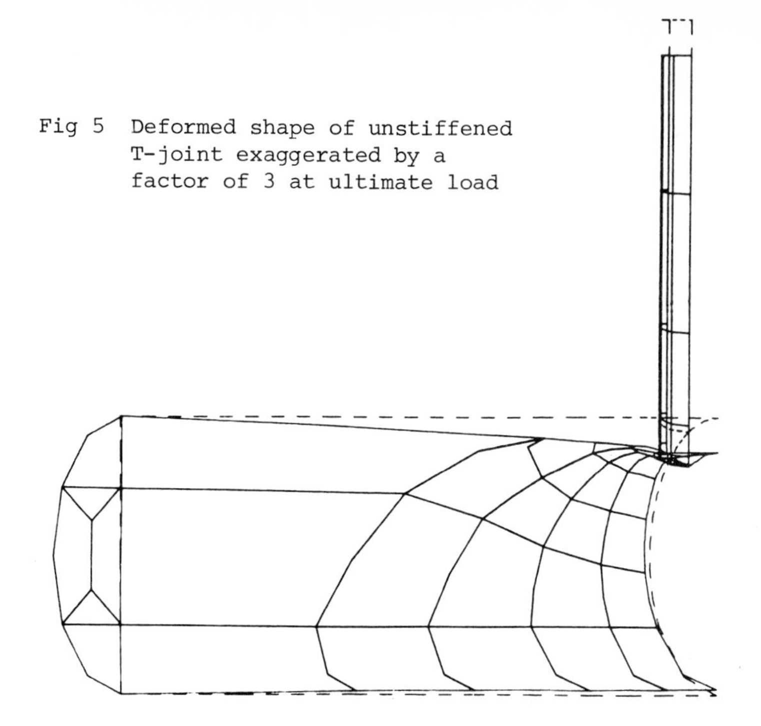

Fig 5 Deformed shape of unstiffened T-joint exaggerated by a factor of 3 at ultimate load

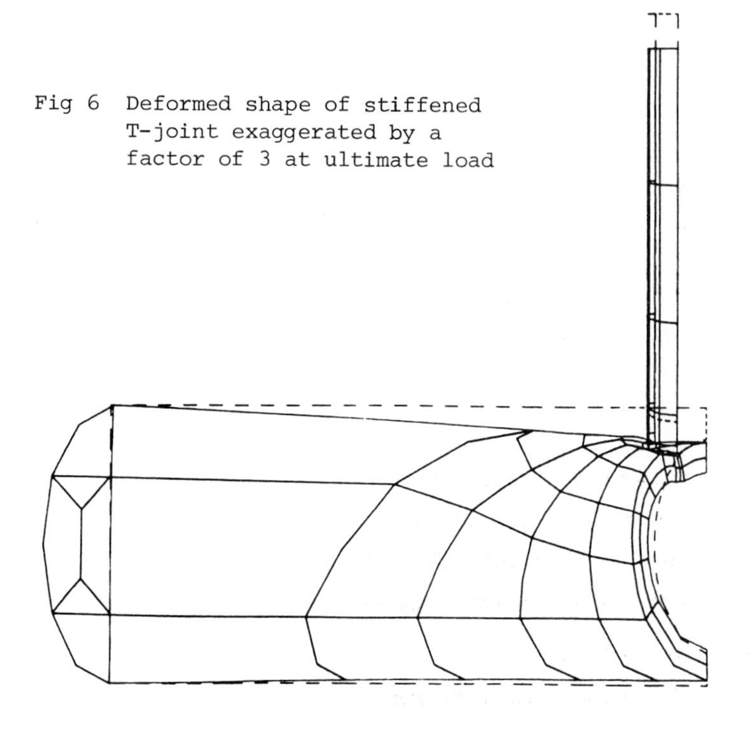

Fig 6 Deformed shape of stiffened T-joint exaggerated by a factor of 3 at ultimate load

WELDED JOINTS IN STEEL STRUCTURAL HOLLOW SECTIONS

N F Yeomans, T W Giddings

Technical Centre, British Steel Corporation,
Tubes Division, Corby, Northants

Synopsis

A method is given for the presentation of design tables for joints between steel structural hollow sections.

Notation

b_e, b_{ep}, b_{eov} = effective widths where

$$b_e = \frac{C}{1.25} \cdot \frac{t_o}{b_o} \cdot \frac{Y_{so}}{Y_{si}} \cdot \frac{t_o}{t_i} \cdot b_i$$

$$b_{ep} = \frac{C}{1.25} \cdot \frac{t_o}{b_o} \cdot b_i$$

$$b_{eov} = \frac{C}{1.25} \cdot \left(\frac{t_i}{b_i}\right)_{ov} \cdot \frac{(Y_{si}\ t_i)_{ov}}{Y_{si}\ t_i} \cdot b_i$$

- C — 13.5 if $Y_{si} < 350$ or 11.5 if $Y_{si} \geq 350$
- g — gap between bracings (negative for overlapping bracings)
- A_q — RHS chord shear area, $2h_o t_o + \alpha b_o t_o$
- N_i — ultimate bracing load for chord failure (deformation)
- N_e — ultimate bracing load for bracing failure
- N_p — ultimate bracing load for chord punching shear failure
- N_s — ultimate bracing load for chord shear failure in the gap between bracings
- Y_{si} — material yield strength
- Y_k — ultimate compressive strength of RHS chord wall for slenderness ratio of $3.46(h_o/t_o - 2)/\sqrt{\sin\theta_1}$
- α — 0 for CHS bracings, $1/\sqrt{(1 + 4g^2/3t_o^2)}$ for RHS bracings
- β — bracing to chord width ratio, b_1/b_o, d_1/d_o, d_1/b_o
- θ_i — angle between bracings and chord

Suffix $i = 0$ - chord

$= 1$ - compression bracing

= *2* - tension bracing

ov = overlapped bracing

All other symbols are shown in Fig 1.

Introduction

Their efficiency, their aesthetic appeal and the technical support given by their producers have rapidly established a major place in the construction market for hot formed steel structural hollow sections (SHS). The foresight of the manufacturers in establishing, early in the development of SHS, the international research organisation CIDECT (Comite International pour le Developpement et l'Etude de la Construction Tubulaire) has considerably aided this success.

SHS have been developed to take advantage of the development and increasing use of welding as an acceptable and often preferable method of structural jointing. Since its inception one of CIDECT's major tasks has been the development and analysis of data on the strength and behaviour of welded connections with the aim of producing recommendations for their efficient and economic design.

There are two main structural hollow section shapes; the circular hollow section (CHS) and the rectangular (including square) hollow section (RHS); this paper gives a brief review of the work that has been carried out on statically loaded welded connections with CHS or RHS chord members and how this work has been used to develop design recommendations.

1. SHS IN LATTICE GIRDER DESIGN

1.1 Joint types

The most common forms of joints used in lattice construction are shown in Fig 1, generally the members can be of CHS or RHS. When a chord member is used it is usual also to use CHS bracing members; however, if the chord is an RHS the bracings can be either CHS or RHS.

1.2 Design

In the majority of braced framework systems the frame is designed assuming the transfer of load by the development of axial forces in the members. In SHS construction, however, this is not usually the case; the chord members are usually continuous and the bracings are welded to the chord, thus there is significant continuity in the frame and stiffness of the joints which results in the generation of bending moments in the members and reduced overall deflections. Nevertheless, provided that the members and the joints have the capacity to redistribute these moments, it is well established that it is perfectly acceptable to analyse the girder on the basis of pin joints, and therefore axial loads. This has been verified on several occasions by testing isolated joints and joints in girders with the resulting joint strengths being comparable (1).

When member centre lines do not node, the resulting moments must be taken into account in the design of the members; however, since the joint's design is based upon experimental data, in which eccentricities of up to $h_o/2$ or $d_o/2$ have been incorporated, these moments have already been taken into account in the joint itself.

2. JOINT DESIGN FORMULAE

2.1 Limit state

Formulae for the design of joints comprising SHS have been developed using a combination of semi-empirical and lower bound theoretical analyses.

The semi-empirical formulae have been developed around the main influencing parameters i.e.

chord thickness ratio, d_o/t_o, b_o/t_o, h_o/t_o

chord/bracing width ratio, d_1/d_o, d_1/b_o, b_1/b_o

Using statistical analyses 95% characteristic strength formulae have been developed based upon test results together with lower bound theoretical formulae for the more simple forms of failure (2).

These formulae are reproduced here as Table 1, for CHS chord joints and Table 2 for RHS chord joints and suggested performance factors to convert them from characteristic strength formulae to design strength are also given.

2.2 Allowable loads

In order to convert the formulae given in Tables 1 and 2 for use in allowable stress design (re BS 449) it is necessary to substitute allowable stress for yield strength and to take account of the performance factors.

Assuming a yield strength of 275 N/mm^2 and rounding to the nearest 5 we have for the main equations:

$$\frac{Y_{so} \text{ or } Y_{si}}{1.5\,\gamma_m\,\gamma_c} = \begin{array}{ll} 165 \text{ N/mm}^2 & \text{if } \gamma_m\,\gamma_c = 1.1 \text{ (characteristic)} \\ 180 \text{ N/mm} & \text{if } \gamma_m\,\gamma_c = 1.0 \text{ (lower bound)} \end{array}$$

and for the shear formulae (which all contain $Y_{so}/\sqrt{3}$)

$$\frac{Y_{so} \text{ or } Y_{si}}{1.5\sqrt{3}\,\gamma_m\,\gamma_c} = 105 \text{ N/mm}^2$$

If the yield strength is not 275 N/mm^2 then the above values should be multiplied by $Y_{so}/275$.

3. SIMPLIFICATIONS FOR DESIGN TABLES

As can be seen from Tables 1 and 2 the formulae for predicting the strength of welded tubular joints are somewhat complicated and it has been found necessary, in order to provide easily followed design tables, to make certain simplifications.

3.1 Joints with CHS chords

Formulae 1, 2 and 4 in Table 1 can, with the exception of the chord end load function, $f(n)$, be related to the chord dimensions d_o and t_o

and the bracing diameter d_1 for fixed values of θ and Y_{so}. Similarly for formula 3 provided a fixed value of $f(d_o/t_o,g)$ is used. Thus design tables with rows representing chord size and thickness and columns representing bracing size can be drawn up, for fixed values of θ and Y_{so}, $f(n)$ and $f(d_o/t_o,g)$. Design loads for different values of θ and Y_{so} can then be calculated by multiplying simply by the ratio of sin θ's and Y_{so}'s. The $f(n)$ can be taken care of by an inset graph and $f(d_o/t_o,g)$ by an inset table, both mounted on the main table.

Thus all parameter combinations can be taken care of in simple look up tables for each of the three types of joint.

3.2 Joints with RHS chords

Developing design tables for joints with RHS chords is more complicated than for those with CHS chords since there are more variables ($0.5 \leq b/h \leq 2.0$) and more than one strength formula for each type of joint, additionally for T, X and Y-joints a sin θ term appears inside the bracketed part of the formula.

3.2.1 T-, Y- and X-joints, $b_1/b_o \leq .85$

The following simplifying assumptions have been made:

1. The $1/\sin\theta$ term in the brackets has been ignored, thus for $\theta < 90^o$ the result will be conservative.
2. The bracings are square or circular, thus if $h_1 > b_1$ the value found will be conservative, but if $h_1/\sin\theta_1 < b_1$ it would be unsafe and the table should not be used.

3.2.2 T-, Y- and X-joints, $b_1/b_o > .85$

There are three different formulae for the prediction of the strength of these joints and at present no design tables have been prepared. This has also been complicated in the case of compression in the bracings by the need to determine a buckling stress, Y_k, for the chord side walls for insertion into the main joint strength formula.

3.2.3 K- and N-joints with a gap between bracings

Formulae N_1, N_2 and N_s (assuming constant yield stress and bracing angle) only vary with b_o, t_o and the periphery of the bracings and can therefore be represented as before for CHS but with the columns relating to effective bracing width $(b_1+b_2+h_1+h_2)/4$. Only occasionally do the punching shear N_p or effective width N_e criteria become critical and these circumstances can be indicated by shading on the effected area of the table. The chord end load function $f(n)$ can, as with CHS chord joints, be given on a graph inset in the table.

Part of the design table for this type of joint with square RHS chords is shown here as Table 3.

3.2.4 K- and N-joints with overlapping bracings

The formulae for overlapped joints are basically simple in that an effective bracing periphery is merely multiplied by the bracing thickness and yield stress; however, the terms used to determine the effective width of the bracing faces are relatively complex.

In order that design tables of a manageable size could be produced various simplifying assumptions, which limit the use of these tables have had to be made. These are:

1. The bracings are of the same width, thickness and yield stress.
2. The bracings have the same yield stress as the chord.
3. The value of the constant C, which varies with yield stress has been given its minimum value.

Thus $$b_e = \frac{C}{1.25} \cdot \frac{t_o}{b_o} \cdot \frac{b_i}{t_i} \cdot t_o$$

$$b_{eov} = \frac{C}{1.25} \cdot t_i$$

where i relates to the bracing member under consideration and $C = 11.5$

4. *CONCLUSIONS*

(1) Possible to present simple design data for many of the practical joint types.

(2) Possible to determine the strength of greater ranges from the basic design equations, the use of which is considerably simplified by the use of microcomputer programs.

References

1. Koning, C.H.M. de and Wardenier, J. 'Tests on welded joints in complete girders made of square hollow sections'. INO-IBBC report No. BI-79-19/0063.4.3.471, CIDECT Report No. 5Q/79/, March 1979.

2. International Institute of Welding. 'Design recommendations for hollow section joints - predominantly statically loaded'. IIW Document No. XV-491-81.

Acknowledgements

The authors wish to express their gratitude to the Tubes Division of the British Steel Corporation for permission to publish this paper.

Table 1 Strength formulae for CHS chord joints

Joint type (see Fig 1)	Strength formulae	Design factor $\gamma_m \gamma_c$
1. T- & Y-joints	$N_1 = \frac{Y_{so} t_o^2}{\sin \theta_1} (3.1 + 15.6 \beta^2) \left(\frac{d_o}{2t_o}\right)^{0.2} f(n)$	1.1
2. X-joints	$N_1 = \frac{Y_{so} t_o^2}{\sin \theta_1} \frac{5.7}{1 - 0.81 \beta} f(n)$	1.1
3. K- & N-joints with gap or overlap	$N_1 = \frac{Y_{so} t_o^2}{\sin \theta_1} (2.3+10.8\beta) f(n^*) \frac{f(d_o,g)}{t_o}$; $N_2 = N_1 \frac{\sin \theta_1}{\sin \theta_2}$	1.1
4. Punching shear (not for overlap joints)	$N_p = \frac{Y_{so} t_o \pi}{\sqrt{3}} d_1 \frac{1 + \sin \theta_1}{2 \sin^2 \theta_1}$	1.0

Functions $f(n) = 1.2 + 0.5n$ but not greater than 1

$f(n^*) = 1 + 0.3(n^*-n^{*2})$ but not greater than 1

where $n = \frac{\text{maximum nominal axial + nominal bending stress in chord at joint}}{\text{chord yield strength}}$

$n^* = \frac{\text{smaller axial chord load}}{\text{chord yield load}}$

n and n^* are negative for compression

$$\frac{f(d_o.g)}{t_o} = \frac{d_o}{2t_o}^{0.2}\left[1 + \frac{0.012(d_o/2t_o)^{1.5}}{\exp(0.39(g/t_o)-0.53)+1}\right]$$

Table 2 Strength formulae for RHS chord joints

Joint type (see Fig 1)	Strength formulae	Design factor $\gamma_m \gamma_c$
1. T-, Y- & X-joints (if $0.85 < \beta < 1.0$ interpolate between values)	$\beta \leq 0.85$ $N_1 = \frac{Y_{so} t_o^2}{\sin \theta_1} \left[\frac{2h_1}{b_o \sin \theta_1} + 4(1-\beta)^{0.5} \right] \frac{1}{1-\beta}$	1.0
	$\beta = 1.0$ lower value of N_1, N_e or N_p below $N_1 = \frac{Y_{ko} t_o}{\sin \theta_1} \left[\frac{h_1}{\sin \theta_1} + 10t_o \right]$ $N_e = Y_{s_1} t_1 (2h_1 - 4t_1 + 2b_e)$ $N_p = \frac{Y_{so} t_o}{\sqrt{3} \sin\theta_1} \left[\frac{2h_1}{\sin \theta_1} + 2b_{ep} \right]$	1.0 but 1.25 if compression in N_1
2. K- & N-joints with gap. (lower value of N_1, N_s, N_e and N_p)	$N_1 = \frac{Y_{so} t_o^2}{\sin \theta_1} 9.8 \frac{b_1+b_2+h_1+h_2}{4b_o} \left(\frac{b_o}{2t_o} \right)^{0.5} f(n)$; $N_2 = N_1 \frac{\sin \theta_1}{\sin \theta_2}$ $N_s = \frac{Y_{so} A_g}{\sqrt{3} \sin \theta_1}$; $N_e = Y_{si} t_i (2h_i - 4t_i + b_i + b_e)$ $N_p = \frac{Y_{so} t_o}{\sqrt{3} \sin \theta_1} \left[\frac{2h_i}{\sin \theta_i} + b_i + b_{ep} \right]$	1.1 on N_1 and N_2 1.0 1.0 on N_s, N_e and N_p
3. K- & N-joints with 100% overlap	$N_i = Y_{si} t_i (2h_i - 4t_i + b_i + b_{eov})$	1.0
4. K- & N- joints with 30% to 99% overlap	$N_i = Y_{si} t_i (2h_i - 4t_i + b_e + b_{eov})$	1.0

Table 3 Typical joint design table (part)

RHS CHORD
K- or N- type joint with gap
between bracings

If $\theta < 90$ multiply by $1/\sin\theta$
If Y_{so} <> 275 multiply by $Y_{so}/275$
If bracings are CHS multiply by $\pi/4$
If chord load is compressive multiply by $f(n)$ from inset graph

Chord			Allowable joint load in kN for effective bracing width b_e of:																		
b_o	h_o	t_o	20	30	40	50	60	70	80	90	100	120	140	150	160	180	200	250	300	350	400
40	40	2.5	14	21	29																
40	40	3.0	19	28	38																
40	40	3.2	21	31	40																
40	40	4.0	29	43	50																
50	50	2.5	13	19	26	32															
50	50	3.0	17	25	34	42															
50	50	3.2	19	28	37	46															
50	50	4.0	26	39	52	63															
50	50	5.0	36	54	70	79															
70	70	3.0		21	28	36	43	50													
70	70	3.6		28	37	47	56	65													
70	70	5.0		46	61	76	92	107													
90	90	3.6			33	41	49	58	66	74											
90	90	5.0			54	67	81	94	109	121											
90	90	6.3			76	95	114	133	152	172											
100	100	4.0			37	46	55	64	73	82	91										
100	100	5.0			51	64	77	89	102	115	128										
100	100	6.3			71	90	108	127	145	163	181										
100	100	8.0			103*	129	155	181	207	233	252										
100	100	10.0			145*	181*	217*	253*	279	301	315										

* Also check N_p and N_e in Table 2

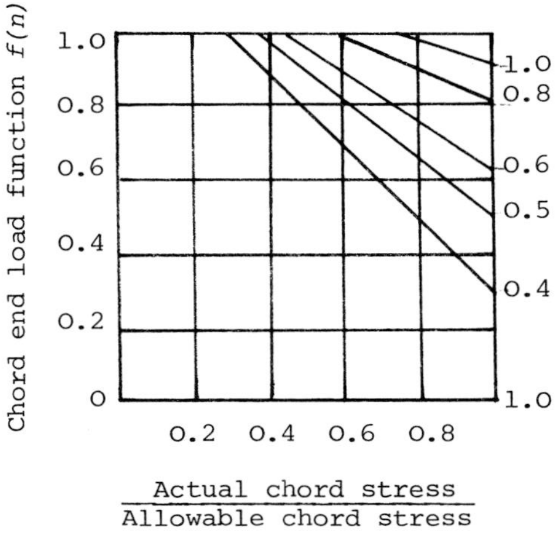

Chord end load function $f(n)$
1.0
0.8
0.6
0.4
0.2
0
0.2 0.4 0.6 0.8
1.0
0.8
0.6
0.5
0.4
1.0
Actual chord stress / Allowable chord stress

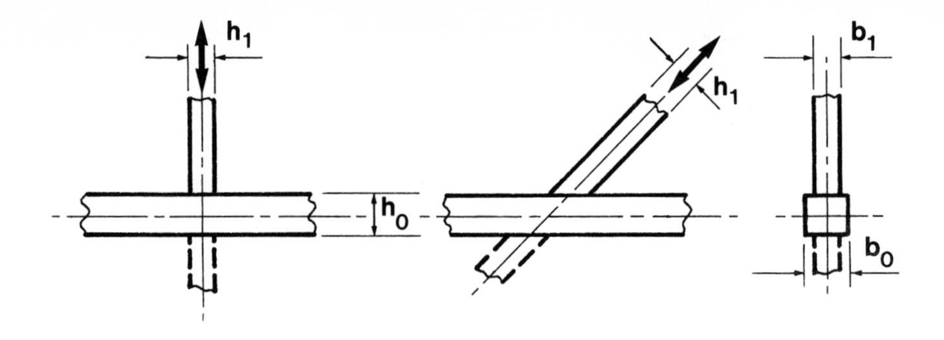

a. T-, Y- & K- joints

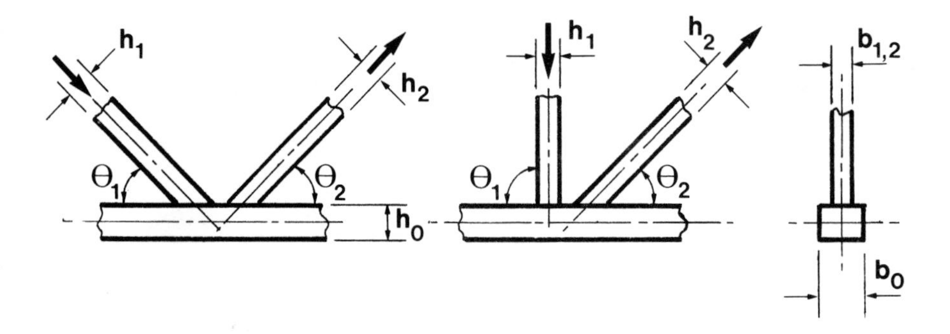

b. K- & N- joints with gap

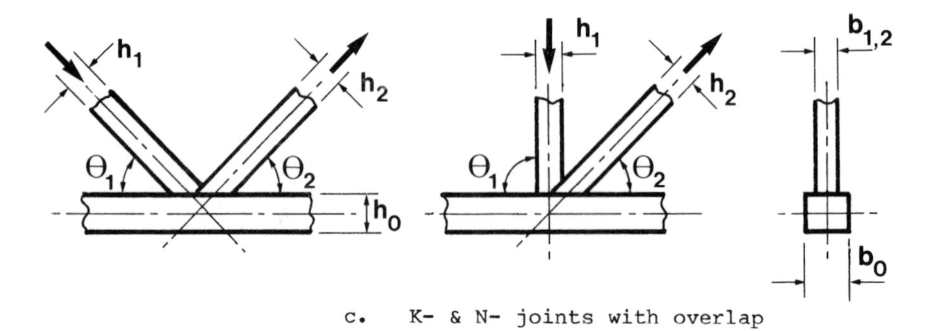

c. K- & N- joints with overlap

Figure 1 Joint Types and Symbols